PYRROLES

Part Two

This is a part of the forty-eighth volume in the series
THE CHEMISTRY OF HETEROCYCLIC COMPOUNDS

THE CHEMISTRY OF HETEROCYCLIC COMPOUNDS

A SERIES OF MONOGRAPHS

EDWARD C. TAYLOR, *Editor*

ARNOLD WEISSBERGER, *Founding Editor*

PYRROLES

Part Two

THE SYNTHESIS, REACTIVITY, AND PHYSICAL PROPERTIES OF SUBSTITUTED PYRROLES

Edited by
R. Alan Jones

School of Chemical Sciences
University of East Anglia
Norwich, U.K.

AN INTERSCIENCE® PUBLICATION

JOHN WILEY & SONS, INC.

NEW YORK · CHICHESTER · BRISBANE · TORONTO · SINGAPORE

In recognition of the importance of preserving what has been
written, it is a policy of John Wiley & Sons, Inc., to have books
of enduring value published in the United States printed on
acid-free paper, and we exert our best efforts to that end.
In Interscience® Publication.

Copyright © 1992 by John Wiley & Sons, Inc.

All rights reserved. Published simultaneously in Canada.

Reproduction or translation of any part of this work
beyond that permitted by Section 107 or 108 of the
1976 United States Copyright Act without the permission
of the copyright owner is unlawful. Requests for
permission or further information should be addressed to
the Permissions Department, John Wiley & Sons, Inc.

Library of Congress Cataloging in Publication Data:

Pyrroles.

  (The chemistry of heterocyclic compounds, 0069-3154: v. 48)
  "An Interscience publication."
  Includes bibliographical references and index.
  Contents: pt. 1. The synthesis and the physical and
chemical aspects of the pyrrole ring.—pt. 2. The
synthesis, reactivity, and physical properties of
substituted pyrroles.
  1. Pyrrole I. Jones, R. Alan (Richard Alan)
QD401.P994 1990   547'.593   89-16553
ISBN 0-471-51306-7 (pt. 2)

Printed in the United States of America

10 9 8 7 6 5 4 3 2 1

Contributors

Enrico Aiello
Istituto Farmacochimico
Facultà di Farmacia
Università di Palermo
Palermo, Italy

Anna Maria Almerico
Istituto Farmacochimico
Facultà di Farmacia
Università di Palermo
Palermo, Italy

Girolamo Cirrincione
Istituto Farmacochimico
Facultà di Farmacia
Università di Palermo
Palermo, Italy

Gaetano Dattolo
Istituto Chimico Farmaceutico
 e Tossicoligico
Facultà di Farmacia
Università di Milano
Milano, Italy

Hamish McNab
Department of Chemistry
University of Edinburgh
Edinburgh, U.K.

Lilian C. Monahan
Department of Chemistry
University of Edinburgh
Edinburgh, U.K.

Trevor P. Toube
Department of Chemistry
Queen Mary and Westfield College
University of London
London, U.K.

Boris A. Trofimov
Institute of Organic Chemistry
Siberian Branch of the U.S.S.R.
 Academy of Sciences
Irkutsk, U.S.S.R.

The Chemistry of Heterocyclic Compounds
Introduction to the Series

The chemistry of heterocyclic compounds is one of the most complex and intriguing branches of organic chemistry. It is of equal interest for its theoretical implications, for the diversity of its synthetic procedures, and for the physiological and industrial significance of heterocyclic compounds.

The Chemistry of Heterocyclic Compounds, published since 1950 under the initial editorship of Arnold Weissberger, and later, until Dr. Weissberger's death in 1984, under our joint editorship, has attempted to make the extraordinarily complex and diverse field of heterocyclic chemistry as organized and readily accessible as possible. Each volume has dealt with syntheses, reactions, properties, structure, physical chemistry, and utility of compounds belonging to a specific ring system or class (e.g., pyridines, thiophenes, pyrimidines, three-membered ring systems). This series has become the basic reference collection for information on heterocyclic compounds.

Many broader aspects of heterocyclic chemistry are recognized as disciplines of general significance that impinge on almost all aspects of modern organic and medicinal chemistry, and for this reason we initiated several years ago a parallel series entitled *General Heterocyclic Chemistry*, which treated such topics as nuclear magnetic resonance, mass spectra, and photochemistry of heterocyclic compounds, the utility of heterocyclic compounds in organic synthesis, and the synthesis of heterocyclic compounds by means of 1,3-dipolar cycloaddition reactions. These volumes were intended to be of interest to all organic and medicinal chemists, as well as to those whose particular concern is heterocyclic chemistry.

It has become increasingly clear that this arbitrary distinction creates as many problems as it solves, and we have therefore elected to discontinue the more recently initiated series *General Heterocyclic Chemistry*, and to publish all forthcoming volumes in the general area of heterocyclic chemistry in *The Chemistry of Heterocyclic Compounds* series.

EDWARD C. TAYLOR

Department of Chemistry
Princeton University
Princeton, New Jersey 08544

Preface

Part 1 of this volume, which is devoted to the chemistry of pyrrole and its derivatives, described the synthesis and the structural, physical, and chemical properties of the 1H-, 2H- and 3H-pyrrole ring systems.

Substituents attached to the π-electron excessive pyrrole ring differ markedly in their properties from those of substituents on other aromatic systems, and this book is the first of a series that will present a survey of the chemistry of the substituents. In some instances, much has been recorded about the physical and chemical characteristics of such substituents; in other cases very little is known or the reports are widely scattered throughout the literature. It is hoped that, by bringing together a review of the chemistry of such substituents, further research may be stimulated.

I am deeply indebted to all of the authors, who have provided the chapters for this book, for their endeavors and for their cooperation in the preparation of the final texts. I must also pay tribute to Professor A. H. (Tony) Jackson, who through his research made a major practical contribution to our present-day understanding of the chemistry of pyrroles. He also gave me significant encouragement in my task as editor and, but for his untimely death, he would have provided authoritative chapters for this and other parts of the series.

Finally, I acknowledge the support and helpful advice of my colleagues at the University of East Anglia and also the support given to me by my wife, without whose continued patience and understanding this book would not have been possible.

R. ALAN JONES

Norwich U.K.
September 1992

Contents

ABBREVIATIONS USED IN THE TEXT, TABLES, AND FORMULAS	xiii
1 ACYLPYRROLES Trevor P. Toube	1
2 VINYLPYRROLES Boris A. Trofimov	131
3 AMINOPYRROLES Girolamo Cirrincione, Anna Maria Almerico, Enrico Aiello, and Gaetano Dattolo	299
4 3-HYDROXYPYRROLES Hamish McNab and Lilian C. Monahan	525
INDEX	617

Abbreviations Used in the Text, Tables, and Formulas

ABN	azobisisobutyronitrile
Ac	acetyl
Am	amyl
ANRORC	assisted nucleophilic ring opening ring closure
Ar	aryl
ax	axial
Bu	butyl
Bt	benztriazol-l-yl
Bz	benzyl
c	cyclo
d	doublet
DABCO	1,4-diazabicyclo [2.2.2]octane
DCC	N,N'-dicyclohexylcarbodiimide
DDQ	2,3-dichloro-5,6-dicyano-1,4-quinone
DMAD	dimethyl acetylenedicarboxylate
DMF	dimethylformamide
DMI	1,3-dimethylimidazolidin-2-one
DMSO	dimethylsulfoxide
DPPA	diphenylphosphoryl azide
EDC	1-ethyl-3-(3-dimethylaminopropyl)carbodiimide hydrochloride
Et	ethyl
eq	equatorial
FVP	flash vacuum pyrolysis
HOMO	highest occupied molecular orbital
HSAB	hard–soft acid–base (theory)
HMPA	hexamethylphosphoric triamide
HMPT	hexamethylphosphorous triamide
i	iso
LCAO SCF	linear combination of atomic orbitals self consistent field.
LUMO	lowest unoccupied molecular orbital
Me	methyl
MNDO	modified neglect of differential overlap (molecular orbital calculations)
n	normal
NBS	N-bromosuccinimide
NCS	N-chlorosuccinimide

NXS	*N*-halosuccinimide
PA	proton affinity
PEG	polyethylene glycol
Ph	phenyl
Pr	propyl
R	alkyl group (unless otherwise stated)
s	singlet
STO	Slater-type orbital
t	tertiary
t	triplet
TEA	triethylamine
Tf	triflate
TFA	trifluoroacetic acid
TFAA	trifluoroacetic acid anhydride
THF	tetrahydrofuran
TMS	trimethylsilyl
Tol	tolyl
Tos	toluenesulfonyl
TOSMIC	*p*-toluenesulfonylmethylisocyanide

PYRROLES

Part Two

This is a part of the forty-eighth volume in the series
THE CHEMISTRY OF HETEROCYCLIC COMPOUNDS

CHAPTER 1

Acylpyrroles

Trevor P. Toube

*Department of Chemistry,
Queen Mary and Westfield College,
Mile End Road,
London E1 4NS U.K.*

1.1. Occurrence	2
1.1.1. From Microbial Sources	2
1.1.1.1. The Lexitropsins	4
1.1.2. From Sponges	6
1.1.3. Higher Plants	8
1.1.4. Other Sources	9
1.1.5. Pharmaceuticals	10
1.2. Reactivity of Acylpyrroles	12
1.3. Reactions of Acyl Groups with Carbon Nucleophiles	13
1.3.1. Reactions with Cyanide Ions	13
1.3.2. Aldol Condensations and Related Reactions	15
1.3.3. Wittig and Related Reactions	24
1.3.4. Knoevenagel and Related Reactions	30
1.3.5. Acylpyrroles as Carbanion Sources	34
1.3.6. The Use of C—C Bond-Forming Reactions in the Preparation of Heterocyclic Rings	36
1.4. Reactions with Nitrogen Nucleophiles	46
1.4.1. Reactions with Amines	46
1.4.2. Reactions with Hydrazines	51
1.4.3. Reactions with Hydroxylamines	53
1.4.4. Amide Formation	54
1.4.5. Other Reactions with Nitrogen Nucleophiles	55
1.4.6. The Use of C—N Bond-Forming Reactions in the Preparation of Heterocyclic Rings	55
1.5. Oxidation	64
1.6. Reduction	67
1.6.1. Reduction Using Lithium Aluminum Hydride	67
1.6.2. Reduction by Boron Hydrides	70
1.6.3. Reduction by Diisobutylaluminum Hydride	72

Pyrroles Part Two: The Synthesis, Reactivity, and Physical Properties of Substituted Pyrroles, Edited by R. Alan Jones.
ISBN 0-471-51306-7 © 1992 John Wiley & Sons, Inc.

1.6.4. Reduction Using Catalysts	73
1.6.5. Other Methods of Reduction	75
1.7. Protecting Groups in Acylpyrroles	76
1.7.1. Protection of C-Acyl Groups	76
1.7.1.1. Aldehydes	76
1.7.1.2. Carboxylic Acids	78
1.7.2. N-Protection	83
1.7.2.1. N-Protection by Acyl Groups	83
1.7.2.2. N-Protection of Acylpyrroles	86
1.8. Decarboxylation and Related Reactions	86
1.8.1. Decarboxylation	87
1.8.2. Cleavage of Aldehyde and Ketone Groups	90
1.9. Cycloaddition Reactions of Acylpyrroles and Their Derivatives	92
1.9.1. Diels–Alder and Other [4 + 2] Cycloadditions	92
1.9.2. Other Cycloadditions	95
1.10. Rearrangement Reactions	96
1.11. Miscellaneous Reactions	99
1.12. Complexes of Acylpyrroles and Their Derivatives	111
1.12.1. Complexes of Acylpyrroles	111
1.12.2. Complexes of Imines	112
1.13. References	115

In Part 1,[1] various methods for the synthesis of acylpyrroles were examined. The main preparative routes to these compounds are total synthesis[1a] or acylation of pyrroles.[1b] The modifications by acyl substituents of the reactivity of the pyrrole ring toward nitration, sulfonation and halogenation,[1c] or oxidation and reduction[1d] are also recorded.

The present chapter begins with a survey of the many acylpyrroles that have been isolated as natural products and then continues with the main classes of reactions of these compounds. The final section deals with the use of acylpyrroles and their derivatives as ligands in the formation of metal complexes.

1.1. OCCURRENCE

1.1.1. From Microbial Sources

Acylpyrroles with structures of various degrees of complexity have been isolated from a variety of microbial sources. Perhaps the simplest is pyrrole-2-carboxylic acid (**1a**) found in cultures of *Marasmiellus ramealis*[2] and *Streptomyces griseoflavus*,[3] among others.[4] The closely related methyl ester **1b** is a trail pheromone for the leaf-cutting ants *Atta texana*[5] and *Acromyrmex octospinosus*.[6] The butterfly pheromone **2** of *Lycorea ceres ceres*[7] has been synthesized.[8] Pyoluterin (**3**),[9] one of a number of halogenated naturally occurring acylpyrroles, has also been made in the laboratory.[10,11]

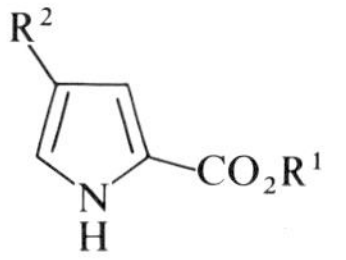

1.1. Occurrence

1
a $R^1 = R^2 = H$
b $R^1 = R^2 = Me$

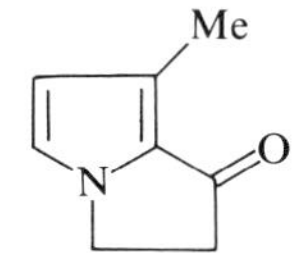

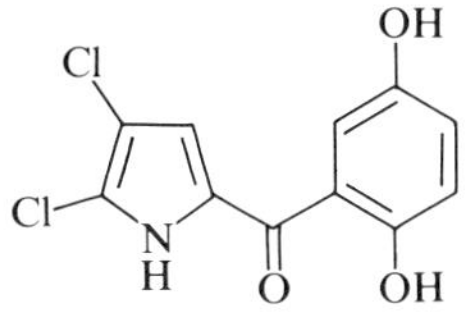

Indanomycin (**4**),[12] which undergoes a number of interesting rearrangements,[13] has a more complex structure, while from *Streptomyces griseovariabilis* come **5a** (chlorbiocin) and **5b**,[14] whereas *S. rishiriensis* produces the related coumermycin **6**.[15] The antibiotic **7** has been isolated from *Nocardia argentinensis*.[16] Calcimycin **8** comes from *S. chartreusensis*,[17,18] and models of the ionophore have been synthesized.[19]

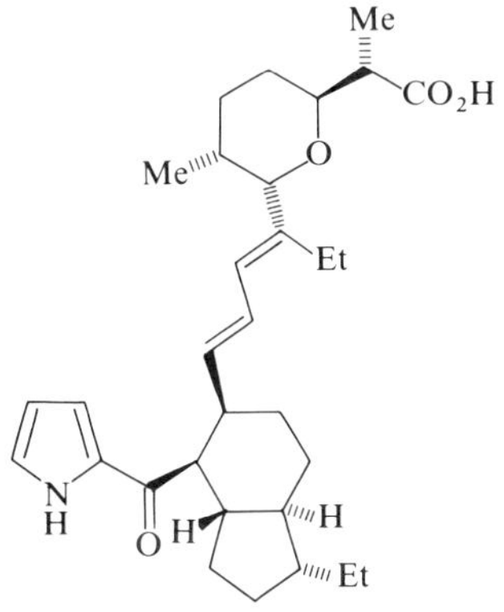

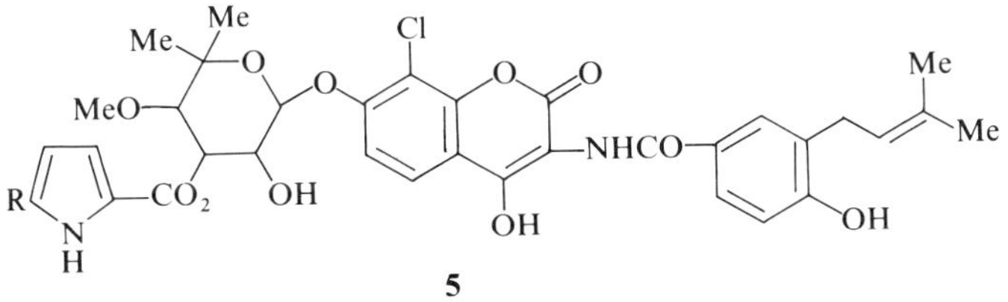

5
a R = H
b R = Me

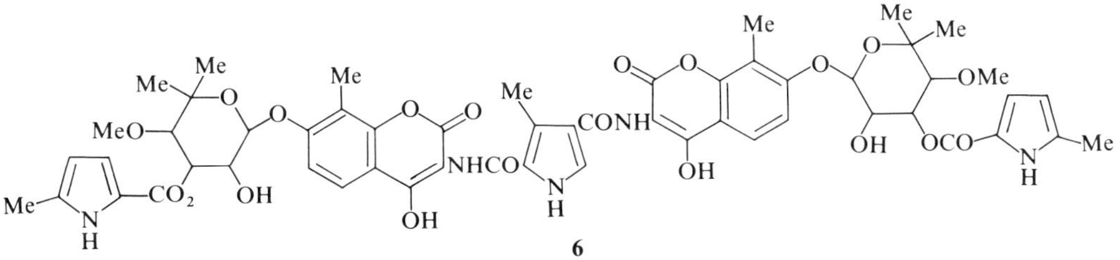

Some of the simple acylpyrroles isolated from microorganisms[20] may be artifacts caused by known rearrangements of β-lactams, for example, the conversions **9** → **10** + **11a**,[21] **12** → **10** + **11b**,[21] and **13** → **14**.[22]

1.1.1.1. The Lexitropsins

This group of antibiotics includes netropsin (also called congocidin) (**15**) from *S. netropsis*,[23] distamycin (**16**) from *S. distallicus*,[24] kikumycin B (**17a**),[25] and anthelvencin A (**17b**).[26] These compounds bind very specifically to an AT-rich minor groove in deoxyribonucleic acid (DNA) and have therefore been much studied. Syntheses following similar routes have been reported for netropsin[27] and distamycin,[27,28] as well as for anthelvencin (and its enantiomer)[29] and dihydrokikumycin.[30] A large number of derivatives and analogs have also been prepared,[31-48] usually by the route described above, in conjunction with standard peptide procedures.

1.1. Occurrence

Among the most interesting analogs that have been synthesized are the 2,5-linked isomers **18**[42] and the dimeric antibiotics **19**,[49] **20**,[50] and **21**.[51] These compounds do not appear to bind as strongly to DNA as do (**15**) and (**16**). An iron(II) complex of distamycin–ethylenediaminetetracetic acid has been made and it has been shown to cleave DNA sequence-specifically.[52]

1.1.2. From Sponges

Marine sponges have proved a fruitful source of acylpyrroles, many of them bearing one or more bromine atoms on the heterocyclic ring. A relatively simple group are the compounds **22**, isolated from *Oscarella lobularis*.[53] Similar substances have been found in *Laxosuberites*,[54] namely, **23**, **24**, and **25**. The metabolite **26** from the sponge *Reniera*[55] had its proposed structure confirmed by synthesis from dimethyl 1-methylpyrrole-3,4-dicarboxylate (**27**) (see Section 1.3.2).

1.1. Occurrence

From *Pseudodaxinyssa cantharella*, a group of carbonyl-containing pyrrolic metabolites **28–32** have been isolated.[56] Several of these compounds are known from other sources; **28a** and **28b** have been found together with **33a** and **33b**,[57] and **30** is found with **33c**.[58] The marine sponge *Agelas flabelliformis* produces **33c**.[59] *Agelas* was the source from which oroidine (**32a**) was first isolated,[60] and *A. oroides* has also been shown to yield **33c**.[61] *Agelas sceptrum* produces oroidine (**32a**), together with a cyclobutane dimeric analog, sceptrin (**34**).[62] The related midpacamides **35a** and **35b** have been obtained[63] from *A. mauritiana*, together with the *N*-methyl derivative of **33b**. *Hymeniacidon* yields **32b**,[64] and another *Agelas* species produces **36**.[65] A further *Agelas* species yields ageline B (**37**).[66]

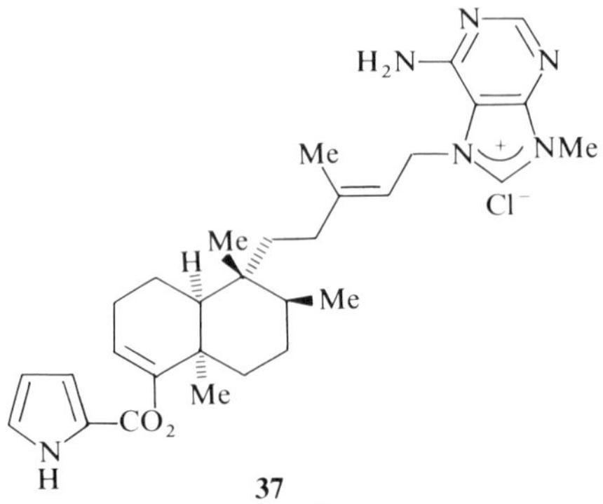

37
ageline B

Oroidine (**32a**) has been synthesized[67] from methyl 4,5-dibromopyrrole-2-carboxylate (**33b**), and dihydrooroidine is an intermediate in the synthesis[68] of yet another sponge metabolite, dibromophakellin (**38**), isolated from *Phakellia flabellata*.[69] The ester **39**, obtained from *Pseudostellaria heterophylla*, has been synthesized.[70]

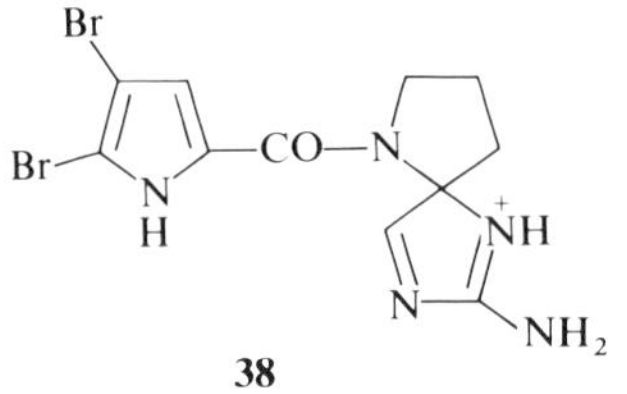

38 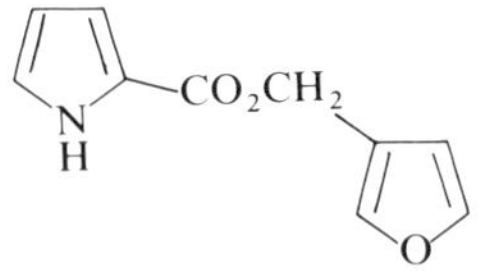   **39**

The soft corals are related to the marine sponges and produce similar pyrrolic compounds. Thus, the formylpyrrole **40**, a close analog of **23**, has been isolated from *Telesto* and its structure has been confirmed by synthesis.[71]

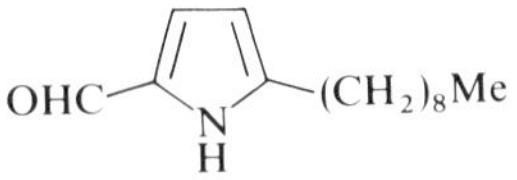

40

1.1.3. Higher Plants

Relatively few acylpyrroles are reported to have been isolated from higher plants. One example is the alkaloid, funebral (**41**),[72] which has been found in *Quararibea funebris*. There is speculation that (**41**) probably lies on the biosynthetic pathway to the related imine, funebrine (**42**).[73] *Achillea ageratifolia* produces a series on N-acylpyrroles **43–47**, together with the corresponding pyrrolidines.[74] The N-pyrrolyl lactone **48** has been shown to be involved in cell cycle control in *Pisum sativum*.[75]

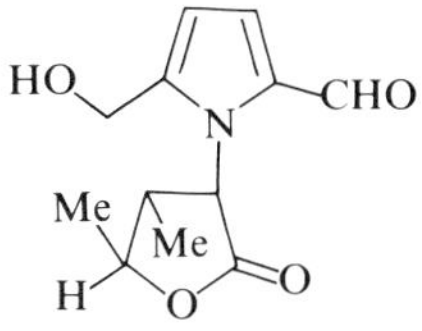

41
funebral

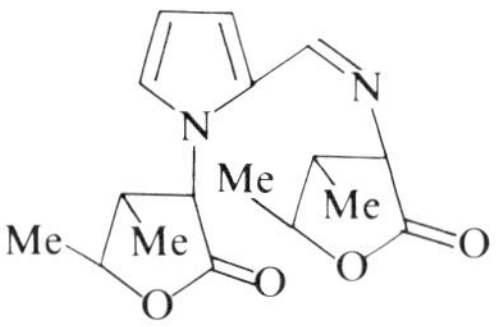

42
funebrine

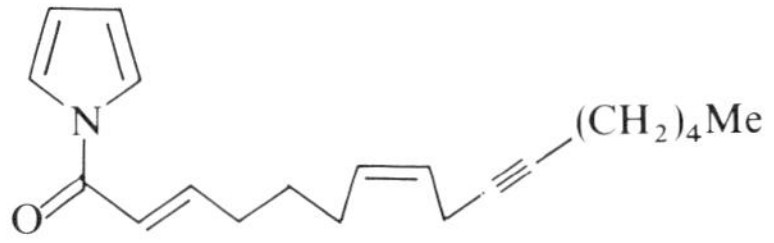

43

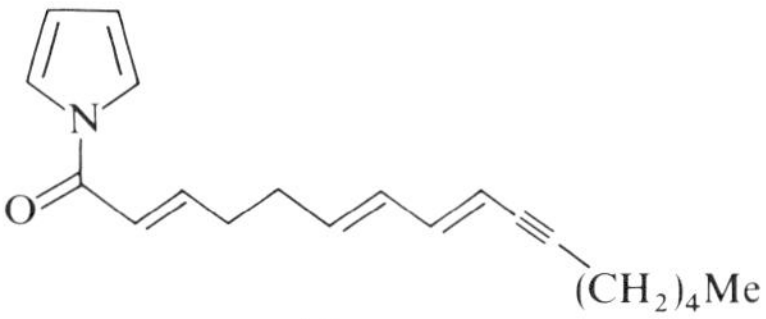

44

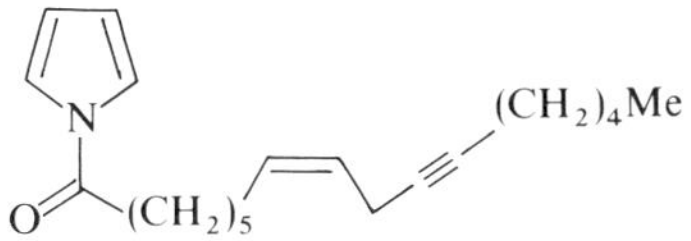

45

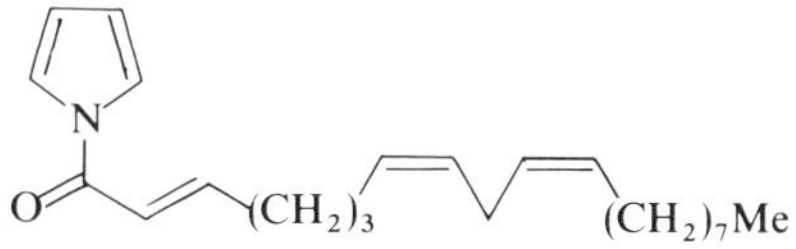

46

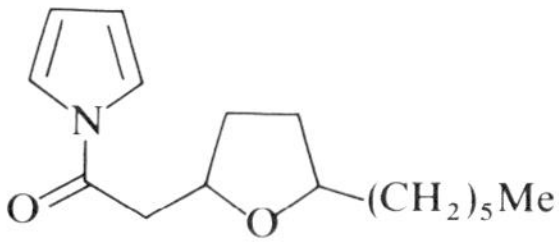

47

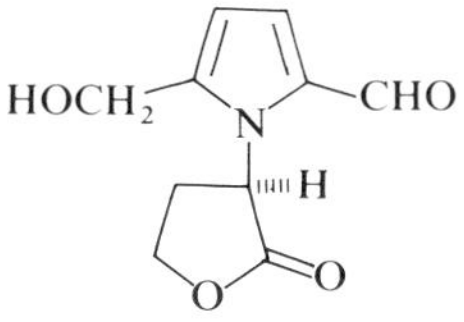

48

1.1.4. Other Sources

Formylpyrroles and other simple acylpyrroles are frequently detected in analyses of the flavor and aroma components of foods as diverse as coffee, soya sauce, and roasted meats. They are probably the products of Maillard reactions; for example, D-glucose and glycine, when heated at 100°C, give 2-formyl-4-methylpyrrole (**49**) and 2-acetylpyrrole (**50a**), together with other products, the yield depending on the pH and reaction times,[76] whereas D-glucose and lysine

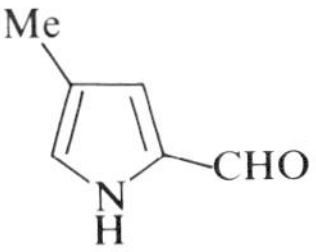

49

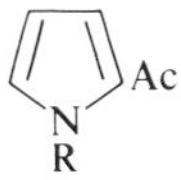

50
a R = H
b R = Et

produce **51** as a Maillard product;[77,78] the structure is established by synthesis.[79] In the presence of sodium carbonate, the Maillard reaction of D-glucose with lysine gives **50a** and **52a** among other products;[80] the corresponding reaction with L-theanine yields **50b** and **52b**.[81] When the aminosugar **53** is heated with base, the pyrrole **54** is obtained in 45% yield.[82]

1.1.5. Pharmaceuticals

A number of acylpyrroles have found applications in the pharmaceutical area. Prominent among these are tolmetin (**55a**)[83] and zomepirac (**55b**). Many analogs, essentially 5-acylpyrrole-2-acetic acid derivatives, have been prepared and patented for their properties as analgesics and antiinflammatories.[84-87] Some derivatives have additional substituents on the ring, notably ethoxycarbonyl groups at C-3[88] and alkoxy[89] or trifluoromethyl[88] groups at C-4.

The amides **56a**,[90] **56b**,[90] and **56c**[91] have also been prepared for pharmacological study, as have the pyrrolyldiketones **57**.[92]

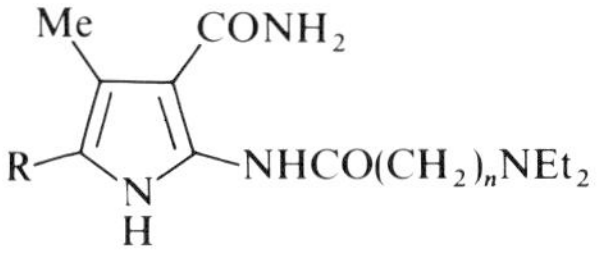

1.1. Occurrence

56
a R = Me, n = 2
b R = Et, n = 2
c R = PhCH$_2$, n = 1

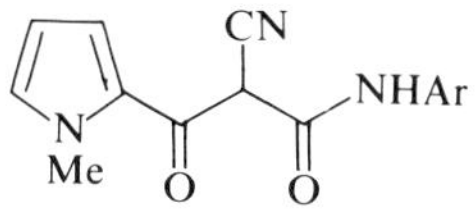

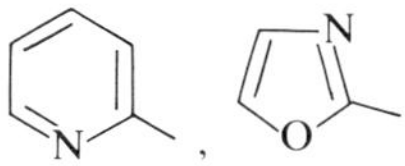

A number of pyrroles with sugar groups (or their analogs) as substituents have been prepared as model pyrrole C-nucleosides. Examples include **58**,[93] **59**,[94] **60** and **61**,[95] **62**,[96] **63**,[97] **64**,[98] **65** and **66**,[99] and **67**.[100] The X-ray structure of **68**[98] has been determined.[101] The reaction of a benzylaminopentalose with dimethyl acetylenedicarboxylate gives **69**.[102] Periodate cleavage[103] of a series of nitropyrrole sugars yields the aldehydes **70** and **71**.[104]

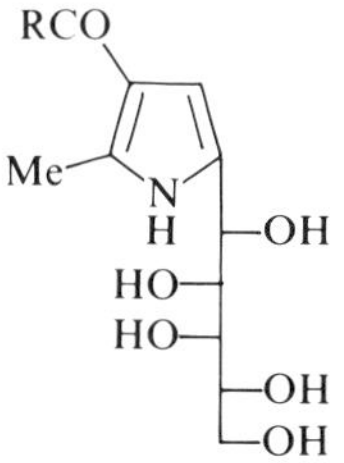

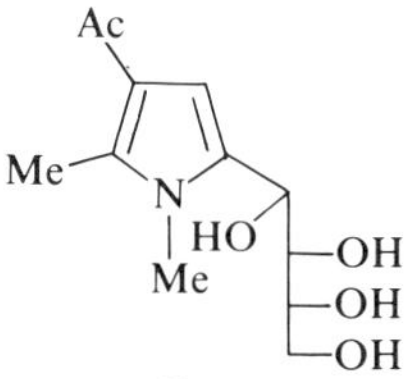

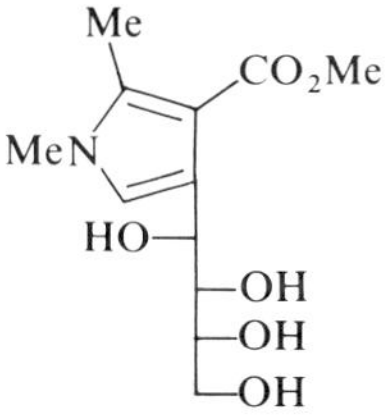

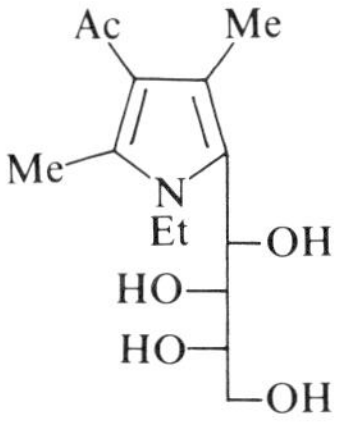

58 R = Me, Ph

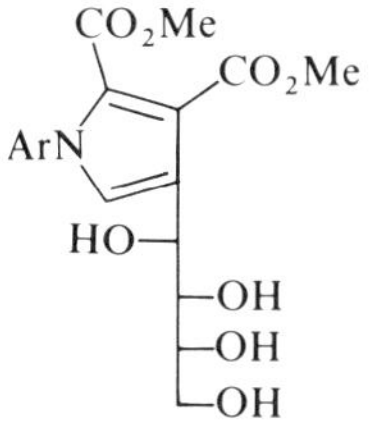

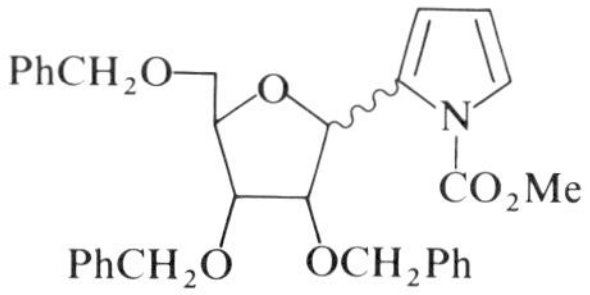

63 23% β, 3% α

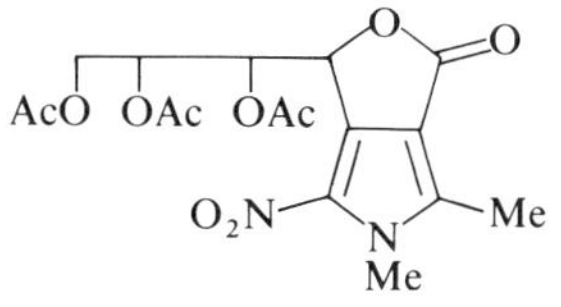

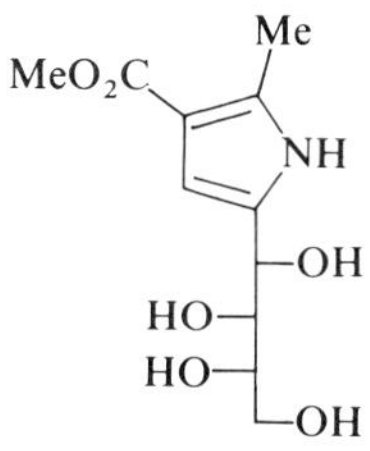

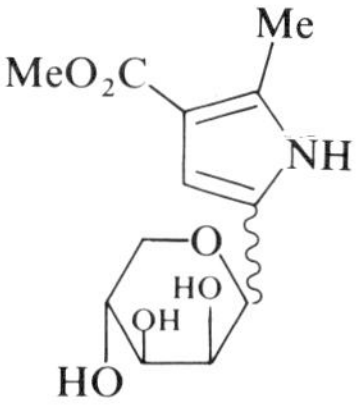

In a related area, the series of 1-(2-pyrrolyl)isoquinoline-2-oxides **72** has been patented as herbicides.[105] The acylpyrroles **73** have been patented as memory-improving drugs.[106]

1.2. REACTIVITY OF ACYLPYRROLES

Pyrrole aldehydes and ketones are generally much less reactive toward nucleophiles than many other aromatic carbonyl compounds.[107–109] It is now accepted[107,110–112] that this low reactivity results from the contribution of the zwitterionic canonical form **74b** to the resonance hybrid of 2-acylpyrroles **74a**, which may thus be regarded as being to some extent vinylogous amides. Much of the evidence for this contention comes from spectroscopic evidence (see Part 1).[1] Similar canonical forms may be invoked to explain the poor reactivity of acyl groups at C-3 on the pyrrole ring.

R = H, Me; R' = H, alkyl

The contribution of the canonical form **74b** also provides an explanation of the relatively high barriers to rotation about the formal ring–aldehyde C—C single bond in 2-formylpyrroles.[107] Evidence for the preferred *syn*-conformation[107] **75** in a range of 3-substituted 2-formylpyrroles (R = H, Cl, OMe, OEt, Me, CN, CO$_2$Me, NO$_2$; R^1 = H, Me, Et) has been provided by the presence of $^5J_{5H,CHO}$ coupling in their ^{1}H NMR spectra,[113–118] although attempts by some authors[115,116] to calculate the *syn* : *anti* ratios of the two rotamers from the absolute size of these coupling constants have been queried[118] in the light of the lack of agreement in the literature over the numerical values of the coupling constants in question.

The exceptionally low reactivity of aldehyde groups adjacent to other acyl functionalities on the pyrrole ring has been ascribed[119] to steric factors, again based on NMR evidence. Dipole moments and infrared spectroscopic measurements have been interpreted as indicating that in the 2-chloro-3-formylpyrrole **76** the oxygen atom of the formyl group is *trans* to chlorine, as shown.[120]

1.3. REACTIONS OF ACYL GROUPS WITH CARBON NUCLEOPHILES

It is perhaps an oversimplification of the art of organic synthesis to claim that at its heart lies the process of attaching one carbon atom to another. It is, however, clear that the formation of new carbon–carbon bonds is a prerequisite for the elaboration of carbon skeleta of increasing complexity. Acyl groups are widely used by organic chemists to expedite such reactions, either by their reactions with carbon nucleophiles or by their ability to stabilize nucleophilic centres on adjacent atoms.

1.3.1. Reactions with Cyanide Ions

2-Formylpyrrole (**77a**) is reported[107] not to form a cyanhydrin. However, ethyl 3-formylpyrrole-1-carboxylate (**78**)[121] does give the acylated cyanhydrin **79** with potassium cyanide and acetic anhydride,[122] as an oil, in 20% yield. Cyanhydrins protected as their silyl ethers, **81a** and **81b**, have been prepared in yields of 94–97% from the *N*-protected aldehydes **80a** and **80b**, respectively, by the action of *tert*-butyldimethylsilyl chloride and either potassium cyanide and zinc iodide in acetonitrile[123] or lithium cyanide in tetrahydrofuran.[124] The compounds **81a** and **81b** are, of course, umpole synthons of the corresponding aldehydes and are therefore of considerable interest.

Cyanide ions are also used generally to initiate the benzoin condensation. 2-Formylpyrrole (**77a**), however, fails to undergo a self-condensation of this type.[107] This failure cannot be ascribed to the acidity of the imine proton, which might be thought to inhibit the formation of **82**, which is also an intermediate in cyanhydrin formation, as the reaction also fails in the case of both 2-formyl-1-methylpyrrole (**77b**)[112] and 3-formyl-1-methylpyrrole (**83**).[125] However, the N-substituted 3-carbethoxy-2-formylpyrroles **84** do undergo a benzoin-type self-condensation, albeit in low yield, to give the acyloins **85**,[126] which are readily oxidized to **86** and can undergo cyclization to give **87**. It is generally accepted that electron-withdrawing substituents attached to the ring increase the susceptibility of acyl groups on pyrroles toward nucleophilic attack, and this may explain the success of the reaction in this case.

2-Formylpyrrole (**77a**) reacts with *p*-anisaldehyde (**88**) in a mixed benzoin condensation to form the acyloin **89**.[122] From the known mechanism of the benzoin condensation and the fact that the carbonyl group in **89** is attached to the heterocyclic ring, not the benzene moiety, it might be concluded that **82** must have been an intermediate in the reaction and that the stability of **82** is therefore greater than has been assumed above. However, **89** may arise tautomerically from its isomer having the carbonyl group conjugated with the benzene ring via a route that does not involve **82**.

It is clear, therefore, that the factors affecting the success or failure of these (and other) reactions of acylpyrroles are finely balanced, and it is not always easy to predict the outcome.

1.3.2. Aldol Condensations and Related Reactions

Reactions in which carbanions attack aldehyde or ketone substituents on pyrrole are well established and, predictably, the formyl group is generally much more reactive than acetyl functionalities.[107]

The 2-formylpyrroles, **77a** and **77b**, react with methyl ketones **90** to yield enones **91** (R^1 = H or Me, R^2 = Me or Ph;[112] R^1 = H, R^2 = Me, Et or Ph,[128,129] or *p*-ClC$_6$H$_4$,[130] or *p*-MeC$_6$H$_4$ or heteroaryl). With acetone, **77a** can also give the double-condensation product **92**.

Condensation of formylpyrroles with carbanions is a general reaction. Thus, **77a** undergoes what is essentially a Stobbe reaction with dimethyl succinate **93** to give **94**,[131] and **77b** reacts in the same way with dimethyl succinate or glutarate to yield **95** (n = 1 or 2).[132,133] Substitution of the ring by electron-withdrawing groups should increase the reactivity of the formyl group toward carbanions, and, predictably, **96** reacts readily with a variety of methyl ketones to produce the enones **97**.[134]

Doebner condensations have been reported for pyrroles bearing nitro groups. Thus, 2-formyl-5-nitropyrrole (**98**) gives the acrylic acid **99**;[135] the 4-nitro analog **100** yields the corresponding product **101**.[136]

The sodium hydride-catalyzed reaction of 3,4-dialdehydes **102** with cyclohexane-1,4-dione in benzene gives the condensation product **103** after 3 h.

1.3. Reactions of Acyl Groups with Carbon Nucleophiles

Prolonged reaction time leads to condensation with a second molecule of the aldehyde to yield the pentacyclic product **104**.[137]

The carbanion derived from nitromethane reacts with 2-formyl-1-methylpyrrole (**77b**) to give the nitro compound **105** (m.p. 102°C) in 72% yield.[138] This reaction is again an entirely general one for formylpyrroles. Thus, the highly substituted aldehydes **106**–**110** all give the corresponding vinyl derivatives **111**,[139] **112**,[139] **113**,[139] **114**,[140] and **115**,[141] whereas the dipyrrolylmethane **116** undergoes condensation at both formyl groups to yield **117**.[142]

Carbanions stabilized by nitrile groups also condense with 2-formylpyrroles; the reaction of **77a** or **77b** with phenylacetonitrile produces **118a** and **118b**, respectively.[112] The rate of the reaction is greater for the 1-methyl derivative, but is significantly slower for both of these compounds, compared with the corresponding reactions of a range of other aryl and heteroaryl aldehydes.[143]

There appears to be only one example of the pyrrole analog of the Perkin reaction,[107] and it involves the condensation of 2-formylpyrrole (**77a**) with 2-pyrrolylacetic acid (**119**) in the presence of triethylamine to yield, after treatment with acidified methanol, the ester **120** (5%).[144] The reaction also produces a large amount of tarry material from which the free acid cannot be isolated.

Condensations between 2-formylpyrrole (**77a**) and a range of oxo derivatives of heterocyclic systems have been reported. With oxazolone **121**, **77a** gives the product **122**, together with its geometric isomer **123**.[145] Reaction with rhodamine (**124**) yields **125**,[146] or **126**,[146,147] while, with hydantoin (**127**), the product is **128**.[146,148] These latter reactions were undertaken as part of vain attempts to synthesize β-(2-pyrrolyl)alanine.[107]

1.3. Reactions of Acyl Groups with Carbon Nucleophiles

The cyclopentadienyl anion reacts in a similar fashion with formylpyrroles. Sodium cyclopentadienyl (**129**) and 2-formylpyrrole (**77a**) yield the cyclopentadiene derivative **130**.[149] The analogous reaction of the aldehyde **107**, which is more reactive than **77a** because of the presence of an ethoxycarbonyl group, with ferrocene (**131**) produces the iron complex **132**.[150]

An extremely important reaction in the synthesis of porphyrins and other linear and cyclic tetrapyrrolic systems is the condensation of 2-formylpyrroles with pyrroles having an unsubstituted 2-position. The first reported reaction of this type was the formation of the dipyrrolylmethene **135** by the condensation between the aldehyde **133** and 2,4-dimethylpyrrole (**134**).[151] This reaction is again a general one. Thus, for example, the aldehyde **110**, activated by the presence of two ethoxycarbonyl groups, condenses readily with ethyl 2,4-dimethylpyrrole-3-carboxylate (**136**) to generate the dipyrrolylmethene **137**.[152]

The reaction has been used in the synthesis of analogs of prodigiosin (**138**).[153] The aldehyde **139** combines with the dipyrrole **140** in the presence of phosphorus oxychloride to produce **141**, which has the tripyrrolic skeleton of prodigiosin, in 59% yield.[154] A similar route to prodigiosin analogs involves the reaction of aldehydes **142** with the dipyrrole **143**. In these cases, it is necessary to ensure that the formyl group reacts with the pyrrole ring in **143** which is unsubstituted on the nitrogen atom, rather than with the *N*-methylated ring to give **144** as the product.[155] The synthetic strategy can be reversed, however, to avoid the danger of ambiguities in the regiochemistry of the reaction, by using the dipyrrole aldehyde **145**, which is then condensed with the 2-unsubstituted pyrroles **146** to yield **144**.[155] The aldehyde **145** was prepared from the dipyrrole **143** by the standard Vilsmeier–Haack procedure (phosphorus oxychloride and dimethylformamide),[113] which in this case gave a 3 : 1 predominance of the desired product, **145** (m.p. 96–98° C), over the isomer (m.p. 90–92°C) bearing the formyl group at the corresponding position of the *N*-methylated ring, **147**.[155]

2-Formylpyrrole (**77a**) reacts with a number of carbanions or reagents equivalent to carbanions. Piperidine catalyzes the condensation with the pyridinium salt **148** to give the pyridinium analog **149** of a styrylpyrrole.[156,157]

2-Formyl-1-methylpyrrole (**77b**) reacts, after activation by treatment with trimethylsilyl trifluoromethylsulfonate, with the silyl enol ether **150** to give the 2,3-bis(trimethylsiloxy)carboxylate **151** in 90% yield, as a 1 : 1 mixture of the *erythro* and *threo* stereoisomers.[158] 2-Formylpyrrole (**77a**) reacts very slowly with the chromium carbonyl complex **152** in the presence of triethylamine and trimethylsilyl chloride to give the condensation product **153** (m.p. 96°C) in 12% yield.[159]

The carbanion, formed from the cysteine derivative **154** by treatment with lithium *tert*-butoxide, adds to 2-formyl-1-methylpyrrole (**77b**) to give the pyrrolylcarbinol **155** (m.p. 106–107°C) in 55% yield.[160]

Tandem one-pot reductive alkylations of pyrrolyl aldehydes and ketones have been reported. The carbonyl compound **156** was reacted with an aryl metallic reagent **157** to produce an intermediate alkoxide **158**, which is then treated with lithium in liquid ammonia in the presence of ammonium chloride to yield the benzylpyrrole **159**.[161,162]

Acylpyrroles

R¹	R²	R³	M	% yield of 159
H	H	p-OMe	Li	98
H	H	H	Li	99
H	H	p-Me	Li	97
H	H	m-OMe	Li	57
H	H	o-OMe	Li	99
H	Me	p-OMe	Li	75
H	Ph	p-OMe	Li	92
H	Ph	p-OMe	MgBr	15
Me	H	p-OMe	MgBr	84
PhCH₂	H	p-OMe	MgBr	83
MeO(CH₂)₂OCH₂	H	p-OMe	MgBr	67
MeO(CH₂)₂OCH₂	H	p-Me	MgBr	72

Although Grignard reagents, which can be regarded as ambient carbanions, generally react with N-unprotected pyrroles in an exchange reaction to yield the 1-pyrrolylmagnesium halide, it is reported that 1-(2-pyrrolyl)propan-1-one (160) reacts with excess ethyl magnesium bromide to give the carbinol 161, while the aldehyde 162 yields the alcohol 163 with excess methyl magnesium bromide.[107] With N-protected acylpyrroles, however, the reaction is somewhat less problematical. The cyclic ketone 164 reacts with methyl magnesium bromide to yield an unstable alcohol, which rapidly dehydrates to the dihydroindole 165.[163] Not unexpectedly, in the case of methyl 2-formylpyrrole-1-carboxylate (166), the reaction between the formyl group and the Grignard reagent 167, formed from the bromoacetal, is faster than the reaction with the ester group to a sufficient extent for the carbinol 168 to be isolated in satisfactory yield.[164] The authors claim this reaction to be general, leading to a range of precursors to substituted indoles via cleavage of the cyclic acetal, followed by cyclization onto the 3-position of the pyrrole ring.

1.3. Reactions of Acyl Groups with Carbon Nucleophiles

The metabolite **26** from the sponge, *Reniera*, has been synthesized using the carbanion generated from the dithianyl derivative of propanal **169**. Treatment of the pyrrole diester **27** with one equivalent of this reagent produces the dithian **170** (m.p. 95–96°C) in 27% yield. Cleavage of the protecting group to give the dione **171** is almost quantitative and, when followed by a sodium hydride-catalyzed intramolecular condensation analogous to the Dieckmann reaction, produces the bicyclic ketone **172** (m.p. 49–50°C) in 30% yield. Methylation with diazomethane gives the natural product **26** in 40% yield.[55]

The Reformatzky reaction may be considered as being closely related to Grignard reactions. Reaction of the *N*-protected ketone **173** with ethyl bromoacetate and zinc gives the condensation product **174** in over 90% yield.[163] The reaction fails with the *N*-unprotected analog.

1.3.3. Wittig and Related Reactions

The Wittig reaction involves the attack on an aldehyde or ketone of the formal carbanion adjacent to a positively charged phosphonium cation, which is sometimes drawn as a carbon–phosphorus double bond and the reaction as a [2 + 2] cycloaddition to the carbonyl group. When the formal carbon–phosphorus double bond of the Wittig reagent is conjugated, as, for example, to a carbonyl function, relatively weak bases can be used in its generation from the phosphonium salt, and the resulting ylid is often stable enough to be isolated; ylids that are not stabilized in this fashion are generally generated and used in situ and require stronger bases, such as alkyl or aryl lithiums, for their formation, but also tend to be more reactive than their stabilized analogs.

A common variant of the Wittig reaction (the Wadsworth–Emmons or Horner reaction) involves the use of phosphonate esters instead of the phosphonium salts as starting materials. In these cases, weaker bases usually suffice for the generation of the ylids.

Aldehyde groups attached to π-excessive aromatic rings would be expected to be relatively unreactive in the Wittig reaction. However, both Wittig and Horner reactions involving both stabilized and more reactive ylids have proved widely successful with both 2-formylpyrrole (**77a**) and its *N*-methyl analog (**77b**).[165] Condensation of the appropriate ylids **175a–175c** with (**77a**) gave the appropriate alkenes **176a–176c**; with (**77b**), the products were the corresponding *N*-methylated derivatives **176d–176f**.[165]

$$Ph_3\overset{+}{P}CHR + 77 \longrightarrow$$

175
a R = Ph
b R = Ac
c R = CO$_2$Et

176
a R^1 = H, R^2 = Ph
b R^1 = H, R^2 = Ac
c R^1 = H, R^2 = CO$_2$Et
d R^1 = Me, R^2 = Ph
e R^1 = Me, R^2 = Ac
f R^1 = Me, R^2 = CO$_2$Et

The acrylic ester **176c** has also been prepared by a number of groups.[166–169] The *E*-isomers are usually the major or exclusive products of these reactions, but *Z*-**176c** has also been isolated.[166,168,169] This reaction is sensitive to pressure, the yield from the reaction in dichloromethane at 1 atm. being only 12%, whereas, at 10 kbar, 63% of the product is isolated.[170] The reaction is also more effective at high temperatures.[168] In the same fashion, the ester **178** has been prepared using the ylid **177**.[167] A series of substituted styrylpyrroles, **180a–180l** have been synthesized from the appropriate ylids **179a–179f**.[171] The products **176a–176f** can also be prepared via the diethyl alkylphosphonates **181a–181c**

1.3. Reactions of Acyl Groups with Carbon Nucleophiles 25

using the Horner procedure.[165,172-174] The reaction with **181c** is sensitive to reaction conditions; the Horner reaction of **181a** with **77b** at 70°C under phase-transfer conditions using C200 in dioxan gives **176f** in 42% yield.[173] On the other hand, the corresponding reaction using 2-formylpyrrole (**77a**) gives no product at all; nor does the reaction in water and dioxan using potassium carbonate as base, but replacement of this salt by cesium carbonate yields 86% of **176c** entirely as the E-isomer.[174] The Wittig reaction also proceeds satisfactorily with formylpyrroles bearing ring substituents; for example, the aldehyde **182**, on treatment with the phosphorane **175c**, gives the acrylate **183**.[175]

$$77a + Ph_3\overset{+}{P}\underset{Ph}{\overset{-}{C}}CO_2Et \longrightarrow 178$$

177

a	$R^1 = H$, $R^2 = OMe$
b	$R^1 = H$, $R^2 = Me$
c	$R^1 = H$, $R^2 = F$
d	$R^1 = H$, $R^2 = Cl$
e	$R^1 = H$, $R^2 = Br$
f	$R^1 = H$, $R^2 = NO_2$
g	$R^1 = Me$, $R^2 = OMe$
h	$R^1 = Me$, $R^2 = Me$
i	$R^1 = Me$, $R^2 = F$
j	$R^1 = Me$, $R^2 = Cl$
k	$R^1 = Me$, $R^2 = Br$
l	$R^1 = Me$, $R^2 = NO_2$

$$77 + Ph_3\overset{+}{P}\overset{-}{C}H\text{–}C_6H_4\text{–}R \longrightarrow 180$$

179
a R = OMe
b R = Me
c R = F
d R = Cl
e R = Br
f R = NO_2

$$(EtO)_2\overset{O}{P}CH_2R$$

181
a R = Ph
b R = Ac
c R = CO_2Et

182 → **183**

2-Formylpyrrole (**77a**), or its sodium salt,[176] react with the ylid **184** to give the dienoate **185**,[167,168,176,177] which can be cyclized to yield the 3H-pyrrolizine **186**.[167] The 3-substituted dienoates **185b**[177] and **185c–185f**[178,179] have also been prepared by either the Wittig or the Horner route (in some case, by

$$Ph_3\overset{+}{P}\overset{-}{C}HCH=CHCO_2Et$$

184

185
a R = H   d R = OMe
b R = Cl   e R = CN
c R = Me   f R = CO_2Et

both reactions), as have the corresponding acrylic esters.[177-179] The 5-nitro-2-pyrrolyldienoate **187** has been synthesized by the Wittig reaction from the corresponding formylpyrrole **98**.[135]

186

187

Reaction of the appropriate pyrrole aldehydes with the ylid generated by treating methyltriphenylphosphonium bromide with sodium hydride yields the vinylpyrroles **188** and **189**.[180] 2-Formylpyrrole (**77a**) reacts with the vinylphosphonium salt **190a** to give the 3H-pyrrolizine **191a**.[181] With the ylid **192**, the dimethylpyrrolizine **191b** is formed;[182] mixtures of the other pyrrolizines result from the cyclisation of the products of the reaction of **77a** with the ylids from allylphosphonium salts **193**.[171] The corresponding reaction of the vinylphosphonium salt **190b** yields **191c**,[183] while the cyclopropylphosphonium salt **194** on the other hand, gives the dihydroindolizine **195**.[183]

188
a R = CHO
b R = Ac
c R = CO$_2$Me

189
a R = CHO
b R = CO$_2$Me

190
a R = H
b R = Me

191
a R^1 = R^2 = H
b R^1 = R^2 = Me
c R^1 = Me, R^2 = H

192

193
a R^1 = R^2 = H
b R^1 = H, R^2 = CO$_2$R
c R^1 = Me, R^2 = H
d R^1 = Me, R^2 = CO$_2$R

194

195

The reaction between 2-formylpyrrole (**77a**) and the ylid **196** produces not only the acrolein derivative **197a**, but also the dienal **198a**, which results from further condensation with **197a**.[177-179] The enals **197b** (X = Cl,[178] Me, OMe, CN, or CO$_2$Et[179]) and the dienals **198b** (X = Cl,[178] Me, OMe, CN, or

1.3. Reactions of Acyl Groups with Carbon Nucleophiles

CO_2Et^{179}) have been synthesized in the same way. The reaction of 2-formyl-1-methylpyrrole (**77b**) with the protected phosphonium salt **199** in the presence of sodium ethoxide gives, after cleavage of the acetal under acidic conditions, the pure *E*-isomer of the acrolein **200** in 60% yield.[184]

201
$R^1 = H, Cl, OMe, Me, CN, CO_2H$
$R^2 = H, Me$
$R^3 = CO_2R, CN, CH=CHCO_2R$

Reactions between the appropriate 3-substituted 2-formylpyrroles and the relevant Wittig or Horner reagents produce the substituted alkenes **201**, together in some cases with their geometric isomers.[177–179]

The products of the reaction of a series of *N*-substituted 2-formylpyrroles with **175a** are shown in Table 1.1.[185] The results of the Wittig reactions of the same aldehydes with a range of heteroaryl phosphonium salts are given in Table 1.2. The ethene (R = Me, Y = S) has also been prepared by the Horner reaction between (**77b**) and the phosphonate **202**; only the *E*-isomer was obtained.[186] The same phosphonate reacts with 2,5-diformyl-1-methylpyrrole (**203a**) to give the bis-adduct **204**.[186] The latter compound and the corresponding tripyrrole **205**, prepared by a similar route, act as semiconductors when doped with iodine and TCNQ.[186]

TABLE 1.1. 1-SUBSTITUTED 2-STYRYLPYRROLES OBTAINED FROM
1-SUBSTITUTED 2-FORMYLPYRROLES[185]

1-Substituent	Yield (%)	(E) : (Z) Ratio	m.p. [b.p.]
H	23–77	—[a]	141–142°C [141–144°C, 1.5 torr][165]
Me	85	3 : 1	71–73°C
NMe_2	66	1 : —	102–103°C
$PhSO_2$	65	1 : —	75–77°C
CH_2OMe	73	1 : —	36–37°C

[a] Dependent on solvent used.

TABLE 1.2. THE WITTIG REACTION BETWEEN HETEROARYL PHOSPHONIUM SALTS AND 2-FORMYLPYRROLES

R	Y	Yield (%)	(E) : (Z) Ratio	m.p. (E)	m.p. (Z)
H	NMe	39	1.2 : 1	118–120°C	Oil
Me	NMe	76	14 : 1	149–150°C	Oil
$PhSO_2$	NMe	80	1 : —	138–141°C	—
NMe	NMe	51	1 : —	(b.p. 115/0.1 torr)	—
H	NCH_2OMe	81	6 : 1	78°C	(b.p. 100°C, 0.1 torr)
H	O	62	2.7 : 1	104°C	72–73°C
H	S	50	2.3 : 1	141–144°C	57–60°C
Me	O	86–90[a]	1 : —	34–36°C	—
Me	S	45–90[a]	1 : —	65–66°C	—
$PhSO_2$	S	53	1 : —	61–62°C	—

[a] Dependent on solvent used.

As part of a synthesis of the antitumor agent CC-1065, the dipyrrolylethene **208** was prepared by the reaction of ethyl 2-formylpyrrole-5-carboxylate **206** with the phosphonium salt **207** using sodium hydride or potassium tert-butoxide as base.[187] In contrast, the condensation of the 2,5-dialdehyde **203a**

1.3. Reactions of Acyl Groups with Carbon Nucleophiles

with the phosphonium salt **209**, using butyllithium as base, was reported to give both the E- and Z-isomers of the retinal analog **210**.[188] 2-Formylpyrrole (**77a**) has also been converted into the prostaglandin analog **211**.[189]

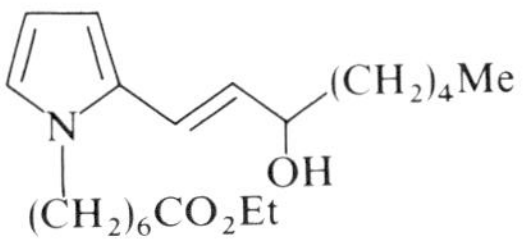

211

Aldehyde groups at the 3-position of the pyrrole ring also react with ylids (e.g., in the preparation of **189**[180]). 3-Formyl-1-methylpyrrole (**212a**), on treatment with **175a**, yields the styryl derivative **213** (m.p. 95°C) in 13% yield.[125] In the course of a synthesis of the antitumor agent CC-1065, the aldehyde **212b** was induced to undergo a Horner reaction with the carbanion generated with sodium hydride from the phosphonate **214** to produce the acrylate **215**, which was subsequently treated with TOSMIC and hexamethyldisilazane to give the dipyrrole **216** (m.p., 152–154°C) in 30% yield.[190] In the synthesis of the coenzyme, methoxatin, 3-formylpyrrole-5-carboxylic ester **217** (R = Me or Et) was reacted with the phosphonium salt **218** and sodium hydride to give the condensation product **219** in 84% yield predominantly (95%) as the E-isomer.[191, 192]

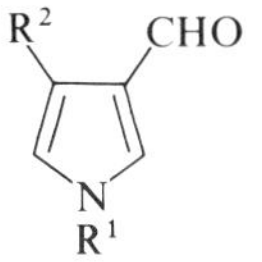

212
a $R^1 = Me, R^2 = H$
b $R^1 = Tos, R^2 = CO_2Me$
c $R^1 = R^2 = H$

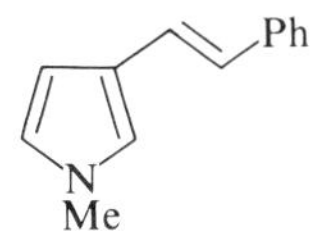

213

214

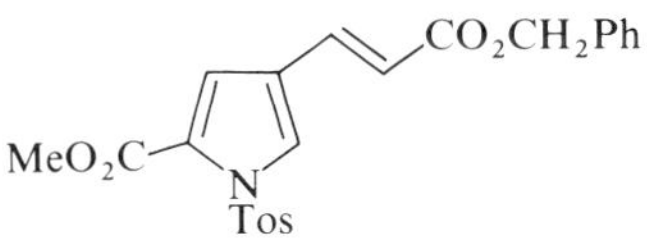

215

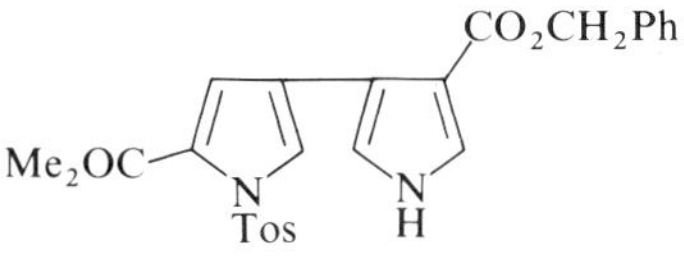

216

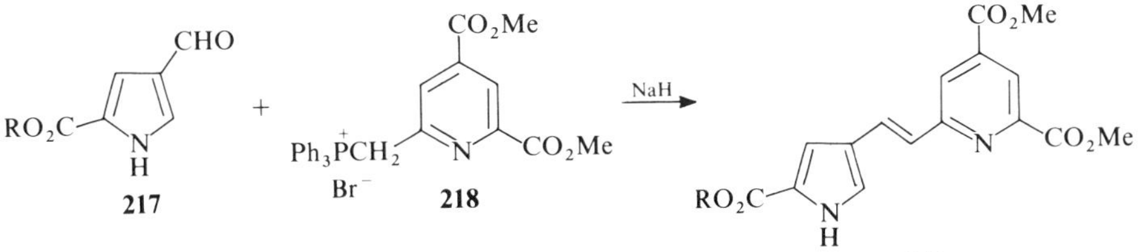

217   **218**   **219**

Pyrrolyl ketones are generally much less reactive than the corresponding aldehydes, and forcing conditions are required before they undergo the Wittig or Horner reaction. Thus, for example, the reaction between 2-acetylpyrrole **220a** and the ylid **175a** gives the expected alkene **221**,[165] while, under standard conditions with **175c**, it gives **222** together with the Z-isomer, but it is necessary to heat the ylid with the ketone in a sealed tube at 120°C in order to produce the acrylates in reasonable yield.[168] The reaction between **220a** and the vinylphosphonium salt **190a** gives the 3H-pyrrolizine **223**. The corresponding reaction with 2-benzoylpyrrole (**220b**) gives only tars,[181,182] although Wittig reactions of 2-benzoylpyrrole (**220b**) with simple alkylphosphoranes have, however, been reported to be successful.[183]

220
a R = Me
b R = Ph

1.3.4. Knoevenagel and Related Reactions

Both 2- and 3-formylpyrroles react with compounds containing methylene groups that can form carbanions stabilized by two adjacent ester groups. Thus, the aldehyde **224** reacts with diethyl malonate in the presence of piperidine to give the acrylic acid **225**.[193] The 3-formylpyrrole **226** reacts in an analogous manner to give **227**,[193] and the reaction also proceeds with 3-formylpyrrole.[125] The dicarboxylic acid **229** may be synthesized from **224** (R = Me),[194] and the ester **230** can equally well be employed in this reaction and has been used in the preparation of **231** from ethyl 2-formylpyrrole-2-carboxylate.[195] It is also possible to use malonic acid in the Knoevenagel condensation with formylpyrroles; 2-formylpyrrole (**77a**) reacts with malonic acid in the presence of aniline to give **228**, which can be cyclized by heating with acetic anhydride to yield **232a**.[196] Ethyl 3-formylpyrrole-1-carboxylate (**78**)[121] condenses with malonic acid in the presence of pyridine and piperidine to give the dicarboxylic acid, which, on basification, loses its protecting group and carbon dioxide to yield the acrylic acid **233**.[122]

1.3. Reactions of Acyl Groups with Carbon Nucleophiles

The reaction conditions for the condensation of **224** (R = Me) with diethyl malonate can also be modified to give the diester **234**,[142] and the corresponding base-catalysed reaction with ethyl cyanoacetate gives the *E*- and *Z*-isomers **235** and **236**.[142] The cyanovinyl ester group introduced in this fashion has been used widely by pyrrole chemists as a protecting group for aldehydes, as it is fairly readily removed by hydrolysis. 3-Formylpyrrole (**212c**) can be protected in this fashion,[197] as can **162**,[139] **224**,[139,141] **237a**,[139] and **237b**.[142] The same reaction on the aldehydes **238a** and **238b** gives rise to cyanovinyl compounds **239a** (m.p. 161–163°C) and **239b** (m.p. 126–127°C), respectively,[198] and the 3-formylpyrrole **240** yields the antiviral compound **241a**. Similarly, malononitrile reacts with the aldehydes **240** to give **241b**,[199] and the aldehydes **238a** and **238b** to give **242a** (m.p. 204–205°C) and **242c** (m.p. 117–118°C), respectively.[198]

With cyclic dicarbonyl compounds, the Knoevenagel reaction proceeds with equal efficiency. Thus, the reaction of the activated aldehyde **243** with Meldrum's acid in the presence of pyridine gives **244** (m.p. 173–175°C) in 74% yield.[200] Subsequent catalytic hydrogenation of the vinyl group over Raney nickel yields **245**, which, after hydrolysis and decarboxylation, followed by reesterification produces **246** (m.p. 103–106°C) in good yield.[200] The aldehyde **247**, which is activated by two ester functions on the pyrrole ring, reacts with a series of derivatives of barbituric acid to give the condensation products **248** and **249** (R = alkyl or aryl groups).[201] With the lactone **250**, 2-formyl-1-methylpyrrole (**77b**), in the absence of either acid or base, produces **251** in 57% yield.[202]

The 3,4-diformylpyrrole **252** reacts with the 3-ketoglutaric ester **253** to yield the cyclic ketone **254**.[203] Thiol and sulfoxide groups also stabilise adjacent carbanions, although not to the same extent as carbonyl groups. Thus, 1,3-bis(methylthio)acetone (**256**) reacts with the dialdehyde **255a** in a similar fashion in the presence of pyridine and morpholine to give the analogous cyclic ketone **257** (m.p. 250°C) in 55% yield;[204] the corresponding bis-sulfoxide **258** condenses with the *N*-methylated dialdehyde **255b** to yield the sulfoxide **259** (m.p. 280–290°C) in an even higher yield (85%).[205] In contrast, the dialdehyde **255c** reacts with more difficulty with a variety of β-diketones **260** to produce the monocondensation derivatives **261** in moderate yields.[206] In most of these reactions, only the *E*-isomers are produced, although in the case where R = CO$_2$Et, the *Z*-isomer could also be isolated. The analogous reaction between **255d** and the diketone **260** gives **262**.[206]

1.3.5. Acylpyrroles as Carbanion Sources

To this point, only the reactions of carbanions with the carbonyl function of acylpyrroles (primarily of pyrrolyl aldehydes) have been considered. However, acylpyrroles, or more specifically, pyrrolyl ketones having one or more protons on the carbon atom adjacent to the carbonyl group, can equally be expected to form carbanions themselves. The condensation of these carbanions with carbonyl groups, either inter- or intramolecularly, forms another powerful set of carbon–carbon bond-forming reactions. The condensations with aldehydes or ketones are essentially aldol reactions and give rise to α,β-unsaturated ketones; with esters, the reactions are analogs of the Claisen condensation reaction and produce β-dicarbonyl compounds.

The aldol condensation between 2-acetylpyrrole **220a** and 2-formylpyrrole (**77a**) in the presence of base leads to the enone **263**.[207] The same ketone **220a** has been used in the preparation of a series of unsaturated compounds **264a** (m.p. 165–167°C), **264b** (m.p. 178°C), and **264c** (m.p. 207–209°C).[208] Pyrroles bearing 3-acetyl substituents also undergo aldol reactions. Thus, **265**[209] condenses with benzaldehyde in the presence of base to give the styryl ketone **266**,[207,210,211] and with a series of substituted benzaldehydes it produces the analogous compounds **267**.[212] N-Acyl groups have also been reported as sources of anions, at least in the intramolecular reaction of 2-formyl-1-phenyl-

1.3. Reactions of Acyl Groups with Carbon Nucleophiles 35

acetylpyrrole, formed by the reaction of 2-formylpyrrole with phenacetyl chloride, in which the benzylic carbanion condenses with the adjacent formyl group to yield the bicyclic compound **268**,[196] the phenyl derivative of **232a** (see Section 1.3.4).

Although the initial product of an aldol reaction is the β-hydroxyketone, almost all of the reported reaction of pyrrolyl ketones lead directly to the α,β-unsaturated ketones under the reaction conditions used. There is, however, one report in which the kinetic enolate of the N-protected ketone **269**, generated by base at $-78°C$, is trapped as the boron enolate **270** by dibutylboryl triflate. The E-isomer constitutes over 95% of the trapped enolate, which, when allowed to react with benzaldehyde, yields the aldol **271**. This compound appears to be entirely in the *erythro* form as shown; there is no evidence for the *threo* isomers in the ^{1}H NMR spectrum.[213]

Claisen condensation reactions of 2-acetylpyrrole (**220a**) have also been reported. With ethyl oxalate and sodium ethoxide, the β-ketopyruvate **272**, which is predominantly in the enolic form, is prepared.[214] This product is unstable in aqueous base and readily reverts to the starting ketone. With ethyl trifluoroacetate, 2-acetylpyrrole (**220a**) produces the β-diketone **273**.[215] Acetyl groups in the 3-position of the pyrrole ring are generally less reactive and may require stronger bases, such as sodamide or sodium hydride, in order to generate their carbanions. The reaction of ethyl 3-acetyl-2,4-dimethylpyrrole-5-carboxylate (**265**) with diethyl carbonate in the presence of sodium hydride leads to the β-ketoester **274** as well as the aroyl malonate **275** in variable yields.[216] The corresponding reaction with ethyl formate fails.[216] However, ethyl formate does react with the cyclic ketone **164** to give the dione **276** (largely in the enolic form).[163]

1.3.6. The Use of C—C Bond-forming Reactions in the Preparation of Heterocyclic Rings

A number of the reactions discussed in earlier parts of Section 1.3 have been used in the elaboration of a wide variety of heterocyclic system.

It has already been shown that 2-formylpyrrole (**77a**) can be converted into the pyrrolizin-3-one **232a** via the dicarboxylic acid **228**.[196] This preparation is, however, difficult to carry out. The intermediate **228** is a sensitive compound that is difficult to isolate. A more convenient synthesis involves the condensation of the formylpyrrole with Meldrum's acid to give **277a**, that is, the protected analog of **228**. Flash-vacuum pyrolysis of **277a** at 600°C and 10^{-3} mm Hg produces the pyrrolizin-3-one **232a** in 88% yield.[217] Under similar conditions, 4-bromo-2-formylpyrrole is converted via **277b** (m.p. 155°C) into the bromopyrrolizinone **232b** (m.p. 59–60°C).[218] An alternative route to the pyrrolizine ring system starts from the anion of 2-formylpyrrole **278**, generated from the free base with benzyltrimethylammonium hydroxide; reaction with but-1-en-3-one yields the keto-aldehyde **279** (68%), which undergoes an ethoxide-catalyzed intramolecular aldol condensation to produce the pyrrolizine **280** (m.p. 60–61°C) in 31% yield. The alternative cyclization via the carbanion formed at the methyl carbon is somewhat more efficient and leads to the 8-membered cyclic ketone **281** (m.p. 55–56°C) in 40% yield.[219] 2-Formyl-3,4-diphenylpyrrole (**282a**) may be converted into its *N*-acetyl derivative **282b** by treatment with acetic anhydride and triethylamine; thermal cyclization of this intermediate produces the pyrrolizinone **283** (m.p. 132–134°C) in 72% yield. In a different approach to the same ring system,[221] the anion **278** is allowed to react with phenyl vinyl sulfone, followed by cyclization of the intermediate sulfone **284** to **285**, which then reacts with a further two molecules of the sulfone to form **286**, or with one molecule of sulfone to provide, after elimination of the equivalent of two molecules of phenylsulfinic acid, the tetracyclic system **287**. Reaction of **285** with one molecule each of **284** and the sulfone furnishes **288**.[221]

277
a R = H
b R = Br

232
a R = H
b R = Br

1.3. Reactions of Acyl Groups with Carbon Nucleophiles

A tricyclic ring compound **290** is isolated from the base-catalyzed intramolecular vinylogous aldol condensation of the ketone **289**.[222]

Reference to the use of sulfur to assist in the stabilization of carbanions has been made in Section 1.3.5. 3,4-Diformyl-1-methylpyrrole (**255b**) can be converted into the sulfur heterocycle **291a**; the ester groups may be hydrolyzed to give the dicarboxylic acid **291b**, which is readily decarboxylated to give the parent ring compound **291c**.[223] The corresponding reaction sequence with 2,3-diformyl-1-methylpyrrole produces the sulfone **292a**, the dicarboxylic acid **292b**, and the parent ring compound **292c**. The bicyclic systems **293a** and **293b** can be synthesized in an analogous manner from the appropriate sulfides.[233] When **292a** is heated, extrusion of sulfur dioxide results in the formation of the indole **294a**. The same product is formed when 2,3-diformyl-1-methylpyrrole is treated with bis(methoxycarbonylmethyl)sulfide, and an analogous reaction occurs with bisphenacyl sulfide to yield the corresponding diketone **294b**.[223] The diester **294a** may also be prepared by reacting the diformylpyrrole with succindinitrile to furnish the dicyanoindole **294c**, which is then converted into **294a** by standard procedures.[223]

255b $\xrightarrow{SO_2(CH_2CO_2Me)_2}$ **291**

a R = CO_2Me
b R = CO_2H
c R = H

292
a R = CO_2Me
b R = CO_2H
c R = H

293
a R = CO_2Me
b R = COPh

294
a R = CO_2Me
b R = COPh
c R = CN

There are several examples in the literature in which aldol condensations of 2-formylpyrroles are used in the synthesis of compounds in which a nitrogen heterocycle is linked to the pyrrole ring. In the simplest cases, the carbanion derived from the nitrogen heterocycle condenses with the aldehyde group. Thus, amide ketals **295–297** react with 2-formylpyrrole (**77a**) to furnish the corresponding racemic carbinols **298–300**, together with their dehydration products **301–303**.[224] Similarly, succinyl hydrazide reacts with 2-formylpyrrole (**77a**) to yield what is essentially the Stobbe product **304**.[225] This α,β-unsaturated amide undergoes conjugate addition with Grignard reagents to give alkylated derivatives **305a**, and with hydrogen cyanide to give the cyano compound **305b**.[225]

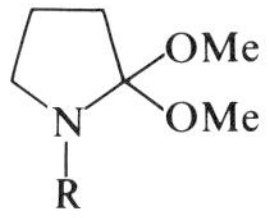

1.3. Reactions of Acyl Groups with Carbon Nucleophiles

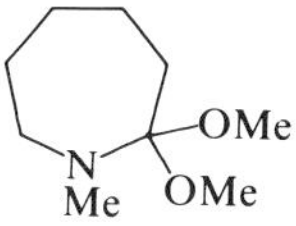

295
- a R = Me
- b R = Bu
- c R = PhCH$_2$

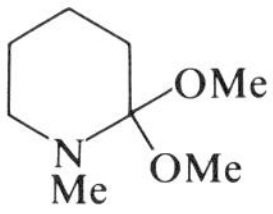

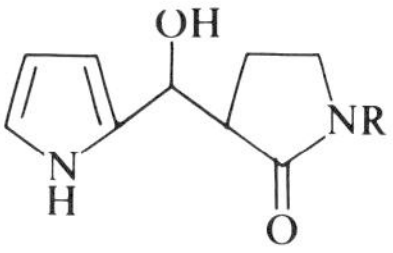

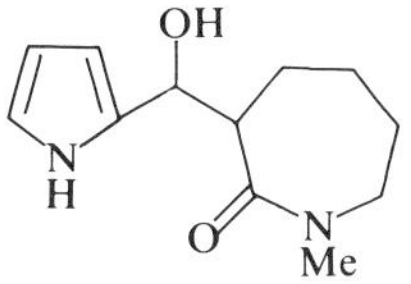

298
- a R = Me
- b R = Bu
- c PhCH$_2$

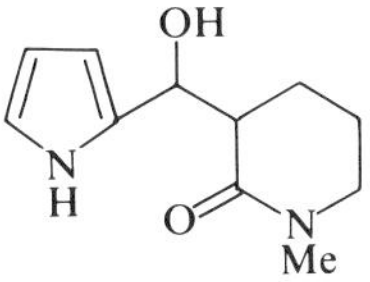

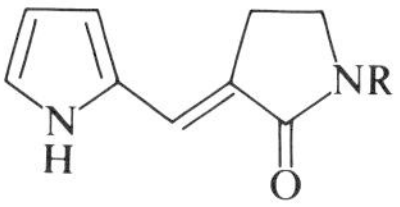

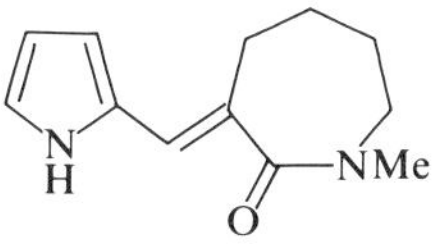

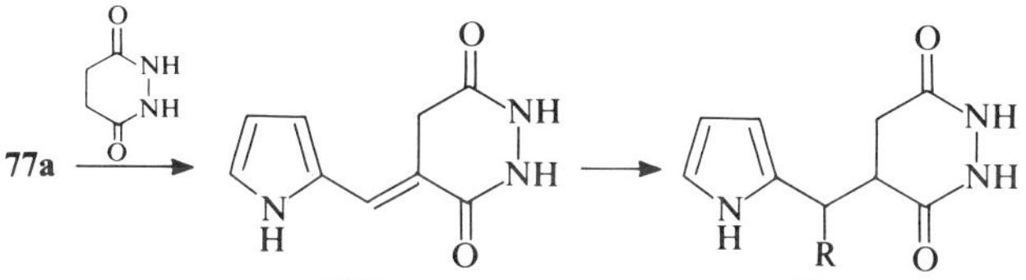

301
- a R = Me
- b R = Bu
- c R = PhCH$_2$

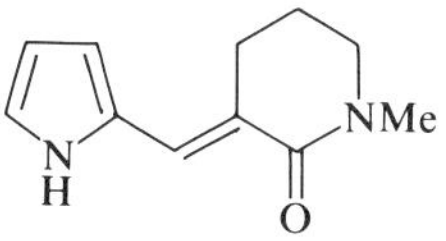

305
- a R = alkyl
- b R = CN

In a more complicated reaction sequence, when the benzocycloheptanone **306** is treated with 2-formyl-1-methylpyrrole (**77b**) in the presence of base, the condensation product **307** can be isolated, and may then be reacted with urea to give **308a** (m.p. 210–212°C) in 83% yield or with thiourea to yield 70% of the sulfur analog **308b** (m.p. 220–222°C).[226]

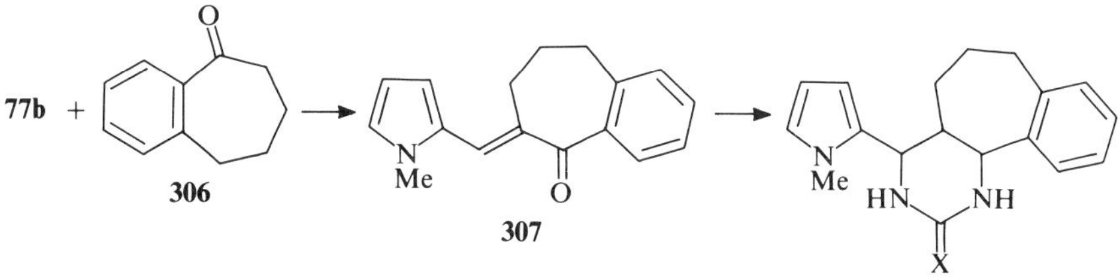

308
- a X = O
- b X = S

With acylglycines **309**, 2-formylpyrrole (**77a**) gives the products **310**, which can be converted in the usual way into the corresponding oxazolones **311**.[227]

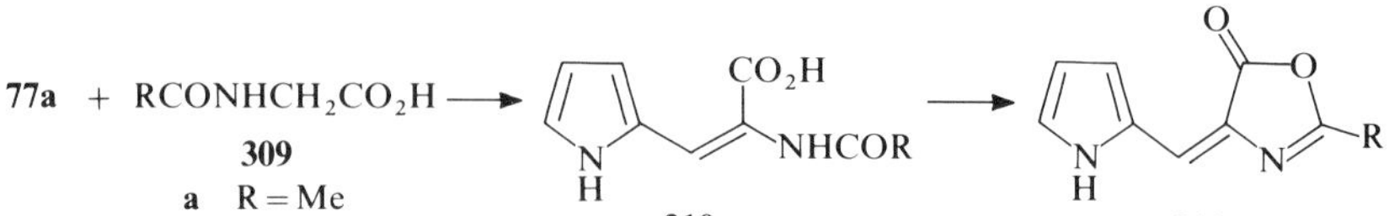

77a + RCONHCH$_2$CO$_2$H ⟶ **310** ⟶ **311**

309
a R = Me
b R = Ph

310
a R = Me
b R = Ph

311
a R = Me
b R = Ph

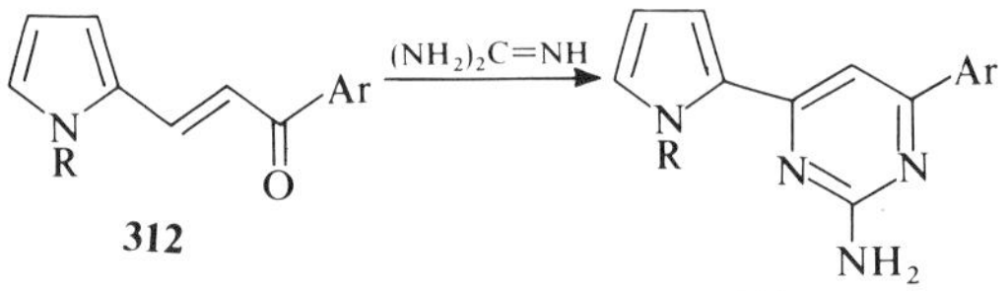

The condensation of either 2-formylpyrrole (**77a**) or its 1-methyl derivative (**77b**) with a series of aryl methyl ketones furnishes the corresponding enones **312**, which, on treatment with guanidine and base, yield the 2-aminopyrimidines **313** (Table 1.3).[228]

TABLE 1.3. PREPARATION OF α,β-UNSATURATED ARYL KETONES **312** AND 2-AMINOPYRIMIDINES **313** FROM 2-FORMYLPYRROLES[228]

Ar	R	Yield (%) of **312**	m.p. (°C)	Yield (%) of **313**	m.p. (°C)
Ph	H	80	136	71	164–166
Ph	Me	83	76	70	121–122
2-Thienyl	H	81	130	82	137
2-Furyl	H	82	105	75	170

The Wittig reaction can furnish similar starting materials.[165] The enone **91** (R = Me)[165,229] prepared by this means, on treatment with the dimethylformamide acetal,[230] gives the product **314** (m.p. 151°C) in 52% yield; this compound undergoes thermal cyclization to the 7-membered ring ketone **315** in 42% yield.[231]

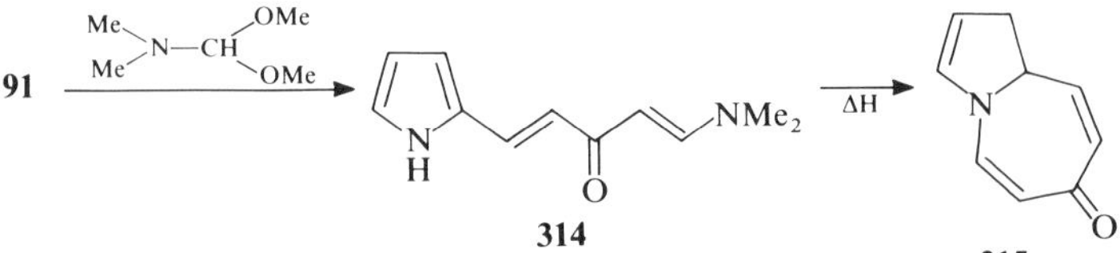

1.3. Reactions of Acyl Groups with Carbon Nucleophiles

Treatment of the vinyl triphenylphosphine derivative **316** with the 2-formylpyrrolyl anion **278** [prepared from (**77a**) by the action of sodium hydride] gives, as intermediates, the ylids **317a** and **317b**, which rapidly cyclize to give the corresponding tautomeric pyrrolizines **318**⇌**319**.[232] If the same reaction is carried out using the sulfoxide **320**, instead of the sulfide **316b**, the intermediate ylid **321** eliminates the sulfur function to give the alkene **322**. The further addition of the pyrrolyl anion **278** to this species leads to the intermediate ylid **323**, which undergoes an intramolecular Wittig condensation to yield the pyrrolizine **324**. Its tautomer **325**, having a reactive methylene position ideally placed for cyclization to the remaining aldehyde group, produces the tetracyclic product **326**.[232]

The use of the disulfone **327** in the reaction with **278** gives the intermediate **328**, which leads to the pyrrolizine **331** by a sequence **328** → **329** → **330** → **331** analogous to that observed with the corresponding triphenylphosphine derivatives. The pyrrolizine **331** coexists with its tautomer **332**, and intramolecular condensation with the second formyl group again yields a tetracyclic product **333**.²³²

Reaction of the ylid **334** with the pyrrole **335** is believed to occur at the unprotected nitrogen atom to yield the intermediate ylid **336**, which then adds to the ester function at C-2 to give the pyrrolizinone **337**. Removal of the conjugated double bond of the ethyl enol ether system by hydrogenation over palladium, followed by elimination of the tetrahydropyranyl group, gives 5,7a-didehydroheliotridin-3-one **338**.²³³

The Wittig reaction between the *N*-protected pyrrolylmethylphosphonium salt **339** and the *N*-protected 2-formylpyrrole **340** in dimethylformamide in the presence of potassium carbonate gives the *E*-alkene **341** (m.p. 149–150°C) in 84% yield. Irradiation of this compound in ethanol in a vessel open to the air produces the tricyclic product **342** (m.p. 145–146°C) in 75% yield.²³⁴

The Dieckmann reaction may also be used to construct heterocyclic rings fused to pyrrole. Thus, the ester **343**, when treated with ethyl acrylate in acetic

1.3. Reactions of Acyl Groups with Carbon Nucleophiles

acid, yields the Michael product **344**, which undergoes a Dieckmann cyclization on treatment with sodium bis(trimethylsilyl)amide to give **345**, as a mixture of diastereomers. Dehydration of this material over palladium-on-carbon yields 30% of the fused system **346** (m.p. 206–208°C).[235] In a similar reaction sequence, **348**, prepared from the diester **347**, may be cyclized to **349** under Dieckmann conditions.[236] An analogous reaction may be used to convert **350** into **351**.[237]

The anion formed by the reaction of sodium in toluene on the ester **352** reacts with ethyl 3-bromopropionate to give the diester **353** in 65% yield. Dieckmaan cyclization of **353** produces the intermediate **354**, which is decarboxylated to yield[238] the butterfly pheromone, danaidone **2**,[7] in 24% yield. The esters **355**, on treatment with ethyl 3-chloropropionate in the presence of sodium hydride, are converted into the corresponding diesters **356**, which cyclize in the presence of sodium bis(trimethylsilyl)amide to yield **357**. Dehydrogenation of **357** by the action of iodine and potassium acetate in ethanol gives the pyrrolopyridines **358** (Table 1.4).[239]

TABLE 1.4. SYNTHESIS OF PYRROLOPYRIDONES **358**[239]

R^1	R^2	Yield (%) of **356**	Yield (%) of **357**	Yield (%) of **358**	m.p.(°C) **358**
$PhCH_2OCH_2$	Me	50	73	85	115–116
$PhCH_2OCH_2$	H	56	58	82	135–136
Me	H	31	75	72	107–109

The diketone **359** is prepared from the anion of 2-acetylpyrrole (**220a**) in a manner analogous to the synthesis of **279** from **278**.[219] The aldol condensation of **359** can proceed in two directions to give the pyrrolizine **360** or the pyrrolocycloheptenone derivative **361**.[219]

1.3. Reactions of Acyl Groups with Carbon Nucleophiles

Heterocycles have also been elaborated by Friedel–Crafts reactions of pyrrolecarboxylic acids. The acid chlorides **362** cyclize under the influence of aluminum trichloride to yield the tricyclic products **363** (Table 1.5).[240] Reaction of the pyrrole-2-carboxylates **160**, **364a**, and **364b** with the styrene epoxides **365**

TABLE 1.5. FRIEDEL–CRAFT CYCLIZATION OF SOME
N-PHENYLETHYLPYRROLES[240]

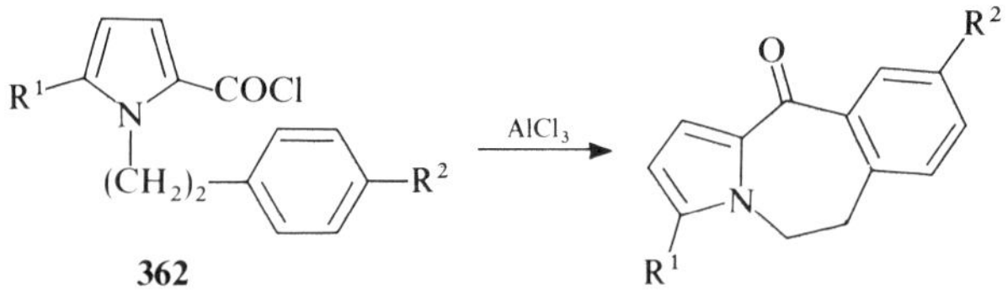

R^1	R^2	m.p. (°C) **362**	m.p. (°C) **363**
H	H	85–86	54–55
H	4-CN	114–115	146–147
H	4-Cl	91–92	100–105
H	4-Br	101–103	101–103
H	4-NO_2	119–120	174–176
H	3,4,5-Br	144–145	169–171
Me	H	Oil	Oil
Me	3,4,5-Br	81–83	160–163

TABLE 1.6. SYNTHESIS OF SOME PYRROLOBENZAZEPINES[241]

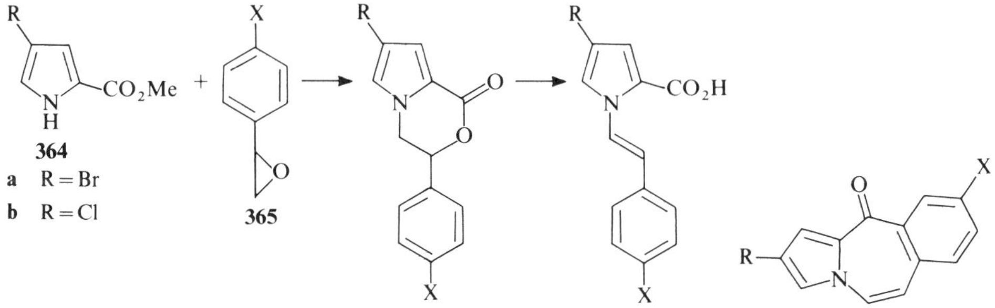

364
a R = Br
b R = Cl

R	X	Yield (%) of **367**	m.p. (°C)	Yield (%) of **368**	m.p. (°C)
H	CN	96	192–194	67	190–191
H	H	84	183–184	75	113–114
H	Cl	87	168–170	63	139–141
H	SCF_3	88	191–193	61	147–148
Br	H	98	198–199	60	210–214
Cl	CN	79	225–227	67	303–307

yields the bicyclic system **366**, which on treatment with potassium *tert*-butoxide, followed by hydrochloric acid, produces the *E*-isomers of the pyrrolylstyrenes **367**. Photolytic isomerization of **367**, followed by conversion into their mixed anhydrides in the presence of tin(IV) chloride lead to the tricyclic pyrrolo-[1,2-*b*]benzazepinones **368** (Table 1.6).[241]

1.4. REACTIONS WITH NITROGEN NUCLEOPHILES

Although, as indicated in earlier sections, the carbonyl groups of pyrrolyl aldehydes and ketones have some vinylogous amide character, the reactions of formyl and acetylpyrroles with amines, hydroxylamines, and hydrazines are, in general, quite normal.

1.4.1. Reactions with Amines

2-Formylpyrrole (**77a**) reacts with ammonia to give a polymeric product,[242] but its reaction with methylamine gives the expected imine **369**,[243] and with arylamines the corresponding compounds **370** are readily formed, unless the benzene ring bears strongly electron-withdrawing substituents.[244,245] Such *N*-arylimines are easily hydrolyzed to the original carbonyl compound.[246] The kinetics of this hydrolysis, which follows a first-order rate law under neutral aqueous conditions, have been studied for **370** (X = H, OMe, NMe$_2$).[247] The phenylimine hydrolyzes most rapidly, nearly 3 times as fast as the anisyl derivative, while the dimethylamino compound reacts even more slowly.[247] In contrast, an examination of the kinetics of the hydrolysis in aqueous methanol of the 1-substituted derivatives **371** shows that the substituents (e.g., *m*-Cl, *m*-Me, *p*-OEt) on the aromatic ring have little effect on the rate of the reaction.[248] A polarographic study of the effect of pH on the electrochemical behavior of the imines **370** (X = H or OH) has been reported.[249] Benzylamine reacts with 2-formylpyrrole (**77a**) to give the condensation product **372a**[244,250] and 2-aminomethylpyrrole yields the analogous imine **372b**. 2-Formyl-1-methylpyrrole (**77b**) reacts in the same way with simple amines.[244]

The corresponding reactions with diamines are not straightforward. With 1,2-diaminoethane, 2-formylpyrrole (**77a**) gives the bis-adduct **373**.[251] However,

1.4. Reactions with Nitrogen Nucleophiles

it has been reported that diethyl 2-formyl-4-methylpyrrole-3,5-dicarboxylate (**96**) produces the imine **374** by condensation with only one of the amino groups.[252] Similarly, 1,4-diaminobenzene reacts with two molecules of (**77a**) to yield **375**,[251] whereas (**96**) is reported to give **376**.[252] With 1,3-diaminobenzene, 2-formylpyrrole (**77a**) again gives a bis-adduct **377**,[251] but it is reported that with 1,2-diaminobenzene either the bis-adduct **378** or the benzimidazole **379** may be formed. With 2-formyl-1-methylpyrrole (**77b**), reaction with bis(N-methylamino)ethane under reflux in benzene gives the nitrogen analog of a cyclic acetal **380** in 70% yield.[253] This cyclic diamino system can be used as a protecting group for formylpyrroles; it is readily removed by quaternization with methyl iodide, followed by simple hydrolysis using water.[253]

The behavior of the 3,4-diformylpyrrole **381** with amines is anomalous. With aliphatic amines, efficient substitution for one of the chlorine atoms is the primary reaction; pyrrolidine and cyclohexylamine give **382** (65%) and **383**

(56%), respectively. In contrast, the reaction with arylamines leads to the monoimines **384** in 70–80% yield, and, when an excess of the amine is used, **385** is obtained in 60–70% yield.[254]

When 2-formylpyrrole (**77a**) is treated with a secondary amine in the presence of an excess of acid, or with the corresponding ammonium perchlorate, iminium salts **386**, identical to the intermediate isolable from the Vilsmeier–Haack reaction,[255] may be isolated.[256] On treatment with triethylamine in nonaqueous solvent, **386** dimerizes to yield **387**. The same dimers are the usual products of the treatment of **77a** with a secondary amine in the presence of an acid catalyst,[257] although the reaction will also proceed in the absence of acid.[258] These cyclic dimers **387** (R = Me) have been employed in an efficient one-pot two-step synthesis of 5-substituted 2-formylpyrroles.[259] Treatment of **387** with *tert*-butyllithium at −15°C in tetrahydrofuran, followed by quenching of the pyrrolyllithium thus formed by an electrophile, leads to the substituted dimer **388**, which is in equilibrium with the azafulvene **389** and is readily hydrolyzed to the aldehyde **390** (Table 1.7).[259] This reaction sequence has been modified, starting from 4-bromo-2-formylpyrrole **391**, to form initially the adduct **392**, which on treatment with an alkyllithium results in lithiation at the 4-position, rather that at C-5. Reaction with an electrophile gives the 4-substituted dimer **393**, and hence, after hydrolysis, the 4-substituted 2-formylpyrrole **394** (Table

1.4. Reactions with Nitrogen Nucleophiles

TABLE 1.7. ONE-POT TWO-STEP SYNTHESIS OF SOME 5-SUBSTITUTED 2-FORMYLPYRROLES[259]

Electrophile	R	Yield (%) of **390**
MeI	Me	91
Me$_2$SO$_4$	Me	75
CH$_3$(CH$_2$)$_7$I	CH$_3$(CH$_2$)$_7$	47
CH$_3$(CH$_2$)$_{15}$I	CH$_3$(CH$_2$)$_{15}$	19
CH$_2$=CHCH$_2$I	CH$_2$=CHCH$_2$	23
i-Pr$_3$SiOTf	i-Pr$_3$Si	68
PhSSPh	PhS	78
DMF	CHO	63
CH$_3$(CH$_2$)$_4$CON(Me)OMe	CH$_3$(CH$_2$)$_4$CO	54
CH$_3$OCOCl	CH$_3$OCO	64

1.8).[260] Treatment of the substituted dimer **393** (R^1 = Me) with *tert*-butyllithium, followed by reaction with an electrophile, has been shown to lead via **395** to 4,5-disubstituted 2-formylpyrroles **396** (Table 1.9).[260]

TABLE 1.8. ONE-POT TWO-STEP SYNTHESIS OF SOME 4-SUBSTITUTED AND 4,5-DISUBSTITUTED 2-FORMYLPYRROLES[260]

Electrophile	R^1	Yield (%) of 394
MeI	Me	87
$CH_3(CH_2)_7I$	$CH_3(CH_2)_7$	58
$CH_2{=}CHCH_2Br$	$CH_2{=}CHCH_2$	60
$i\text{-}Pr_3SiOTf$	$i\text{-}Pr_3Si$	74
PhSSPh	PhS	83
DMF	CHO	78
$CH_3(CH_2)_4CON(Me)OMe$	$CH_3(CH_2)_4CO$	62
CH_3OCOCl	CH_3OCO	71

Electrophile	R^2	Yield (%) of 396
MeI	Me	87
DMF	CHO	69

A similar basic ring system has been obtained on treating the 2-chloromethylpyrrole 397 with aqueous potassium hydroxide; the product 398a is in thermal equilibrium with the azafulvene 398b.[261]

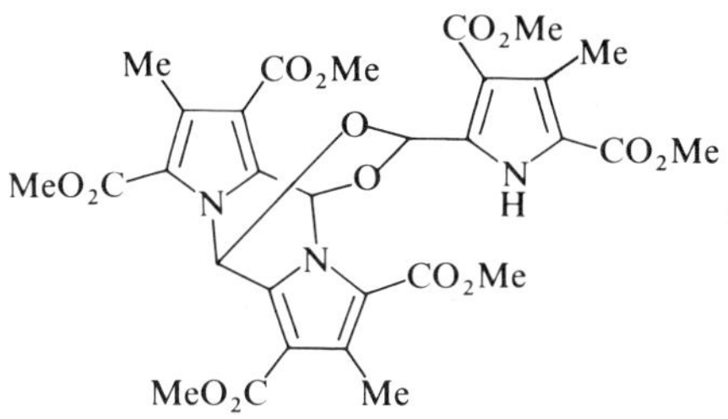

1.4. Reactions with Nitrogen Nucleophiles

398a 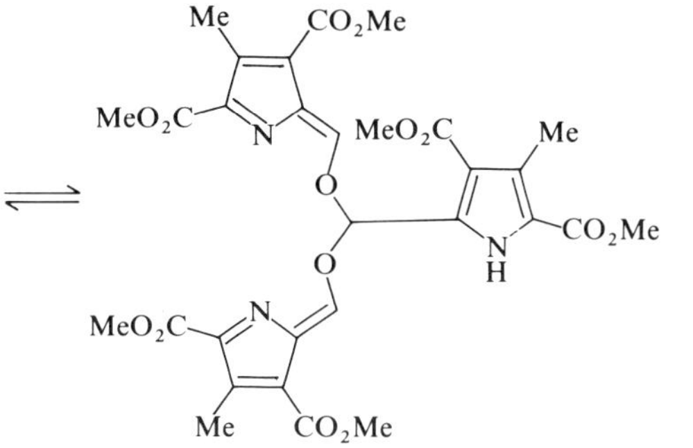 ⇌ **398b**

1.4.2. Reactions with Hydrazines

Pyrrole aldehydes and ketones react normally with hydrazine to form the corresponding hydrazones.[107] The compounds having carbonyl functions in the 2-position are more reactive than the corresponding 3-substituted pyrroles. Some more unusual hydrazines have been employed during recent years. 2-Formylpyrrole (**77a**) has been converted into the hydrazones **399** [R = CO_2Et,[262,263] naphthyl-1-CH_2CO,[262] N-piperidyl (m.p. 175°C),[265] pyrazinyl (m.p. 227°C),[265] 2-pyrimidyl (m.p. 139°C),[265] 2-thiazolyl (m.p. 195°C),[265] and 2-quinolyl (m.p. 148°C)[265]]. Several of these compounds have been used as colorimetric reagents in the quantitative analysis of copper(II) and cobalt(II).[265] Aroylhydrazones **400** have also been prepared.[266] Both 2-formylpyrrole (**77a**) and its N-methyl derivative (**77b**) have been converted into the thiosemicarbazones and semicarbazones **401**.[267] The analogous reaction of 2-formyl-1-methylpyrrole with N-hydroxy-N'-aminoguanidine in the presence of p-toluenesulfonic acid gives the hydrazone **402** as its tosylate salt.[268] 2,5-Diformylpyrrole (**203b**) gives the bishydrazone **404** with N-aminoguanidine.[269] 2-Formylpyrroles **405** have been converted into their hydrazones **406** by conventional means.[270]

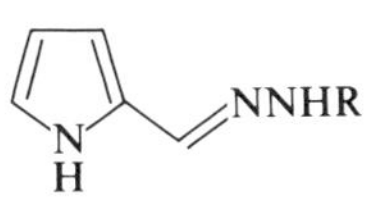

399    **400** 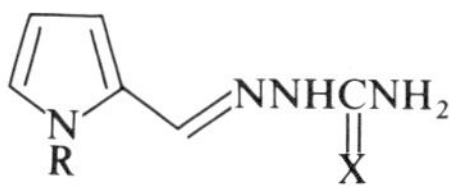   **401**
R = H, Me
X = O, S

77b → **402**

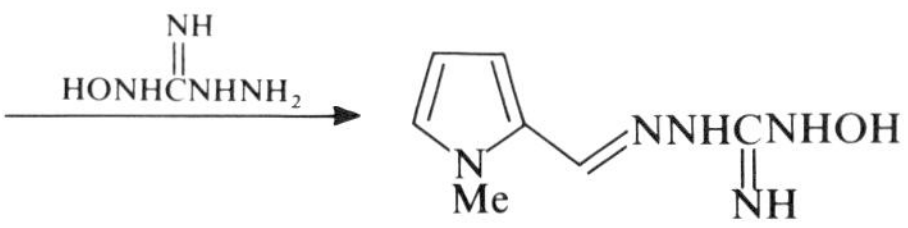

Ar = Ph; 3,4-(MeO)$_2$C$_6$H$_3$;
3,4-CH$_2$O$_2$C$_6$H$_3$; 2-NO$_2$-4,5-(MeO)$_2$C$_6$H$_2$

Pyrrolylhydrazones generally undergo the Wolff–Kishner reaction and may be used in the conversion of acylpyrroles into the corresponding alkyl derivatives.[107] However, treatment of the tosylhydarazone **402** with butyllithium results in the formation of the two cyclopenta[b]pyrroles **408** and **409**.[271]

Pyrrolecarboxylic acids and the corresponding acid chlorides react with hydrazine to give the carbonylhydrazides (e.g. **410**)[263,272–277] which react with aldehydes to form hydrazones[273] and with arylsulfonyl chlorides to yield arylsulfonylhydrazides.[274–276] They have also been converted into acylhydrazides **411**[263] and hence into the heterocyclic compounds **412**.[262] An unusual route to pyrrolylhydrazine involves the condensation of the acid chloride **413** with the 1-aminopyridinium salt **414** to give the nitrogen ylid **415** (m.p. 152–154°C) in 86% yield. Reduction of **415** with sodium borohydride produces the hydrazine **416** (m.p. 127–128°C) in 86% yield.[278]

R = CH$_2$Cl, CHCl$_2$, CCl$_3$, 2-thienyl

1.4.3. Reactions with Hydroxylamines

Pyrrole aldehydes and ketones react in the usual manner with hydroxylamine to form oximes.[107] With electron-withdrawing groups on the pyrrole nucleus, the reaction proceeds readily; the 2-formylpyrroles **405** produce the corresponding oximes in 75–88% yield.[270] In general, reduction of the pyrrolyl oximes with Raney nickel produces the corresponding amines in reasonable yields (e.g., **417** → **418**),[279] and they also undergo the Beckmann rearrangement.[107,280,281] In most cases, (e.g., **419** → **420**) the pyrrolyl group, rather than the alkyl group, migrates, although with less highly substituted rings the reaction conditions lead instead to the hydrolysis of the oxime to the original carbonyl compound.[281] Oximes may also be dehydrated to form the corresponding cyanides; for example, when **421** is treated in a halogenated solvent with triphenylphosphine and triethylamine, it is converted into the nitrile **422** (m.p. 116°C);[282] the yield in 1,2-dibromo-1,1,2,2-tetrachloroethane is 82%, significantly higher than that in carbon tetrachloride. This procedure generally gives better yields than the standard method using acetic anhydride. The 2-formylpyrrole **423a** forms an oxime, **424a**, which is explosive, but which can nevertheless be dehydrated to the cyanide **425a**,[283] and, in the same way, **423b** can be converted into **424b**, which gives the nitrile **425b** in 81% yield on treatment with acetic anhydride. Ethanolic sodium hydroxide solution converts **425b** into the amide **426** (m.p. 102–103°C) in 93% yield.[284] 2-Formylpyrrole (**77a**) can be transformed directly into 2-cyanopyrrole in 88% yield by treatment with hydroxylamine sulfonic acid.[285] The oximes **427** have been converted into compounds **428** and **429**.[286]

1.4.4. Amide Formation

Amides have been prepared in the conventional fashion by the reaction of carboxylic esters with amines; for example, the ester **430** reacts with ammonia to give the corresponding amide **431**.[287] It is, however, usually more efficient to use the acyl halide rather than the ester. The acyl chloride **432** is converted into the amide **433**, and no reaction of the ester groups is observed.[288] In the same way, the acyl chloride **434**, on treatment with dimethylamine, gives the tertiary amide **435**.[289] The reaction between **436** and aniline gives **437**,[290] and **438** has been converted into the amides **439** [R = Ph; naphthyl; 2-, 3- or 4-$NO_2C_6H_4$; halo-C_6H_4; 4-MeC_6H_4; $(CH_2)_2OH$].[291] Anilides can also be obtained directly from the pyrrole carboxylic acids, via the initial reaction with dicyclohexylcarbodi-imide, followed by the appropriate aniline.[292] 2-Formyl-1-methylpyrrole (**77b**) has been converted into the corresponding amide **442** via the silylated cyanohydrin **440**, which with lithium diisopropylamide gives the umpole anion; subsequent reaction with **441** leads to **442** in almost quantitative yield.[293]

1.4. Reactions with Nitrogen Nucleophiles

[Scheme showing compound 440 (N-methylpyrrole with CN and SiMe₃) + Ph₃PONMe₂ (441) → compound 442 (N-methylpyrrole-CONMe₂)]

Amides can also be obtained by the reaction of trichloracetylpyrroles with amines (see Section 1.5).

1.4.5. Other Reactions with Nitrogen Nucleophiles

Reaction of 2-amino-2-methylpropan-1-ol with the acyl chloride **443a** gives the protected acid **444a** (R = H), which on treatment with butyllithium in tetrahydrofuran leads to directed lithiation of the 3-position of the pyrrole ring. Subsequent reaction with benzophenone or benzaldehyde produces the carbinols **444b** [R^1 = H; R^2 = Ph; m.p. 189–190°C (22%); R^1 = H; R^2 = H; m.p. 97–100°C (32%)].[294] The corresponding reaction using the bromo derivative **443b** results in replacement of the bromine atom on reaction with butyllithium leading, after hydrolysis, to **445**.[294]

The aldehyde **446** forms the imine with the silyloxyamine **447**. Subsequent reaction with methyl triflate, followed by vinyl magnesium bromide, gives **448**, an intermediate in the synthesis of a tumor-promoting agent.[295]

[Structures of compounds 443 (a R=H, b R=Br), 444a, 444b, 445, 446, 447, 448]

1.4.6. The Use of C—N Bond-Forming Reactions in the Preparation of Heterocyclic Rings

As was the case with C—C bond-forming reactions (Section 1.3.6), the reactions described in Section 1.4 have been used in the elaboration of heterocyclic systems.

Reaction of the dialdehyde **449a** with hydrazine yields the pyridazine **450a**.[296] The *N*-methylated analog has been obtained in a similar manner, **449b** → **450b**,[297] and the corresponding 3,4-diacetylpyrrole **451** gives the dimethylpyridazine **452**.[296] In all cases, the reaction appears to proceed via the initial formation of the bis-hydrazone. The corresponding reactions of the diester **453** leads to the pyridazinedione **454**.[298] Oxidation of **454** with lead tetraacetate produces **455**, which, like azadicarboxylates, is a good dienophile and reacts with cyclopentadiene to yield the tetracyclic adduct **456**.[298]

In contrast to the reaction with hydrazine, **449a** and **449b** combine with 1,2-diaminoethane to yield the 2 : 2 adduct **457**;[299] in the case where R = H, the adduct prefers to exist as the doubly degenerate 2-azafulvene **458**, which is probably stabilized by intramolecular H-bonding. A polymer is produced with 1,3-diaminopropane while the higher homologs react with **449** to form 1 : 1 adducts, which again, when R = H, prefer to exist as the azafulvenes **460** and not as the bis-imines **459**. The pyrrolotriazacycloundecine **461** results from the reaction of **449a** with bis(2-aminoethyl)amine.[299]

1.4. Reactions with Nitrogen Nucleophiles

The reaction between 2-formyl- and 2-acetylpyrrole-3-carboxylates **462** (R^1 = H or Me) and hydrazine leads in some cases, via the expected hydrazone, to the bicyclic product **463**.[287,300] The analogous reaction with 3-formyl or 3-acetylpyrrole-2-carboxylates **464** (R^1 = H or Me) yields the pyridazinones **465**.[301,302] In contrast, although ethyl 2-formylpyrrole-1-carboxylate **466** reacts with hydrazines to form hydrazones, **467**, treatment with acid is required to effect the cyclization to the pyrrolotriazines **468**.[303] When an acylhydrazine is employed, the resulting triazine **468** (R = acyl) can be hydrolyzed by base to generate the parent heterocycle **468** (R = H).[304] The hydrazone derived from the aldehyde **423a** undergoes an intramolecular nucleophilic displacement reaction of the pyridyl chlorine atom to give the tricyclic product **469** (R = H).[283] The analogous 2-acyl-1-(2-chloropyridyl)pyrroles produce the corresponding 7-membered ring compounds **469** (R = Me or Et).[283]

2-Formylpyrrole (**77a**) reacts with hydrazine hydrate to give the bis-condensation product **470**, which, on heating with sulfur, is converted into the thiadiazole **471**. Electrochemical polymerization of this monomer gives a film for which a conductivity of 9 S cm^{-1} is claimed.[305]

The aldol products **91** (R = Ph, p-ClC$_6$H$_4$, p-MeOC$_6$H$_4$, heteroaryl) react with hydrazines to produce the heterocycles **472** (R^1 = H, Me, Ph),[130] and the β-ketoesters **473** (R = H or alkyl groups: Me to C$_{16}$H$_{33}$) give the corresponding dihydropyrazolones **474** in good yield (R = H, 90%; R = alkyl 42–71%).[306] These compounds are active as antipyretics.

With hydroxylamine, the 3-ethoxycarbonyl ketone **475** forms the cyclic product **476**.[287] The dialdehyde **449b** reacts with glycine to give the pyrrolopyridine **477**.[297]

1.4. Reactions with Nitrogen Nucleophiles

Catalytic hydrogenation of the nitro group in **478** over a palladium catalyst yields the tricyclic product **480**, presumably via the intermediate imine **479**.[307] In the same way, reduction of **481a** with sodium borohydride in the presence of cobalt chloride, followed by treatment of the resulting amino ester **481b** with potassium hydride in tetrahydrofuran yields the pyrrolopyrazine **482**.[308] This reaction was carried out as part of a synthesis of peramine **483**, an insect antifeedant isolated from *Lolium perenne* infected with *Acremonium lolii*.[309] Reduction of the nitro group in **484a** with iron dodecacarbonyl gives the amine **484b**, which can be cyclized by treatment with *p*-toluenesulfonic acid to give **485**, an analog of sibiromycinone.[310] The model compounds **486**, on catalytic reduction of the nitro group, followed by cyclization under the influence of acid, give the analogous products **487** (Table 1.9).[311] If **486** (R^1 = OH, R^2 = Me, R^3 = H) is treated instead with potassium ferricyanide, the dimeric compound **488** is formed.[311] Catalytic reduction of 2-formyl-1-(2-nitrobenzyl)pyrrole (**489**) yields the tricyclic product **491**.[312] The reaction presumably proceeds via the intermediate imine **490**, which is then further reduced. Reduction of the corresponding oxime **492** also gives **491**, and, if Adams' catalyst is employed, the intermediate imine **490** can be isolated.[312]

TABLE 1.9. SYNTHESIS OF SOME SIBROMYCINONE MODEL COMPOUNDS **487** FROM 1-AROYLPYRROLE-2-CARBOXYLIC ESTERS **486**[311]

R^1	R^2	R^3	m.p. (°C) **486**	Yield (%) of **487**	m.p. (°C) **487**
H	H	H	82–83	77	225–226
H	OMe	OMe	173	74	302–303
OH	Me	H	112	83	258–260

In a synthesis of porphobilinogen **496**, the 2-formylpyrrole **493** has been converted into its oxime **494**. Reduction of the oxime, followed by cyclization, gives the cyclic amide **495**, essentially a protected porphobilinogen.[313] The amides **498**, which are prepared by hydrolysis of the corresponding nitriles **497** by phosphoric acid, cyclize to give **499**.[314] The conversion of the nitriles **497** into the cyclic products **499** can be carried out directly by treatment with phosphorus pentoxide and dimethylaminocyclohexane (Table 1.10).[314]

TABLE 1.10. SYNTHESIS OF PYRROLOPYRIMIDINONES **499** FROM 2-ACYLAMINOPYRROLE-3-CARBOXAMIDES **498** AND 2-ACYLAMINO-3-CYANOPYRROLES **497**[314]

R^1	R^2	R^3	R^4	Yield (%) of			m.p. **498**	m.p. (°C) **499**
				497 → 498	497 → 499	498 → 499		
H	Me	Me	H	47	76	61	228–230	—
H	Me	Ph	H	58	75	43	193–195	—
H	Me	Me	PhCH$_2$	73	76	59	259–261	360
H	Me	Me	C(Me)$_3$	67	70	64	243–245	302–303
Ph	Me	Me	Me	62	75	64	216–218	328
Ph	Me	—(CH$_2$)$_4$—			40	62	70	163–165
								302–303
Ph	Ph	Me	Me	55	57	42	218–220	321–322
Ph	CF$_3$	Me	Me	—	65	48	—	299–301

The diesters **500**, on treatment with liquid ammonia, react selectively at the aliphatic ester site to yield the amides **501**, which can be dehydrated to the nitriles **502** by phosphorus oxychloride; these latter compounds cyclize to give pyrrolopyridones **503**.[315] The 1-pyrrolylurea derivative **504** is cyclized by base to yield the bicyclic **505**.[316] The acylaminopyrrole **506**, when heated with a series of substituted anilines in the presence of phosphorus pentoxide and triethylamine hydrochloride, produces the pyrrolopyrimidines **507** (R = 2-Me, 4-Me, 4-Et, 2-Cl, 4-Cl, 2-F, 3-F, 4-F) in 44–66% yields.[317] 1-(3-Thienyl)pyrrole-2-carboxylic acid (**508**) has been converted into the azide **509**, which forms the tricyclic product **510**, presumably via an electrophilic attack of a nitrene intermediate on the thiophene ring.[318]

1.4. Reactions with Nitrogen Nucleophiles

In a manner analogous to the reaction of 2-formylpyrrole (**77a**) with 1,2-diaminobenzenes, described in Section 1.4, the reaction of **77a** with the diaminonaphthalene **511** yields the adduct **512**. Dehydrogenation over palladium-on-carbon, or oxidation with manganese dioxide, gives **513** (m.p. 283–284°C) in 57% yield.[319] The same aldehyde (**77a**) reacts with benzil **514** in the presence of ammonium acetate and acetic acid to produce the compound **515**.[320] An imine, generated from **77a**, and the triphenylphosphine derivative **516**, in a sequence analogous to the Wittig reaction, cyclizes to give the triazole **517** (m.p. 197–198°C) in 33% yield.[321]

The thioether group in the amide **518** is deprotected by treatment with sodium in liquid ammonia to give the thiol **519** (m.p. 125–129°C) in 83% yield. This compound cannot be cyclized thermally, as the only product is the dimeric sulfide. However, hydrolysis of the amide to the corresponding carboxylic acid **519** (m.p. 140–141°C) by hydroxide can be accomplished in 88% yield, and **519** can then be cyclized by treatment with dimethylaminopyridine and dicyclohexylcarbodiimide to give **520** (43%).[286]

1.5. OXIDATION

2-Formylpyrrole (**77a**) has too much of the vinylogous amide character, as discussed in Section 1.2, to give a positive result with Fehling's solution[125] and 3-formylpyrrole (**212c**) is, if anything, even less reactive.[125] However, both compounds are oxidized by Tollen's reagent.[125,322] The N-methylated derivatives (**77b** and **212a**) are not generally oxidized, although treatment of 2-formyl-1-methylpyrroles with silver oxide is reported to give the carboxylic acid.[112] Oxidation to the appropriate carboxylic acids of both 2- and 3-formylpyrroles by aqueous potassium permanganate solution under basic conditions is a general reaction;[107,323] for example, the oxidation by permanganate of the N-protected aldehyde **521** to the corresponding acid **522** goes at room temperature in 70% yield,[324] and the same reagent is used to oxidize the formyl group of **523** to the carboxylic acid **524**; hydrogen peroxide, chromium trioxide, or bromine water give other oxidation products.[325] Oxidation of the dialdehyde **525**[326] with hydrogen peroxide (or, more efficiently, with *m*-chloroperbenzoic acid) gives the Baeyer–Villiger product **526**, which is hydrolyzed by base to yield the dehydropyrrolidone **289**.[222] The dipyrrolyl ketone **527** also undergoes a Baeyer–Villiger oxidation with *tert*-butyloxychloride in chloroform in the presence of ethanol to give the ester **528**.[327] In the absence of an alcohol, the 2-methyl substituent is simultaneously transformed into a chloromethyl group. The same oxidizing agent in 1,2-dichlorethane and in the presence of propan-2-ol converts the benzoylpyrroles **529** into the Baeyer–Villiger products **530a** (m.p. 112–113°C) and **530b** (m.p. 105°C) in 56 and 36% yields, respectively. No trace of the alternative Baeyer–Villiger isomers is detected.[328]

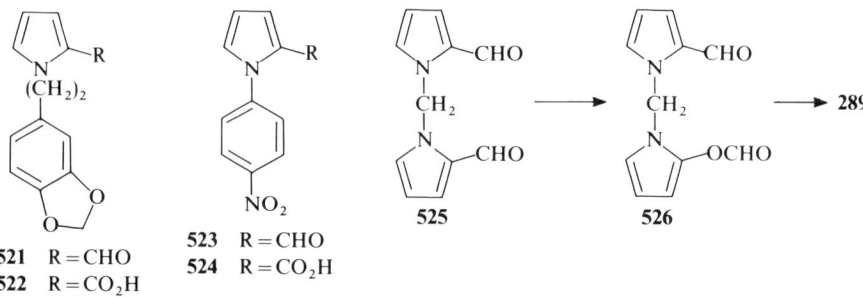

1.5. Oxidation

Oxidation of 2-acetylpyrrole (**220a**) with selenium dioxide gives the glyoxal **531**,[329] while treatment of **220a** with amyl nitrite produces the oxime of **531**.[329] 3-Acetylpyrroles also give α-ketoaldehydes with selenium dioxide, **532a** giving **533a**,[330] **532b** producing **533b**,[330] and **532c** leading to **533c**.[331]

The diol **535**, produced by reduction of the pyrrole α-ketoester **534** with sodium borohydride, is immediately cleaved by periodate oxidation to give 3-formylpyrrole (**212c**) in 63% overall yield; this route is claimed as the best available synthesis of **212c**.[332] Periodate oxidation also cleaves the hydroxy-ketones **536** to 2-formyl-1-methylpyrrole (**77b**) in ~20% yield and gives 2,5-diformyl-1-methylpyrrole (**203a**) in a similar yield from **537**.[333] The same conditions can be used to prepare the 2,4-dialdehyde **539** from **538**.[333]

The haloform reaction of 3-acetyl-1-tosylpyrrole **540** using bromine and aqueous sodium hydroxide is a formal oxidation and leads to 1-tosylpyrrole-3-carboxylic acid (**541**), which is readily deprotected by base to give pyrrole-3-carboxylic acid (**542**) in 60% overall yield; this is claimed to be the best synthesis of **542**.[334] The intermediate 2-trihalomethylketones **543**, which can also be obtained by trihaloacetylation of pyrrole, are often isolated;[335-337] and, with base, they give the corresponding carboxylic acid **544**.[335] The corresponding treatment of 2-trichloroacetylpyrroles **543** with alcohols produces high yields (often > 80%) of the corresponding esters **545**,[338] and the reaction with primary and secondary amines results in the formation of secondary and tertiary amides.[339] 1-Methyl-2-trichloroacetylpyrrole **546**, on reaction with sodium hydroxide, gives the corresponding carboxylic acid **547**. In a one-pot reaction, this may be converted into the acid chloride **548** by treatment with oxalyl chloride and then reacted with pyrrolylmagnesium bromide to give the dipyrrolyl ketone **549** (m.p. 129–131°C) in 45% yield.[340] The 3-trifluoroacetylpyrrole **550** is converted into the corresponding carboxylic acid **551** by treatment with sodium hydroxide in methanol.[341] 3-Trichloroacetylpyrroles (e.g., **552**) are also converted into the carboxylic acids on treatment with base.[335]

R^1	R^2	R^3	R^4
H	H	H	H
Me	H	H	H
H	Me	H	Me
H	Me	CO_2Et	Me
H	Me	COEt	H

R^1	R^2	R^3	R^4	R
H	H	H	H	Me
H	H	H	H	Et
H	H	H	H	CH_2=$CHCH_2$
H	H	H	H	CMe_2=$CHCH_2$
H	CO_2Et	Me	Et	Me
H	CO_2Me	Me	$CHMe_2$	$PhCH_2$
H	Me	H	Me	CCl_3CH_2
H	Me	H	Me	$Cl(CH_2)_2$

546 → **547** → **548** → **549**

550 → **551**   **552**

1.6. REDUCTION

The carbonyl groups of acylpyrroles may be reduced by a variety of hydride reducing agents, by catalytic hydrogenation, or by more indirect means. Catalytic hydrogenation requires careful control as the pyrrole ring is itself susceptible to reduction, especially under acidic conditions (see Part 1). In general, formylpyrroles are more easily reduced than are ketones, which in turn are more susceptible to reduction than are esters or carboxylic acids, although the ease of reduction is often dependent on pH.[107] Reduction may, in the first place, lead to the production of the corresponding alcohols, but these are often unstable under the reaction conditions or during workup and are converted into the azafulvenes **553**, which may undergo further reduction to the pyrrole with a fully saturated sidechain[107] or may polymerize to give materials that until recently were discarded as "tars," but that now attract considerable interest as potential organic conductors.

553

1.6.1. Reduction Using Lithium Aluminum Hydride

The reduction of 2-formylpyrrole (**77a**) by lithium aluminum hydride gives 2-methylpyrrole, whereas 2-formyl-1-methylpyrrole (**77b**) is reduced to the unstable alcohol **554**,[342] supporting the interpretation that complete reduction of acyl groups on pyrrole goes via the azafulvene **553**.[107] More highly alkylated formylpyrroles **555** also undergo complete reduction of the formyl group to the methylpyrrole **556**, when the nitrogen atom is unprotected,[342,343] or to the carbinol **556c** in the case of the N-methylated derivative **555c**.[344] Complete reduction requires the use of excess hydride; if only an equimolar amount is used, polymeric tars result.[345] Pyrrolyl ketones may also be reduced to the corresponding saturated systems.[127,327,343-346] Thus, 3-propionylpyrrole is reduced to 3-propylpyrrole[127] and the acetylpyrroles **555e** and **555g** are reduced to the ethyl derivatives **556e**[344] and **556g**,[344,345] respectively, and benzoylpyrrole **555f** is reduced to the benzyl compound **556f**.[345] 2-Acetyl-, 2-propionyl-, and 2-benzoylpyrrole are reduced to the corresponding saturated compounds, 2-ethyl-, 2-propyl-, and 2-benzylpyrrole, respectively, on treatment with excess lithium aluminum hydride in dry ether,[346] but the unstable carbinols have also been formed.[347] Dipyrrolyl ketones are reduced to the corresponding dipyrrolylmethanes,[127] for example, **527** → **557**.[327] The analogous ketone **558a** is reduced to **558b**,[343] but the N-protected cyclic ketone **559** is converted into the corresponding alcohol **560**.[348]

	R^1	R^2	R^3	R^4	R^5	R^1	R^2	R^3	R^4	R^5
a	H	Me	H	CHO	Me	H	Me	H	Me	Me
b	H	Me	H	Me	CHO	H	Me	H	Me	Me
c	Me	Me	H	CHO	Me	Me	Me	H	CH$_2$OH	Me
d	H	H	Me	Ac	Me	H	H	Me	Et	Me
e	H	Ac	Me	Ac	Me	H	Et	Me	Et	Me
f	H	Me	COPh	Me	H	H	Me	CH$_2$Ph	Me	H
g	H	Me	Ac	Me	H	H	Me	Et	Me	H

558 a X = O
         b X = H$_2$

Pyrrole esters are reduced more slowly than are aldehydes or ketones.[345] The butanoyl substituent of **561** is converted into a butyl group, while the ester group remains intact, giving **562** as the product.[344] Carboxylic esters can be reduced to methyl groups in the absence of N-protection of the pyrrole ring (e.g.,

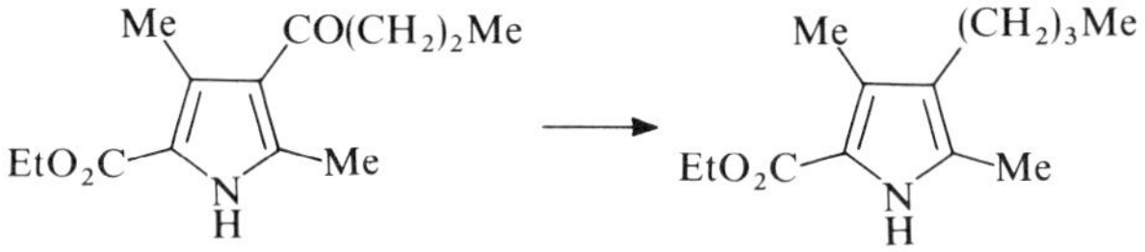

1.6. Reduction

561 → **562**

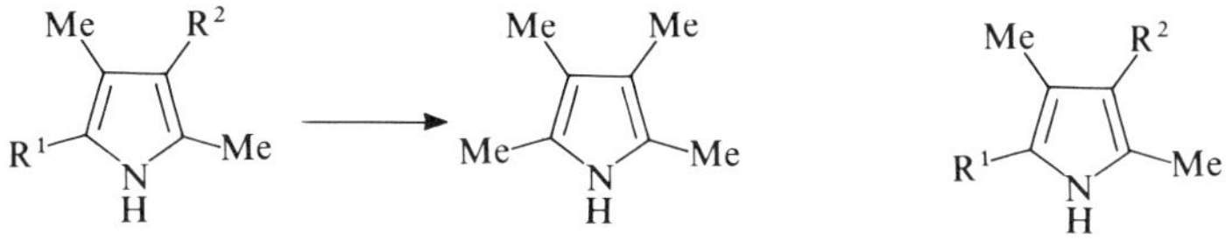

563 → **564**

a $R^1 = CO_2H$, $R^2 = CHO$
b $R^1 = R^2 = CO_2H$
c $R^1 = CO_2H$, $R^2 = CO_2Et$
d $R^1 = R^2 = CO_2Et$

565

a $R^1 = CO_2Et$, $R^2 = Et$
b $R^1 = CO_2H$, $R^2 = Ac$

563 → **564**;[344,345] ethyl 2-methylpyrrole-3-carboxylate is slowly reduced to 2,3-dimethylpyrrole,[345] and the esters **565** give 3-ethyl-2,4,5-trimethylpyrrole.[344] When the pyrrolyl nitrogen atom is protected by an alkoxycarbonyl group (e.g., **566**), it is cleaved during the reduction and the reaction goes to completion to give the alkylpyrrole **567**,[342] but, in the case of methyl 1-methylpyrrole-2-carboxylate, reduction stops at the carbinol stage,[349] as the intermediate azafulvene cannot be formed. With N-arylpyrroles, reduction of the ester groups also stops at the carbinol stage, **568** → **569** (Table 1.11),[350] and reduction of the N-protected diesters **570** also produces the diols **571**.[351]

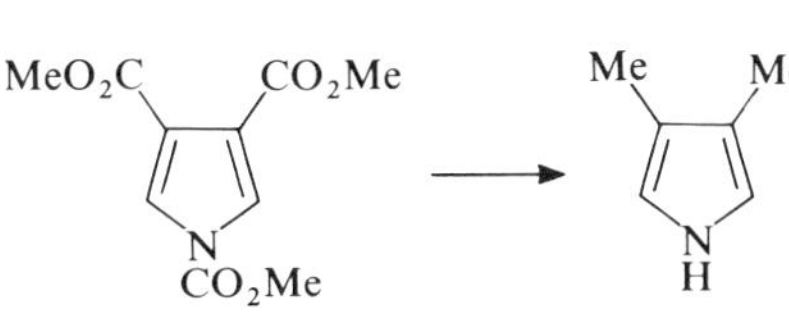

566 → **567**

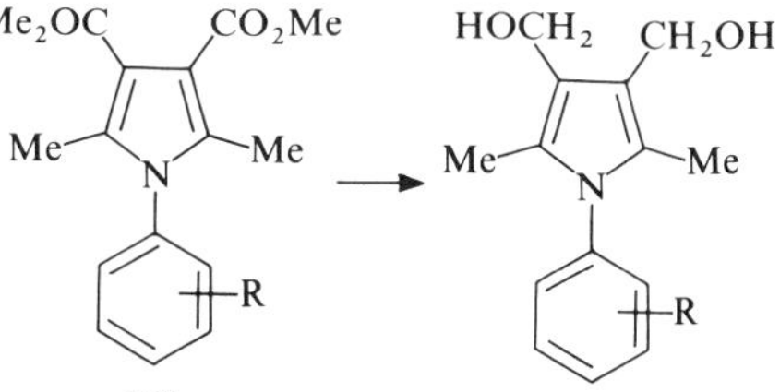

568 → **569**

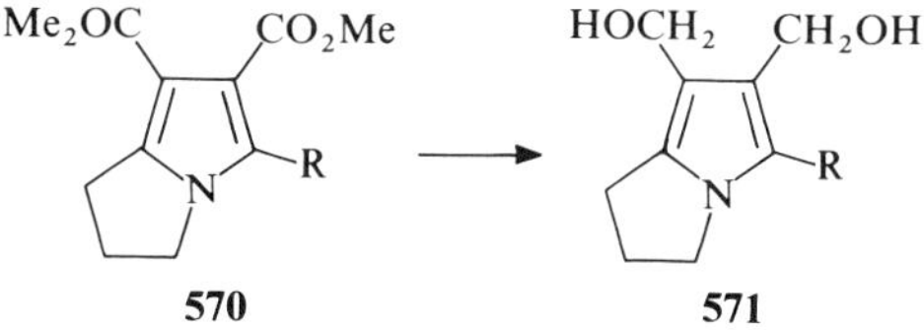

570 → **571**

$R = CF_3$; 3,4-$Cl_2C_6H_3$; 4-ClC_6H_4; Ph; 4-$MeOC_6H_4$; 4-FC_6H_4; Me

TABLE 1.11. REDUCTION OF PYRROLE-3,4-DICARBOXYLATES **568** WITH LITHIUM ALUMINUM HYDRIDE[356]

R	m.p. (°C) of Diester **568**	m.p. (°C) of Dicarbinol **569**
4-OMe	75–76	143–145
4-OEt	129–130	134–136
3,4-OCH$_2$O	131–132	143–145
4-Me	119–120	157–159
4-n-Bu	93–94	Oil
4-F	109–110	156–158
4-Cl	110–111	163–165
4-Br	139–140	164–167
3,4-Br$_2$	83–92	83–84

N-Acyl groups, whether ketones or esters, are generally cleaved by lithium aluminum hydride to the N-unprotected pyrrole.[352] However, it has been reported that, by using a modified technique[353] in which reduction is performed in the presence of aluminum trichloride, N-acetyl groups may be reduced to N-ethyl groups.[107]

1.6.2. Reduction by Boron Hydrides

Reduction of N-protected pyrrole aldehydes and ketones with sodium borohydride generally leads to the unstable carbinols.[342,354] As with the reduction with lithium aluminum hydride, 2-formyl-1-methylpyrrole (**77b**) gives the alcohol,[342] although workup conditions with borohydride are usually less inimical to such unstable carbinols.[355,356] The Maillard product **51**[77] has been prepared from 2,5-diformylpyrrole in a reaction sequence where the borohydride reduces only one of the aldehyde groups of **572**.[79] The 2-formylpyrroles **405** are reduced by sodium borohydride to the corresponding alcohols **573** in yields of 50–85%,[270] but it is interesting that the corresponding reduction of 2-acylpyrroles in the absence of an N-protecting group leads to the corresponding 2-alkylpyrroles, although somewhat forcing conditions may be required.[340,357] 2-Formylpyrrole (**77a**) is not generally reduced under these conditions (Table 1.12).[340,357]

OHC─⟨N⟩─CHO
(CH$_2$)$_2$
CH—NHAc
CO$_2$Me
572

⟶

HOCH$_2$─⟨N⟩─CHO
(CH$_2$)$_2$
CH—NHAc
CO$_2$Me

⟶ **51**

Me$_2$OC──Ar
Me─⟨N⟩─CH$_2$OH
       H
573
Ar = Ph; 3,4-(MeO)$_2$C$_6$H$_3$;
3,4-OCH$_2$OC$_6$H$_3$; 2-NO$_2$-4,5(MeO)$_2$C$_6$H$_2$

TABLE 1.12. REDUCTION OF 2-ACYLPYRROLES WITH SODIUM BOROHYDRIDE

$$\text{pyrrole-COR} \xrightarrow[(CH_3)_3CHOH]{NaBH_4} \text{pyrrole-CH}_2R$$

R	Yield (%) of Alkylpyrrole	m.p. (°C)	Ref.
N-Methylpyrrol-2-yl	65	74–75	340
Ph	99	Oil	357
3,4-(MeO)$_2$C$_6$H$_3$	98	94–95	357
3,4-(PhCH$_2$O)$_2$C$_6$H$_3$	88	54–55	357
CH$_3$(CH$_2$)$_{14}$	98	67–68	357
PhCH$_2$	92	45–47	357
H	13	Oil	357

Reductive amination of 2-formylpyrrole (**77a**) or its N-methyl derivative (**77b**) occurs when they are treated with an amino compound in the presence of sodium borohydride. The reaction involving **574** produces **575a** and **575b**, which can act as blockers of adrenoreceptor sites.[359]

$$77 + H_2N(CH_2)_6NH(CH_2)_2SS(CH_2)_2NH(CH_2)_6NH_2$$
574

↓

pyrrole(R)-CH$_2$NH(CH$_2$)$_6$NH(CH$_2$)$_2$SS(CH$_2$)$_2$NH(CH$_2$)$_6$NHCH$_2$-pyrrole(R)

575
a R = H
b R = Me

With diborane, 2-formylpyrrole (**77a**) gives polymeric material only, and a similar result is obtained with its N-methyl derivative (**77b**).[360] However, formyl and acyl groups in other pyrroles are reduced by diborane, usually to the corresponding alkyl side chains.[327,361] The amide **576** is reduced by diborane to a mixture of the alkane **577a**, the carbinol **577b**, and a boron complex of the amine **577c**.[327] The 3-chloroacetylpyrrole **578** reacts with diborane, whether generated in situ or added after being generated externally, to form the chlorohydrin **579** and the vinyl chloride **580**; **579** is the only product formed when sodium borohydride is used as the reducing agent.[362] If the diborane reduction is performed in the presence of boron trifluoride–ether complex, the chloroethylpyrrole **581** is produced instead.[362] Diborane reacts very slowly, if at all, with pyrrole carboxylic acids or esters.[360,361,363] With methyl pyrrole-2-carboxylate, diborane reduction gives 2-methylpyrrole slowly and in poor yield.[363] Bromination of the ring, however, appears to enhance this reaction dramatically; methyl 3,4,5-triboromopyrrole-2-carboxylate is converted into 3,4,5-tribromo-2-methylpyrrole in 15 min in excellent yield.[363]

1.6.3. Reduction by Diisobutylaluminum Hydride

Reduction of vinylogous pyrrole aldehydes to the corresponding alcohols is most conveniently carried out using diisobutylaluminum hydride.[364,365] This reagent also reduces nitriles to either alcohols or aldehydes, depending on the reaction conditions. In a synthesis[366] of 2,3-diformylpyrrole (**584**),[367,368] 3-cyano-2-formylpyrrole (**582**) is reduced with diisobutylaluminum hydride at 0°C to give the 3-hydroxymethyl compound **583** (m.p. 107–109°C) in 50% yield; oxidation of the latter with manganese dioxide produces **584** in 72% yield.[366] The corresponding reduction of the protected aldehyde **585** with diisobutylaluminum hydride at −30°C gives the aldehyde **586** (m.p. 94–99°C) in 92% yield.[366] An analogous synthesis of an acid-sensitive formylpyrrole involves

1.6. Reduction 73

oxidation with manganese dioxide of the carbinol **588** (the product of lithium aluminum hydride reduction of **587**) to give the aldehyde **589**.[310]

1.6.4. Reduction Using Catalysts

Catalytic reduction may be used to reduce pyrrole aldehydes or ketones to the corresponding alkyl compounds, provided palladium is used as the catalyst. The acetyl substituent in **590** is reduced to the ethyl compound **591**.[369] The use of platinum catalyst, however, usually results in reduction of the ring to give the pyrrolidine, especially under acidic conditions.[370,371] In the case of 2-formyl-pyrroles having no substituent at C-5, an alternative reductive dimerization occurs to form dipyrrolylmethanes, presumably via the acid-catalyzed condensation to the dipyrrolymethene followed by hydrogenation.[371] Catalytic hydrogenation over rhodium-on-carbon at atmospheric pressure of the 1,2-dibenzoylpyrrole **592** gives the pyrrolidine **593** in 63% yield with no reduction of the carbonyl groups.[372] The yield is raised to 84%, when a platinum catalyst is used. The analogous hydrogenation of ethyl 2-benzoylpyrrole-1-carboxylate gives the pyrrolidine (69%) when Adams' catalyst is used, together with 10% yield of the cyclohexoylpyrrolidine.[371] Rhodium-on-alumina catalyzes the reduction of the diester **246** to give the pyrrolidine **594** in 79% yield.[200]

Using Raney nickel as a catalyst, hydrogenation of N-benzoylpyrroles generally results in the cleavage of the benzoyl group; C-benzoyl and other C-acyl groups may be reduced.[373] Care is needed in this reaction, as reduction under pressure may result in the conversion of acylpyrroles into the corresponding pyrrolidines,[375–377] although under suitable conditions only the acyl group is reduced.[279,373,377] However, Raney nickel will not catalyse the reduction of carboxylic acids, nor carboxylate esters, so that acetylpyrroles that also bear alkoxycarbonyl groups may be selectively reduced using this catalyst.[279,373] Pyrrole carboxylates may, in fact, act as catalyst poisons.[279,378] Reduction of the ester group requires copper chromite.[373] N-Ethoxycarbonyl groups may also be reduced using copper chromite.[373,377]

Pyrrole esters may also be reduced by indirect means via conversion into the thioesters, which are susceptible to hydrogenolysis by Raney nickel. The ethyl pyrrolethiocarboxylate **595**, for example, is reduced to the aldehyde **596**, which may be further reduced to the carbinol **597**, although the latter reaction proceeds slowly.[379] The reaction may be interrupted to give formylpyrroles.[380] Use of a more active Raney nickel catalyst (W-5) at higher temperatures and pressure results in the reduction of the thioester **595**, or the aldehyde **596**, to the methyl compound **598**.[379] In contrast, reduction of the thioester **599** produces only the hydroxymethyl derivative **600**.[381]

It has been reported that hydrogenolysis of thioesters over Raney nickel may lead to complete cleavage of the substituent.[104,382] Thus, for example, the thioester group in **601** is cleaved to yield 3-formylpyrrole in 52% yield, although the dialdehyde **602** (m.p. 152°C) is also produced in this reaction.[383] The corresponding 3-acetylpyrrole **604** is obtained in 79% yield by cleavage of the thiocarboxylate group of **603** on treatment with Raney nickel.[382] With a partially deactivated catalyst, the cleavage is suppressed and the hydroxymethyl product **605** (m.p. 128°C) is obtained in 52% yield instead.[382]

1.6.5. Other Methods of Reduction

Pyrrole aldehydes may be reduced to the corresponding alcohols by aluminum amalgam in base or under neutral conditions.[383] With sodium amalgam, 2-formylpyrrole (**77a**) is reported to give the pinacol **606**.[354] A pinacol **607** is also claimed as the major product of the reduction of 2-acetylpyrrole with sodium amalgam; the hydroxyethylpyrrole **608** is also obtained in this reaction.[384] Acylpyrroles do not undergo the pinacol reaction easily.[365] Because of the lability of the pyrrole ring in the presence of acid, reduction by metal amalgams are generally carried out in basic or neutral media. However, the ketones **609** were reduced to the corresponding saturated species **610** using zinc amalgam and hydrochloric acid.[385] It is interesting to note that acylpyrroles of the type **609** ($X = SO_3^- Na^+$) were prepared for electrochemical polymerization to self-doping conducting polypyrrole.

Pyrrole carboxylic acids and their esters do not appear to be reduced by aluminum or sodium amalgams.[107]

Formyl and acetylpyrroles may be reduced indirectly via their hydrazones (see Section 1.4.2) using the Wolff–Kishner procedure.[151,301,386–389] For example, 2-formyl-3,4,5-trimethylpyrrole **133** is reduced to 2,3,4,5-tetramethylpyrrole;[151] and ethyl 2-ethyl-3-formyl-4-methylpyrrole-5-carboxylate, to ethyl 2-ethyl-3,4-dimethylpyrrole-5-carboxylate.[301] A Wolff–Kishner reduction of the acylpyrrole **611** ($n = 7$) to the alkylpyrrole **612** ($n = 7$) was used in the synthesis of 2-formyl-5-n-nonylpyrrole **23** ($n = 8$), which has been isolated from the soft coral *Telesto*.[71] In the same way, the ketones **611** ($n = 2,3,4,5,6,9,11$) were converted into the corresponding alkyl derivatives **612** in yields of 57–93%.[388] The earliest reported use of this procedure is the reduction of **611** ($n = 1$) to **612** ($n = 1$).[389] The semicarbazone of 2-formylpyrrole (**77a**) is decomposed by base to give 2-methylpyrrole.[390]

$$\underset{\mathbf{611}}{\text{pyrrole-CO(CH}_2)_n\text{Me}} \xrightarrow{\text{Wolff–Kishner}} \underset{\mathbf{612}}{\text{pyrrole-(CH}_2)_{n+1}\text{Me}}$$

Pyrrole aldehydes do not normally undergo Cannizzaro reactions. However, the crossed Cannizzaro between 2-formyl-1-methylpyrrole (**77b**) and formaldehyde causes reduction of the formyl group to give the hydroxymethyl derivative **554** in reasonable yield.[355] The formal reduction of the carboxylic acid group in **613** to the acetylpyrrole **614** goes in 87% yield on treatment with methyllithium.[8] In the absence of an *N*-protecting group, however, pyrrole-2- and -3-carboxylic acids do not undergo the corresponding reaction, presumably as a result of the preferential formation of the pyrrolyl anion.[274] The treatment of pyrrole-2-carboxylic acid (**615**) with 50% phosphorous acid, acetic acid, and hydrogen iodide causes partial reduction of the ring to give dehydroproline (**616**) in 76% yield.[391,392]

$$\underset{\mathbf{613}}{\text{pyrrole}} \xrightarrow{\text{MeLi}} \underset{\mathbf{614}}{\text{pyrrole}}$$

$$\underset{\mathbf{615}}{\text{pyrrole-CO}_2\text{H}} \xrightarrow{\text{H}_3\text{PO}_4/\text{HI}} \underset{\mathbf{616}}{\text{dihydropyrrole-CO}_2\text{H}}$$

1.7. PROTECTING GROUPS IN ACYLPYRROLES

1.7.1. Protection of C-Acyl Groups

1.7.1.1. Aldehydes

Formylpyrroles do not form acetals unless other electron-withdrawing groups are substituted on the ring. Treatment of simple formylpyrroles with 1,2- or 1,3-diols in the presence of acid generally results in the formation of the

1.7. Protecting Groups in Acylpyrroles

products of acid-catalyzed dimerization or trimerization of the pyrroles. Formylpyrroles substituted with electron-withdrawing groups do, however, give acetals. Thus, for example, the carboxylic acid **617**, on treatment with 2-methylpentane-2,4-diol, gives the acetal **618**,[393a] and the ester **619** gives the acetal **620** under the same conditions.[393a] The formylpyrrole **619** has also been converted into the hemiacetal **621**.[393a]

A more suitable diol for the preparation of the acetals is 2,2-dimethylpropane-1,3-diol in the presence of *p*-toluenesulfonic acid.[394] The acetal groups are readily cleaved and the aldehyde regenerated by trifluoroacetic acid. Treatment of the monoprotected dialdehyde **586** with trifluoroacetic acid at 0°C, however, results mainly in the transfer of the protecting group from the 2- to the 3-formyl group to give **622**; at higher temperatures, deprotection occurs to give 2,3-diformylpyrrole (**584**) in 74% yield.[366] Propane-1,3-diol has also been used to prepare acetals; for example, **601** has been converted into the acetal **623**.[125] The dicyanovinyl group (see Section 1.3.4) can also be used as a protecting group, as for example, in condensation of **619** with malonitrile to give **624**.[393b] Protection of the formyl group has also been effected by a condensation reaction with ethyl cyanoacetate.[365]

619 ⟶ [structure **624**: pyrrole with MeO, Me, EtO_2C, CN, CN substituents]

1.7.1.2. Carboxylic Acids

It is often useful to be able to protect and deprotect carboxylic acids selectively, especially because free pyrrolecarboxylic acids are subject to decarboxylation (see Section 1.8).

The simplest procedure is conversion into their alkyl esters. The most common method of esterification involving heating the carboxylic acid with the relevant alcohol in the presence of a mineral acid is often accompanied by the formation of polymeric materials.[107] However, the method can be used, as illustrated by the conversion of **625** into the triester **626**, by treatment with methanolic hydrogen chloride.[394] Under the same conditions, the carboxylic acid group adjacent to the 3-ethoxycarbonyl substituent undergoes esterification to give **628** before complete reaction to produce **629**.[394] As an alternative to the reaction under acidic conditions, the silver salt of pyrrole-2-carboxylic acid (**615**), on treatment with methyl iodide, gives methyl pyrrole-2-carboxylate.[395] Treatment of **615** with silver oxide and methyl iodide gives a less pure product, while the reaction with potassium carbonate and methyl iodide leads to a 75% yield of the ester, but only if the potassium carbonate is completely anhydrous; moist potassium carbonate gives rise to methyl 1-methylpyrrole-2-carboxylate.[396] A better route to the methyl ester in 94% yield involves treatment of **615** with diazomethane in ether.[396] Among many other examples, diazomethane has also been used to convert **630** into **631**.[397] The dipyrrolyl-

625 (Me, CO_2Et, EtO_2C, CO_2H) —MeOH/HCl→ **626** (Me, CO_2Et, EtO_2C, CO_2Me)

627 (HO_2C, Me, CO_2Et, CO_2H) —MeOH/HCl→ **628** (HO_2C, Me, CO_2Et, CO_2Me) —MeOH/HCl→ **629** (MeO_2C, Me, CO_2Et, CO_2Me)

1.7. Protecting Groups in Acylpyrroles

$$\text{MeO}_2\text{C}\underset{\underset{H}{N}}{\diagdown\diagup}\text{CO}_2\text{H} \xrightarrow{\text{CH}_2\text{N}_2} \text{Me}_2\text{OC}\underset{\underset{H}{N}}{\diagdown\diagup}\text{CO}_2\text{Me}$$

630   631

632
a   R = H
b   R = CO$_2$H

633
a   R = H
b   R = Me

methane carboxylic acids **632a** and **632b** may be transformed into their respective ethyl esters **633a** and **633b** using diazoethane.[398]

A mixed anhydride method may also be used. For example, 2,4,5-trimethylpyrrole-3-carboxylate and its N-methyl derivative form mixed anhydrides when treated with trifluoroacetic anhydride; the resulting anhydrides have been reacted with the desired alcohol (tropanol, piperidol, or pyrrolidinol) to yield the corresponding esters.[399]

Phase-transfer catalytic esterification, using tetraalkylammonium halides, of the potassium salt of pyrrole-2-carboxylic acid (**615**) with suitable alkylating agents produces the corresponding esters in high yield.[400] For example, at room temperature, diethyl sulfate gives 87% of ethyl pyrrole-2-carboxylate, and, under similar conditions, benzyl bromide gives a similar yield of the benzyl ester; with 1-bromooctane, the reaction requires a temperature of 85°C to yield 93% of the octyl ester.[400]

tert-Butyl esters, of 1-methylpyrrole-2-carboxylic acid, for example, have been prepared by the acid-catalyzed reaction of the carboxylic acid with 2-methylpropene.[401] *tert*-Butyl esters have also been obtained from the reaction of trichloromethylpyrroles with *tert*-butanol and sodium acetate, **634** → **635**.[402]

634   635

R = (CH$_2$)$_2$Cl; CH$_2$CO$_2$Et; (CH$_2$)$_2$CO$_2$Me

The base-catalysed hydrolysis of simple methyl or ethyl esters of pyrrole-2-carboxylic acids is generally more rapid than that of the isomeric pyrrole-3-carboxylates.[151] It is not easy to rationalize this fact completely, although possible stabilization of the tetrahedral intermediate **636** has been invoked.[403] On the other hand, the mesomeric interaction of the 3-ester group with the pyrrole ring is weaker than that of the 2-ester, and the 3-ester might therefore be expected to hydrolyze more readily.[107] The base-catalyzed hydrolysis of the diester **637** yields the 2-carboxylic acid **638**;[404] similarly, it is the 2-ester of **639** that is hydrolyzed to give acid **640**.[405] Hydrolysis of ethyl 3,5-dimethylpyrrole-2-carboxylate is more rapid than that of the isomeric ethyl 2,4-dimethylpyrrole-3-carboxylate,[406] and, generally, hydrolysis of the esters in the 3-position requires longer reaction times.[407] The triester **641** hydrolyzes first to give **642** and then to give the diacid **643**.[394] Anomalously, the analogous methyl ester **644a** undergoes initial hydrolysis of the 2-ester group to produce **644b** before yielding the dicarboxylic acid **644c**.[394]

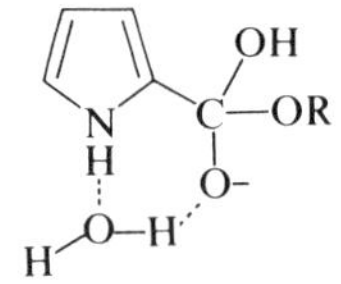

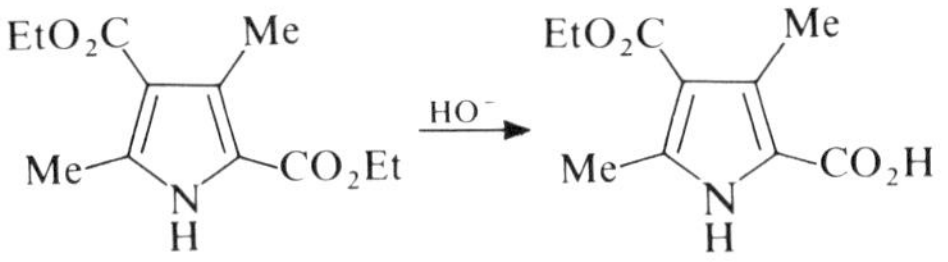

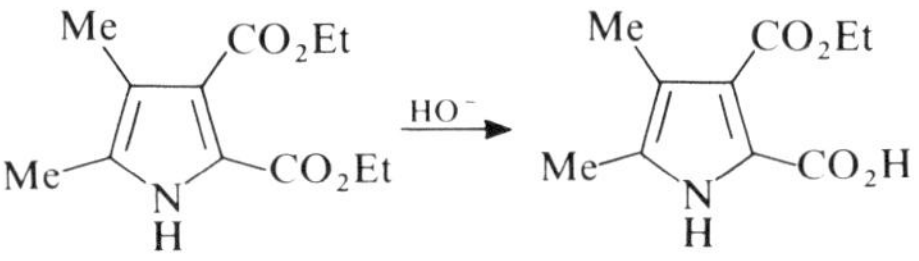

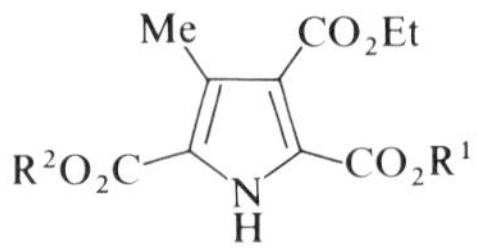

644
a $R^1 = R^2 = Me$
b $R^1 = H, R^2 = Me$
c $R^1 = R^2 = H$

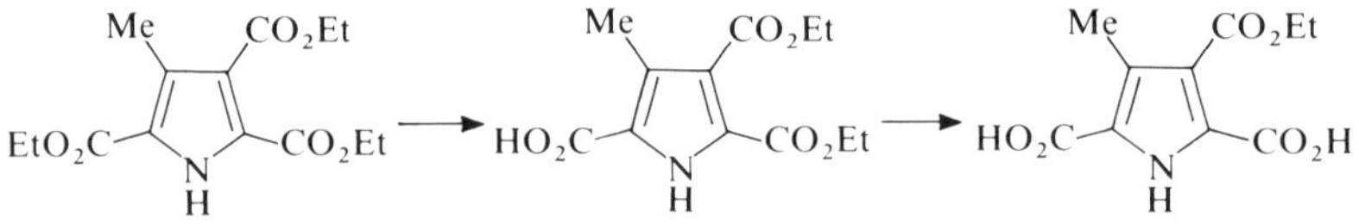

An alternative method of hydrolysis, often used for hindered esters, uses concentrated sulfuric acid at 35–40°C, followed by dilution of the reaction mixture on crushed ice.[408] Interestingly, this method results in a reversal of the order of hydrolysis; esters in the 3-position are selectively hydrolyzed, reflecting a greater stabilization of the intermediate acylium ion at the 3-position.[107] The ester **637**, for example, is hydrolyzed by this technique to the 3-carboxylic acid

1.7. Protecting Groups in Acylpyrroles

645.[409,410] The same result is obtained with analogous esters, **646a** → **647a**[410] and **646b** → **647b**.[411] Hydrolysis of **648** by aqueous ethanol containing hydrochloric acid gives **504**.[316]

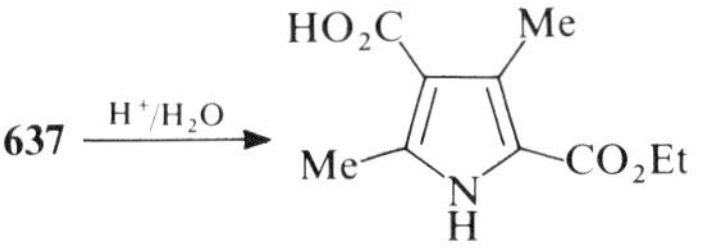

$$637 \xrightarrow{H^+/H_2O} 645$$

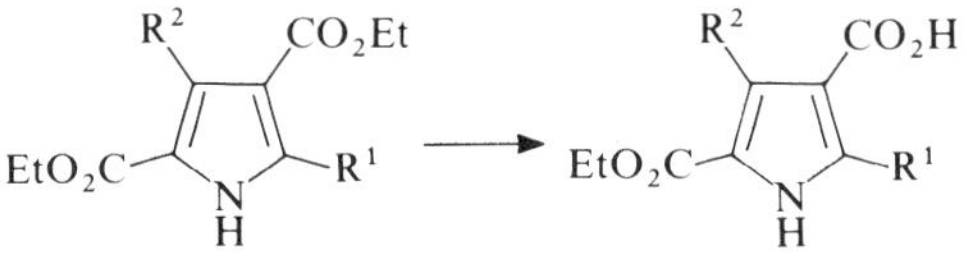

646 → 647

a $R^1 = Me$, $R^2 = Et$
b $R^1 = H$, $R^2 = Me$

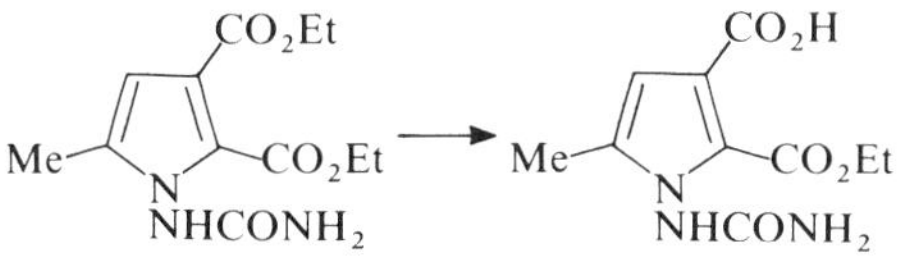

648 → 504

The ester **649a** under basic hydrolytic conditions (EtOH:KOH) gives a product **649b**.[412] This behavior of the β-ketoester side chain contrasts with that of the analogous N-substituted pyrrole, which, under the same conditions, produces the carboxylic acid **650b**.[413] The same hydrolysis of the 3-ester is observed in the corresponding β-ketoester, compounds **651a** and **615b**, and in compounds **652**.[413] The diester **653a** related to the pharmaceutical compounds

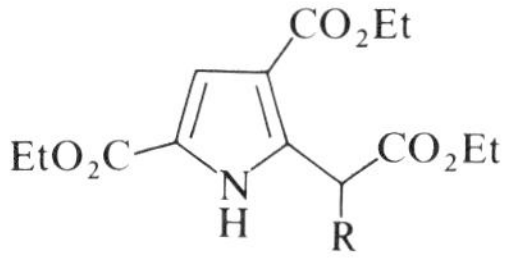

649
a R = Ac
b R = H

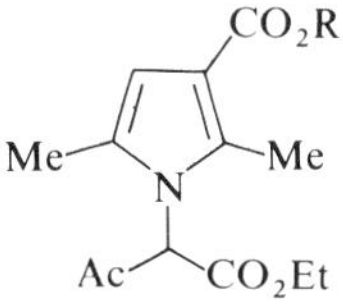

650
a R = Et
b R = H

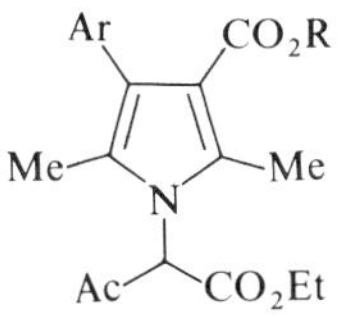

651
a R = Et
b R = H

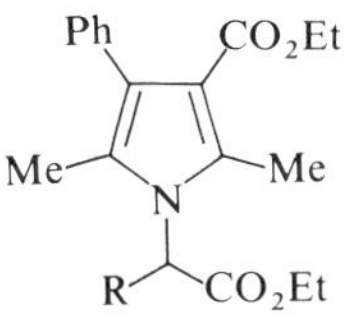

652
R = CN, CO_2Et, $CONH_2$

653
a R = H
b R = 3,4-$Cl_2C_6H_3$

55a and **55b**, on heating at 0°C in concentrated solution of sulfuric acid in ethanol, is hydrolyzed to give the carboxylic acid **654**, which is unstable under the reaction conditions.[414] This behavior contrasts with the claim that the acetate group of **653b** can be selectively hydrolyzed under acidic conditions.[415]

In the synthesis of complex pyrrole-containing natural products, selective protection and deprotection of the carboxylic acid groups is necessary. Thus, aryloxy esters are generally readily hydrolyzed by base, but are relatively stable under acidic conditions. Pentachlorophenoxy esters appear to be especially labile. In contrast, *tert*-butyl esters are rapidly hydrolyzed on treatment with catalytic amounts of acid; *p*-toluenesulfonic acid, sulfuric acid, phosphoric acid, and trifluoroacetic acid are all suitable for this purpose.[416-418] In a third method of deprotection, benzyl esters are cleaved by hydrogenolysis. For example, the esters **655**, on hydrogenation over Raney nickel, yield the carboxylic acids **656**.[276] Benzyl esters have also been cleaved by hydrogenolysis over palladium-on-carbon.[419]

R = Me; CH_2CO_2Me; $(CH_2)_2CO_2Et$

An extremely unusual reaction results when the benzyl esters **657** are treated under Reimer–Tiemann conditions with potassium hydroxide and chloroform in *tert*-butanol; the products **658**, in which the ester group has been replaced by a carboxaldehyde function, are obtained in low yield, presumably via hydrolysis followed by decarboxylative formylation.[420] This reaction also occurs with the corresponding carboxylic acid and ethyl ester of the chloro and bromo derivatives.

R = H, Et, Cl, Br, CHO

Carboxylic acids have also been protected as their oxazoline derivatives, prepared by reacting the acid chlorides with 2-amino-2-methylpropan-1-ol. In this way, for example, **443** has been converted into **444**;[294] and 1-methylpyrrole-2-carboxylic acid, into its oxazoline derivative.[421]

1.7.2. N-Protection

1.7.2.1. N-Protection by Acyl Groups

Earlier in this chapter it has frequently been noted that certain reactions of acylpyrroles fail unless the nitrogen atom of the ring is protected. The acidity of the unprotected group and the concomitant ease with which the proton is lost often inhibits the desired reaction. N-Acyl groups are readily removed by base and are thus widely used as N-protecting groups.

2-Formylpyrrole (**77a**) may be converted readily into its N-acetyl derivative;[422] using N,N-dimethylaminopyridine, triethylamine, and acetic anhydride, the conversion is effected in 80% yield.[423] The same conditions are used for the N-acetylation of **659** to give **660** (R = Me, m.p. 85–87°C, 91% yield; R = CHO, m.p. 172–173°C, 73% yield).[423] Benzoylation of the phenyl 3,4,5-trimethylpyrrole-2-carboxylate using benzoyl chloride in acetic acid gives the N-protected compound in 37% yield.[424]

R = Me, CHO

Removal of N-acyl groups is usually achieved by treatment with base. The kinetics and mechanism of this process have been examined in some detail.[425,426] The reaction sequence appears to be that shown in Scheme 1 for **661** (R = H, Me, Ph, 4-$NO_2C_6H_4$). The participation of the dianion **662** has been invoked to explain the somewhat complex rate equation required to account for the fact that the kinetics are almost first-order at high concentrations of alkali, but second-order in hydroxide in more dilute solutions.[425] In water at 25°C, formyl groups are displaced about 50 times faster than acetyl groups, and the rate for the removal of a formyl group is over two orders of magnitude greater than that for a benzoyl group.[426] With a salicyloyl protecting group (**661**, R = 2-HOC_6H_4), the rate at pH 10 is about three orders of magnitude greater than that for an acetyl group.[425,427] The suggested participation of the phenoxide anion **663** in enhancing the reaction is shown in Scheme 2.[425,427]

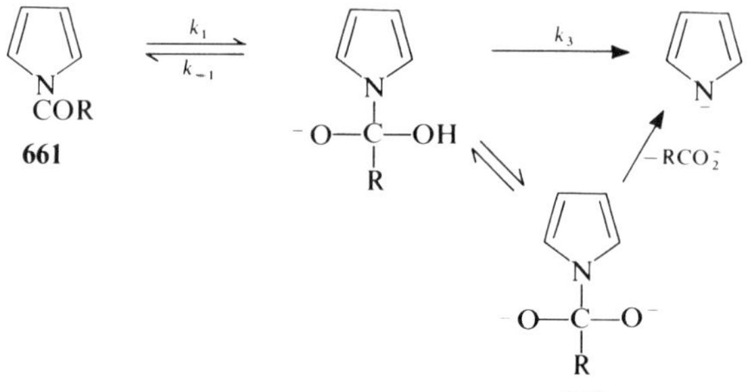

Scheme 1

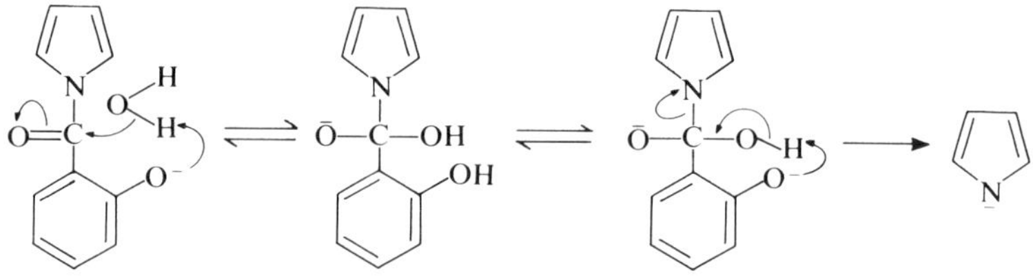

Scheme 2

Hydrolysis of the *N*-benzoylpyrrole diester **664**, on treatment with potassium hydroxide in methanol for 3 h, removes the benzoyl group to yield **665a**, leaving both ester groups intact. After a further 40 h, one of the ester groups has hydrolyzed, giving **665b**; complete hydrolysis to **665c** requires the use of 50% aqueous potassium hydroxide in methanol.[381,428] *N*-Benzoyl protecting groups can also be removed by hydrogenolysis using Raney nickel.[373]

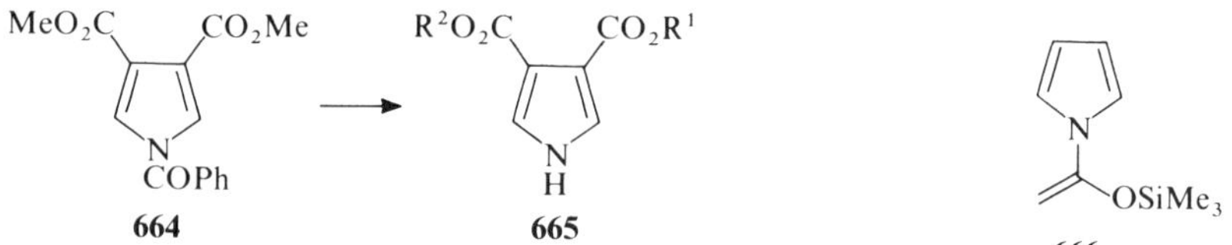

664 **665**
a $R^1 = R^2 = Me$
b $R^1 = H, R^2 = Me$
c $R^1 = R^2 = H$

666

A modification of the common acetyl protecting group is achieved by converting it into its trimethylsilyl enol ether, for example, **666**, by treatment of the *N*-acetyl compound with trimethylsilyl triflate and triethylamine.[429]

The removal of *N*-alkoxycarbonyl groups is also usually carried out using base. Deprotection of ethyl 3-formylpyrrole-1-carboxylate by treatment with sodium hydroxide regenerates 3-formylpyrrole in 36% yield.[274,382,430] *N-tert-*Butoxycarbonyl groups are particularly labile and can be introduced onto the pyrrole system by a variety of means.[431,432] For example, treatment of pyrrole with sodium hydride in tetrahydrofuran, followed by *tert*-butyl phenyl carbonate,[433] gives the *N*-protected species[434] and, in marginally better yield, by using (*tert*-BuOCO)$_2$ and *N,N*-dimethylaminopyridine in acetonitrile.[435] The former method gives 67% of the corresponding protected species of 2,5-dimethylpyrrole,[433] and the latter process has been used to obtain high yields of a range of *N*-*tert*-butyloxycarbonylpyrroles (Table 1.13).[435] 1-*tert*-Butoxycarbonylpyrrole, on treatment with *tert*-butyllithium, gives the *N*-protected 2-lithiated pyrrole,[436] which reacts with a variety of electrophiles to give *tert*-butyl 2-substituted pyrrole-1-carboxylates (2-Ac, 35%; 2-EtCHOH, 47%; 2-PhCO, 45%; 2-PhCHOH, 75%; 2-Me$_3$Si, 76%).[436] The protecting group is removed in moderate to high yield on treatment with sodium methoxide.[436] The *tert*-butoxycarbonyl group may also be removed under acidic conditions by treatment with trifluoroacetic acid.[433] It is also reported that heating *N*-*tert*-butoxycarbonylpyrroles above 150°C, in the absence of acid, base, or solvent, results in clean deprotection of the nitrogen, for example, **667a** → **667b** (99%), **668a** → **668b** (93%), and **669a** → **669b** (92%).[437]

667	668	669
a R = CO$_2$*t*Bu	a R = CO$_2$*t*Bu	a R = CO$_2$*t*Bu
b R = H	b R = H	b R = H

TABLE 1.13. *N*-PROTECTION OF ACYLPYRROLES BY *tert*-BUTYLOXYCARBONYL GROUPS[435]

R^1	R^2	R^3	R^4	Yield (%) of Protected Pyrrole	m.p. (°C)
OEt	H	H	H	85	10–11
CCl$_3$	H	H	H	81	8–9
OEt	H	H	Me	94	28
OEt	Me	H	Me	94	5–6
OEt	Me	CO$_2$Et	Me	81	32–33
OCH$_2$Ph	H	NO$_2$	H	87	101
OCH$_2$Ph	H	NHCO$_2$*t*Bu	H	80	126–127

The lithium salt of pyrrole-1-carboxylic acid has also been used to protect the nitrogen atom in the synthesis of 2-substituted pyrroles. Treatment with butyllithium, followed by the appropriate electrophile and aqueous workup conditions, yields the corresponding 2-substituted pyrroles (2-CO$_2$H, 2-CHO, 2-Tos, 2-PhNCO, Ph$_2$COH).[438]

1.7.2.2. N-Protection of Acylpyrroles

A variety of N-protecting groups have been used for acylpyrroles. The N-p-toluenesulfonyl derivative of pyrrole-3-carboxylic acid **541** is deprotected by treatment with sodium hydroxide,[334] whereas the N-protecting group of 1-triphenylmethylpyrrole-3-carboxylic acid can be removed by sodium in liquid ammonia.[341] Pyrroles may also be protected as their N-(2-chloroethyl) derivatives by treatment with 1,2-dichloroethane and tetra-n-butylammonium iodide; 2-formylpyrrole is converted into its 1-(2-chloroethyl) derivative in 94% yield, and 2-benzoylpyrrole produces the corresponding N-protected compound in 89% yield.[439] Deprotection is achieved by treatment with sodium in acetonitrile, followed by either mercury(II) acetate and sodium borohydride, or hydrochloric acid and sodium acetate; 2-formylpyrrole is regenerated in 85% yield.[439] An analogous protecting group containing sulfur may be used; the N-protected species **670a** and **670b** are prepared in 84 and 87% yields, respectively.[439] Deprotection of **670b** by treatment with sodium hydride or diazabicyclo[4.3.0]non-5-ene regenerates 2-benzoylpyrrole (91%).[439] The trimethylsilylethyl methyl ether function has also been used for N-protection of acylpyrroles. The protecting group is introduced by treating the appropriate pyrrole with sodium hydride and 2-(trimethylsilyl)ethoxymethyl chloride; the N-protected derivatives of 2-substituted pyrroles **671** are prepared in high yield (2-CHO, 948%; 2-Ac, 91%; 2-PhCO, 94%).[440] The protecting group is removed in two stages by treatment with boron trifluoride–diethyl ether complex to give initially 1-hydroxymethylpyrroles, which are then converted by base into the unprotected pyrroles in high yield.[440]

670
(CH$_2$)$_2$SO$_2$Ph
a R = CHO
b R = COPh

671
CH$_2$O(CH$_2$)$_2$SiMe$_3$

1.8. DECARBOXYLATION AND RELATED REACTIONS

The decarboxylation of pyrrolecarboxylic acids is a valuable synthetic tool; the acid (or ester) function is used to block a ring position, which can then be released for subsequent substitution on removal of the acid group.

1.8. Decarboxylation and Related Reactions

1.8.1. Decarboxylation

Many pyrrolecarboxylic acids decompose with the evolution of carbon dioxide on heating at or above their melting points. Pyrrole-3-carboxylic acid decarboxylates at 161°C;[441] and the 2-carboxylic acid, at 192°C.[442] Of the isomeric monoesters, **672a** and **672b**, the former loses carbon dioxide at 210°C;[443] the latter, at 270°C.[411] Decarboxylation may be aided by heating with copper bronze,[151] as, for example, for the decarboxylation 4-methylpyrrole-3-carboxylic acid.[444] Heating with copper chromite, often supported on barium oxide, in the presence of quinoline is also commonly used to bring about decarboxylation. The acids **673a** and **673b** have been decarboxylated in this fashion,[274,445] as have **674** and **675**.[446] The dicarboxylic acid **677**,[447] which is produced by hydrolysis of the diester obtained by thermal decarboxylation of **676** after moistening with glycerol,[279] loses the 2-carboxylic group on heating in water to give **678**.[448] The monoester **679** may be decarboxylated by boiling in water for 7 h, or by treatment with ethanol saturated with hydrogen chloride, to give **680**.[316] It is of interest that the same product is obtained by heating the isomeric monoester **504** with saturated ethanolic hydrogen chloride,[316] presumably via an intramolecular transesterification to **679**, as the 3-carboxylate esters are more difficult to hydrolyze.[107] Heating in water, glycerol, or 2-aminoethanol is often used as a milder alternative to high-temperature pyrolysis. This technique, using glycerol, has been used for the decarboxylation of 4-methylpyrrole-3-carboxylic acid[444] and for the conversion of **681** into **682**.[417] Heating with 2-aminoethanol has been employed in the decarboxylation of

679 → **680**

(structures: ethyl 5-methyl-1-(ureido)pyrrole-2-carboxylic acid-3-carboxylate → corresponding decarboxylated product retaining NHCONH$_2$)

681 → **682**

(structures: 4-carboxy-2-methyl-3-(ethoxycarbonylmethyl)-5-(ethoxycarbonyl)pyrrole → decarboxylated analog)

683a,[449] **683b**,[450] and **684**;[451] **685** gives 3-methylpyrrole presumably via the amide formed on reaction between the 3-carboxylic ester and 2-aminoethanol.[452] Amide groups are displaced less readily than are carboxylic acids.[107]

683 (R group at 4-position, Me at 3, Me at 5, CO$_2$H at 2)
  a R = Ac
  b R = CN

684 (ArCO at 4-position, CO$_2$H at 2)

685 (Me at 4, CO$_2$Et at 3, CO$_2$H at 2)

Esters of pyrrolecarboxylic acids having electron-donating groups on the ring are decarboxylated under fairly mild conditions. Distillation of an alkaline solution of the ester, or even heating under reflux, may suffice. In this manner, ethyl 2-methylpyrrole-5-carboxylate has been converted into 2-methylpyrrole.[453] Similarly, heating ethyl 2,3,4-trimethylpyrrole-5-carboxylate under reflux with aqueous potassium hydroxide gives 2,3,4-trimethylpyrrole,[151,454–456] and 2-ethyl-3,4-dimethylpyrrole has been obtained from the corresponding 5-carboxylic ester in the same fashion.[301] The carboxylic acid **686** undergoes decarboxylation, when heated with acidified methanol, with concomitant nucleophilic substitution of the 5-bromo group by the solvent molecule to give **687**; this reaction does not occur with the corresponding ester.[457]

686 (5-Br, 3-Et, 4-Me, 2-CO$_2$H pyrrole) —MeOH/HCl→ **687** (5-MeO, 3-Et, 4-Me pyrrole)

Pyrrolecarboxylic acids having several electron-withdrawing substituents on the ring are remarkably resistant to decarboxylation even at elevated temperatures.[449,458,459] Thus, for example, the acid **688** cannot be decarboxy-

1.8. Decarboxylation and Related Reactions

lated by any of the common methods.[460] Not surprisingly, the more labile N-ethoxycarbonyl compound **689** is readily converted by heating at 160°C into the decarboxylated species **690** (m.p. 35°C) in 80% yield.[460] It is possible to circumvent the difficulty in decarboxylating resistant carboxylic acids by using an indirect route. The carboxylic acid function may be electrophilically substituted by bromine or iodine; the pyrrolyl bromides and iodides are subject to reductive cleavage using hydrogen over palladium-on-carbon catalyst.[107] In this way, **691a** has been converted into **691c** via the bromide **691b**.[394] The same technique has been used to decarboxylate **692a**[449] and **693a**[461] to give the α-unsubstituted pyrroles **692b** and **693b**, respectively. The dicarboxylic acid **694** may be converted into **695** via the diiodide; this reaction has also been carried out stepwise, each carboxylic acid group being in turn replaced by iodine and the resulting iodide cleaved.[462] As part of a synthesis of prodigiosin **138**, the principal pigment of *Bacillus prodigiosus*, the diester **696a** was partially hydrolyzed by lithium hydroxide in tetrahydrofuran and either water or aqueous methanol to give **696b**; treatment with sodium iodide and iodine in 1,2-dichloroethane gave the diiodopyrrole **697a**, which on hydrogenolysis over a palladium catalyst gave the ester **697b**.[463,464]

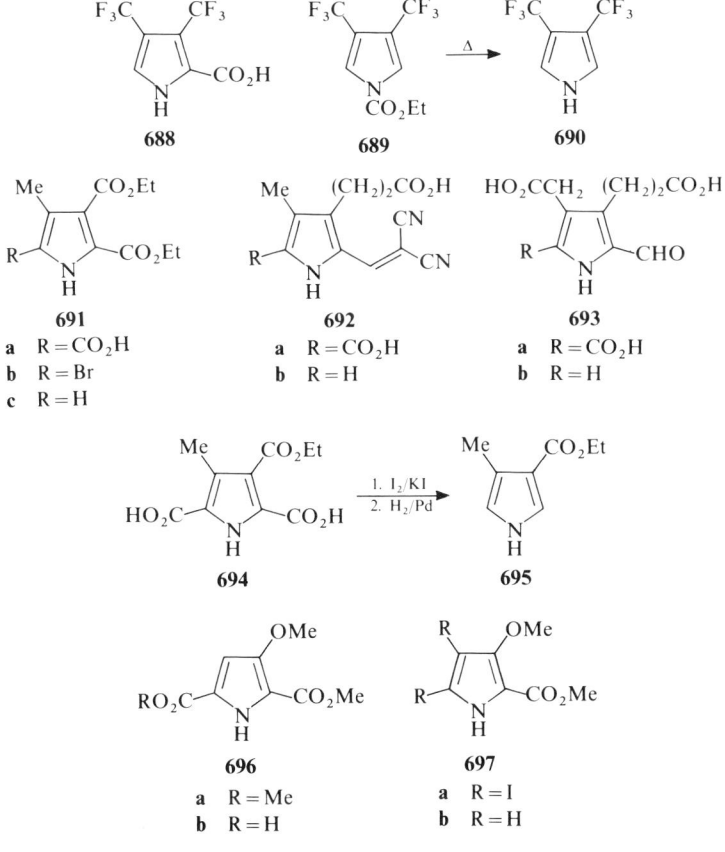

Pyrolysis of pyrrolecarboxylic acids in an attempt to bring about decarboxylation may also lead to condensation to form dimers.[107] Thus, pyrrole-2-carboxylic acids having no substituent on the nitrogen atom may form the pyrocoll **698**; pyrrole-3-carboxylic acids having free 2-positions dimerize to the related quinones **699**. 1-Phenylpyrrole-2,3-dicarboxylic acid (**700**) can be converted into the isomeric quinones **701** and **702** by pyrolysis of its anhydride.[452,465]

1.8.2. Cleavage of Aldehyde and Ketone Groups

It is generally more difficult to remove formyl and acyl groups from the pyrrole ring than it is to decarboxylate the corresponding carboxylic acids. Treatment of formylpyrroles with dilute sulfuric acid is, however, reported to result in cleavage under relatively mild conditions. Acetyl groups may be removed from the pyrrole nucleus by treatment with alkoxide ions, but this process results in the substitution of the acetyl group by alkoxy groups.[389] However, satisfactory techniques have been developed for the removal of acetyl and formyl groups.[467,468] Treatment of the 3-acetyl- or 3-formylpyrroles **703**

TABLE 1.14. DEACYLATION OF SOME FORMYL AND ACETYLPYRROLES

R^1	R^2	R^3	R^4	R^5	Reaction Conditions[a]	Yield (%) of Deacylated Product	Ref.
H	CO_2CH_2Ph	Ac	Me	Me	A	95	467
H	CO_2CH_2Ph	CHO	Me	Me	A	55	467
H	CO_2Et	Ac	H	Me	A	45	467
H	CO_2CH_2Ph	Ac	Me	Me	B	95	467
H	CO_2Et	Ac	H	Me	B	83	467
H	CO_2Et	$COCH_2CO_2Et$	Me	Me	B	35	467
H	Ac	Me	Et	Me	B	35	468
H	Ac	CO_2Et	H	Me	B	85	468
H	CO_2Me	Me	Me	Ac	B	88	468
H	Me	Ac	Et	H	B	40	468
Me	Me	Ac	H	Me	B	68	468
Me	Me	Ac	Ac	Me	B	72	468
Me	Me	Ac	H	Me	B	68	468
H	H	CO_2Et	Ac	Me	B	84	468
H	Me	Ac	Me	NO_2	B (C)	20 (85)	468
H	Ac	Me	CO_2Et	Me	C	77	468

[a] Method A—$HS(CH_2)_2SH/BF_3$; method B—$HO(CH_2)_2OH/p$-TosOH; method C—$HO(CH_2)_2OH/HClO_4$.

with ethane-1,2-dithiol and boron trifluoride leads to cleavage of the acyl group to give **704** (Table 1.14).[467] However, for the compound **705**, which has both a formyl and acetyl group, only the dithioketal **706** is formed under the reaction conditions and no cleavage results.[467] The treatment of **703** with ethane-1,2-dithiol and p-toluenesulfonic acid also results in cleavage of the acyl function to give **704** (Table 1.14), but again only the stable acetal is formed from **705** and the corresponding 2-acetyl compound.[467] Similar conditions have been used to deacetylate several other acetylpyrroles (Table 1.14).[468]

1.9. CYCLOADDITION REACTIONS OF ACYLPYRROLES AND THEIR DERIVATIVES

1.9.1. Diels–Alder and Other [4 + 2] Cycloadditions

Pyrroles undergo [4 + 2] cycloaddition reactions, although the competing Michael addition is often preferred[107] (see Part 1). The [4 + 2] cycloaddition reaction is generally favored when the pyrrole ring is substituted at the 1-position by an electron-withdrawing group. Thus, the reaction between ethyl pyrrole-1-carboxylate (**707**) and diethyl acetylenedicarboxylate yields the cycloadduct **708**,[469] which readily undergoes the retroreaction to give the pyrrole triester **709**.[470–472] Better yields are achieved in this reaction sequence with 1-benzoylpyrrole.[381] Methyl pyrrole-1-carboxylate (**710**) undergoes the Diels–Alder reaction with p-toluenesulfonylacetylene to give the adduct **711** (m.p. 95–96°C) in 60% yield after 24-h reaction at 85°C; the tosyl group may be removed by reduction with sodium amalgam, and the resulting bicyclic compound **712a** has been converted into the trimethylsilyl ester **712b** in 89% yield. Deprotection of the nitrogen atom by heating **712b** under reflux in methanol and dichloromethane for 10 min gives the parent bicyclic system **713** (80%).[473] Ethyl and tert-butyl pyrrole-1-carboxylates react with benzyne to form the adducts **714a**[474] and **714b**,[475] respectively. 1-Benzoylpyrrole reacts with hexafluorobut-2-yne to give the Diels–Alder adduct **715a** in quantitative yield; the analogous reaction of 1-benzoyl-2,5-dimethylpyrrole (**715b**) is equally efficient.[476] Both adducts undergo a retroreaction to give the 3,4-bis(trifluoromethyl)pyrroles **716a** and **716b**, respectively, in high yield.[476] The corresponding reaction of tert-butyl pyrrole-1-carboxylate is slightly less efficient, with the adduct **715c** being produced in 96% yield.[460] The retroreaction to give **689** requires treatment with 2,4,6-trimethylbenzonitrile oxide[460] (see Part 1).

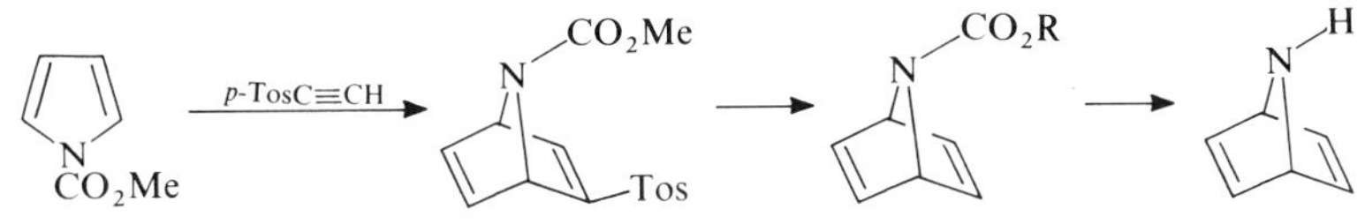

1.9. Cycloaddition Reactions of Acylpyrroles and Their Derivatives

710 → **711** → **712** (a R=Me, b R=SiMe$_3$) → **713**

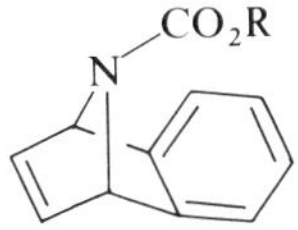

714 a R=Et, b R=tBu

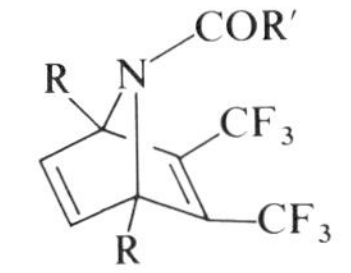

715 a R=H, R'=Ph; b R=Me, R'=Ph; c R=H, R'=tBuO

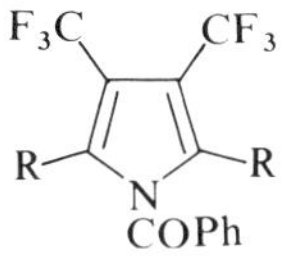

716 R=H, R=Me

Reaction with less reactive dienophiles, such as maleic anhydride or N-substituted maleimides, is ineffective unless performed under pressure. Under such conditions, however, the expected cycloadducts are formed (Table 1.15).[477] Pressure is also required for the cycloaddition reaction between ethyl 3-mercaptopyrrole-1-carboxylates **717** and phenyl vinyl sulfone to give **718a**, with

TABLE 1.15. HIGH-PRESSURE DIELS–ALDER REACTIONS OF 1-ACYLPYRROLES[477]

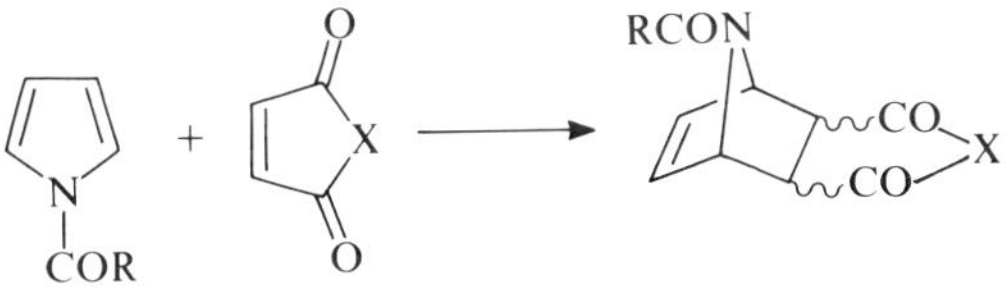

R	X	Yield (%) of endo-Isomer	Yield (%) of exo-Isomer
Ph	NPh	45	64
Ph	O	0[a]	25[a]
Ph	O	20[b]	0[b]
Ph	NMe	66	11
Me	NPh	0	84
Me	O	35	0
Me	NMe	77	0
OEt	NPh	0	46
OEt	O	0	26
OCH$_2$Ph	NPh	90	0
OC$_6$H$_4$Cl-p	NPh	90	0

[a] In chloroform.
[b] In ethyl acetate.

methyl acrylate to produce **718b**, and with *N*-phenylmaleimide to yield **719**.[478] With singlet oxygen, methyl pyrrole-1-carboxylate gives the [4 + 2] cycloadduct **720**, which reacts with tin(II) chloride to give **721**; this product on

TABLE 1.16. PREPARATION OF 2-SUBSTITUTED PYRROLES **722** VIA CYCLOADDITION WITH SINGLET OXYGEN[479]

	2-Substituted Pyrrole **722**	
Nucleophile	2-Substituent	Yield (%)
OSiMe₃ / =⟨Ph	CH₂COPh	57
cyclohexenyl-OSiMe₃	2-methylcyclohexanone	67
CH₂CHOSiMe₃	CH₂CHO	54
CH₂=CHCH=CHOSiMe₃	CH₂CH=CHCHO	62
CH₂=C(OSiMe₃)CH=CH₂	CH₂COCH=CH₂	72
CH₂=C(OSiMe₃)CH=CH(CH₂)₃Me	CH₂COCH=CH(CH₂)₃Me	72
CH₂=C(OSiMe₃)CH=CHPh	CH₂COCH=CHPh	64
OEt / =⟨Me	CH₂COMe	48
N-methylpyrrole	5-methyl-N-methylpyrrole	14
N-CO₂Me pyrrole with 2-CH₂Ac	5-methyl N-CO₂Me pyrrole with 2-CH₂Ac	21

1.9. Cycloaddition Reactions of Acylpyrroles and Their Derivatives 95

treatment with a variety of nucleophiles yields the 2-substituted pyrroles **722** (Table 1.16).[479]

3-Acetyl-1-benzenesulfonylpyrrole (**723**) reacts as a dienophiles in a completely nonregioselective Diels–Alder reaction with 2-methylbutadiene to give equimolar amounts of **724a** and **724b** in 51% overall yield.[480] Cycloadditions between dimethyl acetylenedicarboxylate and the 4-acyl-1-methyl-2-vinylpyrroles **188a** and **188b** and 2-formyl-1-methyl-4-vinylpyrrole (**189a**) give the adducts **725a** (m.p. 144–146°C, 32%), **725b** (m.p. 171–174°C, 50%), and **726** (m.p. 108–112°C, 30%), respectively.[180]

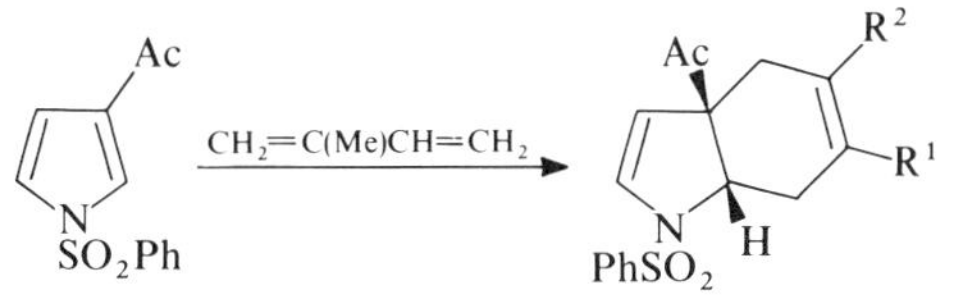

724
a $R^1 = Me, R^2 = H$
b $R^1 = H, R^2 = Me$

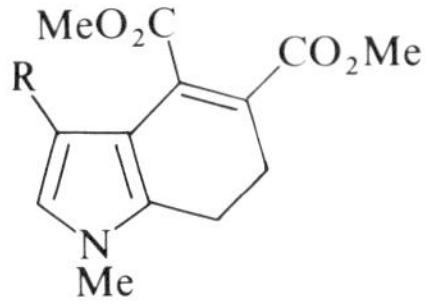

725
a R = CHO
b R = Ac

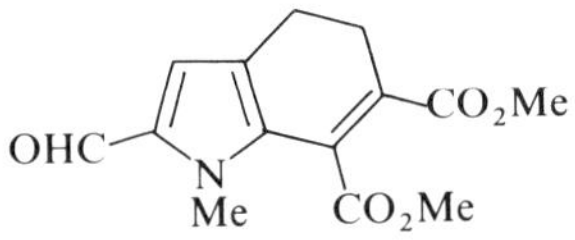

1.9.2. Other Cycloadditions

Methyl pyrrole-1-carboxylate (**710**) reacts with ethoxycarbonyl carbene, generated from ethyl diazoacetate in the presence of copper(I) bromide, to give the cyclopropyl derivatives **727** and **728**.[481] Only simple electrophilic substitu-

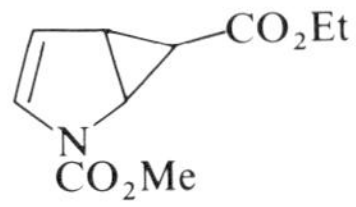

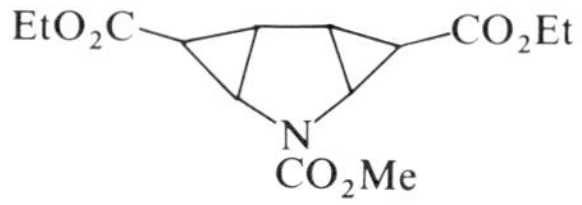

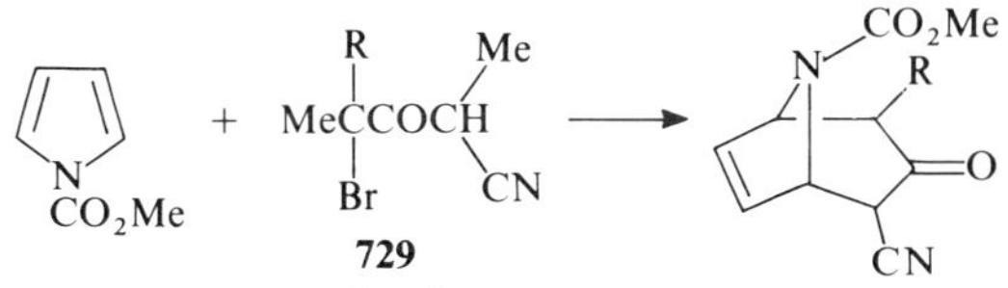

729
a R = H
b R = Me

730
a R = H
b R = Me

tion products are obtained in the absence of acyl substitution on the pyrrolyl nitrogen atom. Methyl pyrrole-1-carboxylate also reacts in [4 + 3] cycloadditions with the β-ketonitriles **729a** and **729b** in the presence of silver oxide to give low yields of the adducts **730a** and **730b**, respectively, as a mixture of stereoisomers, some of which can be isolated.[482]

1.10. REARRANGEMENT REACTIONS

Acylpyrroles undergo a number of rearrangement processes. Some of these reactions are academic curiosities, but many of them have significant applications in synthesis.

When the 2-formyl-1-nitrophenylpyrroles **731** are heated in 80% polyphosphoric acid for several hours, the formyl group migrates to the 3-position to give **732** in 55–60% yield.[483] A similar migration from C-2 to C-3 is observed in the acylpyrroles **733**, which are converted by polyphosphoric or trifluoroacetic acids into **734** (Table 1.17).[484]

TABLE 1.17. ACID-CATALYZED REARRANGEMENT OF 2-ACYLPYRROLES **733** INTO 3-ACYLPYRROLES **734**[484]

R	R'	Acid Catalyst[a]	Yield (%) of **734**	m.p. [b.p.]
Ph	H	P	41	90–93°C
4-MeC$_6$H$_4$	H	P	48	[153–157°C, 0.4 torr]
4-ClC$_6$H$_4$	H	P	36	48–49°C
		T	78	
4-CF$_3$C$_6$H$_4$	H	P	49	[136–142°C, 0.4 torr]
Me	H	P	58	[78–81°C, 0.1 torr]
		T	80	
4-MeC$_6$H$_4$	Me	T	64	119–120°C

[a] P = polyphosphoric acid; T = trifluoroacetic acid.

1.10. Rearrangement Reactions

Hydrolysis under basic conditions of the cyclopropylcarboxylate ester **735**, followed by treatment with trimethylsilylethynyl ethyl ether, gives the anhydride **736**. Reaction of the anhydride with methyllithium in tetrahydrofuran at −78°C, followed by methylation with diazomethane, gives a 4 : 1 mixture of the two acetyl compounds **737** and **738** in 65% overall yield.[485] The ferrocene derivative **740**, produced by heating the acyl chloride **739** with ferrocene in the presence of aluminum trichloride, rearranges on prolonged heating to give the 3-pyrrolyl derivative **741**.[486]

3-Acetylpyrroles, such as **742**, on treatment with thallium(III) nitrate in methanol, undergo an oxidative rearrangement to the 3-pyrrolylacetate **743**;[487,488] this reaction has been used in a synthesis of porphobilinogen (**496**).[488] In this reaction, when the carbonyl carbon atom of the acetyl group in **742** is labeled, the label is also found to be located on the carbonyl group of the ester function in **743**.[489] The reaction proceeds even more efficiently when thallium(III) trinitrate is supported on Montmorrilonite clay.[490] However, overoxidation may occur, with the 2-methyl group being converted into an aldehyde, giving the formylpyrrole **744**.[490] The mechanism of this latter oxidation[491] is probably related to that of the rearrangement reaction. A corresponding process in which 3-chloroacetyl groups are converted into

3-acetic esters has also been reported. For example, treatment of **745** with methyl orthoformate and *p*-toluenesulfonic acid, followed by oxidation with silver nitrate, yields 75% of **746**.[492]

Reaction of the pyrrole carboxylic acid **747** with thionyl chloride in dioxane gives the expected acyl chloride, which reacts with piperidine to give the amide

748. However, if neat thionyl chloride is used instead, the product is **749**, presumably formed via the reaction sequence shown in Scheme 3.[490]

1.11. MISCELLANEOUS REACTIONS

Acylation of the pyrrole ring is often required before certain reactions will take place. The pseudo-Gomberg reaction between pyrroles and substituted anilines in the presence of amyl nitrite to give the 2-arylpyrroles fails if the pyrrolyl nitrogen atom is not protected by an acyl function (Table 1.18). The yields are affected by the choice of solvent and by the relative concentrations of pyrrole and aniline.[494]

TABLE 1.18. PSEUDO-GOMBERG REACTIONS TO YIELD 2-ARYLPYRROLES **750**[494]

R	R'	Yield (%) of 2-Arylpyrrole	m.p.
H	any	—	—
Me	any	—	—
CO_2Et	H	29	Oil
CO_2Et	3-NO_2	75	—
CO_2Et	3-Cl	48	—
CO_2Et	3-Me	—	—
CO_2Et	3-OMe	—	—
CO_2Et	2-NO_2	84	66–67°C
CO_2Et	4-NO_2	80	93–95°C
CO_2Et	4-Cl	52	b.p. 72°C, 0.6 torr
COPh	2-NO_2	75	98–99°C

Treatment of 1-aroylpyrroles **751** with palladium(II) acetate may lead to a variety of products. When the reaction is carried out in acetic acid, the main product is the dimer **752**,[495,496] but, for 1-benzoylpyrrole, the intermediate palladium complex **753** may also cyclize intramolecularly to give the tricyclic product **754**.[496] In the presence of an aromatic hydrocarbon as solvent or cosolvent, arylation of the ring may occur, giving either the 2-arylpyrrole **755** or the 2,5-disubstituted product **756** (Table 1.19).[495–497] Under similar conditions, the dimer **757** is formed in 38% yield from 2-benzoyl-1-methylpyrrole.[495] The intermolecular cycloaddition reaction has also been used in the synthesis of prodigiosin (**138**). The di(1-pyrrolyl)ketone **758**, on treatment with palladium(II) acetate in acetic acid at 80°C, gives the cycloadduct **759**, which can be decarbonylated in 88% yield by lithium methoxide in methanol to give the bipyrrole **760**.[463,464] The di(1-pyrrolyl)methanes **761** (R = H, m.p. 145°C; R = Ph, m.p. 159°C) are obtained in essentially quantitative yield by the simple alkylation of 2-formylpyrrole and 2-benzoylpyrrole with dichloromethane in the presence of aqueous sodium hydroxide, using tetra-n-butylammonium iodide as a phase-transfer catalyst.[498]

Mercury(II) acetate mercuriates the pyrrole carboxylic esters **762** at the unoccupied β-position to give **763**, which can react with methyl acrylate to yield the acrylate esters **764**. The benzyloxycarbonyl protecting group of **764b** can be removed by heating or by treatment with trifluoroacetic acid to produce **764c**.[499] The analogous reaction sequence with **765** leads to 2-pyrrolylacrylic esters **766**.[499] The corresponding reaction of methyl 1,3-dimethylpyrrole-5-carboxylate, which has two unsubstituted positions, is not entirely selective; the reaction product consists of 94% of the 2-pyrrolylacrylate **767** with some of the 4-pyrrolyl isomer, and a trace of the bis-acrylic ester.[499]

TABLE 1.19. PALLADIUM ACETATE-CATALYZED REACTIONS OF 1-ACYLPYRROLES[495-497]

Ar	Ar'	Solvent	Yield (%) of				Ref.
			752	754	755	756	
Ph	Ph	PhH	47[a]	Trace	—	25	497
Ph	4-ClC$_6$H$_4$	Ar'H	90% conversion into **752** and **754**		—	—	497
Ph	4-MeC$_6$H$_4$	Ar'H	90% conversion into **752** and **754**		—	—	497
Ph	—	AcOH	47	30[b]	—	—	495, 496
2,6-Cl$_2$C$_6$H$_3$	Ph	PhH	—	—	—	81	497
2,6-Cl$_2$C$_6$H$_3$	4-ClC$_6$H$_4$	Ar'H	—	—	30	—	497
2,6-Cl$_2$C$_6$H$_3$	4-MeC$_6$H$_4$	Ar'H	—	—	12	36	497
4-MeC$_6$H$_4$	—	AcOH	43[c]	—	—	—	497
1-Naphthyl	—	AcOH	56[d]	—	—	—	497

[a] m.p. 150–151°C.
[b] With 1.0 molar ratio of Pd; only a trace of **754** is obtained with 0.34 moles of Pd.[496]
[c] m.p. 190–192°C.
[d] m.p. 225–227°C.

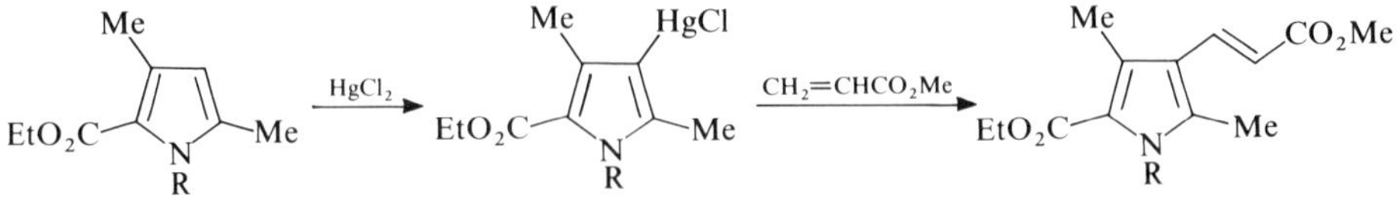

Acylpyrroles

762 → (HgCl$_2$) **763** → (CH$_2$=CHCO$_2$Me) **764**

a R = Me, b R = CO$_2$tBu, c R = H

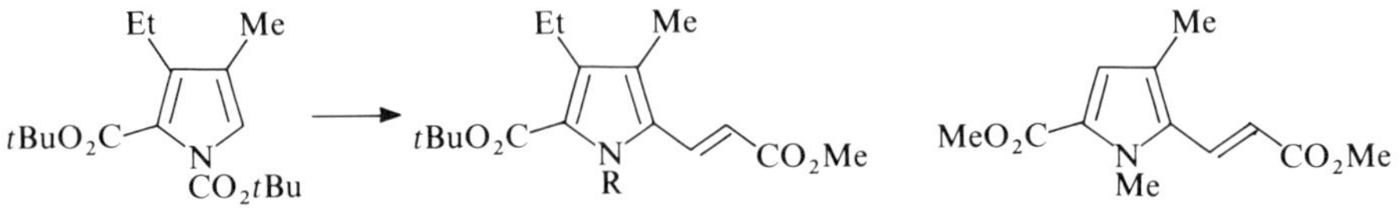

765 → **766**

a R = CO$_2$tBu
b R = H

767

1,2,5-Trisubstituted 3-acetylpyrroles and pyrrole-3-carboxylates are halogenated, as expected, in high yield at the unsubstituted 4-position on treatment with N-halosuccinimide in anhydrous dimethylformamide[500] (Table 1.20) (see also Part 1). Under Mannich conditions, reaction of pyrrole-2-carboxylic esters takes place at the 5-methyl group, **768** (R = Et, Me) → **769**. The presence of the second group in Knorr pyrrole **768c** inhibits this reaction.[501]

TABLE 1.20. HALOGENATION OF 3-ACYLPYRROLES WITH N-HALOSUCCINIMIDES[500]

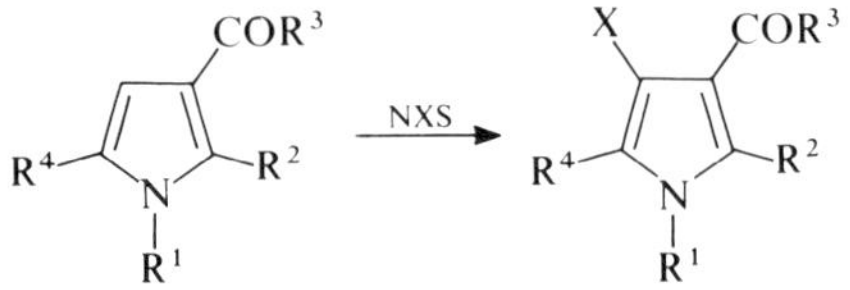

R^1	R^2	R^3	R^4	X	Yield (%)	m.p. (°C)
H	Me	OEt	Ph	Br	87	155
H	Ph	OEt	Ph	Br	92	129
Ph	Ph	OEt	Ph	Br	65	158
H	Me	Me	2-NO$_2$C$_6$H$_4$	Br	94	218
Me	Me	Me	2-NO$_2$C$_6$H$_4$	Br	90	136
Ph	Me	Me	2-NO$_2$C$_6$H$_4$	Br	80	122
H	Me	OEt	Ph	Cl	75	147
H	Ph	OEt	Ph	Cl	88	129
Ph	Ph	OEt	Ph	Cl	91	134
H	Me	Me	2-NO$_2$C$_6$H$_4$	Cl	78	193
Ph	Me	Me	2-NO$_2$C$_6$H$_4$	Cl	65	130
H	Ph	OEt	Ph	I	95	158
H	Me	Me	4-NO$_2$C$_6$H$_4$	I	90	215

1.11. Miscellaneous Reactions

768 R = Et, Me

769
a R' = H
b R' = Me
c R' = CO_2Et

770

Acylation of the pyrrolyl nitrogen atom inhibits the formation of N,N-diacylaminopyrroles analogous to **770** obtained from N-alkylated pyrroles upon treatment with N-chloroimides.[502]

Not unexpectedly, treatment of 2-formyl-1-methylpyrrole with lithium N-methylpiperizide, followed by butyllithium and then methyl iodide, gives 2-formyl-1,5-dimethylpyrrole in 88% yield. However, if lithium N,N,N'-trimethylethylenediamine is used, alkylation takes place on the pyrrolyl N-methyl substituent to give 1-ethyl-2-formylpyrrole in 74% yield.[503]

Perfluorobutanoyl anhydride reacts with 2-formylpyrrole at room temperature to produce 5-heptafluoropropyl-2-formylpyrrole (**771**) (58%);[504] under the same conditions, 2-formyl-5-methylpyrrole produces the 3-heptafluoropropyl derivative **772** in 48% yield. However, at 60°C the reaction fails, with a trace of 2-formyl-5-methyl-3-trifluoromethylpyrrole being formed.

771

772

The intramolecular rearrangement of **773** to **774** occurs in 55–80% yield on treatment with aluminum trichloride.[505] Although intermolecular Friedel–Crafts acylation of 1-methylpyrrole generally yields the 2-acylpyrroles (see Part 1), the reaction with pivaloyl chloride produces the 3-isomer.[506] An intramolecular Friedel–Crafts reaction with aluminum trichloride converts the acid chloride **775** into **776** (45%). If the amide **777** is employed instead of the acid chloride, the cyclic thioester **778** is produced.[507] The carboxyl acid group in **779** acylates the pyrrole ring on treatment with trifluoroacetic acid and trifluoroacetic anhydride to give the quinone **780**. Methylation of the initial cyclization

773
R' = Me, n-Pr, n-Bu, tBuCH$_2$

774

product, using dimethyl sulfate and potassium carbonate, produces **781**.[508] In a similar sequence, starting from the *N*-unprotected pyrrole, the methylation reaction yields a mixture of *N*-methyl analogs of **780** and **781**, while reaction of the initial cyclization product with diazomethane is said to yield only the *N*-methyl analog of **780**.[508]

The diazoketone **782** undergoes an acid-catalyzed cyclization with the elimination of a molecule of nitrogen to give **783**. Treatment of the latter with

1.11. Miscellaneous Reactions

hydrogen sulfide causes extrusion of sulfur to give the indole system **784**; with acetic anhydride, the acetylated indolethiol **785** is formed.[509]

An indole ring system **787** also results from treatment of the β-diketone **786** with methyl iodide,[509] and the 1-(2-pyrrolyl)hexan-2,4-dione **788** cyclizes in the presence of zinc triflate in dichloromethane to the indole **789**;[510] the protected β-ketoaldehyde **790** yields the two indoles **791** and **792** under the influence of magnesium triflate.[510]

Lithiation of N,N-diethyl 1-methylpyrrole-3-carboxamide using butyl-lithium, followed by treatment with 3-phenylmercaptopropenal, produces the adduct **793**; reaction with a further molecule of butyllithium leads to an intermediate, which cyclizes in the presence of moisture and air to lose diethylamine and yield the quinone **794**.[511] The 4-(2-pyrrolyl)but-2-enal **795**[512] has been cyclized to an indole system, which was further elaborated to give **796** as part of an ergot alkaloid synthesis.[513]

Thermal cyclization of the 1-(3-pyridyl)pyrrole-2-carboxylic acid **797** yields the lactone **798**.[514] Reduction of the 1-aryl-2-formylpyrrole **799** with sodium borohydride, or sodium dihydrobis(2-methoxyethoxyl)aluminate, produces the diol **800**, which can be cyclized to the ether **801a**. Inverse addition of the aluminate to **799** produces the acetal **801b** instead.[515]

Treatment of the 2-formylpyrrole **802** with triethyl phosphite, followed by sodium ethoxide, produces only a trace of the expected Horner product **804**; the main product is the phosphonate **803** (m.p. 76°C).[515] Triethyl phosphate and sodium hydride N-alkylate the 2-formylpyrroles **805** to yield the N-ethyl derivatives **806**,[516] while trimethyl phosphate, under the same conditions, converts 2-formylpyrrole into its N-methyl derivative in excellent yield.[516] It has also been noted that, during the Horner reaction, which uses phosphonates (see

1.11. Miscellaneous Reactions

805 R = H, Cl, Me, OMe, CN, CO$_2$Me → **806**

731 → **807** R = 3- or 4-NO$_2$

809 → **810** → **811**

Section 1.3.3), the reaction conditions can lead to N-alkylation of the olefinic products derived from the N-unsubstituted formylpyrroles.[516] Dipropyl phosphate reacts with 1-aryl-2-formylpyrroles **731** to give the phosphorus ester **807** in 70–80% yield; these compounds have bactericidal properties.[517] The ylid **808** reacts with Knorr pyrrole (**809**) to produce the N-substituted pyrrole **810** in 79% yield; this compound gives the normal Wittig product **811** with 4-nitrobenzaldehyde.[518]

Pyrroles bearing carboxylic acid or acid chloride groups on the nitrogen atom may be converted into N-acylpyrroles **812** by treatment with nucleophiles

TABLE 1.21. CONVERSION OF PYRROLE-1-CARBOXYLIC ACIDS AND ACID CHLORIDES INTO 1-ACYLPYRROLES[519]

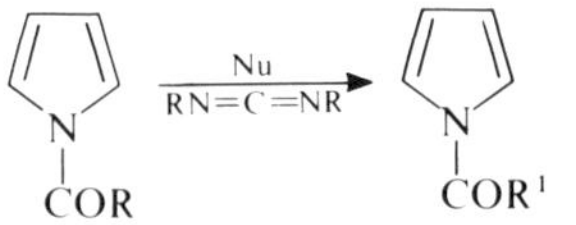

[Scheme: pyrrole-N-COR + Nu (RN=C=NR) → pyrrole-N-COR¹ (812)]

Nucleophile	R¹	Yield (%) of 812 from R = OH	Yield (%) of 812 from R = Cl
N-Na pyrrole-2-CO₂Me	pyrrole(N-H)-2-CO₂Me	89	69
N-Na pyrrole-2-CHO	pyrrole(N-H)-2-CHO	—	63
4-Me-C₆H₄-ONa	4-Me-C₆H₄-O—	87	99
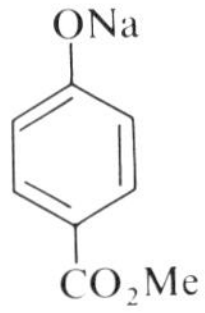 4-(CO₂Me)-C₆H₄-ONa	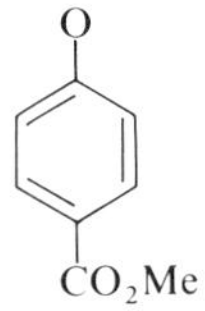 4-(CO₂Me)-C₆H₄-O—	82	88
PhONa	PhO—	74	79
PhNH₂	PhNH—	84	93
CH=CHNH₂	CH₂=CHNH—	94	95

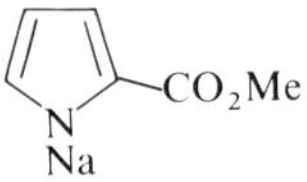

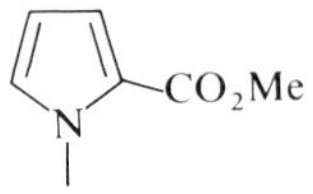

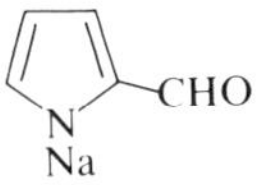

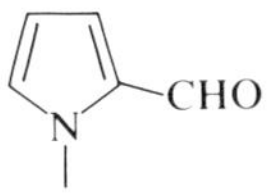

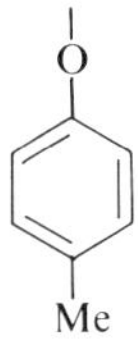

in the presence of diimides (Table 1.21).[519] The anion derived from 2-formylpyrrole reacts with the anhydride **813** to produce the 1-acyl-2-formylpyrrole **814**.[520]

N-Acylpyrroles may be used as acylating agents. Treatment of the 1-acylpyrroles **815** with butyllithium and diisopropylamine in the presence of the

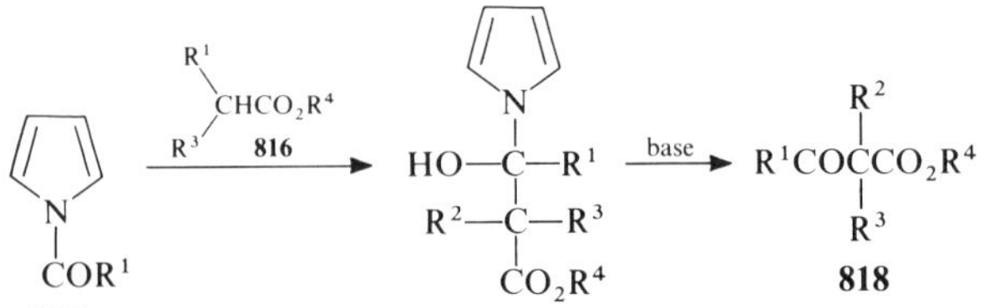

[Scheme: **815** (pyrrole-N-COR¹) + R³-CH(R¹)-CO₂R⁴ (**816**) → HO-C(R¹)-C(R²)(R³)-CO₂R⁴ (**817**) —base→ R¹COC(R²)(R³)CO₂R⁴ (**818**)]

1.11. Miscellaneous Reactions

TABLE 1.22. β-KETOESTERS $R^1COCR^2R^3CO_2R^4$ (**818**) OBTAINED USING N-ACYLPYRROLES AS ACYLATING AGENTS[521]

R^1	R^2	R^3	R^4	Yield (%) of **818**
Me	H	H	t-Bu	82
Me	H	n-Bu	Et	90
Me	Me	Me	Et	90
Et	H	H	Et	100
i-Pr	H	H	Et	92

esters **816** produces the adduct **817** in very high yield; this adduct, on treatment with DABCO in toluene or acetonitrile, or with anhydrous potassium carbonate under totally anhydrous conditions, yields the β-ketoesters **818** (Table 1.22).[521]

The dibutyl acetal of 2-formylpyrrole **819**, when heated over various aluminum catalysts, regenerates the aldehyde, together with the butyl ether **820**, and butyl pyrrole-2-carboxylate **821** in various ratios, depending on the catalyst used and the temperature at which the pyrolysis is carried out (Table 1.23).[522]

TABLE 1.23. PYROLYSIS OF **819**[522]

Catalyst	Temperature (°C)	Conversion (%)	Product ratio		
			77a	**820**	**821**
γ-Alumina	200 ± 5	82	8 :	30 :	62
	250 ± 5	91	16 :	32 :	52
	300 ± 5	97	21 :	40 :	39
	350 ± 5	100	29 :	53 :	18
Aluminum phosphate	150 ± 5	61	47 :	42 :	11
	200 ± 5	73	41 :	46 :	13
	250 ± 5	85	31 :	56 :	13
	300 ± 5	100	22 :	61 :	17
Aluminum sulfate	150 ± 5	72	65 :	0 :	35
	200 ± 5	84	62 :	0 :	38
	250 ± 5	95	59 :	0 :	41
	300 ± 5	96	58 :	0 :	42

The 2-formylpyrrole **822**, when treated with 3-benzyl-5-(2-hydroxyethyl)-4-methyl-1,3-thiazolium chloride and butenone in the presence of triethylamine and anhydrous sodium acetate, gives the γ-diketone **823** in 80% yield. This diketone may be converted into the bipyrrole **824** by the Paal–Knorr reaction with ammonium carbonate.[523] The pyrroles **825** may be dimerized photochemically to give **826**, as a mixture of stereoisomers, by irradiation in acetonitrile (Table 1.24).[524]

TABLE 1.24. PHOTODIMERIZATION OF 3-HYDROXYPYRROLE-2-CARBOXYLIC ESTERS[524]

R^1	R^2	Yield (%)	m.p. (°C)
H	H	45	166
Me	Me	75	179
CH_2=$CHCH_2$	Me	65	95
CH_2=$CH(CH_2)_2$	Me	63	34
—$(CH_2)_4$—		75	145

N-Acylpyrroles have been polymerized by oxidative electrochemical techniques to give conducting polymers,[525] and electrochemical polymerization of C-acylpyrroles has also been used to produce the polymers **827**.[526] Electropolymerization at the anode of a mixture of pyrrole and 2-formylpyrrole in acetic acid and acetonitrile, with tetra-n-butylammonium benzenesulfonate as supporting electrolyte, yields polymers for use in batteries.[527]

827
$n > 4$; R = Ac, CO_2Me

Although pyrrole has no such effect, 1-benzoylpyrrole quenches the photochemical reduction of benzophenone strongly.[528]

1.12. COMPLEXES OF ACYLPYRROLES AND THEIR DERIVATIVES

In previous sections, a number of examples have been given of acylpyrroles or their derivatives being used as ligands. Further cases are presented in this final section of this chapter.

1.12.1. Complexes of Acylpyrroles

Pyrrole-2-carboxylic acid forms tin(IV) complex salts, which can be represented as **828**, on reaction with organotin(IV) oxides[529] or chlorides.[530] 2-Formylpyrrole has been reported to form a bridged acyl cluster complex, which can be represented as **829**, containing three osmium atoms.[531] With pentane-2,4-dione as a second ligand, 2-formylpyrrole also forms complexes of the type **830** [M = Ni(II), Cu(II), or Pt(II)], when treated with the appropriate metal acetates.[532] With the same dione as a second ligand, 2,5-diformylpyrrole forms complexes with nickel, copper, cobalt, iron, palladium, and platinum.[533] Dimeric copper(II) complexes **831** (R = Me or Et) have also been obtained from 2,5-diformylpyrrole.[534] The diketones **832** have been synthesised as ligands for copper(II).[535]

112 Acylpyrroles

2-Formylpyrrole condenses with L-cysteine methyl ester to give the diastereoisomers **833**, which form complexes with rhenium; these compounds are potential enantioselective catalysts in the hydrosilylation of acetophenone,[536] which may also be carried out in up to 55% enatiomeric excess using a rhenium complex catalyst, formed from cyclooctadiene and the pyrrole **834**.[537]

1.12.2. Complexes of Imines

The condensation of acylpyrroles with amines provides a route to imines (Section 1.4.1), and a number of such imines have been used as ligands. The simple methylimine **369** forms complexes with copper and cobalt.[243] Complexes with copper(II) have also been prepared with **835**, obtained from 2-formylpyrrole and either aminoethanol or 3-aminopropan-1-ol.[538] Condensation of the same aldehyde with orthanilic acid, or with taurine, produces **837** or **838**, respectively, both of which have been used as ligands for chromium, manganese, iron, cobalt, nickel, and copper.[539] With valine, (**77a**) gives the ligand **838** (LH$_2$), which forms a complex with molybdenum of formula [MoO$_2$L.3H$_2$O].[540] 2-Formylpyrrole and 2-acetylpyrrole form, with 1-(aminoethyl)benzene, imines of the type **839**, which have been used in the preparation of the rhodium complexes **840** with cyclooctadiene as the second ligand; these complexes act as chiral catalysts, which give a very small enatiomeric excess (1–1.8%) in the hydrosilylation of acetophenone.[541]

1.12. Complexes of Acylpyrroles and Their Derivatives 113

A somewhat more elaborate ligand **842** is prepared from **77a** and the diamine **841**; it forms copper and nickel complexes.[542] The dianionic ligand **843**, which complexes copper(II), nickel(II), and cobalt(II), is prepared from salicylaldehyde, 1,3-diaminopropane, and **77a**.[543] The ligand **844**, obtained from the appropriate tetraamine and 2 equiv of **77a**, forms complexes with iron(III).[544] Other ligands for copper(II) include **845**, prepared by condensation of **77a** with the appropriate diamine.[535] The triamines **846** react with **77a** to give diimines **847**;[545] of particular interest is the related system **848**.[545] These compounds bind copper(II), with binuclear complexes being produced from **848**.[545] The ligands **849** and **850**, synthesized from **77a** and 1,2-diaminoethanol or cis-1,2-diaminocyclohexane, respectively, form a series of zinc(II) complexes, some of which are dimers in which the pyrrole nitrogen atom is involved in binding with the metal.[546]

The conversion of a formyl complex into one containing an imino group may in some cases be carried out in situ. Treatment of the complexes **839** with a primary amine yields the imino complexes **851**.[532] Similarly, 1,2-diaminoethane or 1,3-diaminopropane convert **831** into the copper(II) complexes **852**; with longer-chain diamines, such as 1,4-diaminobutane, 1,5-diaminopentane and 1,6-diaminohexane, the binuclear copper complexes **853** are produced.[534]

The powerful ligand **854** has been synthesized from 2,5-diformylpyrrole; it forms complexes with a wide range of metals: manganese, cobalt, nickel, copper, zinc, palladium, cadmium, tin, lead, and uranium (as uranyl).[547] The same dialdehyde has also been used in the preparation of the macrocyclic ligands **855**, which give complexes containing one or two copper(II) atoms; in the binuclear complexes the metal atoms may be bridged by alkoxy groups, but the pyrrolyl nitrogen atoms are not generally involved in binding to copper.[548] A similar macrocycle **856** has been synthesized as a potential ligand.[549]

Hydrazones (Section 1.4.2) also form metal complexes. The derivatives **399** (R = 2-pyridyl, pyrazinyl, 2-pyrimidinyl, 2-thiazolyl, or 2-quinolinyl) react quantitatively with copper(II) and cobalt(II) and may be used as analytic reagents for the estimation of these metals.[265] The hydrazone **857** forms a nitrate-containing complex of manganese nitrate.[550] The acylhydrazones **858** (R = Me, m.p. 165–166°C; R = OMe, m.p. 200–201°C; R = NH$_2$, m.p. 193–195°C),[551] obtained from 3-acetyl-1-tosylpyrrole, act as ligands for palladium, forming complexes of the type **859** or **860** with a donor ligand (D), such as triphenylarsenic, alkylphosphines, or pyridine.[551]

1.13. REFERENCES

1. R. A. Jones, ed., *The Chemistry of Heterocyclic Compounds*. Vol. 48, *Pyrroles*, Part 1, Wiley, New York, 1990: (a) Chapter 2; (b) Section 3.4; (c) Section 3.5; (d) Section 3.6.
2. M. Y. Jarrah and V. Thaller, *J. Chem. Soc., Perkin Trans. 1*, **1983**, 1719.
3. G. Hofle and H. Wolf, *Liebigs Ann. Chem.*, **1983**, 835.
4. W. A. Corpe, *Appl. Microbiol.*, **11**, 145 (1963).
5. J. H. Tumlinson, R. M. Silverstein, J. C. Moser, R. G. Brownlee, and J. M. Ruth, *Nature*, **234**, 348 (1971).
6. J. H. Cross, J. R. West, R. M. Silverstein, A. R. Justum, and J. M. Cherrette, *J. Chem. Ecol.*, **8**, 1119 (1982); *Chem. Abstr.*, **97**, 124395 (1982).
7. J. M. Meinwald and Y. C. Meinwald, *J. Am. Chem. Soc.*, **88**, 1305 (1966).
8. J. M. Meinwald and H. C. J. Otterheym, *Tetrahedron*, **27**, 3307 (1971).
9. A. J. Birch, P. Hodge, R. W. Rickards, R. Takeda, and T. R. Watson, *J. Chem. Soc.*, **1964**, 2641.
10. D. G. Davies and P. Hodge, *Tetrahedron Lett.*, **1970**, 1673.
11. B. W. Cue, J. P. Dirlam, L. J. Czuba, and W. W. Windisch, *J. Heterocycl. Chem.*, **18**, 191 (1981).
12. J. W. Westley, R. H. Evans, C. -M. Liu, T. Hermann, and J. F. Blount, *J. Am. Chem. Soc.*, **100**, 6784 (1978).
13. M. P. Edwards and S. V. Ley, *J. Chem. Soc., Perkin Trans. 1*, **1984**, 1761.
14. L. N. Lysenkova, M. G. Brazhnikova, V. N. Borisova, G. B. Fedorova, L. M. Rubasheva, and N. P. Potapova, *Antibiotiki (Moscow)*, **25**, 483 (1980); *Chem. Abstr.*, **93**, 219276 (1980).

15. H. Kawaguchi, T. Naito, and H. Tsukiura, *J. Antibiot. (Jpn.), Ser. A*, **18**, 11 (1965); *Chem. Abstr.*, **63**, 16293 (1965).
16. W. D. Celmer, G. N. Chmurny, C. E. Moppett, R. S. Ware, P. C. Watts, and E. B. Whipple, *J. Am. Chem. Soc.*, **102**, 4203 (1980).
17. M. O. Chaney, P. V. Demarco, N. D. Jones, and J. L. Occolowitz, *J. Am. Chem. Soc.*, **96**, 1932 (1974).
18. J. Westley, *Polyether Antibiotics: Naturally Occurring Acid Ionophores*, Marcel Dekker, New York, 1982.
19. M. Prudhomme, G. Dauphin, and G. Jeminet, *J. Chem. Res.*, **1987** (S), 420.
20. T. Momose, T. Tanaka, T. Yokota, N. Nagamoto, and K. Yamada, *Chem. Pharm. Bull.*, **22**, 2224 (1978).
21. J. S. Davies, *Tetrahedron Lett.*, **23**, 5089 (1982).
22. A. G. Brown, D. F. Corbett, A. J. Eglington, and T. T. Howarth, *Tetrahedron*, **39**, 2551 (1983).
23. M. Julia and N. P. Joseph, *Compt. Rend. Hebd. Seances Acad. Sci.*, **257**, 1115 (1963).
24. F. Arcamone, P. G. Orezzi, W. Barbieri, V. Nicoletta, and S. Penco, *Gazz. Chim. Ital.*, **97**, 1097 (1967).
25. T. Takaishi, Y. Sugawara, and M. Suzuki, *Tetrahedron Lett.*, **1972**, 1873.
26. G. W. Probst, M. M. Hoehn, and B. L. Woods, *Antimicrob. Agents Chemother.*, **1965**, 789; *Chem. Abstr.*, **65**, 9689 (1966).
27. J. W. Lown and K. Krowicki, *J. Org. Chem.*, **50**, 3774 (1985).
28. L. Grehn and U. Ragnarsson, *J. Org. Chem.* **46**, 3492 (1981).
29. M. Lee, D. M. Coulter, and J. W. Lown, *J. Org. Chem.*, **53**, 1855 (1988).
30. M. Lee and J. W. Lown, *J. Org. Chem.*, **52**, 5717 (1987).
31. S. Penco, S. Redaetti, and F. Arcamone, *Gazz. Chim. Ital.*, **97**, 1110 (1967).
32. E. N. Glibin, B. V. Tsukerman, and O. F. Ginzburg, *Zh. Org. Khim.*, **13**, 2231 (1977); *Chem. Abstr.*, **88**, 50585 (1979).
33. M. Bialer, B. Yagen, and R. Mechoulam, *J. Med. Chem.*, **23**, 1144 (1980).
34. M. Bialer, B. Yagen, and R. Mechoulam, *J. Heterocycl. Chem.*, **17**, 1797 (1980).
35. P. L. Gendler and H. Rapoport, *J. Med. Chem.*, **24**, 33 (1981).
36. E. Glibin, B. V. Tsukerman, and O. F. Ginzburg, *Zh. Org. Khim.*, **17**, 657 (1981); *Chem. Abstr.*, **95**, 62099 (1981).
37. P. Gendler and H. Rapoport, *J. Med. Chem.*, **24**, 33 (1981).
38. S. L. Grokhovskii, A. L. Zhuze, and B. P. Gottikh, *Biol. Inorg. Khim.*, **8**, 1070 (1982); *Chem. Abstr.*, **97**, 215877 (1982).
39. A. A. Kharlin, S. L. Grokhovskii, A. L. Zhuze, and B. P. Gottikh, *Biol. Inorg. Khim.*, **8**, 1063 (1982); *Chem. Abstr.*, **97**, 216683 (1982).
40. L. Grehn, U. Ragnarsson, B. Eriksson, and B. Oberg, *J. Med. Chem.*, **26**, 1043 (1983).
41. Ger. Pat. 3623880; *Chem. Abstr.*, **106**, 156157 (1987).
42. J. W. Lown, K. Krowicki, J. Balzarini, and E. De Clercq, *J. Med. Chem.*, **29**, 1210 (1986).
43. L. Grehn, U. Ragnarsson, and R. Datema, *Acta Chem. Scand.*, **B40**, 145 (1986).
44. E. Nishiwaki, S. Tanaka, H. Lee, and M. Shibuya, *Heterocycles*, **27**, 1945 (1988).
45. K. Krowicki, J. Balzarini, E. De Clercq, R. A. Newman, and J. W. Lown, *J. Med. Chem.*, **31**, 341 (1988).
46. F. Debart, C. Perigaud, G. Gosselin, D. Mrani, B. Rayner, P. Le Ber, C. Auclair, J. Balzarini, E. De Clercq, C. Paoletti, and J. -L. Imbach, *J. Med. Chem.*, **32**, 1074 (1989).
47. M. Bialer, B. Yagen, R. Mechoulam, and Y. Becker, *J. Pharm. Sci.*, **69**, 1334 (1980).

1.13. References

48. C. Bailly, N. Helberger, and J. P. Henichart, *Colloq. INSERM*, **174**, 579 (1989); *Chem. Abstr.*, **111**, 50009 (1989).
49. S. S. Khorlin, S. L. Grokhovskii, A. L. Zhuze, and B. P. Gottikh, *Bioorg. Khim.*, **8**, 1358 (1982); *Chem. Abstr.*, **98**, 72699 (1983).
50. A. S. Krylov, A. A. Khorlin, S. L. Grokhovskii, A. L. Zhuze, A. S. Zasedatelev, G. V. Gurskii, and B. P. Gottikh, *Dokl. Akad. Nauk SSSR*, **254**, 234 (1980); *Chem. Abstr.*, **94**, 60015 (1981).
51. N. G. Plekhanova, E. N. Glibin, B. V. Tsukerman, and O. F. Ginzburg, *Zh. Org. Khim.*, **19**, 1533 (1983); *Chem. Abstr.*, **99**, 176269 (1983).
52. P. G. Schultz, J. S. Taylor, and P. B. Dervan, *J. Am. Chem. Soc.*, **104**, 6861 (1982).
53. G. Cimino, S. de Stefano, and L. Minale, *Experientia*, **31**, 1387 (1975).
54. D. B. Stierle and D. J. Faulkner, *J. Org. Chem.*, **45**, 4980 (1980).
55. J. M. Frinke and D. J. Faulkner, *J. Am. Chem. Soc.*, **104**, 265 (1982).
56. G. De Nanteuil, A. Ahond, J. Guilhem, C. Poupat, E. Tran Huu Dau, P. Potier, M. Pusset, J. Pusset, and P. Laboute, *Tetrahedron*, **41**, 6019 (1985).
57. F. J. Schmitz, S. P. Gunasekera, V. Lakshmi, and L. M. V. Tillekeratne, *J. Nat. Prod.*, **48**, 47 (1985).
58. N. K. Utkina, S. A. Fedoreev, and O. B. Maksimov, *Khim. Prir. Soedin.*, **1985**, 578; *Chem. Abstr.*, **104**, 145784 (1986).
59. S. P. Gunasekera, S. Cranick, and R. E. Longuey, *J. Nat. Prod.*, **52**, 757 (1989).
60. L. Minale, G. Cimino, S. De Stefano, and G. Sodano, *Prog. Chem. Nat. Prod.*, **33**, 1 (1976).
61. S. L. Forenza, L. Minale, R. Riccio, and E. Faturosso, *J. Chem. Soc., Chem. Commun.*, **1971**, 1129.
62. R. P. Walker, D. J. Faulkner, D. Van Engen, and J. Clardy, *J. Am. Chem. Soc.*, **103**, 6772 (1981).
63. R. Fatni-Afshar and T. M. Allen, *Can. J. Chem.*, **66**, 45 (1988).
64. J. Kobayashi, Y. Ohizumi, H. Nakamura, and Y. Hirata, *Experientia*, **42**, 1176 (1986).
65. H. Nakamura, H. Wu, R. Abe, J. Kobayashi, Y. Ohizume, and Y. Hirata, *Tennen Yuki Kagobutsu Toronkai Koen Yoshishu*, **1983** (26th), 118; *Chem. Abstr.*, **100**, 83029 (1984).
66. R. Capon and D. J. Faulkner, *J. Am. Chem. Soc.*, **106**, 1819 (1984).
67. G. de Nanteuil, A. Ahond, C. Poupet, O. Thoison, and P. Potier, *Bull. Soc. Chim. Fr.*, **1986**, 813.
68. L. H. Foley and G. Buchi, *J. Am. Chem. Soc.*, **104**, 1776 (1982).
69. G. M. Sharma and P. R. Burkholder, *J. Chem. Soc., Chem. Commun.*, **1971**, 151; G. M. Sharma and B. Magdoff-Fairchild, *J. Org. Chem.* **42**, 4118 (1977).
70. M. G. Reinecke and Y.-Y. Zhao, *J. Nat. Prod.*, **51**, 1236 (1988).
71. B. F. Bowden, P. S. Clezy, J. C. Coll, B. N. Ravi, and D. M. Tapiolas, *Aust. J. Chem.*, **37**, 227 (1984).
72. T. M. Zennie, J. M. Cassady, and R. F. Raffauf, *J. Nat. Prod.*, **49**, 695 (1986).
73. R. F. Raffauf, T. M. Zennie, K. D. Chan, and P. W. Le Quesne, *J. Org. Chem.*, **49**, 2714 (1984).
74. H. Greger, C. Zdero, and F. Bohlmann, *Phytochemistry*, **26**, 2289 (1987).
75. D. G. Lynn, K. Jaffe, M. Cornwall, and W. Tramontano, *J. Am. Chem. Soc.*, **109**, 5858 (1987).
76. T. Nyhammar, K. Olsson, and P. A. Pernemalm, *ACS Symp. Ser.*, **1983**, 215.
77. R. Miller, K. Olsson, and P. -A. Pernemalm, *Acta Chem. Scand.*, **B38**, 689 (1984).
78. T. Nakayama, F. Hayase, and H. Kato, *Agric. Biol. Chem.*, **44**, 1201 (1980); *Chem. Abstr.*, **93**, 95648 (1980).
79. R. Miller and K. Olsson, *Acta Chem. Scand.*, **B39**, 717 (1985).
80. H. Kato, T. Nakayama, S. Sugimoto, and F. Hayase, *Agric. Biol. Chem.*, **46**, 2599 (1982); *Chem. Abstr.*, **97**, 214388 (1982).

81. T. Hara, *Nippon Nogei Kagaku Kaishi*, **55**, 1069 (1981); *Chem. Abstr.*, **96**, 102603 (1982).
82. C. R. Hall, T. D. Inch, and N. E. Williams, *J. Chem. Soc., Perkin Trans. 1*, **1983**, 1977.
83. M. Artico, F. Corelli, S. Massa, and G. Stefancich, *J. Heterocycl. Chem.*, **19**, 1493 (1982).
84. Span. Pat. 502650; *Chem. Abstr.*, **97**, 162811 (1982).
85. Jpn. Pat. 82 70865; *Chem. Abstr.*, **97**, 162809 (1982).
86. Eur. Pat. 249236; *Chem. Abstr.*, **108**, 150482 (1988).
87. Eur. Pat. 32048; *Chem. Abstr.*, **95**, 203737 (1981).
88. Br. Pat. 2107304; *Chem. Abstr.*, **99**, 122288 (1983).
89. Eur. Pat. 72013; *Chem. Abstr.*, **99**, 22311 (1983).
90. J. W. Sowell, A. J. Block, M. E. Derrick, J. J. Freeman, J. W. Kosh, P. F. Mubarak, and P. A. Tenthovey, *J. Pharm. Sci.*, **70**, 537 (1981).
91. L. D. Wang, J. W. Sowell, J. J. Freeman, and J. W. Kosh, *J. Pharm. Sci.*, **70**, 699 (1981).
92. U.S. Pat. 4435407; *Chem. Abstr.*, **101**, 23327 (1984).
93. M. Gomez Guillen, J. A. Galbis Perez, P. Areces Bravo, and M. Bueno Martinez, *An. Quim., Ser. C*, **77C**, 278 (1981); *Chem. Abstr.*, **97**, 39287 (1982).
94. E. Pando, C. Martin Madero, and C. Vergara, *An. Quim., Ser. C*, **76C**, 53 (1980); *Chem. Abstr.*, **94**, 157162 (1981).
95. J. Fernandez-Bolanos, J. Fuentes Mota, and I. Robina Ramirez, *An. Quim., Ser. C*, **79C**, 317 (1983); *Chem. Abstr.*, **102**, 62543 (1985).
96. J. A. Galbis Perez, J. C. Palacios Albarran, J. L. Jiminez Requejo, and M. Avalos Gonzalez, *Carbohydr. Res.*, **132**, 153 (1984).
97. I. Maeba, T. Takeuchi, T. Iijima, K. Kitaori, and H. Muramatsu, *J. Chem. Soc., Perkin Trans. 1*, **1989**, 649.
98. J. Fernandez-Bolanos, I. Rabina Ramirez, and J. Fuentes Mota, *An. Quim. Ser. C*, **81C**, 49 (1985); *Chem. Abstr.*, **105**, 24533 (1986).
99. J. A. Galbis Perez, E. Roman Galan, M. A. Arevalo Arevalo, and F. Polo Corrales, *An. Quim., Ser. C*, **82C**, 76 (1986); *Chem. Abstr.*, **106**, 176786 (1987).
100. M.-I. Lim and R. S. Klein, *Tetrahedron Lett.*, **22**, 25 (1981).
101. M. Millan, C. F. Conde, A. Conde, and R. Marquez, *Acta Crystallog.*, **C39**, 120 (1983).
102. E. Pando, M. C. Garcia Bala, and C. Martin Madero, *Affinidad*, **40**, 445 (1983); *Chem. Abstr.*, **100**, 156451 (1984).
103. A. Gomez-Sanchez, F.-J. Hidalgo, and J.-L. Chiara, *Carbohydr. Res.*, **167**, 55 (1987).
104. P. Fournari, M. Farnier, and C. Fournier, *Bull. Soc. Chim. Fr.*, **1972**, 283.
105. Br. Pat. 2194946; *Chem. Abstr.*, **109**, 73320 (1988).
106. Eur. Pat. 226099; *Chem. Abstr.*, **107**, 217488 (1987).
107. R. A. Jones and G. P. Bean, *The Chemistry of Pyrroles*, Academic Press, London, 1977.
108. A. Gossauer, *Die Chemie der Pyrrole*, Springer-Verlag, Berlin, 1974.
109. A. Treibs and H. G. Kolm, *Liebigs Ann. Chem.*, **606**, 166 (1957).
110. R. A. Jones, in A. R. Katritzky and C. W. Rees eds., *Comprehensive Heterocyclic Chemistry*, Vol. 4, Pergamon Press, Oxford, 1984.
111. U. Eisner and R. L. Erskine, *J. Chem. Soc.*, **1958**, 971.
112. W. Herz and J. Brasch, *J. Org. Chem.*, **23**, 1513 (1958).
113. G. J. Karabatsos and F. M. Vane, *J. Am. Chem. Soc.*, **85**, 3886 (1963).
114. R. A. Jones and P. H. Wright, *Tetrahedron Lett.*, **1968**, 5495.
115. M. Farnier and T. Drakenberg, *Tetrahedron Lett.*, **1973**, 429.
116. B. P. Roques and S. Combrisson, *Can. J. Chem.*, **51**, 573 (1973).
117. M. Farnier and T. Drakenberg, *J. Chem. Soc., Perkin Trans. 2*, **1975**, 333.

1.13. References

118. G. S. Coumbarides, J. M. Mercey, and T. P. Toube, *J. Chem. Res.*, **1990**, (S) 151.
119. A. Nizhnik, I. A. Vasilenko, and A. F. Mironov, *Zh. Obshch. Khim.*, **54**, 2326 (1984); *Chem. Abstr.*, **102**, 61704 (1985).
120. S. P. Fradkina, I. T. Kvitko, and I. N. Vasil'eva, *Zh. Org. Khim.*, **19**, 190 (1983); *Chem. Abstr.*, **98**, 197271 (1983).
121. H. Pleininger, R. El-Berins, and R. Hirsch, *Synthesis*, **7**, 422 (1973).
122. I. M. Labouta, P. Jacobsen, P. Thorbek, P. Krogsgaard-Larsen, and H. Hjeds, *Acta Chem. Scand.*, **B36**, 699 (1982).
123. V. H. Rawal, J. A. Rao, and M. P. Cava, *Tetrahedron Lett.*, **26**, 4275 (1985).
124. R. Yoneda, H. Hisakawa, S. Harusawa, and T. Kurihara, *Chem. Pharm. Bull.*, **35**, 3850 (1987).
125. H. J. Anderson and H. Nagy, *Can. J. Chem.*, **50**, 1961 (1970).
126. E. Bisagni, J.-P. Marquet, and J. Andre-Louisfert, *Bull. Soc. Chim. Fr.*, **1968**, 637.
127. T. S. Gardner, E. Wenis, and J. Lee, *J. Org. Chem.*, **23**, 823 (1958).
128. E. Lubrzynska, *J. Chem. Soc.*, **109**, 1118 (1916).
129. S. V. Tsukerman, Y. N. Surov, and V. F. Lavrushin, *Zh. Obshchei Khim.*, **37**, 364 (1967); *Chem. Abstr.*, **67**, 43294 (1967).
130. N. R. El-Rayyes, G. H. Hovakeemian, and H. S. Hmoud, *J. Chem. Eng. Data*, **29**, 225 (1984); *Chem. Abstr.*, **100**, 156537 (1984).
131. N. R. El-Rayyes, G. V. Hovakeemian, and A. F. Samara, *J. Chem. Eng. Data*, **31**, 369 (1986); *Chem. Abstr.*, **105**, 114699 (1986).
132. N. R. El-Rayyes, *J. Prakt. Chem.*, **314**, 915 (1972).
133. N. R. El-Rayyes, *J. Prakt. Chem.*, **315**, 295 (1973).
134. A. Biev, *Dokl. Bolg. Akad. Nauk*, **35**, 1503 (1982); *Chem. Abstr.*, **99**, 22254 (1983).
135. J. H. Lange, W. T. Colwell, and D. W. Henry, *J. Med. Chem.*, **12**, 946 (1969).
136. J. Hrabovsky, J. Kovac, and K. Vagacova, *Coll. Czech. Chem. Commun.*, **51**, 1301 (1986).
137. R. P. Kreher and J. Pfister, *Angew. Chem. Int. Ed. Engl.*, **23**, 914 (1984).
138. J. Bourguignon, G. Le Nard, and G. Queguiner, *Can. J. Chem.*, **63**, 2354 (1985).
139. H. Fischer and B. Weiss, *Ber. Dtsch. Chem. Ges.*, **57**, 602 (1924).
140. E. Bisagni, J. D. Bourzat, and J. Andre-Louisfert, *Tetrahedron*, **26**, 2087 (1970).
141. H. Fischer and M. Neber, *Liebigs Ann. Chem.*, **496**, 1 (1932).
142. H. Fischer and K. Zeile, *Liebigs Ann. Chem.*, **462**, 210 (1928).
143. G. Alberghina, M. E. Amato, S. Fisichella, and D. Pisano, *J. Chem. Soc., Perkin Trans. 2*, **1988**, 295.
144. G. M. Badger, G. E. Lewis, and V. P. Singh, *Aust. J. Chem.*, **20**, 2785 (1967).
145. W. Herz, *J. Am. Chem. Soc.*, **71**, 3982 (1949).
146. W. Herz and C. Dittmer, *J. Am. Chem. Soc.*, **70**, 503 (1948).
147. H. H. Moharram, A. M. Abdel-Fattah, M. M. El-Merzabani, and S. A. Mansour, *Egypt. J. Chem.*, **26**, 301 (1983).
148. D. G. Harvey, *J. Chem. Soc.*, **1950**, 1638.
149. A. Treibs and N. Haberle, *Liebigs Ann. Chem.*, **739**, 220 (1970).
150. A. Treibs and R. Zimmer-Galler, *Chem. Ber.*, **93**, 2539 (1960).
151. H. Fischer and B. Walach, *Liebigs Ann. Chem.*, **450**, 109 (1926).
152. A. H. Corwin and J. S. Andrews, *J. Am. Chem. Soc.*, **58**, 1086 (1935).
153. H. Rapoport and K. G. Holden, *J. Am. Chem. Soc.*, **84**, 635 (1962).
154. D. Eickinger and H. Falk, *Monatsh. Chem.*, **118**, 255 (1987).
155. D. Brown, D. Griffiths, M. E. Rider, and R. C. Smith, *J. Chem. Soc., Perkin Trans. 1*, **1986**, 455.

156. R. N. Castle and C. W. Whittle, *J. Org. Chem.*, **24**, 1189 (1959).
157. F. J. Villani, E. A. Wefer, T. A. Mann, J. Mayer, L. Peer, and A. S. Levy, *J. Heterocycl. Chem.*, **9**, 1203 (1972).
158. T. Oesterle and G. Simchin, *Liebigs Ann. Chem.*, **1987**, 693.
159. R. Aumann and H. Heinen, *Chem. Ber.*, **120**, 537 (1987).
160. D. Seebach and T. Weber, *Helv. Chim. Acta*, **67**, 1650 (1984).
161. D. P. Schumacher and S. S. Hall, *J. Org. Chem.*, **46**, 5060 (1981).
162. S. S. Hall, D. Loebenberg, and D. P. Schumacher, *J. Med. Chem.*, **26**, 469 (1983).
163. W. A. Remers, R. H. Roth, G. J. Gibs, and M. J. Weiss, *J. Org. Chem.*, **36**, 1232 (1971).
164. H. Muratake and M. Natsume, *Heterocycles*, **29**, 783 (1989).
165. R. A. Jones and J. A. Lindner, *Aust. J. Chem.*, **18**, 875 (1965).
166. W. Flitsch and U. Neuman, *Chem. Ber.*, **104**, 2170 (1971).
167. W. Flitsch, B. Muter, and U. Wolf, *Chem. Ber.*, **106**, 1993 (1973).
168. Y. Badar, W. J. S. Lockley, T. P. Toube, B. C. L. Weedon, and L. R. G. Valadon, *J. Chem. Soc., Perkin Trans. 1*, **1973**, 1416.
169. B. A. J. Clark, M. M. S. El-Bakoush, and J. Parrick, *J. Chem. Soc., Perkin Trans. 1*, **1974**, 1531.
170. A. Nonnenmacher, R. Mayer, and H. Pleininger, *Liebigs Ann. Chem.*, **1983**, 2135.
171. R. A. Jones, T. Porjarlieva, and R. J. Head, *Tetrahedron*, **24**, 2013 (1968).
172. E. J. Seus, *J. Heterocycl. Chem.*, **2**, 318 (1965).
173. J. V. Sinisterra, Z. Mouloungui, M. Delmas, and A. Gaset, *Synthesis*, **1985**, 1097.
174. Z. Mouloungui, I. Murengezi, M. Delmas, and A. Gaset, *Synth. Commun.*, **18**, 1241 (1988).
175. F. Corelli, S. Massa, G. Stefanich, A. Mai, M. Artico, S. Panico, and N. Simonetti, *Farmaco, Ed. Sci.*, **42**, 893 (1987).
176. E. E. Schweitzer and K. K. Light, *J. Org. Chem.*, **31**, 870 (1966).
177. F. R. Ahmed and T. P. Toube, *J. Chem. Soc., Perkin Trans. 1*, **1984**, 1577.
178. F. R. Ahmed and T. P. Toube, *J. Chem. Res.*, **1986** (S), 440; (M), 3601.
179. J. M. Mercey, Ph.D. thesis, London University, 1988.
180. E. Gonzalez-Rosende, R. A. Jones, J. Sepuldeva-Arques, and E. Zaballos-Garcia, *Synth. Commun.*, **18**, 1669 (1988).
181. K. E. Schweitzer and K. K. Light, *J. Am. Chem. Soc.*, **86**, 2963 (1964).
182. K. E. Schweitzer and K. K. Light, *J. Org. Chem.*, **31**, 2912 (1966).
183. P. L. Fuchs, *J. Am. Chem. Soc.*, **96**, 1607 (1974).
184. C. W. Spangler and R. K. McCoy, *Synth. Commun.*, **18**, 51 (1988).
185. W. Hinz, R. A. Jones, and T. Anderson, *Synthesis*, **1986**, 620.
186. A. Berlin, S. Bradamante, R. Ferraccioli, G. A. Pagani, and F. Sannicolo, *J. Chem. Soc., Perkin Trans. 1*, **1987**, 2631.
187. V. H. Rawal, R. J. Jones, and M. P. Cava, *J. Org. Chem.*, **52**, 19 (1987).
188. M. Muradin-Szweykowska, A. J. M. Peters, and J. Lugtenburg, *Recl., J. R. Neth. Chem. Soc.*, **103**, 105 (1984).
189. V. A. Dombrovskii, E. V. Gracheva, and P. M. Kochergin, *Khim. Geterosikl. Soedin*, **1986**, 40; *Chem. Abstr.*, **105**, 190716 (1986).
190. P. Carter, S. Fitzjohn, and P. Magnus, *J. Chem. Soc., Chem. Commun.*, **1986**, 1162.
191. J. B. Hendrickson and J. G. de Vries, *J. Org. Chem.*, **50**, 1688 (1985).
192. J. B. Hendrickson and J. G. de Vries, *J. Org. Chem.*, **47**, 1148 (1982).
193. H. Fischer and J. Klarer, *Liebigs Ann. Chem.*, **442**, 1 (1925).
194. W. C. Agosta, *J. Am. Chem. Soc.*, **82**, 2258 (1960).

1.13. References

195. W. A. Davies, A. R. Pinder, and I. G. Morris, *Tetrahedron*, **18**, 405 (1963).
196. W. Flitsch and U. Neuman, *Chem. Ber.*, **104**, 2170 (1971).
197. H. Fischer and H. Wasenegger, *Liebigs Ann. Chem.*, **461**, 277 (1928).
198. S. Gronowitz and R. Kada, *J. Heterocycl. Chem.*, **21**, 1041 (1984).
199. M. V. Mesentseva, I. S. Nikolaeva, A. N. Fomina, and M. I. Akimova, *Khim. Farm. Zh.*, **21**, 1206 (1987); *Chem. Abstr.*, **109**, 6360 (1988).
200. W. W. Turner, *J. Heterocycl. Chem.*, **23**, 327 (1986).
201. A. T. Biev, *Dokl. Bolg. Akad. Nauk*, **35**, 1665 (1982); *Chem. Abstr.*, **98**, 215560 (1983).
202. D. G. Schmidt and H. Zimmer, *J. Heterocycl. Chem.*, **20**, 787 (1983).
203. J. Duflos, D. Letouze, G. Queguiner, and P. Pastour, *Tetrahedron Lett.*, **1973**, 3453.
204. G. Seitz and H.-S. The, *Arch. Pharm.* (*Weinheim*), **316**, 730 (1983).
205. G. Seitz and P. Imming, *Chem.-Ztg.*, **112**, 9 (1988).
206. J. Vorkarpic-Furac and M. Suprina, *Z. Chem.*, **29**, 176 (1989); *Chem. Abstr.*, **112**, 76847 (1990).
207. S. V. Tsukerman, V. P. Izvekov, and V. F. Lavrushin, *Khim. Geterosikl. Soedin*, **1965**, 527; *Chem. Abstr.*, **64**, 676 (1966).
208. W. Wei, Z. Yue, C. Dai, and M. Jiang, *Sci. Sin.* (*Ser. B*), **29**, 113 (1986); *Chem. Abstr.*, **107**, 216961 (1987).
209. H. Fischer, B. Weiss, and M. Schubert, *Ber. Dtsch. Chem. Ges.*, **56**, 1194 (1923).
210. S. V. Tsukerman, V. P. Izvekov, and V. F. Lavrushin, *Khim. Geterosikl. Soedin*, **1967**, 9; *Chem. Abstr.*, **70**, 77079 (1969).
211. S. V. Tsukerman, V. P. Izvekov, and V. F. Lavrushin, *Khim. Geterosikl. Soedin*, **1968**, 823; *Chem. Abstr.*, **71**, 2798 (1969).
212. S. El-Meligny, M. Shaban, and K. M. Ghoneim, *Egypt. J. Pharm. Sci.*, **26**, 59 (1985); *Chem. Abstr.*, **107**, 154188 (1987).
213. D. A. Evans, J. V. Nelson, E. Vogel, and T. R. Taber, *J. Am. Chem. Soc.*, **103**, 3099 (1981).
214. H. Fischer and J. Muller, *Z. Physiol. Chem.*, **132**, 102 (1924).
215. E. M. Larsen and G. A. Terry, *J. Am. Chem. Soc.*, **73**, 500 (1951).
216. T. T. Howarth, A. H. Jackson, J. Judge, G. W. Kenner, and D. J. Newman, *J. Chem. Soc., Perkin Trans. 1*, **1974**, 490.
217. H. McNab, *J. Org. Chem.*, **46**, 2809 (1981).
218. A. J. Blake, H. McNab, and R. Morrison, *J. Chem. Soc., Perkin Trans. 1*, **1988**, 2145.
219. G. Jones and P. M. Radley, *J. Chem. Soc., Perkin Trans. 1*, **1982**, 1123.
220. G. Dannhardt and L. Steindl, *Arch. Pharm.* (*Weinheim*), **319**, 749 (1986).
221. W. Flitsch and W. Lubisch, *Chem. Ber.*, **117**, 1424 (1984).
222. J. J. G. S. van Es, J. H. Hoek, C. Erkelins, and J. Lugtenburg, *Recl. Trav. Chim. Pays-Bas*, **105**, 360 (1986).
223. J. Duflos, G. Dupas, and G. Queguiner, *J. Heterocycl. Chem.*, **20**, 1191 (1983).
224. V. Virmani, J. Singh, P. C. Jain, and N. Anand, *J. Chem. Soc. Pak.*, **1**, 109 (1979); *Chem. Abstr.*, **93**, 46308 (1980).
225. A. M. Islam, K. A. M. El-Bayouki, A. M. Khairy, and H. H. Moharram, *Pharmazie*, **39**, 382 (1984); *Chem. Abstr.*, **102**, 24540 (1985).
226. N. R. El-Rayyes and H. M. Ramadan, *J. Heterocycl. Chem.*, **24**, 589 (1987).
227. C. Cativiela, M. D. Diaz de Villeges, J. I. Garcia, J. A. Mayoral, and E. Menendez, *An. Quim., Ser. C*, **81**, 56 (1985); *Chem. Abstr.*, **105**, 182726 (1986).
228. N. R. El-Rayyes, *J. Heterocycl. Chem.*, **19**, 415 (1982).
229. R. M. Acheson and J. Woolard, *J. Chem. Soc., Perkin Trans. 1*, **1975**, 446.
230. H. Brederick, F. Effenberger, and H. Botsch, *Chem. Ber.*, **97**, 3397 (1964).

231. W. Flitsch and M. Hohenhorst, *Liebigs Ann. Chem.*, **1988**, 276.
232. W. Flitsch and W. Lubisch, *Chem. Ber.*, **115**, 1547 (1982).
233. W. Klose, K. Nikisch, and F. Bohlmann, *Chem. Ber.*, **113**, 2694 (1980).
234. R. Rajeswari, A. A. Adesomojo, and M. P. Cava, *J. Heterocycl. Chem.*, **26**, 557 (1989).
235. E. Toja, J. Kettenring, B. Goldstein, and G. Tarzia, *J. Heterocycl. Chem.*, **23**, 1561 (1986).
236. M. M. Vora, C. S. Yi, and D. Blanton, *Heterocycles*, **16**, 399 (1981).
237. M. M. Vora, C. S. Yi, and D. Blanton, *Heterocycles*, **18**, 507 (1983).
238. E. Roder, H. Wiedenfeld, and T. Bourauel, *Liebigs Ann. Chem.*, **1985**, 1708.
239. A. Toja, G. Tarzia, P. Ferrari, and G. Tuan, *J. Heterocycl. Chem.*, **23**, 1555 (1986).
240. Y. Girard, J. G. Atkinson, P. C. Belanger, J. J. Fuentes, J. Rokach, C. S. Rooney, D. C. Remy, and C. A. Hunt, *J. Org. Chem.*, **48**, 3320 (1983).
241. P. C. Belanger, J. G. Atkinson, C. S. Rooney, S. F. Britcher, and C. D. Remy, *J. Org. Chem.*, **48**, 3234 (1983).
242. K. Yeh and R. H. Baker, *Inorg. Chem.*, **6**, 830 (1967).
243. B. Emmert, K. Diehl, and F. Gollwitzer, *Ber. Dtsch. Chem. Ges.*, **62**, 1733 (1929).
244. R. A. Jones, *Aust. J. Chem.*, **17**, 894 (1964).
245. H. Tanaka and O. Yamauchi, *Chem. Pharm. Bull.*, **9**, 588 (1961).
246. A. Triebs and E. Dietl, *Chem. Ber.*, **94**, 298 (1961).
247. A. A. H. Saeed, *Indian J. Chem.*, **23B**, 92 (1984).
248. H. S. Lyn and H. J. Chae, *Taehan Hwahakhoe Chi*, **27**, 133 (1983); *Chem. Abstr.*, **99**, 52712 (1983).
249. F. Capitan, F. Molina, P. Espinosa, and L. F. Capitan-Valivey, *Bull. Soc. Chim. Belg.*, **94**, 387 (1985).
250. H. Tanaka and O. Yamauchi, *Chem. Pharm. Bull.*, **10**, 435 (1962).
251. P. Pfeiffer, T. Hesse, H. Pfitzer, W. Scholl, and H. Thielert, *J. Prakt. Chem.*, **149**, 217 (1937).
252. A. Bizhev and R. Borisova, *Farmatsiya (Sophia)*, **33**, 1 (1983); *Chem. Abstr.*, **99**, 38416 (1983).
253. A. Carpenter and D. J. Chadwick, *Tetrahedron*, **41**, 3803 (1985).
254. E. A. Panifilova, I. Y. Kvitko, and A. V. El'tsov, *Khim. Geterosikl. Soedin.*, **1981**, 1489; *Chem. Abstr.*, **96**, 122558 (1982).
255. M. A. Kira and A. Bruckner-Wilhelms, *Acta Chim. (Budapest)*, **56**, 47 (1968); *Chem Abstr.*, **69**, 86888 (1968).
256. P. E. Sonnett, *J. Org. Chem.*, **36**, 1005 (1971).
257. H. R. Heusel, *Chem. Ber.*, **99**, 868 (1966).
258. W. Herz and J. Brasch, *J. Org. Chem.*, **23**, 711 (1958).
259. J. M. Muchowski and P. Hess, *Tetrahedron Lett.*, **29**, 777 (1988).
260. J. M. Muchowski and P. Hess, *Tetrahedron Lett.*, **29**, 3215 (1988).
261. J. M. Brittain, R. A. Jones, R. O. Jones, and T. J. King, *J. Chem. Soc., Perkin Trans. 1*, **1981**, 2656.
262. N. R. El-Rayyes and F. M. Al-Kharafi, *Egypt. J. Chem.*, **23**, 151 (1980); *Chem. Abstr.*, **96**, 142575 (1982).
263. J. C. Lancelot, D. Maume, and M. Robba, *J. Heterocycl. Chem.*, **17**, 625 (1980).
264. F. D. Popp, *J. Heterocycl. Chem.*, **21**, 617 (1984).
265. A. A. Schildt and F. H. Case, *Talanta*, **28**, 863 (1981); *Chem. Abstr.*, **96**, 154515 (1982).
266. I. A. Tossidis, *Chem. Chron.*, **12**, 181 (1983); *Chem. Abstr.*, **102**, 6129 (1985).
267. M. P. Ceneviva, M. I. A. Goncalves, F. C. Lacerda de Almeida, and R. J. Giordano, *Rev. Farm. Bioquim. Univ. Sao Paulo*, **21**, 121 (1985); *Chem. Abstr.*, **106**, 138191 (1987).

1.13. References

268. A. T'ang, E. J. Lien, and M. M. C. Lai, *J. Med. Chem.*, **228**, 1103 (1985).
269. L. Garuti, A. Ferranti, G. Giovanninetti, M. Baserga, and A. M. Palenzona, *Farmaco, Ed. Sci.*, **36**, 393 (1981).
270. A. Mukherjee, V. L. Sharma, V. Seth, and A. P. Bhaduri, *Indian J. Chem.*, **27B**, 537 (1988).
271. H. Volz, U. Zirngibl, and B. Messner, *Tetrahedron Lett.*, **1970**, 3593.
272. H. Fischer and M. Hussong, *Liebigs Ann. Chem.*, **492**, 128 (1931).
273. H. Fischer and A. Waibel, *Liebigs Ann. Chem.*, **512**, 195 (1934).
274. M. K. A. Khan, K. J. Morgan, and D. P. Morrey, *Tetrahedron*, **22**, 2095 (1966).
275. H. Rapoport and K. G. Holden, *J. Am. Chem. Soc.*, **82**, 5510 (1960); **84**, 635 (1962).
276. S. F. MacDonald, *J. Chem. Soc.*, **1952**, 4176.
277. A. Treibs and D. Grimm, *Liebigs Ann. Chem.*, **752**, 44 (1971).
278. J. M. Yeung, L. A. Corleto, and E. E. Knaus, *J. Med. Chem.*, **25**, 191 (1982).
279. H. Adkins, I. A. Wolff, A. Pavlic, and E. Hutchinson, *J. Am. Chem. Soc.*, **66**, 1293 (1944).
280. G. Buchi and G. Lukas, *J. Am. Chem. Soc.*, **85**, 647 (1963).
281. V. Sprio, P. Madonia, and R. Caronia, *Ann. Chim. (Italy)*, **49**, 169 (1959).
282. G. Bringmann and S. Schneider, *Synthesis*, **1983**, 139.
283. J.-C. Lancelot, D. Laduree, H. El Kashef, and M. Robba, *Heterocycles*, **23**, 909 (1985).
284. V. Nacci, A. Garofalo, and I. Fiorine, *J. Heterocycl. Chem.*, **22**, 259 (1985).
285. C. Fizet and J. Streith, *Tetrahedron Lett.*, **1974**, 3187.
286. R. Granados, D. Mauleon, and M. Perez, *An. Quim., Ser. C*, **79C**, 275 (1983); *Chem. Abstr.*, **102**, 62011 (1985).
287. H. Fischer, H. Beyer, and E. Zauker, *Liebigs Ann. Chem.*, **486**, 55 (1931).
288. A. Hanck, *Chem. Ber.*, **101**, 2280 (1968).
289. P. Clezy and A. W. Nichol, *Aust. J. Chem.*, **18**, 1977 (1965).
290. H. Fischer and H. Orth, *Liebigs Ann. Chem.*, **489**, 62 (1931).
291. S. El-Meligy, M. Shaban, and M. Ghoneim, *Egypt. J. Pharm. Sci.*, **26**, 51 (1985); *Chem. Abstr.*, **107**, 154187 (1987).
292. D. C. Rustidge, Ph.D. thesis, University of East Anglia, 1977.
293. G. Boche, F. Bosold, and M. Niessner, *Tetrahedron Lett.*, **23**, 3255 (1982).
294. M. E. K. Cartoon and G. W. H. Cheeseman, *J. Organomet. Chem.*, **234**, 123 (1982).
295. A. P. Kozikowski and X.-M. Cheng, *Tetrahedron Lett.*, **26**, 4047 (1985).
296. F. Acar, S. S. Badesha, W. Flitsch, R. Gozogul, O. Inel, S. Inel, R. A. Jones, C. Ogretir, and D. C. Rustidge, *Chim. Acta Turc.*, **9**, 225 (1981); *Chem. Abstr.*, **95**, 169111 (1981).
297. R. Kreher and G. Vogt, *Angew. Chem. Int. Ed. Eng.*, **9**, 955 (1970).
298. B. T. Gilis and J. C. Valentour, *J. Heterocycl. Chem.*, **8**, 13 (1971).
299. S. A. N. Taheri, R. A. Jones, S. S. Badesha, and M. M. Hania, *Tetrahedron*, **45**, 7717 (1989).
300. H. Fischer, E. Sturm, and H. Friedrich, *Liebigs Ann. Chem.*, **461**, 244 (1928).
301. H. Fischer and B. Putzer, *Ber. Dtsch. Chem. Ges.*, **61**, 1068 (1928).
302. H. J. Anderson and S. J. Griffiths, *Can. J. Chem.*, **45**, 2227 (1967).
303. C. Jaureguiberry and M. Roques, *Compt. rend. Hebd. Seances Acad. Sci.*, **274C**, 1703 (1972).
304. J. P. Cress and D. M. Forkey, *J. Chem. Soc., Chem. Commun.*, **1973**, 35.
305. German Pat. 3721534; *Chem. Abstr.*, **109**, 93860 (1988).
306. V. P. Zhestkov, V. G. Voronin, M. L. Suslina, and A. S. Zaks, *Khim.-Farm. Zh.*, **16**, 687 (1982); *Chem. Abstr.*, **97**, 109917 (1982).
307. G. Stefancich, M. Artico, F. Corelli, and S. Massa, *Synthesis*, **1983**, 757.

308. M. A. Brimble and D. D. Rowan, *J. Chem. Soc., Chem. Commun.*, **1988**, 978.
309. D. D. Rowan, M. B. Hunt, and D. L. Gayner, *J. Chem. Soc., Chem. Commun.*, **1986**, 925.
310. F. A. Carey and R. M. Guiliano, *J. Org. Chem.*, **46**, 1366 (1981).
311. R. Singh, P. C. Cain, and N. Anand, *Indian J. Chem.*, **21B**, 225 (1982).
312. M. Artico, G. de Martino, R. Giuliano, S. Massa, and G. C. Porretta, *J. Chem. Soc., Chem. Commun.*, **1969**, 671.
313. K. Faber, H. J. Anderson, C. J. Loader, and A. S. Daley, *Can. J. Chem.*, **67**, 1046 (1984).
314. N. S. Girgis, A. Jorgensen, and E. B. Pedersen, *Synthesis*, **1985**, 101.
315. S. W. Schneller, J.-K. Luo, R. S. Hosmane, and R. H. Durrfeld, *J. Heterocycl. Chem.*, **21**, 1153 (1984).
316. L. Lamartina, O. Migliara, and V. Sprio, *J. Heterocycl. Chem.*, **19**, 1381 (1982).
317. A. Jorgensen, *Heterocycles*, **24**, 997 (1986).
318. Y. Effi, M. C. de Sevricourt, S. Rault, and M. Robba, *Heterocycles*, **16**, 1519 (1981).
319. E. Toja, D. Selva, and P. Schiatti, *J. Med. Chem.*, **27**, 610 (1984).
320. German Pat. 3141063; *Chem. Abstr.*, **99**, 88196 (1983).
321. L. Bruche, L. Garanti, and G. Zecchi, *Synthesis*, **1985**, 304.
322. P. Hodge and R. W. Rickards, *J. Chem. Soc.*, **1963**, 468.
323. R. A. Nicolaus, *Gazz. Chim. Ital.*, **83**, 239 (1953).
324. Jpn. Pat. 58 59961; *Chem. Abstr.*, **99**, 139929 (1983).
325. V. G. Kul'nevich, E. Baum, and T. E. Goldovskaya, *Khim. Geterosikl. Soedin.*, **1982**, 495; *Chem. Abstr.*, **97**, 77207 (1982).
326. J. A. de Groot, R. van der Steen, and J. Lugtenburg, *Rec. Trav. Chim. Pays-Bas*, **101**, 35 (1982).
327. J. A. Ballantyne, A. H. Jackson, G. W. Kenner, and G. McGillivray, *Tetrahedron*, **Suppl. 7**, 241 (1966).
328. G. McGillivray, E. Ten Krooden, and M. Beyers, *S. Afr. J. Chem.*, **39**, 51 (1986).
329. T. Ajello and V. Sprio, *Gazz. Chim. Ital.*, **89**, 2526 (1959).
330. T. Ajello, V. Sprio, and P. Madonia, *Gazz. Chim. Ital.*, **87**, 11 (1957).
331. V. Sprio and P. Madonia, *Gazz. Chim. Ital.*, **87**, 171 (1957).
332. V. J. Demopoulos, *Org. Prep. Proced. Int.*, **18**, 278 (1986).
333. T. Severin and I. Ipach, *Chem. Ber.*, **108**, 1768 (1975).
334. C. Cativiela and J. I. Garcia, *Org. Prep. Proced. Int.*, **18**, 283 (1986).
335. A. Treibs and F. H. Kreuzer, *Liebigs Ann. Chem.*, **721**, 105 (1969).
336. P. E. Sonnett, *J. Med. Chem.*, **15**, 97 (1972).
337. D. M. Bailey, R. E. Johnson, and U. Salvador, *J. Med. Chem.*, **16**, 1298 (1973).
338. J. W. Harbuck and H. Rapoport, *J. Org. Chem.*, **37**, 3618 (1972).
339. D. M. Bailey and R. E. Johnson, *J. Med. Chem.*, **16**, 1300 (1973).
340. J. A. de Groot, R. van der Steen, R. Fokkens, and J. Lugtenburg, *Rec. Trav. Chim. Pays-Bas*, **101**, 219 (1982).
341. D. Chadwick and S. T. Hodgson, *J. Chem. Soc., Perkin Trans. 1*, **1983**, 93.
342. R. M. Silverstein, E. E. Ryskiewicz, C. Willard, and R. C. Koehler, *J. Org. Chem.*, **20**, 668 (1955).
343. R. L. Hinman and S. Theodoropolis, *J. Org. Chem.*, **28**, 3052 (1963).
344. A. Treibs and H. Derra-Scherer, *Liebigs Ann. Chem.*, **589**, 188 (1954).
345. A. Treibs and H. Scherer, *Liebigs Ann. Chem.*, **577**, 139 (1952).
346. W. Herz and C. F. Courtney, *J. Am. Chem. Soc.*, **76**, 576 (1954).
347. R. Rips and N. P. Buu-Hoi, *J. Org. Chem.*, **24**, 372 (1959).
348. E. Laschuvka and R. Huisgen, *Chem. Ber.*, **93**, 81 (1960).

1.13. References

349. F. P. Doyle, M. D. Mehta, G. Sach, and J. L. Person, *J. Chem. Soc.*, **1958**, 4458.
350. W. K. Anderson and P. F. Carey, *J. Med. Chem.*, **20**, 1691 (1977).
351. W. K. Anderson and P. F. Carey, *J. Med. Chem.*, **20**, 813 (1977).
352. V. M. Micovic and M. L. Mihailovic, *J. Org. Chem.*, **18**, 1190 (1953).
353. R. F. Nystrom and C. R. A. Berger, *J. Am. Chem. Soc.*, **80**, 2896 (1958).
354. R. M. Silverstein, E. E. Ryskiewicz, and S. W. Chaiken, *J. Am. Chem. Soc.*, **76**, 4485 (1954).
355. E. E. Ryskiewicz and R. M. Silverstein, *J. Am. Chem. Soc.*, **76**, 5802 (1954).
356. P. E. Sonnett, *J. Heterocycl. Chem.*, **7**, 1101 (1970).
357. R. Greenhouse, C. Ramirez, and J. M. Muchowski, *J. Org. Chem.*, **50**, 2961 (1985).
358. C. F. Hobbs, C. K. McMillin, E. P. Papadopoulos, and C. A. van der Werf, *J. Am. Chem. Soc.*, **84**, 43 (1962).
359. P. Angeli, L. Brasili, E. Brancia, D. Giardina, W. Quaglia, and C. Melchiorre, *J. Med. Chem.*, **28**, 1643 (1985).
360. K. M. Biswas and A. H. Jackson, *J. Chem. Soc. (C)*, **1970**, 1667.
361. K. M. Biswas and A. H. Jackson, *Tetrahedron*, **24**, 1145 (1968).
362. Y. K. Shim, J. Y. Shim, and W. J. Kim, *Bull. Korean Chem. Soc.*, **9**, 410 (1988); *Chem. Abstr.*, **111**, 57455 (1989).
363. P. E. Sonnett, *J. Heterocycl. Chem.*, **9**, 1395 (1972).
364. F. R. Ahmed and T. P. Toube, *J. Chem. Res.*, **1986**, (S), 440, (M), 3601.
365. F. R. Ahmed and T. P. Toube, unpublished work.
366. J. M. Mercey and T. P. Toube, *J. Chem. Res.*, **1988**, (S), 333.
367. M. Farnier and P. Fournari, *Bull. Soc. Chim. Fr.*, **1975**, 2335.
368. C. E. Loader, G. H. Barnett, and H. J. Anderson, *Can. J. Chem.*, **60**, 383 (1982).
369. G. G. Kleinspehn and A. H. Corwin, *J. Am. Chem. Soc.*, **82**, 2750 (1960).
370. R. Renshaw and W. E. Cross, *J. Am. Chem. Soc.*, **61**, 1195 (1939).
371. H. Fischer and A. Stern, *Liebigs Ann. Chem.*, **446**, 229 (1925).
372. H.-P. Kaiser and J. M. Muchowski, *J. Org. Chem.*, **49**, 4203 (1984).
373. S. Umio, *Yakugaku Zasshi*, **79**, 1048 (1959); *Chem. Abstr.*, **54**, 5611 (1960).
374. F. K. Signaigo and H. Adkins, *J. Am. Chem. Soc.*, **58**, 709 (1936).
375. J. L. Rainey and H. Adkins, *J. Am. Chem. Soc.*, **61**, 1104 (1939).
376. R. Renshaw and W. E. Cross, *J. Am. Chem. Soc.*, **61**, 1195 (1939).
377. L. H. Andrews and S. M. McElvain, *J. Am. Chem. Soc.*, **51**, 887 (1929).
378. H. Adkins and H. L. Coonradt, *J. Am. Chem. Soc.*, **63**, 1563 (1941).
379. E. Bullock, T.-S. Chen, and C. E. Loader, *Can. J. Chem.*, **44**, 1007 (1966).
380. E. Bullock, T.-S. Chen, C. E. Loader, and A. E. Wells, *Can. J. Chem.*, **48**, 1651 (1970).
381. J. K. Groves, N. E. Cundasawmy, and H. J. Anderson, *Can. J. Chem.*, **51**, 1089 (1973).
382. C. E. Loader and H. J. Anderson, *Tetrahedron*, **25**, 3879 (1969).
383. H. Fischer and A. Stern, *Liebigs Ann. Chem.*, **446**, 229 (1925).
384. M. Dennstedt and J. Zimmerman, *Ber. Dtsch. Chem. Ges.*, **19**, 2204 (1886).
385. E. E. Havinga, Woten Hoeve, E. W. Meijer, and H. Wynberg, *Chem. Mater.*, **1**, 650 (1989).
386. J. W. Cornforth and M. E. Frith, *J. Chem. Soc.*, **1958**, 1091.
387. P. S. Skell and G. P. Bean, *J. Am. Chem. Soc.*, **84**, 4655 (1962).
388. D. O. Alonso Garrido, G. Buldain, and B. Frydman, *J. Org. Chem.*, **49**, 2619 (1984).
389. L. Knorr and K. Hess, *Ber. Dtsch. Chem. Ges.*, **45**, 2631 (1912).
390. P. A. Cantor, R. Lancaster, and C. A. Vander Werf, *J. Org. Chem.*, **21**, 918 (1956).
391. A. Carbella, P. Garibaldi, G. Jommi, and F. Mauri, *Chem. Ind. (Lond.)*, **1969**, 583.

392. J. W. Scott, A. Focella, A. O. Hengartner, D. R. Parrish, and D. Valentine, *Synth. Commun.*, **10**, 529 (1980).
393. (a) P. S. Clezy, C. J. R. Fookes, D. Y. K. Lau, A. W. Nichol, and G. A. Smythe, *Aust. J. Chem.*, **27**, 357 (1974); (b) R. Chong and P. S. Cleezy, *Aust. J. Chem.*, **20**, 935 (1967).
394. A. H. Corwin and L. Straughn, *J. Am. Chem. Soc.*, **70**, 1416 (1948).
395. F. Blicke and E. Blake, *J. Am. Chem. Soc.*, **52**, 235 (1930).
396. P. Hodge and R. W. Rickards, *J. Chem. Soc.*, **1963**, 2543.
397. P. Hodge and R. W. Rickards, *J. Chem. Soc.*, **1965**, 459.
398. A. Corwin and S. R. Buc, *J. Am. Chem. Soc.*, **66**, 1151 (1944).
399. J. A. Waters, *J. Med. Chem.*, **20**, 1094 (1977).
400. J. Barry, G. Bram, and A. Petit, *Heterocycles*, **23**, 875 (1985).
401. D. J. Chadwick, J. Chambers, G. D. Meakins, and R. L. Snowden, *J. Chem, Soc., Perkin Trans. 1*, **1973**, 1766.
402. K. M. Smith, G. W. Craig, F. Eivazi, and Z. Martynenko, *Synthesis*, **1980**, 493.
403. A. Williams and G. Salvatori, *J. Chem. Soc., Perkin Trans. 2*, **1972**, 883.
404. L. Knorr, *Liebigs Ann. Chem.*, **236**, 290 (1886).
405. O. Piloty and K. Wilke, *Ber. Dtsch. Chem. Ges.*, **45**, 2586 (1912).
406. G. Magnanini, *Ber. Dtsch. Chem. Ges.*, **22**, 35 (1889).
407. H. Fischer and O. Wiedemann, *Z. Physiol. Chem.*, **155**, 57 (1926).
408. H. Fischer and B. Walach, *Ber. Dtsch. Chem. Ges.*, **58**, 2818 (1925).
409. A. H. Corwin and W. M. Quattlebaum, *J. Am. Chem. Soc.*, **58**, 1081 (1936).
410. E. J. Chu and T. C. Chu, *J. Org. Chem.*, **19**, 266 (1954).
411. A. H. Corwin and J. Straughn, *J. Am. Chem. Soc.*, **70**, 2968 (1948).
412. G. Doleschall, P. Seres, L. Parkanyi, G. Toth, A. Almasy, and E. Bihatsi-Karsai, *J. Chem. Soc., Perkin Trans. 1*, **1986**, 927.
413. F. Boberg, K.-H. Garburg, K.-J. Gorlich, E. Pipereit, and M. Ruhr, *J. Heterocycl. Chem.*, **23**, 753 (1986).
414. G. Stahly, E. M. Martlett, and G. E. Nelson, *J. Org. Chem.*, **48**, 4423 (1983).
415. Ger. Pat. 3415321; *Chem. Abstr.*, **104**, 19507 (1986).
416. J. Crook, A. H. Jackson, and G. W. Kenner, *J. Chem. Soc. (C)*, **1971**, 474.
417. A. Treibs and K. H. Hintermeier, *Chem. Ber.*, **87**, 1167 (1954).
418. A. Treibs and W. Ott, *Liebigs Ann. Chem.*, **615**, 137 (1958).
419. T. T. Howarth, A. H. Jackson, and G. W. Kenner, *J. Chem. Soc., Perkin Trans. 1*, **1974**, 502.
420. K. M. Smith, F. W. Bobe, O. M. Minnetian, H. Hope, and M. D. Yanuck, *J. Org. Chem.*, **50**, 790 (1985).
421. D. J. Chadwick, M. V. McKnight, and R. Ngochindo, *J. Chem. Soc., Perkin Trans. 1*, **1982**, 1343.
422. C. F. Candy, R. A. Jones, and P. A. Wright, *J. Chem. Soc. (C)*, **1970**, 2563.
423. K. Nikisch, W. Klose, and F. Bohlmann, *Chem. Ber.*, **113**, 2036 (1980).
424. R. A. Jones and R. L. Laslett, *Aust. J. Chem.*, **17**, 1056 (1964).
425. F. M. Menger and J. A. Donahue, *J. Am. Chem. Soc.*, **95**, 432 (1973).
426. P. Linda, A. Stener, A. Cipiciani, and G. Savelli, *J. Heterocycl. Chem.*, **20**, 247 (1973).
427. C. R. W. A. Smith and D. A. Copeland, *J. Am. Chem. Soc.*, **95**, 3808 (1973).
428. A. Shirazi, R. S. Marianelli, and G. D. Sturgeon, *Inorg. Chim. Acta*, **72**, 5 (1983).
429. V. Frick and G. Simchen, *Liebigs Ann. Chem.*, **1987**, 839.
430. H. Pleininger, *Synthesis*, **1973**, 422.

1.13. References

431. L. A. Carpino and D. E. Barr, *J. Org. Chem.*, **31**, 764 (1966).
432. I. Hasan, E. R. Martinelli, L.-C. C. Lin, F. W. Fowler, and A. B. Levy, *J. Org. Chem.*, **46**, 157 (1981).
433. D. Dhanak and C. B. Reese, *J. Chem. Soc., Perkin Trans. 1*, **1986**, 2181.
434. L. Grehn, U. Ragnarsson, B. Eriksson, and B. Oberg, *J. Med. Chem.*, **26**, 1042 (1983).
435. L. Grehn and U. Ragnarsson, *Angew. Chem. Int. Ed. Eng.*, **23**, 296 (1984).
436. I. Hasan, E. R. Martinelli, L.-C. C. Lin, F. W. Fowler, and A. B. Levy, *J. Org. Chem.*, **46**, 157 (1981).
437. V. H. Rawal and M. P. Cava, *Tetrahedron Lett.*, **26**, 6141 (1985).
438. A. Katritzky and K. Akutagawawa, *Org. Prep. Proced. Int.*, **20**, 585 (1988).
439. C. Gonzalez, R. Greenhouse, and R. Tallabs, *Can. J. Chem.*, **61**, 1697 (1983).
440. J. M. Muchowski and D. R. Solas, *J. Org. Chem.*, **49**, 203 (1984).
441. G. Ciamician, *Monatsh. Chem.*, **1**, 625 (1880).
442. G. Ciamician and P. Silber, *Ber. Dtsch. Chem. Ges.*, **17**, 1150 (1884).
443. M. K. A. Khan and K. J. Morgan, *Tetrahedron*, **21**, 2197 (1965).
444. J. Elguero, R. Jacquier, and B. Schimizu, *Bull. Soc. Chim. Fr.*, **1967**, 2996.
445. P. Pfaffli and C. Tamm, *Helv. Chim. Acta*, **52**, 1911 (1969).
446. H. J. Anderson, J. A. Clase, and C. E. Loader, *Synth. Commun.*, **17**, 401 (1987).
447. G. S. Marks, D. K. Dougall, E. Bullock, and S. F. MacDonald, *J. Am. Chem. Soc.*, **82**, 3183 (1960).
448. S. H. Wilen, D. Shen, J. M. Licata, E. Baldwin, and C. S. Russell, *Heterocycles*, **22**, 1747 (1984).
449. G. M. Badger, R. L. N. Harris, and R. A. Jones, *Aust. J. Chem.*, **17**, 987 (1964).
450. G. M. Badger, R. L. N. Harris, and R. A. Jones, *Aust. J. Chem.*, **17**, 1013 (1964).
451. B. W. Cue and N. Chamberlain, *J. Heterocycl. Chem.*, **18**, 667 (1981).
452. G. M. Badger, R. L. N. Harris, and R. A. Jones, *Aust. J. Chem.*, **17**, 1022 (1964).
453. A. Treibs and R. Schmidt, *Liebigs Ann. Chem.*, **577**, 105 (1952).
454. A. Treibs and R. Zinsmeister, *Chem. Ber.*, **90**, 87 (1957).
455. R. Grigg, A. W. Johnson, and J. W. F. Wasley, *J. Chem. Soc.*, **1963**, 359.
456. G. M. Badger, R. L. N. Harris, and R. A. Jones, *Aust. J. Chem.*, **17**, 1002 (1964).
457. W. Siedel, *Liebigs Ann. Chem.*, **554**, 144 (1943).
458. F. Feist, *Ber. Dtsch. Chem. Ges.*, **35**, 1545 (1902).
459. A. H. Corwin, W. A. Bailey, and P. Viohl, *J. Am. Chem. Soc.*, **64**, 1267 (1942).
460. J. Leroy, D. Cantacuzene, and C. Wakselman, *Synthesis*, **1982**, 813.
461. G. P. Arsenault and S. F. MacDonald, *Can. J. Chem.*, **39**, 2043 (1961).
462. G. Kleinspehn and A. H. Corwin, *J. Am. Chem. Soc.*, **75**, 5295 (1953).
463. D. L. Boger and M. Patel, *Tetrahedron Lett.*, **28**, 2499 (1987).
464. D. L. Boger and M. Patel, *J. Org. Chem.*, **53**, 1405 (1988).
465. M. P. Cava and L. Bravo, *Tetrahedron Lett.*, **1970**, 4631.
466. T. Ajello and S. Giambrone, *Ricerca Sci.*, **23**, 2233 (1953).
467. K. M. Smith, M. Miura, and H. D. Tabba, *J. Org. Chem.*, **48**, 4779 (1983).
468. M. W. Moon and R. A. Wade, *J. Org. Chem.*, **49**, 2663 (1984).
469. R. Kitzing, R. Fuchs, M. Joyeaux, and H. Prinzback, *Helv. Chim. Acta*, **51**, 888 (1968).
470. R. M. Acheson and J. M. Vernon, *J. Chem. Soc.*, **1961**, 457.
471. R. M. Acheson and J. M. Vernon, *J. Chem. Soc.*, **1963**, 1000.
472. N. W. Gabel, *J. Org. Chem.*, **27**, 301 (1962).

473. H.-J. Altenbach, B. Blech, J. A. Marco, and E. Vogel, *Angew. Chem. Int. Ed. Engl.*, **21**, 778 (1982).
474. G. Kaupp, J. Perreten, R. Leute, and H. Prinzbach, *Chem. Ber.*, **103**, 2288 (1970).
475. L. A. Carpino and D. E. Barr, *J. Org. Chem.*, **31**, 764 (1966).
476. R. W. Kaesler and E. LeGoff, *J. Org. Chem.*, **47**, 4779 (1982).
477. M. G. B. Drew, A. V. George, N. S. Isaacs, and H. S. Rzepa, *J. Chem. Soc., Perkin Trans. 1*, **1985**, 1277.
478. J. Keijsers, B. Hams, C. Kruse, and H. Scheeren, *Heterocycles*, **29**, 79 (1989).
479. M. Natsume and H. Muratake, *Tetrahedron Lett.*, **1979**, 3477.
480. E. Wenkert, P. D. R. Moeller, and S. R. Piettre, *J. Am. Chem. Soc.*, **110**, 7188 (1988).
481. F. W. Fowler, *J. Chem. Soc., Chem. Commun.*, **1969**, 1359.
482. B. Fohlisch, R. Herter, and E. Wolf, *J. Chem. Res.*, **1986**, (S), 128; (M) 1201.
483. V. A. Budylin, M. del C. Pina, and Y. G. Bundel, *Khim. Geterosikl. Soedin*, **1984**, 562; *Chem. Abstr.*, **101**, 138295 (1984).
484. J. R. Carson and N. M. Davis, *J. Org. Chem.*, **46**, 839 (1981).
485. T. A. Bryson and G. A. Roth, *Tetrahedron Lett.*, **27**, 3689 (1966).
486. B. E. Maryanoff, S. L. Keeley, and F. J. Fersico, *J. Med. Chem.*, **26**, 226 (1983).
487. G. W. Kenner, K. M. Smith, and F. J. Unsworth, *J. Chem. Soc., Chem. Commun.*, **1973**, 43.
488. G. W. Kenner, J. Rimmer, K. M. Smith, and F. J. Unsworth, *J. Chem. Soc., Perkin Trans. 1*, **1977**, 332.
489. K. M. Smith, Z. Martynenko, and H. D. Tabba, *Tetrahedron Lett.*, **22**, 1291 (1981).
490. A. H. Jackson, K. R. N. Rao, N. S. Ooi, and E. Adelakun, *Tetrahedron Lett.*, **25**, 6049 (1984).
491. A. H. Jackson, K. R. N. Rao, and E. Smeeton, *Tetrahedron Lett.*, **30**, 2673 (1989).
492. T. Wollmann and B. Franck, *Angew. Chem. Int. Ed. Engl.*, **23**, 226 (1984).
493. D. Brown and D. Griffiths, *Synth. Commun.*, **13**, 913 (1983).
494. S. Saeki, T. Hayashi, and M. Hamana, *Chem. Pharm. Bull.*, **32**, 2154 (1984).
495. T. Itahara, *J. Chem. Soc., Chem. Commun.*, **1980**, 49.
496. T. Itahara, *J. Chem. Soc., Chem. Commun.*, **1981**, 254.
497. T. Itahara, *J. Org. Chem.*, **50**, 5272 (1985).
498. C. Gonzalez and R. Greenhouse, *Heterocycles*, **23**, 1127 (1985).
499. J. A. Ganske, R. K. Pandey, M. J. Postick, K. M. Snow, and K. M. Smith, *J. Org. Chem.*, **54**, 4801 (1989).
500. A. M. Almerico and I. D'Asdia, *J. Heterocycl. Chem.*, **19**, 977 (1982).
501. A. Curulli, M. T. Giardi, and G. Sleiter, *Gazz. Chim. Ital.*, **113**, 115 (1983).
502. M. De Rosa, G. C. Nieto, and F. F. Gago, *J. Org. Chem.*, **54**, 5347 (1989).
503. D. L. Comins and M. O. Killpack, *J. Org. Chem.*, **52**, 104 (1987).
504. M. Yoshida, T. Yoshida, M. Kobayashi, and N. Kamigata, *J. Chem. Soc., Perkin Trans. 1*, **1989**, 909.
505. C. W. Jefford, Q. Tang, and J. Boukouvalas, *Tetrahedron Lett.*, **31**, 995 (1990).
506. M. C. Harsanyi and R. K. Norris, *J. Org. Chem.*, **52**, 2209 (1987).
507. V. Nacci, A. Garofalo, and I. Fiorini, *J. Heterocycl. Chem.*, **23**, 769 (1986).
508. Y. Murakami, T. Wanatabe, and H. Ishii, *J. Chem. Soc., Perkin Trans. 1*, **1988**, 3005.
509. R. J. Sundberg, G. S. Hamilton, and J. P. Laurino, *J. Org. Chem.*, **53**, 976 (1988).
510. A. P. Kozikowski, X.-M. Cheng, C.-S. Li, and J. G. Scripko, *Isr. J. Chem.*, **27**, 61 (1986); *Chem. Abstr.*, **108**, 21658 (1988).
511. M. Iwao and T. Kuraishi, *Tetrahedron Lett.*, **26**, 6213 (1985).
512. M. Natsume and H. Muratake, *Tetrahedron Lett.*, **1979**, 3477.

1.13. References

513. M. Natsume, H. Muratake, and Y. Kanda, *Heterocycles*, **16**, 959 (1981).
514. J.-C. Lancelot, D. Laduree, and M. Robba, *Chem. Pharm. Bull.*, **33**, 3122 (1985).
515. G. W. H. Cheeseman, S. A. Eccleshall, and T. Thornton, *J. Heterocycl. Chem.*, **22**, 809 (1985).
516. J. M. Mercey and T. P. Toube, *J. Chem. Res.* **1987**, (S) 62; (M), 680.
517. P. A. Gurevich, V. V. Kiselev, V. V. Moskva, and G. I. Ruzal, *Khim.-Farm. Zh.*, **17**, 51 (1983); *Chem. Abstr.*, **98**, 215679 (1983).
518. H. J. Bestmann, G. Schmid, and D. Sandmeier, *Chem. Ber.*, **113**, 912 (1980).
519. D. L. Boger and M. Patel, *J. Org. Chem.*, **52**, 2319 (1987).
520. D. L. Boger and M. Patel, *J. Org. Chem.*, **52**, 3934 (1987).
521. S. Brandange and B. Rodriguez, *Acta Chem. Scand.*, **B41**, 740 (1987).
522. S. Raja, N. Xavier, and S. J. Arulraj, *Indian J. Chem.*, **27B**, 916 (1988).
523. W. Hinz, R. A. Jones, S. U. Patel, and M.-H. Karatza, *Tetrahedron*, **42**, 3753 (1986).
524. R. Ghaffari-Tabrizi and P. Margaretha, *Helv. Chim. Acta*, **66**, 1902 (1983).
525. Ger. Pat. 3534415; *Chem. Abstr.*, **107**, 59730 (1987).
526. U. S. Pat. 4548696; *Chem. Abstr.*, **104**, 98051 (1986).
527. Ger. Pat. 3420854; *Chem. Abstr.*, **104**, 110395 (1986).
528. R. Bolivar, R. Machado, L. Montero, F. Vargas, and C. Rivas, *J. Photochem.*, **22**, 91 (1983).
529. G. K. Sandhu, N. S. Boparai, and S. S. Sandhu, *Synth. React. Inorg. Met.-Org. Chem.*, **10**, 535 (1980).
530. G. K. Sandhu and S. P. Verma, *Polyhedron*, **6**, 587 (1987).
531. A. J. Arce, Y. De Sanctis, and A. J. Deeming, *J. Organomet. Chem.*, **311**, 371 (1986).
532. P. Mehta and R. K. Mehta, *J. Indian Chem. Soc.*, **61**, 571 (1984).
533. H. A. Tayim and A. S. Salameh, *Polyhedron*, **5**, 687 (1986).
534. H. Adams, N. A. Bailey, D. E. Fenton, S. Moss, C. O. Rodriguez de Barbarin, and G. Jones, *J. Chem. Soc., Dalton Trans.*, **1986**, 693.
535. N. A. Bailey, A. Barrass, D. E. Fenton, M. S. Leal Gonzalez, R. Moody, and C. O. Rodriguez de Barbarin, *J. Chem. Soc., Dalton Trans.*, **1984**, 2741.
536. H. Brunner, R. Becker, and G. Riepl, *Organometallics*, **3**, 1354 (1984).
537. H. Brunner and H. Fisch, *J. Organomet. Chem.*, **335**, 1 (1987).
538. G. Cros and J.-P. Laurent, *Inorg. Chim. Acta*, **154**, 31 (1988).
539. C. P. Gupta, K. G. Sharma, R. P. Mathur, and R. K. Mehta, *Acta Chem. Acad. Sci. Hung.*, **111**, 19 (1982); *Chem. Abstr.*, **98**, 36580 (1983).
540. L. Casella, M. Gulotti, A. Pintar, S. Colonna, and A. Manfredi, *Inorg. Chim. Acta*, **144**, 89 (1988).
541. S. Hietkamp, H. Sommer, and O. Stelzer, *Angew. Chem. Int. Ed. Eng.*, **21**, 376 (1982).
542. H. Adams, N. A. Bailey, I. S. Baird, D. E. Fenton, J.-P. Costes, G. Cros, and J.-P. Laurent, *Inorg. Chim. Acta*, **101**, 7 (1985).
543. P. J. Burke and D. R. McMillin, *J. Chem. Soc., Dalton Trans.*, **1980**, 1794.
544. A. W. Addison and C. G. Wahlgren, *Inorg. Chim. Acta*, **147**, 61 (1988).
545. L. Casella, M. Gulotti, C. Pessina, and A. Pintar, *Gazz. Chim. Ital.*, **116**, 41 (1986).
546. G. C. van Stein, G. van Koten, H. Passenier, O. Steinbeck, and K. Vrieze, *Inorg. Chim. Acta*, **89**, 79 (1984).
547. H. Tayim, A. S. Salameh, and V. S. I. Meri, *Polyhedron*, **5**, 1509 (1986).
548. H. Adams, N. A. Bailey, D. E. Fenton, and S. Moss, *Inorg. Chim. Acta*, **83**, L79 (1984).
549. D. E. Fenton and R. Moody, *J. Chem. Soc., Dalton Trans.*, **1987**, 219.
550. S. Ianelli, M. Nardelli, and C. Pelizzi, *Gazz. Chim. Ital.*, **115**, 375 (1985).
551. M. Nonoyama, *Inorg. Chim. Acta*, **145**, 53 (1988).

CHAPTER 2

Vinylpyrroles

Boris A. Trofimov

*Institute of Organic Chemistry,
Siberian Branch of the U.S.S.R. Academy of Sciences,
Irkutsk, SU 664033, U.S.S.R.*

2.1. Introduction	132
2.2. 1-Vinylpyrroles	134
2.2.1. Synthesis	134
2.2.1.1. Direct One-Pot Synthesis from Ketoximes and Acetylene	134
2.2.1.1.1. Reaction Conditions	134
2.2.1.1.2. Symmetrical Dialkyl Ketoximes	136
2.2.1.1.3. Alkyl Aryl Ketoximes	137
2.2.1.1.4. Alkyl Heteroaryl Ketoximes	143
2.2.1.1.5. Oximes of Cyclic and Heterocyclic Ketones	145
2.2.1.1.6. Functional Ketoximes	151
2.2.1.1.7. Regiospecificity of the Trofimov Reaction	153
2.2.1.1.8. Vinyl Halides and Dihaloalkanes as Acetylene Equivalents	155
2.2.1.2. Vinylation of Pyrroles with Acetylene	157
2.2.1.2.1. Early Vinylation Procedures	158
2.2.1.2.2. Vinylation in Superbase Media	158
2.2.1.2.3. Structural Effects	160
2.2.1.2.4. Evidence for One-Electron Channel	162
2.2.1.3. Other Vinylation Procedures	165
2.2.2. Reactivity	169
2.2.2.1. Introduction	169
2.2.2.2. Hydrogenation	171
2.2.2.3. Protonation	172
2.2.2.4. Acid-Catalyzed Hydrolytic Cleavage Reactions	180
2.2.2.5. Addition of Alcohols and Phenols	183
2.2.2.6. Addition of Thiols	185
2.2.2.7. Hydrosilylation	186
2.2.2.8. Acylation	188
2.2.2.9. Other Reactions	192
2.2.2.10. Polymerization	195
2.2.3. Physical Chemistry	197
2.2.3.1. Introduction	197
2.2.3.2. Dipole Moments	198

Pyrroles Part Two: The Synthesis, Reactivity, and Physical Properties of Substituted Pyrroles, Edited by R. Alan Jones.
ISBN 0-471-51306-7 © 1992 John Wiley & Sons, Inc.

	2.2.3.3. Relative Basicity	201
	2.2.3.4. Electronic Spectra	201
	2.2.3.5. IR Spectra	204
	2.2.3.6. NMR Spectra	204
2.2.4.	Applications	209
	2.2.4.1. Biological Properties	209
	2.2.4.2. Polymers	212
2.3. 2-Vinylpyrroles		213
2.3.1.	Synthesis	213
	2.3.1.1. From 2-Formylpyrroles and CH-Acids	213
	2.3.1.1.1. Reactions with Ketones	213
	2.3.1.1.2. Reactions with Carboxylic Acids, Ester, and Nitriles	215
	2.3.1.1.3. Reactions with Other CH-Acids	217
	2.3.1.2. Via the Wittig Reaction	218
	2.3.1.3. Addition Reactions of Pyrroles with Alkynes	224
	2.3.1.4. From Pyrroles and Carbonyl Compounds	227
	2.3.1.5. Miscellaneous Syntheses	231
2.3.2.	Reactivity	237
	2.3.2.1. [$4\pi + 2\pi$] Cycloaddition Reactions	237
	2.3.2.2. Other Reactions	242
2.4. 3-Vinylpyrroles		248
2.4.1.	Synthesis	248
	2.4.1.1. From 3-Formylpyrroles and CH-Acids	248
	2.4.1.2. From β-Unsubstituted Pyrroles and Carbonyl Compounds	252
	2.4.1.3. Via the Wittig Reaction	252
	2.4.1.4. Addition of 2,5-Disubstituted Pyrroles to Alkynes	253
	2.4.1.5. From 3-(2-Aminoethyl)pyrroles and Related Compounds	254
	2.4.1.6. From Oximes of Ethylenic Ketones and Acetylene	254
	2.4.1.7. Miscellaneous Syntheses	255
2.4.2.	Reactivity	256
	2.4.2.1. [$4\pi + 2\pi$] Cycloaddition Reactions	256
	2.4.2.2. Other Reactions	261
2.5. Physical Properties of C-Vinylpyrroles		264
2.5.1.	Dipole Moments	264
2.5.2.	Electronic Spectra	265
2.5.3.	IR Spectra	269
2.5.4.	NMR Spectra	271
2.6. Acknowledgments		278
2.7. References		278

2.1. INTRODUCTION

The vinylpyrroles (structures **A**, **B**, and **C**) are extensively studied reactive intermediates in the synthesis of diverse compounds of the pyrrole series,

2.1. Introduction

especially condensed heterocycles genetically related to pyrrole. They are also of interest as vinyl monomers, although this aspect is less well developed at present. Fragments of 2-vinylpyrroles (**B**) are widely found in natural molecules having vital importance for living matter, such as porphyrins, chlorophylls, vitamin B_{12}, bile pigments, bilirubin, biliverdin, stercobilin, bilenes, biladienes, bilatrienes, chlorins, phlorins, oxophlorins, corroles, and prodigiosins.

3-Vinylpyrroles structures (**C**) are components of chlorophylls *a*, *b*, and *c*, which are of key importance in photosynthesis, that is, in basic processes of transformation and conservation of solar energy. They are also constituents of myoglobin and hemoglobin, which are responsible for the oxygen transport in mammals.

Until recently, the 1-vinylpyrroles remained the least studied. In the 1970s they were regarded[1,2] as being almost unknown and inaccessible compounds and, at that time, with the exception of well-investigated *N*-vinyl derivatives of indole[3,4] and carbazole,[5-7] which, although containing the pyrrole ring, are still noticeably different from typical pyrroles, other members of this series were reported in only relatively few papers.[5,8-10] The situation suddenly and drastically changed, when a one-pot reaction of acetylene with ketoximes in an alkali metal hydroxide–dimethylsulfoxide (DMSO) system leading to pyrroles and 1-vinylpyrroles in high yields was discovered, and this area of chemistry was systematically developed.[11,12] Previously expensive and exotic 1-vinylpyrroles became the cheapest and most accessible compounds of the pyrrole series. At present, they are being extensively and systematically explored by many scientific teams as promising monomers (their polymers and copolymers with other monomers possess valuable properties), intermediates for fine organic synthesis, and bioactive substances. The 1-vinylpyrroles also present a gratifying area to test and apply modern concepts of theoretical chemistry and reactivity.

Thus, on the borderlines of pyrrole and acetylene chemistry, the chemistry of 1-vinylpyrroles, a promising new research area, is developing rapidly. The results of investigations have been reported in many publications, monographs,[11,13,14] reviews,[12,15-19] and theses.[20-24]

During the investigation of the pyrrole synthesis from ketoximes and acetylene it has been observed[11,12,15,16,18,19] that the superbase KOH/DMSO system greatly facilitates the vinylation of pyrroles with acetylene. This observation has laid the foundation of a new effective method for the vinylation of compounds having an N—H bond, a method principally different from other known procedures in that it is carried out under atmospheric pressure and at moderate temperatures (80–100°C). Currently, this method is recommended for the vinylation of other NH-heterocycles, which are resistant to the action of alkali. Besides its evident usefulness in industry, the procedure may turn out to be indispensable for laboratories having no facilities for work with acetylene under pressure.

The wide range of 1-vinylpyrroles that are now available have made it possible to examine systematically their dipole moments, acid–base properties and IR, UV, and NMR spectra. As a result, there is strong evidence establishing that there is a common π-system in 1-vinylpyrroles that readily transmits weak perturbations along its whole range depending on the conformation of the vinyl group with respect to the plane of the pyrrole ring.

Unlike the chemistry of 1-vinylpyrroles, that of 2- and 3-vinylpyrroles has been well documented in other readily available monographs[1,2] and allows their coverage in this chapter to be presented in less detail with the emphasis mainly on the latest papers.

2.2. 1-VINYLPYRROLES

2.2.1. Synthesis

2.2.1.1. Direct One-Pot Synthesis from Ketoximes and Acetylene

2.2.1.1.1. Reaction Conditions

The most extensively developed method for the preparation of 1-vinylpyrroles is based on the heterocyclization of ketoximes with acetylene in a strong base–DMSO system[25–39] (Scheme 1).

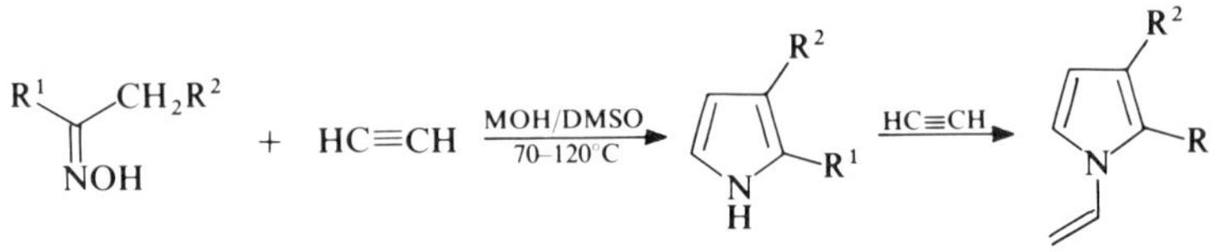

$$R^1\underset{NOH}{\overset{CH_2R^2}{\diagdown\!/}} + HC\equiv CH \xrightarrow[70-120°C]{MOH/DMSO} \underset{H}{\overset{R^2}{\text{pyrrole}}}\!\!-\!R^1 \xrightarrow{HC\equiv CH} \underset{\text{vinyl}}{\overset{R^2}{\text{pyrrole}}}\!\!-\!R^1$$

Scheme 1

The reaction (Scheme 1) is viable for all ketoximes having at least one methylene or methyl group in the α-position to the oxime function as long as no substituents are sensitive to strong bases. The major advantage of this reaction, which has been called[40,41] "the Trofimov reaction," is that for the synthesis of 1-vinylpyrroles it requires no pyrrole precursors and starts from cheap and readily available ketones. Besides, it is possible to synthesize 1-vinylpyrroles stepwise by initial isolation of the intermediate NH-pyrroles and then N-vinylation using other acetylenes, which extends the scope of application of this reaction.

The reaction allows the preparation of diverse 2-, 2,3-, 2,5-, and 2,3,5-substituted 1-vinylpyrroles. The yields depend on the structure of reactants; with simple ketoximes and unsubstituted acetylene yields of 70–95% are attained [25–34,37–39] under optimal conditions. Together with simple alkyl- and aryl-substituted 1-vinylpyrroles, it is also possible to obtain other 1-vinylpyrroles for which even the NH-precursors are accessible only with difficulty by conventional methods, or are unknown, for example, annelated pyrroles from oximes of cyclohexanone,[25–39] suberone,[42–43] tetralone,[44] decalones, or various alkenyl-, furyl-, and thienylketones.[45–49]

2.2. 1-Vinylpyrroles

The formation of 1-vinylpyrroles from ketoximes and acetylene occurs smoothly over the temperature range 70–140°C (usually at 90–100°C).[25–30,50,51] Sometimes, simple heating of the reactants to this temperature initiates a mild exothermal process, which can be regulated by the addition of acetylene under atmospheric pressure. The simple reaction takes 3–5 h on average to complete, but use can also be made of an autoclave, since the reaction times are reduced when under pressure.

The reaction is catalyzed by a superbase couple (a strong base–DMSO) and, as a rule, alkali metal hydroxides are used. The dependence of the catalytic activity of the system on the nature of the cation has been studied in the reactions of cyclohexanone and acetophenone oximes with acetylene.[51,52] In general, the reaction rate is increased with increasing atomic number of the cation, but the activity maximum is observed with potassium ($Rb^+ < Cs^+ < K^+ > Na^+ > Li^+$).[38,50,52] This sequence, although valid for many oximes of aliphatic and alicyclic ketones, is far from absolute and can change depending on the reaction conditions and the type of ketoxime. Alkali metal hydroxides differ not only in their activity but also in selectivity. Thus, LiOH selectively catalyzes the formation of the pyrrole ring from alkyl aryl ketoximes,[52–55] but it is almost inactive in N-vinylation of the pyrrole ring. In contrast, with alicyclic ketoximes, LiOH is totally ineffective in both the ring formation and N-vinylation steps.[51] However, in this case, formation of the pyrrole ring is selectively accelerated by the use of rubidium and tetrabutylammonium hydroxides.[50,51] A decrease in the yield of 1-vinyl-2-phenylpyrrole, when RbOH and CsOH are used as catalysts, is a result of an extensive resinification.[52] The effect of the KOH concentration[52] on the yield of 3-methyl-2-phenyl-1-vinylpyrrole from propiophenone oxime and acetylene (100°C, 3 h) is

KOH, %	10	15	20	30	100
Yield, %	23	39	50	76	68

Even a slight decrease in the reaction temperature (to 80°C) leads to a sharp reduction in the yield of 3-methyl-2-phenyl-1-vinylpyrrole, while when the reaction is conducted at 100°C using 20% KOH, as the catalyst, the yield increases progressively from 23% after 30 min to 57% after 3 h.[52]

In general, the reaction rate increases with an increase in the base concentration;[51,52] good preparative results are achieved with an overstochiometric excess (ca. tenfold) of the base with respect to ketoxime.[12] However, the optimal ketoxime : alkaline ratio is, as a rule, close to equimolar.[25–30,50,52–54,56] This effect is clearly demonstrated by the conversion of cyclohexanone oxime to 1-vinyl-4,5,6,7-tetrahydroindole.[51] At a moderate temperature (100°C), an increase in the KOH concentration (up to the equimolar ratio relative to the initial oxime) leads to an increase in the yield of 1-vinyl-4,5,6,7-tetrahydroindole. Under more severe conditions (120°C), the by-processes accelerated as well, which then results in an inverse dependence of the yield of 1-vinyl-4,5,6,7-tetrahydroindole on the base concentration.[51]

The reaction is effective only in highly polar nonhydroxylic solvents, which, in pairs with strong bases, form superbase media; DMSO is the best one of all media studied.[51,52] Compared with DMSO, analogous systems, such as hexamethylphosphoamide (HMPA) or sulfolane, appear to produce far less active catalytic systems, although alternative procedures, which make use of them, have not been well developed and are of less preparative value. The reaction does not occur in ethers, alcohols, or hydrocarbons.[12,14]

The reaction rate is approximately directly related to the acetylene pressure. The synthesis of 1-vinyl-4,5,6,7-tetrahydroindole from cyclohexanone oxime and acetylene carried out on a 5- or 25-liter scale under a pressure of up to 1.5 bar at 110°C with 0.4 M potassium hydroxide produces between 50 and 100 g of pyrrole per hour from each liter of reaction mixture.[57] Thus, the reaction is technologically feasible and provides a basis for the further development of the bench-scale production of 1-vinylpyrroles. Further studies indicate that with equimolar concentration of potassium hydroxide and cyclohexanone oxime and an acetylene pressure of 0.3–1.5 bar, it is possible to obtain 1-vinyl-4,5,6,7-tetrahydroindole in yields of up to 83% with 99% purity without heating as a result of the exothermal nature of the process. It was later established that high yields ($\sim 80\%$) of 1-vinylpyrroles could be obtained under atmospheric pressure of acetylene at 93–97°C, when a higher base concentration (up to 50% of the ketoxime mass) is employed over a longer reaction time.[14]

2.2.1.1.2. Symmetrical Dialkyl Ketoximes

The reaction of acetylene with oximes of symmetrical dialkyl ketones gives, together with 2-alkyl- and 2,3-dialkylpyrroles, their 1-vinyl derivatives (72–95%).[31,56–63]

The reaction is generally conducted in an autoclave at 120–140°C with a large excess acetylene and a 0.1 to 0.3 : 1 mass ratio of potassium hydroxide to the ketoxime. The reaction is complete within 1–3 h when the DMSO : ketoxime volume ratio is between 1 : 8 and 1 : 16. When the alkaline content is increased to 1–2 mol per 1 mol of ketoxime, the reaction proceeds at a considerably higher rate and a satisfactory yield (50%) is attained after only 3 h even at atmospheric pressure (flow system) and a lower temperature (93–97°C).

The reaction of acetoxime with acetylene generally gives a low yield (7–12%) of 2-methyl-1-vinylpyrrole under the standard conditions. A systematic study of the effect of acetylene pressure and base concentration[56] has established that under optimum conditions (3 h, 130°C in an autoclave at 12 bar, 30% KOH, oxime : DMSO ratio 1 : 15) the yield of 2-methyl-1-vinylpyrrole can be increased to 45%, or, if acetylene is passed continuously through the reaction mixture under atmospheric pressure (oxime : DMSO ratio 1 : 10, 120–140°C, 21 h) and 40% of potassium hydroxide is added in batches every 5 or 6 h, the yield is improved to 54%. However, when the concentration of potassium hydroxide is increased to 50% and introduced in one batch, all of the acetoxime is consumed in 4 h, but the yield of 2-methyl-1-vinylpyrrole drops sharply to

24%, while a further decrease in the yield to 7% is noted when the concentration of potassium hydroxide is increased to 100% (97°C, 14 h). No by-products, such as 1-butyn-3-ol, O-vinylacetoxime, and pyridines,[14] were detected, although considerable amounts of an undistillable tar-like residue suggest that 2-methyl-1-vinylpyrrole is more reactive than other 1-vinylpyrroles and is more readily resinified, resulting in the significant drop in yield.[56] It is also possible, however, that high yields of 2-methyl-1-vinylpyrrole cannot be achieved due to autocondensations of acetoxime, analogous to those furnishing pyridines in an aqueous medium.[14]

2.2.1.1.3. Alkyl Aryl Ketoximes

A number of alkyl phenyl ketoximes have been successfully used in the synthesis of a series of 3-alkyl(or phenyl)-2-phenyl-1-vinylpyrroles, together with 3-alkyl(or phenyl)-2-phenylpyrroles (Scheme 2). The reaction proceeds readily at 100°C (3 h) in DMSO in the presence of 20–30% potassium hydroxide under an initial pressure of 10–14 bar of acetylene. With a deficiency of acetylene, the corresponding 2-phenylpyrrole is the main product,[54] whereas, with excess acetylene, the reaction proceeds further to form 2-phenyl-1-vinylpyrroles. The structure of the alkyl phenyl ketoximes does not affect the yield of vinylpyrroles significantly, although a tendency toward increased yield with long-chain alkyl derivatives has been observed.

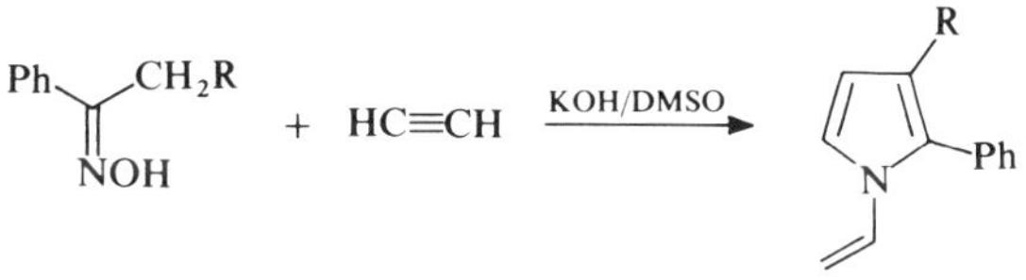

$R = H$, Me, Et, n-Pr, n-Bu, n-Am, n-C_9H_{19}, Ph

Scheme 2

The synthesis has been extended[53] to methyl aryl ketoximes for the successful preparation of 2-aryl-1-vinylpyrroles.

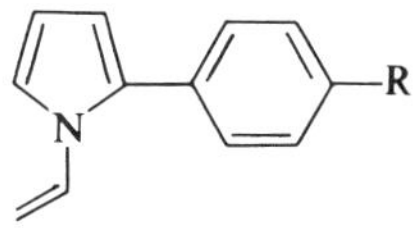

$R = H$, Et, Cl, Br, MeO, PhO

Normally, the reaction proceeds smoothly at 100°C in the presence of 30% (of the ketoxime mass) KOH in an autoclave under an initial acetylene pressure of 8–16 bar. The maximum pressure, developed when the reaction mixture is at its operational temperature, is 20–30 bar. The yield of 2-aryl-1-vinylpyrroles, obtained when a twofold excess of acetylene is used, is dependent on the substituent on the benzene ring. The yield of 4-bromophenyl-1-vinylpyrrole is reduced, as a result of solvolysis of the bromo group by the strong base, while the pyrrole could not be obtained from the oxime of nitroacetophenone, because of its instability in the superbase medium.

The isolation from natural products of new antibiotics related to the arylpyrroles and the increasing interest to this class of compounds[1,2,64,65] has stimulated further investigation[66] of the reaction of alkyl aryl ketoximes with acetylene in MOH/DMSO systems (M = Li, K) for the synthesis of new 2-aryl-1-vinylpyrroles (see Table 2.1). As can be seen from the table, an increase in the chain length of alkyl group produces an increase in the yield of the 1-vinylpyrroles, as a result of the increased basicity of the pyrrolic precursor that makes it more susceptible to nucleophilic attack by acetylene. Milder reaction conditions[67] allow simplification of the purification of pyrroles with the isolation of the intermediate O-vinyloximes[68] and 5-hydroxy-4-5-dihydro-3H-pyrroles.[69] Thus, use of an equimolar amount of potassium hydroxide and the ketoxime with a tenfold excess of DMSO makes it possible to lower the reaction temperature to 50–60°C, which results in 89% overall yield of the 3-alkyl-2-aryl-1-vinylpyrroles and their nonvinylated precursors. The overall yields and the ratio of the two pyrroles are not noticeably affected by the structure of the ketoxime (Table 2.2). Thus, 2-phenylpyrrole is formed in ∼ 10% yield, while the pyrrole from n-hexyl phenyl ketoxime was isolated in 83% yield, together with the corresponding 1-vinylpyrrole (6%). In general, as the length of the alkyl group is increased the overall yield of the pyrroles first increases (to C_4–C_5) and then drops, to 52–70%. The dependence on the alkyl substituent appears to be related not only to electronic effects, as indicated above, but maybe also to a change in the hydrophobic–hydrophilic balance of the starting and final compounds and the degree of their solubility in the reaction mixture. At the same time, there is a tendency toward an inhibition of the heterocyclization step with increasing electronegativity of the alkyl group. Methyl phenyl ketoxime, for example, forms the O-vinylation product (22%), which is not completely converted into the pyrrole under the reaction conditions (50°C, 3 h), the

TABLE 2.1. 3-SUBSTITUTED 2-ARYL-1-VINYLPYRROLES

Ar	R	Yield (%)	b.p. (°C/mm Hg) [m.p.]	d_4^{20}	n_D^{20}
Ph	i-Pr	70	116/1–2	0.9986	1.5741
Ph	C_6H_{13}	90	158/2–3	0.9700	1.5615
Ph	C_7H_{15}	71	160/1–2	0.9636	1.5551
Ph	C_8H_{17}	82	166–167/1	0.9575	1.5500
4-MeC_6H_4	H	71	104/2	1.0250	1.6070
4-i-PrC_6H_4	H	58	113/1–2	1.0050	1.5849
4-t-BuC_6H_4	H	72	123/1–2	0.9989	1.5221
4-EtC_6H_4	Me	80	129–130/10	1.0041	
$PhCH_2$	Ph	48	[50–51]		

TABLE 2.2. YIELD OF 3-SUBSTITUTED
2-PHENYLPYRROLES OBTAINED FROM KETOXIMES
AND ACETYLENE[a]

	Yield (%)	
R^2	$R^1 = H$	$R^1 = CH_2{=}CH$
H	10	Traces
Me	36	Traces
Et	38	Traces
n-Pr	47	13
i-Pr	26 ± 2[b]	4
n-Bu	63	20
n-Am	83	6
n-C_6H_{13}	44 ± 3[b]	8
n-C_7H_{15}	48	9
n-C_8H_{17}	49	17
n-C_9H_{19}	57	13

[a] Initial acetylene pressure 12–14 bar.
[b] From two runs.

1-vinylpyrrole is totally absent. It is interesting that similar results were obtained with methyl 4-chlorophenyl ketoxime, the nucleophilicity of which is weaker compared with those of methyl phenyl ketoxime, due to the electronegative chloro group on the benzene ring.

2-(4-Phenylphenyl)-1-vinylpyrrole (**1**) is the major product (68%) from the reaction with 4-phenylacetophenone oxime while 3-isopropyl-2-(4-methoxyphenyl)-1-vinylpyrrole (**2**) and 2-(4-chlorophenyl)-3-isopropyl-1-vinylpyrrole (**3**) are obtained in good yield ($\sim 75\%$) under similar reaction conditions.[70]

Whereas 2-(4-ethylthiophenyl)- and 2-(4-phenylthiophenyl)pyrroles are readily prepared in $\sim 50\%$ yield from 4-ethylthio- and 4-phenylthioacetophenone oximes (atmospheric pressure, 96°C, 5 h, equimolar oxime : KOH ratio), the 1-vinyl derivatives (**4, 5**) are formed in a yield as low as 1%.[70] The vinyl derivatives can be obtained, however, by direct vinylation of pyrroles with acetylene at atmospheric pressure (120°C, 5 h, KOH : DMSO molar ratio 1 : 5).[70]

The condensation of 4-fluoroacetophenone oxime with acetylene (90°C, initial acetylene pressure 12 bar) is complicated by the formation of 4-fluoroacetophenone (24%) as well as by solvolysis of fluoro group in both the oxime and the resultant N-unsubstituted pyrrole. The 2-(4-fluorophenyl)-1-vinylpyrrole (**6**) is also formed in the mixture, but is better synthesized (70% yield) by vinylation of the pyrrole precursor (atmospheric pressure, 125–135°C, pyrrole : KOH molar ratio 1 : 1.4).

The yield of 2-(4-bromophenyl)-1-vinylpyrrole (**7**) from 4-bromoacetophenone oxime is improved by lower reaction temperature (90°C), while at 60°C, in addition to the expected product, 2-(4-vinyloxyphenyl)-1-vinylpyrrole (**8**) is isolated in 2% yield.[70] The latter product obviously arises from vinylation of an intermediate hydroxyphenyl derivative.

The same pyrrole (**8**) is obtained in 27% from 4-hydroxyacetophenone oxime;[71] its nonvinylated precursor is detected in trace (0.1%) amounts.[71]

As a result of a side reaction with DMSO, Z-1-(2-methylthiovinyl)pyrrole (**9**) is also formed in trace quantities in this synthesis[71,72]

2-(4-Aminophenyl)-1-vinylpyrrole (**10**) and 2-(3-aminophenyl)1-vinylpyrrole (**11**) were obtained (70–90°C, autoclave, equimolar oxime : KOH ratio) in

2.2. 1-Vinylpyrroles

yields 30 and 19%, respectively. With a twofold decrease in the amount of potassium hydroxide, other conditions being equal, the yield of the 1-vinylpyrrole (**10**) drops to 14%. A 20°C increase in the reaction temperature also reduces the yield of **10** and enhances resinification.[71]

10 4-NH_2
11 3-NH_2

The reaction of 4-nitroacetophenone oxime fails to give the expected pyrroles and is recovered unchanged when equimolar amounts of the oxime and potassium hydroxide are used. At 100°C, with a oxime : KOH 1 : 2 ratio, partial elimination of the nitro group with the formation of 2-phenylpyrrole (74%) and its 1-vinyl derivative is observed.[7,73] Similarly, the condensation of 4-dimethylaminoacetophenone oxime with acetylene at atmospheric pressure (100°C, 5 h, oxime : KOH molar ratio 1 : 3) yields, not only the expected 1H- and 1-vinylpyrroles (2–4%), but also 2-phenylpyrrole and its 1-vinyl derivative.[73]

X = NO_2, $N(Me)_2$; R = H, CH_2=CH

As in the case of the 4-hydroxyacetophenone oxime (see above), small amounts (0.05–0.13% yield) of the Z-1-(2-methylthiovinyl)pyrroles (**9, 12–14**), which have been characterized by their ^1H and ^{13}C NMR and mass spectra, have been isolated.[72]

9 R = 4-CH_2=$CHOC_6H_4$
12 R = Me
13 R = 4-ClC_6H_4
14 R = 4-HOC_6H_4

Further kinetic studies[74] on the effect of the alkali metal cation, the solvent and added salts (e.g., Cs_2CO_3, CsF, RbCl), as well as linear and macrocyclic polyethers to the KOH/DMSO catalytic system on the formation of the pyrroles and their subsequent N-vinylation, were undertaken. All of the reactions were carried out at atmospheric pressure and at a temperature of

$96 \pm 2°C$, and the kinetic data confirm the formation of NH-pyrroles via the intermediate O-vinyloximes, 5-hydroxy-1-pyrrolines, and 3H-pyrroles.

The activities of alkali metal hydroxides follow the order Li ≪ Na < K < Rb < Cs. The lower than expected ring-closure rate in the presence of rubidium hydroxide, compared with potassium hydroxide, is explained by a higher moisture content of the former. The hydration of the hydroxides is obviously important. Thus, cesium hydroxide, despite its hydration, is active, while lithium hydroxide, which is not hydrated, is a weaker catalyst than sodium hydroxide. Under the conditions studied, lithium hydroxide is predictably[14] inactive for the vinylation stage.

Although rubidium and cesium hydroxides are most active in catalyzing the conversion of acetophenone oxime into pyrroles, the addition of salts of these metals to the KOH/DMSO system fails to increase catalytic activity of the latter to any great extent. The catalytic effect of the added salts is more pronounced at the N-vinylation stage, where a twofold increase in the yields is observed.

The basicity of the KOH/DMSO system has been enhanced on the addition of dibenzo-18-crown-6, diethyleneglycol diethyl ether, diethyleneglycol divinyl ether, and polyethylene glycol (PEG) with a resultant increase in the rate of formation of the pyrrole and its N-vinyl derivatives.

DMSO possesses a specific catalytic effect and is much better than HMPA, 1-methyl-2-pyrrolidone, and tetramethylurea. Hydrazine hydrate and DMF show no catalytic effect.

The reaction of ketoximes with acetylene has been successfully extended to oximes of condensed aromatic ketones such as 1-and 2-acetylnaphthalenes.[61]

M = Li, K; R = H, CH$_2$=CH

The optimum reaction temperature is 90°C when, for example, with an acetylene pressure of 12–16 bar, the oxime of 1-acetylnaphthalene forms a mixture of 2-(1-naphthyl)pyrrole and 2-(1-naphthyl)-1-vinylpyrrole in 15 and

TABLE 2.3. THE EFFECT OF TEMPERATURE ON THE PRODUCTS OF THE HIGH-PRESSURE REACTION OF 2-ACETYLNAPHTHALENE OXIME WITH ACETYLENE

	Yield (%)	
Temperature (°C)	2-(2-Naphthyl)pyrrole	2-(2-Naphthyl)-1-vinylpyrrole
60	Traces	0
70	33	Traces
90	13	49
100	0	36

48% yield, respectively.[61] The effect of temperature on the corresponding reaction of 2-acetylnaphthalene is shown in Table 2.3.[61]

Naphthylpyrroles can be obtained in good yield at atmospheric pressure. At 90–100°C, 2-acetylnaphthalene oxime is converted in 64% yield into 2-(2-naphthyl)pyrrole, while, at 110–120°C, the product mixture comprises 2-(2-naphthyl)pyrrole and 2-(2-naphthyl)-1-vinylpyrrole in a ratio of 1 : 2 in an overall isolated yield of 49%.[61]

It has been shown that dioximes also react with acetylene;[48] 1,4-bis(1-vinyl-2-pyrrolyl)benzene (**15**) has been prepared in 30% yield from diacetylbenzene dioxime in a one-pot reaction.

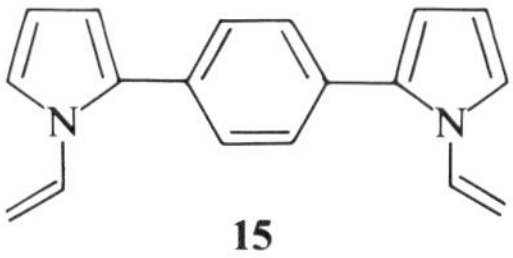

15

2.2.1.1.4. Alkyl Heteroaryl Ketoximes

The preparation of 3-alkyl-2-(2-furyl)-1-vinylpyrroles (**16a–f**) from alkyl furyl ketoximes[45,46,62] (Scheme 3) occurs under the typical reaction conditions (100°C, ketoxime : KOH molar ratio 5 : 3, excess acetylene under initial pressure 14–16 bar).

M = Li, K
a   R = H   d   R = n-Pr
b   R = Me   e   R = i-Pr
c   R = Et   f   R = n-Bu

Scheme 3

Methyl furyl ketoxime gives a mixture of 2-(2-furyl)-1-vinylpyrrole (**16a**) and corresponding NH-pyrrole in a ratio of 4 : 1 in 71% yield (55 and 16%, respectively),[62] while ethyl furyl ketoxime readily produces 2-(2-furyl)-3-methyl-1-vinylpyrrole (**16b**) (57%) and the corresponding NH-pyrrole (8%). The best general conditions for the exclusive synthesis of the vinylpyrroles (50–85%) (**16a–f**) requires the higher temperature (110–130°C) and a 1 : 1 ketoxime : KOH ratio. Under these conditions, for example, ethyl furyl ketoxime is converted into **16b** in a 65% yield. At 130–140°C, alkyl furyl ketoximes readily form 1-vinylpyrroles at atmospheric pressure as well, although it is necessary to increase the reaction time to 6–8 h. A mixture of 2-(2-furyl)-3-methylpyrroles (19%) and 2-(2-furyl)-3-methyl-1-vinylpyrrole (**16b**) (33%) was also obtained from ethyl furyl ketoxime under these conditions at 100°C over 6 h.[62]

The 2-(2-thienyl)-1-vinylpyrroles (**17a–f**)[47] have been obtained from methyl 2-thienyl ketoximes. The reaction (Scheme 4) occurs smoothly in an autoclave under an initial acetylene pressure of 8 bar at 100–130°C.

M = Li, K; **a** R = H, **b** Me, **c** Et, **d** i-Pr, **e** n-Bu, **f** n-C$_5$H$_{11}$

Scheme 4

The vinylpyrroles (Table 2.4) can also be obtained with acetylene at atmospheric pressure.[75]

TABLE 2.4. 3-ALKYL-2-(2-THIENY)-1-VINYLPYRROLES

Pyrrole	Yield (%)	b.p. (°C at 1 mm Hg)	d_4^{20}	n_D^{20}
2-(2-Thienyl)-1-vinylpyrrole (**17a**)	62	110–111	1.0526	1.6350
3-Methyl-2-(2-thienyl)-1-vinylpyrrole (**17b**)	67	113–115	1.0812	1.6078
3-Ethyl-2-(2-thienyl)-1-vinylpyrrole (**17c**)	70	116–118	1.0450	1.5981
3-Propyl-2-(2-thienyl)-1-vinylpyrrole (**17d**)	53	135–136	1.0447	1.5823
3-Butyl-2-(2-thienyl)-1-vinylpyrrole (**17e**)	62	144–146	1.0206	1.5805
3-Hexyl-2-(2-thienyl)-1-vinylpyrrole (**17f**)	50	153–157	1.0247	1.5698

The ratio of pyrroles and their 1-vinyl derivatives can be controlled, as with other analogous reactions, by a suitable choice of reaction conditions. Thus, 2-(2-thienyl)-1-vinylpyrrole (**17a**) is most conveniently prepared in an autoclave at an elevated acetylene pressure and at 120°C in the presence of a 1 : 3–1 : 6 ratio of potassium hydroxide to ketoxime over a period of 3 h. Longer reaction times (5 h) and higher temperatures (140–150°C) result in resinification and reduced yields.

The analogous condensation of 2-acetyl-1-methylpyrrole oxime with acetylene at 70°C leads to 1-methyl-2-(2-pyrrolyl)pyrrole (35%) and its 1-vinyl derivative (**18**) (7%), together with recovered 2-acetyl-1-methylpyrrole (32%) and polymer.[76] At 90°C, the N-vinylated derivative is the only isolated product (35–40%), but resinification is considerably more extensive.

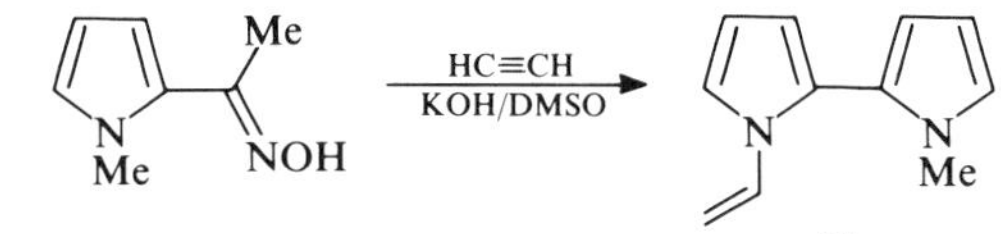

2.2. 1-Vinylpyrroles

[Scheme showing pyrrole-Me-NOH reacting with HC≡CH / KOH/DMSO to give compound **18**]

Acylindole oximes with acetylene (DMSO, 100–105°C, atmospheric pressure, ketoxime : KOH molar ratio 1 : 1.5, 0.5–5 h) produce the (1-vinyl-2-pyrrolyl)indoles.[77–81]

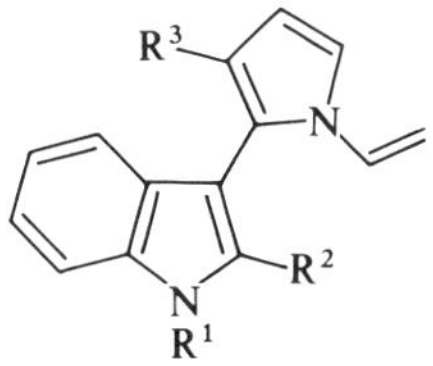

19
R¹ = H, Me
R² = H, Me
R³ = H, Me, i-Pr, Ph

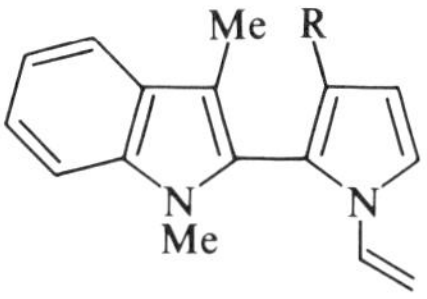

20
R = H, Me, Ph

The methyl group at the 2-position of 2-methyl-3-(2-pyrrolyl)indole increased the rate of vinylation to give **19** in 30–40% yield. The increased rate is possibly caused by distortion of coplanarity of the molecule, such that the pyrrolyl nitrogen atom becomes more accessible to attack by the acetylene. Additionally, nucleophilicity of the pyrrolyl anion is enhanced by the electron-donating effect of the methyl group. With 2-acetylindole oximes, the vinylation of the intermediate 2-(2-pyrrolyl)indoles is even faster and the 2-(1-vinyl-2-pyrrolyl)indoles (**20**) can be prepared in yields of up to 70%.[78]

2.2.1.1.5. Oximes of Cyclic and Heterocyclic Ketones

1-Vinyl-4,5,6,7-tetrahydroindole (**21a**) was the first member of the pyrrole series obtained from ketoximes and acetylene.[14] When the reaction is carried out under pressure, the yields of **21a** and the N-unsubstituted precursor are 93 and 74–81%, respectively. Under atmospheric or a slight excess pressure (1.2–1.5 bar) the yields are reduced to 90 and 50%. 7-Methyl-1-vinyl-4,5,6,7-tetrahydroindole (**21b**) has been synthesized (80%) from 2-methylcyclohexanone oxime.

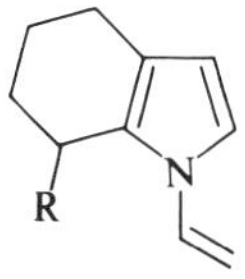

21
a R = H
b R = Me

Recently,[82] the effect of the ring size of cycloalkanone oximes (22) on the rate of their ring closure to pyrroles under the action of acetylene in the KOH/DMSO system (86°C, atmospheric pressure) to give the 1-vinylpyrrole (23, $n = 4$–8) has been studied. It was found that there was little effect on either the rate of the ring closure or the N-vinylation. The largest effect of the ring size was observed at the ring-closure step; the cyclopentanone oxime ($n = 1$) fails to produce the pyrrole.

$$(CH_2)_n \underset{=NOH}{\overset{R^1}{\diagup}} \xrightarrow{HC\equiv CH} (CH_2)_n \underset{\underset{R^2}{N}}{\overset{R^1}{\diagup}}$$

22

$R^1 = H$ ($n = 1$–4)
$R^1 = $ 5-Me, 7-Me ($n = 2$)
$R^2 = H$, $CH_2=CH$

23

The conversion of the oxime of 2,2,6,6-tetramethyl-4-piperidone (24a) and its N-oxidized derivatives (24b,c) into new pyrrolo[3,2-c]piperidines has also been reported.[83]

$$\underset{Me}{\overset{Me}{\diagdown}}\underset{Me}{\overset{Me}{\diagup}} X-N \diagup =NOH \xrightarrow[KOH/DMSO]{HC\equiv CH} \underset{Me}{\overset{X}{\diagdown}}\underset{Me}{\overset{Me\;Me}{\diagup N}}$$

24
a $X = H$ c $X = O^\bullet$
b $X = OH$ d $X = OCH=CH_2$

25

At an elevated temperature (105°C) and with excess acetylene, both oximes (24a and 24b) are converted into 25a as the KOH/DMSO system can behave as a reducing system for the N-oxidized derivative. Additionally, O-vinylation of the 25b takes place to form 4,4,6,6-tetramethyl-1-vinyl-5-vinyloxy-4,5,6,7-tetrahydro-5-azaindole (25d). It is not possible to obtain the 1-vinylazaindole (25c) having a nitroxyl radical from the oximes (24b,c). In both cases, at 50–65°C, only the corresponding N-substituted pyrrole ($X = O^\bullet$) is formed. With rising temperature (95°C) the oxime (24c) is converted into 1-vinylazaindole (25a) having no radical center. This may be regarded as an indication that the vinylation of pyrroles in the KOH/DMSO system involves a one-electron stage (see Section 2.2.1.2.4).

As far as the preparative aspects are concerned, the best yield (90%) of the 1-vinylazaindole (25a) was achieved with a 1 : 3 : 43 molar ratio of the oxime : KOH : DMSO (105°C, 8 h). The same reaction with acetylene under pressure (14 bar, 100°C, 3 h, equimolar oxime : KOH ratio) is accompanied by considerable resinification, which decreases the yield of 25a to 38%. Analogously, the oximes (**26a–c**), when treated with acetylene (KOH/DMSO, atmospheric pressure, 90–100°C, 4–5 h), form the 1-vinylazaindoles (**27a–c**) (Table 2.5).[84]

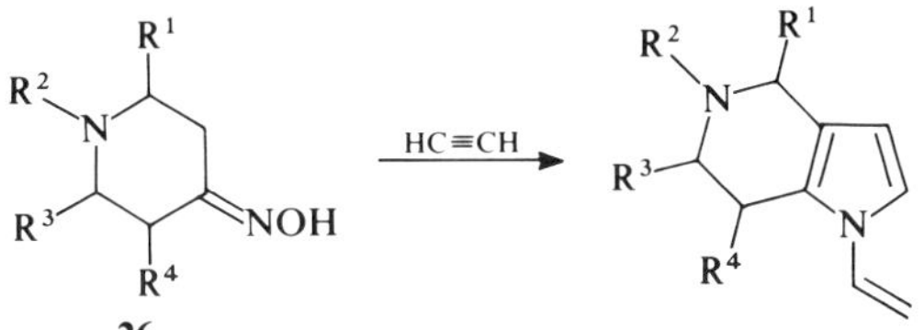

	R^1	R^2	R^3	R^4
a	Me	H	H	Me
b	H	Me	H	Me
c	Ph	Me	Ph	Me

Transformation of the substituted piperidine-4-one oximes (**28**) and their bicyclic derivatives, decahydroquinolines (**29–32**), to pyrroles (Table 2.6) has also been studied[85] (Scheme 5).

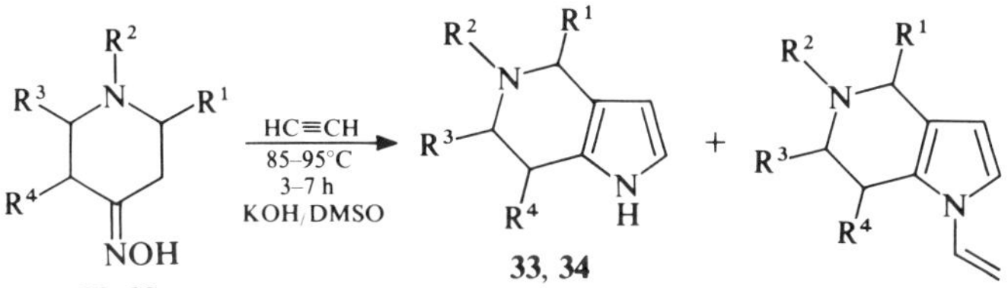

a equatorial R^4 **b** axial R^4 **c** equatorial R^1

Compound	R^1	R^2	R^3	R^4
28, 33a, b, 35a, b	Me	Me	H	Me
29, 36a, b	Me	H	$(CH_2)_4$	
30	Me$_{eq}$	H	$(CH_2)_4$	
31, 34a, b, 37	2-Fur$_{ax}$	H	$(CH_2)_4$	
32, 34c	2-Fur$_{eq}$	H	$(CH_2)_4$	

Scheme 5

It has been confirmed experimentally that initially formed pyrrolopiperidine (**33a**) (Table 2.6) is *N*-vinylated to produce the corresponding 1-vinyl-4,5,7-trimethylpyrrolo[3,2-*c*]piperidine (**35a**) having the unchanged skeleton of the molecule. The large *gem*- and *trans*-coupling values for the 6-axial-H protons and the large *gem*- and small *cis*-coupling for the 6-equatorial-H proton in the ^{1}H NMR spectra of the pyrrolopiperidines (**33a**) and (**35a**) are indicative of an equatorial orientation of the 7-methyl group. In the isomers **33b** and **35b**, the 7-methyl group is axial. In an inert atmosphere, the isomer pairs **33a** and **33b**,

TABLE 2.5. CONDENSED 1-VINYLPYRROLES FROM CYCLIC KETOXIMES

Compound		Yield (%)	b.p. (°C/mm Hg) [m.p.]	d_4^{20}	n_D^{20}	Ref.
1-Vinyl-4,5,6,7-tetrahydroindole		97	85–86/3	1.0010	1.5562	50
1-Vinyl-4,5,6,7,8-pentahydro-1-H-cyclohepta[b]pyrrole		93	73–74/1	0.9337	1.5560	14
Vinyl-4,5,6,7,9-hexahydro-1-cycloocta[b]pyrrole		~100	64.5–66/3·10⁻²	0.9809	1.5512	82
1-Vinyl-4,5,6,7,8,9,10,11,12,13-decahydro-1-H-cyclododeca[b]pyrrole		~100	114–116/5·10⁻²		1.5472	82

Compound	Structure	Yield	Other	n	Ref
4,5,7-Trimethyl-1-vinyl-pyrrolo[3,2-c]piperidine, (4-Me_eq and 4-M_ax isomers)			2885		
4,4,6,6-Tetramethyl-1-vinypyrrolo[3,2-c]piperidine		0.9800	90	1.5303	83 [25–27]
4-Methyl-1-vinyl-6,7-tetramethylenepyrrolo[3,2-c]piperidine (4-Me_eq and 4-M_ax isomers)			55 / 45	(4-Me_ax)[a] / (4-Me_ax)[a]	85
4-(2-Furyl)-1-vinyl-6,7-tetra-methylenepyrrolo[3,2-c]piperidine (4-Fur_eq isomer)					85

[a] Axial and equatorial isomers identified by ^{1}H NMR spectroscopy.

TABLE 2.6. RING-CLOSURE PRODUCTS FROM OXIMES 28–32

Starting Oximes	Yield	
	NH-Pyrrole	1-Vinylpyrrole
28[a]	69% (33a), 18% (33b)	6% (35a)
28[b]	4% (33a), 4% (33b)	28% (35a), 22% (35b)
29	Not observed	55% (36a), 45% (36b)
30	Not observed	Not observed
31	18% (34a), 14% (34b)	31% (37)
32	3% (34c)	Not observed

[a] Atmospheric pressure of acetylene, 85°C, 1.4% H_2O in DMSO.
[b] In autoclave, 12 bar, 85–95°C, 0.2% H_2O in DMSO.

35a and **35b** are stable and do not transform into each other; it is therefore most likely that their ratio (Table 2.6) is predetermined at the stage of prototropic isomerization of the intermediate O-vinyloxime[14] (Scheme 6), which can undergo reversible prototropic isomerization with the hydroxylamines **39** and **40**. As the double bond of **40** is attached directly to the substituent R^4 (the 7-position of the final product), it can cause changes in configuration of this moiety during the inverse reaction leading to the O-vinyloxime (**38**).

Even more rigorous reaction conditions lead to a sharp decrease of stereo- and regioselectivity of the sequence with the appearance of the regioisomeric pyrrolopyrimidine (**41**), which is consistent with Scheme 6.[85]

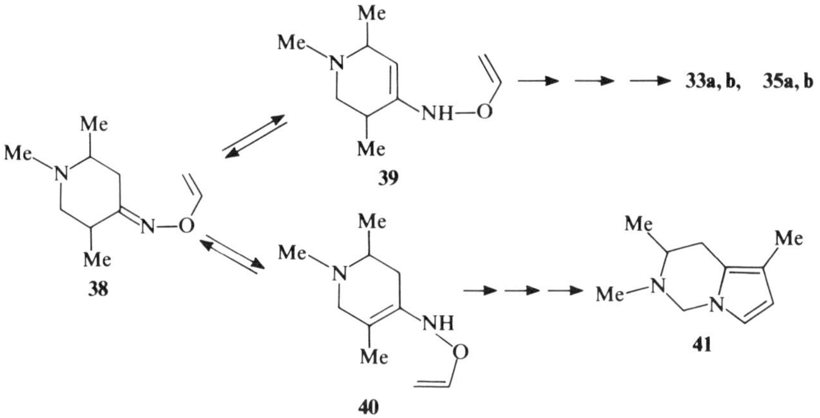

Scheme 6

Each oxime of 2-methyl- and 2-(2-furyl)-4-oxodecahydroquinolines (**29–32**) exists as two stereoisomers with either an axial or equatorial arrangement of the substituent,[85] which, with a *trans*-junction of the bicyclic system, cannot interconvert. The hydrogen atom attached to the nitrogen atom can also take a mobile axial or equatorial configuration. The oxime (**29**) is quantitatively converted into a mixture of the configurational isomers of the 1-vinyl-6,7-tetramethylenepyrrolo[3,2-*c*]piperidine (**36a,b**). However, because of complex ^{1}H NMR spectra, it has not been possible to establish unambiguously whether these isomers are formed as a result of the stabilization of mobile structures by the pyrrole ring, or because of a change in the configuration of the 6-membered ring junction. From a consideration of the ease with which the configuration of the 7-methyl group changes during the synthesis of the pyrroles **33a,b** and **35a,b**, the latter alternative (Scheme 7) is preferable.

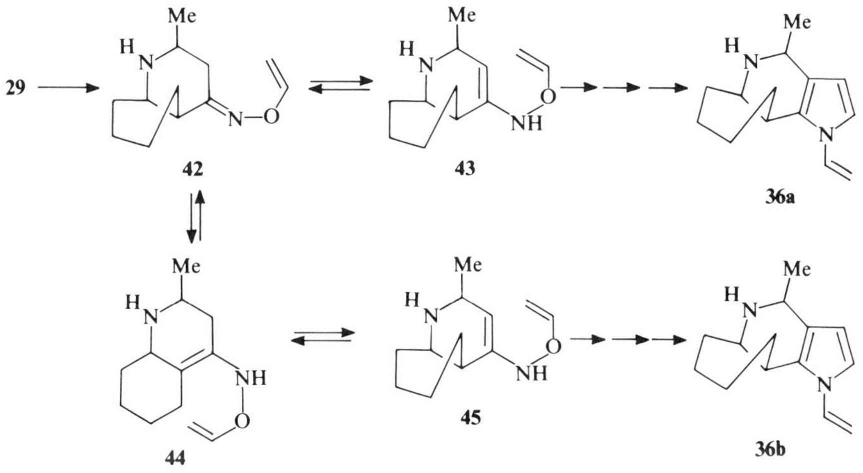

Scheme 7

The reversible prototropic isomerization of the intermediate **42** leads to the hydroxylamines **43** and **44**, which can isomerize to give the hydroxylamine **45** having the cyclohexane ring in a boat conformation (*cis*-junction, 6-H_{ax}–7-H_{eq}). Intermediates **43** and **45** produce the 1-vinylpyrroles **36a** and **36b**, respectively, via a [3,3] sigmatropic shift.

The hindered pyrrole formation in the presence of equatorial substituent (Me, 2-furyl) on the decahydroquinolin-3-one oximes (Table 2.6) may be connected with steric hindrance to the [3,3] sigmatropic shift, which is created by substituents located in the piperidone ring plane of intermediates **43** and **45**. Consequently, ring-closure reactions on piperidin-4-one oximes invariably results in the formation of pyrrolo[3,2-*c*]piperidines having axial and equatorial 7-substituents, while, with rigid ketoximes (**29–32**), the equatorial substituent at the 2-position precludes the reaction.

2.2.1.1.6. Functional Ketoximes

The 1-ethoxy-2-propanone oxime (**46**) reacts anomalously with excess acetylene to give 2-methyl-1-vinylpyrrole (**47**) in a low yield (up to 15%), which is sometimes higher than that of the expected 2-ethoxymethyl-1-vinylpyrrole (**48**).[86]

This contrasts with the reaction of other alkyl methyl ketoximes from which the pyrrole ring is formed via the methylene group of the alkyl group (see Section 2.2.1.1.1). In one case, the second expected isomer, 3-ethoxy-2-methyl-1-vinylpyrrole (**49**),[86] was detected as a trace impurity in the ^{1}H NMR spectrum of a reaction mixture.

In the reaction with excess acetylene in an autoclave at 100°C, the 1-hydroxy-2-methyl-3-butanone oxime (**50**) forms two anomalous products:[86] 2,3-dimethyl-1-vinylpyrrole (**51**)[87] and 2-(2-propenyl)-1-vinylpyrrole (**52**) in 20 and 80% yields, respectively.

The vinyl ether of the oxime (**50**), 2-methyl-1-vinyloxy-3-butanone oxime (**53**) with excess acetylene, also gives the 1-vinylpyrrole (**52**), which polymerizes readily upon distillation. None of the pyrrole (**51**) is formed in this case. In an unusual reaction, 2-hydroxy-2-methylpentanone oxime (**54**) reacts with acetylene to form the 2-methyl-1-vinylpyrrole (**47**) in ~ 20% yield.

A one-pot conversion of 3-butenyl methyl ketoxime **55** into 2-methyl-3-(1-propenyl)-1-vinylpyrrole (**56**) and other pyrroles has also been developed[28] (see Section 2.4.1.6).

2.2. 1-Vinylpyrroles

[Scheme: compound **55** (CH2=CHCH2CH2C(Me)=NOH) + HC≡CH, KOH/DMSO → compound **56** (1-vinyl-2-methyl-3-propenyl pyrrole)]

The 2-cyclopropyl-1-vinylpyrroles (**58a,b**), a poorly explored family of pyrroles that provides new opportunities in the search for biologically active compounds, have been obtained by a one-pot procedure[88] from the alkyl cyclopropyl ketoximes (**57a,b**) and acetylene (KOH, DMSO, 80–95°C, 5–6 h, atmospheric pressure) in unoptimized yields of 7–64%.

[Scheme: **57** (cyclopropyl-C(=NOH)-CHR-) + HC≡CH → **58**]

a R = H
b R = n-Pr

The low yield (10%) of **58a** is possibly due to the competing annelation via the CH group of the cyclopropane ring, which could lead to the unstable 3H-pyrrole (**59**).

[Scheme: **57a** + HC≡CH → [**59** spirocyclopropane 3H-pyrrole] → tars]

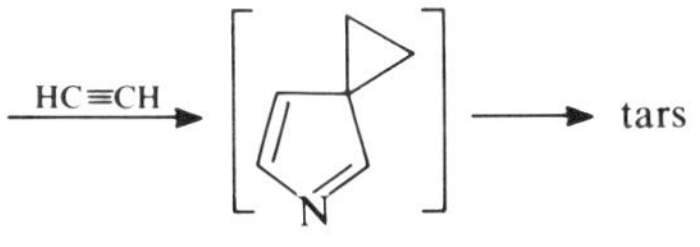

2.2.1.1.7. Regiospecificity of the Trofimov Reaction

Two isomeric 1-vinylpyrroles can be obtained from the reaction of acetylene with unsymmetrical dialkyl ketoximes.[58]

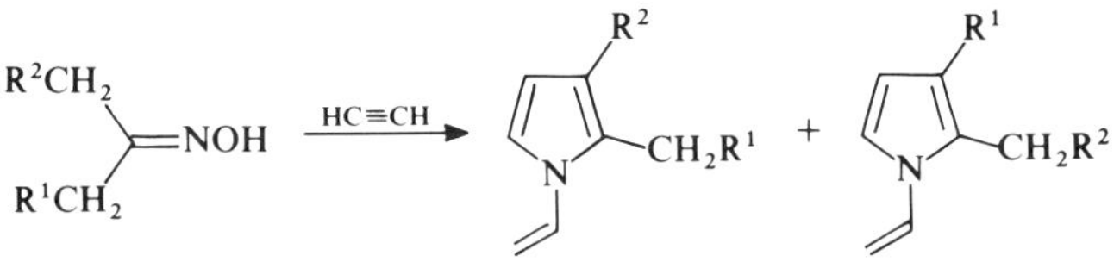

[Scheme: R^2CH_2-C(=NOH)-CH_2R^1 + HC≡CH → 1-vinyl-3-R^2-2-CH_2R^1-pyrrole + 1-vinyl-3-R^1-2-CH_2R^2-pyrrole]

R^1 = H, alkyl; R^2 = alkyl

However, at 120°C, methyl alkyl ketoximes generally react at the methylene group of the alkyl group independently of its structure to give the only isomer. As the temperature is increased, the reaction loses its regiospecificity and at

140°C, pyrroles resulting from reaction at the methyl group are also produced in yields reaching 20–50% of the product mixture, which permits preparative separation of the two isomers. Unfortunately, the overall yield is reduced at 140°C. An exception is to be found with methyl isopropyl ketoxime, which reacts exclusively at the methyl group over the whole temperature range. Only the 2-isopropyl-1-vinylpyrrole (**60**) and its nonvinylated precursor are isolated in 15 and 10% yields, respectively, and the 3H-pyrrole (**61**) appears to be unstable under the reaction conditions.

Although benzyl methyl and benzyl ethyl ketoximes may lead to the two structural isomers **62a,b** and **63a,b** (together with their nonvinylated precursor),[60] only the 2-methyl-3-phenyl-1-vinylpyrrole (**62a**) is isolated from the potassium hydroxide-catalyzed reaction of benzyl methyl ketoxime with excess acetylene at 100°C. Even at 120°C the reaction mixture contains none of the isomeric 2-benzyl-1-vinylpyrrole (**63a**).

An analogous result was obtained when the reaction was carried out in a LiOH/DMSO system at 120°C, which catalyzes the vinylation stage less effectively. Under these conditions, the 1-vinylpyrrole (**62a**) and its nonvinylated precursor are formed in almost equal proportions (1 : 1.2) and isomers of the type **63** are absent.

In contrast, benzyl ethyl ketoxime reacts with an excess of acetylene at 100°C to give the two isomeric 1-vinylpyrroles (**62b** and **63b**), and, as the temperature is raised, the proportion of **62b** decreases from 85% at 100°C to 45% at 150°C.

2.2.1.1.8. Vinyl Halides and Dihaloalkanes as Acetylene Equivalents

It has been shown[89-95] that the reaction of ketoximes with vinyl halides at 80–130°C in the presence of alkali metal hydroxides and DMSO leads to 1H- and 1-vinylpyrroles (Scheme 8). By varying the reaction conditions, it is possible to obtain selectively either the pyrroles (R^3 = H) (~40%) or the 1-vinylpyrroles (up to 35%).

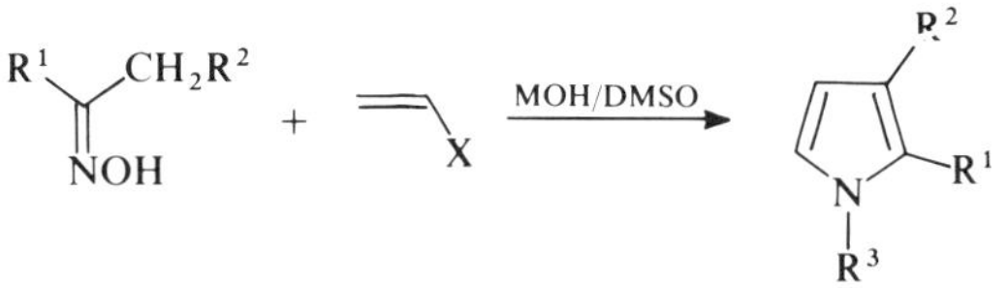

$$R^1\text{C}(=\text{NOH})\text{CH}_2R^2 + \text{CH}_2=\text{CHX} \xrightarrow{\text{MOH/DMSO}} \text{pyrrole}$$

X = Cl, Br; M = Na, K;
R^1 = alkyl, R^2 = H, alkyl; R^3 = H, $CH_2=CH$

Scheme 8

4,5,6,7-Tetrahydroindole and its 1-vinyl derivative can be obtained by this route using vinyl chloride and a potassium hydroxide : oxime ratio of 2–6 : 1 at 90–140°C.[90,91]

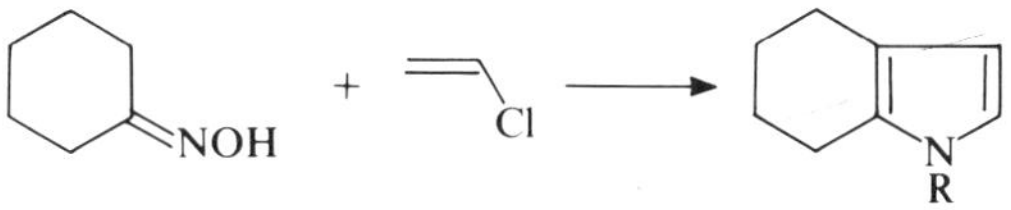

R = H, $CH_2=CH$

Using this procedure, with a five- to sixfold excess of potassium hydroxide at 120°C, 2-arylpyrroles and their 1-vinyl derivatives have been synthesized[93-95]

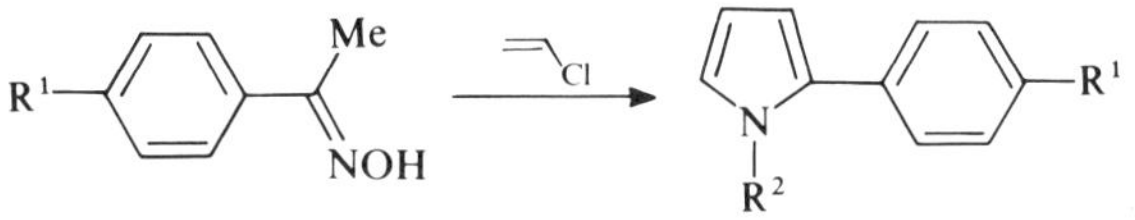

% yield

R^1	R^2 = H	R^2 = CH_2=CH
H	45	7
Me	39	13
Et	41	11
i-Pr	43	11
t-Bu	41	14
EtS	45	7
n-PrS	40	8
i-PrS	42	6
n-BuS	43	6
i-BuS	43	12
t-BuS	36	8
PhS	19	—

Scheme 9

and, similarly, 2-aryl-3-methylpyrroles have been obtained[96] from aryl ethyl ketoximes in good yields (45–51%), although the yields of the corresponding 1-vinyl derivatives are low (4–6%) (Scheme 9).

This alternative pyrrole synthesis from ketoximes has allowed a one-pot preparation of 1-vinyl-4,5-dihydro[*g*]indole (**64**) and its nonvinylated precursor from the readily accessible α-tetralone oxime.[92] The overall yield of 4,5-dihydro[*g*]indole and **64** is 50%.

The preparation of pyrroles and 1-vinylpyrroles in ~30% yield by the reaction of ketoximes with dihaloethanes in the presence of excess alkali metal hydroxides in DMSO has been reported.[97,98] For optimum yields, the ketoxime : 1,2-dihaloethane : potassium hydroxide ratio should be 1 : (2–3) : (7–10) at 100–120°C with a ketoxime : DMSO ratio of 1 : 10. Using this procedure, a synthesis of 4,5,6,7-tetrahydroindole and its vinyl derivative from cyclohexanone oxime and 1,2-dihaloethanes has been realized[99] with the optimum yield (60%) attained with a molar ratio of cyclohexanone oxime : dichloroethane : KOH : DMSO of 1 : (1–2) : 7 : 10 at 115°C. With an increase in relative amounts of dihaloethane and alkali, the yield of 1-vinyl-4,5,6,7-tetrahydroindole increases.

The synthesis of 2,3-dialkyl-1-vinylpyrroles has been shown[100] to be sensitive to a change in temperature, reaction time, KOH concentration, and the general reaction procedures.

2.2. 1-Vinylpyrroles

Aryl methyl ketoximes are converted into 2-arylpyrroles (up to 70%) by reaction with 1,2-dichloroethane; the simultaneous formation of the 1-vinylpyrroles does not exceed 12%.[101]

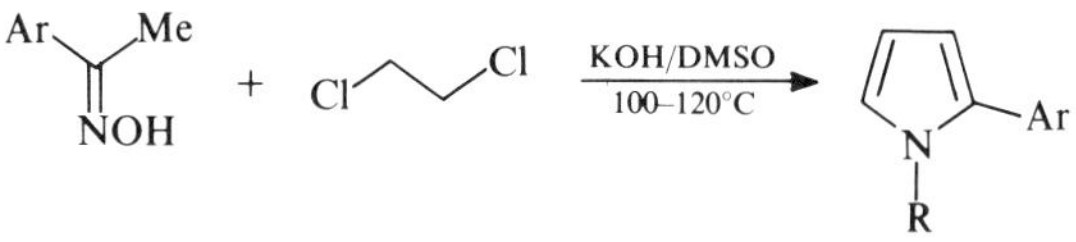

Ar = Ph, 4-MeC$_6$H$_4$, 4-MeOC$_6$H$_4$, 4-ClC$_6$H$_4$
R = H, CH$_2$=CH

In the preparation of the pyrroles and their 1-vinyl derivatives (**65–67**) use has been made of vinyl chloride and 1,2-dichloroethane, as well as acetylene.[102] However, although the yield of the pyrrole (**66**) amounted to 62% when acetylene was used, it did not exceed 6% from vinyl chloride.

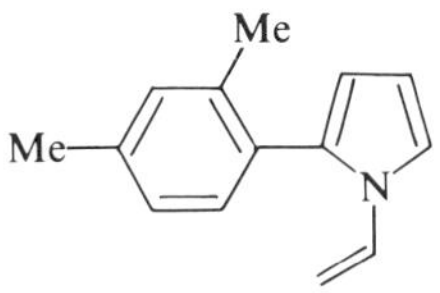

65 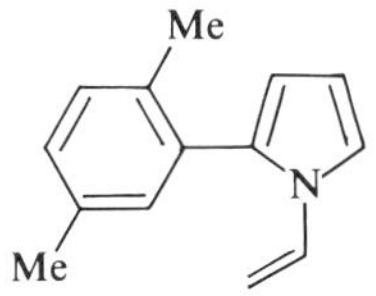   **66** 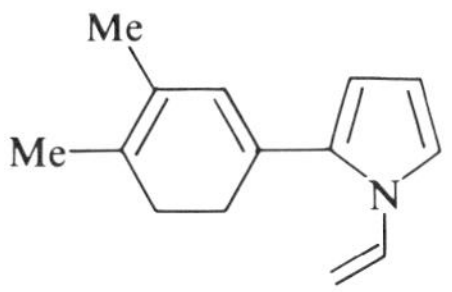   **67**

2-(2-Furyl)- and 2-(2-thienyl)pyrroles in yields to 54 and 20%, respectively, and their 1-vinyl derivatives have been synthesized using 1,2-dichloroethane with a suspension of MOH (M = Li, Na, K) in DMSO at 100–135°C.[103]

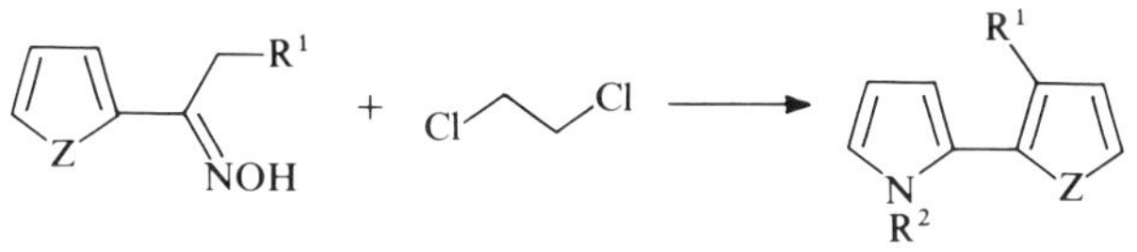

R^2 = H, CH$_2$=CH

Z	R^1
O	H
O	Me
O	Et
S	H
S	Me
S	Et
S	n-Bu

In spite of the lower yields, use of vinyl chloride or 1,2-dichloroethane can be more advantageous than acetylene under pressure to those without relevant experience to handle it.

2.2.1.2. Vinylation of Pyrroles with Acetylene

Interest in the vinylation of nitrogen-containing heterocycles has been encouraged by their use as monomers and synthons. Thus, they are finding ever-increasing use in the manufacture of plastics, and synthetic fibers; in radio engineering and medicine (e.g., Refs. 6,7,105,106); and in the production of semiconducting films[107] and electrophotographic materials.[108,109]

2.2.1.2.1. Early Vinylation Procedures

The traditional methods for the synthesis of N-vinyl nitrogen heterocycles have been reviewed,[6,7,110,111] and involve either dehydration of β-hydroxyethyl derivatives,[4,9,10,112–114] the dehydrohalogenation of haloethyl derivatives,[115,116] or the vinylation of NH-heterocycles with acetylene,[3,5–7] vinyl chloride,[5,117,118] or vinyl ethers.[119] All these processes have been used for carbazole, and, in addition, N-alkenylcarbazole has been synthesized by the isomerization of N-allyl derivatives[120] and by catalytic decomposition of N-(1-alkoxyalkyl)carbazoles.[121] Systematic work on the vinylation of diverse NH-heterocycles, including indoles, using acetylene under pressure, has been carried out by Skvortsova et al.[3,122,123]

Until recently, direct vinylation with acetylene under pressure at 160–200°C was the most widely accepted procedure.[3,124,125]

It is of interest to note that, although N-vinylcarbazole was first prepared in 1924 without acetylene,[115] its commercial use for polymers came only after the development of the acetylenic vinylation procedure. Commercial production of monomeric N-vinylcarbazole and its polymers dates back to the first years of World War II when, for military purposes, Germany speeded up large-scale work with acetylene under pressure.[5–7] However, the use of pressure has restricted the introduction of these methods in industrial processes.

Pyrrole and its benzo derivatives have also been subjected to vinylation in the gas phase under normal pressure.[128,129] However, this procedure requires high reaction temperatures (250–320°C), and the yields are considerably lower than those obtained by liquid-phase vinylation.

2.2.1.2.2. Vinylation in Superbase Media

In systems of the KOH/DMSO type that can be regarded as superbase media,[11,15,16,19,128] there is extra activation of the NH-heterocycle anions with a simultaneous increase in their concentration, as well as activation of acetylene. These effects lead to a substantial increase in the reaction rate and reduce the need for excessive acetylene pressure.

The first systematic study[129] of the vinylation of a series of pyrroles in superbase media resulted in 1-vinylpyrroles being prepared in up to 97% yields.

R^1 = alkyl, aryl
R^2 = H, alkyl, aryl

2.2. 1-Vinylpyrroles

The reaction proceeds most effectively in the presence of 30% potassium hydroxide in aprotic polar solvents (DMSO, sulfolane, HMPA). The use of DMSO allows the reaction temperature to be reduced to 80–100°C, which is about 100°C lower than that in the classical vinylation of NH-heterocycles, and allows for the use of low-pressure acetylene (1–1.5 bar). Higher concentrations of alkali in the DMSO (60–100%) at 95–100°C improve the yields to 75–95%.[130]

The method[129,130] used for the vinylation of pyrroles has been extended to the use of phenylacetylene and has permitted the successful vinylation of indole and carbazole.[131] The reaction proceeds regio- and stereospecifically under mild conditions (100°C, 20% KOH) to give the E-isomers of 1-(2-phenylvinyl)indole (**68**) and 9-(2-phenylvinyl)carbazole (**69**) in 80 and 97% yields, respectively. 1-Vinyl-4,5,6,7-tetrahydroindole has also been produced by this method,[132] and a commercial production of N-vinylcarbazole has been launched.[133–135]

Evidence for the high activity of the KOH/DMSO catalytic system has been provided by vinylation of 2-furyl- and 2-thienylpyrroles, 4,5-dihydrobenzo[g]indole, and 1,2,3,4-tetrahydro-γ-carbolines[136] using acetylene under atmospheric pressure. Vinylation to produce (**70a–c**) in almost quantitative yields (96–99%) makes use of a fivefold molar excess of potassium hydroxide (Table 2.7). At the higher reaction temperature of 120°C, the vinylation time is reduced, but the yield falls by 9–16% as a result of side processes.

TABLE 2.7. VINYLATION CONDITIONS AND YIELDS OF THE 1-VINYLPYRROLES (**70a–c**)

Reaction Temperature (°C)	Reaction Time (h)	Yield (%)		
		70a	**70b**	**70c**
110	4.5	98		
110	5		98.5	
110	6			96
120	3	86		
120	4		82	87

Use of an autoclave with an initial pressure of 10–14 bar results in reduced yields (< 50%) of the vinylpyrroles (**70a–c**), as a result of resinification and

160   Vinylpyrroles

70
a Z=O
b Z=S
c Z=CH=CH

71, 72
R=H, Me

experimental difficulties. Consequently, vinylation under atmospheric pressure proves to be not only more safe, but also more effective.

The rate of the formation of 1-vinylpyrroles increases in the sequence **70c** < **70b** < **70a**, corresponding to the NH-acidity of the pyrroles.[14] The nucleophilicity of the anions, however, decreases in the same order. Thus, the observed order in the reactivity of pyrroles is not compatible with the nucleophilic character of the reaction, but supports the observation, cited earlier, that the rate of the base-catalyzed addition of azoles to acetylene increases with their acidity[137] and is compatible with formation of ion pairs in which the more acidic azoles form looser ion pairs and should be more prone to the insertion of acetylene into the $N^-\ldots K^+$ bond. A similar argument applies to the vinylation of 4,5-dihydrobenzo[g]indole, which, possessing a lower acidity than 2-phenylpyrrole (pK_a 22.2 and 21.6, respectively[14]), adds more reluctantly to acetylene. In this case, in order to obtain the vinyl derivative (**64**) in 91% yield, it is necessary to perform the reaction for 8 h at 120°C with the addition of a threefold molar excess of the base in portions to reduce resinification. At 95–110°C, compressed acetylene does not increase the yield of the 9-vinylcarbolines (**71, 72**), and when, used at 120°C, it even leads to a 15% decrease in yield.[136]

2.2.1.2.3. Structural Effects

The ease of vinylation,[136,137] as assessed by the yield of N-vinyl derivatives, can give a distorted picture. Evaluation of the effect of substituents on the rate of the vinylation of pyrroles[138] correlates the kinetic data for the vinylation of a number of pyrroles (**73–83**) with the known pK_a or relative acidity values,[14,139] as well as relating these values with substituent constants.

Figures 2.1 and 2.2 show the kinetic curves for the variation in the yield of the 1-vinylpyrroles with time and indicate that the rate of vinylation increases in the order of decreasing acidity (increasing pK_a value).[139]

The kinetic curves for vinylation of the p-substituted phenylpyrroles (**78–81**) (Fig. 2.2) indicates that the rate drops with the electron-withdrawing power of the substituents, which control the acidity of the pyrroles. However, there are some deviations: 2-benzyl-3-phenyl- (**76**), 2,3-diphenyl- (**75**), 4-phenylphenyl- (**80**), and 2-(2-thienyl)-(**83**)-pyrroles are more readily vinylated than it could be expected from their acidity values. All these compounds react more readily than 2-phenylpyrrole (**73**), but possess higher acidities. The deviations may be

2.2. 1-Vinylpyrroles

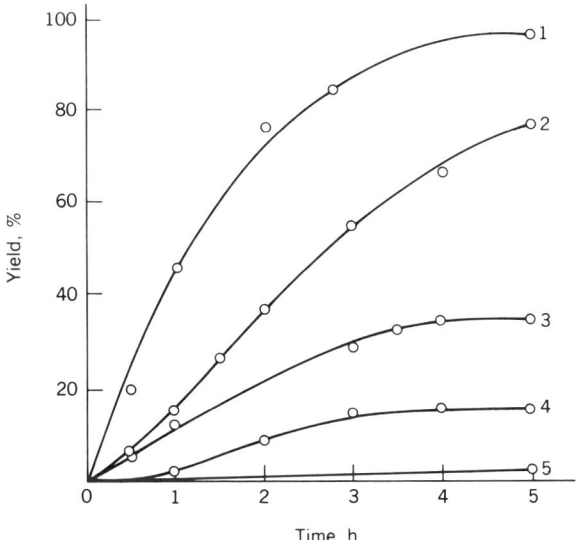

	R¹	R²	R³
73	Ph	H	H
74	Ph	n-C_7H_{15}	H
75	Ph	Ph	H
76	CH_2Ph	Ph	H
77	Ph	H	Ph
78	4-EtC_6H_4	H	H
79	4-$MeOC_6H_4$	H	H
80	4-PhC_6H_4	H	H
81	4-ClC_6H_4	H	H
82	2-Furyl	H	H
83	2-Thienyl	H	H

rationalized by the greater polarizability and lower ionization potential of the pyrroles **76**, **75**, **80**, and **83**, which results in an increased nucleophilicity with a simultaneous decrease in the basicity. This effect is well illustrated by the data for 2-phenyl- (**73**), 2-(2-thienyl)-(**83**), and 2-(2-furyl)pyrrole (**82**) (Fig. 2.2). Although the acidity of pyrroles increases, the nucleophilicity of corresponding anions falls in the sequence **73**–**83**–**82**, such that the reaction rate all increases as a result of the decreasing ionization potentials[140] [**73** (7.61 eV) > **83**

Figure 2.1. Yield–time relationship for the vinylation of pyrroles by acetylene (100°C, DMSO, [KOH] 1.5 M, [pyrrole] 0.5 M):[138] (1) 2-benzyl-3-phenylpyrrole; (2) 3-heptyl-2-phenylpyrrole; (3) 2,3-diphenylpyrrole; (4) 2-phenylpyrrole; (5) 2,5-diphenylpyrrole.

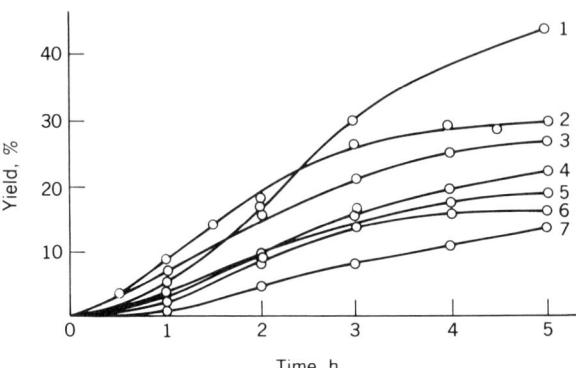

Figure 2.2. Yield–time relationship in the vinylation of 2-aryl(heteroaryl)pyrroles by acetylene (100°C, DMSO, [KOH] 1.5 M):[138] (1) 2-(4-ethylphenyl)pyrrole; (2) 2-(4-methoxyphenyl)pyrrole; (3) 2-(2-furyl)pyrrole; (4) 2-(4-phenylphenyl)pyrrole; (5) 2-phenylpyrrole; (6) 2-(4-chlorophenyl)-pyrrole.

(7.41 eV) > **82** (7.29 eV)]. Consequently, although the anions of the pyrroles **83** and **82** are less basic than those of the pyrrole **73**, they are more nucleophilic. The very low rate of vinylation of the 2,5-diphenylpyrrole (**77**) can be explained in terms of steric effects.

Nearly all kinetic curves have a clearly defined S-form (Figs. 2.1 and 2.2) that is typical for autocatalytic reactions.[138] Autocatalysis should be intrinsic for all vinylation reactions in aprotic and, especially, superbase media, as the protogenic substrate (1H-pyrrole in this case) in being consumed results in an increase in both the nucleophilicity of the remaining anions (due to lowering solvation) and the degree of ionization of the remaining substrate molecules (due to the increasing basicity of medium). Simultaneously, the chemical activity of acetylene is enhanced by complexation with partially dehydrated molecules of the catalyzing base (KOH).[14] A further factor, which may be responsible for the S-shape of the kinetic curves, may be an one-electron channel of nucleophilic addition to the triple bond (see Section 2.2.1.2.4), which may follow a chain mechanism and be inhibited by impurities active with respect to radical and ion–radical intermediates.

2.2.1.2.4. Evidence for One-Electron Channel

The synthesis of azaindoles from oximes of the piperidine series[83] and the effect of the pyrrole structure on the rate of vinylation[138] reveals facts that suggest the presence of a one-electron channel for the nucleophilic addition of pyrroles to the triple bond. In order to substantiate this hypothesis, the effect of compounds active toward radical and ion–radical intermediates (radical process inhibitors) on the rate of vinylation of 2-phenylpyrrole and 4,5,6,7-tetrahydroindole has been studied.[141]

The kinetic curves (Figs. 2.3 and 2.4) indicate that the superbase-catalyzed addition of pyrroles to acetylene is hindered by the radical inhibitors: benzophenone, hydroquinone, and nitroxyl radicals such as 2,2,5,6-tetramethyl-

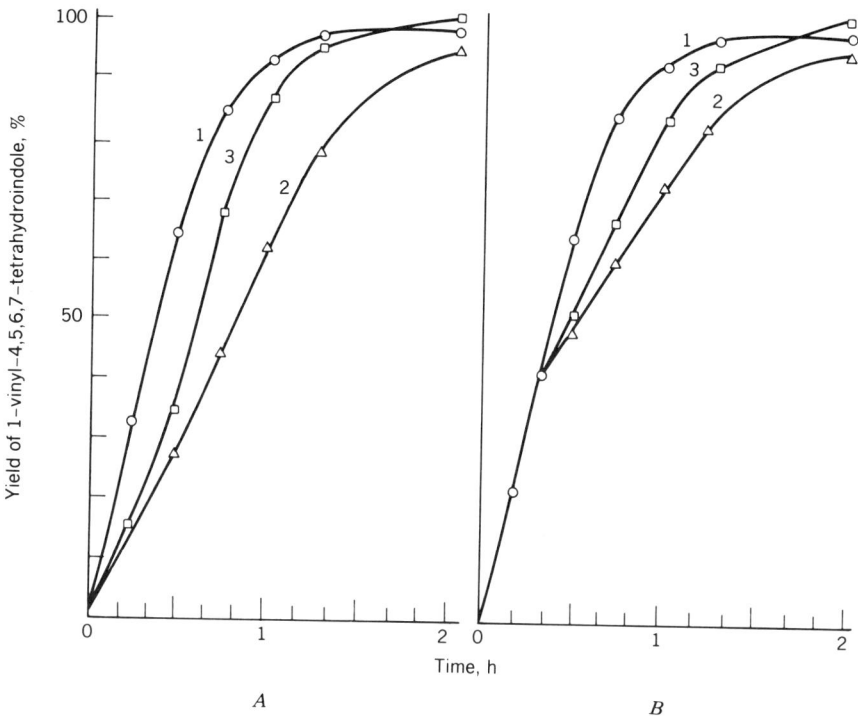

Figure 2.3. Effect of radical inhibitors on the vinylation of 4,5,6,7-tetrahydroindole:[144] (1) none; (2) hydroquinone; (3) 2,2,5,5-tetraethyl-4-phenyl-3-imidazolin-1-oxyl-3-oxide; (A) inhibitors and reactants mixed simultaneously; (B) inhibitors introduced 20 min after the reaction starts.

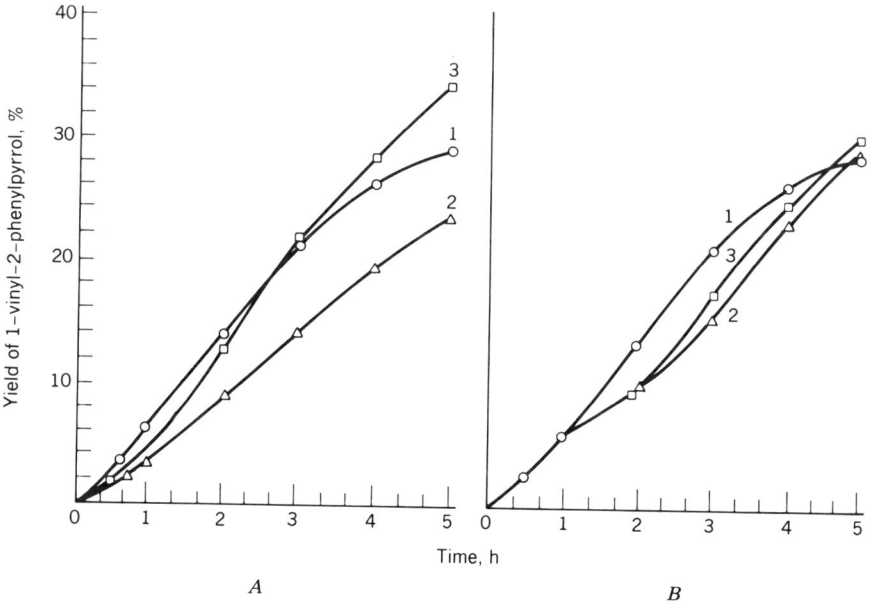

Figure 2.4. Effect of radical inhibitors on the vinylation of 2-phenylpyrrole:[141] (1) none; (2) hydroquinone; (3) 2,2,5,5-tetramethyl-4-phenyl-3-imidazolin-1-oxyl-3-oxide; (A) inhibitors and reactants mixed simultaneously; (B) inhibitors introduced 1 h after the reaction starts.

4-phenyl-3-imidazolyl-1-oxyl-3-oxide (**84**) and 2,2,5,6-tetramethyl-4-cyano-3-imidazolyl-1-oxyl-3-oxide (**85**).

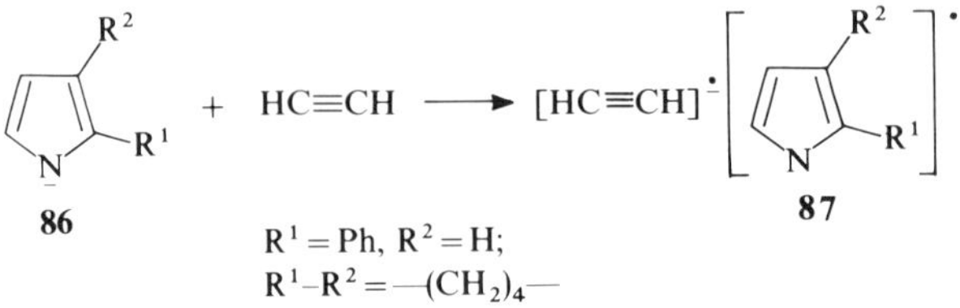

$$\underset{\mathbf{86}}{\text{[pyrrole with } R^2, R^1\text{]}} + HC{\equiv}CH \longrightarrow [HC{\equiv}CH]^{\cdot -} \left[\underset{\mathbf{87}}{\text{[pyrrole with } R^2, R^1\text{]}} \right]^{\cdot}$$

$R^1 = Ph, R^2 = H;$
$R^1-R^2 = -(CH_2)_4-$

The largest inhibiting effect is that of hydroquinone, and the lowest effect is displayed by benzophenone and the nitroxyls **84** and **85**; the radical trap activity of the latter is decreased with time and results from deterioration of the inhibitor through its reduction by the vinylation system.[83]

The vinylation reaction with acetylene is commonly regarded as a typical ionic nucleophilic addition to the triple bond,[142] and in numerous examples of this reaction involving attack by O-, N-, and S-nucleophiles, the one-electron transfer process has not been observed, except for the interaction of KSCN and KI with propiolates and acetylenedicarboxylates.[143] Consequently, the evidence indicating the formation of ion–radical intermediates in the vinylation of pyrroles with acetylene in the superbase KOH/DMSO system is particularly significant.

ESR measurements conducted during the vinylation of 4,5,6,7-tetrahydroindole by either acetylene or phenyl acetylene in KOH/DMSO have confirmed the formation of radical species.[144]

The electron transfer from the pyrrolic anion (**86**) to acetylene activated by the superbase medium[16] also leads to aminyl radicals of type **87**, which are unstable and, therefore, detected only by use of spin traps,[145] [most frequently, 2-methyl-2-nitrosopropane (t-BuNȮ)]. However, in some cases, the spin trap can be reduced by anions, if the difference in their redox potentials is not less than 0.2 V.[146,147] In the 4,5,6,7-tetrahydroindole/KOH/DMSO/t-BuNȮ system in the absence of acetylene, an ESR signal (Fig. 2.5) with hyperfine interaction $^3N \times {}^3N \times {}^3H$ and constants 15.25, 1.05, and 1.05 Oe, which can be assigned to the spin adduct (**88**) (Scheme 10), has been observed.

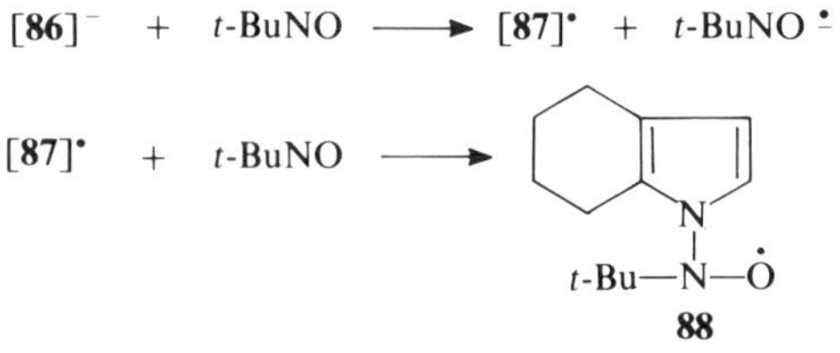

[**86**]⁻ + t-BuNO ⟶ [**87**]˙ + t-BuNO˙⁻

[**87**]˙ + t-BuNO ⟶ **88** (tetrahydroindole with t-Bu—N—Ȯ substituent)

Scheme 10

2.2. 1-Vinylpyrroles

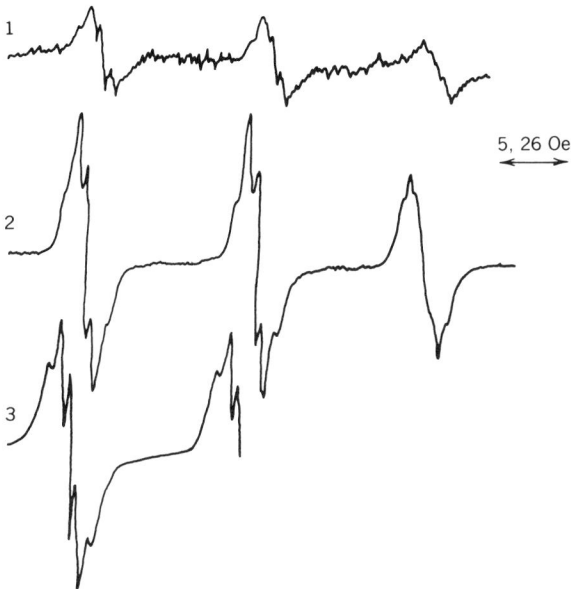

Figure 2.5. ESR spectrum of the spin adduct of 4,5,6,7-tetrahydroindolyl radical with 2-methyl-2-nitrosopropane: (1) in the system KOH/DMSO; (2) in the system KOH/DMSO/acetylene; (3) computer-simulated spectrum.

The addition of acetylene or phenylacetylene to the system increases the intensity of this signal by more than one order (Fig. 2.5), indicating that a competing reaction occurs, which is responsible for the sharp increase in the yield of the aminyl radical (**87**) captured by the spin trap. The electron transfer from the anion (**86**) to acetylene appears to be the source of this process, since, when the acetylene flow is stopped, the ESR signal intensity is reduced practically to the initial value.

The absence of signals in the ESR spectra[148] from the spin adducts of vinyl radicals shows that the radical ion $[HC{\equiv}CR]^{\pm}$ interacts with the tetrahydroindolyl (**87**) at a rate exceeding that of protonation. Additionally, on vinylation of 4,5,6,7-tetrahydroindole with acetylene, the ESR spectra of the reaction mixture exhibits a weak singlet ($\Delta H = 6$–7 Oe),[144] the intensity of which increases as the reaction proceeds. The g-factor of the signal corresponds to that of a free electron, and the parameters are typical of polyacetylene.[149]

2.2.1.3. Other Vinylation Procedures

The substituted 1-vinylpyrroles (**93**) have been prepared[150] via the Horner reaction of aldehydes with diethyl 1-pyrrolylmethanephosphonate and its aryl-substituted derivatives (**91**), obtained by reaction of the α-aminomethanephosphonates (**89**) with the 2,5-diethoxytetrahydrofuran (**90**)[150] (Scheme 11). The intermediate β-hydroxyphosphonate (**92**) can be isolated in some cases.

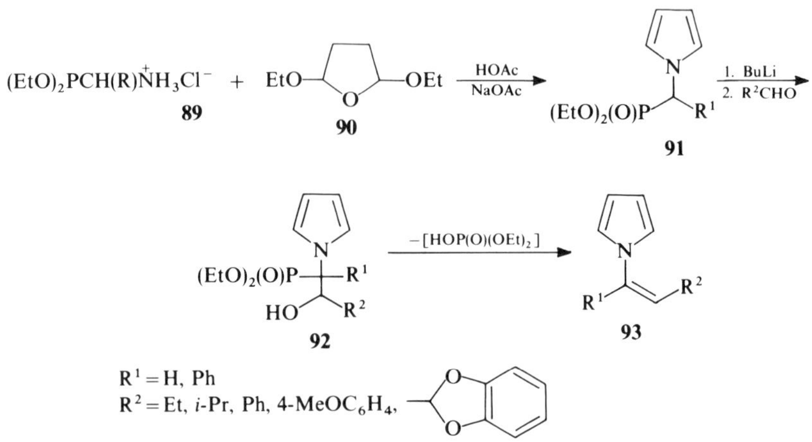

Scheme 11

1-Styrylpyrrole (**93**, R^1 = H, R^2 = Ph) is formed only as *E*-isomer, whereas other 1-vinylpyrroles (**93**) are obtained as mixtures of the *E*- and *Z*-isomers (~4 : 1 with aliphatic and 3 : 2 with aromatic aldehydes) (Table 2.8).

TABLE 2.8. 1-VINYLPYRROLES (**93**)

R^1	R^2	Yield (%)	b.p. (°C) (mg Hg)	m.p. (°C)
H	Ph	52	160–165 (0.01)	98–100
Ph	Ph	82		48–50
Ph	4-MeOC$_6$H$_4$	72	148–152 (0.01)^a	
Ph	Et	66	69–70 (0.01)^b	
Ph	*i*-Pr	57	74–75 (0.01)^c	

^{a} n_D^{24} 1.6523.
^{b} n_D^{27} 1.5641.
^{c} n_D^{25} 1.5448.

A competing process is observed in the reaction of the phosphonate (**91**) (R = Ph) with ketones. With ethyl methyl ketone, for example, a mixture of the 1-vinylpyrrole (**94**) and the α-pyrrol-1-yl-α-ethyl-α-phenylmethanediethylphosphonate (**95**) is formed in a 2 : 3 ratio with an overall yield of 38%.

The 1-vinylpyrroles (**99**), having several electron-withdrawing substituents attached to the vinyl group, have been prepared by reaction of pyrrolylpotassium with the substituted ethoxyethylenes (**96**) (Scheme 12).[151]

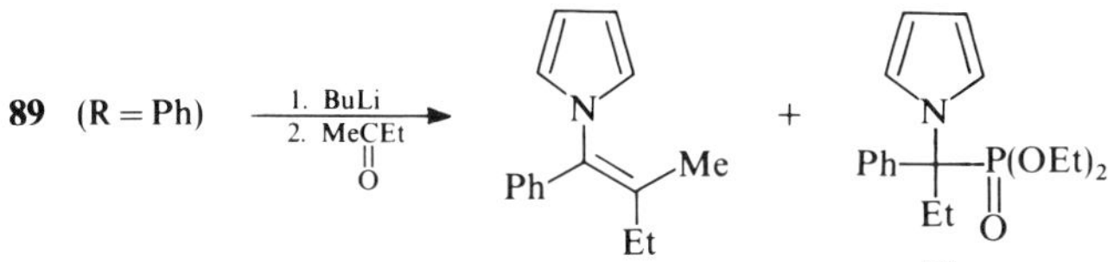

2.2. 1-Vinylpyrroles

Scheme 12

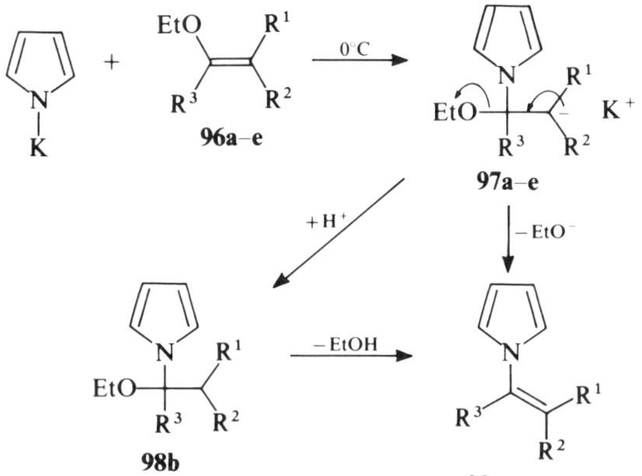

	R^1	R^2	R^3
a	CN	CN	H
b	CO_2Et	CO_2Et	H
c	CN	CO_2Et	H
d	CO_2Et	CN	H
e	CN	CN	Ph

The intermediate salts (**97a–e**) generally produce the 1-vinylpyrroles (**99a,c–e**) directly and do not form the ethers (**98a–e**). The exception is found in the reaction of 1,1-di(ethoxycarbonyl)2-ethoxyethene (**96b**), which with pyrrolylpotassium at 80°C produces the adduct (**98b**). Alkaline hydrolysis of **98b** yields the 1-(2,2-dicarboxyvinyl)pyrrole (**101**) via the dicarboxylate (**100**). Decarboxylation of **101** in refluxing toluene or pyridine produces E-3-pyrrol-1-ylpropenic acid (**102**) (Scheme 13).

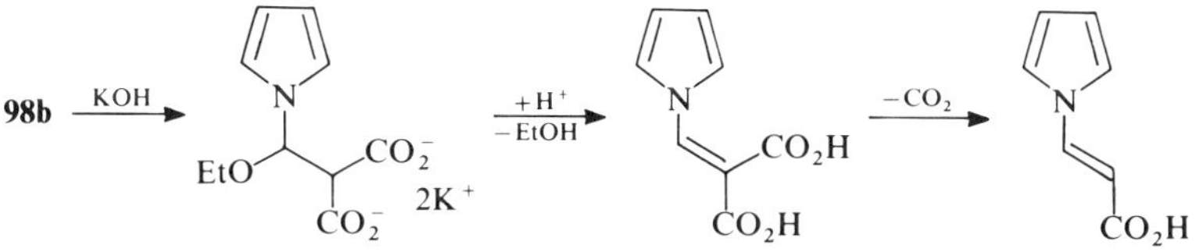

Scheme 13

Recently,[152] the 1-(1-phenylvinyl)pyrrole (**103**) has been obtained (74% yield) by methylenation of 1-benzoylpyrrole with Tebbe's reagent ($Cp_2TiCH_2ClAlMe_2$).[153-155] The reaction has also been utilized for the methylenation of other 1-acylpyrroles, and 1-vinylindoles, for example, 1-isopropenylindole (**104**), 1-isopropenyl-3-phenylindole (**105**), 1-(1-phenylvinyl)-3-phenylindole (**106**), 1-isopropenyl-7-azaindole (**107**), and 1-isopropenyl-2-methyl-3-(4-fluorophenyl)-4,7-diazaindole (**108**) have been obtained in 51-68% yield. This reaction provides a convenient route to 1-vinylpyrroles with various substituents at the α-position of the vinyl group, which are inaccessible by other methods.

A synthesis of 1-vinylpyrroles (35%) by the base-catalyzed decomposition of the 3-(2-hydroxyethyl)-1,3-oxazolidines (**109**) has been reported.[156]

R^1, R^2 = H, alkyl, aryl;
R^1-R^2 = $(CH_2)_4$, $(CH_2)_5$;
M = Na, K

1-Vinylpyrrole has been synthesized[157] by thermal cleavage of the 1-(1-acetoxyethyl)pyrrole (**110**) and, in a reaction similar to those described in earlier sections, the *N*-vinyldipyrrolylmethane (**111**) has been isolated in small yield

(0.08%) from the reaction of acetoxime with acetylene; the major products are 2-methyl-1-vinylpyrrole and 2-methylpyrrole.[158]

The acid-catalyzed condensation of 2-formylfuran with the 3-ethyl-2,4-dimethylpyrrole (**112**) initially forms the azafulvenium salt (**113**),[159,160] which undergoes an unusual nucleophilic ring-opening reaction to produce **114**.[158]

2.2.2. Reactivity

2.2.2.1. Introduction

1-Vinylpyrroles are highly reactive compounds. Conjugation of the vinyl group with the aromatic ring results in the vinyl substituents being more susceptible to electrophilic attack than are normal alkenes or 2- and 3-vinylpyrroles, while the pyrrolyl α-position is less reactive than is normal for other pyrrole systems.

Consequently, the behavior of 1-vinylpyrroles is frequently determined by competition at these two centers. The high π-electron density of the pyrrole ring and its capability of stabilizing the positive charge favor the formation of charge-transfer complexes, which precede or accompany the reactions of 1-vinylpyrroles with electron-deficient reactants.

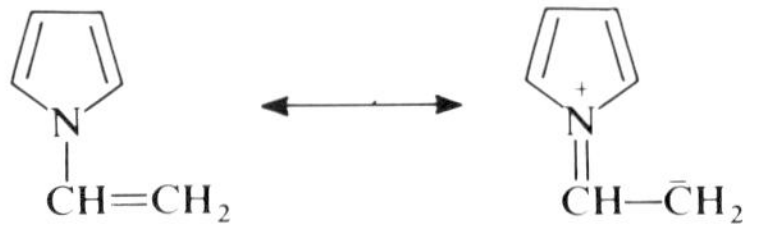

The high sensitivity of N-vinyl compounds to oxygen is well known, and 1-vinylpyrroles are no exception, although it has been noted[6] that, unlike vinylamines and vinylamides, N-vinylcarbazole is not oxidized by air.

Until recently, few chemical properties of 1-vinylpyrroles had been reported, and their reactivity could be surmised only from the known reactivities of N-vinylcarbazole[5-7, 161-168] and 1-vinylindole[3, 169-175] which are not typically representative of pyrrole derivatives. However, 1-vinylpyrroles have now been shown to react additively with alcohols,[176] thiols,[177] and hydrosilanes[178] (Scheme 14).

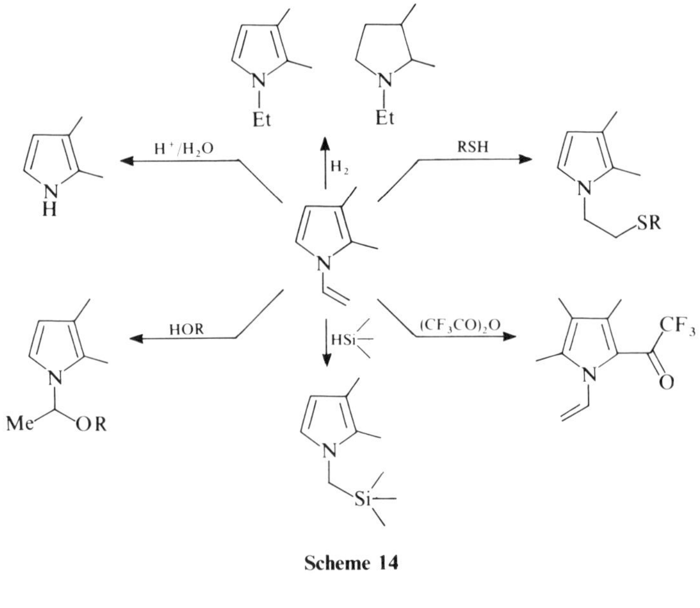

Scheme 14

Procedures for selective[179] and total[180] hydrogenation of 1-vinylpyrroles have also been developed. Hydrolysis reactions[181] and the formation of charge-transfer complexes with bromine, iodine, and alkyl bromides[182] have been described. The acid-catalyzed dimerization of 1-vinylpyrrole to give 2-(1-pyrrolyl-1-ethyl)-1-vinylpyrroles has been studied in detail.[183]

In contrast, electrophilic substitution of the pyrrole ring of 1-vinylpyrroles (for example, trifluoroacetylation[184-189]) with retention of the 1-vinyl group,

2.2.2.2. Hydrogenation

Selective hydrogenation of 1-vinylpyrroles over Raney nickel to yield 1-ethyl derivatives has been reported[179] (Table 2.9).

TABLE 2.9. 1-ETHYLPYRROLES

R^1	R^2	Yield (%)	b.p. (°C/mm Hg) m.p.	d_4^{20}	n_D^{20}
Ph	H	93	114/4–5	1.0175	1.5795
Ph	Me	85	94/1	1.0169	1.5730
Ph	Et	81	135–136/2	0.9931	1.5595
Ph	n-Pr	80	126/1	0.9846	1.5535
Ph	i-Pr	85	66–67/1	0.9760	1.5480
—(CH$_2$)$_4$—		90	66–67/1	0.9783	1.5200

The reduction is normally conducted at atmospheric pressure, but hydrogenation under pressure reduces the reaction time approximately by a factor of 10. Thus, the reaction can be employed for the preparation of 1-ethylpyrroles along with conventional methods.[190,191]

Catalytic hydrogenation of the 1-isopropenyl derivative (**108**) (5% Pd/C, EtOH, 2.8 bar, ambient temperature, 2 h) leads to the 1-isopropyl compound (**115**) as the only product in 95% yield,[152] while hydrogenation[192] of the 2-trifluoroacetyl-1-vinylpyrroles (**116**) over Raney nickel reduces not only the vinyl group but also the trifluoroacetyl group, yielding 1-ethyl-2-(1-hydroxy-2,2,2-trifluoroethyl)pyrroles (**117**) (Table 2.10).

172 Vinylpyrroles

TABLE 2.10. 1-ETHYL-2-(2,2,2-TRIFLUORO-
1-HYDROXYETHYL)PYRROLES (117) OBTAINED BY
HYDROGENATION OF 2-TRIFLUOROACETYL-1-VINYLPYRROLES

R	Yield (%)	b.p. (°C/mm Hg)	m.p. (°C)
Me	72	81–82/1	56
Ph	96	140/3–4	88–89
4-EtOC$_6$H$_4$	94	162–164	56–71
4-MeOC$_6$H$_4$	95	162–164/3	70–71

1-Vinyl-4,5,6,7-tetrahydroindole is a source of the relatively inaccessible 1-ethyloctahydroindole (96% yield)[180] by catalytic hydrogenation over Raney nickel at 140°C. Hydrogenation at 120°C gives a mixture of 1-ethyloctahydro- and 1-ethyltetrahydroindoles. Hydrogenolysis is insignificant. Hydrogenation of 2-phenyl- and 3-alkyl-2-phenyl-1-vinylpyrroles derivatives under similar conditions gives a complex mixture of products.

2.2.2.3. Protonation

Many of the questions concerning the specific behavior of 1-vinylpyrroles in acid media—with the possibility of generating 1-vinylpyrrolium cations, their detection, and studies of their stability and chemical and physical properties—are still unanswered.

1-Vinylpyrrolium cations have been generated[24] in situ during NMR-spectral measurements under the conditions that minimize the possibility of their intermolecular reactions with unprotonated species. The spectral data provide evidence of protonation at the ring α-position (118) or 1-vinyl group (119) of

R^1 = H, alkyl, aryl, heteroaryl; R^2 = H, alkyl;
A = Cl, Br, CF$_3$CO$_2$, SO$_3$F

2.2. 1-Vinylpyrroles

1-vinylpyrroles, as well as simultaneous reaction at both sites (**120**). At $-80°C$, protonation occurs predominantly at the α-position of the pyrrole ring with retention of the double bond irrespective of the nature of acid used or of ring substituents.

In the formation of the cations (**118**) the ring proton signals undergo large (1–1.5 ppm) downfield shifts compared with unprotonated molecules, which is compatible with the distribution of the positive charge on the ring. The vicinal coupling constant values, $^3J_{AC}$ and $^3J_{BC}$, of the vinyl group (15 and 9 Hz, respectively) do not change relative to those of unprotonated 1-vinylpyrroles, whereas the geminal constant $^2J_{AB}$ increases (0.8–3 Hz) reflecting a decrease in conjugation of the vinyl group with the pyrrole ring.

The cations (**118a**), generated at $-50°C$ by reaction of 1-vinylpyrroles with fluorosulfonic acid, have been studied by ^{13}C NMR spectroscopy.[124,193,194] In accord with the ^1H NMR spectral data, the difference in chemical shifts of carbon atoms of the vinyl group in the cations (**118a**) and the initial 1-vinylpyrroles (110–115 and 95–99 ppm, respectively) indicates a substantially reduced electron density at these atoms in the protonated species. The small chemical shift, compared with that of ethylene (123.3 ppm), suggests a residual conjugation of the lone electron pair of the pyrrole nitrogen atom even in the cationic species.

118a

A linear correlation between the C_β, H_A, and H_B chemical shifts has been observed:[24]

$$\delta C^+_\beta = 40.9 + 12.8(\pm 4.0) \, \delta H^+_A, \quad r = 0.95, \quad s_o = 0.04, \quad n = 9$$

$$\delta C^+_\beta = 20.6 + 14.9(\pm 1.8) \, \delta H^+_B, \quad r = 0.99, \quad s_o = 0.02, \quad n = 9$$

There is also a satisfactory correlation between δC_β of 1-vinylpyrrolium cations and their unprotonated precursors of the form

$$\delta C^+_\beta = 39.4 + 0.76(\pm 0.14) \, \delta C^\circ_\beta, \quad r = 0.98, \quad s_o = 0.05, \quad n = 10$$

The points corresponding to 2-(2-furyl)- and 2-(5-methyl-2-furyl)-1-vinylpyrrole deviate significantly from this correlation.

The ^{13}C NMR spectra of 3-alkyl-substituted cations are distinguished by a slightly greater shielding of the C_β and C_5 atoms (0.5–2.0 ppm), compared with those for the 3-unsubstituted compounds. The origin of this upfield shift lies in the inductive effect of the alkyl substituent polarizing the π-system.

The sums of the chemical shifts ($\sum \delta C^+_i$) of the ring C_2, C_3, and C_4 atoms of the cationic species, whose hybridization does not change on protonation, have been compared[24] with corresponding values for the neutral molecules ($\sum \delta C^\circ_i$). The difference $\sum \delta C^+_i - \sum \delta C^\circ_i$ reflects the total deshielding of nuclei in the ring carbon moiety on protonation. For 2,3-dialkyl-1-vinylpyrroles this value is 120 $\pm$ 4 ppm, while, for 2-aryl- and 2-heteroarylpyrroles, it is equal to 109 $\pm$ 2 and 95 $\pm$ 4 ppm, respectively, indicating an increase in the degree of delocalization of the charge on the protonated system in the order 2-alkyl < 2-aryl < 2-heteroaryl. Further evidence has been deduced from the total ^{13}C nuclei deshielding (for aromatic and heteroaromatic substituents). The increase in the total deshielding in the series 2-phenyl < 2-thienyl < 2-furyl (5.1, 33.0, and 44.0 ppm, respectively) reflects the very strong π–π interaction between the rings of 2-furylpyrrole derivatives and accounts for the lack of correlation cited above. In this case, the resultant conjugation of the pyrrolyl lone pair of electrons with the vinyl group is enhanced.

The regiospecificity of the protonation of 1-vinylpyrroles with hydrogen halides depends on temperature.[24,195] At -80°C, the α-protonated pyrrole species (**118**) is formed exclusively. At -40°C, addition of the second molecule of the hydrogen halide on the vinyl group produces **120**. Also, at -40°C, reaction with an equimolar amount of the acid leads to the 1-(1-haloethyl)pyrroles (**119**). These effects result from an interaction of kinetic and thermodynamic factors. Thus, the formation of **118** is brought about by a primary (kinetic) protonation at low temperature (-80°C). With increasing temperature (-40°C), **118** is converted into the thermodynamically more stable 1-(1-haloethyl)pyrrole (**119**), which is then protonated to form the cation (**120**).

The isopropyl group in 1-(1-chloroethyl)-3-isopropyl-2-phenyl pyrrolium chloride (**121**) and analogous compounds is a diastereotropic tracer.[195] Because of the presence of the chiral α-chloroethyl group, the methyl signals are anisochronous. Coalescence of the signals has been assumed to indicate the interconversion of enantiomeric forms (**A** and **B**), which involves the deprotonation of the pyrrolium moiety and ionization of the C—Cl bond, which is catalyzed by free hydrogen chloride. At high concentration of hydrogen chloride, the proton exchange between free base and conjugated acid (**121**) predominates.

The proton exchange between HCl and the 5-CH$_2$ is confirmed by broadening of its signals simultaneously with the coalescence of the methyl group signals.

2.2. 1-Vinylpyrroles

Further elevation of the sample temperature above the coalescence temperature results in a broadening of the H_4 signal, suggesting a significant but low participation of the β-protonated species (**122**).

Intermolecular proton exchange has been studied with the 1-(1-chloroethyl)pyrrolium chlorides **123** and **124**. In the cation (**123**) the exchange has been shown to involve only the 3- and 5-positions, whereas, for the phenyl substituted cation (**124**), the exchange involves the 3-, 4-, and 5-positions.

The protonation of 2-(2-furyl)-1-vinylpyrroles (**125**) has been investigated by ^{1}H NMR spectroscopy.[24,194,196,197] At $-80°C$, it is the 5-position of the pyrrole ring that is protonated irrespective of the nature of acid. At the higher temperature of $-30°C$, reaction with hydrogen halides leads to the formation of a mixture of pyrrolium (**126**) and furanium (**127**) ions with simultaneous addition of the acid to the vinyl group.

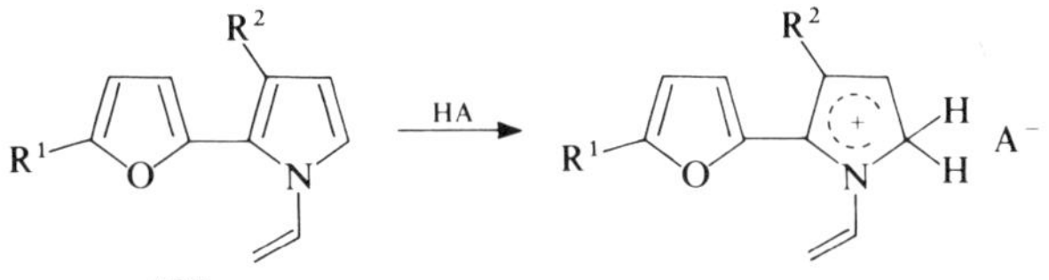

125
$R^1 = H$, Me; $R^2 = H$, Me;
$A = Cl$, Br, CF_3CO_2, SO_3F

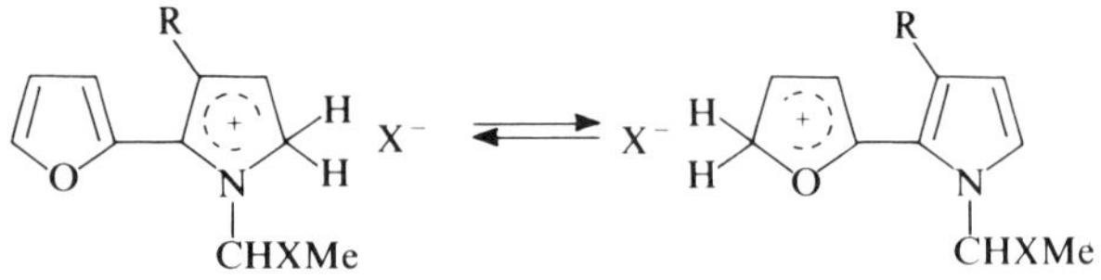

126   **127**
$X = Cl$, Br

It has been postulated that at $-80°C$ kinetic protonation leads to the nonequilibrium system with the pyrrolium ions prevailing. With rising temperature, equilibrium is achieved and the relative concentrations of isomeric pyrrolium and furanium ions correspond to their energies.[197,198] With an excess of hydrogen bromide, protonation is incomplete and, if the temperature is allowed to increase to 0°C, 4-bromo-4,5-dihydro-2-(2-pyrrolyl)-3H-furanium bromides (**129**) result from protonation of the brominated dihydrofuran intermediate (**128**). The presence of aluminum bromide, which binds to the bromide ion to form the complex anion $Al_2Br_7^-$, impedes this reaction, while, by contrast, the addition of tetramethylammonium bromide accelerates the formation of

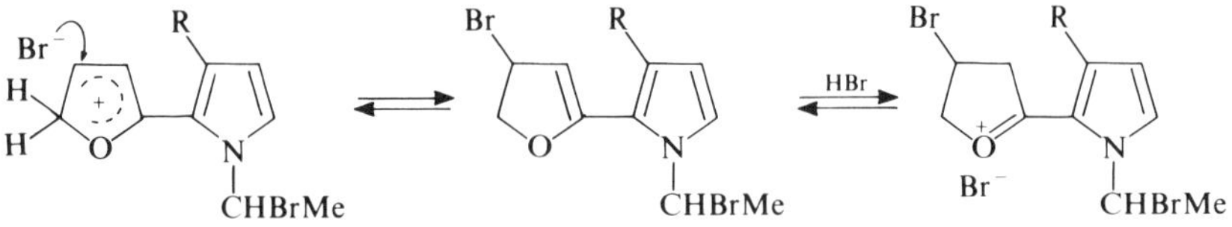

128   **129**

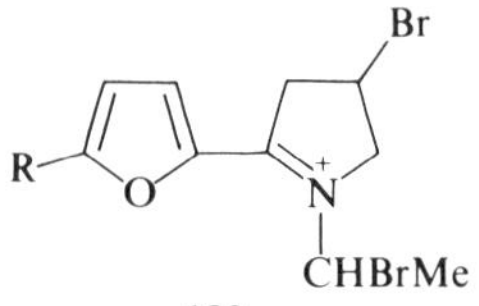

130

129. With the less nucleophilic chloride ion, no transformations of this kind occur.

A further increase in temperature above $\sim 10°C$ leads to a transformation of the dihydrofuranium cations into 4-bromo-4,5-dihydro-3-(2-furyl)-3H-pyrrolium bromides (**130**) with retroaromatization of the furan ring.

The transformation of dihydrofuranium cations to dihydropyrrolium ions (**129** → **130**) has been interpreted[24,197] in terms of the greater thermodynamic stability of the protonated pyrrole ring system, compared with the protonated furan system, in spite of kinetically more favorable initial addition of the hydrogen bromide on the less aromatic furan ring.

At higher temperatures (20°C), pyrrolium ions (**118**) react with nonprotonated systems to yield the dimeric cations (**133**), the structure of which has been proved by ^{1}H NMR spectroscopy.[24,193] The postulated mechanism requires proton transfer (**118** → **133**) (seemingly intermolecular) from the 5-position of the protonated pyrrole ring to the β-vinyl carbon atom and subsequent electrophilic attack at the 5-position of the unprotonated pyrrole to form the dimeric cation (**133**) (Scheme 15).

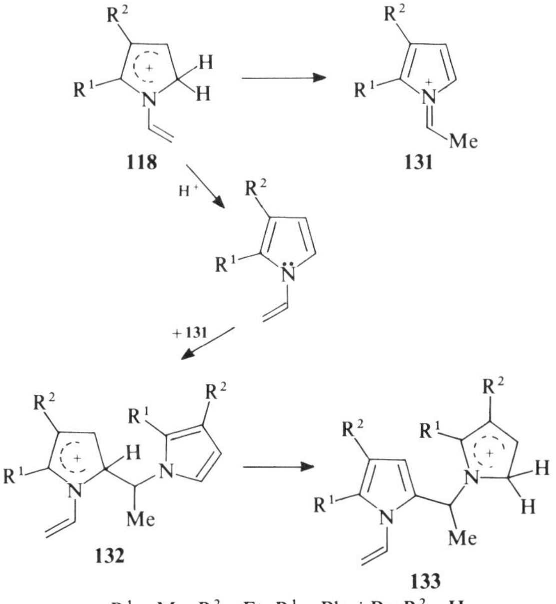

$R^1 = Me, R^2 = Et; R^1 = Ph, i-Pr; R^2 = H$

Scheme 15

The rate of the dimerization is significantly higher for the 2-phenyl derivative ($R^1 = Ph$), indicating the lower stability of the initial cation (**118**).

Although electrophilic substitution, rather than addition, is typical for the pyrrole and its derivatives, several 2-substituted-1-vinylpyrroles have been observed to react with excess hydrogen bromide at -40 to $0°C$ to yield

4-bromo-4,5-dihydro-2-substituted 3H-pyrrolium ions (**135**) via the intermediate enamine (**134**).[24,197] Predictably, the corresponding reaction with hydrogen chloride does not occur.

At $-70°C$, 2-t-butyl-1-vinylpyrrole (**136**) forms the stable β-protonated derivative (**137**) with the superacid system $HSO_3F/SbF_5/SO_2FCl$ with retention of the double bond,[24] while, under the same conditions, 2-phenyl-1-vinylpyrroles are diprotonated (**138 → 139**).[24] The formation of the dication is aided by delocalization of the charge by the phenyl ring. When the dication is heated at 30°C for 1 h, irreversible ring closure leads to the monocation (**140**), from which, on neutralization, the bicyclic system (**141**) has been isolated in 7% yield.[24]

Protonation of the 1,4-bis(1-vinyl-2-pyrrolyl)benzene (**142**) with the superacid $HSO_3F/SbF_5/SO_2FCl$ at $-70°C$ affords the dication (**143**).[199] The ^1H NMR spectrum confirms the presence of the α-protonated pyrrole system with signals at 5.54 (5-CH_2), 7.42 (d, 3-H), and 8.25 ppm (d, 4-H) with $^3J_{(H_3,H_4)} = 6.0$ Hz. The vinyl group on this ring produces signals at 7.36 (H_C), 6.35 (H_A) and 5.83 ppm

2.2. 1-Vinylpyrroles

(H_B). The β-protonated pyrrole nucleus gives rise to signals at 4.75 (4-CH_2), 7.30 (d, 3-H), and 7.71 ppm (d, 5-H) with $^4J_{(H_3,H_5)} = 3.9$ Hz, and the protons of the vinyl group associated with this ring resonate at 7.36 (H_C) and 6.13 ppm (H_A, H_B).

Although α-protonation of the pyrrole ring is generally a thermodynamically controlled process,[200] the formation of the cation (143) is readily understood from a consideration of the resonance stabilization of the monocation (144a–c) in which the 3- and 5-positions of the nonprotonated ring are deactivated to electrophilic attack. In contrast, reaction of (142) with hydrogen chloride and hydrogen bromide results in addition of the halogen acid to the vinyl groups and protonation of both pyrrole rings at the α-position to form the dication (145).[199]

α-Protonation of the two rings is confirmed by the presence of a single set of signals at 5.64 (5-CH_2), 7.44 (d, 3-H), and 8.49 ppm (d, 4-H) with $^2J_{(H_3,H_4)} = 5.8$ Hz. The protons of the $CHBrCH_3$ group resonate at 6.61 (q) and at 2.40 ppm (d). The difference in behavior of 142 with the halogen acids is probably caused by steric hindrance to coplanarity of the molecule, created by initial formation of the bulky 1-haloethyl substituents.

180   Vinylpyrroles

The formation of **146** from the reaction of **142** with hydrogen bromide at −70°C is the first reaction stage, which subsequently leads, at higher temperatures (−40°C), to the addition of second molecule of hydrogen bromide to each pyrrole ring to form the dication (**146**), the structure of which is confirmed by ^{1}H NMR spectroscopy.[197]

Variable temperature ^{1}H NMR measurements of the dication (**146**) indicate an equilibrium with the monocationic species.[197]

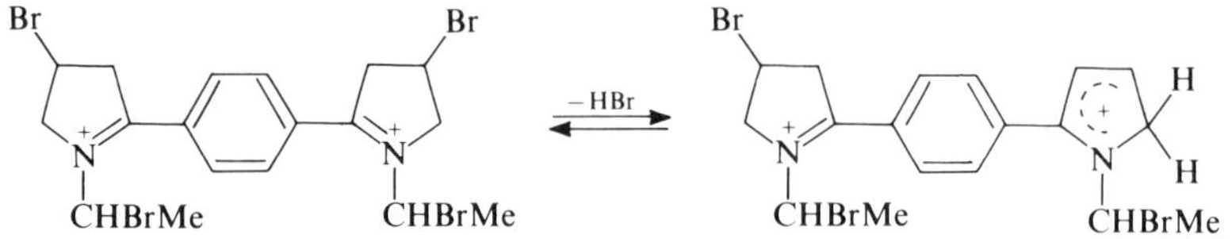

2.2.2.4. Acid-Catalyzed Hydrolytic Cleavage Reactions

Although 1-vinylpyrroles might be expected to undergo hydrolytic cleavage analogous to 1-vinylindole,[201] 9-alkenylcarbazoles,[6,161,164,202] N-vinyl phenothiazine,[163] and N-vinyllactams,[203] (e.g., Scheme 16), little information has been available until quite recently.

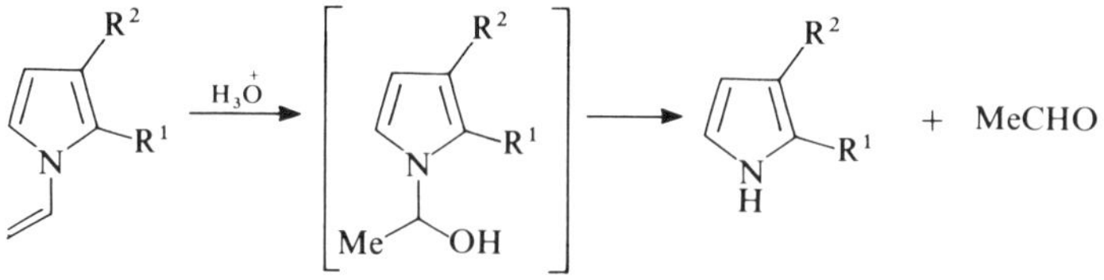

R^1 = Me, Ph; R^2 = H, Me; R^1–R^2 = $(CH_2)_4$

Scheme 16

It has been shown[181] that hydrolytic cleavage of 1-vinylpyrroles can be realized in dilute acidic solutions if the released acetaldehyde can be removed, for example, by reaction with hydroxylamine. Under normal conditions, however, complex acid-catalysed condensation reactions between the pyrroles and the aldehyde take place (see Part 1, Chapter 3). As a result, mixtures of deeply colored oligomers, the composition of which depend on the conditions of hydrolysis and isolation, are formed. Physical data[1,2,204] have led[181] to the identification of 2-(1-hydroxyethyl)pyrroles (**147**), 1,1-di(2-pyrrolyl)ethanes (**148**), 2-pyrrolylpyrrolines (**149**) [together with pyrrolinone (**150**)], and pyrrolidinone fragments (**151**) in the reaction mixtures.

The intermediate carbimmonium ion (**152**) plays a key role in the formation of the bipyrrolic systems (**148, 153,** and **154**) and of the polymers.

2.2. 1-Vinylpyrroles

147 **148** **149** **150** **151** **152** **153** **154**

As expected, 1-vinylpyrroles with ring alkyl substituents are the least stable under acidic conditions[181] and are readily cleaved by diluted acids at room temperature. Thus, for example, only 65% of 1-vinyl-4,5,6,7-tetrahydroindole is recoverable from a solution in 1% aqueous hydrochloric acid after 1 h at room temperature. Above 65°C, the compound is rapidly and completely converted into an orange-red resin. The IR spectrum of the resin indicates that dominant features of the products include conjugated double bonds, together with carbonyl, NH— and OH— groups, while the ^{1}H NMR spectrum provides evidence for the CH—CH$_3$ groups of fragments **147**, **148**, **153**, **154** (3.6–4.3 ppm and 1.5–1.7 ppm), as well as weak signals for 1-vinyl (5.0–5.2 ppm) and pyrrole protons (5.6–6.3 ppm). Broadened signals are indicative of the polymeric nature of the products, and a well-developed polyconjugated chain is confirmed by a narrow intense signal in the ESR spectrum ($\sim 1.10^{18}$ spin/g, $\Delta H = 0.8$–1.0 mT).

Hydrolytic cleavage of the vinyl group to yield 4,5,6,7-tetrahydroindole (5%) has been observed[181] only when the reaction was conducted with an excess of acidified NH$_2$OH·HCl. Evidence for the active participation of the 1-vinyl group in the hydrolytic cleavage is provided by the stability of the analogous 1-alkyl compounds under similar conditions. The processes are also greatly influenced by oxidative transformations leading to structures of the types **150** and **151**, as indicated by the fast conversion of 1-vinyl-4,5,6,7-tetrahydroindole into polymeric material on heating with 33% hydrogen peroxide (see Part 1, Chapter 3).

2-Phenyl-1-vinylpyrroles are noticeably more stable than 2-alkyl- and 2,4-dialkyl-1-vinylpyrroles but even then will eventually react under acidic conditions at elevated temperatures to give products of the types **147, 148, 150, 151, 153**, and **154**, as established by infrared and ^{1}H NMR spectroscopy.[14,181]

The 1-vinylpyrroles (**93**) having alkyl or aryl substituents attached to the vinyl group are stable in 2N hydrochloric acid. However, prolonged heating in the acid results in their complete destruction.[150]

Comparison of the analytical data for the products of hydrolytic reactions of 2-phenyl-1-vinylpyrrole and 3-methyl-2-phenyl-1-vinylpyrrole leads to the conclusion that, under similar reaction conditions, the latter compounds are subject to a greater oxidative process. This observation is in agreement with the well established fact that the introduction of alkyl substituents not only facilitates protonation but also accelerates the oxidation and polycondensation of the pyrrole ring.[1,2,204]

Spectral data indicate that, at a lower concentrations (0.8%) of hydrochloric acid, the product contains the partial structure (**155**),[181] in addition to the fragments **147–154**.

155

Additionally, the degree of the hydrolytic reaction and the character of the products are affected by the nature of acid.[181] Thus, reaction of 2-phenyl-1-vinylpyrrole in 0.5% sulfuric acid at 96°C produces a product in which the presence of neither the pyrrole nor benzene rings can be detected and spectral evidence indicates a high degree of polycondensation and oxidation. Under similar conditions, 2-phenyl-1-vinylpyrrole is virtually unaffected by 2% acetic acid, while 12-day contact of the pyrrole with 7% acetic acid at room temperature produces trace quantities of 2-phenylpyrrole.

It is known[161] that, at room temperature, iron(III) nitrate either catalyzes the decomposition of N-vinylcarbazole in aqueous methanol (1 : 9) to carbazole and acetaldehyde or, at high concentration, induces its dimerization to trans-1,2-dicarbazolylcyclobutane. Under similar conditions, 2-phenyl-1-vinylpyrrole undergoes almost no hydrolysis, and the yield of 2-phenylpyrrole is ~10%. Instead, methanol adds to the vinyl group to form 1-(1-methoxyethyl)-2-phenylpyrrole (**156**) in high yield.[181] The corresponding reaction in n-butanol produces the 1-butoxyethyl derivative.

R = Me, n-Bu   156

Hydrolytic cleavage to yield 2-phenylpyrrole in 53% yield is only accomplished in a dilute aqueous–dioxane solution of hydrochloric hydroxylamine at 80°C.[181]

$$\text{[1-vinyl-2-phenylpyrrole]} \xrightarrow[\text{H}_2\text{O}]{\text{HO}\overset{+}{\text{N}}\text{H}_3\text{Cl}^-} \text{[2-phenylpyrrole]} + \text{MeCH=NOH} + \text{HCl}$$

2.2.2.5. Addition of Alcohols and Phenols

The acid-catalyzed reaction of 1-vinylindole[205–207] and of N-vinylcarbazole[208–210] with alcohols yields the corresponding N-(α-alkoxyethyl) derivatives. Under similar conditions, in the case of 1-vinylindole, the polymerization reaction competes with the acid-catalyzed addition reaction and considerably reduces the yield of adducts.[205] Alcohols and phenols have also been reported to add to N-vinylpyrrolidone[203,211] according to the Markovnikov rule. However, the structure of the adduct with phenol appears to be unusual and has been suggested to involve the electrophilic attack on the aromatic ring instead of on the hydroxyl group.

The electrophilic addition of a range of alcohols to 1-vinyl-2-phenylpyrrole and 1-vinyl-4,5,6,7-tetrahydroindole to yield 1-(1-alkoxyethyl)pyrroles (157) has been reported[176] (Table 2.11).

Protic and aprotic acids, either alone or in combination, have been used as catalysts. Best results have been achieved with iron(III) nitrate, although silver, chromium(III), and cobalt(II) nitrates do not catalyze the reaction. The rate of the addition reaction falls with an increase in the donating power of the alkyl

TABLE 2.11. 1-(1-ALKOXYETHYL)PYRROLES (157)

R^1	R^2	R^3	Yield (%)	b.p. (°C/mm Hg)	d_4^{20}	n_D^{20}
Ph	H	Me	76	98/1	1.0475	1.5640
Ph	H	Et	55	104/2	1.0302	1.5490
Ph	H	n-Pr	57	122–123/3	1.0159	1.5456
Ph	H	i-Pr	47	102/1	1.0055	1.5452
Ph	H	n-Bu	52	126–130/5	1.0109	1.5550
—(CH$_2$)$_4$—		n-Bu	55	93–94/1	0.9582	1.4960
—(CH$_2$)$_4$—		n-Am	51	110–112/1	0.9568	1.4930

group (R^3), which is in agreement with the electrophilic nature of the process. The character and shape of the kinetic curves obtained at 76–78°C indicate the reversibility of the reaction; the addition reaction is incomplete and the yield decreases with time, as a result of polymerization of 1-vinylpyrrole. An analogous equilibrium has been observed earlier with 1-alkenyl- and N-(1-alkoxyalkyl)carbazoles.[208] The 1-(1-alkoxyethyl)pyrroles (157) are not prone to disproportionation and alcoholysis; when the reaction is carried out with excess alcohol, neither the corresponding dialkyl acetal nor any noticeable amount of NH-pyrroles is observed.

In the presence of radical initiators, such as azobisisobutyronitrile (ABN) and carbon tetrachloride, alcohols have been reported to add to 1-vinylcarbazole to form α-adducts.[208] The corresponding reaction with substituted 1-vinylpyrroles has been examined,[212] and 1-(1-alkoxyethyl)pyrroles were found to be the major products (50–70%). In the absence of carbon tetrachloride, ABN fails to catalyze the reaction with 1-vinylcarbazole[208] and 2-phenyl-1-vinylpyrrole.[212] Similarly, no catalytic effect is exerted by carbon tetrachloride alone at 75–80°C, or by an ABN/CHCl$_3$ system. However, at 100°C, carbon tetrachloride catalyzes a 5% conversion of 2-phenyl-1-vinylpyrrole. This observation supports the conclusion[208] that the reaction is electrophilic in nature and is catalyzed by traces of hydrochloric acid and Cl$_2$, the products of transformation involving chlorine atoms, which are also generated by the ABN in carbon tetrachloride.

The adducts of 1-vinylpyrroles with acetylenic alcohols, such as 158,[213,214] 159, 160,[215] and 161,[214] are of interest as potential biologically active substances.

2.2. 1-Vinylpyrroles

Noncatalytic and acid-catalyzed (CF_3COOH) addition of phenols to 1-vinyl-4,5,6,7-tetrahydroindole afford a range of 1-(1-aroxyethyl)-4,5,6,7-tetrahydroindoles in yields of up to 60%.[216] The yields of adducts decrease as a result of oligomerization of the vinyl system, with increased acidity of the phenol and in the presence of trifluoroacetic acid.

2.2.2.6. Addition of Thiols

The addition of alkanethiols to 1-vinylpyrroles with and without free-radical initiators has been investigated.[177] It is well established that with radical initiators, as well as under noncatalytical conditions, cyclic enamines[216] add mercaptans according to the Markovnikov rule. In contrast, N-vinylpyrroles and related systems react with hydrogen sulfide and with thiols in an anti-Markovnikov manner[3,5-7] to yield the β-adducts.[217] ABN-catalyzed free-radical addition of thiols to 1-vinylindole and 1-vinylimidazoles gives a mixture containing the products of α- and β-addition.[218]

A convenient synthetic route to the novel 1-(2-alkylthioethyl)pyrroles (**162**), by the addition of alkanethiols to 1-vinylpyrroles, has been developed[177] and

TABLE 2.12. 1-(2-ALKYLTHIOETHYL)PYRROLES (**162**)

R^1	R^2	R^3	Yield (%)	b.p. (°C at 1–2 mm Hg)	d_4^{20}	n_D^{20}
—$(CH_2)_4$—		Et	83	111–112	1.0402	1.5475
—$(CH_2)_4$—		n-Pr	82	128–129	1.0313	1.5410
—$(CH_2)_4$—		i-Pr	77	116–117	1.0166	1.5390
—$(CH_2)_4$—		n-Bu	89	150–151	1.0336	1.5350
—$(CH_2)_4$—		i-Bu	75	142–143	1.0321	1.5330
—$(CH_2)_3CH(Me)$—		Et	80	125–126	1.0632	1.5375
—$(CH_2)_3CH(Me)$—		n-Pr	82	128–129	0.9991	1.5340
—$(CH_2)_3CH(Me)$—		i-Pr	76	110–111	0.9844	1.5335
—$(CH_2)_3CH(Me)$—		n-Bu	80	139–140	1.0014	1.5290
—$(CH_2)_3CH(Me)$—		i-Bu	73	128–129	1.0128	1.5280
Me	n-Pr	n-Bu	87	141–142	0.9488	1.5100
Me	Me	n-Bu	99	103–104	0.9645	1.5153
Me	n-Am	n-Bu	81	138–139	0.9378	1.5048
Ph	H	Et	93	144–145	1.0766	1.5955
Ph	H	n-Pr	94	158–160	1.0550	1.5826
Ph	H	n-Bu	70	166–167	1.0450	1.5757

the orientation of the addition reaction has been clarified. When heated to 70–80°C, 1-vinylpyrroles react with alkanethiols in the presence or absence of a radical initiator to produce only the β-adducts (**162**) (Table 2.12).

With ABN, the preparative yields of **162** are 73–99%. Under comparable conditions without initiator, the yield of adducts drops to 20–30%.[177] No side products or polymers were detected. In the presence of the ABN/CCl$_4$ system, thiols also react with 1-vinylpyrroles to produce the β-adducts; for example, 1-(2-alkylthioethyl)-2-phenylpyrroles are obtained in 50–54% yields.[212]

The free-radical thiylation of the 5-trifluoroacetyl-1-vinylpyrroles (**163**) occurs without complication to form only the β-adducts (**164**).[219]

R^1 = Me, Ph, 4-EtC$_6$H$_4$, 4-MeOC$_6$H$_4$; R^2 = H, Me; R^3 = Et, n-Pr

Under free-radical initiation, thiophenols also selectively add to 1-vinylpyrroles to form 1-(2-arylthioethyl)pyrroles (**165**).[220] However, under similar conditions, but without initiators, up to 80% of the 1-(1-arylthioethyl)pyrroles (**166**) are formed with ∼ 20% of the β-adducts (**165**).

R^1 = Ph, R^2 = H;
R^1–R^2 = —(CH$_2$)$_4$—;
Ar = Ph, 4-FC$_6$H$_4$

The thiolation of a mixture of 1-vinyl-4,5,6,7-tetrahydroindole in the presence of 4,5,6,7-tetrahydroindole and cyclohexanone with and without ABN leads selectively to the α-adducts [**166** R^1, R^2 = (CH$_2$)$_4$].[220] 4-Nitrothiophenol does not add to 1-vinyl-4,5,6,7-tetrahydroindole; at room temperature an exothermic reaction results in polymerization.[220] Thus, unlike alkanethiols, which add to 1-vinylpyrroles exclusively by the radical mechanism to form the β-adducts, thiophenols, because of their increased acidity, show a tendency to electrophilic addition leading to the α-adducts.

2.2.2.7. Hydrosilylation

Silicon-containing nitrogen heterocycles are often biologically active and have been extensively explored as starting materials for the synthesis of drugs.[221] 1-[3-(Trialkoxysilyl)propyl]pyrroles, obtained by a multistage synthesis,[222] have also been reported to improve the adhesion of polypyrrole film to *n*-type silicon photoanodes, and it is conceivable that the same properties would be found with 1-[2-(trialkoxysilyl)ethyl]pyrroles, which could be obtained by direct addition of trialkoxysilanes to 1-vinylpyrroles.

Hydrosilylation of a series of 1-vinylpyrroles with trialkylsilanes, catalyzed by $H_2PtCl_6 \cdot 6H_2O$ at 80–100°C, regiospecifically gives the β-adducts (Table 2.13) in 55–93% yield.[223,224] High yields are favored by aprotic solvents (THF, acetone, dioxane) and prolonged reaction times.

TABLE 2.13. HYDROSILYLATION OF 1-VINYLPYRROLES TO GIVE 1-[2-(TRIALKYLSILYL)ETHYL]PYRROLES

R^1	R^2	R^3	Yield (%)	b.p. (°C mm Hg)	d_4^{20}	n_D^{20}
—(CH$_2$)$_4$—		Me(Et)$_2$	93	152–153/5	0.9588	1.5103
—(CH$_2$)$_4$—		Et$_3$	82	165–166/3.5	0.9418	1.5116
—(CH$_2$)$_4$—		Me(*n*-Pr)$_2$	84	196–197/13	0.9264	1.5033
—(CH$_2$)$_4$—		(*n*-Pr)$_3$	86	186–187/3	0.9321	1.5026
—(CH$_2$)$_4$—		Me(*n*-Bu)$_2$	93	190–191/3	0.9243	1.4987
—(CH$_2$)$_4$—		(Me$_3$SiO)$_2$Et	55	160–162/2	0.9622	1.4680
Ph	H	Me(Et)$_2$	76	150–151/1	0.9644	1.5449
Ph	H	Et$_3$	79	168–169/1	0.9656	1.5430
Ph	H	(*n*-Pr)$_3$	68	186–188/4	0.9451	1.5310
Ph	Me	Et$_3$	79	180–182/2	0.9704	1.5365
Ph	Me	(*n*-Pr)$_3$	79	188–189/3	0.9499	1.5288
Me	Et	Et$_3$	77	150–152/6	0.8461	1.4873
Me	Et	Me(*n*-Pr)$_2$	80	150–152/4	0.8840	1.4824

The yields also depend on the hydrosilane used. Contrary to expectations, an attempt[224] to add trichloro-, alkyldichloro-, and triethoxysilanes to 1-vinylpyrroles was unsuccessful and the starting 1-vinylpyrroles were completely recovered, although even a trace of trimethylchlorosilane readily converts the vinylpyrroles into their dimers and polymers.[225] (see Section 2.2.2.10). Only in the reaction of methylchlorosilane with 2-(2-thienyl)-1-vinylpyrrole was the formation of the expected adduct (~7%) detected by ^{1}H NMR spectroscopy.[224]

The rhodium complex $(Ph_3P)_3RhCl$, which readily catalyzes the addition of triethylsilane to 1-vinylpyrroles (50–60% yields of the adducts),[224] also appears to be inactive with trichloro-, alkyldichloro-, and trialkylsilanes. No products of the hydrosilylation of 1-vinylpyrroles with the above silanes were formed when the addition was initiated by UV irradiation.[224]

Triethylsilane also reacts with 2-aryl- and 2-heteroaryl-1-vinylpyrroles with varying success at 140°C using either H_2PtCl_6 or the rhodium complex $(Ph_3P)RhCl$, as a catalyst[224] (Table 2.14).

TABLE 2.14. 1-[2-(TRIETHYLSILYL)ETHYL]-2-HETARYL(PHENYL)PYRROLES

X	R^1	R^2	Yield (%)	b.p. (°C/mm Hg) [m.p.]	d_4^{20}	n_D^{20}
O	H	H	40	140–143/1	0.9531	1.5195
S	H	H	70	165–167/4–5	1.0232	1.5525
S	Me	H	65	192–196/4	1.0030	1.5420
S	Et	H	44	193–196/4	1.0690	1.5400
S	H	COCF$_3$	7^a	[32–33]		
HC=CH	H	H	43	169–170/1	0.9656	1.5430
HC=CH	Me	H	43	180–182/2	0.9704	1.5365
HC=CH	H	COCF$_3$	7^a	[44–46]		

^{a} Also prepared by trifluoroacetylation of the corresponding 1-[2-triethylsilyl)ethyl]pyrroles.[186]

The thiophene ring clearly facilitates the reaction. Evidently, the 2-thienyl substituent coordinates to the platinum ions to form a more active catalytic species. This postulate is supported by the observed decrease in the yield from 70 to 40%, when the bulk of the 3-substituent is increased, thereby inhibiting coordination.[14] 1-[2-(Trialkylsilyl)ethyl]-5-trifluoroacetylpyrroles are better synthesized by acylation of 1-[(2-trialkylsilyl)ethyl]pyrroles.[14]

2.2.2.8. Acylation

Investigations[184–187] of the trifluoroacetylation have been undertaken in order to assess the relative activity of two most probable nucleophilic centers of the 1-vinylpyrrole system and to develop preparative methods[186] for the synthesis of trifluoroacetylpyrroles.

2.2. 1-Vinylpyrroles

Normally, pyrroles undergo acylation at the α-position or, when it is occupied, at the β-position[226-228] (see Part 1, Section 3.4). A number of cases are known[3] where electrophiles attack 1-vinylindoles at the vinyl group, rather than on the pyrrole moiety. This observation is compatible with data concerning both the ease of acylation of enamines[229] and the electrophilic substitution on the N-vinyl group of N-vinylamides[230] on trifluoroacetylation. However, in the presence of pyridine, trifluoroacetic anhydride readily, and exclusively, acylates the α-position of 1-vinylpyrroles[184-187] in 73–96% yield (Table 2.15).

TABLE 2.15. TRIFLUOROACETYLPYRROLES

R^1	R^2	R^3	Yield (%)	b.p. (°C/mm Hg) [m.p.]	d_4^{20}	n_D^{20}
Me	H	$CH_2=CH$	82	73–74/3	1.2590	1.5110
Me	Me	$CH_2=CH$	76	92–93/4 [37]		
—$(CH_2)_4$—		$CH_2=CH$	73	122–123/3	1.2675	1.5460
Ph	H	$CH_2=CH$	88	132/1.5	1.2582	1.5883
Ph	Me	$CH_2=CH$	93	136/1	1.2400	1.5810
4-EtC_6H_4	H	$CH_2=CH$	93	138–140/1	1.2126	1.5732
4-MeOC_6H_4	H	$CH_2=CH$	96	158–163/1 [68–70]		
4-BrC_6H_4	H	$CH_2=CH$	76	[90–93]		
Ph	H	H	83	[158–159]		
Ph	H	Et	98	122/3	1.2572	1.5600
Ph	H	$(CH_2)_2SEt$	77	164/3–4	1.2419	1.5720
Ph	H	$(CH_2)_2SiEt_3$	97	166/1 [44–46]		1.5371

For a qualitative comparison with 1-vinylpyrroles, the trifluoroacylation of pyrroles having other substituents at the nitrogen atom, such as H, Et, $(CH_2)_2SEt$, $(CH_2)_2SiEt_3$, has been studied.[187] No noticeable difference in the reaction course has been observed. It is apparent that conjugation of the N-vinyl group with the pyrrole ring is not as strong as analogous conjugation in vinyl ethers, sulfides, and enamines and in N-vinylamides, which are readily trifluoroacylated at the β-position of the vinyl group;[230] i.e., of two alternative intermediate cations A and B, cation A is the more stable.

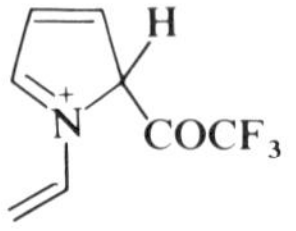

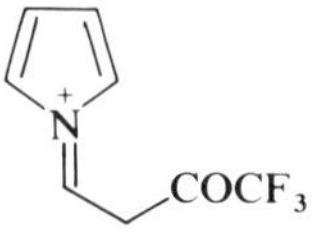

The observed preference for acylation of the ring is somewhat unexpected in view of the fact that the vinyl group of 1-vinylindole[3] and, especially, of 1-alkenylcarbazoles[231] is very prone to electrophilic addition.

Trifluoroacetylation of various pyrrole systems has been extensively studied[1, 2, 232–234] (see also Part 1), and the reaction has frequently been used to compare the reactivities of pyrrole, furan, and thiophene.[227, 228, 235] The relative rates of the trifluoroacetylation of these heterocycles with trifluoroacetic anhydride are $5.3 \times 10^7 : 1.4 \times 10^2 : 1$, as a result of the greater capability of the pyrrole ring to stabilize the positive charge of the cationic σ-complexes.[236] On this basis, it is to be expected that 2-(2-furyl)- and 2-(2-thienyl)-1-vinylpyrroles (**167, 168**) would be acylated only at the 5-position of the pyrrole nucleus (Scheme 17). This is found to be true for the trifluoroacylation of 2-(2-furyl)- and 2-(2-thienyl)pyrroles,[237, 238] but, unexpectedly, 2-(2-furyl)-1-vinylpyrrole **167** behaves in quite a different manner, for while 2-(2-thienyl)-1-vinylpyrrole (**168**) is trifluoroacetylated to produce **170** exclusively, **167** yields **169** and **171** in a 9 : 1 ratio.

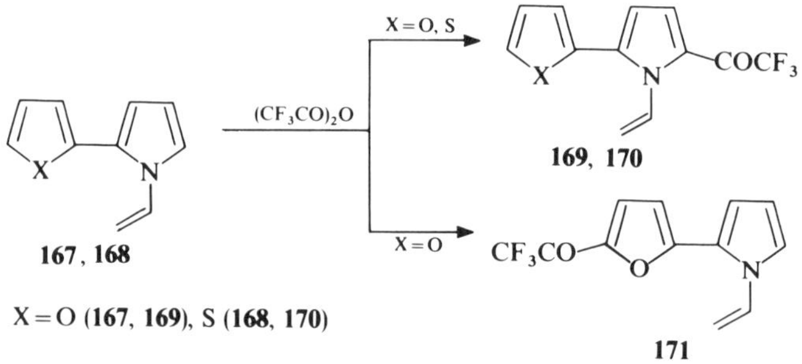

Scheme 17

The factors that are important in the protonation of 2-(2-furyl)-1-vinylpyrrole (see Section 2.2.2.3) are also important in the acylation reaction. Interestingly, 2-(2-thienyl)-1-vinylpyrrole (**168**), on trifluoroacetylation, gives better preparative results than does its nonvinylated precursor (53 and 71%, respectively), and the overall yield of the trifluoroacetylation products of 2-(2-furyl)-1-vinylpyrrole (**169 + 171**) is increased from 50 to 63% by the use of a twofold excess of the anhydride, while the relative amount of **171** is increased to 38% (Table 2.16).[237]

2.2. 1-Vinylpyrroles   191

TABLE 2.16. 1-VINYL-2-(2-HETEROARYL)-5-TRIFLUOROACETYLPYRROLES

Pyrrole	Yield (%)	b.p. (°C/mm Hg)
169	25	124–130/3
170	71	131/4
171	38	124–130/3

The absence of the secondary acylation products is compatible with the strong deactivating effect of the trifluoroacetyl substituent, which is transmitted to the α-position of the furan and thiophene rings.

Trifluoroacetylation of 2-phenyl-1-vinylpyrrole also produces 2-phenyl-5-trifluoroacetyl-1-vinylpyrrole, although the rate of reaction is noticeably lower than that for the heteroaryl derivatives. The maximum yield (58%) is achieved with a twofold excess of the anhydride.[239] Similarly, 3-alkyl-2-(2-phenyl)-1-vinylpyrroles yield the 5-trifluoroacetyl derivatives[240] and, in contrast with 2-(2-furyl)-1-vinylpyrrole,[237] trifluoroacetylation of 2-(2-furyl)-3-methyl-1-vinylpyrrole also produces the 5-trifluoroacetylpyrrole (34%).

Hydrogenation and thiolation of 2-trifluoroacetyl-1-vinylpyrroles, which confirm the previously established regularities,[14, 192, 219] and their phosphorylation[241] have been reported.[239, 240] Aliphatic alcohols ROH (R = Me, Et, n-Bu) do not add to 2-phenyl-5-trifluoroacetyl-1-vinylpyrrole under electrophilic conditions.[239]

Cleavage of 2-trifluoroacetyl-1-vinylpyrroles under basic conditions leads to the corresponding acids (**172**) (75–88%); however, formation of the expected intramolecular adduct (**173**) has not been observed.[242]

R^1 = Ph, 4-MeC$_6$H$_4$, 4-EtC$_6$H$_4$, 4-MeOC$_6$H$_4$, 4-ClC$_6$H$_4$, 2-thienyl; R^2 = H
R^1–R^2 = —(CH$_2$)$_4$—

2.2.2.9. Other Reactions

The phosphorylation of 1-vinylpyrroles, 1-vinylindole, and 9-vinylcarbazole has been examined.[243-245]

The phosphorylation product of the reaction of 1-vinyl-4,5,6,7-tetrahydroindole with phosphorus pentachloride has been identified as 6,7,8,9-tetrahydroindolo-[1,2-α]-1,1,3-trichloro-4,1-azaphospholanium hexaphosphate (**174**) on the basis of a signal at 66.6 ppm in its ^{31}P NMR spectrum. Reaction of **174** with sulfur dioxide produces 6,7,8,9-(tetrahydroindolo-[1,2-α]-oxo-1,3-dichloroazaphospholane (**175**).

The formation of pyrrolo[1,2-*a*]-1,1,3-trichloro-4,1-azaphospholanium hexachlorophosphates (**176**), as confirmed by their ^{31}P NMR spectra, is general for a range of 1-vinylpyrroles. On treatment with sulfur dioxide, the salts (**176**) are transformed into the dichloro-4,1-azaphospholanes (**177**).[243]

$R^1 = $ Me, Ph; $R^2 = $ H, Ph

As the nucleophilicity of the 1-vinylpyrroles decreases, the phosphorus-containing 1-vinylpyrroles (**178**) are formed, in addition to the bicyclic system (**176**). Thus, 2-trifluoroacetyl-1-vinyl-4,5,6,7-tetrahydroindole (**180**) forms the product of the electrophilic substitution on the vinyl group (**181**).[241] Subsequent treatment of **178** and **181** with sulfur dioxide yields **179**, **182**, and **183**.[241] Similar transformations have been performed with 2-methyl- and 2-phenyl-5-trifluoroacetyl-1-vinylpyrroles,[241] and the structures have been established by IR and ^{31}P NMR spectroscopy. The phosphonyl chloride (**179**) has been characterized (Scheme 18).

It is unlikely that the azaphospholanes **174** and **176** are formed from a primary phosphorylation of the 1-vinylpyrroles at the *N*-vinyl group. The attack of phosphorus pentachloride at the α-position of the pyrrole nucleus is more

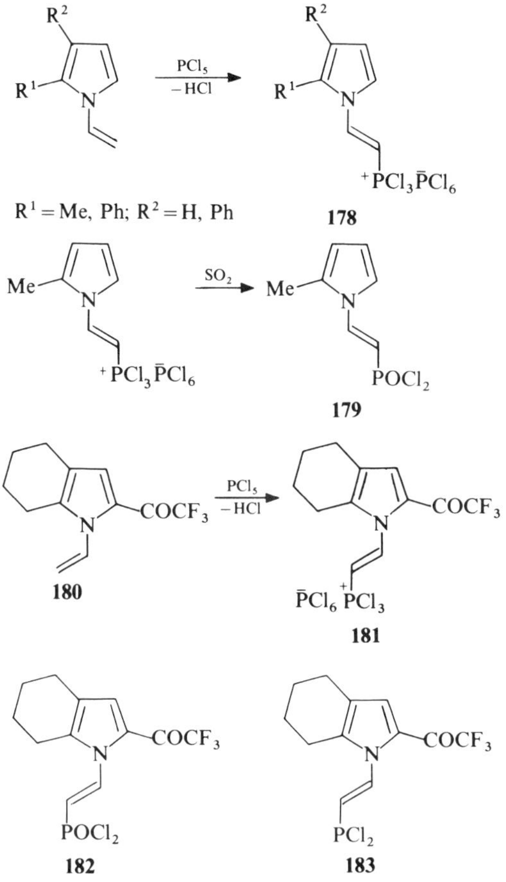

2.2. 1-Vinylpyrroles

Scheme 18

probable. The intermediate (**184**) forms the salts (**174** and **176**) by a subsequent intramolecular electrophilic attack at the vinyl group.

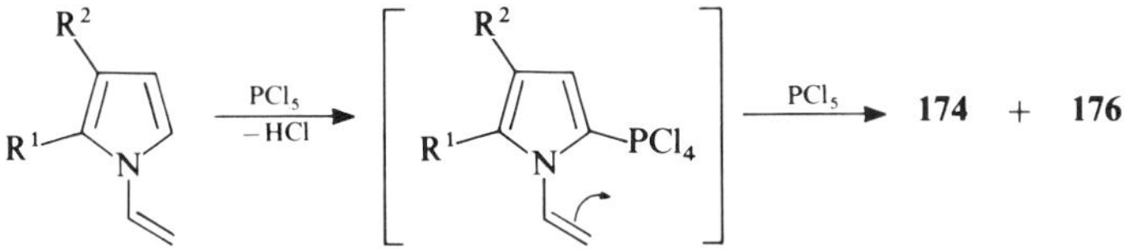

When the compounds **175** and **177** are treated with triethylamine, elimination of hydrogen chloride results in the formation of pyrrolo[1,2-*a*]-1-oxo-1-chloro-4,1-azaphosphol-2-enes (**185**), as confirmed by their ^{31}P, ^{1}H, and ^{13}C NMR spectra.[244]

In contrast with the reaction of the vinylpyrroles, phosphorylation of N-vinylindole and N-vinylcarbazole affords the products exclusively on the vinyl group (e.g., **186**).[243] The subsequent conversion of **186** into **187** has been described.[243]

1-(2,2,3,3-Tetracyano-1-cyclobutyl)pyrroles (**188**), prepared by the [π + π]-cycloaddition of tetracyanoethene to 1-vinylpyrroles, form 3-methyl- or (2,3-dimethyl)-5,6-dihydro-5-methoxy-7,8-dicyanoindolizines (**189**) in almost quantitative yield,[246, 247] when refluxed in methanol. It is of interest that, under comparable conditions, 1-vinylindole and 9-vinylcarbazole are transformed into the corresponding 3,4,4-tricyano-1,3-butadien-1-yl derivatives.[248, 249]

2-Phenyl-1-vinylpyrrole undergoes the normal acid-catalyzed reaction with formaldehyde (see Part 1, Section 3.3) to form dipyrrolylmethanes without affecting the vinyl groups.[250]

In the presence of Pd/γ-Al_2O_3 or Cr_2O_3/γ-Al_2O_3 modified with rare-earth elements and potassium oxide, mixtures of 1-vinyl-4,5,6,7-tetrahydroindole and

4,5,6,7-tetrahydroindole are transformed into 1-ethyl-4,5,6,7-tetrahydroindole, indole, and 2-ethylindole in ratios depending on the reaction conditions and the catalyst.[251]

2.2.2.10. Polymerization

1-Vinylpyrrole polymers may be used for the preparation of semiconductors and photosensitive materials, diverse charge-transfer complexes, synthetic dyes, and pigments analogous to natural ones. The photoconductive properties of doped modifications of poly-N-vinylcarbazole[252, 253] contribute much to its increasing use in electrophotography, thermoplastic records, and holography, as well as in electronics.

In spite of the potential practical value of poly-1-vinylpyrroles, their application has, until recently, been limited. General discussion of the polymerization of 1-vinyl monomers have been presented in monographs,[6, 7] reviews,[3, 5, 254, 255] and numerous theses.[21, 231, 256] However, the data on the polymerization of simple 1-vinylpyrroles occupy no more than a few paragraphs, although one thesis[21] is devoted to the polymerization of 1-vinylpyrroles. Most of the literature on the polymerization of 1-vinylpyrrole has been summarized in a recent review.[17] In contrast, the polymerization of 1-vinylindole has been studied in detail by Skvortsova et al.,[3, 122, 170–172, 257–260] and significant research has been conducted on polyvinylcarbazole[261–267] and its derivatives[268–272] and on N-vinylphenothiazine,[273–275] particularly by Lopatinsky and Sirotkina.[270–272, 276–280]

The charge-transfer polymerization of the 1-vinyl derivatives of indole and pyrrole illustrates their relative reactivities, as well as the difficulty encountered in its interpretation. Chloranil and 2,5-dichloro-p-benzoquinone initiate the polymerization of 1-vinylindole after a prolonged induction period, although DDQ is a more effective initiator.[4] In contrast, under the same conditions, 1-vinylpyrrole yields only traces of polymer after a prolonged reaction time (36 h), and it is apparent that the ease of polymerization initiated by electron acceptors is in the order 9-vinylcarbazole > 1-vinylindole > 1-vinylpyrrole.[4]

The polymerization of 1-vinylpyrrole occurs on mixing with excess CCl_4. The IR spectrum of the polymer is analogous with that of poly-N-vinylpyrrole obtained by cationic polymerization in the presence of boron trifluoride etherate, although the mechanism of CCl_4-induced polymerization is far from being understood in spite of considerable research in this area.[255]

Copolymerization of 1-vinylindole and 1-vinylpyrrole in the presence of DDQ[4] leads to only traces of the copolymer; the molar fraction of 1-vinylpyrrole is only 0.32.

In general, 1-vinylpyrroles undergo polymerization in the presence of organic electron-acceptors via charge-transfer complexes,[3, 4, 182] under the influence of typical radical initiators,[21, 281–294] and in the presence of protic or aprotic acids via a cationic mechanism.[4] Quantum-chemical calculations suggest that

1-vinylpyrroles should also polymerize by an anionic mechanism;[295, 296] however no reliable experimental evidence has been reported.

Unlike poly-N-vinylcarbazole, the polymers of 1-vinylpyrroles (and 1-vinylindoles) can be readily modified by a suitable choice of substituted pyrroles[297] to prepare cross-linked electroconductive materials of polypyrrole type.[298]

Treatment of 1-vinyl-4,5,6,7-tetrahydroindole with hydrogen chloride produces, in addition to the polymer, the dimer (**191**) in 20% yield, as a result of electrophilic attack of the cation (**190**) on the neutral pyrrole.[183]

The dimer (**191**) has also been isolated in 35% yield during an attempted dichlorocarbenylation of 1-vinyl-4,5,6,7-tetrahydroindole under phase-transfer catalytic conditions.[299]

Other 1-vinylpyrroles undergo dimerization in a similar way on the action of diverse Brönsted and Lewis acids.[300] The previously unknown alternating oligomerization of 1-vinylpyrroles involving the double bond and the pyrrole ring has been studied in detail for 1-vinyl-4,5,6,7-tetrahydroindole.[301–303] Oligomers of type **192** with molecular mass of 1000–2000 have been isolated (~ 35% yield) and characterized,[303] the yield of the dimer (**191**) amounts to 65–68%.[301–303]

Polymerization of 1-vinylpyrroles in the presence of basic reagents has been studied recently. Although apparently stable in superbases, even at elevated temperatures, 1-vinylpyrroles (1-vinyl-4,5,6,7-terahydroindole, 2-phenyl-1-vinylpyrrole, 2,3-dimethyl-1-vinylpyrrole) undergo oligomerization with sodium metal at 190°C.[304, 305] The reaction involves both the vinyl group and the endocyclic double bond of the pyrrole ring to furnish the oligomers of the type **193** containing short polyene chains as a result of elimination of the pyrrole rings (47%, M = 900–1200).

$$-\!\!\mathopen{[}CH_2\!\!-\!\!CH\mathclose{]}_{\!l}\!-\!\!\mathopen{[}CH\!\!-\!\!CH\mathclose{]}_{\!m}\!-\!\!\mathopen{[}CH\!\!=\!\!CH\mathclose{]}_{\!n}\!-$$

193

$R^1\text{--}R^2 = -\!(CH_2)_4\!-, \; R^1 = Ph, \; R^2 = H; \; R^1 = R^2 = Me$

2.2.3. Physical Chemistry

2.2.3.1. Introduction

1-Vinylpyrroles have been extensively studied by various physicochemical methods.[20, 35, 36, 43, 46, 188, 189, 306–333] In particular, the coplanarity of the vinylpyrroles system and the degree of conjugation between the vinyl group and the ring have been examined in detail.

In summary, it has been shown that the conjugative effect is weakened by the bulky substituents at the 2-position, which distort the molecular coplanarity, and a marked temperature dependence of the dipole moments of 1-vinylpyrrole and its alkyl derivatives supports the existence of nonplanar conformation.[316] The decrease in conjugation due to a distortion in coplanarity with increasing bulk of the 2-substituent is also evident from the ultraviolet absorbtion spectra.

The integrated intensities of bands of the vinyl group stretching vibrations in the IR spectra of N-vinylazoles, including those of 1-vinylpyrrole, have been shown to be linearly related to the π-electron polarity of the N-vinyl group, and the resonance constants of the azole rings have been calculated.[320]

The ^{1}H, ^{13}C, and ^{14}N NMR spectra of 1-vinylpyrroles have been used to measure the influence of steric and electronic effects of substituents on the conjugation of the pyrrole ring with the vinyl group;[20, 35, 36, 46, 188, 307, 309–311, 318] a dependence of the ^{13}C chemical shifts of ring carbon atoms and CNDO/2 calculated total charges has been observed.[20, 307, 311] Long-range ^{1}H–^{1}H, ^{1}H–^{19}F, and ^{1}H–^{13}C spin–spin coupling for a series of 1-vinylpyrroles provides a basis for the determination of the preferred conformation of the vinyl group.[46, 188, 307, 310–325, 331–333]

Photoelectron spectra and quantum-chemical calculations[140, 312, 328, 330] confirm the preference of a nonplanar conformation for 2-substituted 1-vinylpyrroles (in contrast with the data for 9-vinyl- and 9-alkenylcarbazoles[329]).

The total energy of the 2-methyl-1-vinylpyrrole molecule has been calculated by CNDO/2 for dihedral angles (φ) between the vinyl group and the ring of 0° (planar *anti* conformation) to 180° (planar *syn* conformation). The minimum energy is found at 0° with a second minimum corresponding to a nonplanar *gauche* rotamer (φ 120–130°). The difference in energies between these two stable rotamers indicates a population ratio of 3 : 1, which is in agreement with the observed average conformation angle of 35°.[20, 307, 311]

2.2.3.2. Dipole Moments

From a theoretical aspect, the principal points of interest are (1) how the vinyl group on the nitrogen atom affects the electron distribution in the pyrrole ring (and, consequently, the molecular polarity) and (2) the way in which the polarity of the 1-vinylpyrrole molecule depends on mutual orientation of its fragments.

Dipole moments for a series of 2- and 3-substituted 1-vinylpyrroles have been measured (Table 2.17).[316] The 1-vinyl group causes an almost twofold decrease in polarity of the molecule ($\Delta\mu = 0.74$ D), compared with the *N*-unsubstituted system (1.82 and 1.08 D for pyrrole and 1-vinylpyrrole, respectively). This observation is unexpected, as a 1-phenyl group causes a significantly smaller depolarization ($\Delta\mu = 0.46$ D). Since the dipole moment of pyrrole is directed from the nitrogen atom into the ring,[2] it is evident that the 1-vinyl group produces an inverse displacement of electrons, thereby reducing the "aromaticity" of the pyrrole nucleus. The overall electron withdrawal results from a joint σ-inductive and π-conjugative effect of the 1-substituent. The induction effects of vinyl and phenyl groups are almost equivalent ($\sigma^* = 0.65$ and 0.60, respectively).[337] It follows that the π-conjugation effect of the 1-vinyl group is greater than that of a 1-phenyl group. This difference between the conjugation of the vinyl and phenyl groups with the pyrrole ring is likely to reflect the difference in the coplanarity of molecules.

$\mu = 1.82$ D $\quad\mu = 1.08$ D $\quad\mu = 1.36$ D
$\Delta\mu = 0.74$ D $\quad\Delta\mu = 0.46$ D

Bulky 2-substituents further decrease the $\Delta\mu$ value, indicating a steric inhibition to coplanarity of the 1-vinylpyrrole system.[316] A similar effect has been observed with 2,5-dimethylpyrrole (2.08 D) and its 1-phenyl derivative (2.0 D) having very close dipole moments.[2]

Predictably, the replacement of the 1-vinyl group by an ethyl group increases the dipole moment of the molecule relative to both the corresponding 1-vinylpyrroles and the *N*-unsubstituted pyrroles, as a result of the electron-donating inductive effect of the ethyl group. In contrast, alkyl substituents at the 3-position of the pyrrole ring produce a depolarizing effect.[316]

2.2. 1-Vinylpyrroles

TABLE 2.17. DIPOLE MOMENTS OF PYRROLES AND 1-VINYLPYRROLES (25°C, BENZENE)[316]

R^1	R^2	R^3	$\mu(D)$
H	H	H	1.82[a]
H	Me	H	1.99
H	t-Bu	H	1.95[a]
			2.20
H	Ph	H	1.61
			1.66[a]
H	2-Furyl	H	1.58[b]
H	2-Thienyl	H	1.89
			1.82[c]
H	—(CH$_2$)$_4$—		1.73
Et	H	H	2.04[d]
Et	Me	H	2.31
Et	—(CH$_2$)$_4$—		1.96
CH$_2$=CH	H	H	1.08
CH$_2$=CH	Me	H	1.50
CH$_2$=CH	t-Bu	H	1.64
CH$_2$=CH	n-Bu	n-Pr	1.20
CH$_2$=CH	Me	Et	1.36
CH$_2$=CH	—(CH$_2$)$_4$—		1.37
CH$_2$=CH	Ph	H	1.30
CH$_2$=CH	2-Furyl	H	1.42
CH$_2$=CH	2-Thienyl	H	1.18
CH$_2$=CH	Ph	n-C$_6$H$_{13}$	1.10
CH$_2$=CH	PhCH$_2$	Ph	1.38
CH$_2$=CH	2-Furyl	Me	1.22
CH$_2$=CH	2-Thienyl	n-Pr	1.19
Ph	H	H	1.36[a]

[a] Ref. 334.
[b] Ref. 2.
[c] Ref. 335.
[d] Ref. 336.

The dipole moments of 2-(2-furyl)- and (2-thienyl)pyrroles, which have planar (or near-planar) forms, are governed by the ratio of the *syn* or *anti* conformations of the rings.

X = O, S

Quantum-chemical calculations (ab initio, STO-3G) and dipole moments[335] indicate that these systems exist predominantly as planar *syn* conformers. Consequently, if the μ' values of 0.72 and 0.54 D are taken for the 2-furyl and 2-thienyl systems, respectively,[335] one could expect that, in the absence of ring interaction, the dipole moments of furyl- and thienylpyrroles would be close to 1.10 and 1.28 D. Experimental values show a considerable π-donating effect of both 1-thienyl and 1-furyl rings toward the pyrrole ring, which contribute to the total polarity of systems. In contrast, aryl substituents at the 2-position of 1-vinylpyrrole have an opposite effect on the dipole moment of the vinylpyrrole, compared with the corresponding effects on the N-unsubstituted pyrrole systems. This observation is consistent with an increased nonplanarity of the 1-vinyl and 2-aryl groups.

The polarity of 1-vinylpyrrole and its derivatives without bulky substituents depends considerably on the temperature of measurement; $\Delta\mu$ can vary between 0.11 and 0.16 D over a temperature range of -50 to $50°C$. This effect corresponds to a redistribution of the population of planar and nonplanar conformations and/or to a change in the degree of coplanarity of the system.[316] Unambiguous evidence for this rationale comes from the constancy of the dipole moment for 1-ethyl-2-methylpyrrole over a temperature range from -50 to $+80°C$. The temperature-independent dipole moment of 2-*tert*-butyl-1-vinylpyrrole has been explained in terms of its nonplanar conformation with a high rotational energy barrier.[316]

Inspection of the shape of the μ–T curves shows the greatest gradient for 1-vinylpyrrole between 50 and 100°C, whereas for its derivatives it is located near 0°C (Fig. 2.6). This interesting observation indicates that, at room temper-

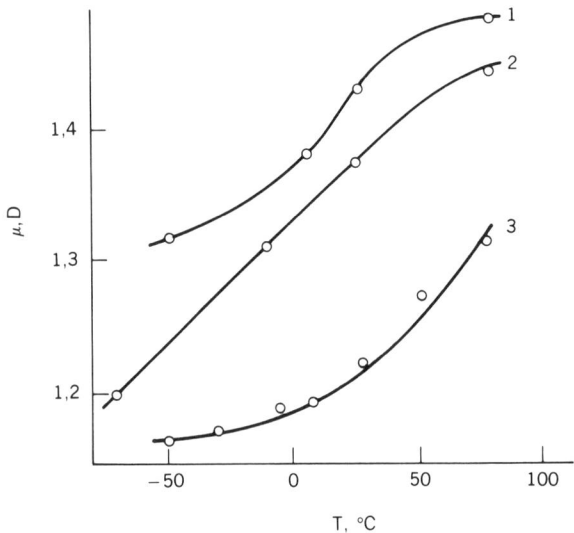

Figure 2.6. Temperature dependence of the dipole moments of 1-vinyl-pyrroles:[316] (1) 1-vinyl-pyrrole; (2) 1-vinyl-4,5,6,7-tetrahydroindole; (3) 2-methyl-1-vinylpyrrole.

ature, 1-vinylpyrrole is essentially planar, while the 2-alkyl- and 2,3-dialkyl derivatives exist as nonplanar conformers.

The dipole moments of a series of 2-trifluoroacetyl-1-vinylpyrroles have been used[314,321,323,326] to evaluate the electronic interactions between the 1-vinyl group and acyl, aryl, and heteroaryl substituents. Comparison with data for the corresponding 1-unsubstituted and 1-ethylpyrroles indicates strong conjugative effects.

2.2.3.3. Relative Basicity

The relative basicities of a series of 1-vinylpyrroles have been measured[308] using the frequency shift (Δv_{OH}) of the v O–H vibration of phenol, caused by the π-hydrogen bonding with the pyrrole (Table 2.18).

The pyrroles investigated form only weak π-hydrogen bonds (Δv_{OH} = 90–150 cm^{-1}) with phenol, indicating that they are less basic than aliphatic ethers[338] and sulfides.[339] 2-Phenyl- and 1-(2-alkylthioethyl)pyrroles produce additional bands (Δv_{OH} = 55 and 250 cm^{-1}, respectively) corresponding to H-bonding of the phenol with the benzene ring and the sulfur atom. It is evident from Table 2.18 that the 1-vinyl group noticeably reduces the basicity of pyrroles, compared with that of 1-alkylsubstituted derivatives, while 2- and 3-alkyl substituents increase the ring basicity. Not unexpectedly, there is no direct relationship between Δv_{OH} and the electron-donating abilities (Taft constants) of the alkyl groups as they would be influenced by the steric and hyperconjugation mechanisms as well.

The replacement of a methyl group at the 2-position by the phenyl group produces the negative contribution to basicity. However, instead of a further decrease in basicity, when a second phenyl group is introduced at the 3-position, the opposite effect is observed. Similar anomalies have also been noted in the UV[43] and IR[306] spectra of 2,3-diphenyl-1-vinylpyrrole. Although "through" conjugation of the vinyl group with both phenyl substituents, and particularly the 3-phenyl group, is possible, the enhanced basicity of this pyrrole, compared with 2-phenyl-1-vinylpyrroles, may result from a buttressing effect of the substituents and steric inhibition of coplanarity.

2.2.3.4. Electronic Spectra

The UV spectra of 2- and 3-alkyl-1-vinylpyrroles, as well as those of annelated derivatives, have two absorption bands at 200–206 and 247–254 nm (Table 2.19).[43] Alkylpyrroles are known[340,341] to have a single medium intensity (log ε 3.30–3.90) absorption maximum in the 208–218 nm region, which, on the basis of quantum-chemical calculations,[342,343] has been assigned to the π–π* transition. The introduction of a 1-vinyl substituent produces a longer wavelength band at 247–254 nm and a hypsochromic shift (8–12 nm) of the shorter wavelength band. These changes may be rationalized in terms of a division of electron energy levels with the possibility of new π–π* transitions.

TABLE 2.18. RELATIVE BASICITIES OF 1-VINYLPYRROLES AND RELATED COMPOUNDS

R^1	R^2	R^3	$\Delta\nu_{OH}$ (cm^{-1})	R^1	R^2	R^3	$\Delta\nu_{OH}$ (cm^{-1})
CH$_2$=CH	H	H	90	CH$_2$=CH	n-C$_6$H$_{13}$	H	121
CH$_2$=CH	Me	H	110	CH$_2$=CH	—(CH$_2$)$_4$—		120
CH$_2$=CH	Me	Me	117	CH=CH	—(CH$_2$)$_3$CH(CH$_3$)—		120
CH$_2$=CH	Me	n-Pr	120	CH$_2$=CH	—(CH$_2$)$_5$—		124
CH$_2$=CH	Me	i-Pr	120	CH$_2$=CH	Ph	H	100
CH$_2$=CH	Me	n-Am	120	CH$_2$=CH	Ph	Me	107
CH$_2$=CH	Et	H	119	CH$_2$=CH	Ph	Ph	120
CH$_2$=CH	Et	Me	126	CH$_2$=CH	Ph	H	135
CH$_2$=CH	n-Bu	H	120	Et	Ph	Me	143
CH$_2$=CH	i-Bu	H	118	Et	Ph	n-Pr	145
CH$_2$=CH	t-Bu	H	120	(CH$_2$)$_2$SEt	—(CH$_2$)$_4$—		150
CH$_2$=CH	n-Bu	n-Pr	125	(CH$_2$)$_2$SBu-n	—(CH$_2$)$_4$—		150

TABLE 2.19. UV SPECTRA OF 1-VINYLPYRROLES

		In Cyclohexane		In Ethanol	
R^1	R^2	λ_{max} (nm)	log ε	λ_{max} (nm)	log ε
Me	H	202	4.00		
		248	4.11		
n-Bu	H	202	4.08	200	4.08
		252	4.12	250	4.09
t-Bu	H	201	4.07	201	4.04
		250	4.11	249	4.06
t-Bu	Me	202	4.09		
		247	4.00		
n-C$_6$H$_{13}$	H	205	4.11		
		252	4.15		
Me	Me	202	4.08	200	4.08
		251	4.13	249	4.13
Et	Me	202	4.09	202	4.03
		254	4.10	254	3.99
n-Bu	n-Pr	201	4.09	200	4.09
		254	4.09	252	4.09
—(CH$_2$)$_4$—		203	4.09	202	4.05
		254	4.11	251	4.07
—(CH$_2$)$_3$CH(Me)—		204	4.20	202	4.17
		254	4.78	252	3.14
—(CH$_2$)$_5$—		206	4.78		
		251	4.09		
Ph	Me	200	4.30	200	4.28
		254	4.10	254	4.11
		270	4.13	268	4.12
Ph	Ph	202	4.51		
		223	4.34		
		295	4.57		
		308	4.53		
1-Ethyl-2-phenyl-3-methylpyrrole		203	4.21		
		218	4.01		
		283	3.97		

The introduction of phenyl substituents in the 2- and 3-positions of the pyrrole nucleus gives rise to a number of new very intense bands and a strong hyperchromic effect in the region of first short-wavelength maximum.

Comparison of the λ_{max} and log ε values for 2-substituted 1-vinylpyrroles shows that, with an increase in the total electron-donating power of the 1- and 2-

substituents, the intensity of the short wavelength band increases, while the long wavelength band is sensitive to the steric effect of the 2-substituent leading to a decrease in intensity and a simultaneous hypsochromic shift of the maximum. Predictably, the spectra of 2-phenyl-1-vinylpyrroles are typical of an extended conjugated system involving the benzene and pyrrole nuclei and the double bond; the spectrum of 3-methyl-2-phenyl-1-vinylpyrrole, for example, has two long wavelength absorption bands at 254 and 270 nm (log ε 4.10 and 4.13), indicating a further allowed $\pi-\pi^*$ transition. The introduction of a second phenyl group at the 3-position increases the conjugation further with the resultant bathochromic shifts of long wavelength bands (30–40 nm) and an increase in their intensity.

The solvent effects (Table 2.19) can be rationalized in terms of hydrogen bonding of the pyrrole system with ethanol with the consequent reduction in conjugation of the ring with the vinyl group. This effect is manifested by weak (1–3 nm) hypsochromic shifts, accompanied by a relatively large hyperchromic effect ($\sim 10\%$).

A quantum-chemical study of the UV spectra of 2-arylpyrroles and their 1-vinyl derivatives has been reported.[217] The first electron transitions (in π-electron approximation) for pyrroles and some other azoles, as well as for their 1-vinyl derivatives, have been calculated[344] and assigned to absorption bands. An agreement in the π-MO (molecular orbital) population found by semi-empirical and nonempirical calculations has been observed. The N-vinyl group perturbs the π-electron system of heterocycle with the pyrrolyl nitrogen atom acting as an effective channel.[344]

2.2.3.5. IR Spectra

The characteristic pyrrole and vinyl IR absorption bands of 1-vinylpyrrole have been established by comparison with the spectra of 1-unsubstituted pyrroles.[306] Five bands of high or moderate intensity, which are assignable to the vinyl group, are to be found at 584 ± 4 (m, HC=CH wagging vibrations); 859 ± 5 (s, CH_2=wagging vibrations); 946 ± 6 (s, HC=CH twisting vibrations) 1585 ± 5 (m), and 1642 ± 2 cm^{-1} (s, C=C stretching vibrations). The band near 859 cm^{-1} frequently appears as two absorption bands or as a broad single band. Absorption bands at 708 ± 4 (s, C—H deformation), which in many cases is split into two or three bands over a ~ 10-cm^{-1} range, one or two bands over a 10–20 cm^{-1} range near 1380 (s), a band at 1490 ± 4 (s) and two weak bands near 1540 cm^{-1} (ring stretching vibrations), are characteristic of pyrrole rings.[1,2,345] A band near 1300 cm^{-1} is characteristic of the C—N stretching vibrations of the 1-substituent.[345] Other absorption bands can be assigned to the substituents.

The frequency and integrated intensity of the $v_{C=C}$ vibration near 1642 cm^{-1} is not significantly influenced by alkyl substituents on the pyrrole ring.[306] It is noteworthy that even the introduction of a phenyl group at the 2-position of the ring has virtually no effect on the intensity of the $v_{C=C}$ absorption band, confirming the nonplanarity of the 1-vinylpyrrole system. However, a sharp decrease in the intensity of the band near 1642 cm^{-1} of 2,3-diphenyl-1-vinylpyrrole shows that the 3-phenyl substituent appreciably changes the electronic distribution of the 1-vinylpyrrole skeleton.

The differences in the calculated π-electron charges on the vinyl α- and β-carbon atoms of 1-vinylazoles have been compared with the observed integrated intensities. A satisfactory linear relationship has been found[320] for all compounds with a sterically fixed *syn* (*s-cis*) conformation or with symmetric coplanar conformations of the vinyl group with respect to the heterocycle, and the π-conjugation for 1-vinylpyrrole and 1-vinylindole has been estimated to be higher than that in other 1-vinylazoles.[320]

2.2.3.6. NMR Spectra

Hydrogen-1 and carbon-13 NMR spectra of 1-vinylpyrroles have been described in detail,[25-30,36] and systematic studies[20,35,46,58,188,189,201,307,309-311,318,325,331-333] have provided unique information concerning the electronic and conformational structure of 1-vinylpyrroles.

Analysis of the ^{1}H and ^{13}C NMR spectra of alkyl-1-vinylpyrroles and 1-vinyltetrahydroindoles[35] shows that the proton chemical shifts and coupling constants (Table 2.20) are in good agreement with the corresponding values known for vinylamines and pyrroles,[346,347] while the ^{13}C chemical shifts (Table 2.21) are compatible with calculated values.[347] In both the ^{1}H and ^{13}C NMR spectral studies of 2-aryl-1-vinylpyrroles, the vinyl group has been found to strengthen the relative transmission of resonance effects due to the increased conjugation in the system.[14,20,309]

Conformational analysis indicates that, although the *s-trans* conformation is preferred,[35] the *s-cis* conformer exists in significant amounts.

The ^{1}H NMR spectra of alkyl-1-vinylpyrroles have been examined[35,58] for the assessment of the sensitivity of their π-systems to the effect of alkyl substituents in the 2- and 3-positions (Table 2.20). A systematically lower shielding of the vinyl H$_2$ and H$_3$ protons and a higher one of the H$_4$ protons of 2-alkyl-1-vinylpyrroles, compared with 2,3-dialkyl derivatives, as well as a clearly expressed sensitivity of these signals to the structural change in R^1 and R^2 situated as far from these protons as four or five bond distances, have been found. A somewhat paradoxal picture is observed: the electron-donating substituents decrease the electron density in the pyrrole nucleus and increase it on the most remote atoms of the vinyl group. These long-range effects are, undoubtedly, a result of interaction of the π-electron systems of the molecule through the nitrogen atom.

The H$_2$ and H$_3$ chemical shifts increase and the H$_4$ chemical shifts decrease in going from 2-methyl derivatives to those with more bulky substituents in the 2-position (Table 2.20). This is in agreement with earlier inference on related

TABLE 2.20. ¹H CHEMICAL SHIFTS (ppm) OF REPRESENTATIVE 1-VINYLPYRROLES[a]

R^1	R^2		H_1	H_2	H_3	H_4	H_5
Et	H	b	6.61	4.40	4.84	5.89	6.77
		c	6.73	4.46	4.90	5.75	6.70
Me	Me	b	6.56	4.35	4.79	5.92	6.71
		c	6.71	4.39	4.84	5.80	6.70
Et	Me	b	6.66	4.35	4.81	5.91	6.69
Me	n-Pr	c	6.72	4.42	4.89	5.85	6.70
n-Bu	H	c	6.72	4.48	4.94	5.73	6.70
Me	i-Pr	c	6.73	4.42	4.87	5.90	6.70
t-Bu	H	c	6.74	4.52	4.94	5.74	6.70
Me	n-Am	c	6.72	4.42	4.87	5.87	6.70
n-C₆H₁₃	H	c	6.73	4.49	4.94	5.72	6.70
n-Bu	Me	c	6.73	4.44	4.87	5.65	6.88
Et	n-Pr	c	6.74	4.45	4.86	5.67	6.89
n-Bu	n-Pr	b	6.75	4.40	4.85	5.96	6.75
—(CH₂)₄—		b	6.55	4.33	4.78	5.88	6.66
—(CH₂)₃CH(Me)—		b	6.66	4.36	4.81	5.89	6.70

[a] In this series the J values are relatively constant: $^3J_{1,2}$ 8.8–9.0; $^3J_{1,3}$ 15.5–16.0; $^2J_{2,3}$ 0.8; $^3J_{4,5}$ 2.5–3.1 Hz.
[b] Neat, 100 MHz (Ref. 35).
[c] 10% CCl₄, 80 MHz (Ref. 58).

heterovinyl systems [348–351] and implies steric inhibition of p–π conjugation in the N-vinyl group by neighbouring alkyl substituents. Similar effects have been observed in vinyl ethers and sulfides,[348] and alkoxy-[350] and alkylthiobenzenes.[351] The results[58] confirm the existence of a nonplanar conformation of 1-vinylpyrroles.

A dependence of ^{13}C chemical shifts for alkyl-1-vinylpyrroles on the alkyl substituents R^1 and R^2 provides information on the conjugation of the N-vinyl group and its relative orientation with respect to the pyrrole ring.[352] As with ^1H NMR spectra, the greatest sensitivity to the alkyl substituents is displayed by the chemical shifts of the C_5 and C_β atoms. The sensitivity of ^{13}C chemical shifts to the change in charge is much higher in 1-vinylpyrroles than in the parent 1-unsubstituted compounds.[14,20,311]

The electron density at the C_3 and C_4 atoms of 1-vinylpyrrole is appreciably lower compared with that for the N-unsubstituted pyrrole (δ increases by

TABLE 2.21. ^{13}C CHEMICAL SHIFTS (ppm) OF ALKYL-1-VINYLPYRROLES

R^1	R^2	C$_2$	C$_3$	C$_4$	C$_5$	C$_\alpha$	C$_\beta$
H	H	118.2[a]	110.1	110.1	118.2	132.8	95.9
Me	H	127.7	108.4	109.3	115.5	130.3	96.5
Et	H	134.1	106.9	109.6	115.93	130.5	97.0
t-Bu	H	140.4	106.6	108.4	118.2	133.4	98.5
Et	Me	130.7	115.3	111.7	114.5	130.2	95.5
n-Bu	n-Pr	128.5	121.3	110.4	114.7	130.9	95.4
—(CH$_2$)$_4$—		127.2	118.9	109.6	114.5	130.4	94.7
—(CH$_2$)$_3$CH(Me)—		131.6	118.0	109.3	114.9	130.4	94.7

[a] ^{13}C Chemical shifts of N-unsubstituted pyrrole are 118.5, 108.2, 108.2, and 118.5 ppm.

3.88 ppm), while there is a slight increase in the charge on the C$_2$ and C$_5$ atoms and the chemical shift is diminished by ~0.64 ppm.

The effect of the 2-alkyl groups on the ^{13}C chemical shift of the ring carbon atoms differs from their effect on the vinyl C$_\beta$ atoms (Table 2.21). In agreement with the increase in their inductive electron-donating effect (H < Me < Et < t-Bu), the C$_3$, C$_4$, and C$_5$ atoms are shielded (except for C$_5$, when R^1 = t-C$_4$H$_9$). In contrast, the vinyl C$_\beta$ atom is deshielded. This suggests that a progressive weakening of the π-conjugation of the 1-vinyl group with the ring occurs with the increasing inductive effect of the 2-alkyl group. Thus, in 2-ethyl-1-vinylpyrrole, the C$_5$ and C$_\alpha$ nuclei are deshielded by 0.42 and 0.22 ppm, compared with the corresponding methyl derivative, while for the tert-butyl derivative, the deshielding increases unevenly and amounts to 2.47 and 2.86 ppm. This latter observation indicates that, with increased steric interaction, the nonplanar rotamer becomes important.

In the series of alkyl-1-vinylpyrroles, the chemical shifts of C$_\beta$ and vinylic β-proton (H$_2$), which is trans with respect to the C$_\alpha$—N bond, show an approximately linear relationship with the steric constants of 2-substituents:[20, 307]

$$\delta C_\beta = 96.5 - 1.0(\pm 0.08) E_s^\circ, \quad r = 0.97, \quad s = 0.24$$

$$\delta H_2 = 4.5 - 0.04(\pm 0.008) E_s^\circ, \quad r = 0.96, \quad s = 0.015$$

Comparison of the ^{13}C chemical shifts of 2-alkyl- and 2,3-dialkyl-1-vinylpyrroles indicates that the effect of the 3-alkyl group is entirely inductive, producing a deshielding shift on the C_4 atom, but a shielding shift on the C_5 and C_β atoms.

This is also supported by the decreased negative value of geminal coupling constant, $^2J_{23}$ (from -1.2 to 0 Hz), which serves as a reliable qualitative criterion of a change in the electron density on the β-carbon in structurally similar compounds of the vinyl series.[20] An alternative probe, which allows quantitative characterization of the conformational structure of 1-vinylpyrroles, is the long-range coupling constant between the vinyl β-trans-proton and the proton in the 3-position of the pyrrole ring.[14,307]

The Overhauser effect has been used as a measure of the dihedral angles (ϕ) between the vinyl group and the pyrrole ring [1-vinylpyrrole (0°), 2-tert-butyl-1-vinylpyrrole (55°), and 2-methyl-5-phenyl-1-vinylpyrrole (90°)] and a linear dependence of the geminal coupling constant $^2J_{23}$ on $\cos^2 \phi$ has been observed, which indicates the calculated values to be close to real.[14,20,307,310]

Long-range coupling between the fluorine nuclei and the proton at C_4 in the ^{1}H and ^{19}F NMR spectra of 2-substituted 5-trifluoroacetylpyrroles and their 1-vinyl derivatives provides information concerning the conformational structure of these compounds with respect to both the C_α—N and C_5—$COCF_3$ bonds.[188,189] The size of the long-range coupling constant for 2-furylpyrroles, compared with that of their 1-vinyl derivatives, indicates that the introduction of the vinyl group reduces the coplanarity of the heterocycles by their steric interaction.[46]

Nitrogen-14 chemical shifts are sensitive to the conjugative effects between the vinyl group and the pyrrole nucleus of 1-vinylpyrroles. With sterically hindered derivatives and weakened π-conjugation, shielding of the annular nitrogen atom is increased.[311,318] No allowance has been made for the inductive effect of the alkyl groups, which make a much lower contribution to the shielding of the nitrogen nuclei.[353,354]

TABLE 2.22. ^{14}N CHEMICAL SHIFTS (ppm)[a] OF 1-VINYLPYRROLES AND 1-METHYLPYRROLE

	δ^{14}N
1-Vinylpyrrole	206.4[b]
	205.6[c]
2-Methyl-1-vinylpyrrole	208.5[b]
	207.5[c]
1-Vinyl-4,5,6,7-tetrahydroindole	216.9[b]
	212.9[c]
1-Methylpyrrole	231.4[b]
	231.6[c]

[a] Referred to neat nitromethane, upfield shift corresponds to deshielding.
[b] Neat.
[c] In acetone (2.0 M solution).

2.2. 1-Vinylpyrroles

The annular nitrogen atoms of 1-vinylpyrroles are considerably deshielded compared with those in 1-methylpyrrole (Table 2.22).

Quantum-chemical data (INDO/S) provide evidence for the increased shielding of the nitrogen atom in 1-vinylpyrroles as the vinyl group is rotated out of the pyrrole ring plane.[318]

2.2.4. Applications

2.2.4.1. Biological Properties

The toxicity of several 1-vinylpyrroles has been evaluated (Table 2.23).

TABLE 2.23. TOXICITY OF 1-VINYLPYRROLES AND THEIR DERIVATIVES

Pyrrole	LD_{50} (mg kg^{-1})
2-Methyl-1-vinyl-pyrrole	800
2,3-Dimethyl-1-vinylpyrrole	1250[a]
3-Ethyl-2-methyl-1-vinylpyrrole	800[b]
2-Ethyl-3-methyl-1-vinylpyrrole	50
2-n-Butyl-3-n-propyl-1-vinylpyrrole	30
1-Vinyl-4,5,6,7-tetrahydroindole	600
7-Methyl-1-vinyl-4,5,6,7,-tetrahydroindole	25
2-Phenyl-1-vinylpyrrole	250[a]
3-Methyl-2-phenyl-1-vinylpyrrole	>800[c]
3-Ethyl-2-phenyl-1-vinylpyrrole	2000
2-Phenyl-3-n-propyl-1-vinylpyrrole	400[a]
2-Phenyl-3-isopropyl-1-vinylpyrrole	300[a]
3-n-Butyl-2-phenyl-1-vinylpyrrole	600
3-n-Amyl-2-phenyl-1-vinylpyrrole	800
3-n-Hexyl-2-phenyl-1-vinylpyrrole	>800[d]
3-n-Nonyl-2-phenyl-1-vinylpyrrole	3000
2-(4-Tolyl)-1-vinyl-pyrrole	3000
2-(4-Ethylphenyl)-1-vinylpyrrole	600
2-(4-Chlorophenyl)-1-vinylpyrrole	300
2-Trifluoroacetyl-1-vinyl-4,5,6,7-tetrahydroindole	>800[e]
5-Trifluoroacetyl-2-phenyl-1-vinylpyrrole	1700
5-Fluoroacetyl-3-methyl-2-phenyl-1-vinylpyrrole	500

[a] For comparison, the toxicity of the related 1-alkyl derivatives were 1-ethyl-2-phenylpyrrole 600 mg kg^{-1}; 1-ethyl-3-methyl-2-phenylpyrrole 400 mg kg^{-1}; 1-ethyl-2-phenyl-3-n-propylpyrrole 1300 mg kg^{-1}; 1-ethyl-2-phenyl-3-isopropylpyrrole 600 mg kg^{-1}; 1-(1-methoxyethyl)-2-phenylpyrrole 1500 mg kg^{-1}; 2-phenyl-1-(1-propargyloxyethyl)pyrrole 800 mg kg^{-1}, depression at 600–800 mg kg^{-1}, hyperexcitability, locomotor activity at 200–400 mg kg^{-1}; 1-(2-n-butylthioethyl)-2,3-dimethylpyrrole 1900 mg kg^{-1}.

[b] Depression, decreased mobility.

[c] No signs of intoxication.

[d] At dosage 400, 800 mg kg^{-1} locomotor excitement, hyperexcitability to irritants.

[e] Depression, dyspnea.

The general effect indicates that certain 1-vinylpyrroles are of interest as precursors for neurostimulators; 2-phenyl-1-(1-propargyloxyethyl)pyrrole derived from 2-phenyl-1-vinylpyrrole, for example, stimulates motor activity, increases excitation, and induces a prespasmodic state (tail tremor, Straub phenomenon).

Ichthyotoxicity examinations of a series of 2-alkyl- and 2,3-dialkyl-1-vinylpyrroles and their derivatives have been carried out.[14] It has been found that only 1-(2-alkylthioethyl)-2,3-dialkylpyrroles possess any ichthyotoxicity, although the activity is inferior to that of known ichthyocides. Other alkylated 1-vinylpyrroles are practically harmless under the experimental conditions. However, in contrast, 2-aryl-1-vinylpyrroles and related derivatives have a clearly defined ichthyotoxicity (Table 2.24).

TABLE 2.24. ACUTE ICHTHYOTOXICITY OF 2-ARYL-1-VINYLPYRROLES AND THEIR DERIVATIVES[14]

R^1	R^2	R^3	Exposure Time (h)	Lethality (%)
CH_2=CH	Ph	H	120	100
CH_2=CH	4-EtC$_6$H$_4$	H	48	100
CH_2=CH	4-MeOC$_6$H$_4$	H	120	100
CH_2=CH	Ph	n-Am	120	70
Et	Ph	H	48	100
EtS(CH$_2$)$_2$	Ph	H	24	100

Replacement of the 1-vinyl group by an alkyl group increases the ichthyotoxicity; the toxicity of 1-(2-ethylthioethyl)-2-phenylpyrrole, for example, is 5 times greater than that of 2-phenyl-1-vinylpyrrole. A similar increase in ichthyotoxicity of the 1-(2-alkylthioethyl) derivatives, compared with the 1-vinylpyrroles, has also been noted for alkylpyrroles. The sharp increase in ichthyotoxicity is accompanied by a sharp decrease in the toxicity for warm-blooded animals (Table 2.24). The ease of oxidative hydrolytic degradation of the pyrrole moiety, which should render such compounds ecologically acceptable, suggests that 1-(2-alkylthioethyl)-2-arylpyrroles are promising starting materials for the commercial production of ichthyotoxic preparations.

The activity of the series of 1-vinylpyrroles and some of their derivatives with respect to 11 strains of bacterial and fungal species has been determined.[192] The minimal microbial growth-inhibiting concentrations of the pyrroles are

1–125 µg ml^{-1} for staphylcoccus, 125–1000 µg ml^{-1} for colon bacillus, and 62.5–1000 µg ml^{-1} for typhoid and dysentery pathogens. Blue pus bacillus and *Proteus* are sensitive to the vinylpyrroles at doses of 250–1000 and 250–500 µg ml^{-1}, respectively. A slightly higher sensitivity is observed with anthracoid bacillus (7.8–62.5 µg ml^{-1}) and yeast-like fungi of the *Candida* genus (fungistatic doses range from 1.98 to 125 µg ml^{-1}). 1-Vinyl-4,5,6,7-tetrahydroindole shows a higher activity relative to staphylococcus and *Candida* fungi. An especially high activity against the latter and anthracoid bacilli is observed with 2-phenyl-1-vinylpyrrole.

Antiviral activity of certain 4,5,6,7-tetrahydroindole and octahydroindole derivatives against influenza virus, classical fowl plague virus, paragrippe, Venezuela equine encephalomyelitis, adenovirus, herpes simplex (12-strain), dermovaccina, and coliphages has been reported.[374,375] The reduced indoles exhibited different toxicity relative to cell culture; the highest tolerant concentration spans 25–800 µg ml^{-1}. 1-Vinyl-4,5,6,7-tetrahydroindole shows moderate activity with respect to the influenza virus, while 1-(2-triethylsilylethyl)-4,5,6,7-tetrahydroindole shows moderate activity with respect to the influenza virus, while 1-(2-triethylsilyethyl)-4,5,6,7-tetrahydroindole is moderately active against herpes virus and displays only slight activity toward adenovirus and vaccine.

The antiviral activity of some 1-vinylpyrroles and their adducts with *n*-butylmercaptans has been evaluated.[14] 2-Methyl-, 2-methyl-3-*n*-propyl-, and 3-isopropyl-2-methyl-1-vinylpyrroles showed activity against herpes virus, and 3-*n*-amyl-2-methyl-1-vinylpyrrole is slightly active against influenza virus. A moderate activity against herpes virus is observed with 1-ethyl-2-phenyl-3-*n*-propylpyrroles and 1-(2-ethylthioethyl)-2-phenylpyrrole, the latter also being moderately active against pox virus.

A systematic research of the toxicity and the repellent and attractive properties of the series of 1-vinylpyrroles and their derivatives with respect to forest pests has been reported.[376] 2-Methyl-3-phenyl-1-vinylpyrrole, 3-*n*-hexyl-2-phenyl-1-vinylpyrrole, and 2-phenyl-5-trifluoroacetyl-1-vinylpyrrole have been shown to have an attractive effect. In particular, 3-*n*-amyl-2-methyl-1-vinylpyrrole, and 3-*n*-nonyl-2-phenyl-1-vinylpyrrole are attractants for bark beetle and weevil, respectively.

1-Vinylpyrroles and several derivatives have also been tested as contact insecticides against bark beetle and long-horn beetle larvae.[14]

2-Phenyl-1-vinylpyrrole, 3-*n*-hexyl-2-phenyl-1-vinylpyrrole, 3-*n*-octyl-2-phenyl-1-vinylpyrrole, 2-(4-chlorophenyl)-1-vinylpyrrole, 2-phenyl-1-(2-propargyloxyethyl)pyrrole, 5-trifluoroacetyl-2-phenyl-1-vinylpyrrole, 3-methyl-2-phenyl-1-vinylpyrrole, and 2-(4-methoxyphenyl)-1-vinylpyrrole have been found to be active insecticides against house fly, tick, rice weevil, and beet aphid.[14]

A number of 2-aryl-1-vinylpyrroles and their derivatives have been tested as herbicides and plant growth regulators[14] for oats, soya, radish, sunflower, and beet. The highest herbicidal effect was produced by 2-phenyl-1-vinylpyrrole, 3-*n*-hexyl-2-phenyl-1-vinylpyrrole, 3-*n*-nonyl-2-phenyl-1-vinylpyrrole,

2-phenyl-5-trifluoroacetyl-1-vinylpyrrole, 3-methyl-2-phenyl-5-trifluoroacetyl-1-vinylpyrrole, 2-(4-methoxyphenyl)-5-trifluoroacetyl-1-vinylpyrrole, and 2-(4-methoxyphenyl)-5-trifluoroacetyl-1-vinylpyrrole. Growth-regulating activity was clearly displayed by 3-hexyl-2-phenyl-1-vinylpyrrole, 2-(4-chlorophenyl)-1-vinylpyrrole, 2-phenyl-1-(1-propargyloxyethyl)pyrrole, and 3-methyl-2-phenyl-5-trifluoroacetyl-1-vinylpyrrole.

A study of the effect of some 2-aryl-1-vinylpyrroles on root rot (helmintosporiosis) of oat, powdery mildew of cucumbers, late blight of tomatoes, and rust on wheat stalk has shown that 2-(4-chlorophenyl)-1-vinylpyrrole produces a twofold decrease in the lesion rate for root rot, compared with that for controls, while the field germination rate remains unchanged.[14]

A number of 2-aryl-1-vinylpyrroles and related compounds have been tested for activity against bacteria and fungal mycelium that cause various diseases of plants, and against organisms destroying nonmetallic materials.[14] The highest activity was found with 2-phenyl-1-vinylpyrrole and 2-(4-chlorophenyl)-1-vinylpyrrole, while 2-phenyl-1-vinylpyrrole, 3-n-octyl-2-phenyl-1-vinylpyrrole, 2-phenyl-1-(1-propargyloxyethyl)pyrrole, 2-phenyl-5-trifluoroacetyl-1-vinylpyrrole, and 2-(4-methoxyphenyl)-5-trifluoroacetyl-1-vinylpyrrole possess weak nematocidal activity. A substantial activity against nematodes was exhibited by 2-(4-chlorophenyl)-1-vinylpyrrole.

Investigations into the antithrombosic and nociceptive activity of a series of 1-vinylpyrroles have shown[377] that the 2-aryl derivatives have maximum activity. Some 1-vinyl derivatives of pyrrolecarboxylic acids and their sodium salts show a distinct anticonvulsive (corazol) effect.[242] These compounds exert no inhibiting effect on the gram-negative colon flora.

2.2.4.2. Polymers

The valuable technological properties of polymers of N-vinylcarbazole are well established. It is well known[6] that the addition of small amounts of N-vinylcarbazole to alkene monomers considerably improves the mechanical and thermal properties of the polymers. For example, polyisobutylene with 2% N-vinylcarbazole is as resilient as natural rubber and has a thermal stability exceeding that of polyisobutylene and rubber,[6] while the strong highly flexible copolymer of N-vinylcarbazole with ethylene is employed as an interlayer for shatterproof glass. The N-vinylcarbazole homopolymer is a better insulator than polystyrene and has found particular use in high-frequency electrical engineering.[6].

Over the last two decades, the photosemiconductive properties of polyvinylcarbazole have been described in many publications and patents (e.g., Refs. 355–373 and references cited therein).

To date, the practical applications of simple 1-vinylpyrroles, as monomers, are almost unexploited.

Poly-1-vinylpyrroles, with the exception of poly-2-(2-thienyl)-1-vinylpyrrole, are extremely soluble in organic solvents and form transparent homogeneous films from their solutions in benzene or dioxane. The copolymer of 1-vinyl-4,5,6,7-tetrahydroindole with an epoxyvinyl ether produces a film, the strength and elasticity of which is improved by the presence of ether groups.[14] Sensitizers, such as 2,4,7-trinitrofluorenone, can be added to the polymer without loss of its physical properties and, when charged to a potential of 120–190 V, possess electrophotosensitivity.[14,378] Photosensitive copolymers of 1-vinyl-4,5,6,7-tetrahydroindole with 2-(vinyloxy)ethylamine have also been reported.[293]

Radical polymerization of 1-vinylpyrroles and copolymerization with other vinyl monomers give polymers that can be cross-linked with opening of the pyrrole rings to form electroconductive polymer nets.[298]

The copolymers of 1-vinyl-4,5,6,7-tetrahydroindole with N-vinylpyrrolidone have an appreciable antisilicotic effect; i.e., they protect cell membranes from silica and enhance the level of immune protection. Copolymers with epoxyvinyl ethers effectively protect plants from xylophages, fungi, and diseases, while retaining a low toxicity to warm-blooded organisms (approximately 2500–3000 mg kg^{-1}).[292,379,380]

Water-soluble[294] and spin-labeled[381] copolymers of 1-vinylpyrroles have been obtained. The copolymer of 1-vinyl-4,5,6,7-tetrahydroindole with N-vinylpyrrolidone (M $\sim 1.5 \times 10^4$) has been suggested as a highly selective foamer for lead–zinc ore floatation.[290]

2.3. 2-VINYLPYRROLES

2.3.1. Synthesis

2.3.1.1. From 2-Formylpyrroles and CH-Acids

It is well established that the most general synthesis of 2-vinylpyrroles is the base-catalyzed condensation of 2-formylpyrroles with activated methylene (or methyl) groups. This condensation reaction has been extensively reviewed[1,2] (see also Chapter 1).

$$\text{pyrrole-CHO} \xrightarrow[-H_2O]{CH_2XY} \text{pyrrole-CH=CXY}$$

X, Y = COR, CO$_2$R, CN, NO$_2$, etc.

2.3.1.1.1. Reactions with Ketones

2-Formylpyrroles undergo a normal base-catalyzed aldol-type condensation with aliphatic, aromatic, and heteroaromatic ketones to yield the correspondingly substituted 2-vinylpyrroles (Scheme 19).[1,2,382-393] In the simplest case, 2-formylpyrrole reacts with acetone to yield 1-(2-pyrrolyl)-but-1-en-3-one (**194a**) and 1,4-di(2-pyrrolyl)-but-1-en-3-one (**194b**).[2,382,383]

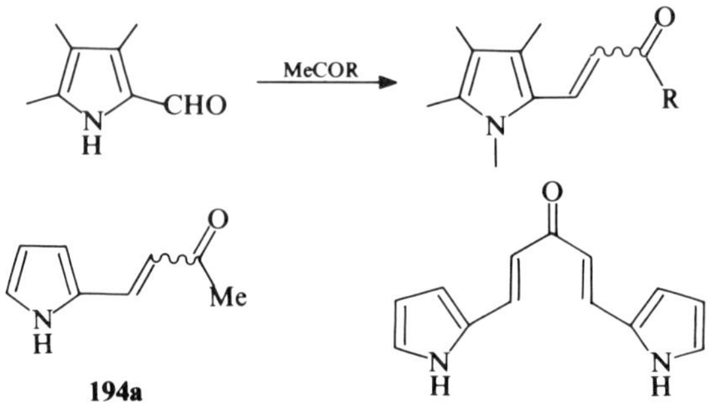

Scheme 19

The reaction of 2-formylpyrrole with 2-(dimethylamino)ethylphenyl ketone leads to the benzoyl-3H-pyrrolizines (**196** and **197**) via the 2-vinylpyrrole (**195**).[2,394]

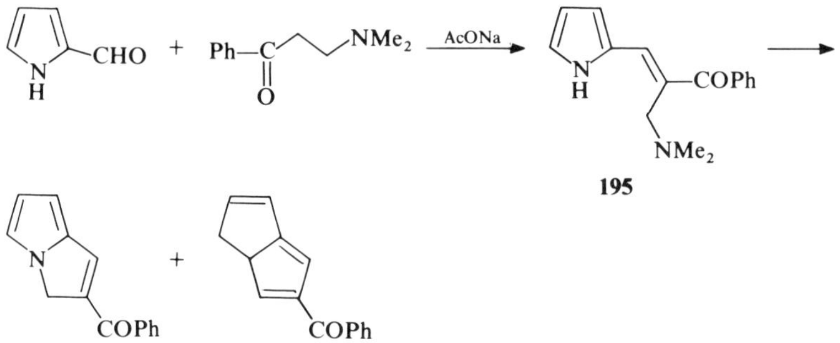

Condensation of 2-formylpyrrole with 2-acetylpyrrole produces 1,3-di(2-pyrrolyl)-prop-1-en-3-one (**198**) in moderate yield.[384,385]

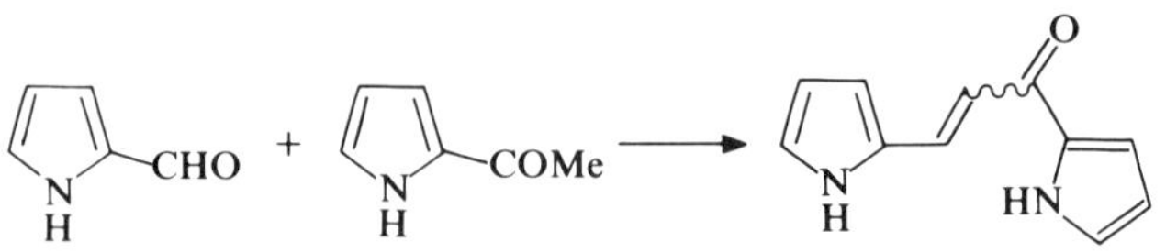

Alkali-sensitive nitropyrrole-2-aldehydes have been reported to condense with 1-indanone and 1-tetralone in the presence of 35% orthophosphoric acid to form pharmacologically interesting derivatives.[2,393]

2.3.1.1.2. Reactions with Carboxylic Acids, Esters, and Nitriles

The Perkin condensation of 2-formylpyrrole with 2-pyrrolylacetic acid gave a tarry product from which can be isolated methyl Z-2,3-di(2-pyrrolyl)acrylate (**199**) (5% yield).[395]

2-Formylpyrroles readily undergo the Knoevenagel condensation yielding 2-vinylpyrroles,[2,391,392,396–404] for example, 2-vinylpyrroles (**201**), have been prepared from 2-formylpyrroles (**200**) and malonic ester. Similarly, the 2-formylpyrrole (**202**) condenses with alkyl hydrogen malonate to form the pyrrolylacrylic ester (**203**) (see Ref. 2 and references cited therein).

Virtually without exception, all pyrrole- (or indole)-2-aldehydes react with malonic acid, malonates, succinates, cyanoacetates, and malonitrile in the presence of bases to furnish the corresponding 2-vinylpyrroles (or indoles) having the ester, nitrile, or anhydride substituents in the vinyl moiety.[2,405–412]

Reactions of formylpyrroles with malononitrile[413–416] or cyanoacetic esters[417,418] are often employed as a method for protection of the aldehyde function in the synthesis of porphyrins. The aldehyde group is readily regenerated on treatment with aqueous NaOH, but, unfortunately, the cyanovinyl and acrylic ester groups strongly deactivate the pyrrole ring toward electrophilic attack and, consequently, hamper the formation of the intermediate dipyrrolyl-methanes.[2]

1-Methyl-5-methylthio-pyrrole-2-aldehyde has been condensed with malononitrile and with methyl cyanoacetate to give 2-cyano-3-(1-methyl-5-methylthio-2-pyrrolyl)acrylonitrile (**204**a) and methyl 2-cyano-3-(1-methyl-5-

methylthio-2-pyrrolyl)acrylate (**204b**). Oxidation of **204a,b** furnishes the sulfones (**205a,b**). The analogous 5-phenylthio derivatives have also been prepared.[419]

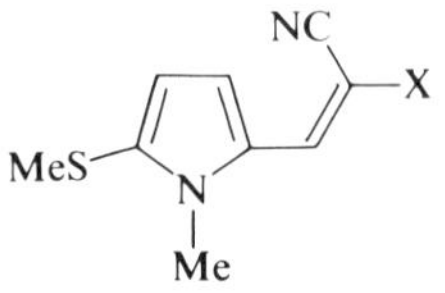

204
205 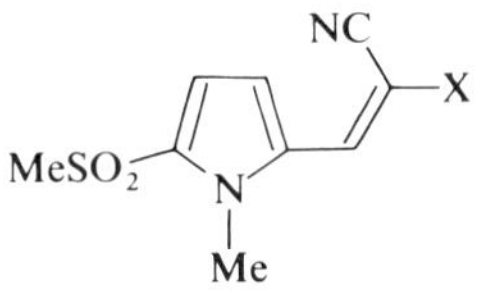
a X = CN
b X = CO_2Me

2-Formyl-1-methylpyrrole (**206**) gives excellent yields of 2-(2-aryl-2-cyanovinyl)pyrroles (**207**) on treatment with the arylacetonitriles under the catalytic action of benzyltrimethylammonium hydroxide,[420] and the Stobbe reaction of (**206**) with succinates or glutarates leads to the expected 2-vinylpyrroles (**208**) in good yield.[2,421,422]

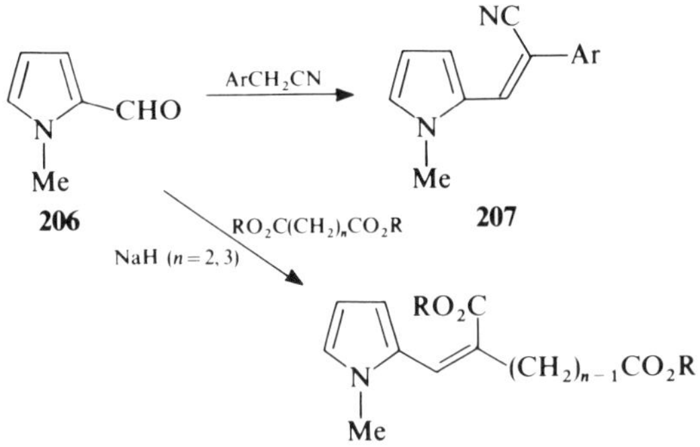

5-Ethoxycarbonyl-3H-pyrrolizine (**209a**) and its isomer (**209b̄**) are formed in the base-catalyzed reaction of 2-formylpyrrole with ethyl acrylate via the initial formation of the 1-pyrrolylpropanoic ester intermediate.[394]

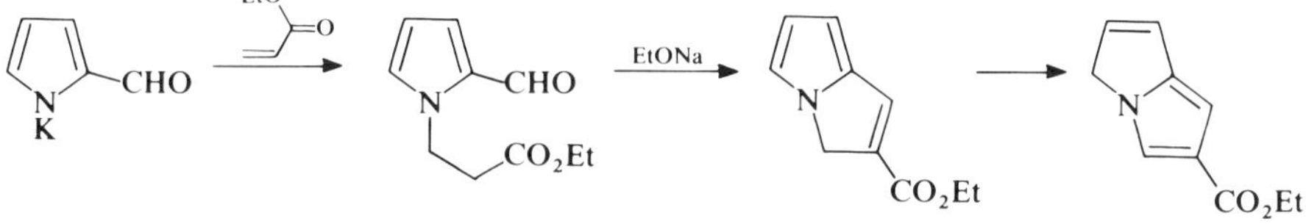

The sodium salt of 2-formylpyrrole reacts with phenylacetyl chloride to furnish 2-phenylpyrrolizin-3-one (**211**) (see Ref. 2 and references cited therein), probably via the intermediate acylation product (**210**).

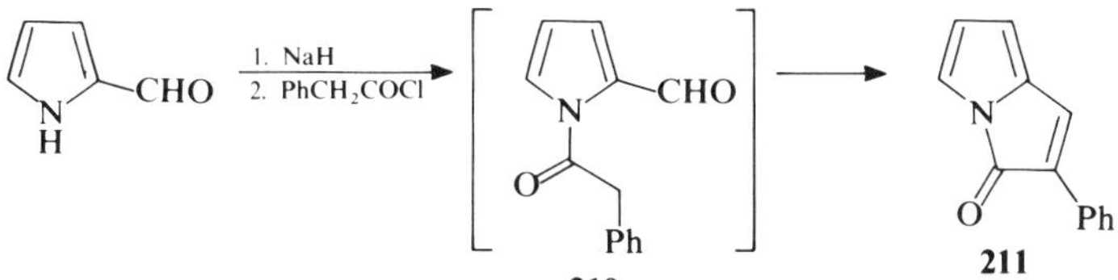

210 **211**

2.3.1.1.3. Reaction with Other CH-Acids

Two isomers (*E* and *Z*) of the azlactone (**212**), 2-phenyl-4-(2-pyrrolylmethylene)-5-oxazalone, have been isolated from the reaction of 2-formylpyrrole with hippuric acid.[423] Hydrolysis of the azlactones yields the expected acrylic acid (**213**) (*E*- and *Z*-isomers) and 4-phenylpyrrolo[1,2-*c*]pyrimidine-2-carboxylic acid (**214**). The corresponding reaction of 2-formylpyrrole with hydantoin (**215a**)[407,424,425] and rhodamine (**215b**)[424] leads to the expected condensation products (**216a,b**).

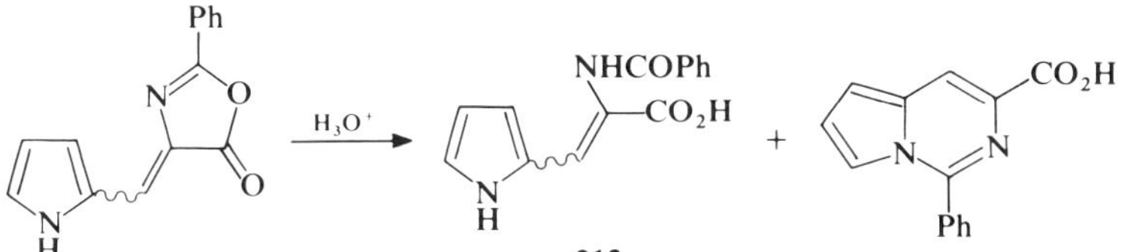

212 **213** **214**

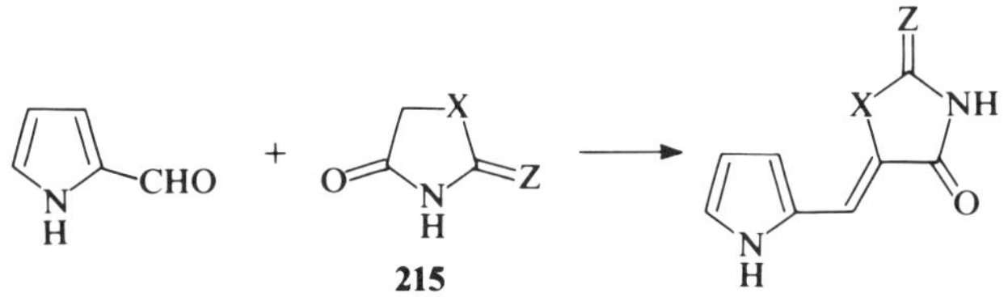

215 **216**

a X = NH, Z = O
b X = Z = S

Condensation of 2-formylpyrroles with barbituric acid and isopropylidene malonate (Meldrum's acid) produce **217** and **218**, respectively.[400,426,427]

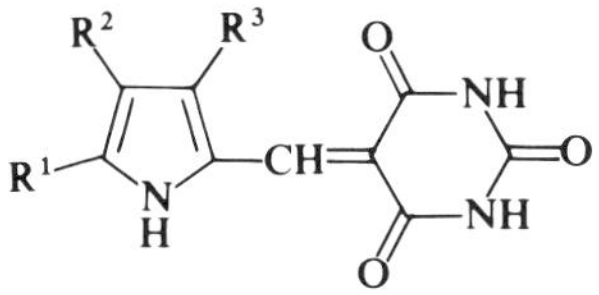

217
$R^1 = R^3 = H$, Me
$R^2 = H$, Me, CO_2Et

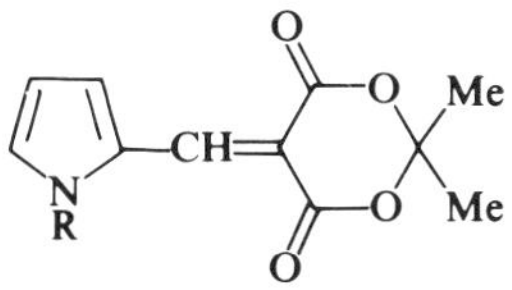

218
R = H, Me

The base-catalyzed condensation of 2-formylpyrroles with Δ^3-pyrroline-2-ones leads to the yellow 2,2'-dipyrrolylmethen-5-ones, e.g., neoxanthobilirubin acid (**219**) containing the 2-vinylpyrrole moiety.[1,2,428] These compounds are important intermediates in the total synthesis of bile pigments.

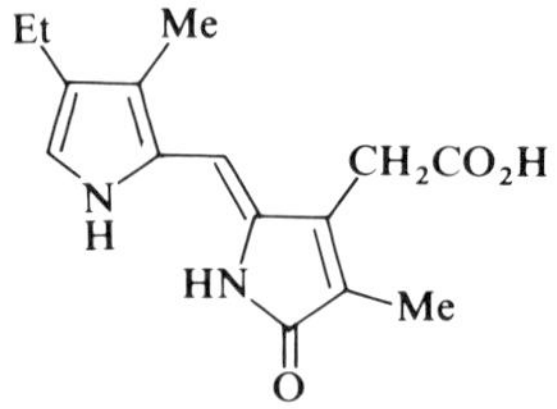

219

Formylpyrroles (and indoles) have also been reported to react with nitromethane[429-431] and cyclopentadiene[432] to yield corresponding vinylpyrroles (and indoles).

A number of heterocyclic analogs of 2-styrylpyrroles (e.g., **220** and **221**) have been synthesized by the condensation of formylpyrroles with heterocyclic quaternary salts.[433-437]

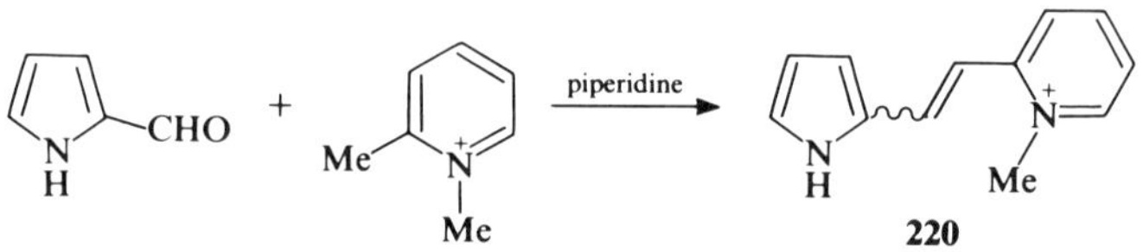

220

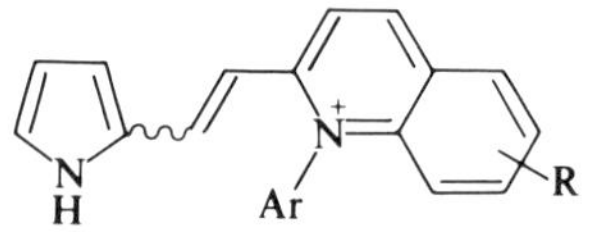

221

R = H, 6-Me, 6-MeO, 5,6-benzo,
Ar = Ph, 4-MeC$_6$H$_4$, 4-MeOC$_6$H$_4$,
2-ClC$_6$H$_4$, α-C$_{10}$H$_7$, β-C$_{10}$H$_7$

2.3.1.2. Via the Wittig Reaction

The Wittig reaction has to be fairly efficient in conversion of formyl- and acylpyrroles to corresponding vinylpyrroles.[1,2,394,403,438-458]

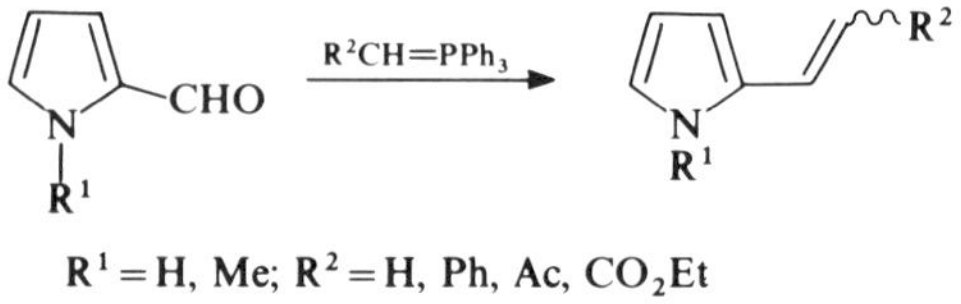

R^1 = H, Me; R^2 = H, Ph, Ac, CO$_2$Et

2.3. 2-Vinylpyrroles

2-Formylpyrrole and 2-formyl-1-methylpyrrole, for example, react readily with both highly reactive phosphoranes and with resonance-stabilized phosphoranes to give the expected 2-vinylpyrroles.[441] Under similar conditions, 2-acetylpyrrole reacts more slowly and only with the reactive benzylidenetriphenylphosphorane and methylenetriphenylphosphorane.[2,441] The successful application of the reaction with other acylpyrroles requires vigorous conditions; 2-acetylpyrrole, for example, yields the E- and Z-isomers of the acrylic ester (**222**) in a ratio of about 2 : 1 only after 12 h at 120°C (see Ref. 2 and references cited therein). Initially, only the E-isomers of the corresponding substituted 2-vinylpyrroles were isolated, possibly as a result of isomerization of the Z-isomers during the isolation procedure. Later, the products of the Wittig reactions were shown to consist of the E- and Z-isomers, although the former predominated.[2,403,450-453]

R^1 = H, Me; R^2 = H; R^3 = CO_2Et

It is also possible to synthesize both the E- and Z-isomers of 2-styrylpyrrole.[2,441,448] In these reactions, although sodium ethoxide may generally be used to produce the phosphoranes from the phosphonium salts, it is imperative that sodium hydride be employed in the preparation of 2-(4-substituted styryl)pyrroles, in order to obviate the unfavorable equilibrium between the pyrrole and the phosphorane. The E-2-(2-pyrrolyl)acrylic esters and E-2-styrylpyrroles have also been prepared by the Wadsworth-Emmons version of the Wittig reaction using alkyldiethylphosphonates.[2,441,442]

3H-Pyrrolizine (**223a**) is obtained from 2-formylpyrrole and vinyltriphenylphosphonium salt in yields of up to 87%.[445] The same reaction with 2-acetylpyrrole affords the methyl derivative (**223b**), but only tars are produced in the reaction with 2-benzoylpyrrole.[2,443-445]

a R=H
b R=Me

3H-Pyrrolizine (3H-pyrrolo[1,2-a]pyrrole) is fundamental to numerous alkaloids and certain pheromones. Hence, its synthesis and that of its derivatives is of considerable importance.[1] A series of 3H-pyrrolizines have been prepared from 2-formylpyrrole by the Wittig reaction with allyl-, methallyl-, and crotyltriphenylphosphonium salts.[2,446] With the exception of the reaction with the crotyl compounds, four isomers (**225–228**) are usually formed. However, in some cases (e.g., R^1 = Ph, R^2 = H) the intermediate 1-(2-pyrrolyl)-but-1,3-diene (**224**) fails to ring-close to yield the pyrrolizine even at elevated temperatures.[444]

In contrast, the reaction of isopropenyltriphenylphosphonium salts with 2-formylpyrrole yields only the 2-methyl-3H-pyrrolizine (**229**), while the same reaction with the 1-ethoxycarbonylcyclopropyltriphenylphosphonium salt provides a route to the dihydroindolizine (**230**).[448]

α-(2-Pyrrolylmethylidene)-γ-butyrolactone is obtained from the reaction of triphenylphosphoranylidene-γ-butyrolactone and 2-formylpyrrole.[452]

The two main pigments of the fungus imperfectus, *Wallemia sebi*, have been identified as 2,10-dimethyl-5-hydroxy-11-(2-pyrrolyl)-undeca-4,6,8,10-tetraen-3-one (**231**) and 2,2-dimethyl-5-[5-methyl-6-(2-pyrrolyl)hexa-1,3,5-trienyl]furan-3(2H)-one (**232**).[454] As models for their synthesis, a series of related 2-vinylpyrroles (**233–241**) have been prepared by Wittig reactions of the appropriate phosphoranes with formylpyrroles.[454]

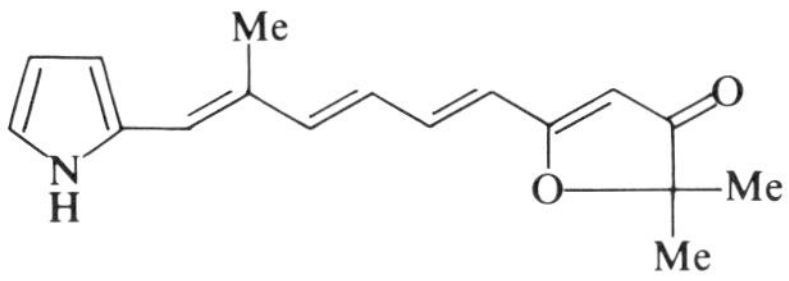

2.3. 2-Vinylpyrroles

232

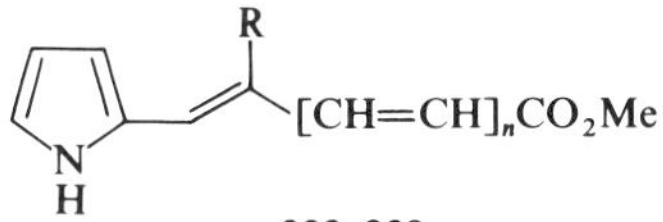

233–239

Compound	n	R
233	0	H
234	0	Me
235	1	H
236	1	Me
237	2	H
238	2	Me
239	3	Me

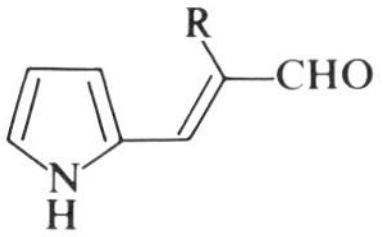

240 R = H
241 R = Me

As noted above, the condensation of 2-formylpyrrole with methoxycarbonylmethylenetriphenylphosphorane (**242**) in benzene gives the enoate (**233**). Under the same conditions the reaction of 2-formylpyrrole with α-methoxycarbonylethylidenetriphenylphosphorane (**243**) or with 3-methoxycarbonylprop-2-enylidenetriphenylphosphorane (**244**) affords the α-methyl derivative (**234**) or dienoate (**235**). The trienoate (**237**) was synthesized by the analogous condensa-

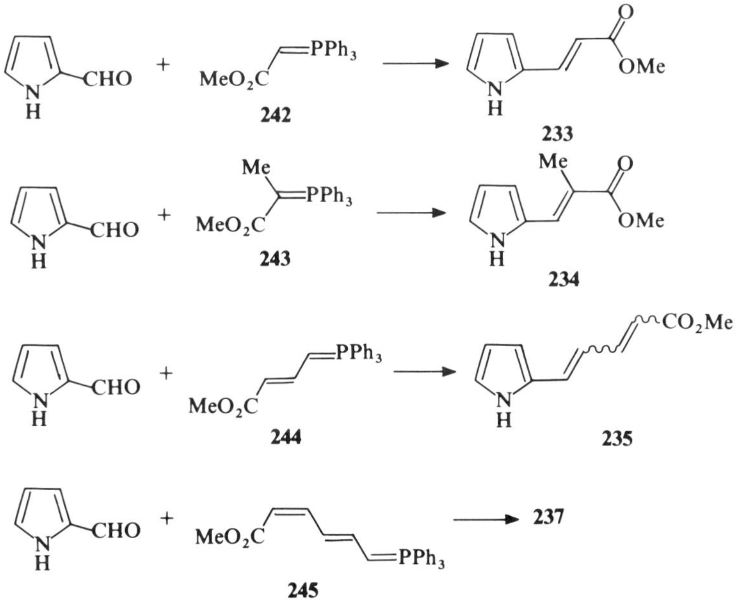

222 Vinylpyrroles

tion of 2-formylpyrrole with the freshly prepared phosphorane (**245**), obtained from 6-methoxycarbonylpenta-2,4-dienyltriphenylphosphonium bromide.

Treatment of 2-formylpyrrole with α-formylethylidenetriphenylphosphorane gives 2-methyl-3-(2-pyrrolyl)-prop-2-enal (**241**), and the subsequent condensation of the aldehyde (**241**) with the phosphorane (**244**) yields the pyrrole (**238**). The reaction between **241** and phosphorane (**242**) gives the E-isomer of the dienoate (**236**) in good yield.

The tetraenoate (**239**) has been synthesized by the same route from the aldehyde (**241**) and the phosphorane derived from 5-methoxycarbonylpenta-2,4-dienyltriphenylphosphonium bromide (**247**).[454]

The heterocyclic analogs of prostaglandin, d,l-2-[E-3-hydroxyoct-1-enyl)-1-(6-ethoxycarbonylhexyl)pyrrole (**249a**), and the indole (**249b**) have been obtained by the Wittig condensation of 2-formylpyrrole (or indole), followed by alkylation and reduction of the carbonyl group.[455]

a $R^1 = R^2 = H$
b $R^1-R^2 = -CH=CH-CH=CH-$

2.3. 2-Vinylpyrroles

The Wittig–Horner condensation of 2-formyl-1-methylpyrrole with the diethyl-2-thienylmethylphosphonate (**250**) in THF affords the E-1-(1-methyl-2-pyrrolyl)-2-thienylethene (**251**) in good yield.[449,456] The analogous condensation of 2,5-diformyl-1-methylpyrrole (**252**) with the phosphonate (**250**) gives 2,5-divinylpyrrole (**253**).[456]

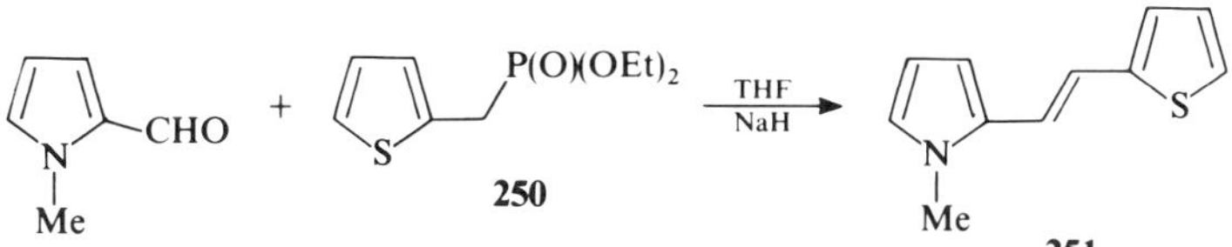

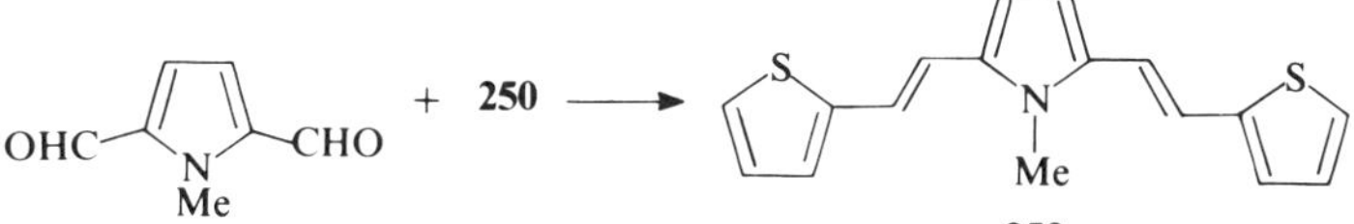

Conversely, 2,5-bis[2-(1-methyl-2-pyrrolylethenyl)]thiophene (**255**) has been obtained by condensation of the 2-formyl-1-methylpyrrole with the thiophene-bisphosphonate (**254**).[456]

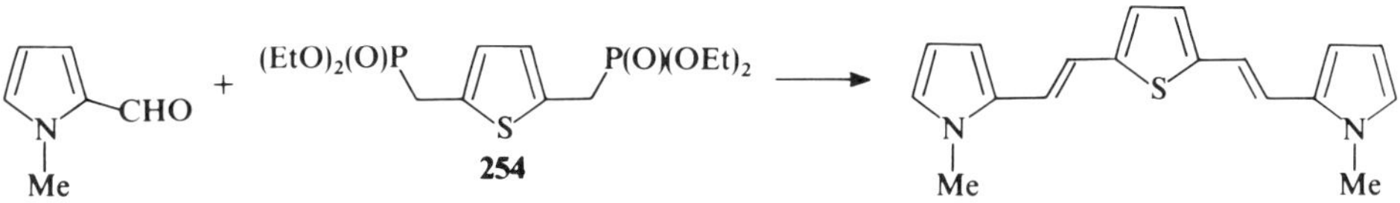

In order to accomplish the condensation of the pyrrolylmethylphosphonate (**256**) or the bis-phosphonate (**258**) with 2-formyl-1-methylpyrrole or the di-

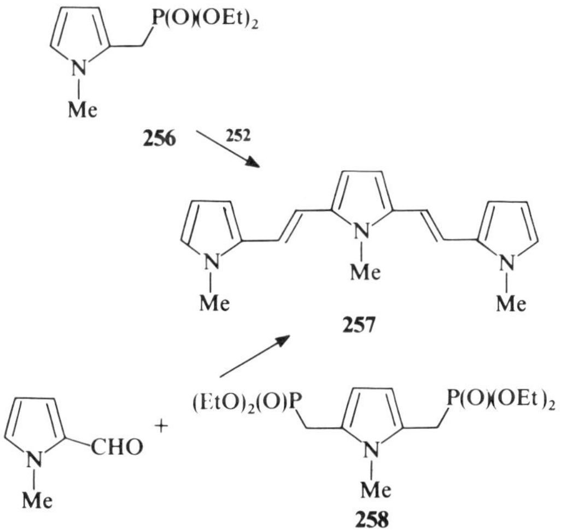

Scheme 20

aldehyde (**252**) to obtain 1-methyl-2,5-bis[2-(1-methyl-2-pyrrolyl)ethenyl]-pyrrole (**257**) in reasonable yields, a considerable excess of both the phosphonates and the base (NaH) has to be employed (Scheme 20).[456]

An intramolecular Wittig process leads to the formation of 2-vinylindole (**259**) in 38% yield.[457]

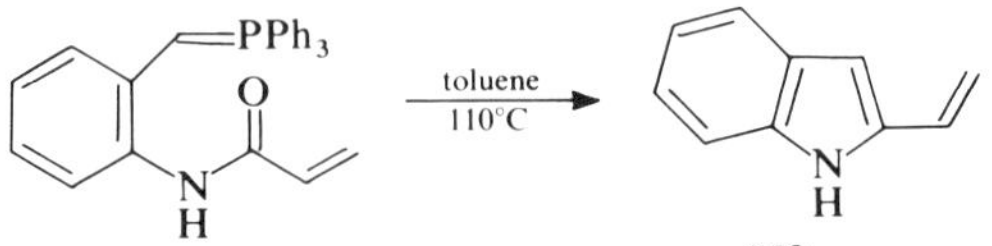

2.3.1.3. Addition Reaction of Pyrroles with Alkynes

Pyrroles undergo either Michael-type addition reactions to electrophilic acetylenes to form substituted 2-vinylpyrroles (**260**) or [$4\pi + 2\pi$] cycloaddition reactions leading to 7-azabicyclo[2,2,1]hepta-2,5-dienes (**261**). The ratio of the two reaction products is dependent on the reaction conditions, the solvent, and the nature of the substituents on both the pyrrole ring and the alkyne.[1,2,458-476] A full discussion of these reactions is to be found in Ref. 2 and in Part 1, Chapter 3.

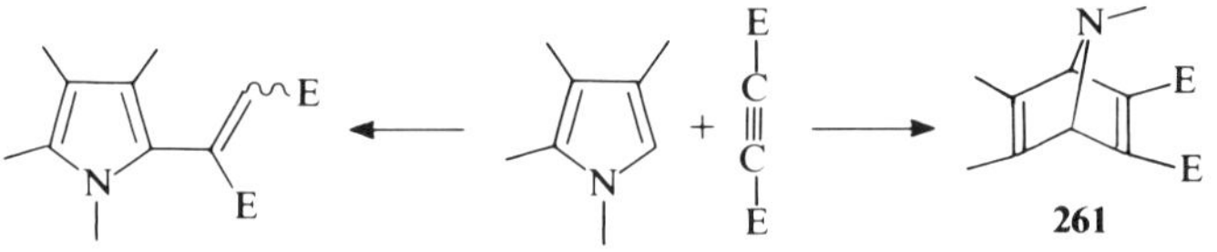

Although the structure of the adduct was not thoroughly established,[460] the reaction of pyrrole with dimethyl acetylenedicarboxylate (DMAD) was first reported to give only a 1 : 1 Michael adduct at the α-position in spite of the also probable N-vinylation. Reinvestigation of the reaction of pyrrole with DMAD has revealed that, when the reaction is carried out at ambient temperature (4 days), not only are the E- and Z-isomers of the 2-vinylpyrrole (**262**) formed in 25 and 42%, respectively, but also the 1-vinylpyrrole (**263**) in 6% yield. However, when the reactants are refluxed in ether, the 1 : 2 adduct (**264a**) is produced in

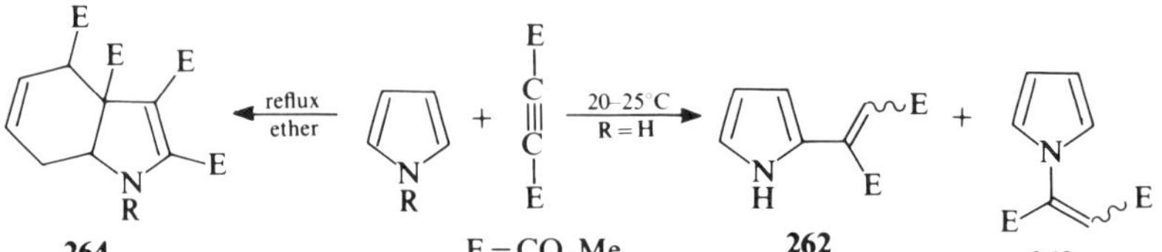

264
a R=H
b R=Me
c R=CH$_2$Ph

E=CO$_2$Me

Scheme 21

2.3. 2-Vinylpyrroles

6–10% yield (Scheme 21).[463] In contrast, under similar conditions, 1-methyl- and 1-benzylpyrroles add to DMAD to afford only the 1 : 2 adducts (**264b,c**) in 36–80% yields.[463,474] The products are derived from the initially formed cycloadduct (**261**) and further addition to a second molecule of the acetylene.[2,463,466,474]

The addition reaction of alkylpyrroles with DMAD has been discussed in detail.[460] Generally, only the C-vinylpyrroles are isolated.[471,472] Both the E- and Z-isomers may be obtained, although it is reported that the reaction of 2,3,4-trimethylpyrrole yields only one isomer and the single 1 : 1 adduct of 1,2-dimethylpyrrole, and DMAD has been assigned the E-configuration.[460] 1-Methoxycarbonylpyrrole reacts with DMAD in the presence of aluminum trichloride to produce both E- and Z-isomers of the expected 2-vinylpyrrole; the kinetically favored Z-isomer is gradually converted into the E-form.[473]

A number of 1-substituted pyrroles have been shown to react with DMAD in the presence of acetic acid to yield 2-vinylpyrroles as the major products.[466] Detailed spectral analysis has provided evidence for the assignment of the E- and Z-isomers.[466]

The addition of pyrroles to DMAD under high pressure (15 kbar)[468] has led in the case of pyrrole to the 2- and 1-vinylpyrroles (**262**, **263**) in 18 and 3% yields, respectively; the E : Z ratio for **263** is 1 : 8. With 1-methylpyrrole, the 1 : 2 cycloadduct (**264**) has been obtained quantitatively, and, under similar conditions, 1-acetylpyrrole afforded only the primary Diels–Alder adduct of type **261** (83%), compared with the yield of 45% reported[475] from the reaction under atmospheric pressure. It is noteworthy that 2,5-dimethylpyrrole gives a 1 : 3 ratio of the corresponding E- and Z-3-vinylpyrroles in 20% overall yield under high-pressure conditions, indicating that these conditions promote addition at the β-position of 2,5-disubstituted pyrroles; predictably, 1,2-dimethylindole reacts with DMAD of the β-position.[475] In general, the high-pressure reaction conditions promote C-vinylation of pyrroles without N-substituents, while those having electron-withdrawing substituents on the nitrogen atom produce the [4π + 2π] cycloadducts.[468]

Some 2-vinylpyrroles, despite their lower nucleophilicity, can add to DMAD in a Michael fashion. Thus, from the reaction of 1-methyl-2-vinylpyrrole (**265**) with DMAD at 20°C, the E- and Z-isomers of the adduct (**266**) were isolated, while at 80°C, the [4π + 2π] adduct, 1-methyl-6,7-dihydroindole-4,5-dicarboxylic acid, was formed. The cycloaddition reactions also occurred with 1-phenyl-2-vinylpyrrole.[467]

2-Phenyl-4H-furo[3,2-b]pyrrole (**257a**) and its 4-methyl derivative (**257b**) add to DMAD in acetonitrile at ambient temperature to give the Michael adducts

(**258a,b**) in 31 and 27% yields, respectively, whereas the corresponding reaction with 4-acetyl-2-arylfuro[3,2-b]pyrrole produces a substituted benzo[b]-furan.[469]

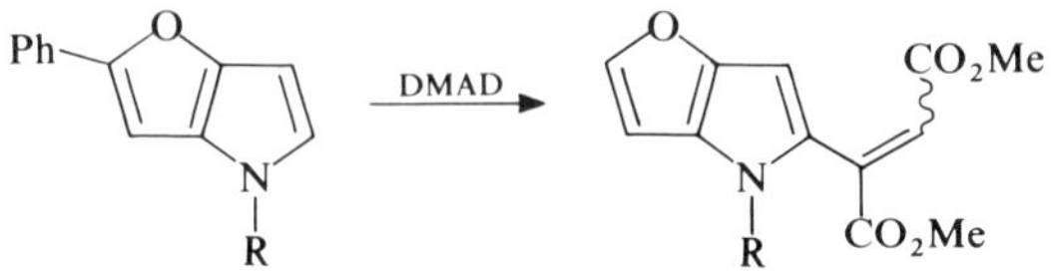

2-Amino-3-cyanopyrroles (e.g., **269**) react with DMAD to give, in addition to the expected *N*- and *C*-5-vinylated species (**270** and **271**), the unusual 1 : 2 adduct (**272**), which can be further transformed into isoindole or isoquinoline derivatives, depending on the reaction conditions (Scheme 22).[470] When conducted in acetic acid, the same reaction, as well as that with methyl propynoate, affords 5-indolone derivatives.[470]

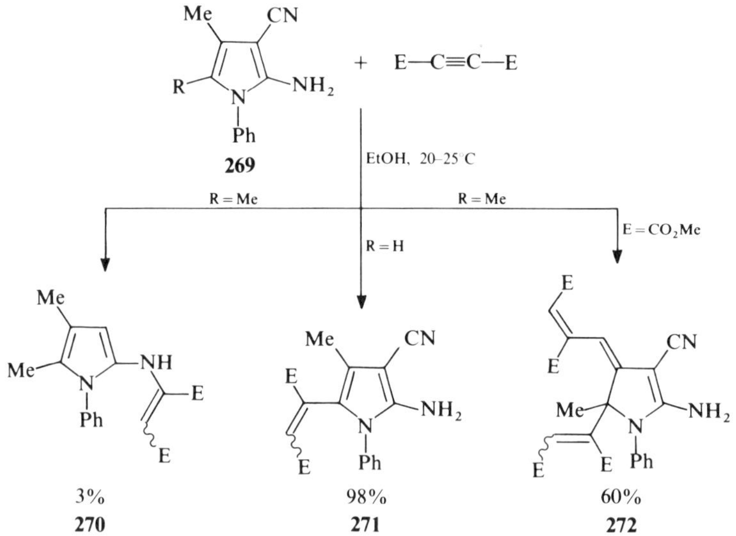

Scheme 22

In the presence of $Rh_4(CO)_{12}$, 1-methylpyrrole adds to diphenylacetylene to give a 31% yield of the *E*- and *Z*-isomers of 1-(1-methyl-2-pyrrolyl)-1,2-diphenylethene (**273**), which were characterized by ^1H NMR spectroscopy.[465]

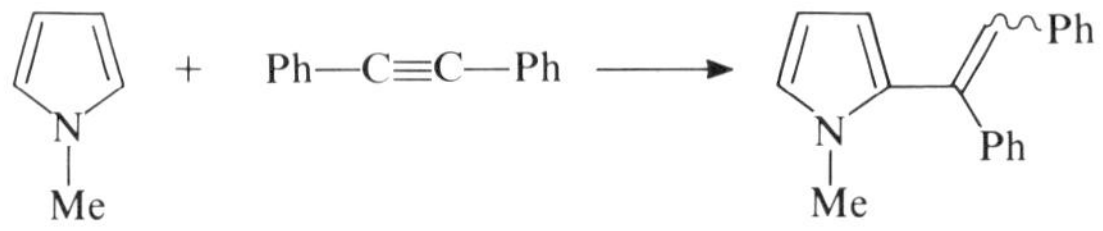

2.3.1.4. From Pyrroles and Carbonyl Compounds

The acid-catalyzed condensation of pyrroles with 2-formyl- and 2-acylpyrroles leads to di(2-pyrrolyl)methenes (**274**) or their salts.[477–492] This is an important procedure in the synthesis of porphyrins[484–490] and prodigiosins (see Ref. 492 and references cited therein). The methenes derived from long-chain alkyl ketones (**274**, R = R′CH$_2$) undergo isomerization to 1,1-di(2-pyrrolyl)alkenes (**275**).[480–488]

In the course of determination[453] of the structure of prodigiosin (**276**), several related compounds containing the prodigiosin nucleus (**277**) have been isolated from natural sources, particularly from a restricted group of eubacteria and actinomycetes.[492]

Prodigiosin has considerable antibacterial and antifungal activity, but its high toxicity has precluded its use as a therapeutic agent. Recently,[492] the synthesis of three novel series of N-substituted prodigiosins, some members of which have an interesting pharmacological action but are devoid of the cytotoxic properties of natural series, has been reported. The majority of prodigiosins (**280a–f**) have been prepared by the acid-catalyzed condensation of either bipyrrole (e.g., **278**) with a 2-formylpyrrole (**279**) or a 5-formyl-2,2-bipyrrole (e.g., **281**) with an α-unsubstituted pyrrole (**282**) (Scheme 23).[492]

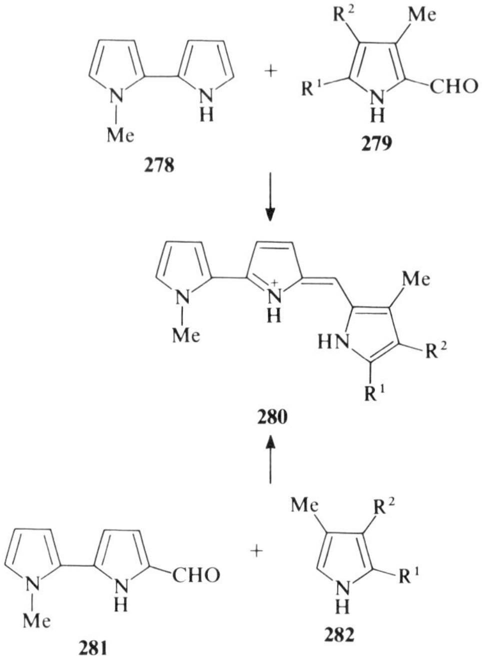

228   Vinylpyrroles

	R^1	R^2
a	n-C$_6$H$_{13}$	Me
b	n-C$_6$H$_{13}$	Br
c	Me	(CH$_2$)$_2$NHCONH-n-C$_{10}$H$_{21}$
d	Me	(CH$_2$)$_2$NHCO$_2$Bu
e	Me	Ph
f	Me	n-C$_6$H$_{13}$

Scheme 23

In contrast to the N-unsubstituted series, the salts of 1-methylprodigiosenes exist as E/Z mixtures, although only Z-isomers have been detected for the neutral species.[492] Dipyrrolyltrimethine dyes of the type **283** containing a 2-vinylpyrrole moiety are prepared by the reaction between pyrroles and α,β-unsaturated aldehydes or ketones, followed by oxidation.[1,479,490,492-494] These peculiar 2-vinylpyrroles have also been obtained by the reaction of pyrroles with 1,3-dialdehydes or diketones,[1,479,495,496] with acyl chlorides[494] or ketones in the presence of orthoformates.[497]

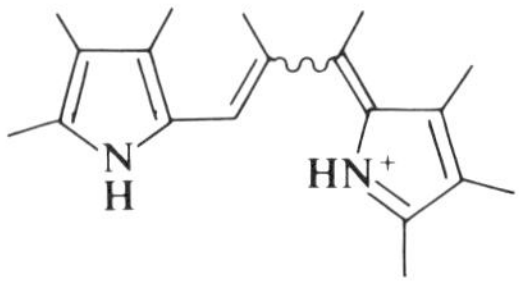

2.3. 2-Vinylpyrroles

Acylpyrroles, prepared from pyrroles and acyl halides, are often contaminated with colored side products, which are presumed to be cyanine dyes of the type **284** produced by autocondensation of the acylpyrroles and their subsequent acylation.[494]

Benzoylation of potassium 3,4,5-trimethylpyrrole-2-carboxylate (**285**) by benzoyl chloride in glacial acetic acid under reflux gives a 44% yield of 3,4,5,3′,4′,5′-hexamethyl-s-phenyldipyrrolylmethene (**286**) and 37% yield of the expected 2-benzoylpyrrole (**287**).[498] Pyrrolylmethene salts have also been obtained by acetylation of pyrroles with acetic anhydride in the presence of boron trifluoride.[499]

Alkylsubstituted pyrroles condense with ω-(N-methylanilino)propenal (**288a**) or -pentadienal (**288b**) under acidic conditions to yield polymethine dyes (**289a,b**), which on treatment with base afford the corresponding conjugated aldehydes.[500]

The reaction of α-unsubstituted pyrroles with squaric acid (3,4-dihydroxycyclobutenedione) (**290**), or cyclobutanediones, proceeds in an unexpected manner to give dipyrrolylcyclobutene derivatives (e.g., **291**) and the tripyrrolylcyclobutene (**292**).[501-505]

The highly colored 2,2′-dipyrrolylmethenes [6-(2-pyrrolyl)-1-azafulvenes or 2-(2-pyrrolylmethylene)-2H-pyrroles] (**274**) and their salts are the best characterized examples of the azafulvene system, and the long-standing wealth of knowledge appertaining to this system is well established[506] (see Refs. 1 and 2 and references cited therein). Recently, there has also been an upsurge in interest in the chemistry of other simple azafulvenes.[2]

Nucleophilic attack on the azafulvene (**293**) by pyrrolin-2-ones in the presence of sodium methoxide leads to the elimination of the 6-dimethylamino group and the formation of the rather specific 2-vinylpyrroles (**294**). The 6-dimethylamino group can also be displaced by carbanions to furnish 2-vinylpyrroles (**295**).[507]

A facile, general, two-step synthesis of 2-vinylindoles (**300**) starting from the ethyl 2-methylindole-3-carboxylates (**296**) has been reported (Scheme 24).[508]

The reaction of the N-protected indole (**297**) with excess lithium diisopropylamide at − 78°C easily forms the 2-methyl anion. Nucleophilic attack by the anion on either an aldehyde or ketone leads to the pyrano[3,4]indoles (**298**), which, under the action of base, produce the 2-vinylindole-3-carboxylic acids (**299**). Decarboxylation of **299** in refluxing bromobenzene affords the 2-vinylindoles (**300**) in excellent overall yields.

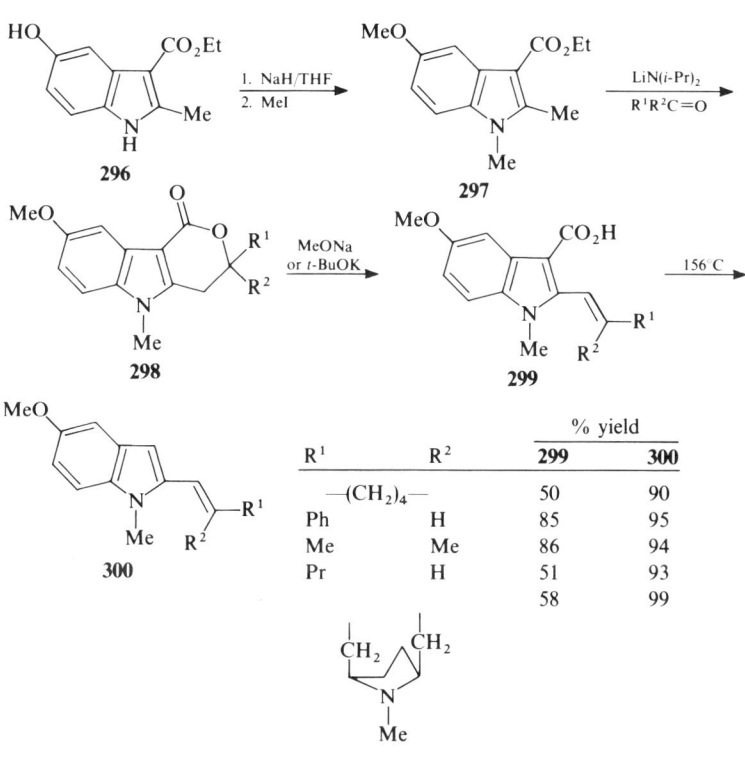

R¹	R²	% yield 299	300
—(CH₂)₄—		50	90
Ph	H	85	95
Me	Me	86	94
Pr	H	51	93
(piperidyl, N-Me)		58	99

Scheme 24

2.3.1.5. Miscellaneous Syntheses

2-Vinylpyrrole can be prepared by catalytic dehydrogenation of 2-ethylpyrrole in the gas phase.[403,441,509,510] The formation of 2-vinylpyrrole, together with 2-methyl- and 2-ethylpyrroles, was noticed during the pyrolysis of 1-n-butylpyrrole.[511]

The dehydrogenation of 3,4,7,8-tetrahydro-8b,8c-diazocyclopent[*fg*]acenaphthylene (**301**) leads to a product that, on the basis of its NMR spectrum[512] and a crystallographic analysis,[513] has been assigned the resonance-stabilized octahydro-8b,8c-diazocyclopent[*fg*]acenaphthylene structure (**302**).

2-(1-Hydroxyalkyl)pyrroles (**303**), prepared from 2-formyl or acylpyrroles by Grignard reactions[514,515] or by reduction with LiAlH$_4$,[514] have been converted into the vinylpyrroles (**304**) in yields of 22–52% by thermal elimination of the water molecule over alumina.[510,514]

During purification on silica gel, the alcohol (**305**) dehydrates to the 1-(4-methoxyphenyl)-1-(2-pyrrolyl)ethene (**306**).[516]

The aromatic 1,4,8,11-bisimino[14]annulene (**309**) and its precursor (**308**), both containing 2-vinylpyrrole structural units, have been synthesized from the diketone (**307**) by reduction with sodium borohydride, followed by a stepwise dehydration of the resulting alcohol via the corresponding tosylate. Treatment of **308** with potassium *tert*-butoxide in DMSO cleaves the N—N bond.[517]

A specific synthesis of 2-vinylpyrrole is based on the ring closure of polyfunctional compounds. Thus, the vinylacetylenic oxiranes (**310**), for example, react with amines at ambient temperature to afford the aminoalcohols (**311**) which, on being heated in quinoline, cyclize to the 2-vinylpyrroles (**312a,b**) in ~76% yield

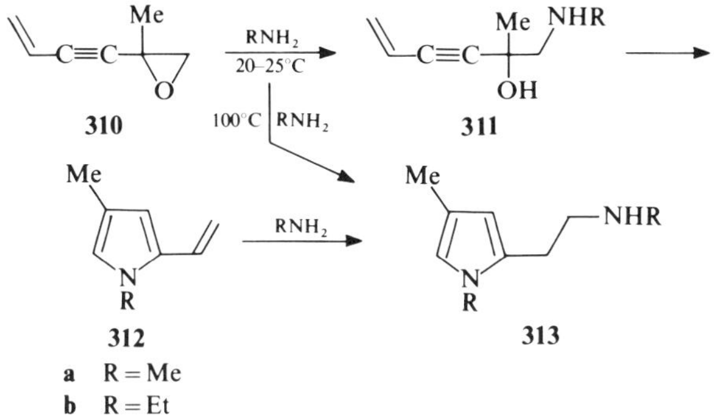

2.3. 2-Vinylpyrroles

Scheme 25

(Scheme 25),[518,519] with the following properties: **312a**, b.p. 49°C at 1.5 mm Hg, n_D^{20} 1.538, d_4^{20} 0.9254; **312b**, b.p. 35–36°C at 1 mm Hg, n_D^{20} 1.5324, d_4^{20} 0.917. Both vinylpyrroles polymerize over a storage period of several weeks to form brittle, glass-like polymers.[519] An attempted one-pot synthesis of the 2-vinylpyrroles (**312a,b**) directly from the oxiranes (**310**) and amines led to the aminopyrroles (**313**), as a result of the simultaneous addition of the excess amine to the double bond.[518,519]

1-Amino-2-vinylpyrroles (**315**) are synthesized by a general reaction of arylhydrazines with the 1,3-dienyl ketoethers (**314**) (see Ref. 2 and references cited therein).

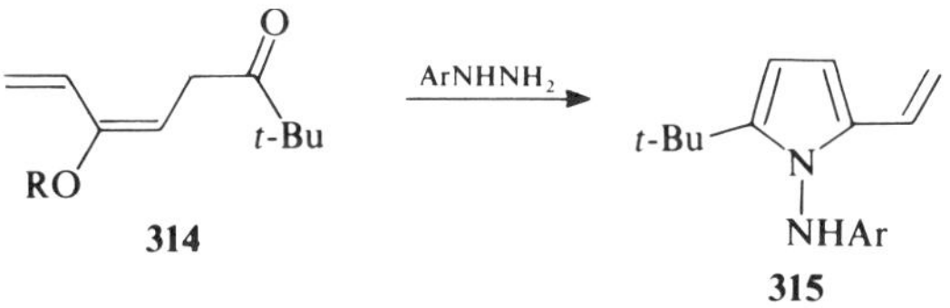

The dialkenylpyrroles (**318**) are formed in one operation by a base-induced regiospecific cycloaddition of 1-tosylalk-1-enyl isocyanides (**316**) to 1,3-dienic ketones and esters (**317**) (Scheme 26).[520] In most cases, pyrroles of the type **318a** were converted, without isolation, into 1-methyl (**318b,d–h**) or 1-acetyl (**318c**) derivatives (Table 2.25).

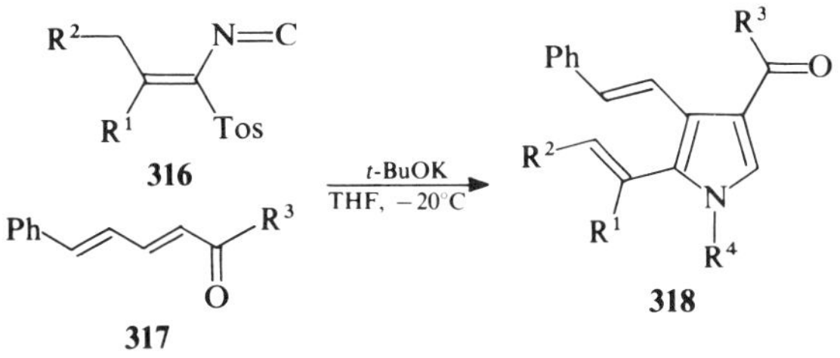

Scheme 26

TABLE 2.25. 2,3-DIALKENYLPYRROLES (318a–h)

Pyrrole	R^1	R^2	R^3	R^4	Yield (%)	m.p. (°C)
a	—(CH$_2$)$_4$—		Ph	H	91	208–209
b	—(CH$_2$)$_4$—		Ph	Me	91	150–151
c	—(CH$_2$)$_4$—		Ph	Ac	82	168–170
d	—(CH$_2$)$_4$—		MeO	Me	90	81–83
e	—(CH$_2$)$_3$—		Ph	Me	84	141–142
f	Me	H	Ph	Me	90	236–237
g	—(CH$_2$)$_5$—		Me	Me	87	109–110
h	—(CH$_2$)$_5$—		Ph	Me	82	164–165

Z-2-(Buta-1,3-dienyl)pyrrole (**320**) has been obtained in ~50% yield by pyrolysis of the 7-azidocycloocta-1,2,5-triene (**319**),[521] and thermal rearrangement of E,Z-mixtures of 2-(buta-1,3-dienyl)-2H-azirines (**321a–c**) afforded 2-vinylpyrroles (**322a–c**). The total isolated yield of **322a** is 58% with an E : Z-isomer ratio of approximately 6.25 : 1. Similarly, the E-isomer of **322c** predominates.[522]

The addition of 2,3-dimethylbut-2-ene to 2-benzoyl-1-methylpyrrole proceeds via the n–π* triplex state to yield the adduct (**323**), which extrudes acetone to give the 2-vinylpyrrole (**324**).[523]

Ultraviolet irradiation of pyrrole in excess benzene gives the adduct (**325**) together with a larger amount of **326**.[524]

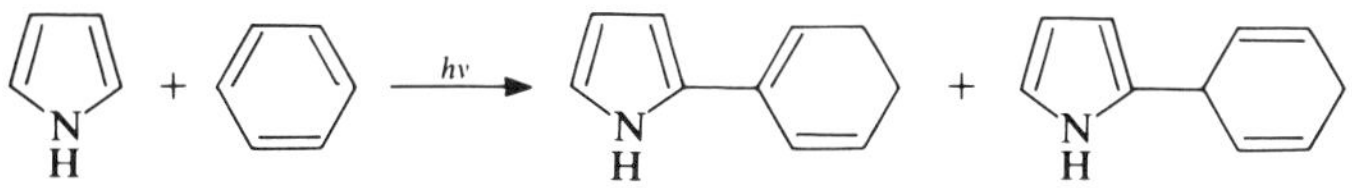

2.3. 2-Vinylpyrroles

325 **326**

Z-4-(Phenylsulfinyl)-4-penten-2-one (**327**) undergoes a versatile acid-catalyzed addition–elimination with pyrrole and 1-methylpyrrole to give the 2-vinylpyrroles (**328a,b**) in 98 and 65% yields, respectively.[525]

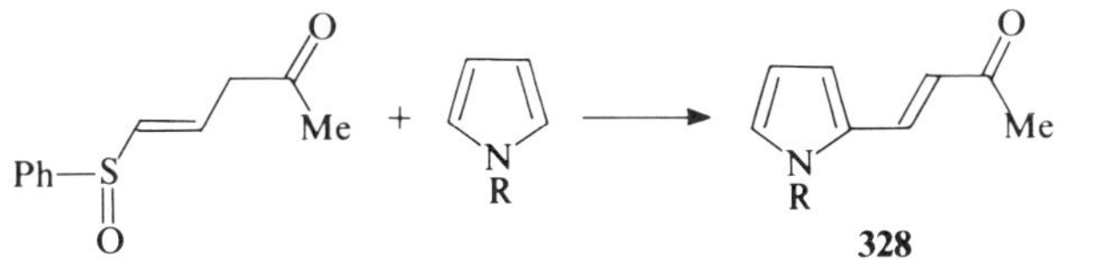

327

a   R = H
b   R = Me

328

An E,Z-mixture of the enediol dimethyl ether (**329**) has been obtained[526] by the route shown in Scheme 27 as a part of the strategy for the synthesis of antitumor agent CC-1065.

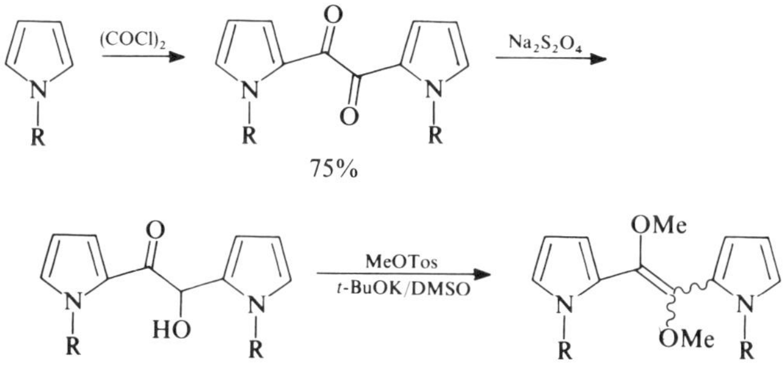

75%

99%   86%
                      329

$R = Me_3Si(CH_2)_2OCH_2$

Scheme 27

E-2-(2-Phenylsulfinylvinyl)pyrrole (**332**), suitable for the synthesis of 4-acylindoles, can be prepared by the addition of the phenylsulfinylcarbanion to the 6-dimethylamino-1-azafulvene (**330**), followed by base-catalyzed (e.g., $NaBH_4/i$-PrOH) elimination of dimethylamine from the resulted aminosulfoxide (**331**).[527]

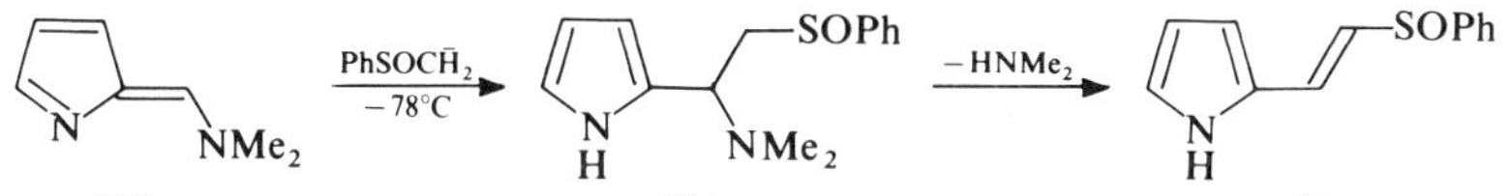

330   **331**   **332**

Trialkylpyrroles unsubstituted at the 5-position (e.g., **333**) react with bromine to yield the di(2-pyrrolyl)methene (**334**).[528–530] The corresponding chloro derivative is isolated from the reaction of **333** and sulfuryl chloride.[531]

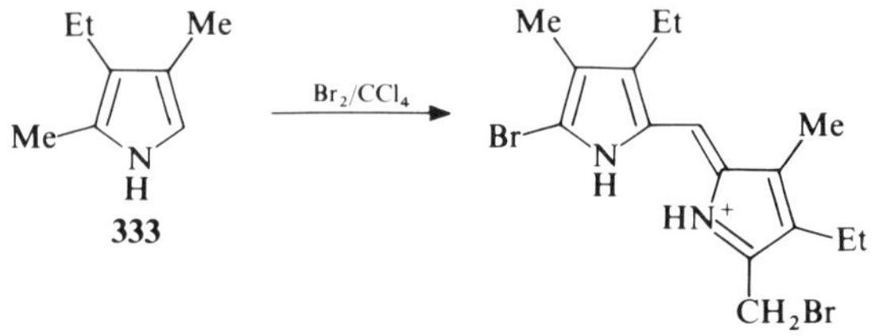

333 **334**

Under alkaline conditions the pyridinium salt (**335**) affords 1,2-(2-pyrrolyl)ethene (**336**) in 14% yield.[532]

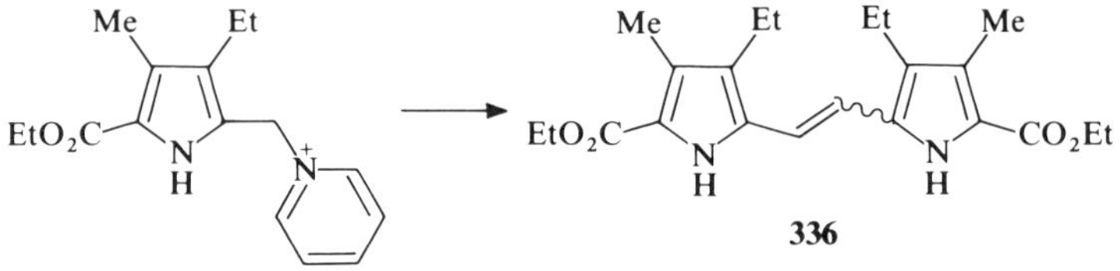

335 **336**

The condensation of 1-methyl-2-nitropyrrole (**337**) with 1-methyl-2-(cyanomethyl)pyrrole (**338**) in the presence of sodium methoxide leads to the oxime (**339**) having a 2-vinylpyrrole moiety.[533]

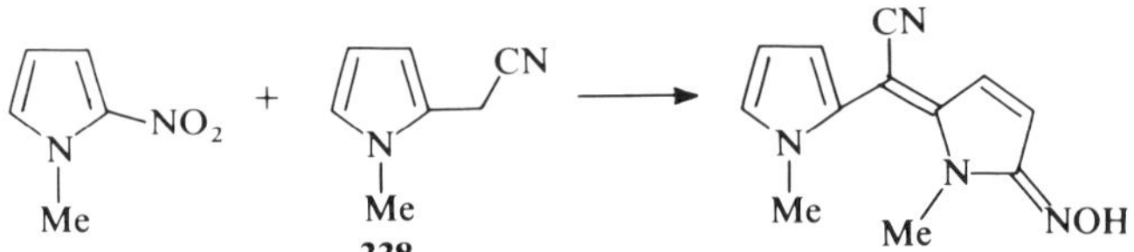

337 **338** **339**

Alkaline hydrolysis of the azetidinones (**340a–c**) results in 2-(1,1-diphenylvinyl)pyrrole (**341**) in 46% yield, together with diphenylacetic acid, 2-iminomethylpyrroles, and arylamines.[534]

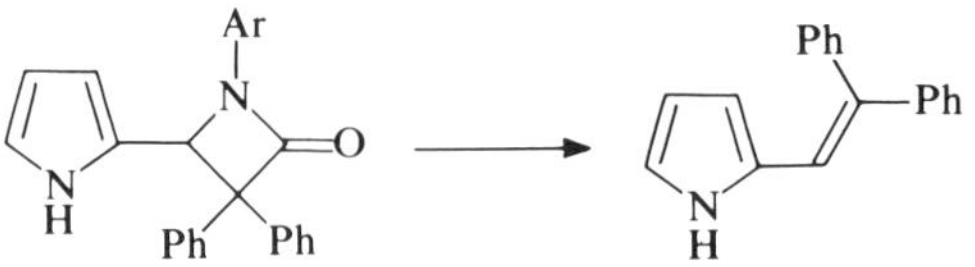

340 **341**

a Ar = Ph
b Ar = 4-MeC_6H_4
c Ar = 4-$MeOC_6H_4$

2.3.2. Reactivity

2.3.2.1. [4π + 2π] Cycloaddition Reactions

Although pyrroles commonly react with dienophiles by a Michael addition or a [4π + 2π] cycloaddition across the 2,5-positions,[2] simple Huckel MO calculations suggest that the [4π + 2π] cycloaddition with C-vinylpyrroles should give dihydrotetrahydroindoles.[535] Similar reactions with vinylfurans[536] and vinylthiophenes,[537] as well as the construction of carbazoles via vinylindoles,[538] are known.

This somewhat atypical starting point for the preparation of indoles had not been developed in any detail until the work of Jones et al.[535] due to the assumed inaccessibility and/or instability of the requisite vinylpyrroles,[539] although the sequential reaction of α-unsubstituted pyrroles with two molecules of π-electron-deficient alkynes to yield indoles via the in situ formed vinylpyrroles had already been described.[540,541]

The facile cycloaddition reaction of 2-vinylpyrroles, as well as 3-vinylpyrroles, with a range of dienophiles was shown[535] to produce tetrahydro- (**343**) and dihydroindoles (**344**) in a two-step process; the second step was a [1,3]-sigmatropic hydrogen shift in the intermediates **342** and **344** to restore the aromaticity (Scheme 28).

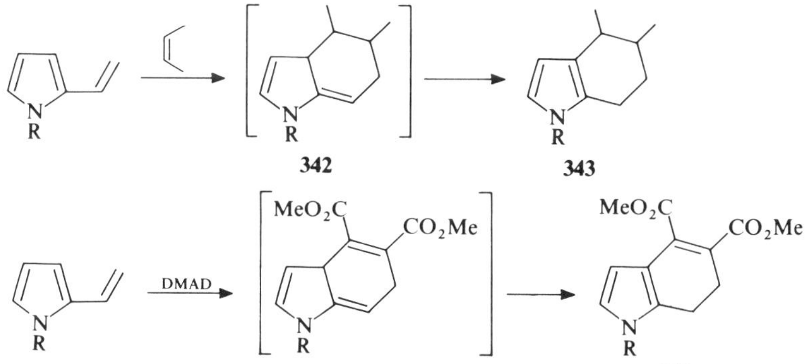

Scheme 28

Maleic anhydride reacts with 2-vinylpyrroles in $CHCl_3$ (or benzene) at ambient temperature very rapidly, although the reactions with dimethyl maleate, E- and Z-dicyanoethenes, acrylonitrile, and ethyl acrylate proceed slowly even under more harsh conditions. Reactivity patterns for these cycloadditions are shown in Table 2.26.

In each case the tetrahydroindoles (**343**) were isolated in 54–81% yield. Interestingly, all attempts to convert them into the indoles with quinones or Pd/C failed. However, the dihydroindoles (**345**), formed in 55–75% yields, were

TABLE 2.26. REACTIVITY OF 1-METHYL-2-VINYLPYRROLE WITH DIENOPHILES IN CDCl$_3$ AT 20°C

Dienophile	$10^5 k_2$ (liters mol^{-1} s^{-1})	$\Delta H^\ddagger_{20\cdot C}$ (kJ mol^{-1})	$\Delta S^\ddagger_{20\cdot C}$ (J deg^{-1} mol^{-1})
Maleic anhydride	>15000		
DMAD	3.23	55.8	−46
Malonitrile	2.22	50.8	−66
Fumaronitrile	0.87	54.6	−61
Dimethyl maleate	~0.30		
Methyl acrylate	~0.32		
Acrylonitrile	~0.31		

readily dehydrogenated to the corresponding indoles, when treated with 2,3-dichloro-5,6-dicyanoquinone.[535]

Methyl propiolate reacts with 2-vinylpyrroles to form the 1-substituted indole-4,7-dicarboxylic esters (**348**) through a further [4π + 2π] cycloaddition of a second molecule of the acetylene to the intermediate 6,7-dihydroindole-4-carboxylic ester (**346**), followed by a retro-Diels–Alder extrusion of ethene (Scheme 29).[535] The intermediates **346** and **347** were not isolated. The NMR monitoring of the reaction showed signals attributable to the intermediate **346** to be of low intensity and not well resolved, suggesting that all of the reaction steps had similar rates.[535]

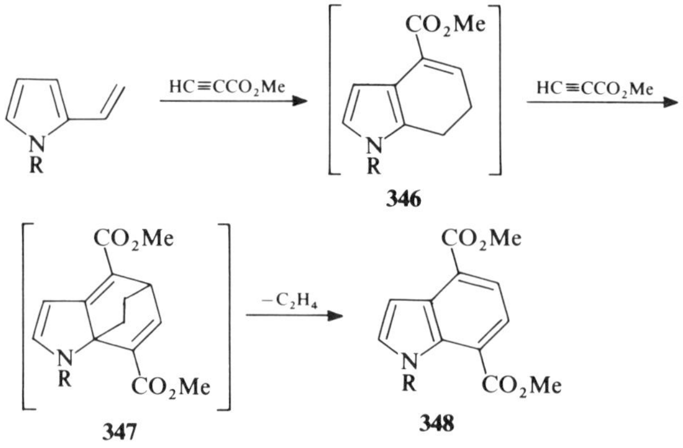

Scheme 29

2-Vinylpyrrole itself failed to give isolable [4π + 2π] cycloadducts with the common dienophiles, and the presence of polymers derived from the Michael additions were spectroscopically detected in the product mixtures.[535] 1-Methyl-2-styrylpyrrole, ethyl E-2-(1-methyl-2-pyrrolyl)acrylate, and 1-(1-methyl-2-pyrrolyl)-E-but-1-en-3-one afforded unstable products with maleic anhydride

2.3. 2-Vinylpyrroles

(80°C) and reacted very slowly and incompletely with dienophiles, in accord with their calculated HOMO energy levels.[535]

The reaction of DMAD with 1-methyl-2-vinylpyrrole is temperature-dependent. At 80°C, the predominant reaction is the [4π + 2π] cycloaddition to yield 1-methyl-6,7-dihydroindole-4,5-dicarboxylate (**345**) (R = Me), whereas at ambient temperature the corresponding Michael *E*- and *Z*-adducts are also formed[542] (see Section 2.3.1.3). 1-Phenyl-2-vinylpyrrole reacts with DMAD to give only the [4π + 2π] cycloadduct (**345**) (R = Ph) over a wide temperature range.[542]

The requirement of the higher temperatures to effect the preferential formation of the [4π + 2π] cycloadducts arises from the necessity to overcome the energy difference between the unreactive *transoid* (**349a**) and the reactive *cisoid* (**349b**) conformations of 2-vinylpyrroles.[543,544] The ratio of the two conformations depends to a large extent on the steric interaction between the substituents R^1 and R^2.

The work[543] has confirmed further the important role of steric requirements in the reaction of the 2-vinylpyrroles (**349**) with DMAD (Table 2.27). Yet, the rationalization[543] of the data presented in Table 2.27 neglects electronic (inductive and conjugative) effects, which should also contribute to the ratio and yield of the products **350** and **351**.

TABLE 2.27. REACTIONS OF 2-VINYLPYRROLES (**349**) WITH DMAD IN $CHCl_3$[543]

R^1	R^2	Reaction Temperature (°C)	Time	Yield (%)		
				E-(**350**)	Z-(**350**)	(**351**)
Me	H	20	4 days	9	20	21
		60	3 h			70
Me	Me	20	3 days	22	17	
		60	3 h	30	41	
Me	*t*-Bu	20	20 days	26	18	
		60	6 h	39	16	
Me	Ph	20	7 days	35	27	38
		60	9 h	7	10	32
Ph	H[a]	20	4 days			65
Ph	Me	20	10 days			25[b]
		60	5 days			39

[a] From Ref. 553.
[b] Dimethyl 7-methyl-1-phenylindole-4,5-dicarboxylate (4%) also formed.

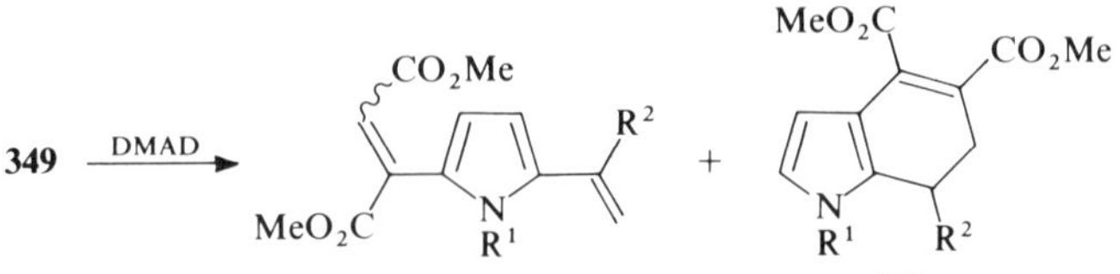

Vinylpyrroles

349 —DMAD→ 350 + 351

Steric interaction between the 1-methyl and the substituent R on the vinyl group is assumed to prevent the diene system from adopting a coplanar *cisoid* conformation and, thereby, inhibits the [4π + 2π] cycloaddition and favors the Michael addition affording 353. The data given in Table 2.28 are generally in accord with this assumption, although the result for the phenyl derivative (352d) where the [4π + 2π] adduct (354d) is the only reaction product is not totally compatible with a steric effect.

TABLE 2.28. REACTIONS OF 2-VINYLINDOLES (352a–d) WITH DMAD IN CHCl$_3$[553]

R	Reaction Temperature (°C)	Time (days)	Yield (%) E-(353)	Z-(353)	(354)
H	62	3			8[a]
Me	20	11	9	7	<2[a]
	62	3	8	6	<2[a]
t-Bu	20	20	17	14	
	62	14	14		30
	20[b]	46	63		
	65[b]	4	39[c]		
Ph	62	8			37[a]
	65[b]	9			30[d]

[a] Polymer formed.
[b] In MeOH.
[c] 35% of the carbazole also formed.
[d] 26% of the carbazole also formed.

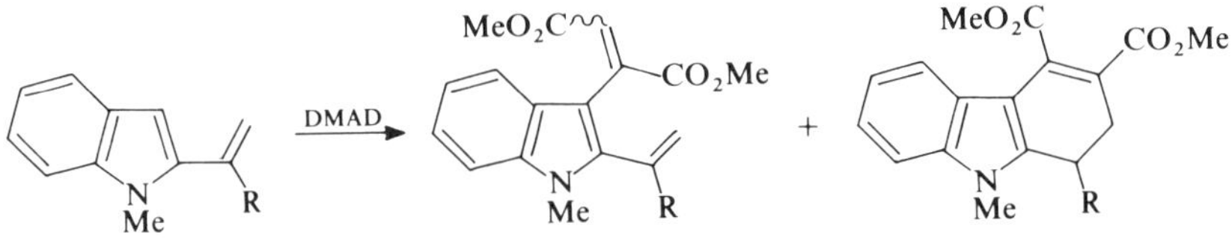

352 —DMAD→ 353 + 354

352
a R = H
b R = Me
c R = t-Bu
d R = Ph

2-Vinylindoles, substituted on the vinyl group, have been shown to behave more generally as the dienophile in [$4\pi + 2\pi$] cycloaddition reactions,[545–549] although with N-phenylmaleimide under harsh conditions and with maleic anhydride, these compounds act as the 4π component.[550–552] The overall reactions of the 1-methyl-2-vinylindoles (352a–d) with DMAD are considerably slower than those of 2-vinylpyrroles.[553]

Recently,[527] the [$4\pi + 2\pi$] cycloaddition of 1-triisopropylsilyl-E-2-(2-phenylsulfinylvinyl)pyrrole (355) with π-electron-deficient acetylenes, followed by N-desilylation of the cycloadducts (357) with fluoride ion (Scheme 30), has been shown to provide a new and viable route to the 4-acylindoles (358) (Table 2.29), which are of considerable interest as starting materials for the synthesis of a variety of important natural products and medicinal agents. Monitoring of this reaction by TLC showed no evidence for the dihydroindole intermediate (356), which presumably aromatized spontaneously under the reaction conditions.[527]

TABLE 2.29. 4-SUBSTITUTED INDOLES 357 AND 358

		Yield (%)	
R^1	R^2	357	358
CO_2Me	OMe	90	56
H	OEt	91	88
Ph	OEt	22	74
H	H	84	84

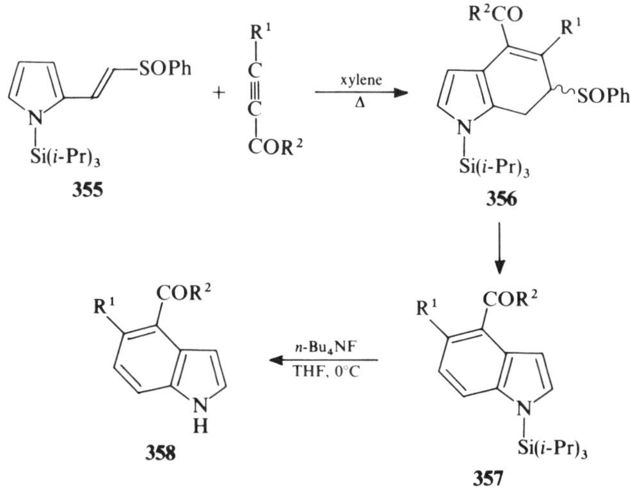

Scheme 30

It is of interest that 1-methyl-2-vinylpyrrole and 2-vinylthiophene display remarkable differences in their reactivity and regioselectivity on the reaction with methyl propiolate in which the 1-methylindole-4,7-dicarboxylate (**359**) and the benzo[*b*]thiophene-4,6-dicarboxylate (**360**) are respectively formed.[554] Also, 1-methyl-2-(1-propenyl)pyrrole reacts with DMAD under reflux in ethanol to give the expected Diels–Alder and the Michael adducts, while 2-(1-propenyl)thiophene produces a 1 : 2 adduct under somewhat more vigorous conditions, as a result of a subsequent ene reaction on the initially formed [$4\pi + 2\pi$] cycloadduct.[554]

In a new approach to synthesis of psoralen and azapsoralen, the vinylpyrrole (**361**) with the propargyloxy group has been subjected to a thermal ring-closure reaction at 150°C to give the Diels–Alder adduct (**362**), which underwent aromatization with 5,6-dicyano-1,4-benzoquinone or with Pd/C.[525]

2-Vinylindole and its donor- and acceptor-substituted derivatives react with dimethyl 1,2,4,5-tetrazine-3,6-dicarboxylate (**363**) to form selectively the (2-indolyl)-1,4-dihydropyridazines (e.g., **364**), as well as heterocyclic annelated pyridazines (**365**–**367**) (Scheme 31).[555]

The [$4\pi + 2\pi$] cycloadditions of 2-vinylindoles with electron-deficient dienophiles (e.g., propynoates, acrylonitrile, α-chloroacrylonitrile, methyl acrylate, 1-penten-3-one) proceed with variable regioselectivities to furnish functionalized carbazole derivatives.[555]

2.3.2.2. Other Reactions

In view of their potential instability, relatively few reactions of vinylpyrroles, other than those described in Section 2.3.2.1, have been systematically studied.

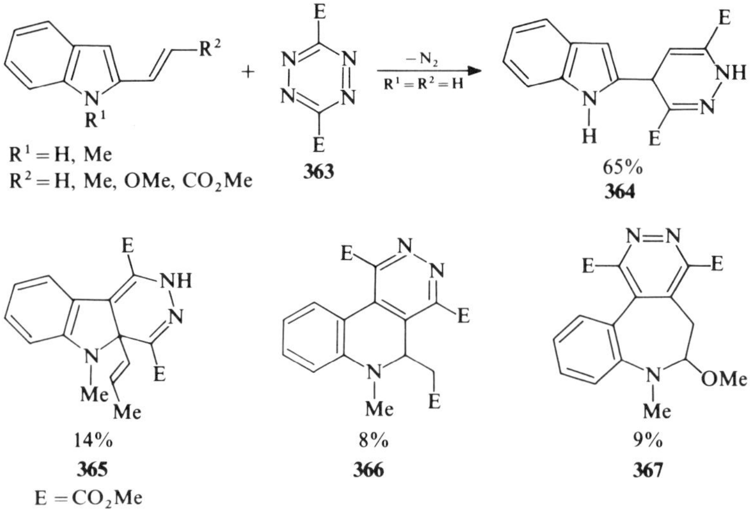

2.3. 2-Vinylpyrroles   243

Scheme 31

R¹ = H, Me
R² = H, Me, OMe, CO₂Me

363

364 65%

365 14%

366 8%

367 9%

E = CO₂Me

One may conclude from available information that, as expected, the vinyl group resembles to some extent that of styrene being sensitive toward radicals, when unsubstituted, and sluggishly active toward nucleophiles. Its reactivity with electrophiles is masked by the instability of the pyrrole system in the presence of acids. As already indicated, 2-vinylpyrroles, particularly those having no substituents on the vinyl group, are prone to polymerization.[1,2,509,518,519]

N-Sustituted 2-vinylpyrroles tend to be more stable, and, for example, 1-ethyl-4-methyl-2-vinylpyrrole (**368**) has been hydrogenated over a platinum catalyst to yield the corresponding 2-ethylpyrrole (70%)[519] and, not very surprisingly, it also undergoes a nucleophilic addition with ethylamine at 100°C to give 1-ethyl-2-[2-(ethylamino)ethyl]-4-methylpyrrole (**369**).[518,519]

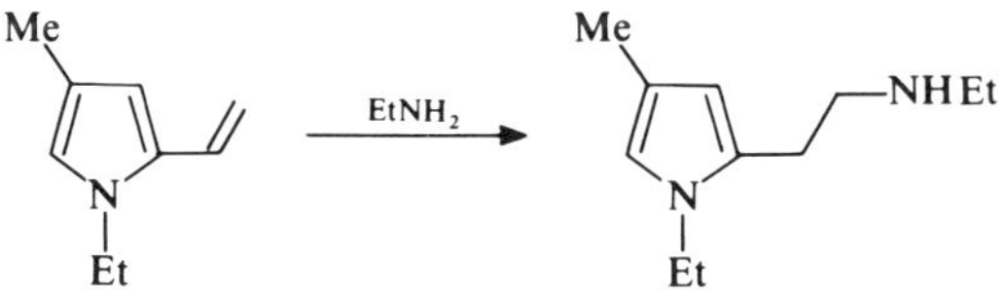

368 → EtNH₂ → 369

The 1-methyl-2-vinylpyrroles (**370**) react with carbenes generated by catalytic decomposition of diazo compounds to form both the cyclopropane adducts (**371**) and the products of the carbene insertion into C—H bonds of the pyrrole ring (**372**)[557] (Table 2.30)

In the reactions of the 2-vinylpyrroles (**370**) with carbenes derived from methyl diazoacetate in the presence of CuCl at 20–50°C, it has been shown that

R¹ = H, Me; R² = H, Me;
X = H, CO₂Me; Y = H, CO₂Me

TABLE 2.30. FORMATION OF THE PYRROLES 371 AND 372[557]

				Yield (%)	
R¹	R²	X	Y	371	373
H	H	H	H	18	6
H	H	H	CO$_2$Me	55	33
H	H	CO$_2$Me	CO$_2$Me	16	60
H	Me	H	H	6	15
H	Me	H	CO$_2$Me	29	44
H	Me	CO$_2$Me	CO$_2$Me	7	74
Me	Me	H	H	8	18
Me	Me	H	CO$_2$Me	20	48
Me	Me	CO$_2$Me	CO$_2$Me	4	78

one (or two) methyl substituents at the β-position of the vinyl group decrease its relative reactivity with the carbene by a factor of 1.8 (or 2.7) and produces an approximately twofold increase in the reactivity of the pyrrole ring.[558]

The regioselective 1,3-dipolar addition reactions of 2-vinylpyrroles with $\text{Ph}\bar{\text{N}}-\overset{+}{\text{N}}\equiv\text{CPh}$ and $\text{MeC}\equiv\overset{+}{\text{N}}-\bar{\text{O}}$ to give the adducts **373** and **374** have been reported.[559] Similar adducts are obtained from the reaction of 1-t-butyl-3-vinylpyrrole with the 1,3-dipolar reagents (see Section 2.3.2.2).

R = Me, Ph

Pyrrole analogs of stilbene, which are readily available via Wittig reaction (see Section 2.3.1.2), cyclize photochemically into the tricyclic ring structure necessary for an antitumor antibiotic CC-1065 and its analogs (Table 2.31).[560–562] The O$_2$ (air)-mediated photocyclizations of these 2-vinylpyrroles

TABLE 2.31. PHOTOCYCLIZATION OF PYRROLE ANALOGS OF STILBENE

Pyrrole	Product	Yield (%)
		50[a]
		82[b]
		75[b] 75[c]
		84[b]
		80[b] 38[c]
		85[b] 75–80[c]
		78[c]
		85[b] 59[c]

[a] $1.5 \cdot 10^{-3}$ M of styrylpyrrole (E + Z) in triethylamine, a RPR 3500-Å lamp, air, 7.75 h.[562]
[b] $2 \cdot 10^{-2}$ M of the vinylpyrrole in refluxing acetonitrile, a 450-W medium pressure Hanovia lamp, under nitrogen, 5% Pd/C, ~ 16 h.[561]
[c] Pyrex, in cylohexane, air.[560]

proceed satisfactorily on a 1 molar scale,[560] whereas yields drop considerably with 1–5-g amounts of reactants due to oxidative destruction of the indole-like products. Several other oxidants (e.g., air/I_2, I_2, $CuCl_2$) were also found to be unsuitable.[561] When photolysis of 2-styrylpyrrole is carried out with iodine in an oxygen-free system, the tetrasubstituted mesoporphyrin is formed.[562] Success is achieved[561] when the 2-vinylpyrroles (Table 2.31) are irradiated in the presence of a catalytic amount of Pd/C. Under these conditions with 1,2-di-(1-methyl-2-pyrrolyl)ethene, the corresponding ethane is formed as a result of transfer hydrogen of the starting material. A similar photoreduction leading to 1-phenyl-2-pyrrolylethane has been observed, when 2-styrylpyrrole was photocyclized in oxygen-free solutions containing triethylamine.[562] By including a more powerful hydrogen acceptor, nitrobenzene, in the reaction mixture, the competing hydrogenation can be circumvented.[561] Chromatographic purification of the photocyclization products is simplified by using a combination of p-nitrobenzene acid and triethylamine. In all the cases[561] the reaction is remarkably clean and proceeds in good yield. The 1-(2-furanyl)-2-(2-pyrrolyl)ethenes cyclize at a much slower rate than do other 2-aryl- or 2-(heteroaryl)vinylpyrroles.[561]

In further developing the approach to the synthesis of the antitumor agent CC-1065, the enediol dimethyl ether (**329**) was photocyclized under anaerobic conditions to afford **375** in 82% yield.[526]

R = MeSi(CH$_2$)$_2$OCH$_2$

Photochemistry of 2-pyrrolylacrylic esters gives a mixture of the isomeric cyclobutane dicarboxylic esters.[563]

2-Vinylpyrroles of the type **376** have been shown to undergo an intramolecular acylation under the action of acetic anhydride[2, 564–566] to furnish either unsubstituted 3H-pyrrolizin-3-one[560] or its derivatives (e.g., **377**[565] and **378**[566]).

Solvolysis of the lactam (**378**) by ethanolic sodium ethoxide leads to the Z-acrylic ester (**379**).[566]

The 2-vinylindoles (**380a,b**) are formylated with (EtO)$_2$ĊHB̄F$_4$ and acylated with (EtO)$_2$ĊMeB̄F$_4$ at the free 3-position to afford 3-formyl- (**381a,b**) and 3-acetylindoles (**381c**) in which the vinyl group is retained intact (Table 2.32).[567]

The vinyl group of **382** also remains intact in the reaction with diethyl azodicarboxylate (**383**). The reaction proceeds cleanly and in high yield at the free 5-position to form the Michael adducts (**384**).[568] In the case of the 2-isopropenylpyrrole (**382**, R = Me), an ene reaction gives the 2-vinylpyrrole (**385**)

2.3. 2-Vinylpyrroles

TABLE 2.32. SYNTHESIS OF THE INDOLES (381)

Indole	R^1	R^2	Reaction Temperature (°C)	Yield (%)
381a	H	H	−78 to −40	55
381b	Me	H	−30 to 20	70
381c	Me	Me	−10 to 20	13

in 18% yield.[568] When the 5-position of 2-vinylpyrroles is blocked or sterically hindered, pyrrolo[3,2-c]pyridazines or dihydropyrrolyl-1,3,4-oxadiazines are formed.[568]

The 1-methyl-2-vinylindoles (**386**) generally react readily at 20°C with 4-phenyl-1,2,4-triazoline-3,5-dione (**387**) to afford 1-(1-methyl-2-vinylindol-3-yl)-4-phenyl-1,2,4-triazolidine-3,5-diones (**388**) in high yield.[569] However, the parent compound (**386**, R = H) produces only polymeric material.[569]

2.4. 3-VINYLPYRROLES

3-Vinylpyrroles are commonly acknowledged as important building blocks in the preparation of vinyl-substituted porphyrins (proto-, pento-, and spirographisporphyrins) and bile pigments (bilirubins, biliverdins, etc).[1,2,570]

2.4.1. Synthesis

Reactions leading to 3-vinylpyrroles are, for the most part, the same as those employed for the synthesis of 2-vinylpyrroles, i.e., condensation of 3-formylpyrroles with CH-acids, or β-unsubstituted pyrroles with carbonyl compounds, the Wittig reaction with 3-formylpyrroles, and the Michael addition to activated acetylenes.

2.4.1.1. From 3-Formylpyrroles and CH-acids

Application of the Knoevenagel reaction has been found to be successful not only with 2-formylpyrroles but also for the 3-isomers.[1,2] Condensation of

2.4. 3-Vinylpyrroles

3-formylpyrroles with malonic esters catalyzed by an amine yields the expected 3-[2,2-di(carboxyvinyl)]pyrroles at low temperature but, under more vigorous conditions, partial decarboxylation takes place to afford β-(3-pyrrolyl)acrylic acids (e.g., **389**).[1,2,429]

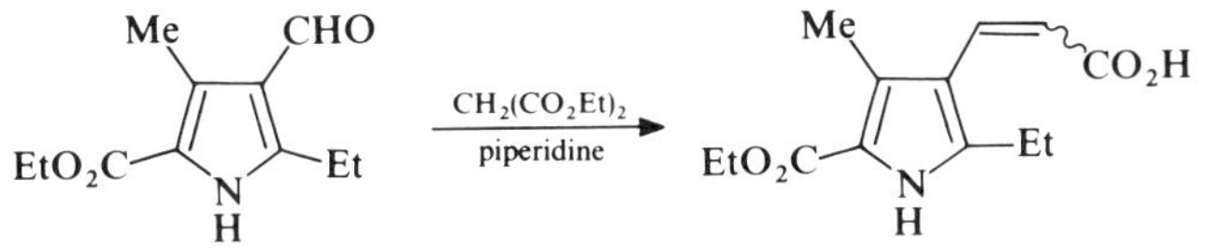

Catalytic hydrogenation of the acrylic acids is a common procedure for the preparation of β-(3-pyrrolyl)propionic acids related to some naturally occurring pyrroles, particularly for those that cannot be synthesized by a total ring synthesis (see Part 1, Chapter 2). 3,4-Diformylpyrroles condense with 1,3- or 1,2-bifunctional CH-acids to give the bicyclic systems[1,2,571–573] (e.g., **390a–d**) (Scheme 32).

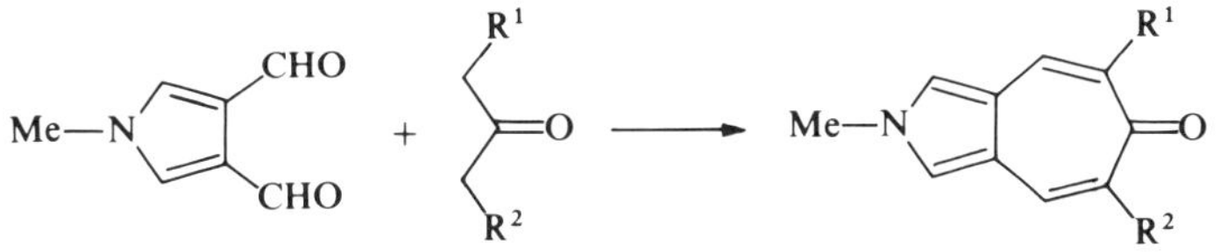

390	R^1	R^2	% yield
a	Me	Me	91
b	H	Me	64
c	H	H	36
d	CO_2Me	CO_2Me	60

Scheme 32

The base-catalyzed condensation of 3-formylpyrroles with nitromethane readily gives 3-(2-nitrovinyl)pyrroles (e.g., **391**),[574] which, when reduced to the β-aminoethyl derivatives, provide a convenient route to 3-vinylpyrroles.[414,575–577]

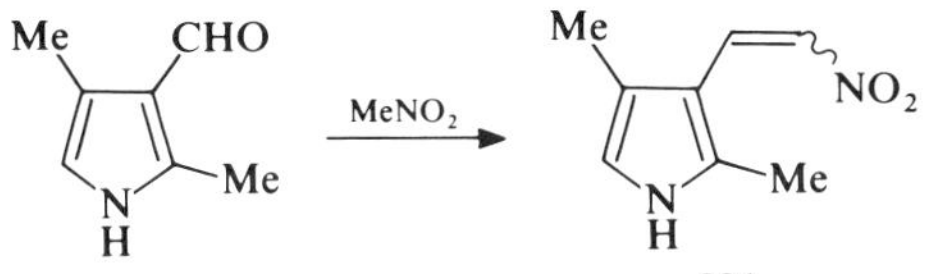

Representative 3-vinylpyrroles, essentially derived from 3-formylpyrroles by condensation with various CH-acids, are listed in Table 2.33.

TABLE 2.33. REPRESENTATIVE 3-VINYLPYRROLES[429]

Pyrrole	m.p. (°C)
3-vinyl-4-methyl-5-methyl-2-(EtO$_2$C)-pyrrole	112
3-vinyl-4-methyl-5-methyl-2-(HO$_2$C)-pyrrole	101–102
3-vinyl-4-ethyl-5-methyl-2-(EtO$_2$C)-pyrrole	79
3-(CH=CHCl)-4-Me-2-(EtO$_2$C)-5-(CO$_2$H)-pyrrole	241
3-(CH=CHBr)-4-Me-5-Me-2-(EtO$_2$C)-pyrrole	158
3-(CH=CHBr)-4-Me-2-(EtO$_2$C)-5-CHO-pyrrole	140
3-(CH=CHNO$_2$)-4-Me-5-Me-2-(EtO$_2$C)-pyrrole	231–232
3-(CH=CHCN)-4-Me-2,5-diMe-pyrrole	154
3-(CH=C(CN)$_2$)-4-Me-2-Me-pyrrole	148

2.4. 3-Vinylpyrroles

TABLE 2.33. Continued

Pyrrole	m.p. (°C)
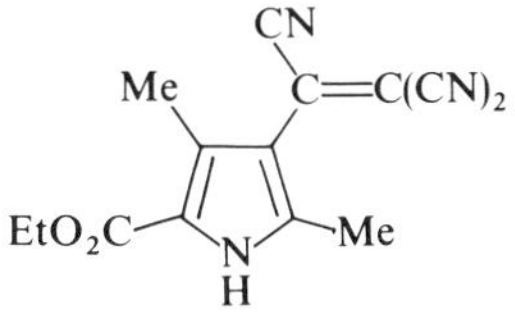 4-Me, 3-[C(CN)=C(CN)$_2$], 5-Me, 2-CO$_2$Et (EtO$_2$C on 5), NH pyrrole	163
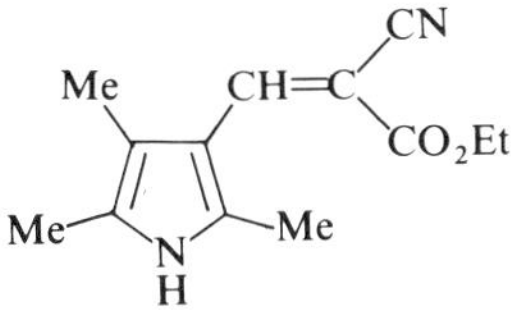 3,4-Me, 3-[CH=C(CN)(CO$_2$Et)], 2,5-Me, NH pyrrole	174
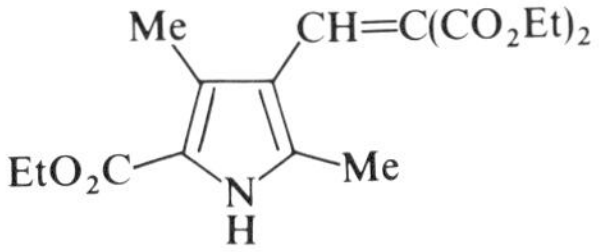 4-Me, 3-[CH=C(CO$_2$Et)$_2$], 5-CO$_2$Et, 2-Me, NH pyrrole	119–120

Physiologically active 3-(2-nitrovinyl)indoles (**393**) have also been obtained in near-quantitative yields by condensation of the corresponding aldehydes (**392**)

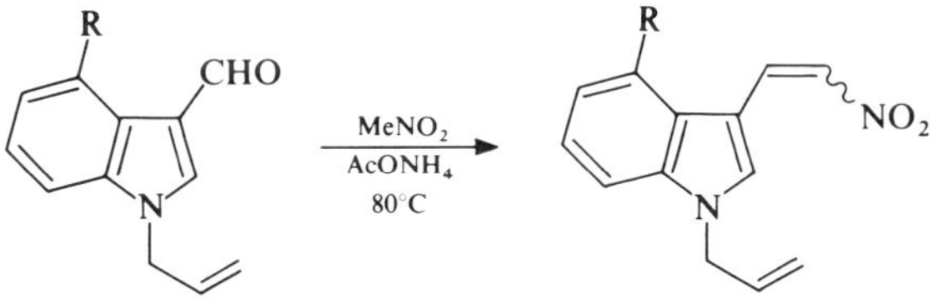

392 → (MeNO$_2$, AcONH$_4$, 80 °C) → **393**

R = CH$_2$=CHCH$_2$, CH$_2$—CHCH$_2$ (epoxide), CH$_2$—CHCH$_2$ (dioxolane),

HOCH$_2$CH(OH)CH$_2$, AcOCH$_2$CH(OH)CH$_2$,

t-BuNHCH$_2$CH(OH)CH$_2$

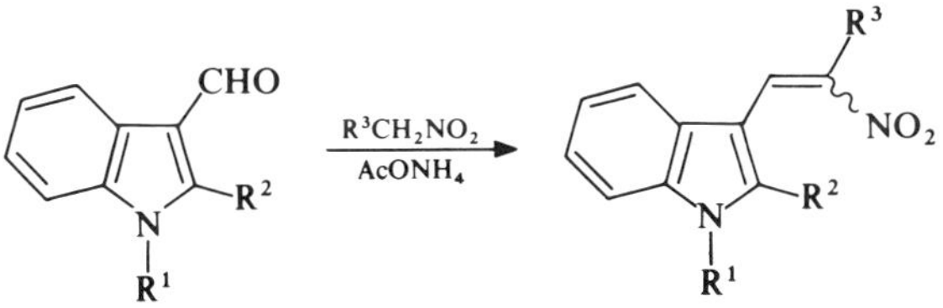

Indole-3-CHO with R^1 on N, R^2 on C2 → (R^{3}CH$_2$NO$_2$, AcONH$_4$) → **394**

R^1 = H, Me; R^2 = H, Me, Ph; R^3 = H, Me

with nitromethane.[578] Likewise, a wide range of the 3-(2-nitrovinyl)indoles (**394**) possessing antifungal activity have been synthesized in moderate (30%) to high (96%) yields.[579]

2,5-Dimethyl-3-formyl-1-phenylpyrrole undergoes condensation with 2-alkyl-3-methylisoquinolinium iodides (**395**) in the presence of piperidine to furnish high yields (86–96%) of isoquinolinopyrrolocarbocyanines (**396**).[580]

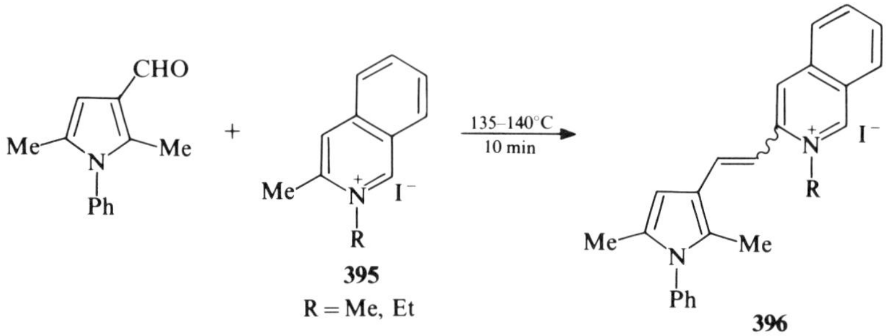

395
R = Me, Et

396

2.4.1.2. From β-Unsubstituted Pyrroles and Carbonyl Compounds

2,5-Disubstituted and 2,3,5-trisubstituted pyrroles react with formic acid at 100°C in the presence of hydrogen bromide to form dipyrrolylmethenes (e.g., **397**).[581]

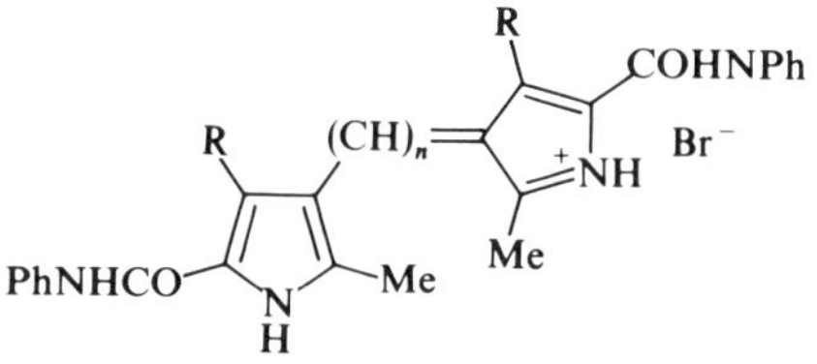

397
R = H, Me   n = 1,3

2.4.1.3. Via the Wittig Reaction

The report[582] that only a 13% yield of 3-styrylpyrrole was obtained from the Wittig reaction with 3-formyl-1-methylpyrrole is thought[2] to be surprising, as no difficulty was reported in the Wittig synthesis of other 3-vinylpyrroles and 2-(3-pyrrolyl) acrylic esters from 1-*tert*-butyl-3-formylpyrrole and several C-substituted 3-formylpyrroles.[2]

The Wittig reaction with 3-formylindole (**398**) to form the 2,3-divinylindole (**399**) has been successfully employed in the synthesis of hyellazole (**400**), a naturally occurring carbazole isolated from the blue-green alga *Hyella caespitosa* (Scheme 33).[583]

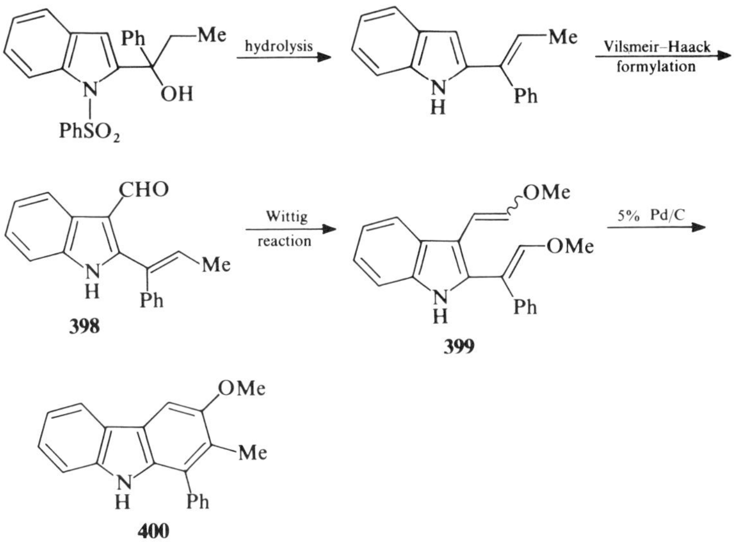

2.4. 3-Vinylpyrroles   253

Scheme 33

Recently,[584] 1-benzenesulfonyl-3-formylindole has also been converted to the corresponding 3-vinylindole by the Wittig reaction with methyltriphenylphosphonium bromide and BuLi.

2.4.1.4. Addition of 2,5-Disubstituted Pyrroles to Alkynes

When both α-positions of a pyrrole are substituted, the pyrrole may react under special conditions with electron-deficient acetylenes at the β-position. Thus, although no adducts have been detected in the reaction between 2,5-dimethylpyrrole (**401**) and DMAD under reflux in CH_2Cl_2 over a period of 30 h, when the reaction was conducted at high pressure (15 kbar, CH_2Cl_2, 40°C, 20 h), the 3-vinylpyrrole (**402**) with an $E : Z$ isomer ratio of 1 : 2 was obtained in a 20% overall yield. Column chromatography of the crude product on silica gel gave the E- and Z-isomers in 5 and 11% yields, respectively.[468] Not unexpectedly, 1,2-dimethylindole has also been reported[468] to react with DMAD to yield the Michael adduct at the 3-position.

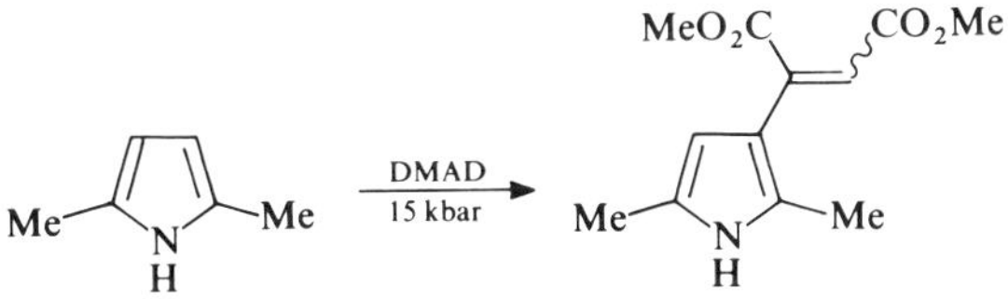

2.4.1.5. From 3-(2-Aminoethyl)pyrroles and Related Compounds

Because of their instability, the 3-vinylpyrroles cannot be used directly in the synthesis of naturally occurring porphyrins.[1,429] Instead, a variety of 3-(2-aminoethyl)pyrroles and related systems (**403**) have been proposed as synthetic equivalents,[570,576,585–594] from which the 3-vinyl moiety can be generated, when the construction of the porphyrin nucleus is complete.

$R = EtO_2CNH$ (Ref. 582), H_2N (Ref. 583), Et_2N (Refs. 584–587), AcNH (Refs. 588, 589), AcO (Ref. 590)

For example, 1,3,5,7-tetramethyl-2,4,6,8-tetra(2-diethylaminoethyl)porphyrin, prepared from 2,4-dimethyl-3-(2-diethylaminoethyl)-5-ethoxycarbonylpyrrole, affords 1,3,5,7-tetramethyl-2,4,6,8-tetravinylporphyrin on Hofmann degradation of the salt obtained by initial treatment with methyl iodide.[590] Similarly, diacetamidoethylporphyrin has been prepared from an acetamidoethylpyrrolic intermediate and then subjected to hydrolysis and Hofmann degradation to give the corresponding divinylporphyrins.[593]

In the case of 3-(2-acetoxyethyl) derivatives, the 3-vinyl group is generated by hydrolysis, followed by conversion of the alcohol to the chloro compound, which undergoes base-catalyzed dehydrohalogenation.[593]

2.4.1.6. From Oximes of Ethylenic Ketones and Acetylene

A new synthetic approach to 3-vinylpyrroles based on the reaction of alkenyl alkyl ketoximes with acetylene in the presence of a KOH/DMSO system (see Section 2.2.1.1), has been developed recently.[49,595] The example of the one-pot transformation of but-3-enyl methyl ketoxime (**404**) into 2-methyl-3-(2-propenyl)pyrrole (**405**) and 2-methyl-3-(1-propenyl)pyrrole (**406**) (E- and Z-isomers) displays two important features: (1) the reaction can be either stopped

E : Z ratio = 7:3

2.4. 3-Vinylpyrroles

selectively at the stage of the pyrrole ring formation with no N-vinylation or prototropic isomerization of the alkenyl group and (2) a complete sequence to yield a 1-vinylpyrrole with a shift of the alkenyl double bond into conjugation with the pyrrole ring.[595]

The pyrrole having the isomerized alkenyl group (**407**) is obtained when the reaction temperature does not exceed 100°C, the heating time is less than one hour (< 1 h), and there is no large excess of acetylene in the closed reaction vessel under a pressure of 3.5–7 bar. The synthesis of the pyrrole (**405**) is also successful when acetylene is passed through the stirring reaction mixture at atmospheric pressure. Optimum yields are attained if the reaction is continued until traces of other pyrroles (**406–408**) become detectable.

At a higher temperature (110°C) with an approximately tenfold excess of acetylene under a pressure of 10–14 bar, both isomerization and vinylation are completed within 3 h, yielding (**406**) ($E : Z$ ratio = 7 : 3), as the only isolable product in a yield of 80%. The reaction proceeds regiospecifically up to 100°C leading only to the introduction of but-3-enyl group in the pyrrole ring.[595]

2.4.1.7. Miscellaneous Syntheses

The attempted Vilsmeir–Haack formylation of the pyrrole (**409**) α-position results in the formation of the salt (**410**).[596]

Dihydroindoles of the type **411**, which are essentially 3-vinylpyrroles, are [4π + 2π] cycloadducts of DMAD to 2-vinylpyrroles[1,2,543] (see Section 2.3.2.1).

On the action of sodium malononitrile at ~50°C, the substituted 2,5-diazapentalene (**412**) undergoes partial ring opening to yield the 3,4-divinylpyrrole (**413**).[597]

2.4.2. Reactivity

2.4.2.1. [4π + 2π] Cycloaddition Reactions

Like 2-vinylpyrroles, 3-vinylpyrroles react with π-electron-deficient dienophiles in a [4π + 2π] manner to give the corresponding dihydro- or tetrahydroindoles.[2,535,542,543,597] The reaction is temperature-dependent and sensitive to steric constraint.

The reaction of 1-*tert*-butyl-3-vinylpyrrole (**414**) with maleic anhydride in chloroform has been reported[535] to be extremely rapid (20°C, 5 min), affording 1-*tert*-butyl-4,5,6,7-tetrahydroindole-Z-6,7-dicarboxylic anhydride (**415**) in a yield of 75%. The corresponding reaction with DMAD conducted under reflux in chloroform in the presence of hydroquinone furnishes dimethyl 1-*tert*-butyl-4,5-dihydroindole-6,7-dicarboxylate (**416**) in 55% yield.[535]

The reaction of **414** with methyl propiolate is slow even under reflux in benzene and can be stopped at the 1 : 1 adduct stage to produce methyl 1-*tert*-butyl-4,5-dihydroindole-7-carboxylate (**417**) in 44% yield (based on reacted

2.4. 3-Vinylpyrroles

vinylpyrrole). This contrasts with the analogous reaction of 2-vinylpyrroles, which, under similar conditions, give only the indole-4,7-carboxylic esters directly. Further reaction of **417** with a second molecule of methyl propiolate yields a mixture of dimethyl indole-4,7-dicarboxylate (**418a**) and its 1-*tert*-butyl derivative (**418b**).[535]

Although there is no apparent large difference in the reactivity of 1-alkyl-2- and 3-vinylpyrroles (when R = H) toward DMAD at ~ 65°C,[535] it was found that the 3-vinylpyrroles bearing electron-withdrawing substituents (**419**) required more vigorous conditions to yield the dihydroindoles (**420**) (30–72%) than were needed for the corresponding reaction of the analogous 2-vinyl isomers.[534] This is consistent with the inhibiting cross-conjugative effect expected of the electron-withdrawing substituent R on the concerted $[4\pi + 2\pi]$ cycloaddition of the π-excessive system of the 3-vinylpyrroles (**419**) with DMAD.[534] The electron-withdrawing substituents inhibited the Michael addition in all the cases, and only the 4,5-dihydroindoles (**422**) were isolated.

R = CHO, CO_2Me, COMe, NO_2

Extremely vigorous conditions (120°C, 48 h) were required to effect the cycloaddition of 2-cyano-1-methyl-4-isopropenylpyrrole (**421**) with DMAD; the reaction produced dimethyl 2-cyano-1,4-dimethylindole-6,7-dicarboxylate (**422**) directly in 48% yield.[598]

As a result of bond fixation within the indole system, the mesomeric interaction between the 3-vinyl group and the π-electron excessive heteroatomic system is stronger than that for the 2-vinylindoles.[553] Like 2-vinylindoles, 3-vinylindoles are heterocyclic dienes and, as such, represent synthetically attractive building blocks for the regio- and stereocontrolled [b] annelation of the indole skeleton[555] (for a recent review, see Ref. 599 and references cited therein). As a consequence of their 4π-donor reactivity, C-vinylindoles (E_{HOMO} ca. -7 to -8 eV)[555,584] participate in $[4\pi + 2\pi]$ cycloaddition reactions having inverse electron demand, i.e., with electron-deficient dienes. The cycloaddition of 3-vinylindoles (**423**) with dimethyl-1,2,4,5-tetrazine-3,6-dicarboxylate (**363**) under mild conditions (Scheme 34) leads exclusively to new annelated and functionalized pyridazines (**424**) with potential pharmacological interest as antihypertensive or antithrombotic drugs (see Ref. 555 and references cited therein). Within the limits of the 400-MHz ^{1}H NMR, only tautomers (**424**) were detected.

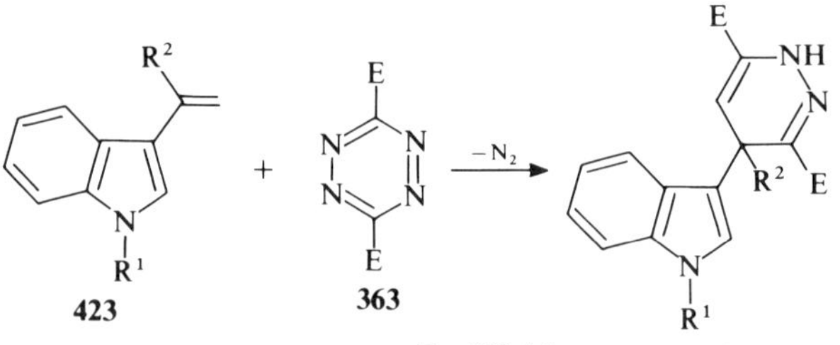

	R^1	R^2	% yield of **424**
a	Me	H	85
b	H	Indol-3-yl	86
c	Me	Me	67

Scheme 34

In an analogous reaction, the 3-vinylindole (**425**) yields the 4-(3-indolyl)pyridazine (**426**) as the only stable adduct in 45% yield.[555]

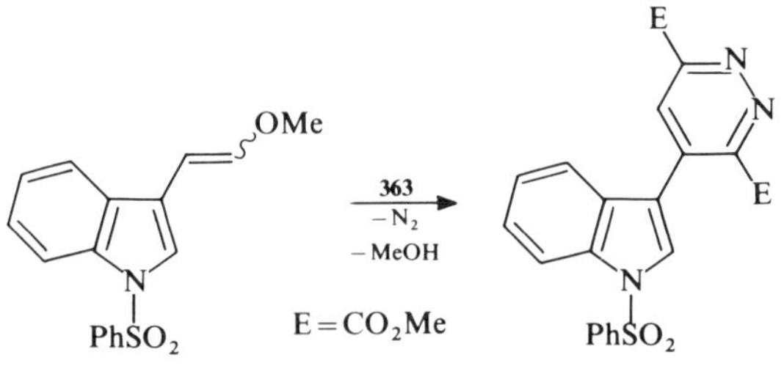

The 3-vinylindole (**427**) undergoes cycloaddition with **363** to furnish 4(3)-(3-indolyl)-1,4-dihydropyridazines (**429a,b**), which were separated by flash chromatography in 49 and 13% yields, respectively (Scheme 35). The two isomers are

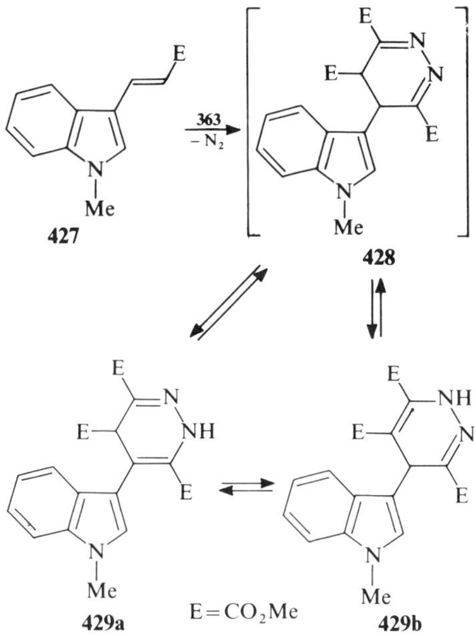

2.4. 3-Vinylpyrroles

Scheme 35

thermodynamically stable and do not equilibrate in solution within the limits of the 400-MHz ^{1}H NMR detection.[555] The reaction is considered to proceed via [$4\pi + 2\pi$] cycloaddition and extrusion of nitrogen to form the intermediate **428**, which is subsequently stabilized by [1,3]-prototropic shifts. Modified neglect of differential orbital (MNDO) calculations for 3-vinylindole predict a LUMO (diene) HOMO(dienophile)-controlled cycloaddition, with the (HOMO) energy value being -8.16 eV.[555]

The reaction of 3-vinylindoles with 4-phenyl-1,2,4-triazoline-3,5-dione leads to adducts of the type **430–432**, whereas the corresponding reaction of 2-vinylindoles gives adducts of the type **433**.[600]

In a reaction analogous to that for the vinylpyrrole (**361**), the 3-vinylindole (**434**) undergoes an intramolecular thermal cyclization to furnish the adduct (**435**), which is oxidatively converted into the carbazole.[601]

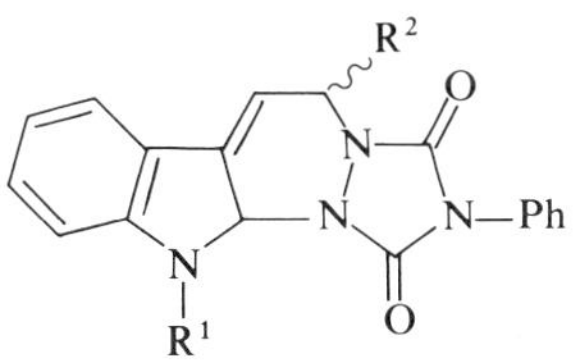

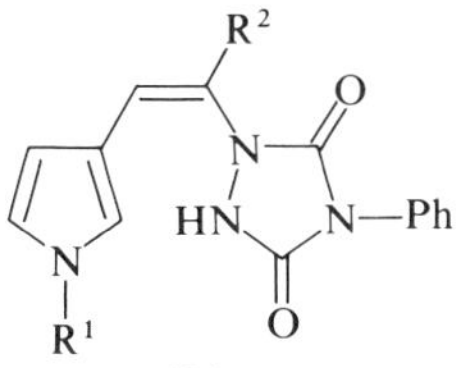

In contrast, 1-benzenesulfonyl-3-(pent-1-enyl)indole (**436**) ($E : Z$ ratio $= 8 : 10$) in toluene undergoes both [$4\pi + 2\pi$] cycloaddition and Michael

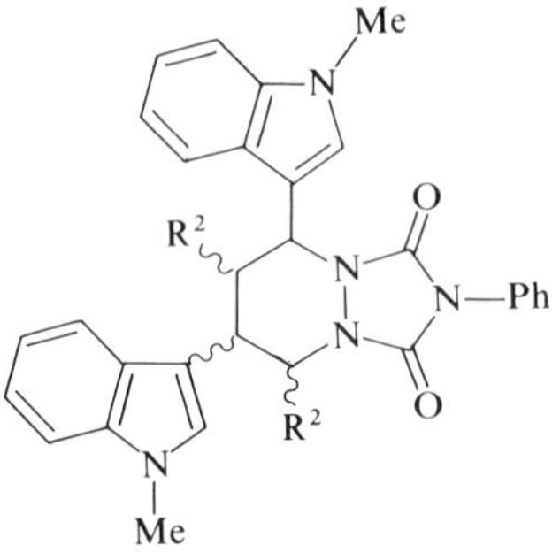

432

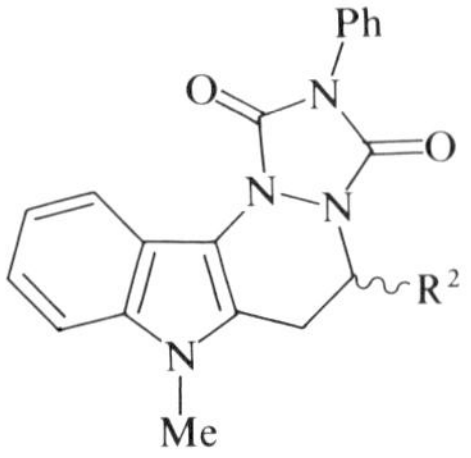

433

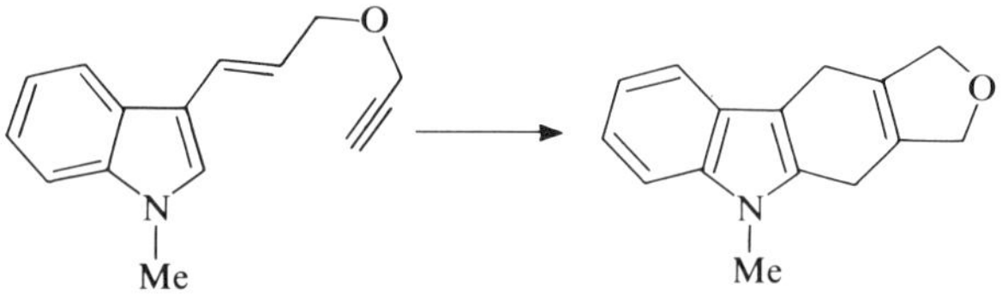

434 **435**

addition with *N*-phenylmaleimide in the presence of $AlCl_3$ to give the isomers (**437a, b**) and the 3-vinylindole (**438**) in 38.5 and 23% yield (Scheme 36).[602] In addition to the ^1H NMR structural assignment for all the compounds, the structure of the isomer (**437b**) has been established by X-ray analysis.[602]

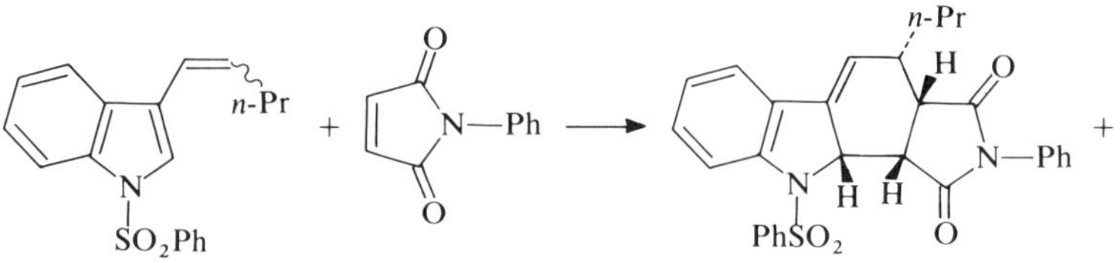

436 **437a**

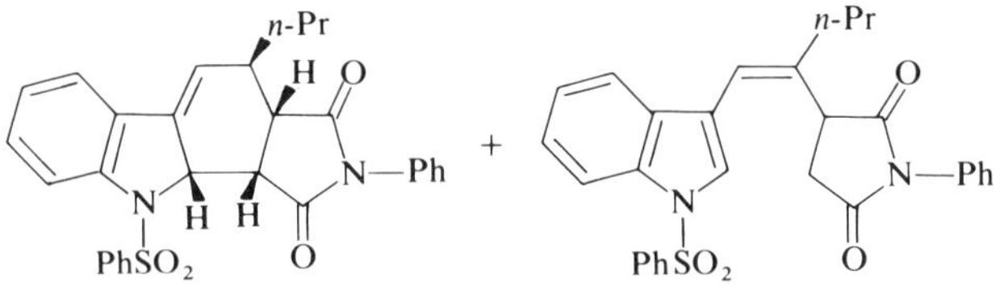

437b **438**

Scheme 36

The cycloaddition reactions of 6-oxo(2*H*)cyclohepta(*c*)pyrroles (**390**) with DMAD afford the 1 : 2 adducts (**439**)[569] via an initial reaction across the 2,5-positions of the pyrrole ring; the vinyl bonds are not involved in the cycloaddition reaction (Scheme 37). The kinetics of the reaction has been investigated by

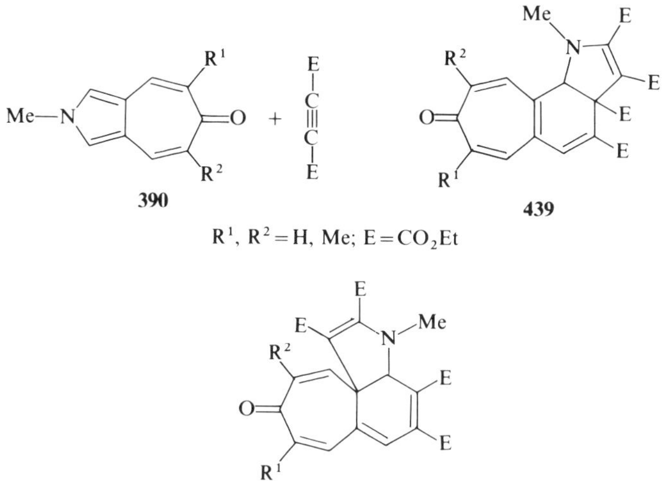

2.4. 3-Vinylpyrroles

Scheme 37

$R^1, R^2 = H, Me; E = CO_2Et$

^{1}H NMR spectroscopy, and an intermediate 1 : 2 adduct (**440**) has been detected and, in the case of $R^1 = R^2 = H$, it has been isolated.[569]

In contrast to 2-vinylpyrroles (see Section 2.3.2.2), 3-vinylpyrroles react with diethyl azodicarboxylate in an extremely complex manner and produce many unstable products in low yield.[569]

2.4.2.2. Other Reactions

The 1,3-dipolar cycloaddition of diphenylnitrilimine with 1-*tert*-butyl-3-vinylpyrrole proceeds rapidly at 80°C to give the 5-(1-*tert*-butyl-3-pyrrolyl)-4,5-dihydropyrazole (**441**) in high yield to the complete exclusion of the 4-isomer.[559] The unequivocal identification of the adduct (**441**) was achieved by comparison of its spectra (IR, NMR) and properties with those of the product obtained from the reaction of 1-phenyl-3-(1-*tert*-butyl-3-pyrrolyl)prop-2-en-1-one (**442**) with phenylhydrazine. The corresponding 1,3-dipolar cycloaddition reaction with methyl nitrile oxide was also regioselective and gave 3-methyl-5-(1-*tert*-butyl-3-pyrrolyl)-4,5-dihydroisoxazole (**443**).[555]

The observed regioselectivity to afford exclusively the isomers **441** and **443** confirms that the cycloadditions are HOMO-(alkene)-controlled as predicted by MO calculations of orbital energies.[559]

The cycloadduct (**444**) was isolated in low yield from the reaction of 1-*tert*-butyl-3-vinylpyrrole with *C*-phenyl-*N*-phenylnitrone.[559] The orientation of this cycloaddition is compatible with MO calculations of the orbital coefficients for a LUMO (dipole)-controlled cycloaddition with "large–large" and "small–small" orbital interactions. Secondary orbital effects producing favorable bonding overlaps appear to govern the cycloaddition such that only one stereo isomer (**444**) is formed.[559]

The 3-styrylpyrroles (**445a, b**), obtained by the Wittig reaction from corresponding 3-formylpyrroles (94 and 90%, respectively), photocyclize into benzo[*g*]indoles (**446a, b**) in small yields (~ 4%, after laborious chromatography workup).[562] In contrast, the 3-vinylindole (**447**), on irradiation in the presence of Pd/C under air-free conditions, gave a high yield (88%) of the photocyclization product (**448**).[561]

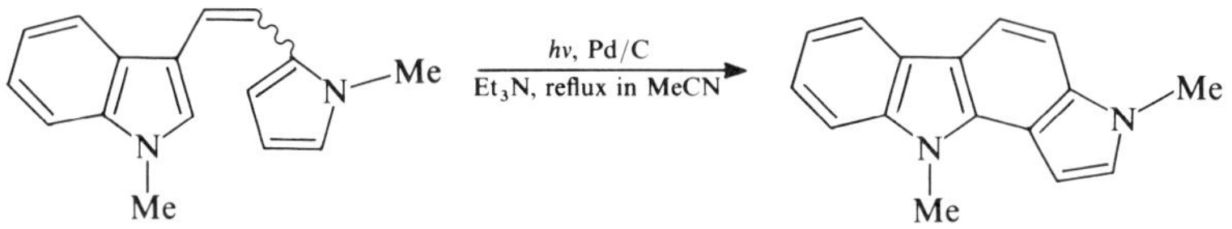

2.4. 3-Vinylpyrroles

A new synthetic approach to the dihydroindoles (**450**) based on 2,3-dialkenylpyrroles (**449**) has been developed (Scheme 38).[520] The electrocyclic and subsequent prototropic shift can be effected either thermally at ~216°C or photochemically.

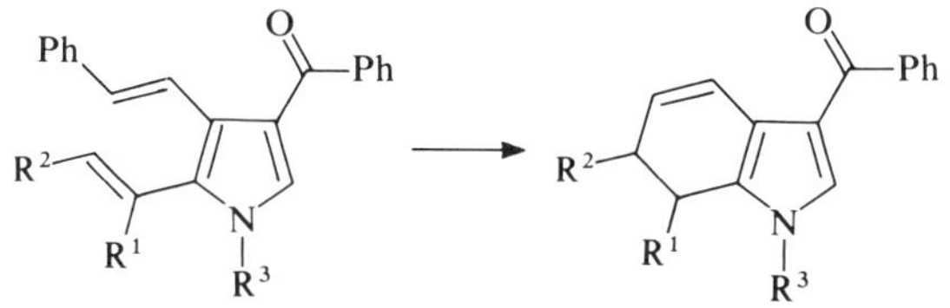

R^1	R^2	R^3	% yield of **450**
—(CH$_2$)$_4$—		H	94
—(CH$_2$)$_4$—		Me	96
—(CH$_2$)$_3$—		Me	92
Me	H	Me	87 (6,7- and 4,5-dihydroisomers, ~1 : 1)

Scheme 38

The 3-vinylindoles (**451a, b**) are formylated with (EtO)$_2$ĊHB̄F$_4$ at −50 to −20°C, to give the formyl derivatives (**452a, b**) (Scheme 39).[567] The corresponding reaction with 3-vinylpyrroles has not been observed.

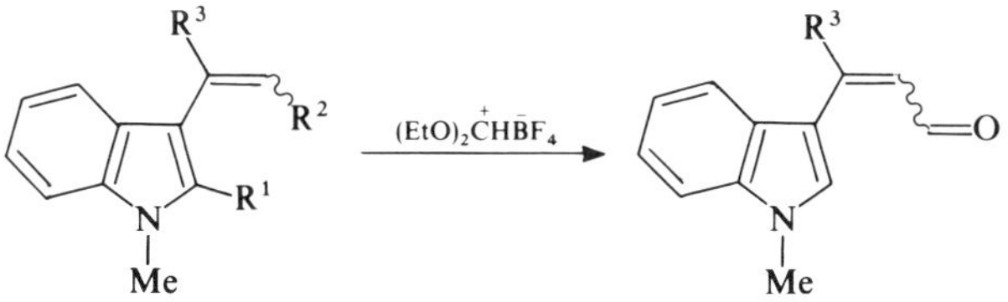

	R^1	R^2	R^3	% yield of **452**
a	H	H	Ph	25 (E : Z ratio = 8 : 2)
b	Me	H	1,2-Dimethyl-3-indolyl	18

Scheme 39

The 3-vinylpyrrole (**453**) reacts with Ehrlich's reagent in an unexpected manner at the vinyl group to furnish a blue pyrrolic dye (**454**).[482]

Catalytic hydrogenation, reduction with hydrogen iodide, bromination, hydrochlorination, oxidation with chromium trioxide, and the acid-catalyzed addition of methanol to ethyl 2,4-dimethyl-3-vinylpyrrole-5-carboxylate are well established. The analogous reduction (either catalytically or with hydrogen iodide), debromination to form the dimethoxy derivative (MeOH/Ag or AgCN), bromination, and reactions with sulfuryl chloride have also been described.[429]

2.5. PHYSICAL PROPERTIES OF *C*-VINYLPYRROLES

Unlike *N*-vinylpyrroles (see Section 2.2.2), the physical properties of *C*-vinylpyrroles, although known for a much longer time, have received virtually no systematic study, obviously as a result of the lack of relatively simple (model) representatives of the series.

Most of the known *C*-vinylpyrroles are heavily functionalized on both the pyrrole ring and the vinyl group with strongly interacting substituents (carboxy, cyano, nitro, acyl, etc.), which often drastically change the vinylpyrrole entity. Most spectral data (UV, IR, NMR) for *C*-vinylpyrroles are concerned with identification purposes and are scattered over a large number of publications that refer to the synthesis of such compounds. When combined, these data, obtained with different instruments, under different conditions, and over a long period of time, hardly ensure valid comparisons and generalizations.

2.5.1. Dipole Moments

Dipole moments of several 2-(2-aroylvinyl)pyrroles and 1,3-di(2-pyrrolyl)-propenone have been measured (Table 2.34).[391] According to their IR spectra, all the compounds were *E*-isomers in the *s-cis* conformation of the double bond relative to the carbonyl group. Additionally, comparison of the measured dipole moments with those calculated for *syn* and *anti* arrangements of the nitrogen atom with the carbonyl group indicated a preference for the *anti–s-cis* form for most of the vinylpyrroles (Table 2.34), except for 1,3-di(2-pyrrolyl)propenone, which exists in the *syn–anti–s-cis* conformation, and 2-[2-(4-nitrobenzoyl)-

2.5. Physical Properties of C-Vinylpyrroles

TABLE 2.34. DIPOLE MOMENTS OF
E-2-(2-AROYLVINYL)PYRROLES[391]

R	X	μ^a (D)
H	H	3.90
H	Me	4.12
H	MeO	4.21
H	Me$_2$N	5.13
H	Cl	4.02
H	Br	4.13
H	NO$_2$	5.32
Me	H	3.14
(pyrrole-CH=CH-C(O)-pyrrole)		2.93

^{a} At 25°C, in dioxane.

vinyl]pyrrole having an experimental dipole value nearer to that of the *syn* form. The polarity of these compounds are determined largely by the carbonyl group, which is sensitive to substitution not only on the benzene ring but also on the pyrrole ring. Introduction of electron-donating substituents (alkyl) at the α-position of the pyrrolyl ring depolarizes the molecule by 0.76D, seemingly due to a reduction of the positive charge on the carbonyl carbon atom demonstrating the ability of the 2-vinylpyrrole system to transmit π-electronic effects. In contrast, both electron-donating and electron-withdrawing substituents on the benzene ring enhance the polarity of the molecules; this may result from conformation changes of the system.[391]

2.5.2. Electronic Spectra

The UV spectrum of 2-vinylpyrrole consists of one absorption maximum at 274 nm (log ε 4.16), which changes position insignificantly on α- or β-substitution (or both) by alkyl groups, although the intensity of this absorption is decreased (log ε 3.0) (Table 2.35). This absorption band is ~ 60–65 nm to longer wavelength and approximately of the same order of intensity as those for pyrrole and 2-alkylpyrroles,[1,2] implying formation of a mutual conjugated system between the pyrrolyl ring and the vinyl group.

Elongation of the conjugated chain by β-substitution of the vinyl group with a second vinyl group, phenyl ring, or carbonyl group leads to a further

TABLE 2.35. ELECTRONIC SPECTRA OF SOME 2-VINYLPYRROLES

Pyrrole		λ_{max} (nm)	log ε	Ref.
2-vinyl		274	4.16	441
2-(propenyl)-Me		270[a]	3.05	514
2-(1-methyl-CH₂OH vinyl)		271		454
2-butadienyl		241	3.77	521[c]
		309.5	4.32	
		243[a]	3.02	514[b]
2-styryl	E	236	4.03	1[b]
		333	4.49	
	Z	229	4.05	441
		330	4.09	
2-(CH=CH-CHO)		350[a]		454[b]
2-(C(Me)=CH-CHO)		345[a]	4.48	454[b]
2-(CH=CH-COMe)		208	3.76	1
		353	4.37	
		348		454[b]
2-(CH=CH-COPh)		262[a]	3.97	1[b]
		386[a]	4.39	
	E	265	3.93	388[b]
		391	4.37	
2-(CH=CH-CO₂Me)		331	4.41	454[b]

TABLE 2.35. Continued

Pyrrole		λ_{max} (nm)	$\log \varepsilon$	Ref.
(pyrrole-C(Me)=CH-COMe)	E	325	441	454[b]
	Z	335	4.18	
(pyrrole-CH=CH-CO$_2$Et)	E	325	3.73	564
	Z	330	4.31	
(Ph-pyrrole-CH=CH-CO$_2$Me)	E	368	4.57	522
	Z	376	4.48	
(pyrrole-CH=C(Ph)-CO$_2$Et)	E	245	3.75	564[b]
		335	4.38	
	Z	239	3.86	
		343	4.35	
(Et-pyrrole-C(CO$_2$Me)=CH-CO$_2$Me)	Maleate	244	3.64	542
		350	3.59	
	Fumarate	381	3.29	
(Me-pyrrole-C(CO$_2$Me)=CH-CO$_2$Me)	Maleate	350	3.37	542
	Fumarate	382	3.49	
(vinyl-pyrrole-C(CO$_2$Me)=CH-CO$_2$Me)	Maleate	277	3.80	542
		366	4.35	
	Fumarate	282	3.88	
		378	2.08	
(isopropenyl-pyrrole(Me)-C(CO$_2$Me)=CH-CO$_2$Me)	Maleate	238	3.96	543
		355	4.21	
	Fumarate	266	3.71	
		390	3.21	
(t-Bu-isopropenyl-pyrrole-C(CO$_2$Me)=CH-CO$_2$Me)	Maleate	346	4.21	543
	Fumarate	380	3.03	

[a] Unknown configuration.
[b] In ethanol.
[c] In n-hexane.

bathochromic shift to 310–390 nm (log ε 3.6–4.6) and, in some cases, to the appearance of a shorter-wavelength maximum in the region of 210–280 nm (log ε 3.6–4.0) (Table 2.35). A similar trend is to be found in the electronic spectra of 2-pyrrolylpolyenes (Table 2.36): 2-(buta-1,3-dienyl)pyrroles absorb at 37–39 nm longer wavelength and with higher intensities than those of the corresponding 2-vinylpyrroles (cf. Tables 2.35 and 2.36), and each additional double bond added to the polyene system produces a "red" shift of 26–29 nm accompanied by an increased intensity (Table 2.36).

TABLE 2.36. PRINCIPAL ABSORPTION MAXIMA IN ELECTRON SPECTRA OF SOME 2-PYRROLYLPOLYENES[454]

Pyrrole (R)		λ_{max} (nm)[a]	log ε
(CH=CH)$_2$CO$_2$Me	E + Z	368	
(CH=CH)$_2$CH$_2$OH	E + Z	317[b]	4.54
CH=C(Me)CH=CHCO$_2$Me	E	364	4.53
	Z	368	4.53
CH=C(Me)CH=CHCH$_2$OH	E	312[b]	
	Z	319[b]	
(CH=CH)$_3$CO$_2$Me		394	4.65
(CH=CH)$_3$CH$_2$OH		342[b]	
CH=C(Me)(CH=CH)$_2$CO$_2$Me		393	4.61
CH=C(Me)(CH=CH)$_2$CH$_2$OH		340[b]	
CH=C(Me)(CH=CH)$_3$CO$_2$Me		409	4.64
CH=C(Me)(CH=CH)$_3$CH$_2$OH		366[b]	
CH=C(Me)CH=CHCOMe		382	
CH=C(Me)CH=CHC(OH)Me		312[c]	

[a] In ethanol.
[b] The absorption maximum of the crude reduction (LiAlH$_4$) product, in benzene.
[c] Sodium borohydride used for reduction.

E-2-(2-Methoxycarbonyl-1-methylvinyl)pyrrole has a molar extinction coefficient similar to that of the unmethylated analog, but the absorption maximum is at a shorter wavelength (Table 2.35). This is attributable to a steric interaction between the methyl group and the adjacent pyrrolyl hydrogen atoms. The strain can be relieved by the enoate system twisting out of the plane of the ring with a consequent decrease in the overlap of the pyrrolyl π-electrons with those of the vinyl group. The corresponding Z-isomer absorbs at a longer wavelength, but with a lower intensity.

The C-methyl groups on the side chain of the dienoates have little effect on the molar extinction coefficient, and the auxochromic effect of the groups on the position of the maximum appears to be exactly balanced, or even reversed, by

2.5. Physical Properties of C-Vinylpyrroles

steric effects (Table 2.36).[454] Similarly, comparison of the maxima for the trienoates (Table 2.36) indicates that the C-methyl groups have an insignificant influence on the wavelength and intensity of the absorption.

The position of the maxima for the highly unstable polyenic alcohols (Table 2.36), prepared by reduction of the corresponding esters,[454] are 40–70 nm to shorter wavelength than those of the parent enoates. A similar effect is observed on reduction of the dienic ketones to secondary alcohols.

2.5.3. IR Spectra

The C=C stretching vibration of the vinyl groups of a series of 2-vinylpyrroles (Table 2.37) ranges from 1570 to 1655 cm^{-1} depending on the structure of the both ring and vinyl substituents. For 2-vinylpyrroles, unsubstituted on the vinyl group, the highest frequency is found for 1-*tert*-butyl-2-vinylpyrrole, which may have a weaker conjugation between the double bond and the pyrrolyl ring due to distortion of coplanarity similar to that noted for 2-*tert*-butyl-1-vinylpyrrole (see Section 2.23).

TABLE 2.37. VINYL ABSORPTION MAXIMA IN IR SPECTRA OF 2-VINYLPYRROLES

Pyrrole	$v_{C=C}$ (cm^{-1})	δCH (cm^{-1})	Ref.
(pyrrole, NH)		985	
(Me-pyrrole, N-Me(Et))	1618	894–902 982	519
(pyrrole, N-Me)	1625		542
(pyrrole, N-Ph)	1625		542
(pyrrole, N-*t*-Bu)	1635		542
(pyrrole-CH=CHMe, NH)	1655^a		514

TABLE 2.37. Continued

Pyrrole		$\nu_{C=C}$ (cm^{-1})	δCH (cm^{-1})	Ref.
2-(CH=CH)Ph pyrrole		1640^a		514
2-(CH=CH-CH=CH$_2$) pyrrole		1617 1588	995 (E) 902 893^c	441^b
2-(CH=CH)CHO pyrrole		1625	968	454^d
2-(CH=CH)COMe pyrrole		1615 1595	960	454
2-(CH=CH)CO$_2$Me pyrrole		1630	970	454
2-(CH=CH)CO$_2$Et pyrrole	E Z	1620 1600	975	564
2-(CH=CH)CH=CHCO$_2$Me pyrrole	E + Z	1615	990	454
2-(CH=CH)(CH=CH)$_2$CO$_2$Me pyrrole		1595	995 882	454
2-(C(Me)=CH)CO$_2$Me pyrrole	E Z	1625 1605	910	454
2-(CH=C(Me))CH=CHCO$_2$Me pyrrole	E Z	 1615	985 882 990 960 880	454
2-(CH=C(Me))(CH=CH)$_2$CO$_2$Me pyrrole		1595^a	980 882	454

TABLE 2.37. Continued

Pyrrole	$v_{C=C}$ (cm^{-1})	δCH (cm^{-1})	Ref.
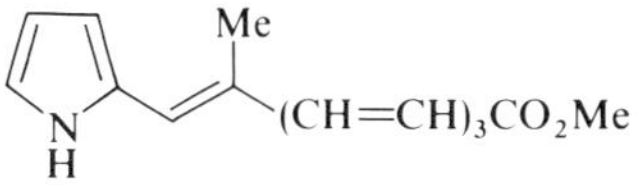 [pyrrole with Me and (CH=CH)₃CO₂Me substituents]	1593^a	980 882	454
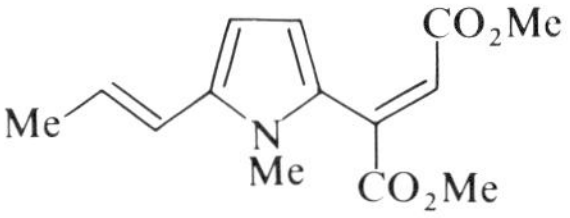 [pyrrole with Me, CO₂Me, CO₂Me substituents]	1612	970	554
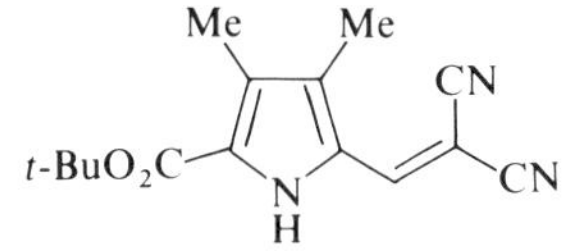 [pyrrole with Me, Me, CN, CN, t-BuO₂C substituents]	1580 1570		417

^{a} Unknown configuration.
^{b} Measured as a KBr disk.
^{c} Overtone at 1800 cm^{-1}.
^{d} Measured in Nujol.

As a rule, the $v_{C=C}$ frequency is lowered on the introduction of substituents, which are capable of conjugation, into both the pyrrole ring and the vinyl group. Elongation of the conjugated chain from methyl 2-pyrrolylacrylate to methyl 2-(2-pyrrolyl)nona-2,4,6,8-tetraenoate results in a regular decrease in frequency from 1630 to 1593 cm^{-1} (Table 2.37). Generally, the Z-isomers absorb at lower frequencies.

A varying number of vinyl CH deformation frequencies have been reported (Table 2.37).

2.5.4. NMR Spectra

The 2- and 3-vinylpyrroles (**455** and **456**) with strong electron-withdrawing substituents are characterized[598] by three sets of ^{1}H NMR signals (doublets of

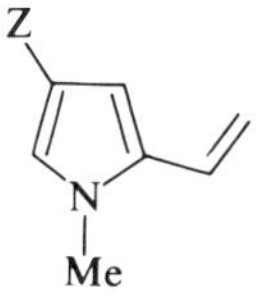

455 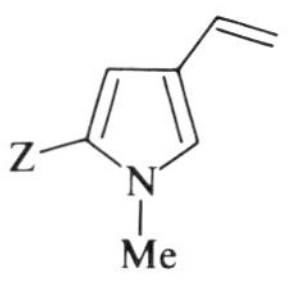 **456**

Z = CHO, COMe, CO₂Me, CN, NO₂

TABLE 2.38. ¹H NMR SPECTRA OF SOME 2-VINYLPYRROLES

Pyrrole	Pyrrolic Protons				Vinyl Protons		Ref.
	H_3	H_4	H_5	H_α	H_β		
(pyrrole, NH, α, β)	5.98 t	6.33 m	6.62 Brs	6.63 dd	4.97 d 4.46 d		558[a]
	5.90 m	6.17 dd	6.36 t	6.46 dd (17.5, 11.2 Hz)	5.33 dd (17.5, 1.6 Hz) 4.91 dd (11.2, 1.6 Hz)		542[b]
(N-t-Bu)	6.05 m	6.31 m	6.49 m	6.55 dd	5.42 dd 4.50 dd		535[c]
(N-)	6.07 m	6.15 m	6.60 m	6.45 dd	5.24 dd 4.80 dd		535[c]
(N-)	6.16 t	6.38 t	6.46 m	6.40 dd	5.37 dd 4.88 dd		535[c]
(N-Ph)	6.12 dd	6.42 dd	6.62 dd	6.33 dd (17.5, 11.2 Hz)	5.34 dd (17.5, 1.7 Hz) 4.86 dd (11.0, 1.7 Hz)		542[b]
(Me, N-Me)	6.06 t	6.18 m	6.63 m	6.30 d	5.62 dd		558[a]

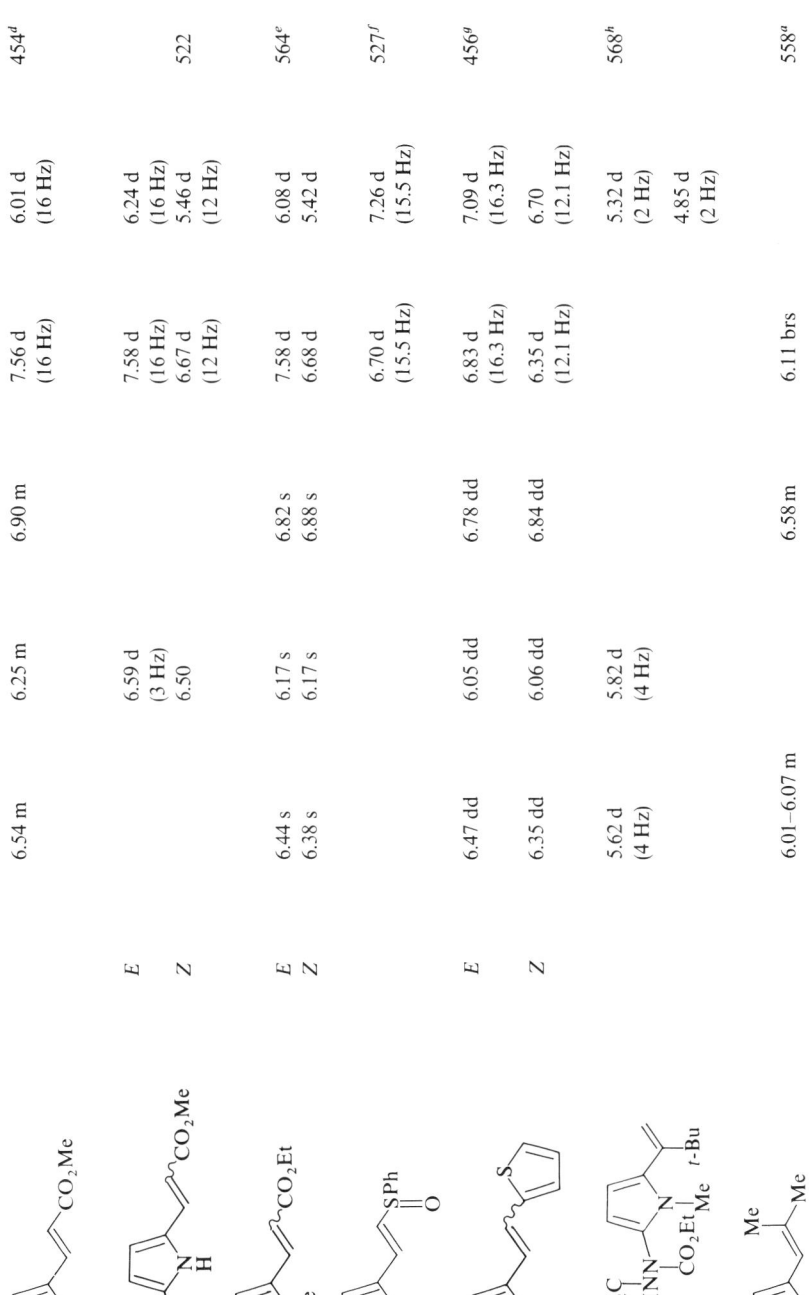

Structure						
pyrrole-CH=CH-CO₂Me		6.54 m	6.25 m	6.90 m	7.56 d (16 Hz)	6.01 d (16 Hz)	454[d]
5-Ph-pyrrole-CH=CH-CO₂Me	E		6.59 d (3 Hz)		7.58 d (16 Hz)	6.24 d (16 Hz)	522
	Z		6.50		6.67 d (12 Hz)	5.46 d (12 Hz)	
N-Me-pyrrole-CH=CH-CO₂Et	E	6.44 s	6.17 s	6.82 s	7.58 d	6.08 d	564[e]
	Z	6.38 s	6.17 s	6.88 s	6.68 d	5.42 d	
pyrrole-CH=CH-S(O)Ph	E	6.47 dd	6.05 dd	6.78 dd	6.70 d (15.5 Hz)	7.26 d (15.5 Hz)	527[f]
pyrrole-CH=CH-thiophene		6.35 dd	6.06 dd	6.84 dd	6.83 d (16.3 Hz)	7.09 d (16.3 Hz)	456[g]
					6.35 d (12.1 Hz)	6.70 (12.1 Hz)	
EtO₂C-HN-N(CO₂Et)-pyrrole(t-Bu, vinyl)		5.62 d (4 Hz)	5.82 d (4 Hz)			5.32 d (2 Hz)	568[h]
						4.85 d (2 Hz)	
N-Me-pyrrole-CH=CMe₂		6.01–6.07 m		6.58 m	6.11 brs		558[a]

273

TABLE 2.38. Continued

Pyrrole		Pyrrolic Protons			Vinyl Protons		Ref.
		H_3	H_4	H_5	H_α	H_β	
![pyrrole with Me, CO2Me vinyl]	E	6.59 m	6.35 m	6.95 m	7.57 brs		454[d]
	Z	6.40 m	6.25 m	6.75 brm	6.94 m		
![pyrrole with CO2Et vinyl]	E	6.48 m	6.15 m	6.79 m		5.90 q	564[e]
	Z	6.52 m	6.18 m	6.92 m		5.50 q	
![pyrrole with Me, CO2Me, CO2Me]	E	5.81 dq (4.0, 0.8 Hz)	6.19 d (4.0 Hz)			5.73 s	542[b]
	Z	5.7 dq (3.6, 0.8 Hz)	5.92 d (3.6 Hz)			6.68 s	

[a] At 250 MHz, in $(CD_3)_2CO$, $CDCl_3$, C_6D_6.
[b] At 100 MHz, in CCl_4 (0.2 M).
[c] At 100 MHz, in $CDCl_3$ (1.0 M).
[d] At 60 MHz in $CDCl_3$.
[e] In CS_2.
[f] In $CDCl_3$.
[g] In DMSO.
[h] At 60 and 100 MHz, in $CDCl_3$ (~30%).

TABLE 2.39. ^{13}C CHEMICAL SHIFTS OF SOME 2-VINYLPYRROLES

Pyrrole	C_2	C_3	C_4	C_5	C_α	C_β	Ref.
1-Me-2-vinylpyrrole	132.0	107.7	106.5	123.1	125.4	110.8	568[a]
1-Ph-2-vinylpyrrole	132.2	109.2	107.2	123.0	126.3	111.1	568[a]
EtO₂C-HNN / t-Bu / CO₂Et, Me	131.6	106.1	103.3	128.6	150.7	116.2	542[b]
Et-pyrrole-CH=C(CO₂Me)₂ (E)	141.0	109.6	107.1	143.3	126.7	115.6	568[c]
(Z)	135.5	111.4	105.0	138.2	124.6	127.8	
vinyl-pyrrole-CH=C(CO₂Et)₂ (E)	140.4	111.9	107.8	139.3	124.9, 128.3	115.8, 115.8	568[a]
(Z)	135.0	112.3	106.4	134.8	125.5, 129.1	112.7, 129.3	

275

TABLE 2.39. Continued

Pyrrole	C_2	C_3	C_4	C_5	C_α	C_β	Ref.
Me-Et pyrrole-CN/CO$_2$Me	124.0	141.0	121.0	128.7	139.2	88.0	418[d]
Me-Et-I pyrrole-CN/CO$_2$Me	128.0	140.2	127.3	84.7	137.1	89.7	418[d]

[a] In CDCl$_3$ (~ 30%).
[b] In CDCl$_3$.
[c] In CCl$_4$ (0.2 M).
[d] At 100.6 MHz, in CDCl$_3$.

doublets) at 5.04 ± 0.15, 5.44 ± 0.15, and 6.46 ± 0.10 ppm, which are assignable to the vinyl group, while the pyrrolyl ring protons resonate at 6.65 ± 0.5 and 7.04 ± 0.15 ppm.

Chemical shifts ^{1}H NMR and ^{13}C NMR and, in some cases, coupling constants of a wide range of substituted 2-vinylpyrroles are recorded in Tables 2.38 and 2.39.

During their studies of the ^{13}C NMR spectra of the products from the Michael addition of DMAD with α-unsubstituted pyrroles, Jones et al.[542] observed that there was an excellent correlation between the value of the $^3J_{CO,H}$

TABLE 2.40. $^3J_{C,H}$ COUPLING CONSTANTS OF THE VINYL GROUP OF 2-VINYLPYRROLES[603]

					$^3J_{C,H}$ (Hz)	
R^1	R^1	R^3	R^4	R^5	cis	trans
H	H	H	H	CO_2Me	6.6	
Me	H	H	H	CO_2Me	6.7	
H	H	H	CO_2Me	CO_2Me	7.4	12.2
Me	H	H	CO_2Me	CO_2Me	7.6	11.7
H	H	H	CN	CO_2Et	6.4	13.2
Me	H	H	CN	CO_2Et	6.6	12.8
H	H	CO_2Et	CN	H	6.9	
Me	Me	CO_2Me	H	CO_2Me		12.2
Me	Me	CO_2Me	CO_2Me	H	6.7	
Me	Et	CO_2Me	H	CO_2Me		12.0
Me	Et	CO_2Me	CO_2Me	H	7.0	
Me	CH_2=CH	CO_2Me	H	CO_2Me		12.1
Me	CH_2=CH	CO_2Me	CO_2Me	H	6.9	
Me	CH_2=C(Ph)	CO_2Me	H	CO_2Me		12.1
Me	CH_2=C(Ph)	CO_2Me	CO_2Me	H	6.8	
H	H	CO_2Et	H	Ph		12.7
H	H	CO_2Et	Ph	H	7.5	
H	H	CO_2Et	H	2-Thienyl		12.1
H	H	CO_2Et	2-Thienyl	H	6.5	
H	H	CO_2Et	H	2-Furyl		12.5
H	H	CO_2Et	2-Furyl	H	7.4	
H	H	H	H	COMe	6.4	
Me	H	H	H	COMe	6.0	
Me	H	H	COMe	COMe	7.1	9.8
H	H	H	H	CN	8.7	
Me	H	H	H	CN	8.6	
H	H	H	CN	CN	7.8	13.2
Me	H	H	CN	CN	7.2	12.9

vicinal coupling (the α-$^{13}CO_2R$ interacts with the β-alkenic proton) and the configuration of the but-2-en-1,4-dioates, irrespective of the nature of the pyrrole ring. Later,[603] they studied ^{13}C NMR spectra of numerous E- and Z-unsaturated esters, ketones, and nitrile, including a number of 2-vinylpyrrole derivatives of the type shown in Table 2.40, and noted that the values of $^3J_{C,H}$ for the two configurational isomers lie within two distinct and well separated and narrow ranges 7.1 ± 0.5 and 11.4 ± 1.1 for the *cis* and *trans* arrangements, respectively, and that substituents have little effect on this parameter. Thus, the measurement of the $^3J_{C,H}$ values from the ^{13}C NMR spectra provides an unequivocal assignment of the configuration of the 2-pyrrolyl substituted α,β-unsaturated esters, ketone and nitrile, even when only one of the isomers is available for analysis.

2.6. Acknowledgments

This chapter would never have seen the light without the skillful and devoted activity of my associates A. I. Mikhaleva, E. A. Petrova, R. N. Nesterenko, O. V. Petrova, O. A. Tarasova, and V. N. Salaurov.

2.7. REFERENCES

1. A. Gossauer, *Die Chemie der Pyrrole*, Springer-Verlag, Berlin, 1974.
2. R. A. Jones and G. P. Bean, *The Chemistry of Pyrroles*, Academic Press, London, 1977.
3. M. F. Shostakovsky, G. G. Skvortsova, and E. S. Domnina, *Usp. Khim.*, **38**, 892 (1969); *Chem. Abstr.*, **71**, 38673 (1969).
4. H. Nomori, M. Hatano, and S. Kambara, *Polym. Lett.*, **4**, 623 (1966).
5. W. Reppe, *Liebigs Ann. Chem.*, **601**, 128 (1956).
6. P. Vatsulik, *Chemistry of Monomers* (in Russian), Inostrannaya Literatura, Moscow, 1960.
7. S. A. Miller, *Acetylene. Its Properties, Manufacture and Uses*, Vol. 2, Benn, London, 1966.
8. U. S. Pat. 3047583 (1962); *Chem. Abstr.*, **58**, 509 (1963).
9. G. Cooper, W. Irwin, and D. L. Wheeler, *Tetrahedron Lett.*, **1971**, 4321.
10. W. Irwin and D. L. Wheeler, *Tetrahedron*, **28**, 1113 (1972).
11. B. A. Trofimov, *Heteroatomic Derivatives of Acetylene. New Polyfunctional Monomers, Reagents and Intermediates* (in Russian), Nauka, Moscow, 1981.
12. B. A. Trofimov and A. I. Mikhaleva, *Khim. Geterotsikl. Soedin.*, **1980**, 1299; *Chem. Abstr.*, **94**, 47025 (1981).
13. B. A. Trofimov, N. I. Golovanova, N. I. Shergina, A. I. Mikhaleva, S. E. Korostova, A. N. Vasil'ev, and I. L. Anisimova, *Atlas of Spectra of Aromatic and Heterocyclic Compounds* (in Russian), Vol. 21, Institute of Organic Chemistry Sib. Div. Akad. Nauk SSSR, Novosibirsk, 1981.
14. B. A. Trofimov and A. I. Mikhaleva, *N-Vinylpyrroles* (in Russian), Nauka, Novosibirsk, 1984; *Chem. Abstr.*, **102**, 203864 (1985).
15. B. A. Trofimov, *Usp. Khim.*, **50**, 249 (1981).
16. B. A. Trofimov, *Zh. Org. Khim.*, **22**, 1991 (1985); *Chem. Abstr.*, **94**, 207925 (1981).

2.7. References

17. B. A. Trofimov, A. I. Mikhaleva, and L. V. Morozova, *Usp. Khim.*, **54**, 1034 (1985); *Chem. Abstr.*, **103**, 123936 (1985).
18. B. A. Trofimov, *Usp. Khim.*, **58**, 1703 (1989).
19. B. A. Trofimov, *Z. Chem.*, **26**, 41 (1986).
20. M. V. Sigalov, thesis for Candidate of Chemical Sciences, Irkutsk, 1979.
21. T. A. Tandura, thesis for Candidate of Chemical Sciences, Irkutsk, 1980.
22. L. N. Sobenina, thesis for Candidate of Chemical Sciences, Irkutsk, 1983.
23. V. V. Druzhinina, thesis for Candidate of Chemical Sciences, Moscow, 1983.
24. E. Yu. Shmidt, thesis for Candidate of Chemical Sciences, Irkutsk, 1988.
25. USSR Inventor's Certificate 455596 (1978), *Official Gazette*, **968**, 1 (1978).
26. Br. Pat. 1463228 (1975); *Chem. Abstr.*, **87**, 53074 (1977).
27. Ger. Pat. 2543850 (1975); *Chem. Abstr.*, **87**, 53074 (1977).
28. U.S. Pat. 4077975 (1975); *Chem. Abstr.* **87**, 53074 (1977).
29. Jpn. Pat. 1090993 (1975), *Izobret. v SSSR i za Rubezhom*, 125 (1982); *Chem. Abstr.*, **81**, 8415 (1974).
30. B. A. Trofimov, A. S. Atavin, A. I. Mikhaleva, G. A. Kalabin, and E. G. Chebotareva, *Zh. Org. Khim.*, **9**, 2205 (1973); *Chem. Abstr.*, **80**, 3694481 (1973).
31. USSR Inventor's Certificate 463666 (1975), *Byull. Izobret.*, N 10 (1975); *Chem. Abstr.*, **89**, 28092 (1975).
32. B. A. Trofimov, A. S. Atavin, A. I. Mikhaleva, S. E. Korostova, E. G. Chebotareva, and L. N. Balabanova, in "The Conference of Young Specialists of the Aniline Dye Industry," *Abstracts in Russian*, p. 14, Moscow, 1974.
33. B. A. Trofimov, A. I. Mikhaleva, G. A. Kalabin, and A. S. Atavin, in "The Fourth All-Union Colloquium on the Chemistry and Pharmacology of Indole Compounds," *Abstracts in Russian*, p. 24, Kishinev, 1975.
34. B. A. Trofimov, A. I. Mikhaleva, S. E. Korostova, and L. N. Balabanova, in "Vth International Congress on Heterocyclic Chemistry," Abstracts, p. 63, Ljubljana, 1975.
35. B. A. Trofimov, G. A. Kalabin, A. S. Atavin, A. I. Mikhaleva, and E. G. Chebotareva, *Khim. Geterotsikl. Soedin.* **1975**, 360; *Chem. Abstr.*, **82**, 169630 (1975).
36. B. A. Trofimov, G. A. Kalabin, A. S. Atavin, A. I. Mikhaleva, and E. G. Chebotareva, in "IInd Symposium on Chemistry and Technology of Heterocyclic Compounds of Fuel Fossils," *Abstracts in Russian*, p. 55, Donetsk, 1973.
37. B. A. Trofimov, in "VIth All-Union Scientific Conference on the Chemistry of Acetylene and Its Derivatives", *Abstracts in Russian*, part 1, p. 3, Baku, 1979.
38. A. I. Mikhaleva, B. A. Trofimov, and A. N. Vasil'ev, ibid., Part 2, p. 42.
39. B. A. Trofimov, in "All-Union Conference on the Chemistry of Heterocyclic Compounds," *Abstracts in Russian*, Vol. 1, p. 47, Riga, 1979.
40. Ya. P. Stradyn, *Khim. Geterotsikl. Soedin.*, **1979**, 1567.
41. Ya. P. Stradyn, ibid., **1981**, 1412.
42. USSR Inventor's Certificate 518493 (1976); *Chem. Abstr.*, **86**, 29620 (1977).
43. B. A. Trofimov, N. I. Golovanova, A. I. Mikhaleva, S. E. Korostova, and A. S. Atavin, *Khim. Geterotsikl. Soedin.*, **1975**, 1225; *Chem. Abstr.*, **84**, 29938 (1976).
44. B. A. Trofimov, A. I. Mikhaleva, and R. N. Nesterenko, *Zh. Org. Khim.*, **14**, 1119 (1978); *Chem. Abstr.*, **89**, 163334 (1978).
45. B. A. Trofimov, A. I. Mikhaleva, R. I. Polovnikova, R. N. Nesterenko, and F. G. Trofimova, ibid., p. 2628; *Chem. Abstr.*, **90**, 137609 (1979).
46. B. A. Trofimov, V. K. Voronov, A. I. Mikhaleva, R. I. Polovnikova, R. N. Nesterenko, and M. V. Sigalov, *Izv. Akad. Nauk SSSR. Ser. Khim.*, **1979**, 2372; *Chem. Abstr.*, **92**, 110423 (1980).

47. B. A. Trofimov, A. I. Mikhaleva, R. N. Nesterenko, A. N. Vasil'ev, A. S. Nakhmanovich, and M. G. Voronkov, *Khim. Geterotsikl. Soedin.*, **1977**, 1136; *Chem. Abstr.*, **87**, 201232 (1977).
48. S. E. Korostova and A. I. Mikhaleva, *Zh. Org. Khim.*, **18**, 2620 (1982); *Chem. Abstr.*, **98**, 107107 (1983).
49. A. I. Mikhaleva, M. V. Sigalov, and G. A. Kalabin, *Tetrahedron Lett.*, **23**, 5063 (1982).
50. A. I. Mikhaleva, B. A. Trofimov, and A. N. Vasil'ev, *Khim. Geterotsikl. Soedin.*, **1979**, 197; *Chem. Abstr.*, **91**, 39240 (1979).
51. A. I. Mikhaleva, A. N. Vasil'ev, and B. A. Trofimov, *Zh. Org. Khim.*, **17**, 1977 (1981); *Chem. Abstr.*, **95**, 203682 (1981).
52. B. A. Trofimov, S. E. Korostova, L. N. Balabanova, and A. I. Mikhaleva, *ibid.*, **14**, 1733 (1978); *Chem. Abstr.*, **89**, 197255 (1978).
53. B. A. Trofimov, S. E. Korostova, L. N. Balabanova, and A. I. Mikhaleva, *Khim. Geterotsikl. Soedin.*, **1978**, 489; *Chem. Abstr.*, **89**, 42986 (1978).
54. B. A. Trofimov, S. E. Korostova, L. N. Balabanova, and A. I. Mikhaleva, *Zh. Org. Khim.*, **14**, 2182 (1978); *Chem. Abstr.*, **90**, 71976 (1979).
55. USSR Inventor's Certificate 601282 (1978); *Chem. Abstr.*, **89**, 24141 (1978).
56. A. I. Mikhaleva, B. A. Trofimov, and A. N. Vasil'ev, *Zh. Org. Khim.*, **15**, 602 (1979); *Chem. Abstr.*, **91**, 56737 (1979).
57. A. I. Mikhaleva, B. A. Trofimov, A. N. Vasil'ev, P. M. Alkhimenkov, I. S. Zhikharev, L. I. Sineva, and N. P. Vasil'ev, *Izv. Sibirsk. Otdeleniya Akad. Nauk. SSSR. Ser. Khim. Nauk*, 150 (1981); *Chem. Abstr.*, **95**, 6964 (1981).
58. B. A. Trofimov, A. I. Mikhaleva, A. N. Vasil'ev, and M. V. Sigalov, *Khim. Geterotsikl. Soedin.*, **1978**, 54; *Chem. Abstr.* **88**, 169214 (1978).
59. A. I. Mikhaleva, I. A. Aliev, R. N. Nesterenko, and G. A. Kalabin, *Zh. Org. Khim.*, **18**, 2229 (1982); *Chem. Abstr.*, **98**, 71862 (1983).
60. B. A.Trofimov, S. E. Korostova, L. N. Sobenina, A. I. Mikhaleva, M. V. Sigalov, and A. S. Atavin, *Khim. Geterotsikl. Soedin.*, **1982**, 193; *Chem. Abstr.*, **96**, 181092 (1982).
61. S. E. Korostova, B. A. Trofimov, L. N. Sobenina, A. I. Mikhaleva, and M. V. Singalov, *ibid.*, **1982**, 1351; *Chem. Abstr.*, **98**, 89108 (1983).
62. B. A. Trofimov, A. I. Mikhaleva, R. I. Polovnikova, S. E. Korostova, R. N. Nesterenko, N. I. Golovanova, and V. K. Voronov, *ibid.*, **1981**, 1058; *Chem. Abstr.*, **96**, 19898 (1982).
63. USSR Inventor's Certificate 694504 (1979); *Chem. Abstr.*, **92**, 94235 (1980).
64. M. Kaneda, S. Nakamura, and N. Ezaki, *J. Antibiotics*, **34**, 1366 (1981).
65. N. Ezaki, T. Shomura, M. Koyama, T. Niwa, M. Kojima, S. Inouye, T. Ito, and T. Niida, *ibid.*, **34**, 1363 (1981).
66. S. E. Korostova, L. N. Sobenina, R. N. Nesterenko, I. A. Aliev, and A. I. Mikhaleva, *Zh. Org. Khim.*, **20**, 1960 (1984); *Chem. Abstr.*, **102**, 131860 (1985).
67. S. E. Korostova, A. I. Mikhaleva, L. N. Sobenina, S. G. Shevchenko, and V. V. Shcherbakov, *Khim. Geterotsikl. Soedin.*, **1985**, 1501; *Chem. Abstr.*, **105**, 114862 (1986).
68. B. A. Trofimov, A. I. Mikhaleva, A. N. Vasil'ev, and M. V. Sigalov, *Izv. Akad. Nauk SSSR. Ser. Khim.*, **1979**, 695; *Chem. Abstr.*, **91**, 38874 (1979).
69. B. A. Trofimov, S. E. Korostova, A. I. Mikhaleva, L. N. Sobenina, V. V. Shcherbakov, and M. V. Sigalov, *Khim. Geterotsikl. Soedin.*, **1983**, 276; *Chem. Abstr.*, **98**, 215436 (1983).
70. S. E. Korostova, S. G. Shevchenko, E. Yu. Shmidt, I. A. Aliev, I. M. Lazarev, and M. V. Sigalov, *ibid.*, (1992).
71. S. E. Korostova, S. G. Shevchenko, and M. V. Sigalov, *ibid.*, in press (1991).
72. S. E. Korostova, A. I. Mikhaleva, S. G. Shevchenko, M. V. Sigalov, and V. Yu. Vitkovsky, *Sulfur Lett.*, **5**, 39 (1986).
73. S. E. Korostova, S. G. Shevchenko, and A. I. Mikhaleva, *Khim. Geterotsikl. Soedin.*, **1989**, 1693.

74. S. E. Korostova, S. G. Shevchenko, E. A. Polubentsev, A. I. Mikhaleva, and B. A. Trofimov, ibid., **1989**, 770; *Chem. Abstr.*, **112**, 76094 (1990).
75. S. E. Korostova, A. I. Mikhaleva, R. N. Nesterenko, N. V. Maznaya, V. K. Voronov, and N. M. Borodina, *Zh. Org. Khim.*, **21**, 406 (1985); *Chem. Abstr.*, **103**, 53894 (1985).
76. S. E. Korostova, S. G. Shevchenko, M. V. Sigalov, and N. I. Golovanova, *Khim. Geterotsikl. Soedin.*, 460 (1991).
77. M. A. Yurovskaya, V. V. Druzhinina, and Yu. G. Bundel', in "IIIrd All-Union Conference on Chemistry of Heterocyclic Compounds," *Abstracts in Russian*, p. 83, Rostov on Don, 1983.
78. V. V. Druzhinina, abstracts of the thesis for Candidate of Chemical Science, Moscow, 1983.
79. M. A. Yurovskaya, V. V. Druzhinina, E. V. Snetkova, and Yu. G. Bundel', *Khim. Geterotsikl. Soedin.*, **1983**, 356; *Chem. Abstr.*, **99**, 5475 (1983).
80. M. A. Yurovskaya, V. V. Druzhinina, M. A. Tyurekhodzhaeva, and Yu. G. Bundel', ibid., **1984**, 69; *Chem. Abstr.*, 100, 191693 (1984).
81. M. A. Tyurekhodzhaeva, in "IIIrd All Union Conference on Chemistry of Chemicals," *Abstracts in Russian*, Part 1, p. 93, Ashkhabad, 1989.
82. A. M. Vasil'tsov, E. A. Polubentsev, A. I. Mikhaleva, and B. A. Trofimov, *Izv. Akad. Nauk SSSR. Ser. Khim.*, **1990**, 864
83. B. A. Trofimov, A. B. Shapiro, R. N. Nesterenko, A. I. Mikhaleva, G. A. Kalabin, N. I. Golovanova, I. V. Yakovleva, and S. E. Korostova, *Khim. Geterotsikl. Soedin.*, **1988**, 350; *Chem. Abstr.*, **110**, 38923 (1989).
84. T. N. Borisova, A. V. Varlamov, N. D. Sergeeva, A. T. Soldatenkov, O. V. Zvolinsky, A. A. Astakhov, and N. S. Prostakov, ibid., **1987**, 937; *Chem. Abstr.*, **108**, 167344 (1988).
85. B. A. Trofimov, A. M. Vasil'tsov, A. I. Mikhaleva, G. A. Kalabin, V.V. Shcherbakov, R. N. Nesterenko, E. A. Polubentsev, and K. D. Praliev, ibid in press (1991).
86. B. A. Trofimov, E. B. Oleinikova, M. G. Sigalov, Yu. M. Skvortsov, and A. I. Mikhaleva, *Zh. Org. Khim.*, **16**, 410 (1980); *Chem. Abstr.*, **93**, 46303 (1980).
87. B. A. Trofimov, E. B. Oleinikova, Yu. M. Skvortsov, and A. I. Mikhaleva, *Izv. Akad. Nauk SSSR. Ser. Khim.*, 2426 (1978); *Chem. Abstr.*, **90**, 71989 (1979).
88. I. N. Domnin, A. I. Mikhaleva, R. N. Nesterenko, S. E. Korostova, and M. V. Sigalov, *Zh. Org. Khim.*, **24**, 1788 (1988); *Chem. Abstr.*, **110**, 154085 (1989).
89. B. A. Trofimov and A. I. Mikhaleva, ibid., **16**, 672 (1980); *Chem. Abstr.*, **93**, 46307 (1980).
90. USSR Inventor's Certificate 840038 (1981); *Chem. Abstr.*, **93**, 150444 (1981).
91. I. A. Aliev, A. I. Mikhaleva, S. Kh. Bairamova, and E. I. Akhmedov, *Zh. Org. Khim.*, **22**, 489 (1986); *Chem. Abstr.*, **106**, 102019 (1987).
92. I. A. Aliev, A. I. Mikhaleva, and M. V. Sigalov, in "XVIth Conference on Chemistry and Technology of Organic Compounds of Sulfur and Sulfur-Containing Oils," *Abstracts in Russian*, p. 135, Riga, 1984.
93. I. A. Aliev, D. T. Almamedova, B. R. Gasanov, and A. I. Mikhaleva, *Khim. Geterotsikl. Soedin.*, **1984**, 1359; *Chem. Abstr.* **102**, 24416 (1985).
94. I. A. Aliev, A. I. Mikhaleva, and M. V. Sigalov, *Sulfur Lett.*, **2**, 55 (1984).
95. I. A. Aliev, B. R. Gasanov, N. I. Golovanova, and A. I. Mikhaleva, *Khim. Geterotsikl. Soedin.*, **1987**, 1486; *Chem. Abstr.*, **109**, 22790 (1988).
96. I. A. Aliev, S. N. Zeinalova, B. R. Gasanov, A. I. Mikhaleva, and S. E. Korostova, *Zh. Org. Khim.*, **24**, 2436 (1988); *Chem. Abstr.*, **110**, 212536 (1989).
97. B. A. Trofimov and A. I. Mikhaleva, *Izv. Akad. Nauk SSSR. Ser. Khim.*, **1979**, 2840, *Chem. Abstr.*, **92**, 94175 (1980).
98. O. Eisenstein, G. Procter, and J. D. Dunits, *Helv. Chim. Acta*, **61**, 2538 (1978).
99. A. I. Mikhaleva, B. A. Trofimov, A. N. Vasil'ev, G. A. Komarova, and V. I. Skorobogatova, *Khim. Geterotsikl. Soedin.*, **1982**. 1202; *Chem. Abstr.*, **98**, 16538 (1983).

100. B. A. Trofimov, A. I. Mikhaleva, A. N. Vasil'ev, S. E. Korostova, and S. G. Shevchenko, ibid., **1985**, 1501; *Chem. Abstr.*, **103**, 53901 (1985).

101. S. E. Korostova, A. I Mikhaleva, L. N. Sobenina, S. G. Shevchenko, and R. I. Polovnikova, *Zh. Org. Khim.*, **22**, 492 (1986); *Chem. Abstr.*, **106**, 84316 (1987).

102. I. A. Aliev, S. E. Korostova, A. I. Mikhaleva, S. G. Shevchenko, S. N. Zeinalova, L. N. Sobenina, M. V. Sigalov, and B. R. Gasanov, *Khim. Geterotsikl. Soedin.*, in press (1991).

103. S. E. Korostova, R. N. Nesterenko, A. I. Mikhaleva, R. I. Polovnikova, and N. I. Golovanova, ibid., **1989**, 901; *Chem. Abstr.*, **112**, 178530 (1990).

104. B. A. Trofimov and A. I. Mikhaleva, in "The All-Union Conference on Synthesis and Reactivity of Organic Sulfur Compounds," *Abstracts in Russian*, p. 24, Tbilisi, 1989.

105. W. Reppe, *Polyvinylpyrrolidon*, Weinheim, 1954.

106. U. N. Musaev, S. Kh. Nasirova, and T. A. Azimov, *Khim.-farm. Zh.*, **1974**, 36; *Chem. Abstr.*, **82**, 98685 (1975).

107. Jpn. Pat. 47-69278 (1981); *Ref. Zh. Khim.* 2T233П (1982).

108. Eng. Pat. 1348437 (1974); *Ref. Zh. Khim.* 5H313П (1975).

109. Jpn. Pat. 49-28455 (1974); *Ref. Zh. Khim.* 5H315П (1975).

110. W. Reppe, *Neue Entwicklungen auf dem Gebiet der Chemie des Acetylens und Kohlenoxyds*, J. Springer, Berlin–Gottingen–Heidelberg, 1949.

111. J. W. Copenhaver and M. H. Bigelow, *Acetylene and Carbon Monoxide Chemistry*, Reinhold, New York, 1949.

112. H. Otsuki, I. Okano, and T. Tadeda, *J. Soc. Chem. Ind. Jpn.*, **49**, 169 (1946).

113. Jpn. Pat. 174356 (1946); *Chem. Abstr.*, **44**, 1544 (1950).

114. U.S. Pat, 2426465 (1947); *Chem. Abstr.*, **42**, 224 (1948).

115. O. Hisao, M. Tadashi, and A. Kiichi, *J. Pharm. Soc. Jpn.*, **98**, 165 (1978).

116. G. S. Kolesnikov, *Synthesis of Vinyl Derivatives of Aromatic and Heteroaromatic Compounds* (in Russian), Izd. Akad. Nauk SSSR, Moscow, 1960; *Chem. Abstr.*, **54**, 18546 (1960).

117. Br Pat. 470077 (1937); *Chem. Abstr.*, **32**, 591 (1938).

118. Ger. Pat. 646995 (1937): *Chem. Abstr.*, **31**, 6528 (1937).

119. V. D. Filimonov, E. E. Sirotkina, N. I. Gaibel, and V. I. Kulachenko, *Zh. Org. Khim.*, **10**, 1790 (1974); *Chem. Abstr.*, **81**, 169884 (1974).

120. V. D. Filimonov, S. G. Gorbachev, and E. E. Sirotkina, *Khim. Geterotsikl. Soedin.*, **1980**, 340; *Chem. Abstr.*, **93**, 71458 (1980).

121. V. D. Filimonov, V. A. Anfinogenov, and E. E. Sirotkina, *Zh. Org. Khim.*, **14**, 2607 (1978); *Chem. Abstr.*, **90**, 103763 (1979).

122. E. S. Domnina, G. G. Skvortsova, N. P. Glazkova, and M. F. Shostakovsky, *Khim. Geterotsikl. Soedin.*, **1966**, 390; *Chem. Abstr.*, **65**, 12294 (1966).

123. USSR Inventor's Certificate 18116 (1966); *Byull. Izobret.*, N 9 (1966).

124. Ger. Pat. 618120 (1935); *Chem. Abstr.*, **30**, 110 (1936).

125. Jpn. Pat. 49-9467 (1974); *Ref. Zh. Khim.*, 5H169П (1975).

126. Ger. Pat. 651734 (1937); *Chem. Abstr.*, **32**, 1716 (1938).

127. E. Matsui, *J. Soc. Chem. Ind. Jpn.*, **45**, 1192 (1942).

128. A. M. Vasil'tsov, B. A. Trofimov, and S. V. Amosova, *Izv. Akad. Nauk SSSR. Ser. Khim.*, **1987**, 1785; *Chem. Abstr.*, **107**, 184724 (1987).

129. B. A. Trofimov, A. I. Mikhaleva, S. E. Korostova, A. N. Vasil'ev, and L. N. Balabanova, *Khim. Geterotsikl. Soedin.*, **1977**, 213; *Chem. Abstr.*, **87**, 22934 (1977).

130. A. I. Mikhaleva, B. A. Trofimov, S. E. Korostova, A. N. Vasil'ev, and L. N. Balabanova, *Izv. Sibirsk. Otdeleniya Akad. Nauk SSSR. Ser. Khim. Nauk*, p. 105, 1979; *Chem. Abstr.*, **91**, 39237 (1979).

131. V. D. Filimonov, *Khim. Geterotsikl. Soedin.*, **1981**, 207; *Chem. Abstr.*, **95**, 6043 (1981).

132. B. A. Trofimov, *Adv. Heterogel, Chem.*, **51**, 177 (1990).

133. B. A. Trofimov, A. I. Mikhaleva, R. N. Nesterenko and E. E. Sirotkina, in "Ist Conference on Chemicals of North Caucasus Region," *Abstracts in Russian*, p. 208, Makhach-Kala, 1988.

134. B. A. Trofimov, A. I. Mikhaleva, R. N. Nesterenko, and E. E. Sirotkina, in "Regional Conference of Siberia and Far East," *Abstracts in Russian*, p. 112, Krasnoyarsk, 1989.

135. B. A. Trofimov, A. I. Mikhaleva, and R. N. Nesterenko, in "Consumers and Producers of Speciality Chemicals. The Fair of Ideas," *Abstracts in Russian*, p. 18, Dilizhan, Armenia, 1991.

136. B. A. Trofimov, R. N. Nesterenko, A. I. Mikhaleva, A. B. Shapiro, I. A. Aliev, I. V. Yakovleva, and G. A. Kalabin, *Khim. Geterotsikl. Soedin.*, **1986**, 481; *Chem. Abstr.*, **106**, 156308 (1987).

137. G. G. Skvortsova, E. S. Domnina, N. P. Glazkova, and L. N. Makhno, in "IIIrd All-Union Conference on Chemistry of Acetylene," *Abstracts in Russian*, p. 115, Nauka, Moscow, 1972; *Chem. Abstr.*, **79**, 42414 (1973).

138. B. A. Trofimov, S. E. Korostova, S. G. Shevchenko, E. A. Polubentsev, and A. I. Mikhaleva, *Zh. Org. Khim.*, **26**, 1110 (1990).

139. B. A. Trofimov, A. I. Shatenstein, E. S. Petrov, M. M. Terekhova, N. I. Golovanova, A. I. Mikhaleva, S. E. Korostova, and A. N. Vasil'ev, *Khim. Geterotsikl. Soedin.*, **1980**, 632; *Chem. Abstr.*, **93**, 167406 (1980).

140. A. F. Ermikov, V. K. Turchaninov, K. B. Petrushenko, A. I. Vokin, S. E. Korostova, L. A. Ostroukhova, and Yu. L. Frolov, *Zh. Obshch. Khim.*, **58**, 450 (1988); *Chem. Abstr.*, **110**, 114125 (1989).

141. B. A. Trofimov, S. E. Korostova, S. G. Shevchenko, and A. I. Mikhaleva, *Zh. Org. Khim.*, **26**, 940 (1990).

142. J. I. Dickstein and S. I. Miller, in S. Patai, ed., *The Chemistry of the Carbon–Carbon Triple Bond*, Part 2, Wiley, New York, 1978.

143. G. F. Dvorko and E. A. Shilov, *Teoretich. i Eksperim. Khimiya*, **3**, 606 (1967); *Chem. Abstr.*, **68**, 95200 (1968).

144. B. A. Trofimov, T. I. Vakul's kaya, S. E. Korostova, S. G. Shevchenko, and A. I. Mikhaleva, *Izv. Akad. Nauk SSSR. Ser. Khim.*, **1990**, 142.

145. V. E. Zubarev, V. N. Belevsky, and L. T. Bugaenko, *Usp. Khim.*, **48**, 1361 (1979); *Chem. Abstr.*, **91**, 156840 (1979).

146. I. M. Sosonkin, V. N. Belevsky, G. N. Strogov, A. N. Domarev, and S. R. Jarkov, *Zh. Org. Khim.*, **18**, 1504 (1982); *Chem. Abstr.*, **97**, 197638 (1982).

147. V. N. Babin, V. V. Gumenyuk, S. P. Solodovnikov, and Yu. A. Belousov, *Z. Naturforsch*, **36b**, 400 (1981).

148. C. M. Cammaggi, R. J. Holman, and M. J. Perkins, *J. Chem. Soc., Perkin Trans. 2*, **1972**, 501.

149. B. R. Weinberger, E. Ehrenfreund, and A. Pron, *J. Chem. Phys.*, **72**, 4749 (1980).

150. H. Gross, S. Beisert, and B. Costisella, *J. Prakt. Chim.*, **323**, 877 (1981).

151. R. Neidlein and G. Jeromin, *Chem. Ber.*, **115**, 714 (1982).

152. S. Wattanasin and F. G. Kathawala, *Synthetic Commun.*, **19**, 2659 (1989).

153. F. N. Tebbe, G. W. Parshall, and G. S. Reddy, *J. Am. Chem. Soc.*, **100**, 3611 (1978).

154. S. H. Pine, R. Zahler, D. A. Evans, and R. H. Grubbs, ibid., **102**, 3271 (1980).

155. S. H. Pine, R. J. Pettit, G. D. Geib, S. G. Gruz, C. H. Gallego, T. Tijerina, and R. D. Pine, *J. Org. Chem.*, **50**, 1212 (1985).

156. B. F. Kukharev, V. K. Stankevich, and V. A. Kukhareva, in "IVth All-Union Symposium on Heterogeneous Catalysis in Chemistry of Heterocyclic Compounds," *Abstracts in Russian*, p. 47, Riga, 1987.

157. O. S. Attaryan, G. V. Asratyan, E. G. Derbinyan, and S. T. Matsoyan, *Zh. Org. Khim.*, **24**, 1339 (1988); *Chem. Abstr.*, **110**, 38831 (1989).

158. M. Strell, A. Kolojanoff, and L. Brem-Rupp, *Chem. Ber.*, **87**, 1019 (1954).
159. H. Fisher and A. Schormiller, *Liebigs Ann. Chem.*, **482**, 232 (1930).
160. P. Piutti, *Gazzetta*, **66**, 265 (1936).
161. S. McKinly, J. V. Crawford, and C. -H. Wang, *J. Org. Chem.*, **31**, 1963 (1966).
162. J. Pielichowski and M. Olszanska, *Pol. J. Chem.*, **52**, 1089 (1978).
163. L. I. Trotsenko, C. N. Kurov, V. K. Turchaninov, E. N. Suslova, L. L. Gaintseva, and G. G. Skvortsova, *Zh. Obshch. Khim.*, **49**, 904 (1979); *Chem. Abstr.*, **91**, 55825 (1979).
164. V. D. Filimonov, V. A. Anfinogenov, and E. E. Sirotkina, *Zh. Org. Khim.*, **14**, 2550 (1978); *Chem. Abstr.*, **90**, 102998 (1979).
165. J. Pielichowski and J. Kyzio, *Monatsh. Chem.*, **105**, 1306 (1974).
166. A. Gandini and S. Prieto, *J. Polym. Sci., Polym. Lett. Ed.*, **15**, 337 (1977).
167. V. A. Anfinogenov, V. D. Filimonov, and E. E. Sirotkina, *Zh. Org. Khim.*, **14**, 1723 (1978); *Chem. Abstr.*, **90**, 38764 (1979).
168. V. A. Anfinogenov, V. D. Filimonov, and E. E. Sirotkina, ibid., **14**, 1687 (1978); *Chem. Abstr.*, **89**, 196558 (1978).
169. G. G. Skvortsova, B. V. Trzhtsinskaya, N. I. Chipanina, L. F. Teterina, and N. M. Deriglazov., *Khim. Geterotsikl. Soedin.*, **1980**, 766; *Chem. Abstr.*, **93**, 186082 (1980).
170. G. G. Skvortsova, E. S. Domnina, and Yu. L. Frolov, ibid., **1968**, 673; *Chem. Abstr.*, **70**, 11465 (1969).
171. M. F. Shostakovsky, G. G. Skvortsova, E. S. Domnina, Yu. L. Frolov, N. P. Glazkova, and G. A. Kravchenko, ibid., **1968**, 1025; *Chem. Abstr.*, **70**, 68050 (1969).
172. E. S. Domnina, L. P. Makhno, and N. N. Chipanina, *Izv. Akad. Nauk SSSR. Ser. Khim.*, **1974**, 1837; *Chem. Abstr.*, **83**, 59535 (1975).
173. E. S. Domnina, G. G. Skvortsova, and L. P. Makhno, in "Chemistry and Pharmacology of Indole Compounds," *Abstracts in Russian*, p. 3, Kiev, 1975.
174. M. G. Voronkov, G. G. Skvortsova, E. S. Domnina, Yu. N. Ivlev, N. F. Chernov, N. N. Chipanina, V. K. Voronov, and D. D. Taryashinova, *Zh. Obshch. Khim.*, **46**, 311 (1976); *Chem. Abstr.*, **84**, 164918 (1976).
175. V. D. Ogorodnikov and I. G. Orlov, *Izv. Akad. Nauk SSSR. Ser. Khim.*, **1976**, 759; *Chem. Abstr.*, **85**, 63453 (1976).
176. B. A. Trofimov, S. E. Korostova, L. N. Sobenina, B. V. Trzhtsinskaya, A. I. Mikhaleva, and M. V. Sigalov, *Zh. Org. Khim.*, **16**, 1964 (1980); *Chem. Abstr.*, **94**, 121214 (1981).
177. A. I. Mikhaleva, S. E. Korostova, A. N. Vasil'ev, L. N. Balabanova, N. P. Sokolnikova, and B. A. Trofimov, *Khim. Geterotsikl. Soedin.*, **1977**, 1636; *Chem. Abstr.*, **88**, 120916 (1978).
178. V. B. Pukhnarevich, L. I. Kopylova, S. E. Korostova, A. I. Mikhaleva, L. N. Balabanova, A. N. Vasil'ev, M. V. Sigalov, B. A. Trofimov, and M. G. Voronkov, *Zh. Obshch. Khim.*, **49**, 166 (1979); *Chem. Abstr.*, **90**, 187032 (1979).
179. B. A. Trofimov, S. E. Korostova, A. I. Mikhaleva, L. N. Balabanova, and A. N. Vasil'ev, *Khim. Geterotsikl. Soedin.*, **1977**, 215; *Chem. Abstr.*, **87**, 22935 (1977).
180. B. A. Trofimov, S. E. Korostova, L. N. Balabanova, A. I. Mikhaleva, and A. N. Vasil'ev, ibid., **1978**, 347; *Chem. Abstr.*, **89**, 42978 (1978).
181. B. A. Trofimov, S. E. Korostova, A. I. Mikhaleva, L. N. Sobenina, and A. N. Vasil'ev, ibid., **1982**, 1631.
182. G. G. Skvortsova, E. S. Domnina, and N. I. Glazkova, in "IIIrd All-Union Conference on Charge-Transfer Complexes and Ion–Radical Salts," *Abstracts in Russian*, p. 140, Riga, 1976.
183. B. A. Trofimov, A. I. Mikhaleva, L. V. Morozova, A. N. Vasil'ev, and M. V. Sigalov, *Khim. Geterotsikl. Soedin.*, **1983**, 269; *Chem. Abstr.*, **98**, 215442 (1983).
184. B. A. Trofimov, A. I. Mikhaleva, G. A. Kalabin, A. N. Vasil'ev, and M. V. Sigalov, *Izv. Akad. Nauk SSSR. Ser. Khim.*, **1977**, 2639; *Chem. Abstr.*, **88**, 74258 (1978).

185. B. A. Trofimov, A. I. Mikhaleva, S. E. Korostova, A. N. Vasil'ev, L. N. Balabanova, M. V. Sigalov, and R. N. Polovnikova, in "IIIrd All-Union Conference on Chemistry of Fluororganic Compounds," *Abstracts in Russian*, p. 131, Odessa, 1978.
186. USSR Inventor's Certificate 698981, *Byull. Izobret.*, N 43 (1979).
187. B. A. Trofimov, A. I. Mikhaleva, S. E. Korostova, L. N. Sobenina, A. N. Vasil'ev, and L. V. Balashenko, *Zh. Org. Khim.*, **15**, 2042 (1979); *Chem. Abstr.*, **92**, 76216 (1980).
188. B. A. Trofimov, M. V. Sigalov, T. N. Tandura, G. A. Kalabin, and S. E. Korostova, *Izv. Akad. Nauk SSSR. Ser. Khim.*, **1979**, 1122; *Chem. Abstr.*, **91**, 74121 (1979).
189. M. V. Sigalov, B. A. Trofimov, G. A. Kalabin, and A. I. Mikhaleva, in "The All-Union Conference on NMR Spectroscopy of Heavy Nuclei of Elementorganic Compounds," *Abstracts in Russian*, p. 71, Irkutsk, 1979.
190. H. Fischer and H. Orth, *Die Chemie des Pyrrols*, Vol. 1, Akad. Verlags. Leipzig, 1934, p. 45.
191. H. Heaney, and S. V. Ley, *J. Chem. Soc., Perkin Trans.* 2, **1973**, 499.
192. B. A. Trofimov, A. I. Mikhaleva, A. I. Belyavsky, N. A. Volyanskaya, S. E. Korostova, L. N. Balabanova, A. N. Vasil'ev, and Ju. L. Volyansky, *Khim.-farm. Zh.*, **1981**, 25; *Chem. Abstr.*, **95**, 24705 (1981).
193. M. V. Sigalov, E. Yu. Shmidt, and B. A. Trofimov, *Izv. Akad. Nauk SSSR. Ser. Khim.*, **1987**, 1146; *Chem. Abstr.*, **108**, 111703 (1988).
194. M. V. Sigalov, E. Yu. Shmidt, and B. A. Trofimov, ibid., **1987**, 2136; *Chem. Abstr.*, **103**, 130916 (1988).
195. M. V. Sigalov, E. Yu. Shmidt, and B. A. Trofimov, *Khim. Geterotsikl. Soedin.*, **1988**, 334; *Chem. Abstr.*, **110**, 38401 (1989).
196. M. V. Sigalov, E. Yu. Shmidt, A. I. Mikhaleva, S. E. Korostova, L. M. Lazarev, and B. A. Trofimov, ibid., in press (1992).
197. M. V. Sigalov, E. Yu. Shmidt, A. B. Trofimov, and B. A. Trofimov, ibid., **1989**, 1343; *Chem. Abstr.*, **112**, 177929 (1990).
198. A. B. Trofimov, B. A. Trofimov, N. M. Vitkovskaya, and M. V. Sigalov, ibid., 746 (1991).
199. M. V. Sigalov, E. Yu. Shmidt, S. E. Korostova, and B. A. Trofimov, ibid., in press (1992).
200. E. B. Whipple, J. Chiang, and R. L. Hinmen, *J. Am. Chem. Soc.*, **85**, 26 (1963).
201. M. F. Shostakovsky, G. G. Skvortsova, E. S. Domnina, and N. P. Glazkova, *Zh. Prikladn. Khim.*, **38**, 2602 (1965); *Chem. Abstr.*, **64**, 6602 (1966).
202. V. D. Filimonov, V. A. Anfinogenov, and E. E. Sirotkina, in "IInd All-Union Symposium on Organic Synthesis," *Abstracts in Russian*, p. 148, Moscow, 1976.
203. F. P. Sidel'kovskaya, *The Chemistry of N-Vinylpyrrolidone and Its Polymers* (in Russian), Nauka, Moscow, 1970; *Chem. Abstr.*, **74**, 100438 (1971).
204. J. Joule and G. Smith, *Principles of Chemistry of Heterocyclic Compounds*, Russian Translation, Mir, Moscow, 1975.
205. G. G. Skvortsova, E. S. Domnina, and N. P. Glazkova, *Khim. Geterotsikl. Soedin.*, **1969**, 255; *Chem. Abstr.*, **71**, 21968 (1969).
206. USSR Inventor's Certificate 367096 (1973), *Byull. Izobret.*, N 8 (1973); *Chem. Abstr.*, **79**, 5258 (1973).
207. G. G. Skvortsova, B. V. Trzhtsinskaya, L. A. Usov, *Khim.-farm. Zh.*, **1975**, 16; *Chem. Abstr.*, **84**, 43752 (1976).
208. V. D. Filimonov, V. A. Anfinogenov, and E. E. Sirotkina, *Khim. Geterotsikl. Soedin.*, **1979**, 497; *Chem. Abstr.*, **91**, 74404 (1979).
209. B. D. Filimonov, E. E. Sirotkina, and N. A. Tsekhanovskaya, *Zh. Org. Khim.*, **15**, 174 (1979); *Chem. Abstr.*, **90**, 186715 (1979).
210. V. P. Lopatinsky, Yu. P. Shekhirev, and E. E. Sirotkina, *Khim. Geterotsikl. Soedin.*, **1966**, 398; *Chem. Abstr.*, **65**, 12158 (1966).

211. U.S. Pat. 3823160 (1974); *Chem. Abstr.*, **81**, 135943 (1974).
212. S. E. Korostova, A. I. Mikhaleva, B. A. Trofimov, L. N. Sobenina, and A. N. Vasil'ev, *Zh. Org. Khim.*, **18**, 525 (1982); *Chem. Abstr.*, **96**, 217630 (1982).
213. B. A. Trofimov, S. E. Korostova, L. N. Sobenina, R. N. Nesterenko, and A. I. Mikhaleva, in "IInd Republic Scientific–Technical Conference on Chemistry and Technology of Acetals," *Abstracts in Russian*, p. 75, Ufa, 1980.
214. B. A. Trofimov, S. E. Korostova, and A. I. Mikhaleva, in "The Conference on Chemistry and Technology of Acetals and Its Heteroanalogs," *Abstracts in Russian*, p. 60, Ufa, 1981.
215. S. E. Korostova, A. I. Mikhaleva, S. G. Shevchenko, V. V. Sherbakov, R. N. Nesterenko, M. V. Sigalov, and B. A. Trofimov, *Zh. Org. Khim.*, **22**, 2489 (1986); *Chem. Abstr.*, **107**, 197975 (1987).
216. M. V. Markova, A. I. Mikhaleva, M. V. Sigalov, L. V. Morozova, A. I. Aliev, and B. A. Trofimov, *Khim. Geterotsikl. Soedin.*, **1989**, 604; *Chem. Abstr.*, **112**, 76866 (1990).
217. Ger. Pat. 624622 (1936); *Chem. Abstr.*, **30**, 4875 (1936).
218. G. G. Skvortsova, N. P. Glazkova, E. S. Domnina, and V. K. Voronov, *Khim. Geterotsikl. Soedin.*, **1970**, 167; *Chem. Abstr.*, **72**, 121437 (1970).
219. S. E. Korostova, L. N. Sobenina, A. I. Mikhaleva, and B. A. Trofimov, in "XVth Scientific Session on Chemistry and Technology of Organic Sulfur Compounds and Sulfur-Containing Oils," *Abstracts in Russian*, p. 129, Ufa, 1979.
220. I. A. Aliev, A. I. Mikhaleva, and B. G. Gasanov, *Khim. Geterotsikl. Soedin.*, **1990**, 750.
221. M. G. Voronkov, G. I. Zelchan, and E. Ya. Lukevits, *Silicon and Life* (in Russian), Zinaitne, Riga, 1978; *Chem. Abstr.*, **90**, 34092 (1979).
222. R. A. Simon, A. J. Ricco, and M. S. Wrighton, *J. Am. Chem. Soc.*, **104**, 2032 (1982).
223. B. A. Trofimov, in "IVth All-Union Conference on Chemistry of Nitrogen-Containing Heterocyclic Compounds," *Abstracts in Russian*, p. 7, Novosibirsk, 1987.
224. L. I. Kopylova, S. E. Korostova, L. N. Sobenina, R. N. Nesterenko, M. V. Sigalov, A. I. Mikhaleva, B. A. Trofimov, and M. G. Voronkov, *Zh. Obshch. Khim.*, **51**, 1778 (1981); *Chem. Abstr.*, **95**, 204039 (1981).
225. M. V. Sigalov, E. Yu. Shmidt, and B. A. Trofimov, in "IVth All-Union Conference on Chemistry of Nitrogen-Containing Heterocyclic Compounds," *Abstracts in Russian*, p. 70, Novosibirsk, 1987.
226. W. D. Cooper, *J. Org. Chem.*, **23**, 1382 (1958).
227. S. Clementi and G. Marino, *J. Chem. Soc., Perkin Trans. 2*, **1972**, 71.
228. S. Clementi and G. Marino, *Tetrahedron*, **25**, 4599 (1969).
229. J. Shmushkovich, in R. A. Raphael, E. C. Taylor, H. Wynberg, eds., *Advances in Organic Chemistry. Methods and Results*, Vol. 4, Interscience, New York, 1960.
230. M. Hojo, R. Masuda, Jo. Kokuryo, H. Shioda, and S. Matsuo, *Chem. Lett.*, 499 (1976).
231. S. G. Gorbachev, abstracts of the thesis for Candidate of Chemical Sciences, Tomsk, 1978.
232. N. K. Genkina, V. N. Eraskina, and N. N. Suvorov, *Zh. Org. Khim.*, **16**, 2154 (1980); *Chem. Abstr.*, **94**, 65419 (1981).
233. J. Kimie and H. Terukijo, *Bull. Chem. Soc. Jpn.*, **49**, 1363 (1976).
234. R. K. Mackie, S. Mhatre, and J. M. Tedder, *J. Fluor. Chem.*, **10**, 437 (1977).
235. G. Marino, *Khim. Geterotsikl. Soedin.*, **1973**, 579; *Chem. Abstr.*, **79**, 52418 (1973).
236. L. I. Belenky, ibid., **1980**, 1587; *Chem. Abstr.*, **94**, 120328 (1981).
237. B. A. Trofimov, S. E. Korostova, A. I. Mikhaleva, R. N. Nesterenko, M. V. Sigalov, V. K. Voronov, and R. I. Polovnikova, *Zh. Org. Khim.*, **18**, 894 (1982); *Chem. Abstr.*, **97**, 55630 (1982).
238. S. E. Korostova, A. I. Mikhaleva, R. N. Nesterenko, R. I. Polovnikova, V. K. Voronov, and B. A. Trofimov, in "IVth All-Union Conference on Chemistry of Fluororganic Compounds," *Abstracts in Russian*, p. 218, Tashkent, 1982.

239. S. E. Korostova, A. I. Mikhaleva, L. N. Sobenina, S. G. Shevchenko, M. V. Sigalov, I. M. Karataeva, and B. A. Trofimov, *Khim. Geterotsikl. Soedin.*, **1989**, 48; *Chem. Abstr.*, **111**, 97023 (1989).
240. S. E. Korostova, R. N. Nesterenko, A. I. Mikhaleva, S. G. Shevchenko, G. A. Kalabin, and R. I. Polovnikova, ibid., 337 (1991).
241. V. G. Rozinov, G. A. Pensionerova, A. I. Mikhaleva, A. N. Vasil'ev, and V. I. Donskikh, in "IVth All-Union Conference on Chemistry of Fluororganic Compounds," *Abstracts in Russian*, p. 220, Tashkent, 1982.
242. L. N. Sobenina, M. P. Sergeeva, A. I. Mikhaleva, M. V. Sigalov, S. E. Korostova, N. I. Golovanova, V. N. Salaurov, E. V. Bakhareva, and N. N. Vasil'eva, *Khim. Geterotsikl. Soedin.*, **1990**, 612.
243. V. G. Rozinov, G. A. Pensionerova, V. I. Donskikh, L. M. Sergienko, S. E. Korostova, A. I. Mikhaleva, and G. V. Dolgushin, *Zh. Obshch. Khim.*, **56**, 790 (1986); *Chem. Abstr.*, **106**, 50302 (1987).
244. V. I. Donskikh, L. B. Krivdin, V. G. Rozinov, and G. A. Pensionerova, in "IInd All-Union Conference on NMR Spectroscopy of Heavy Nuclei of Elementorganic Compounds," *Abstracts in Russian*, p. 70, Irkutsk, 1983.
245. V. G. Rozinov, G. A. Pensionerova, V. I. Donskikh, L. M. Sergienko, O. V. Petrova, A. V. Kalabina, and A. I. Mikhaleva, *Zh. Obshch. Khim.*, **54**, 2241 (1984); *Chem. Abstr.*, **102**, 149358 (1985).
246. A. G. Gorshkov, E. S. Domnina, A. I. Mikhaleva, and G. G. Skvortsova, *Khim. Geterotsikl. Soedin.*, **1985**, 848; *Chem. Abstr.*, **104**, 50749 (1986).
247. USSR Inventor's Certificate 1199757 (1985); *Chem. Abstr.*, **105**, 190926 (1986).
248. J. K. Williams, D. W. Willey, and B. C. McKusich, *J. Am. Chem. Soc.*, **84**, 2216 (1962).
249. A. G. Gorshkov, E. S. Domnina, V. K. Turchaninov, M. F. Larin, and G. G. Skvortsova, *Khim. Geterotsikl. Soedin.*, **1983**, 951; *Chem. Abstr.*, **99**, 139690 (1983).
250. S. E. Korostova, A. I. Mikhaleva, and M. V. Sigalov, *Zh. Org. Khim.*, **23**, 448 (1987); *Chem. Abstr.*, **108**, 5802 (1988).
251. M. A. Ryashentseva, Vu. B. Vol'kenshtein, V. M. Polosin, A. I. Mikhaleva, and R. N. Nesterenko, *Izv. Akad. Nauk SSSR. Ser. Khim.*, 1417 (1991).
252. H. Hoegl, *J. Phys. Chem.*, **69**, 755 (1965).
253. L. I. Boguslavsky and A. V. Vannikov, *Organic Semiconductors and Biopolymers. Conductivity and Physicochemical Properties* (in Russian), Nauka, Moscow, 1968; *Chem. Abstr.*, **71**, 54849 (1969).
254. M. Biswas and D. Chakravarty, *J. Macromol. Sci.*, **C7**, 189 (1972).
255. M. Biswas, *J. Macromol. Sci., Rev. Macromol. Chem.*, **C14**, 1 (1976).
256. S. S. Rogacheva, abstracts of the thesis for Candidate of Chemical Sciences, Tomsk, 1981.
257. E. S. Domnina, thesis for Candidate of Chemical Sciences, Irkutsk, 1967.
258. E. S. Domnina and G. G. Skvortsova, *Vysokomolek. Soedin.*, **8**, 1268 (1966); *Chem. Abstr.*, **65**, 15525 (1966).
259. USSR Inventor's Certificate 384834 (1974), *Ref. Zh. Khim.* 16C421π (1974); *Chem. Abstr.*, **80**, 37623 (1974).
260. M. F. Shostakovsky, G. G. Skvortsova, and E. S. Domnina, *Vysokomolek. Soedin.*, **9**, 2161 (1967); *Chem. Abstr.*, **68**, 3210 (1968).
261. L. P. Ellinger, *Adv. Macromol. Chem.*, **1**, 169 (1968).
262. H. Scott, G. A. Miller, and M. M. Labes, *Tetrahedron Lett.*, **1963**, 1073.
263. N. G. Gaylord, *J. Polym. Sci., D*, **4**, 183 (1970).
264. M. Biswas and D. Chakravarty, *Bull. Chem. Soc. Jpn.*, **43**, 1904 (1970).
265. M. Biswas and D. Chakravarty, *J. Polym. Sci., Polym. Chem.*, **11**, 7 (1973).

266. M. Biswas and D. Chakravarty, ibid., **12**, 1337 (1974).
267. M. Biswas and P. Kamannarayana, ibid., **13**, 2035 (1975).
268. K. Okamoto, M. Yamada, and A. Haya, *Macromolecules*, **9**, 645 (1976).
269. E. Chiellim, R. Solaro, and A. Ledwith, *Macromol. Chem.*, **178**, 701 (1977).
270. V. D. Filimonov, S. G. Gorbachev, and E. E. Sirotkina, *Vysokomolek. Soedin.*, **20A**, 2726 (1978); *Chem. Abstr.*, **90**, 122122 (1979).
271. V. A. Rodionov and V. D. Filimonov, *Zh. Org. Khim.*, **18**, 1094 (1982); *Chem. Abstr.*, **97**, 72213 (1982).
272. V. A. Rodionov, thesis for candidate of Chemical Sciences, Tomsk, 1982.
273. A. G. Gorshkov, V. K. Turchaninov, G. N. Kurov, and G. G. Skvortsova, *Zh. Org. Khim.*, **15**, 767 (1979); *Chem. Abstr.*, **91**, 91575 (1979).
274. K. B. Petrushenko, V. K. Turchaninov, A. I. Vokin, A. G. Gorshkov, and Ju. L. Frolov, *Izv. Akad. Nauk SSSR. Ser. Khim.*, **1979**, 2839; *Chem. Abstr.*, **92**, 146028 (1980)
275. A. G. Gorshkov, V. K. Turchaninov, G. N. Kurov, and G. G. Skvortsova. *Zh. Org. Khim.*, **16**, 2368 (1980); *Chem. Abstr.*, **95**, 24952 (1981).
276. V. P. Lopatinsky, E. E. Sirotkina, M. M. Anosova, L. G. Tikhonova, and S. E. Pavlov, *Izv. Tomsk. Politekhn. Instituta*, **126**, 58 (1964); *Chem. Abstr.*, **63**, 18007 (1965).
277. V. P. Lopatinsky and E. E. Sirotkina, ibid., **126**, 62 (1964); *Chem. Abstr.*, **63**, 18007 (1965).
278. M. V. Kurik, V. S. Manzhara, S. S. Rogacheva, and E. E. Sirotkina, *Zh. Nauchn. i Prikladn. Fotografii i Kinematografii*, **26**, 4 (1981); *Chem. Abstr.*, **94**, 148291 (1981).
279. USSR Inventor's Certificate 441264 (1974); *Byull. Izobret.*, N 32 (1974); *Chem. Abstr.*, **82**, 86898 (1975).
280. U.S. Pat. 3987011 (1976); *Chem. Abstr.*, **86**, 30298 (1977).
281. USSR Inventor's Certificate 539901 (1976); *Byull. Izobret.*, N 47 (1976); *Chem. Abstr.*, **86**, 122016 (1977).
282. USSR Inventor's Certificate 653269 (1979); *Byull. Izobret.*, N 11 (1979); *Chem. Abstr.*, **91**, 5895 (1979).
283. T. T. Minakova, L. V. Morozova, A. I. Mikhaleva, V. K. Stankevich, and B. A. Trofimov, *Zh. Prikladn. Khim.*, **51**, 2123 (1978); *Chem. Abstr.*, **89**, 216123 (1978).
284. B. A. Trofimov, T. T. Minakova, T. A. Tandura, A. I. Mikhaleva, and S. E. Korostova, *Vysokomolek. Soedin.*, **22**, 103 (1980); *Chem. Abstr.*, **93**, 72352 (1980).
285. B. A. Trofimov, T. T. Minakova, T. A. Tandura, and A. I. Mikhaleva, *Izv. Sibirsk. Otdeleniya Akad. Nauk SSSR. Ser. Khim. Nauk*, **1979**, 162; *Chem. Abstr.*, **92**, 111369 (1980).
286. B. A. Trofimov, T. T. Minakova, and T. A. Tandura, *J. Polym. Sci., Polym. Chem. Ed.*, **18**, 1547 (1980).
287. B. A. Trofimov, T. T. Minakova, L. V. Morozova, T. A. Tandura, T. I. Vakul'skaya, and A. I. Mikhaleva, *Vysokomolek. Soedin.*, **22B**, 803 (1980); *Chem. Abstr.*, **94**, 103890 (1981).
288. B. A. Trofimov, T. T. Minakova, T. A. Tandura, and A. I. Mikhaleva, *Deposited Doc.*, VINITI 4243-80; *Chem. Abstr.*, **96**, 123424 (1982).
289. T. T. Minakova, T. A. Tandura, A. I. Mikhaleva, and B. A. Trofimov, *Deposited Doc.*, VINITI 4530-80; *Chem. Abstr.*, **96**, 105014 (1982).
290. USSR Inventor's Certificate 923624 (1982); *Chem. Abstr.*, **97**, 201221 (1982).
291. V. A. Kruglova, G. A. Izykenova, A. V. Kalabina, B. A. Trofimov, and A. I. Mikhaleva, *Vysokomolek. Soedin.*, **24B**, 691 (1982); *Chem. Abstr.*, **98**, 17085 (1983).
292. L. V. Morozova, A. I. Mikhaleva, M. V. Markova, and B. A. Trofimov, in "Ist All-Union Conference on Chemistry, Biochemistry, and Pharmacology of Indole," *Abstracts in Russian*, p. 105, Tbilisi, 1986.
293. L. V. Morozova, A. I. Mikhaleva, and G. F. Myachina, *Zh. Prikladn. Khim.*, **60**, 1193 (1987); *Chem. Abstr.*, **107**, 106130 (1987).

294. L. V. Morozova, A. I. Mikhaleva, M. V. Markova, and B. A. Trofimov, in "IInd All-Union Conference on Water-Soluble Polymers and Its Application," *Abstracts in Russian*, p. 46, Irkutsk, 1987.

295. B. A. Erusalimsky, *Ionic Polymerization of Polar Monomers* (in Russian), Nauka, Leningrad, 1970.

296. A. Rembaum, A. M. Herman, and R. Haack, *J. Polym. Sci.*, **B5**, 407 (1967).

297. Y. Oshiro, Y. Shirata, and H. Mikawa, *Polym. J.*, **6**, 364 (1974).

298. C. I. Simonescu, M. Grigoras, B. Comanita, I. Diaconu, B. A. Trofimov, and I. I. Negulesku, in *Polymeric Materials: Science and Engineering Proceedings of the ACS Division of Polymeric Materials. Science and Engineering*," **62**, American Chemical Society (ACS), Boston, 1990, p. 416.

299. M. G. Voronkov, V. G. Kozyrev, M. V. Sigalov, B. A. Trofimov, and A. I. Mikhaleva, *Khim. Geterotsikl. Soedin.*, **1984**, 420; *Chem. Abstr.*, **100**, 209578 (1984).

300. L. V. Morozova, A. I. Mikhaleva, and M. V. Sigalov, *Izv. Sibirsk. Otdeleniya Akad. Nauk SSSR. Ser. Khim. Nauk*, **1986**, 128; *Chem. Abstr.*, **100**, 209578 (1984).

301. B. A. Trofimov, L. N. Morozova, A. I. Mikhaleva, and M. V. Markova, in "Ist All-Union Conference on Chemistry, Biochemistry, and Pharmacology of Indole," *Abstracts in Russian*, p. 79, Tbilisi, 1986.

302. L. V. Morozova and A. I. Mikhaleva, *Khim. Geterotsikl. Soedin.*, **1987**, 479; *Chem. Abstr.*, **108**, 131495 (1988).

303. B. A. Trofimov, L. V. Morozova, M. V. Sigalov, A. I. Mikhaleva, and M. V. Markova, *Makromol. Chem.*, **188**, 2251 (1987); *Chem. Abstr.*, **107**, 237394 (1987).

304. B. A. Trofimov, L. V. Morozova, A. I. Mikhaleva, E. I. Brodskaya, M. V. Markova, S. E. Korostova, N. I. Golovanova, and D. D. Taryashinova, *Khim. Geterotsikl. Soedin.*, **1989**, 1420; *Chem. Abstr.*, **112**, 235923 (1990).

305. B. A. Trofimov, L. V. Morozova, E. I. Brodskaya, A. I. Mikhaleva, M. V. Markova, and D. D. Taryashinova, *Vysokomolek. Soedin.*, **31**, 897 (1989); *Chem. Abstr.*, **112**, 235921 (1990).

306. B. A. Trofimov, N. I. Golovanova, A. I. Mikhaleva, S. E. Korostova, A. N. Vasil'ev, and L. N. Balabanova, *Khim. Geterotsikl. Soedin.*, **1977**, 910; *Chem. Abstr.*, **87**, 133751 (1977).

307. M. V. Sigalov, B. A. Trofimov, A. I. Mikhaleva, and G. A. Kalabin, *Tetrahedron*, **37**, 3051 (1981).

308. B. A. Trofimov, N. I. Golovanova, A. I. Mikhaleva, S. E. Korostova, A. N. Vasil'ev, and L. N. Balabanova, *Khim. Geterotsikl. Soedin.*, **1977**, 915; *Chem. Abstr.*, **87**, 151567 (1977).

309. B. A. Trofimov, M. V. Sigalov, V. M. Bzhezovsky, G. A. Kalabin, S. E. Korostova, A. I. Mikhaleva, and L. N. Balabanova, *ibid.*, **1978**, 768; *Chem. Abstr.*, **89**, 107165 (1978).

310. M. V. Sigalov, G. A. Kalabin, A. I. Mikhaleva, and B. A. Trofimov, *ibid.*, **1980**, 328; *Chem. Abstr.*, **93**, 94698 (1980).

311. M. V. Sigalov, B. A. Shainyan, G. A. Kalabin, A. I. Mikhaleva, and B. A. Trofimov, *ibid.*, **1980**, 627; *Chem. Abstr.*, **93**, 167489 (1980).

312. A. V. Afonin, M. V. Sigalov, S. E. Korostova, S. A. Aliev, A. V. Vashchenko, and B. A. Trofimov, *Magnetic. Res. Chem.*, **28**, 580 (1990).

313. V. B. Modonov, T. N. Aksamentova, M. V. Sigalov, A. I. Mikhaleva, S. E. Korostova, and B. A. Trofimov, in "Vth All-Union Symposium on Molecular Interaction and Molecular Conformations," *Abstracts in Russian*, p. 187, Alma-Ata, 1980.

314. V. B. Modonov, B. A. Trofimov, and T. N. Aksamentova, in "IIIrd All-Union Conference on, Electric Properties of Molecules" *Abstracts in Russian*, p. 126, Kazan', 1982.

315. T. V. Kashik, C. M. Ponomareva, N. I. Golovanova, A. I. Mikhaleva, S. E. Korostova, R. N. Nesterenko, and B. A. Trofimov, in "IVth All-Union Conference on Chemistry of Fluororganic Compounds," *Abstracts in Russian*, p. 219, Tashkent, 1982.

316. B. A. Trofimov, V. B. Modonov, T. N. Aksamentova, A. I. Mikhaleva, S. E. Korostova, A. N.

Vasil'ev, L. N. Sobenina, and R. N. Nesterenko, *Zn. Obshch. Khim.*, **53**, 1867 (1963); *Chem. Abstr.*, **100**, 5421 (1984).

317. N. I. Golovanova, S. E. Korostova, L. N. Sobenina, A. I. Mikhaleva, B. A. Trofimov, and Yu. L. Frolov, *Zh. Org. Khim.*, **19**, 1294 (1983); *Chem. Abstr.*, **99**, 113213 (1983).

318. M. Witanovsky, S. Biernat, L. Stefanik, B. A. Trofimov, A. I. Mikhaleva, and G. A. Webb, *Bull. Acad. Pol. Sci. Ser. Sci. Chem.*, **29**, 17 (1982); *Chem. Abstr.*, **97**, 109450 (1982).

319. M. V. Sigalov, G. A. Kalabin, A. G. Proidakov, E. S. Domnina, and G. G. Skvortsova, *Izv. Akad. Nauk SSSR. Ser. Khim.*, **1981**, 2676; *Chem. Abstr.*, **96**, 103504 (1982).

320. Yu. L. Frolov, N. I. Chipanina, F. S. Lur'e, D. D. Taryashinova, E. S. Domnina, and G. G. Skvortsova, ibid. **1980**, 1562; *Chem. Abstr.*, **93**, 220267 (1980).

321. V. B. Modonov, B. A. Trofimov, A. I. Mikhaleva, and R. N. Nesterenko, in "XVIth Conference on Chemistry and Technology of Organic Sulfur Compounds and Sulfur-containing Oils," *Abstracts in Russian*, p. 286, Riga, 1984.

322. N. I. Golovanova, B. A. Trofimov, A. I. Mikhaleva, and S. E. Korostova, *Khim. Geterotsikl. Soedin.*, **1984**, 772; *Chem. Abstr.*, **101**, 110197 (1984).

323. V. B. Modonov, A. I. Mikhaleva, S. E. Korostova, and B. A. Trofimov, in "Vth All-Union Conference on Chemistry of Fluororganic Compounds," *Abstracts in Russian*, p. 144, Zvenigorod, 1986.

324. N. I. Golovanova, S. E. Korostova, A. I. Mikhaleva, and R. N. Nesterenko, *Izv. Sibirsk. Otdeleniya Akad. Nauk SSSR. Ser. Khim. Nauk*, **1986**, 122; *Chem. Abstr.*, **107**, 216811 (1987).

325. A. V. Afonin, V. K. Voronov, A. I. Mikhaleva, R. N. Nesterenko, R. I. Polovnikova, N. O. Saldobol, and B. A. Trofimov, *Izv. Akad. Nauk SSSR. Ser. Khim.*, **1987**, 184; *Chem. Abstr.*, **107**, 197977 (1987).

326. B. A. Trofimov, V. B. Modonov, T. N. Aksamentova, A. I. Mikhaleva, S. E. Korostova, L. N. Sobenina, R. N. Nesterenko, and R. I. Polovnikova, ibid., **1987**, 1322; *Chem. Abstr.*, **108**, 166904 (1988).

327. A. V. Afonin, M. V. Sigalov, V. K. Voronov, E. Yu. Shmidt, and B. A. Trofimov, ibid., **1987**, 1418; *Chem. Abstr.*, **108**, 111458 (1988).

328. K. B. Petrushenko, A. I. Vokin, V. K. Turchaninov, and S. E. Korostova, ibid., **1988**, 41; *Chem. Abstr.*, **109**, 189731 (1988).

329. V. D. Filimonov, V. A. Anfinogenov, and S. G. Gorbachev, *Khim. Geterotsikl. Soedin.*, **1982**, 1640; *Chem. Abstr.*, **98**, 82583 (1983).

330. V. K. Turchaninov, A. F. Ermikov, S. E. Korostova, and V. A. Shagun, ibid., **59**, 791 (1989); *Chem. Abstr.*, **111**, 231565 (1989).

331. A. V. Afonin, M. V. Sigalov, B. A. Trofimov, A. I. Mikhaleva, and I. A. Aliev, *Izv. Akad. Nauk SSSR. Ser. Khim.*, 1031 (1991).

332. V. V. Keiko, V. F. Sidorkin, V. A. Shagun, M. V. Sigalov, and B. A. Trofimov, *Zh. Obshch. Khim.*, **60**, 1871 (1990).

333. A. V. Afonin, M. A. Sigalov, S. E. Korostova, V. K. Voronov, and I. A. Aliev, ibid., **1988**, 2765; *Chem. Abstr.*, **111**, 56696 (1989).

334. C. W. N. Cumper and J. W. M. Wood, *J. Chem. Soc.*, **1971**, 1811.

335. V. Galasso, L. Klasinec, A. Sabljic, N. Trinajstic, G. C. Pappalardo, and W. Steglich, *J. Chem. Soc., Perkin Trans. 2*, **1981**, 127.

336. D. M. Bertin, M. Farnier, and F. M. Lumbroso, *Compt. Rend.*, **274C**, 462 (1972).

337. V. A. Palm, *Principles of Quantitative Theory of Organic Chemistry* (in Russian), Khimiya, Leningrad, 1977.

338. B. A. Trofimov, N. I. Shergina, and S. E. Korostova, in *Reactivity of Organic Compounds* (in Russian), Vol. 8, Tartu, 1971, p. 1047.

339. B. A. Trofimov, N. I. Shergina, and E. I. Kositsyna, ibid., Vol. 10, 1973, p. 766.

2.7. References

340. E. S. Stern and C. J. Timmons, *Electronic Absorption Spectroscopy in Organic Chemistry*, Arnold, London, 1970.
341. G. F. Bol'shakov, V. S. Vatago, and F. B. Agrest, *Ultraviolet Spectra of Heteroaromatic Compounds* (in Russian), Khimiya, Leningrad, 1969.
342. N. Solony and F. W. Birss, *Can. J. Chem.*, **43**, 1569 (1965).
343. L. I. Lagutskaya, *Teoretich. i. Eksperim. Khimiya*, **9**, 159 (1973); *Chem. Abstr.*, **79**, 31285 (1973).
344. Yu. L. Frolov, V. B. Mantsivoda, and N. I. Chipanina, *Zh. Prikladn. Spektr.*, **24**, 734 (1967).
345. A. Kreutzberger and P. Kaller, *J. Phys. Chem.*, **65**, 624 (1961).
346. C. Pascual, I. Meier, and W. Simon, *Helv. Chim. Acta*, **49**, 164 (1966).
347. T. T. Clerc, E. Pretsch, and S. Sternhell, ^{13}C *Kernresonanzspectroskopie*, Akad. Verlag, Frankfurt, 1973.
348. B. A. Trofimov, G. A. Kalabin, and V. M. Bzhezovsky, in *Reactivity of Organic Compounds* (in Russian), Vol. 11, Tartu, 1974, p. 365.
349. V. M. Bzhezovsky, thesis for Candidate of Chemical Sciences, Irkutsk, 1977.
350. V. M. Bzhezovsky, G. A. Kalabin, I. A. Aliev, B. A. Trofimov, M. A. Shakhgel'diev, and A. M. Kuliev, *Izv. Akad. Nauk SSSR. Ser. Khim.*, **1976**, 1999; *Chem. Abstr.*, **86**, 54836 (1977).
351. V. M. Bzhezovsky, V. A. Pestunovich, G. A. Kalabin, I. A. Aliev, I. D. Kalikhman, M. A. Shakhgeldiev, A. N. Kuliev, and M. G. Voronkov, ibid., **1976**, 2004; *Chem. Abstr.*, **86**, 54837 (1977).
352. B. A. Trofimov, M. V. Sigalov, V. M. Bzhezovsky, A. I. Mikhaleva, and A. N. Vasil'ev, *Khim. Geterotsikl. Soedin.*, **1978**, 350; *Chem. Abstr.*, **89**, 23315 (1978).
353. G. A. Webb and M. Witanowsky, in *Nitrogen NMR*, Plenum Press, London–New York, 1973, p. 12.
354. W. Witanowsky, L. Stefaniak, and G. A. Webb, *Ann. Rep. NMR Spectra.*, **7**, 118 (1977).
355. K. A. K. Ebrahem, G. A. Webb, and M. Witanowsky, *Org. Magn. Res.*, **11**, 27 (1978).
356. K. Okamoto, A. Itaya, and S. Kusabayashi, *Polymer. J.*, **7**, 622 (1975).
357. M. Ikeda, H. Sato, and K. Morimoto, *Photogr. Sci. Eng.*, **19**, 60 (1975).
358. T. Kiyashi, E. Takamichi, and H. Masahiro, *Makromol. Chem.*, **176**, 3025 (1975).
359. R. Blyumbergas, V. Gaidyalis, and R. Kavalyunas, in *Application Polymeric Materials in National Economy* (in Russian), Vol. 1, Vilnius, 1975, p. 103.
360. Yu. A. Cherkasov, A. D. Lopatko, S. M. Borodkina, and T. V. Cheltsova, *Zh. Nauchn. i Prikladn. Fotografii i Kinematografii*, **20**, 370 (1975); *Chem. Abstr.*, **84**, 372806 (1976).
361. K. Okamoto, N. Oda, A. Ita, and S. Kusabayashi, *Chem. Phys. Lett.*, **35**, 483 (1975).
362. A. N. Faidysh and V. N. Yashchuk, *Dokl. Akad. Nauk of Ukrain. SSR, A*, **1976**, 1127; *Chem. Abstr.*, **86**, 131975 (1977).
363. M. Yokoyama, T. Tamamura, T. Nakano, and H. Mikawa, *J. Chem. Phys.*, **65**, 272 (1976).
364. R. D. Burhart, *Macromolecules*, **9**, 234 (1976).
365. J. H. Slovik, *J. Appl. Phys.*, **47**, 2982 (1976).
366. O. A. Ponomarev and Z. M. Sabirov, in *Chemistry and Physical Chemistry of Macromolecule Compounds* (in Russian), Ufa, 1976, p. 82.
367. M. Yokoyama, Y. Endo, and H. Mikawa, *Bull. Chem. Soc. Jpn.*, **49**, 1538 (1976).
368. K. Okamoto, A. Itaya, and S. Kubayashi, *J. Polym. Sci., Polym. Phys. Ed.*, **14**, 869 (1976).
369. V. G. Uss, S. N. Uss, and J. Sidaravicius, *Litovsk. Fiz. Sbornik*, **16**, 747 (1976); *Chem. Abstr.*, **86**, 113655 (1977).
370. K. Okamoto, N. Oda, A. Itaya, and S. Kusabayashi, *Bull. Chem. Soc. Jpn.*, **49**, 1415 (1976).
371. D. M. Chang, S. Gromelski, and R. Rupp. *Am. Chem, Soc., Polym. Prep.*, **17**, 627 (1976).
372. P. M. Borsenberger and A. I. Ateya, *J. Appl. Phys.*, **49**, 4035 (1978).

373. R. C. Penwell, B. N. Ganguly, and T. W. Smith, *J. Polym. Sci., Macromol. Rev.*, **13**, 63 (1978).
374. S. V. Zhavrid, M. N. Shashikhina, and A. I. Mikhaleva, in "Vth All-Union Colloquium on Chemistry, Biochemistry, and Pharmacology of Indole Derivatives," *Abstracts in Russian*, p. 165, Tbilisi, 1981.
375. S. V. Zhavrid, M. N. Shashikhina, A. I. Mikhaleva, and B. A. Trofimov, *Khim. -farm. Zh.*, **1983**, 25; *Chem. Abstr.*, **99**, 70513 (1983).
376. G. I. Massel', A. S. Rozhkov, A. I. Mikhaleva, A. N. Vasil'ev, and B. A. Trofimov, *Izv. Vuzov., Lesnoi Zh.*, **1984**, 394; *Chem. Abstr.*, **101**, 34503 (1984).
377. B. A. Trofimov, A. I. Mikhaleva, L. N. Sobenina, S. E. Korostova, M. P. Sergeeva, V. B. Kasimirovskaya, S. B. Seredenin, T. Ya. Pushechkina, G. I. Nikiforova, E. B. Bakhareva, N. N. Vasil'eva, and Yu. A. Blednov, in "The All-Union Seminar on Chemistry of Physiological Active Compounds," *Abstracts in Russian*, p. 237, Chernogolovka, 1989.
378. L. V. Morozova, A. I. Mikhaleva, S. E. Korostova, and I. A. Filimonova, in "The All-Union Conference on Organic Semiconductors," *Abstracts in Russian*, p. 305, Tbilisi, 1983.
379. L. V. Morozova, A. I. Mikhaleva, M. V. Markova, and B. A. Trofimov, in "The All-Union Seminar on Chemistry of Physiological Active Compounds," *Abstracts in Russian*, p. 172, Chernogolovka, 1989.
380. L. V. Morozova, A. I. Mikhaleva, M. V. Markova, and B. A. Trofimov, in "The Conference on Biological Active Polymers and Polymeric Reagents for the Plant-Growing," *Abstracts in Russian*, p. 33, Nalchik, 1988.
381. B. A. Trofimov, A. I. Mikhaleva, R. N. Nesterenko, G. A. Kalabin, A. B. Shapiro, and I. B. Yakovleva, in "VIIth All-Union Conference on the Chemistry of Acetylene," *Abstracts in Russian*, p. 110, Erevan, 1984.
382. W. Herz and J. Brasch, *J. Org. Chem.*, **23**, 1513 (1958).
383. E. Lubrzynka, *J. Chem. Soc.*, **109**, 1118 (1916).
384. B. Bolarevic, M. Dezelic, and V. Milovic, *Glasnik Drustva Hemicara Technol., Bosne Herzegovine*, **12**, 111 (1963), *Chem. Abstr.*, **63**, 18006 (1965).
385. S. V. Tsukerman, V. P. Izvekov, and V. F. Lavrushin, *Khim. Geterotsikl. Soedin.*, **1965**, 527; *Chem. Abstr.*, **64**, 676 (1966).
386. S. V. Tsukerman, V. P. Izvekov, and V. F. Lavrushin, ibid., **1966**, 387; *Chem. Abstr.*, **65**, 12159 (1966).
387. S. V. Tsukerman, J. N. Surov, and V. F. Lavrushin, *Zh. Obshch. Khim.*, **37**, 364 (1967); *Chem. Abstr.*, **67**, 43294 (1967).
388. S. V. Tsukerman, V. P. Izvekov, and V. F. Lavrushin, *Khim. Geterotsikl. Soedin.*, **9** (1967); *Chem. Abstr.*, **70**, 77079 (1969).
389. F. Le Coffic, A. Gouyette, and J. E. Faye, *Compt. Rend.*, **276C**, 1327 (1973).
390. J. F. Jimerez and A. V. Roneero, *Anales Real Soc. Espan, Fis. Guim.*, **45B**, 1591 (1949), *Chem. Abstr.*, **46**, 7088 (1952).
391. S. V. Tsukerman, V. P. Izvekov, and V. F. Lavrushin, *Zh. Fiz. Khim.*, **42**, 2159 (1968); *Chem. Abstr.*, **70**, 57025 (1969).
392. J. Duflos, D. Letouzi, G. Gueguiner, and P. Pastour, *J. Heterocycl. Chem.*, **10**, 1083 (1973).
393. R. E. Albrecht and S. Schroder, *Liebigs Ann. Chem.*, **736**, 110 (1970).
394. W. Flitsch and R. Heidhues, *Chem. Ber.*, **101**, 3343 (1968).
395. G. M. Badger, G. E. Lewis, and U. P. Singh, *Austral. J. Chem.*, **20**, 2785 (1967).
396. H. Fisher and C. D. Nenitzescu, *Liebigs Ann. Chem.*, **429**, 175 (1924).
397. H. Fisher and J. Klarer, ibid., **442**, 1 (1925).
398. I. Ju. Grishin, N. M. Przhijalgovskaya, Ju. M. Chunaev, V. F. Mandjkov, L. N. Kurkovskaya, and N. N. Suvorov, *Khim. Geterotsikl. Soedin.*, **1989**, 907.

2.7. References

399. F. P. Doule, M. D. Mehta, G. S. Sach, and J. L. Person, *J. Chem. Soc.*, **1958**, 4458.
400. T. S. Gardner, E. Wenis, and J. Lee, *J. Org. Chem.*, **23**, 823 (1958).
401. H. Fisher and M. Neber, *Liebigs Ann. Chem.*, **496**, 23 (1932).
402. W. A. Davies, A. R. Pinder, and I. G. Morris, *Tetrahedron*, **18**, 405 (1962).
403. P. Fournari and J. Tironflef, *Bull. Soc. Chim. Fr.*, **1963**, 486.
404. W. Kuster, E. Brudi, and G. Koppenhofer., *Bull. Soc. Chim. Fr.* **58**, 1014 (1925).
405. H. Fischer and K. Zeile, *Liebigs Ann. Chem.*, **462**, 210 (1928).
406. T. T. Howarth, A. H. Jackson, J. Judge, G. W. Kenner, and D. J. Newman, *J. Chem. Soc., Perkin Trans. 1*, **1974**, 490.
407. F. G. Gonzales and J. M. R. Gonzales, *Anales. Real. Espan. Fis. Guim.*, **47B**, 549 (1951); *Chem. Abstr.*, **46**, 10149 (1952).
408. H. Fischer and B. Weiss, *Chem. Ber.*, **57**, 605 (1924).
409. H. Fischer and H. Wasenegger, *Liebigs Ann. Chemie.*, **461**, 277 (1928).
410. H. Fischer and M. Neber, ibid., **496**, 1 (1932).
411. I. Yu. Grishin, abstracts of the thesis for Candidate of Chemical Sciences, Moscow, 1990.
412. P. Fourari and J. Tirouflet, *Bull. Soc. Chim. Fr.*, **1963**, 484.
413. G. M. Badger, R. L. N. Harris, and R. A. Jones, *Austral. J. Chem.*, **17**, 987 (1964).
414. R. B. Woodward, W. A. Ayer, J. M. Beaton, F. Bickeihaupt, R. Bonnet, P. Buchschacher, G. L. Closs, H. Dutler, J. Hannah, F. P. Hanck, S. Ito, A. Langemann, E. Golf, W. Leimgruber, W. Lwowski, J. Sauer, Z. Valenta, and H. Volz, *J. Am. Chem. Soc.*, **82**, 3800 (1962).
415. G. M. Badger, R. L. N. Harris, and R. A. Jones, *Austral. J. Chem.*, **17**, 1002 (1964).
416. J. L. Davis, *J. Chem. Soc.*, **1968**, 1392.
417. J. B. Paine III, R. B. Woodward, and D. Dolphin, *J. Org. Chem.*, **41**, 2826 (1976).
418. J. B. Paine III and D. Dolphin, ibid., **53**, 2787 (1988).
419. S. Gronowitz and R. Kada, *J. Heterocycl. Chem.*, **21**, 1041 (1984).
420. W. Herz and J. Brasch, *J. Org. Chem.*, **23**, 711 (1958).
421. N. R. El-Rayyes, *J. Prakt. Chem.*, **314**, 915 (1972).
422. N. R. El-Rayyes, ibid., **315**, 295 (1973).
423. W. Herz, *J. Am. Chem. Soc.*, **71**, 3982 (1949).
424. W. Herz and C. Dittmer, *J. Am. Chem. Soc.*, **70**, 503 (1948).
425. D. G. Harvey, *J. Chem. Soc.*, **1950**, 1638.
426. A. Treibs and R. Zimmer-Galler, *Chem. Ber.*, **93**, 2547 (1960).
427. M. Dezelic and B. Bobarevic, *Croat. Chem. Acta.* **34**, 71 (1961); *Chem. Abstr.*, **58**, 4563 (1963).
428. H. Falk, K. Grubmayr, U. Herzig, and O. Hofer, *Tetrahedron Lett.*, **1975**, 559.
429. H. Fischer and H. Orth, *Die Chemie des Pyrrols*, Vol. 1, Akad. Verlags. Leipzig, 1934, p. 224.
430. W. Kutscher and O. Klamerth, *Hoppe-Seyler's Z. Physiol. Chem.*, **286**, 190 (1951).
431. J. Murakami, H. Takahashi, J. Nakazawa, M. Koshimizu, T. Watanabe, and J. Jokoyama, *Tetrahedron Lett.*, **30**, 2099 (1989).
432. A. Treibs and N. Haberle, *Liebigs Ann. Chem.*, **739**, 220 (1970).
433. R. N. Castle and C. W. Whihtele, *J. Org. Chem.*, **24**, 1189 (1959).
434. F. J. Villani, E. A. Wefer, T. A. Mann, J. Mayer, L. Peel, and A. S. Levy, *J. Heterocycl. Chem.*, **9**, 1203 (1972).
435. Ger. Pat. 2010490, *Chem. Abstr.* **74**, 13137 (1971).
436. G. T. Pilyugin, A. P. Rudko, and I. N. Chernyuk, *Khim. Geterotsikl. Soedin.*, **1967**, 868; *Chem. Abstr.*, **68**, 70158 (1968).
437. G. T. Pilyugin and Ya. O. Gorychok, ibid., **1967**, 122.

438. W. A. Remers and M. J. Weiss, *J. Am. Chem. Soc.*, **87**, 5262 (1965).
439. W. A. Remers, R. H. Roth, G. J. Gibs, and M. J. Weiss, *J. Org. Chem.*, **36**, 1232 (1971).
440. E. E. Schweizer and K. K. Light, *J. Am. Chem. Soc.*, **86**, 2963 (1964).
441. R. A. Jones and J. A. Linder, *Austral. J. Chem.*, **18**, 875 (1965).
442. E. J. Seus, *J. Heterocycl. Chem.*, **2**, 318 (1965).
443. E. E. Schweizer and K. K. Light, *J. Org. Chem.*, **31**, 870 (1966).
444. E. E. Schweizer and K. K. Light, ibid., **31**, 2912 (1966).
445. E. E. Schweizer, A. T. Wehman, and D. M. Nycz, ibid., **38**, 1538 (1973).
446. R. A. Jones, J. Poljarlieva, and R. J. Head, *Tetrahedron*, **24**, 2013 (1968).
447. C. G. Olerberger, A. Wartman, and J. C. Salamone, *Org. Prep. Proced.*, **1**, 117 (1969).
448. P. L. Fuchs, *J. Am. Chem. Soc.*, **96**, 1607 (1974).
449. W. Hinz, R. A. Jones, and T. Anderson, *Synthesis*, **8**, 620 (1986).
450. V. M. Shostakovsky, A. U. Musaev, and A. V. Vasil'vitzky, in "The All-Union Conference on the Chemistry of Unsaturated Compounds in Commemoration of A. M. Butlerov," *Abstracts in Russian*, Part 1, p. 77. Kazan, 1986.
451. W. Flitsch, B. Muter, and U. Wolf, *Chem. Ber.*, **106**, 1993 (1973).
452. D. C. Lankin, M. R. Scalise, J. C. Schmidt, and Y. Zimmer, *J. Heterocycl. Chem.*, **11**, 631 (1974).
453. B. A. J. Clark, M. M. S. El-Bakoush, and J. Parrick, *J. Chem. Soc. Perkin Trans. 1*, **1974**, 1531.
454. Y. Badar, W. J. S. Lockley, T. P. Toube, B. C. L. Weedon, and L. R. G. Valadon, ibid., **1973**, 1416.
455. V. A. Dombrovsky, E. V. Gracheva, and P. M. Kochergin, *Khim. Geterotsikl. Soedin.*, **1986**, 40; *Chem. Abstr.*, **105**, 190716 (1986).
456. A. Berlin, S. Bradamante, R. Ferracceioli, G. A. Pagani, and F. Sannicolo, *J. Chem. Soc., Perkin Trans. 1*, **1987**, 2631.
457. U. Pindur and M. Eitel, *Helv. Chem. Acta*, **71**, **1988**, 1060.
458. C. W. Bird and G. W. H. Cheeseman, in A. R. Katritzky and C. W. Rees, Eds., *Comprehensive Heterocyclic Chemistry*, Vol. 4, Pergamon Press, Oxford, 1984, pp. 39–89.
459. R. M. Acheson and J. M. Vernon, *J. Chem. Soc.*, **1961**, 457.
460. R. M. Acheson and J. M. Vernon, ibid., **1963**, 1008.
461. N. W. Gabel, *J. Org. Chem.*, **27**, 301 (1962).
462. J. Saner, *Angew. Chem., Int. Ed.*, 6, 16 (1967).
463. C. K. Lee, C. S. Hahn, and W. E. Noland, *J. Org. Chem.*, **43**, 3727 (1978).
464. B. H. Lipshutz, *Chem. Rev.*, **86**, 795 (1986).
465. P. Hong, B. -R. Cho, and H. Yamazaki, *Chem. Lett.*, **1980**, 507.
466. W. E. Noland and C. K. Lee, *J. Org. Chem.*, **45**, 4573 (1980).
467. R. P. Rosenthal, Ph.D. thesis, University of East Anglia, 1975.
468. H. Kotsuki, Y. Mori, H. Nishizawa, M. Ochi, and K. Matsuoka, *Heterocycles*, **19**, 1915 (1982).
469. E. Krabovicova, A. Krutosikova, J. Kovac, and M. Dandarova, *Coll. Czech. Chem. Commun.*, **51**, 1455 (1986).
470. K. Eger, G. Folkers, M. Frey, W. Zimmermann, and G. Koop-Kirfel, *Synthetic Commun.*, **18**, 632 (1988).
471. O. Diels, K. Alder, and D. Wimter, *Liebigs Ann. Chem.*, **486**, 211 (1931).
472. O. Diels, K. Alder, H. Winckler, and E. Petersen, ibid., **498**, 1 (1932).
473. R. C. Bansal, A. W. McCulloch, and A. G. McInnes, *Can. J. Chem.*, **47**, 2391 (1969).
474. R. M. Acheson and J. M. Vernon, *J. Chem. Soc.*, **1962**, 1148.
475. R. Kitzing, R. Fuchs, M. Joyeun, and H. Prinzbach, *Helv. Chim. Acta*, **51**, 888 (1968).

2.7. References

476. O. Diels, K. Alder, and W. Lubbert, *Liebigs Ann. Chem.*, **490**, 277 (1931).
477. A. H. Corwin and J. S. Andrews, *J. Am. Chem. Soc.*, **58**, 1086 (1935).
478. H. Fisher and H. Hofelmann, *Hoppe-Seyler's, Z. Physiol. Chem.*, **251**, 218 (1938).
479. A. Treibs and K. Herrmann, *Liebigs Ann. Chem.*, **592**, 1 (1955).
480. A. Treibs, K. Herrmann, E. Meisser, and A. Kuhn, ibid., **602**, 153 (1957).
481. A. Treibs and F. Reitsam, ibid., **611**, 194 (1958).
482. A. Treibs and F. Reitsam, ibid., **611**, 205 (1958).
483. A. Treibs and W. Seifert, ibid., **612**, 242 (1958).
484. L. D. Miroschnichenko, E. I. Filippovich, R. P. Evstigneeva, and N. A. Preobrazhensky, *Dokl. Akad. Nauk SSSR*, **134**, 1100 (1960); *Chem. Abstr.*, **55**, 8377 (1961).
485. E. I. Filippovich, R. P. Evstigneeva, and N. A. Preobrazhensky, *Zh. Obshch. Khim.*, **30**, 3253 (1960); *Chem. Abstr.*, **55**, 21095 (1961).
486. E. I. Filippovich, R. P. Evstigneeva, and N. A. Preobrazhensky, ibid., **31**, 2968 (1961); *Chem. Abstr.*, **57**, 2174 (1962).
487. L. D. Miroshnichenko, R. P. Evstigneeva, E. I. Filippovich, and N. A. Preobrazhensky, ibid., **31**, 2975 (1961); *Chem. Abstr.*, **57**, 2175 (1962).
488. L. D. Miroshnichenko and R. P. Evstigneeva, *Izv. Akad. Nauk SSSR. Ser. Fiz.*, **27**, 50 (1963); *Chem. Abstr.*, **59**, 5111 (1963).
489. R. L. N. Harris, A. W. Johnson, and I. T. Kay, *Quart. Rev.*, **20**, 211 (1966).
490. K. M. Smith, ibid., **25**, 31 (1971).
491. J. R. Sabin, *Int. J. Quant. Chem.*, **2**, 31 (1968).
492. D. Brown, D. Griffiths, M. E. Rider, and R. C. Smith, *J. Chem. Soc., Perkin Trans. 1*, **1986**, 455.
493. H. Rapoport and K. G. Holden, *J. Am. Chem. Soc.*, **84**, 635 (1962).
494. A. Treibs and K. Hintermeier, *Liebigs Ann. Chem.*, **592**, 11 (1955).
495. W. Füler, dissertation, Munich, 1949.
496. A. Treibs and E. Hermann, *Liebigs. Ann. Chem.*, **589**, 207 (1954).
497. A. H. Cook and J. R. Majer, *J. Chem. Soc.*, **1944**, 482.
498. R. A. Jones and R. L. Laslett, *Austral. J. Chem.*, **17**, 1056 (1964).
499. A. Treibs and F. -H. Krenzer, *Liebigs Ann. Chem.*, **718**, 208 (1968).
500. M. Strell and F. Kreis. *Chem. Ber.*, **87** 1011 (1954).
501. A. Treibs and K. Jacob, *Angew. Chem. Int. Ed.*, **4**, 694 (1965).
502. A. Treibs and K. Jacob, *Liebigs Ann. Chem.*, **699**, 153 (1966).
503. A. Treibs and K. Jacob, ibid., **712**, 123 (1968).
504. H. -E. Sprenger and W. Ziegenbein, *Angew. Chem.*, **80**, 541 (1968).
505. A. Treibs, K. Jacob, and R. Tribollet, *Liebigs Ann. Chem.*, **741**, 101 (1970).
506. H. Fischer and H. Orth, *Die Chemie des Pyrrols*, Vol. II, Akad. Verlags, Leipzig, 1939.
507. H. Von Dobeneck, T. Messerschmitt, E. Brunnerand, and U. Wunderer, *Liebigs Ann. Chem.*, **751**, 40 (1971).
508. J. E. Macor, M. E. Newman, and K. Ryan, *Tetrahedron Lett.*, **30**, 2509 (1989).
509. U.S. Pat. 2393132 (1946); *Chem. Abstr.*, **40**, 2472 (1946).
510. G. S. Kolesnikov, *Synthesis of Vinyl Derivatives of Aromatic and Heterocyclic Compounds* (in Russian), the USSR Academay of Science, Moscow, 1960.
511. I. A. Jacobson and H. B. Jensen, *J. Phys. Chem.*, **66**, 1245 (1962).
512. W. W. Pandler and E. A. Stephan, *J. Am. Chem. Soc.*, **92**, 4468 (1970).
513. J. L. Afwood, D. C. Hrncir, C. Wong, and W. W. Pandler, ibid., **96**, 6132 (1974).
514. W. Herz and C. F. Courtney, ibid., **76**, 576 (1954).

515. M. T. Cox, A. N. Jackson, and G. W. Kenner, *J. Chem. Soc. (C)*, **1974**, 1971.
516. D. P. Schumacher and S. S. Hall, *J. Org. Chem.*, **46**, 5060 (1981).
517. W. Flitsch and H. Peters, *Tetrahedron Lett.*, **1975**, 1461.
518. F. Ya. Perveev, E. M. Vekshina, and L. N. Surenkova, *Zh. Obshch. Khim.*, **27**, 1526 (1957); *Chem. Abstr.*, **52**, 3767 (1958).
519. F. Ya. Perveev and E. M. Kuznetsova, ibid., **28**, 2360 (1958); *Chem. Abstr.*, **53**, 3190 (1959).
520. J. Moskal, R. Van Stralen, D. Postma, and A. Van Lausen, *Tetrahedron Lett.*, **27**, 2173 (1986).
521. M. Kroner, *Chem. Ber.*, **100**, 3162 (1967).
522. K. Isomura, T. Tanaka, and H. Taniguchi, *Chem. Lett.*, **1977**, 397.
523. T. S. Canfrell, *J. Org. Chem.*, **39**, 2242 (1974).
524. M. Bellas, D. Bruce-Smith, and A. Gilbert, *J. Chem. Soc., Chem. Commun.*, **1967**, 263.
525. K. Hayakawa, M. Yodo, S. Ohsuki, and K. Kanematsu, *J. Am. Chem. Soc.*, **106**, 6735 (1984).
526. V. H. Rawal and M. P. Cava, ibid., **108**, 2110 (1986).
527. J. M. Muchowski and M. E. Scheller, *Tetrahedron Lett.*, **28**, 3453 (1987).
528. E. T. Tarlon, S. F. MacDonald, and E. Baltazzi, *J. Am. Chem. Soc.*, **82**, 4389 (1960).
529. H. Fisher and H. Orth, *Die Chemie des Pyrrols*, Voll. II, Akad. Verlags, Leipzig, 1939, p. 71.
530. K. M. Smith, *Tetrahedron Lett.*, **1971**, 2325.
531. H. Fiscer, E. Sturm, and H. Friedrich, *Liebigs. Ann. Chem.*, **461**, 244 (1928).
532. G. Hayes, A. H. Jackson, J. M. Judge, and G. W. Kenner, *J. Chem. Soc.*, **1965**, 4384.
533. P. Fournari and T. Marey, *Bull. Soc. Chim. Fr.*, **1968**, 3223.
534. A. K. Uradhyaya and K. N. Mehrotra, *J. Chem. Soc., Perkin Trans 2*, **1988**, 957.
535. R. A. Jones, M. T. P. Mariott, W. P. Rosenthal, and J. Sepulveda Arques, *J. Org. Chem.*, **45**, 4515 (1980).
536. R. Paul, *Bull. Soc. Chim. Fr.*, **1943**, 163.
537. J. F. Scully and E. V. Brown, *J. Am. Chem. Soc.*, **75**, 6329 (1953).
538. W. E. Noland and S. R. Wann, *J. Org. Chem.*, **44**, 4402 (1979).
539. R. S. Hosmane, S. P. Hiremath, and S. W. Schneller, *J. Chem. Soc., Perkin Trans. 1*, **1973**, 2450.
540. R. M. Acheson and J. Woolard, ibid., **1975**, 446.
541. C. K. Lee, Ph.D. thesis, University of Minnesota, Minneapolis, 1976; *Diss. Abstr. Int. B*, **38**, 1210 (1977); *Chem. Abstr.*, **88**, 120912 (1978).
542. R. A. Jones and J. Sepulveda Arques, *Tetrahedron*, **37**, 1597 (1981).
543. R. A. Jones, T. Aznar Saliente, and J. Sepulveda Arques, *J. Chem. Soc., Perkin Trans. 1*, **1984**, 2541.
544. W. E. Noland, C. K. Lee, S. K. Bae, B. J. Chung, C. S. Hahn, and K. J. Kim, *J. Org. Chem.*, **48**, 2488 (1983).
545. F. E. Ziegler, E. B. Spitzner, and C. K. Wilkins, *J. Org. Chem.*, **36**, 1759 (1971).
546. R. J. Sundberg and J. D. Bloom, *Tetrahedron Lett.*, **1978**, 5157.
547. R. J. Sundberg and J. D. Bloom, *J. Org. Chem.*, **45**, 3382 (1980).
548. R. J. Sundberg and J. D. Bloom, ibid., **46**, 4836 (981).
549. C. Marazanj, J. L. Fourrey, and B. C. Das, *J. Chem. Soc., Chem. Commun.*, **1981**, 37.
550. D. Beck and K. Schenker, *Helv. Chim. Acta*, **51**, 260 (1968).
551. N. S. Narismhan, R. S. Kururkar, and D. D. Dhavali, *Ind. J. Chem. Sect. B*, **22B**, 1004 (1983).
552. G. W. B. Reed, P. T. C. Cheng, and S. McLean, *Can. J. Chem.*, **72**, 419 (1982).
553. R. A. Jones and P. M. Fresneda, *Tetrahedron*, **40**, 4837 (1984).
554. C. K. Lee and Gu Mi Ahn, *J. Heterocycl. Chem.*, **26**, 397 (1989).

2.7. References

555. U. Pindur, L. Pfeuffer, and M. -H. Kim, *Helv. Chim. Acta*, **72**, 65 (1989).
556. U. Pindur, M. Eitel, and E. Abdoust-Haushang, *Heterocycles*, **29**, 11 (1989).
557. V. M. Shostakovsky, A. E. Vasilvitsky, A. U. Musaev, and O. M. Nefedov, in "IVth All-Union Conference on Chemistry of Nitrogen-Containing Heterocyclic Compounds," *Abstracts in Russian*, p. 88, Novosibirsk, 1987.
558. V. M. Shostakovsky, A. U. Musaev, A. E. Vasilvitsky, A. M. Guliev, and O. M. Nefedov, *Izv. Akad. Nauk. SSSR. Ser. Khim.*, **1989**, 715; *Chem. Abstr.*, **111**, 214353 (1989).
559. R. A. Jones and M. T. P. Marriott, *Heterocycles*, **14**, 185 (1980).
560. V. H. Rawai and M. P. Cava, *J. Chem. Soc., Chem. Commun.*, **1984**, 1526.
561. V. H. Rawai, R. J. Jones, and M. P. Cava, *Tetrahedron Lett.*, **26**, 2423 (1985).
562. R. A. Jones and J. J. Thomas, unpublished work.
563. R. A. Jones and W. Hinz, unpublished work.
564. W. Flitsch and U. Newmann, *Chem. Ber.*, **104**, 2170 (1971).
565. F. Micheel and W. Kempel, ibid., **69**, 1990 (1936).
566. W. C. Agosta, *J. Am. Chem. Soc.*, **82**, 2258 (1960).
567. C. Flo and U. Pindur, *Liebigs Ann. Chem.*, **1988**, 923.
568. T. Aznar Saliente, R. A. Jones, R. T. Sanchis Llorca, and J. Sepulveda Arques, *J. Chem. Res. (S)*, **1985**, 12; *(M)* **1985**, 232–240.
569. R. T. Sanchis Llorca, J. Sepulveda Arques, E. Zaballos Garcia, and R. A. Jones, *Heterocycles*, **26**, 401 (1987).
570. A. H. Jackson and K. H. Smith, in J. Apsimon, Ed., *The Total Synthesis of Natural Products*, Vol. 1, Wiley, New York, 1973, pp. 143–278.
571. R. Kreher and G. Vogt, *Angew. Chem.*, **82**, 958 (1970).
572. J. Duflos, D. Lefonze, G. Queguiner, and P. Pasfour, *Tetrahedron Lett.*, **1973**, 3453.
573. J. Duflos and G. Queguiner, *Tetrahedron*, **41**, 3303 (1985).
574. E. Bisagni, J. D. Bourzat, and J. Andre-Louisfert, ibid., **26**, 2087 (1970).
575. P. Bamfield, R. Grigg, R. W. Kenyon, and A. W. Johnson, *J. Chem. Soc., Chem. Commun.*, **1967**, 1029.
576. P. Bamfield, R. Grigg, A. W. Johnson, and R. W. Kenyon, *J. Chem. Soc. (C)*, **1968**, 1259.
577. R. Grigg, A. W. Johnson, and M. Roche, ibid., **1970**, 1928.
578. M. Somei, E. Iwasa, and F. Yamada, *Kissey. Yakuhin Koge*, N 61, 211107 (1988); *Zh. Khim.*, 4058II (1989).
579. L. Canoira, R. J. Gonzalo, J. B. Subirals, J. A. Escario, I. Jimenez, and A. R Martinez-Fernandez, *Eur. J. Med. Chem.*, **24**, 39 (1989).
580. L. G. S. Brooker and F. L. White, *J. Am. Chem. Soc.*, **73**, 1094 (1951).
581. A. Treibs and W. Ort, *Liebigs Ann. Chem.*, **577**, 119 (1952).
582. H. J. Anderson and H. Nagy, *Can. J. Chem.*, **50**, 1961 (1972).
583. S. Kano, E. Sugino, S. Shibuya, and S. Hibino, *J. Org. Chem.*, **46**, 3856 (1981).
584. U. Pindur and L. Pfeuffer, *Monatsh. Chem.*, **120**, 157 (1989).
585. R. P. Carr, P. J. Crook, A. H. Jackson, and G. W. Kenner, *J. Chem. Soc., Chem. Commun.*, **1967**, 1027.
586. A. Huni and F. Frank, *Hoppe-Seyler's Z. Physiol. Chem.*, **282**, 244 (1947).
587. R. B. Woodward, *Angew. Chem.*, **72**, 651 (1960).
588. A. M. Fargali, R. P. Evstigneeva, I. N. Handii, and N. A. Preobrazhensky, *Zh. Obshch. Khim.*, **34**, 893 (1964); *Chem. Abstr.*, **60**, 15812 (1964).
589. A. M. Fargali, R. P. Evstigneeva, and N. A. Preobrazhensky, ibid., **34**, 898 (1964); *Chem. Abstr.*, **60**, 15812 (1964).

590. R. D. Evstigneeva, A. M. Fargali, T. A. Lubkova, I. N. Handii, and N. A. Preobrazhensky, in *Sintez Prirodn. Soedin., Ikh Analogov i Fragmentov*, Nauka, Moscow-Leningrad, 1965, p. 216.
591. J. B. Paine III, in D. Dolphin, Ed., *The Porphyrins*, Vol. 1, *Structure and Synthesis*, Part A, Academic Press, London, 1978, pp. 101–234.
592. P. A. Burbidge, G. L. Collier, A. H. Jackson, and G. W. Kenner, *J. Chem. Soc.* (*B*), **1967**, 930.
593. G. L. Collier, A. H. Jackson, and G. W. Kenner, *J. Chem. Soc.* (*C*), **1967**, 66.
594. R. P. Carr, A. H. Jackson, G. W. Kenner, and G. S. Sach, *J. Chem. Soc.* (*C*), **1971**, 487.
595. B. A. Trofimov, L. N. Sobenina, and A. I. Mikhaleva, *Itogi Nauki i Tekhniki., Ser. Org. Khim.*, Vol. 7, VINITI, Moscow, 1987.
596. F. Schierle, H. Reinhard, N. Dieter, E. Lippacher, and H. von Dolenec, *Liebigs Ann. Chem.*, **715**, 90 (1968).
597. F. Closs and R. Gompper, *Angew. Chem.*, **99**, 564 (1987).
598. E. Gonzalez Rosende, R. A. Jones, J. Sepulveda Arques, and E. Zaballos Garcia, *Synthetic Commun.*, **18**, 1669 (1988).
599. U. Pindur, *Heterocycles*, **27**, 1253 (1988).
600. M. -H. Kim and U. Pindur, *Arch. Pharm.*, **322**, 739 (1989).
601. R. Adam and U. Pindur, ibid., **322**, 740 (1989).
602. L. Pfeuffer, U. Pindur, H. -J. Sattler, W. Massa, and C. Frenzen, *Monatsh. Chem.*, **119**, 1289 (1988).
603. B. Gregory, W. Hinz, R. A. Jones, and J. Sepulveda Arques, *J. Chem. Res.* (*S*), 311; (*M*) 2801–2821 (1984).

CHAPTER 3

Aminopyrroles

Girolamo Cirrincione, Anna Maria Almerico, and Enrico Aiello

Istituto Farmacochimico,
Facoltà di Farmacia,
Università di Palermo,
Palermo, Italy

Gaetano Dattolo

Istituto Chimico Farmaceutico e Tossicologico,
Facoltà di Farmacia,
Università di Milano,
Milano, Italy

3.1. Introduction 301
3.2. Synthesis. 301
    3.2.1. Natural Products and Related Compounds. 301
        3.2.1.1. Netropsin or Congocidine 302
        3.2.1.2. Distamycin A 305
        3.2.1.3. Kikumycins 310
        3.2.1.4. Anthelvencin A 314
    3.2.2. Direct Ring Synthesis 315
        3.2.2.1. 1-Aminopyrroles. 315
            3.2.2.1.1. From Four-Carbon Units 315
            3.2.2.1.2. From Ring Transformations 322
            3.2.2.1.3. From 2 + 2 Carbon Units 323
        3.2.2.2. 2-Aminopyrroles. 328
            3.2.2.2.1. From Four-Carbon Units 328
            3.2.2.2.2. From Ring Transformations 332
            3.2.2.2.3. From 3 + 1 Carbon Units 334
            3.2.2.2.4. From 2 + 2 Carbon Units 335
            3.2.2.2.5. From 2 + 1 + 1 Carbon Units 339
        3.2.2.3. 3-Aminopyrroles. 339
            3.2.2.3.1. From Four-Carbon Units 340
            3.2.2.3.2. From Ring Transformations 340
            3.2.2.3.3. From 3 + 1 Carbon Units 342
            3.2.2.3.4. From 2 + 2 Carbon Units 343

Pyrroles Part Two: The Synthesis, Reactivity, and Physical Properties of Substituted Pyrroles, Edited by R. Alan Jones.
ISBN 0-471-51306-7 © 1992 John Wiley & Sons, Inc.

	3.2.2.4. Diaminopyrroles.	345
	3.2.2.4.1. 1,2-Diaminopyrroles.	345
	3.2.2.4.2. 2,4-Diaminopyrroles.	346
	3.2.2.4.3. 2,5-Diaminopyrroles.	347
	3.2.2.5. Triaminopyrroles	347
3.2.3. Conversion of Functional Groups		348
	3.2.3.1. 1-Aminopyrroles	348
	3.2.3.2. 2-Aminopyrroles	349
	3.2.3.3. 3-Aminopyrroles	351
	3.2.3.4. Diaminopyrroles	354
3.3. Structure and Physical Properties		355
3.3.1. Theoretical Methods		356
3.3.2. X-Ray Diffraction		363
3.3.3. ^{1}H NMR Spectra		365
3.3.4. ^{13}C NMR Spectra		403
3.3.5. UV Spectra		424
3.3.6. IR Spectra		435
3.3.7. Mass Spectra		460
3.3.8. Other Physical Measurements		463
3.3.9. Thermodynamic Aspects		463
3.3.10. Tautomeric and Basicity Studies		471
	3.3.10.1. 1-Aminopyrroles	471
	3.3.10.2. 2-Aminopyrroles	472
	3.3.10.3. 3-Aminopyrroles	474
	3.3.10.4. Diaminopyrroles	477
3.4. Reactivity		478
3.4.1. Thermolysis and Photolysis Reactions		478
	3.4.1.1. 1-Aminopyrroles	478
	3.4.1.2. 2-Aminopyrroles	481
	3.4.1.3. Diaminopyrroles	481
3.4.2. Reactions with Electrophiles		482
	3.4.2.1. 1-Aminopyrroles	482
	3.4.2.2. 2-Aminopyrroles	486
	3.4.2.3. 3-Aminopyrroles	489
	3.4.2.4. Diaminopyrroles	494
	3.4.2.5. Triaminopyrroles	494
3.4.3. Reactions with Oxidizing Agents		495
	3.4.3.1. 1-Aminopyrroles	495
	3.4.3.2. 2-Aminopyrroles	496
	3.4.3.3. 3-Aminopyrroles	497
3.4.4. Reactions with Nucleophiles		498
	3.4.4.1. 1-Aminopyrroles	499
	3.4.4.2. 2-Aminopyrroles	499
	3.4.4.3. 3-Aminopyrroles	502
	3.4.4.4. Diaminopyrroles	503
3.4.5. Reactions with Reducing Agents		503
	3.4.5.1. 1-Aminopyrroles	503
	3.4.5.2. 2-Aminopyrroles	504
	3.4.5.3. 3-Aminopyrroles	504
3.4.6. Cycloaddition Reactions		505
3.5. Applications		508
3.5.1. Biological and Medical Uses		508
3.5.2. As Synthons for Drugs		512

3.6. Appendix	513
3.7. Acknowledgments	514
3.8. References	514

3.1. INTRODUCTION

Aminopyrroles have been extensively studied, as is testified by the high number (>350) of papers in literature dealing with their synthesis, identification, and reactivity. However, this does not mean that everything is already known of the reactivity of aminopyrroles. In fact, more than half of the reports deal mainly with preparation and structural determination.

More problems were encountered during the preparation and/or isolation of aminopyrroles than in the determination of their structure; for example, simple substituted aminopyrroles are unknown with the exception of 1-amino and 2,5-diamino derivatives. This is due to the fact that the presence of an electron-donating group on an already electron-rich nucleus makes this class of compounds rather unstable, unless strong electron-withdrawing substituents are present on the ring. Therefore, reactivity studies on aminopyrroles are limited to those compounds that have a high degree of stability. However, a clear difference in the properties and behavior is evident between the *N*- and *C*-amino derivatives.

3.2. SYNTHESIS

Synthetic approaches to aminopyrroles can be divided into three sections. The first section (Section 3.2.1) relates to naturally occurring aminopyrrole derivatives, the total synthesis of which is reviewed regardless of the position of the amino group in the nucleus. This section also reports the synthesis of related compounds. The second section (Section 3.2.2) deals with the synthesis of the pyrrole nucleus with concomitant introduction of the amino group. This approach represents nearly all the routes for the preparation of 1-aminopyrroles, but has also been widely employed in the synthesis of 2-aminopyrroles, although it is less important for the synthesis of 3-aminopyrroles, which are better obtained by the conversion of other functional groups, as described in the third section (Section 3.2.3).

3.2.1. Natural Products and Related Compounds

Natural products containing aminopyrrole moieties represent an important class of oligopeptide antibiotics, with a wide spectrum of biological activities. Since their first isolation in the 1950s, the interest in this class of compounds has increased, as testified by the large number of publications dealing with their chemistry in recent years. Evaluation of their pharmaceutical activity and the interaction of these antibiotics with biologically interesting molecules has resulted in several thousand reports. In this section we review the chemistry of four natural products, reporting data on their isolation and structure determination, syntheses, and the preparation of analogs.

3.2.1.1. Netropsin or Congocidine

The antibiotic netropsin was isolated in 1951 from culture of an Actinomycete, *Streptomyces netropsis*, which was isolated from a soil sample.[1] The first publication reported its biological activity and gave the empirical formula, $C_{32}H_{48}N_{18}O_4$, together with data on its alkaline degradation. A year later an antibiotic that was active against gram-positive(+) and gram-negative(−) bacteria was isolated from culture of *Streptomyces chromagens* and *S. umbifaciens*. The gross formula $C_{10}H_{21}N_5O_3$ was attributed to the compound and, as a result of its remarkable activity against the *Trypanosoma congolense*, it was given the name *congocidine*.[2] The antibiotics sinanomycin[3] and "T-1384"[4] were isolated in 1956 and 1957, respectively. However, all these compounds were later shown to be identical, and the discrepancy between the empirical formulas was due to the high hygroscopicity of the antibiotic. More accurate analysis led to a gross formula $C_{18}H_{26}N_{10}O_3$, and studies on the basic hydrolysis of netropsin,[5,6] congocidine,[7] and antibiotic T-1384,[4] leading to the products shown in Scheme 1, resulted in the single structure **1** being postulated for all four systems.

Support for this structure came from the independent synthesis of **3**, **5a**, **5b**, and **6**, which had been obtained from degradation of antibiotic T-1384.[8] Nevertheless, at this stage doubts existed regarding the position of the amidino moiety,[4] and structure **2** was later proposed, based on the isolation of **8** from the acid degradation and also on an independent total synthesis.[9,10] The total synthesis involved the condensation of the pyrrole **9** with β-aminopropionitrile to give **10**, which, on reduction of the nitro group and condensation with a second molecule of **9**, gave **11**. Subsequent reaction with dry hydrochloric acid followed by dry ammonia (Pinner sequence) and reduction of the nitro group led to the amidine **8**, which by reaction with guanidinoacetic acid (**7**) gave **2** (Scheme 2).

In the latest total synthesis, the introduction of the oligopeptide end groups has been achieved in the reverse order and improved procedures in several steps, which avoided chromatographic purification, have resulted in higher yields.[11]

Because of the biological importance of this antibiotic (see Section 3.5.1), several derivatives of type **15**, analogous of congocidine **2**, have been prepared.[12,13] The synthesis can be achieved either by condensation of **12** with the carboxy compound to give the intermediates **14**, which undergoes a Pinner sequence similar to the first total synthesis of the natural product, or by

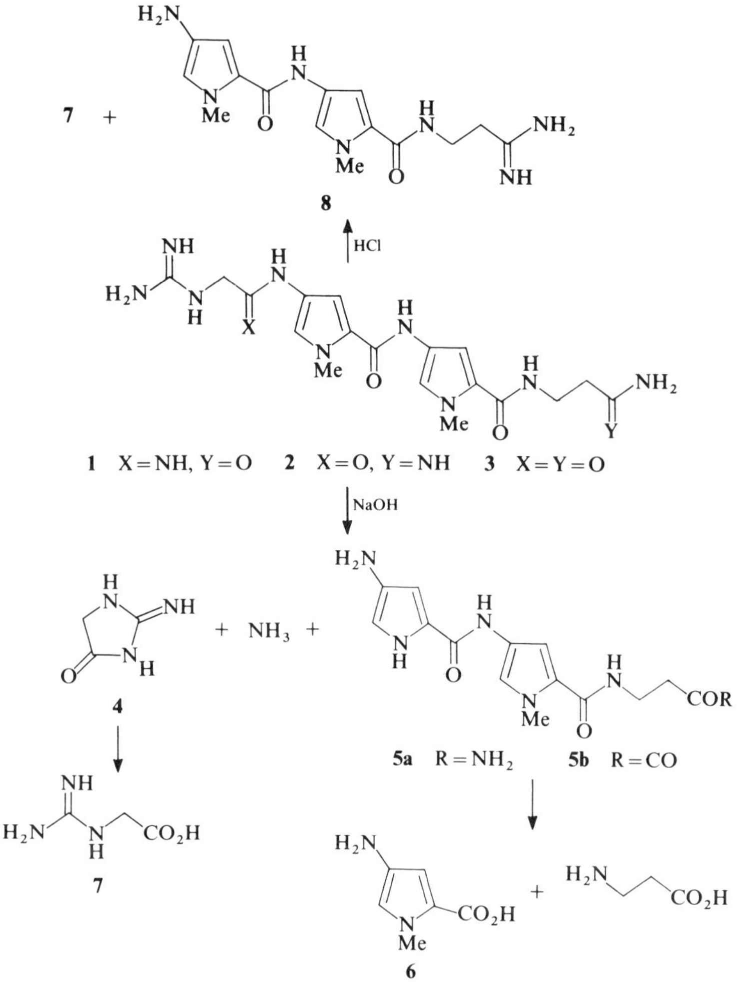

Scheme 1

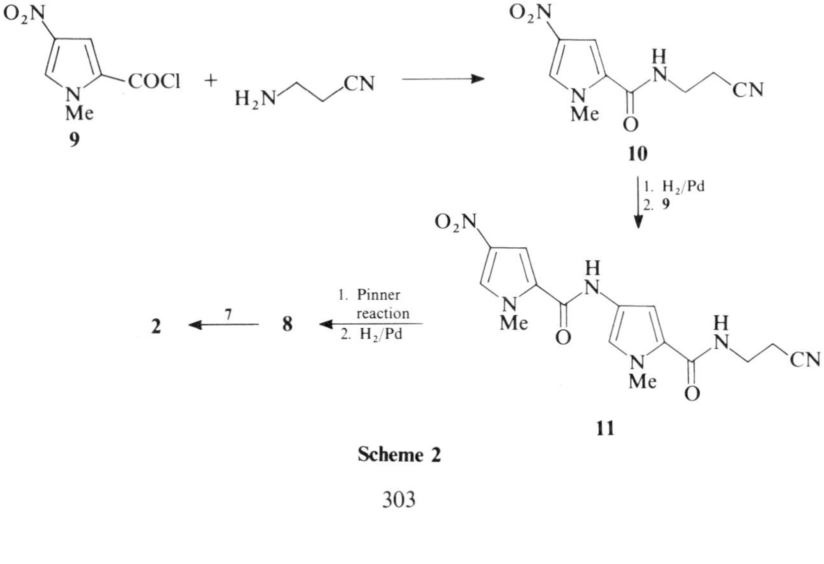

Scheme 2

preliminary formation of the amidine moiety **13** followed by condensation with the carboxy derivative. The condensation reactions were carried out in the presence of base when the carboxy compound was an acyl chloride (X = Cl), whereas dicyclohexylcarbodiimide (DCC) was used in case of a carboxylic acid (X = OH) (Scheme 3).

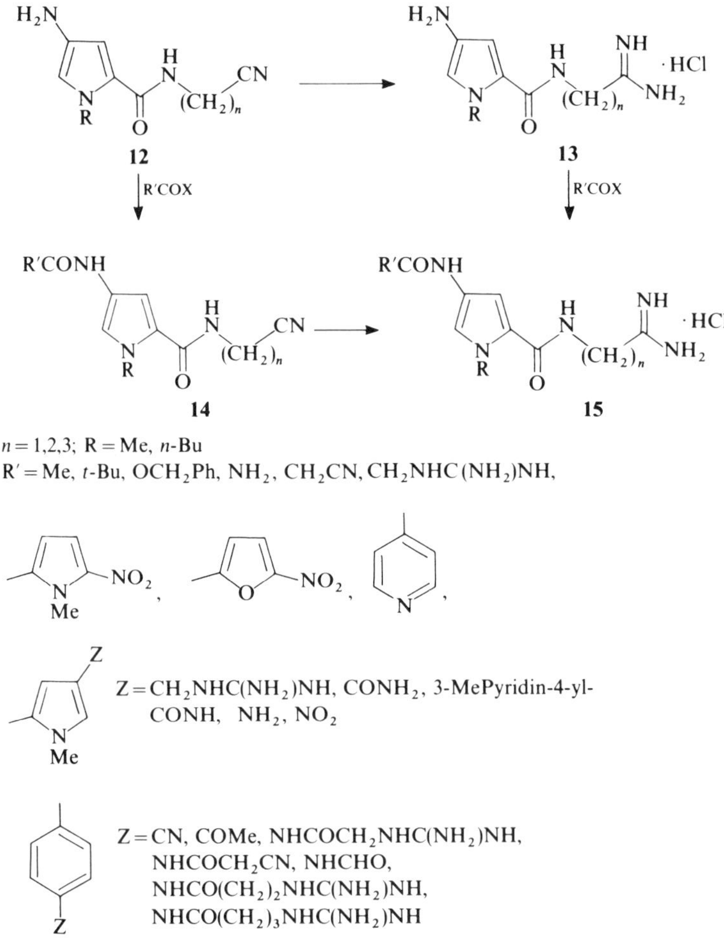

$n = 1,2,3$; R = Me, n-Bu
R' = Me, t-Bu, OCH$_2$Ph, NH$_2$, CH$_2$CN, CH$_2$NHC(NH$_2$)NH,

Z = CH$_2$NHC(NH$_2$)NH, CONH$_2$, 3-MePyridin-4-yl-
CONH, NH$_2$, NO$_2$

Z = CN, COMe, NHCOCH$_2$NHC(NH$_2$)NH,
NHCOCH$_2$CN, NHCHO,
NHCO(CH$_2$)$_2$NHC(NH$_2$)NH,
NHCO(CH$_2$)$_3$NHC(NH$_2$)NH

Scheme 3

Monopeptide and dipeptide analogs of the congocidine **2** where m- and p-disubstituted benzenes, 2,5-disubstituted thiophenes, and 3,5-disubstituted pyridines replace the N-methylpyrrole moieties have also been synthesized following synthetic pathways similar to that used for the total synthesis of the natural product.[14–16] Compounds in which a peptide (NH—CO) group is replaced by an urea moiety have also synthesized.[16]

3.2.1.2. Distamycin A

In 1964, the antibiotic distamycin A, which was shown to have physical and chemical properties similar to those of netropsin, was isolated from the mycelium of the *Streptomyces distallicus*.[17] The molecular formula of the antibiotic was established to be $C_{22}H_{27}N_9O_4$, and degradation studies showed pathways parallel to those of netropsin and led to compounds **17–20**, which can be considered homologs of those obtained from netropsin (Scheme 4).

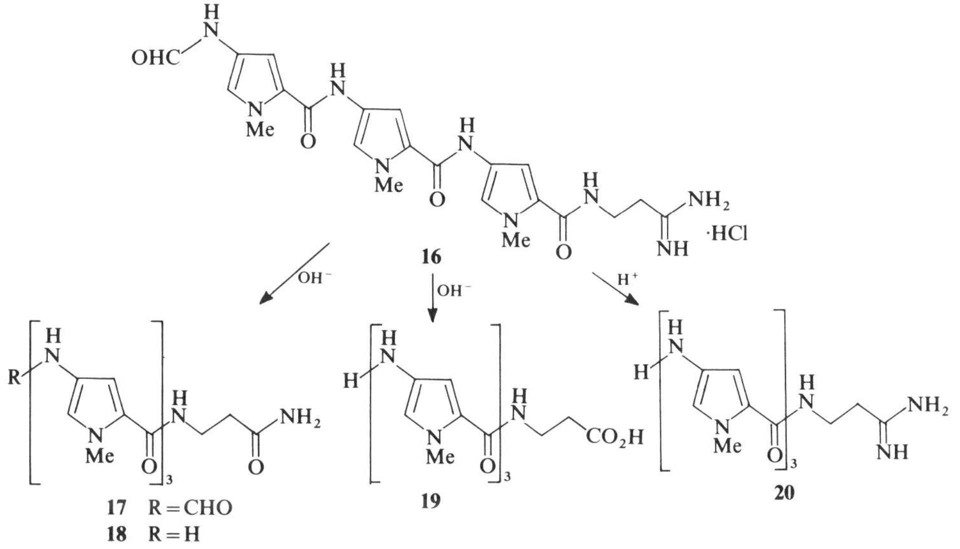

Scheme 4

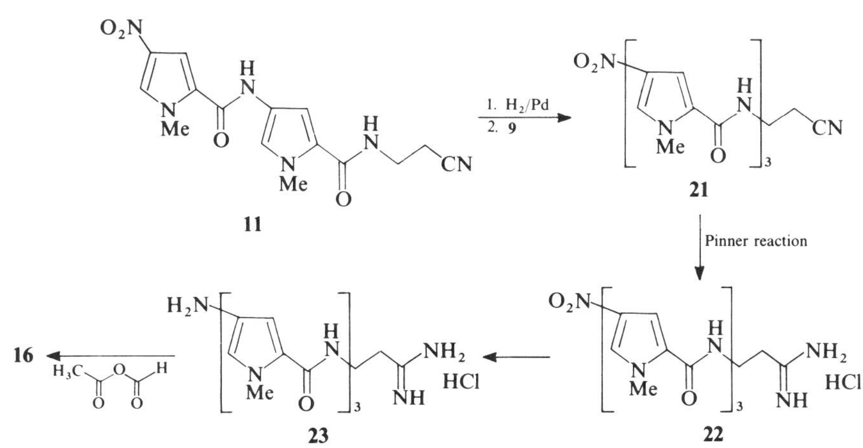

Scheme 5

Structure **16** was assigned to distamycin A[17,18] and was supported by a total synthesis (Scheme 5) that was identical to that used for the synthesis of netropsin up to the precursor **11**.

Reduction of the nitro group and condensation with **9** gave the intermediate **21** containing one more pyrrole unit, which undergoes a Pinner sequence to yield the imino derivative **22**. Subsequent reduction of the nitro group, followed by reaction with formylacetic anhydride, gave distamycin A (**16**)[19] in 21% overall yield. Although the mode of its biological action was unknown, modifications to the natural product were undertaken. Thus, derivatives **24a–j** with modified side chains R and X were prepared by acylation of the aminoamidines of type **24** (R = H) (Scheme 6).

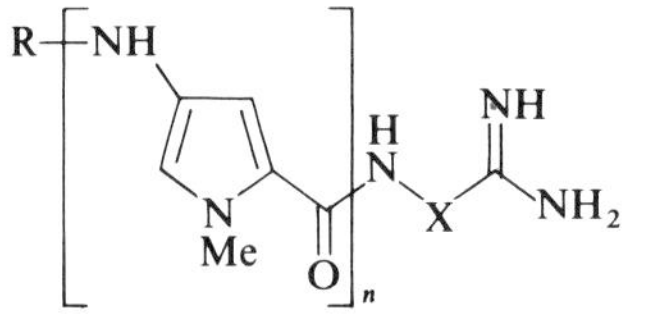

24

a	R=COCH$_3$, $n=3$, X=(CH$_2$)$_2$
b	R=c-C$_5$H$_9$—(CH$_2$)—CO, $n=3$, X=(CH$_2$)$_2$
c	R=CHO, $n=3$, X=CH$_2$
d	R=CHO, $n=3$, X=(CH$_2$)$_3$
e	R=CHO, $n=3$, X=p-C$_6$H$_4$
f	R=CHO, $n=1$, X=(CH$_2$)$_2$
g	R=CHO, $n=2$, X=(CH$_2$)$_2$
h	R=CHO, $n=4$, X=(CH$_2$)$_2$
i	R=CHO, $n=5$, X=(CH$_2$)$_2$
j	R=CHO, $n=6$, X=(CH$_2$)$_2$
k	R=CHO, $n=3$, X=CH(CH$_3$)CH$_2$
l	R=CHO, $n=3$, X=o-C$_6$H$_4$
m	R=CHO, $n=3$, X=m-C$_6$H$_4$

Scheme 6

Preparation of compound **24b** required a modification of the original synthesis. Intermediate **21** was reduced to the corresponding amine **25**, which was condensed with the chloride of the β-cyclopentylpropionic acid to give **26**, and subsequent conversion of the cyano group into the amidine function afforded **24b** (Scheme 7).

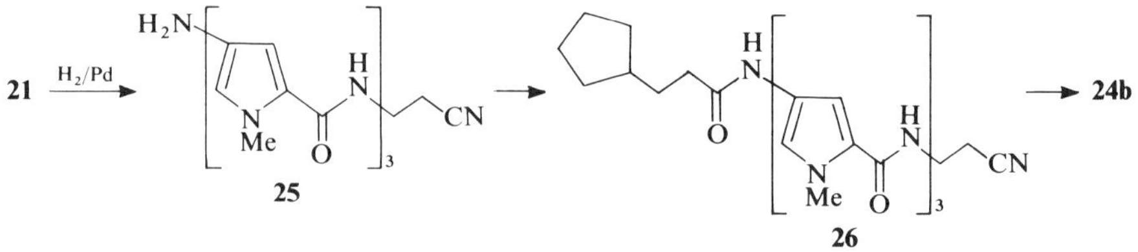

Scheme 7

3.2. Synthesis

Compound **24c** was also synthesized by a modification of the original pathway; the transformation of the cyano group into the amidine moiety was achieved on the nitrocyano intermediate having two pyrrole units. No modification of the original pathway was necessary for the preparation of **24d** and **24e**,[20] and derivatives **24f–j** with different number of pyrrole units ($n = 1,2,4,5,6$) were also synthesized following the original route with the appropriate number of successive reduction–condensation processes used in the sequence **10 → 11**.[21]

The separate synthesis of each side-chain homolog limited the ease with which the structural requirements for the pharmacological activity could be obtained, and it was several years later before a new synthesis, outlined in Scheme 8, was developed starting from the commercially available derivative **27**.[22] The key intermediate was the nitroacid **28**, which can serve as starting material for several side-chain homologs through condensation with the suitable aminonitrile in presence of DCC and 1-hydroxybenzotriazole. Thus, by this

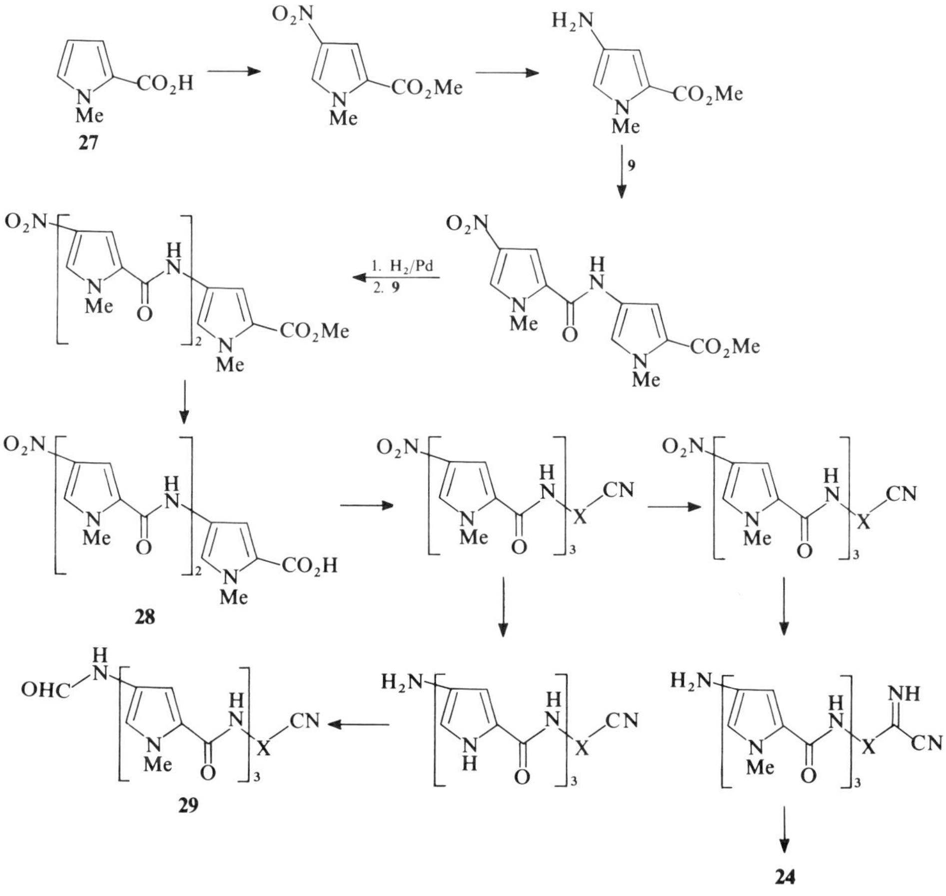

Scheme 8

route were synthesized derivatives **24c** and **24k,** in which replacement of the β-aminoacid derived moiety with an α-aminoacid derived portion was observed. Compounds **24e, 24l,** and **24m** were prepared in order to establish the role of electronic effects on the antiviral activity, while derivatives **29** [X = —$(CH_2)_2$—; —$CH(Me)CH_2$—] were synthesized to verify the requirements of an amidine side-chain for the pharmacological activity (see Section 3.5.1).[23]

By a similar route, the intermediate **28** was used in the synthesis of a tripyrrole homolog of congocidine having two types of side-chain.[13]

In order to determine whether the location of the peptide bond in the pyrrole ring was important for pharmacological activity, isomers of distamycin A and congocidine in which the peptide bonds were moved to the 2- and 5-positions of the pyrrole ring were prepared. The synthetic pathway starting from 1-methyl-5-nitropyrrole-2-carboxylic acid **30**, shown in Scheme 9, is similar to the first total synthesis of distamycin A,[24] and the key intermediate is the nitroamidine derivative **33**, which serves as precursor for the synthesis of the distamycin and the tripyrrole congocidine isomers. Reduction of **33** to **34** and subsequent condensation with formic acid or with guanidinoacetic acid gave **35** (R = CHO) and **35** [R = $COCH_2NHC(=NH)NH_2$], respectively.

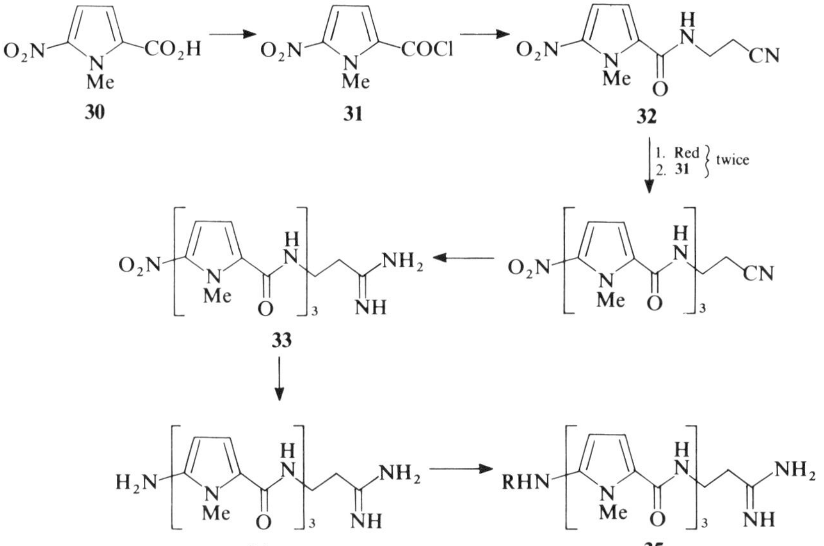

Scheme 9

Compound **32** served as starting material for the 2,5-disubstituted monopyrrole derivatives of distamycin and congocidine, via formation of the amidino group, reduction, and successive condensation with formic acid and guanidinoacetic acid.[24]

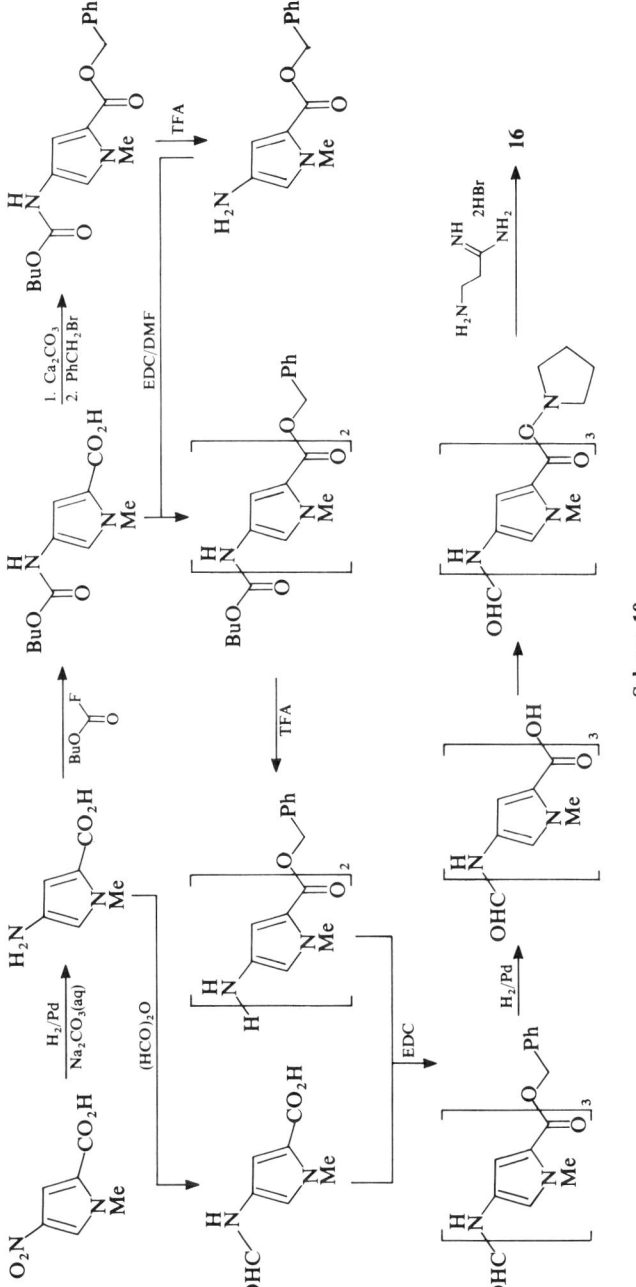

Scheme 10

Later, a new synthesis, outlined in Scheme 10, used the direct application of stable versatile intermediates derived from the unstable monomeric 4-amino-1-methylpyrrole-2-carboxylic acid as well as the use of the preformed aminoamidine side-chain. The synthesis relies on the use of mild and efficient reagents and procedures generally used in the synthesis of peptides, and it has the advantage that strong reagents, such as hydrogen chloride and ammonia, are eliminated from the steps involving pyrroles, thereby allowing the insertion of sensitive fragments, such as unsaturated and/or sulfur-containing aminoamidine residues, into the molecule.[25] Following this synthetic approach, analogs with the amidine nitrogen substituted with one or two methyl groups, derivatives where one or all of the methyl groups were replaced by ethyl or alkyl groups, and the pyrrole rings were unsubstituted at nitrogen were also prepared.[26, 27]

The latest total synthesis of distamycin A involves only a slight modification of the reaction sequences outlined in Schemes 5 and 8, but a sagacious choice of the reagents and reaction conditions have improved the yields and facilitated the workup of the various steps. In particular, the yields for the N-formylation reaction, which has been the least satisfactory step in all the synthetic procedures (10–20%), have been improved to 71% by the use of N-formylimidazole.[11]

Distamycin analogs in which each of three rings is fully methylated have been reported. Such a structural modification, resulting in extraordinarily electron-rich pyrrole rings, required a new synthetic approach,[28] shown in Scheme 11, in which the key step was the formation of an amide bond between a 3-aminopyrrole of type **36** and 1-hydroxybenzotriazole (BtOH) active ester of type **37**.

Reaction of the pyrrole-2-carboxylic acid and the aminopyrroles under the usual conditions resulted in low yields, since the acid was hindered and it was a poor electrophile and acid-sensitive, while the amine is an unstable, hindered weak nucleophile. The amidine side-chain was introduced as an intact unit since the Pinner sequence to transform a nitrile into an amidine moiety led to decomposition products. Evidently the polymethylpyrrole ring is too electron-rich to survive strongly acidic treatment.

The interest in this class of oligopeptides is evident from more recent reports in which many other distamycin analogs have been synthesized.[29–33]

3.2.1.3. Kikumycins

In 1965, two new basic antiviral antibiotics, kikumycin A and B, were isolated from a culture filtrate of *Streptomyces phaeochromogenes*.[34] The originally reported empirical formulas proved to be incorrect, and the correct structures **38a** and **38b** were assigned in 1972 on the basis of the hydrolysis and mass-spectrometric data.[35,36] In 1987, two new oligopeptide antibiotics, TAN-868 A and B, were isolated from the culture broth of *Streptomyces idiomorphus*. TAN-868 A was a new compound to which was assigned the structure **38c**, while TAN-868 B was found to be identical with kikumycin A (**38a**)[37, 38] (Scheme 12).

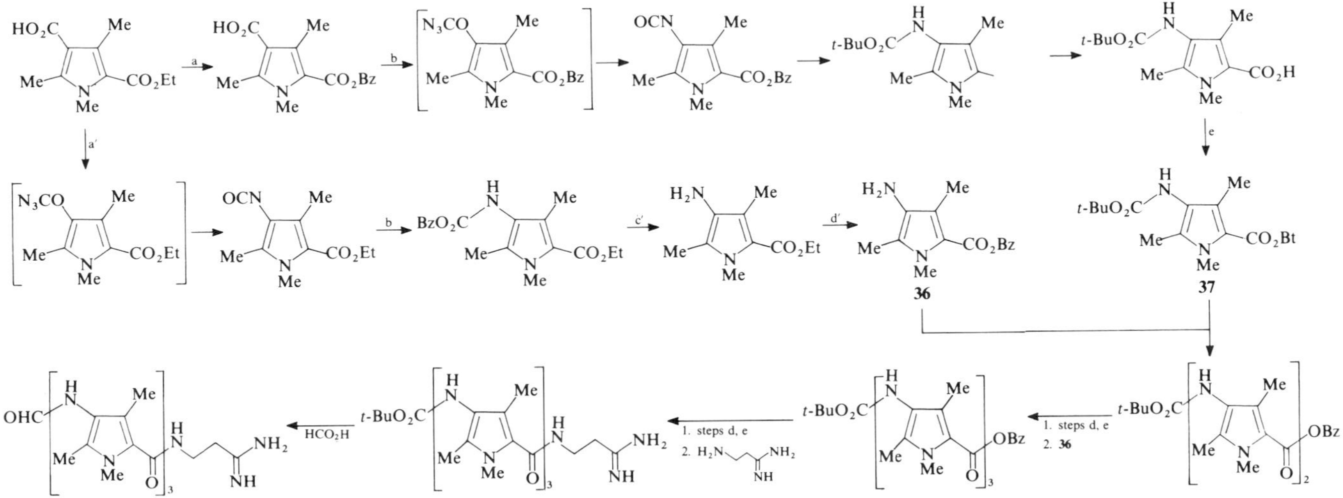

Scheme 11

a BzOH/NaH; b DPPA/MeCN/TEA; c BuOH/CuCl; d H$_2$/Pd; e (1) BtOH/CH$_2$Cl$_2$/TEA, (2) 2-chloro-1-methylpyridinium iodide; a' DPPA/MeCN/TEA; b' BzOH/Na; c' H$_2$/Pd; d' BzOH/Na
Bz = benzyl; t-Bu = t-butyl; BtOH = 1-hydroxybenzotriazole; DPPA = diphenylphosphoryl azide

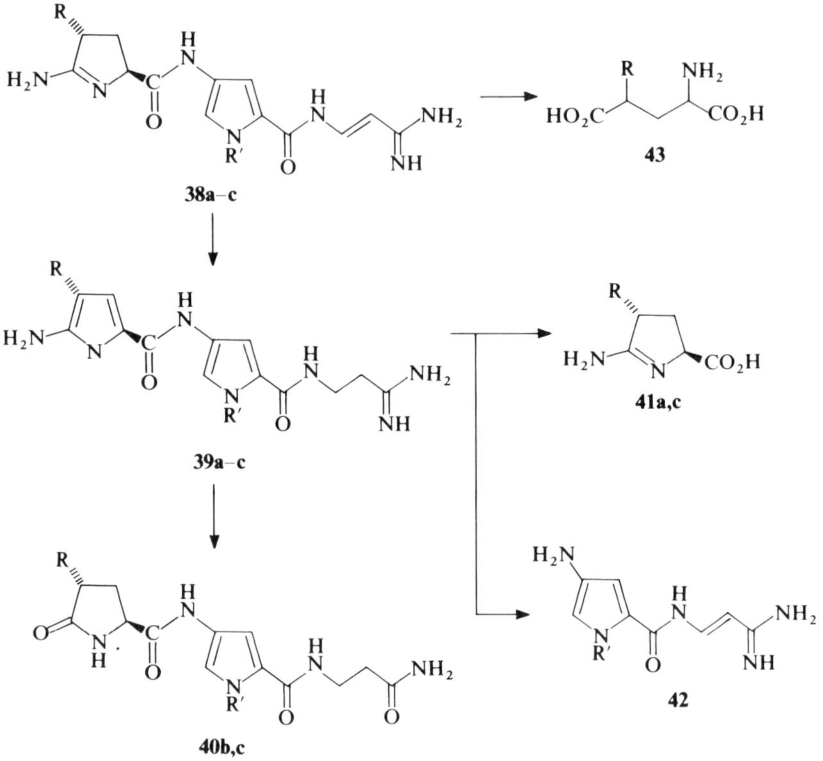

Scheme 12

a R = R' = H
b R = H, R' = Me
c R = OH, R' = H

An important role in the determination of the structure was played by the degradation studies. Thus, catalytic hydrogenation of compounds **38a–c** gave the corresponding dihydro derivatives **39a–c**. Compounds **39b,c** on mild alkaline hydrolysis afforded the deaminated compounds **40b** and **40c**, respectively. Acid hydrolysis in 2N hydrochloric acid of dihydro compounds **39a,c** afforded proline analogs **41a,c**, together with the amino derivative **42**, which was common to both compounds. The above data confirmed the sequence of three moieties and structures **38a–c**. Hydrolysis of compounds **38a–c** in refluxing 6N hydrochloric acid established the absolute configurations. Thus, compounds **38a,b** gave glutamic acid **43** (R = H), originating from the proline moiety, whose absolute configuration was determined to be the L-form, while compound **38c** gave the 4-hydroxyglutamic acid **43** (R = OH), the molar rotations of which in water and in hydrochloric acid were in agreement with those of [2S,4R]-isomer.[35,37]

The only total synthesis of this group of antibiotics refers to the two enantiomeric forms of dihydrokikumycin B **39b**,[39] as shown in Scheme 13. The

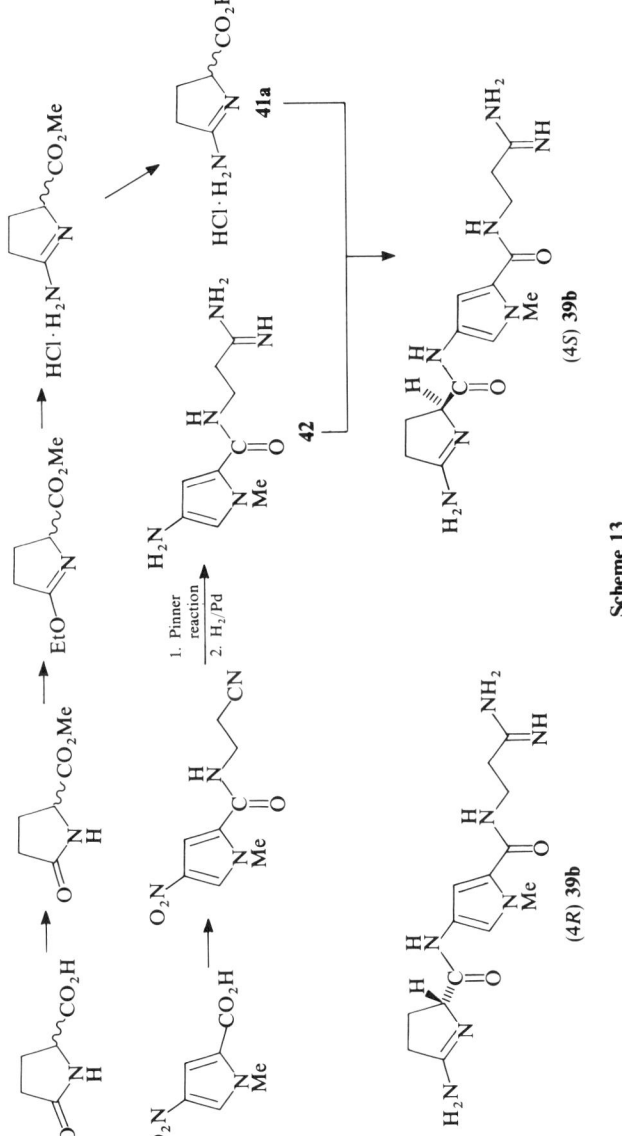

Scheme 13

2-amino-1-pyrroline moiety **41a** and the pyrrole unit **42** were prepared separately, and subsequent condensation of the two fragments furnished the carbon skeleton of **39b**. Both antipodal forms of the 2-amino-1-pyrroline precursors (4S)-4 and (4R)-4 were prepared from the corresponding (S)- and (R)-pyroglutamic acids.

3.2.1.4. Anthelvencin A

Anthelvencin A, an antibiotic with anthelmintic activity, was isolated from culture filtrates of a strain of *Streptomyces venezuelae*, and its structure (**45**) was

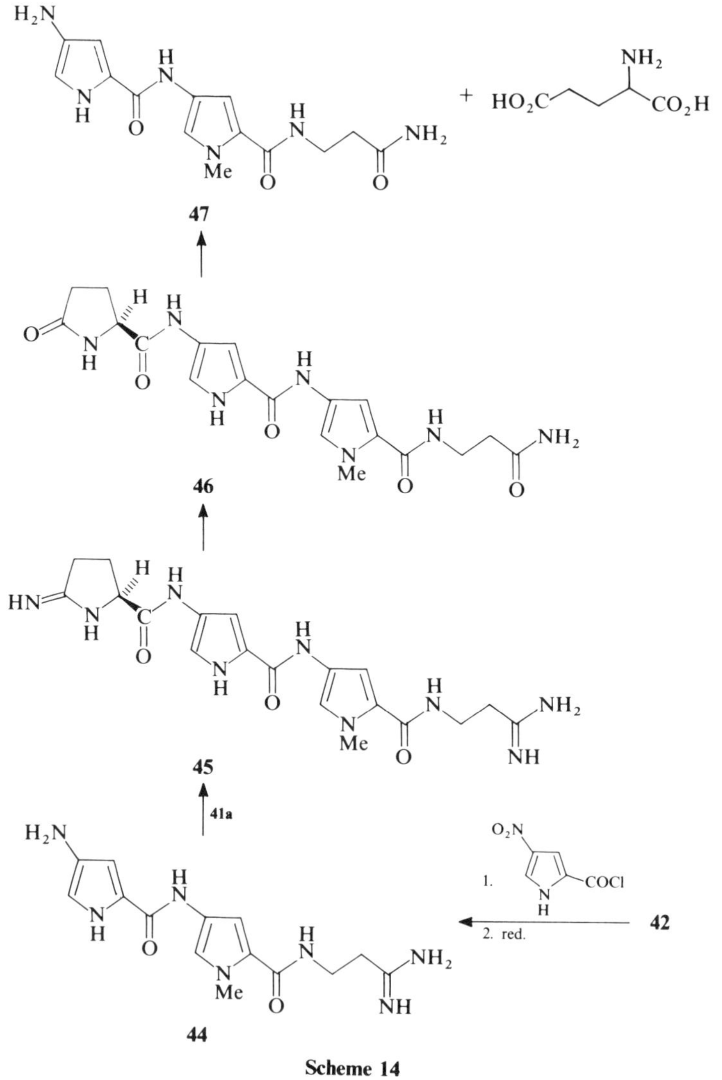

Scheme 14

established by the relationship of the hydrolytic degradation products with those obtained from netropsin.[40] On mild alkaline hydrolysis, **45** gave the amide **46**, which, on acid hydrolysis in 1N hydrochloric acid, gave glutamic acid and an amphoteric compound whose properties were comparable to those reported for compound **5a** obtained from netropsin and to which was assigned structure **47**. Further treatment with acid or alkali gave β-alanine, as the sole identified product.

Verification of the assigned structure by total synthesis came in 1988 when both enantiomeric forms (4S)-(+) and (4R)-(−)-anthelvencin A were prepared. The synthetic pathway is shown in Scheme 14 and involves the use of the key intermediate **42**, which had already been prepared in the synthesis of kikumycin A.[41]

3.2.2. Direct Ring Synthesis

In this section, beside the obvious distinction in the synthesis of 1-, 2-, 3-amino and the polyamino derivatives, the synthetic methods are classified with respect to the carbon units of the starting materials.

3.2.2.1. 1-Aminopyrroles

The most common method used for the preparation of 1-aminopyrroles starts from four-carbon units, and some of these reactions have been improperly reviewed in connection with the synthesis of pyridazines.[42] This is understandable when it is realized that, from the same reactants, slight differences in the experimental conditions can give rise to 5- and 6-membered ring systems, and it also accounts for the discrepancy in reports from different authors. In this section we also refer to reactions that lead exclusively to pyridazines as some of the reactions are not reasonably expected, and we believe that this information will be useful to chemists approaching the synthesis of 1-aminopyrroles.

3.2.2.1.1. From Four-Carbon Units

The reaction of 1,4-dicarbonyl compounds and hydrazines or hydrazides has been known since 1885[43] and, since then, has been widely investigated. Several studies were carried out in the first half of this century, but because of the variety of products that can be formed, the nature of the isolated compounds, usually only the main one, was controversial. It has been shown that from such reactions, 1-aminopyrroles, dihydropyridazines, and sometimes mono- or bishydrazones and bipyrroles can be formed (Scheme 15).

However, in spite of the experimental difficulties and taking into account some doubtful chemical evidence and some unsubstantiated structures proposed in early reports (see, for example, Ref. 44, in which 1,2,4-triazacyclo-2,4,8-octatriene was proposed for what actually was a 1-guanidinopyrrole), it is

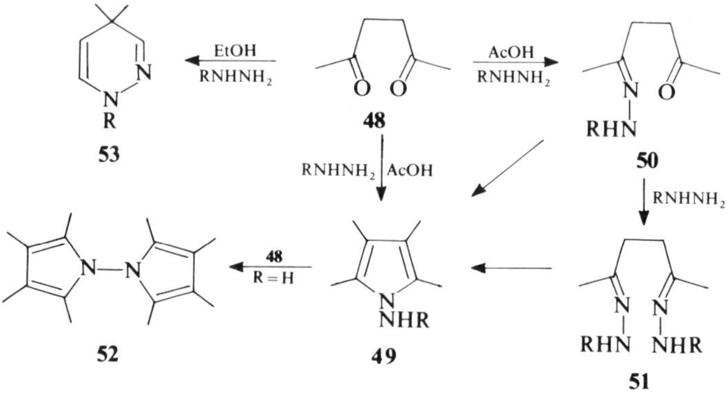

Scheme 15

possible to outline some general features of these reactions. Usually reactions carried out in acetic acid afford 1-aminopyrroles **49** as the sole product; they can also be obtained from either the monohydrazones **50** or the bis-hydrazones **51**, which are formed, and isolated in some cases, during the condensation reaction. When an excess of the dicarbonyl compounds **48** is employed, derivatives **49** undergo further condensation to yield bipyrroles **52**.

When the reaction is carried out in ethanol, dihydropyridazines **53** and/or 1-aminopyrroles **49** can be obtained, depending on the nature of the 1,4-dicarbonyl compounds and the hydrazines. However, in the case of hydrazides and sometimes in the case of phenylhydrazines, only substituted 1-aminopyrroles are obtained, even in neutral conditions. This is evidently a result of the reduced nucleophilicity of nitrogen-2. In contrast, when hindered dicarbonyl compounds are employed, dihydropyridazines are always obtained, even when the reactions are carried out under acidic conditions. It is supposed, therefore, that the first step of the reaction is the formation of a hydrazone of type **50**, and depending on the competitive nucleophilicity of the two nitrogen atoms the hydrazone gives rise to either the 1-aminopyrrole (nitrogen-1 attack) or the 1,2-diazine derivative (nitrogen-2 attack).

The reaction of succinaldehyde **54** ($R^1 = R^2 = H$) in acetic acid or DMF with hydrazides of type **55**, obtained from benzoic or substituted benzoic acids or heterocyclic acids (R = pyridin-4-oyl, thiophen-2-oyl, furan-2-oyl), leads to the corresponding *N*-substituted 1-amidopyrroles **56** (R = COaryl, COheteroaryl)[45,46] (Scheme 16). Although acetonylacetone **54** ($R^1 = R^2 = Me$)

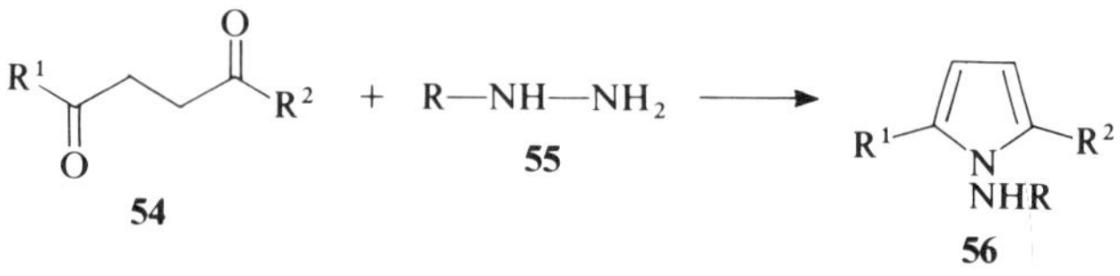

Scheme 16

reacts with phenylhydrazine or substituted phenylhydrazines in acetic acid to give 1-anilinopyrroles **56** (R = aryl),[47–51] the same reactants in acetic acid–ethanol or in pyridine lead to the monoarylhydrazones as main products, together with **56**.[49,50] N-Substituted 1-arylaminopyrroles have been obtained in a similar manner.[50]

Arylhydrazides[45,46,50,52–57] and heteroarylhydrazides[57–60] give the corresponding aroyl and heteroaroyl 1-aminopyrroles of type **56** ($R^1 = R^2 = Me$), when reacted with acetonylacetone under a variety of conditions (no solvent, benzene–ethanol, DMF, acetic acid).

Thiosemicarbazide and aminoguanidine need the stronger hydrobromic acid to form the expected pyrroles, whereas, in acetic acid, only the bis-semicarbazide derivatives were isolated.[61]

1-Diphenylphosphinylamino-2,5-dimethylpyrrole was obtained from the reaction of **54** ($R^1 = R^2 = Me$) and diphenylphosphinylacid hydrazide in both ethanol and acetic acid.[62]

N-Substituted 1-aminopyrroles have been obtained from octane-3,6-dione **54** ($R^1 = R^2 = Et$),[63] the reaction of the diethyl ester of α,α'-dihydroxymuconic acid **54** ($R^1 = R^2 = COOEt$), and hydrazine hydrate in acetic acid led to the corresponding 1-aminopyrrole **56** (R = H, $R^1 = R^2 = COOEt$).[64] However, in contrast, the reaction of 1,2-dibenzoylethane **54** ($R^1 = R^2 = Ph$) and hydrazine in ethanol failed to give 1-aminopyrrole, and 3,6-diphenylpyridazine was isolated.[65] The same compound was also obtained from the reaction of **54** ($R^1 = R^2 = Ph$) with 1-benzenesulfonylhydrazide, as a result of the initial hydrolysis of the sulfonyl residue.[66] The pyridazine was also obtained from the asymmetric 1,4-diketone **54** ($R^1 = Ph; R^2 = Me$) in ethanol,[65] whereas the reaction of **54** ($R^1 = R^2 = Ph$) and thiosemicarbazide in the presence of mineral acid afforded the corresponding thioamidopyrrole **56** (R = $CSNH_2$).[61] Higher yields of the pyrrole were also observed when ethyl *tert*-butylcarbazate was used.[66]

Aryl[67] and heteroaryl hydrazides[59] react with 1,2-dibenzoylethane **54** ($R^1 = R^2 = Ph$) under a variety of conditions to give the corresponding N-substituted 1-aminopyrroles, and a similar reactivity was shown by asymmetric diketones **54** ($R^1 = Me; R^2$ = alkyl, aryl, heteroaryl) with aryl[55] or heteroaryl hydrazides;[55,58,59] the asymmetric 1,4-diketones also react with semicarbazide to give 1-ureidopyrroles under acidic conditions,[68,69] but the reaction of phenacylacetone **54** ($R^1 = Me; R^2 = Ph$) and semicarbazide in refluxing aqueous ethanol afforded the dihydropyridazine **53**, while at room temperature only the corresponding mono semicarbazone was isolated.[69]

1-Ureido-2,3,4-trimethyl- and 1-ureido-2,3,5-trimethylpyrroles have been prepared from the corresponding ketoaldehyde or diketone and semicarbazide in methanol or acetic acid,[69,70] and the 2,3,5-substituted 1-aminopyrrole **58a** (R = H) was obtained from the condensation of hydrazine with 3-carbethoxy-2,5-hexanedione **57a** in acetic acid or acetic acid/ethanol.[65,71–74] When the reaction is carried out in ethanol, the formation of dihydropyridazine was observed with, in one case, the corresponding 1-aminopyrrole[71,74] (Scheme 17). Under similar experimental conditions, the corresponding reactivity was demonstrated by

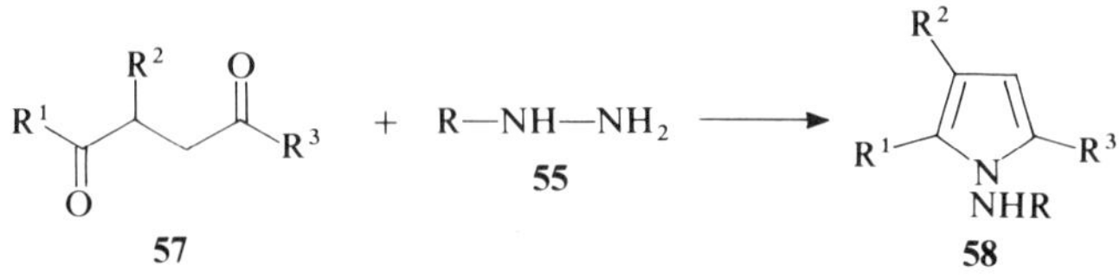

Scheme 17

a $R^1 = R^3 = Me$; $R^2 = CO_2Et$
b $R^1 = R^3 = Ph$; $R^2 = CO_2Et$
c $R^1 = Me$; $R^2 = CO_2Et$; $R^3 = Ph$
d $R^1 = R^2 = R^3 = Ph$
e $R^1 = R^2 = Ph$; $R^3 = C_{10}H_7$

ethyl 2,3-dibenzoylpropionate (**57b**) and ethyl 2-acetyl-3-benzoylpropionate (**57c**).[65] The reaction of the diketoesters **57a–c** with semicarbazide always leads to the corresponding 1-ureidopyrroles **58** (R = $CONH_2$), although in an early report the pyridazine structure was erroneously proposed.[69,72,76,77]

Bulky diketones **57d,e** react with hydrazines **55** (R = H, Ph) and hydrazide **55** (R = COPh) in acetic acid to give dihydropyridazine derivatives.[65,78–80] The nature of the products is controversial, and several authors have proposed the 1-aminopyrrole structures[81,82] for the products. However, because of steric factors, the formation of the pyridazine nucleus appears to be more reasonable.

Fully substituted 1-aminopyrroles of type **60** (R = H) have been prepared from diketones of type **59** (R′ = Me, COOEt) and hydrazine in acetic acid[65,72,83–86] (Scheme 18). When an excess of diketones is used, further reaction with pyrroles **60** (R = H) leads to bipyrroles of type **52**.[84,85] On the other hand, when the reaction was carried out in ethanol, the dihydropyridazines of type **53** were isolated as main products.[65,87,88]

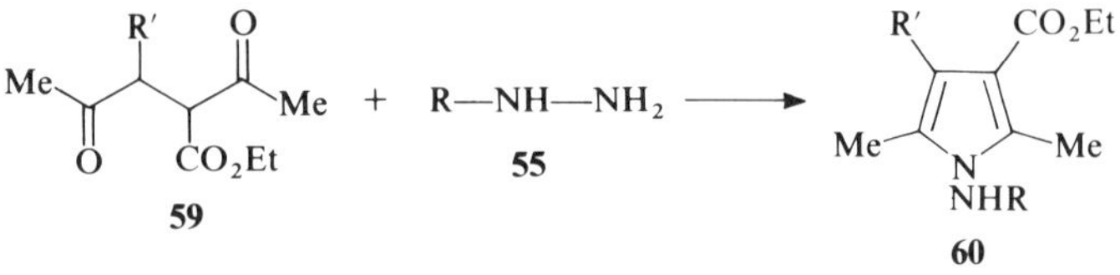

Scheme 18

1,4-Diketone **59** (R′ = COOEt) reacts with phenylhydrazine and substituted phenylhydrazines under acidic conditions to give 1-anilinopyrroles **60** (R = Ph, aryl; R′ = COOEt),[49,89,90] although in some reports the product has been incorrectly assigned as a pyridazine.[43,91] The diketone **59** (R′ = COOEt) also reacts under acidic conditions or in ethanol with other hydrazides, affording 1-amidopyrroles.[83,85,92–96]

An analogous reactivity with phenylhydrazine has also been shown by ethyl α,β-dibenzoylsuccinate.[97]

1-Ureidopyrroles of type **60** (R = $CONH_2$; R′ = Me, COOEt) were obtained in good yields from semicarbazide and the corresponding diketones in acetic acid.[85,98,99] The same 1-ureidopyrroles can also be isolated under neutral or alkaline conditions.[72]

In the case of hydrazides of dicarboxylic acids, the reaction with 1,4-diketones in acetic acid leads to bis-amidopyrroles **61**[50,100,101] (Scheme 19).

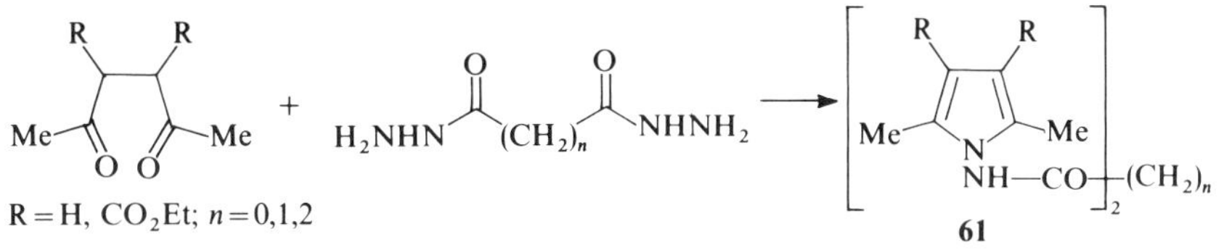

Scheme 19

A peculiar type of reactivity is to be found with bis-1,4-dicarbonyl compounds. Tetraacetyl derivatives **62** and **63**, when reacted with hydrazine in acetic acid, exhibited simultaneously the two different reaction pathways giving rise to condensed 1-aminopyrrole and pyridazine nuclei (derivative **64**)[102,103] (Scheme 20).

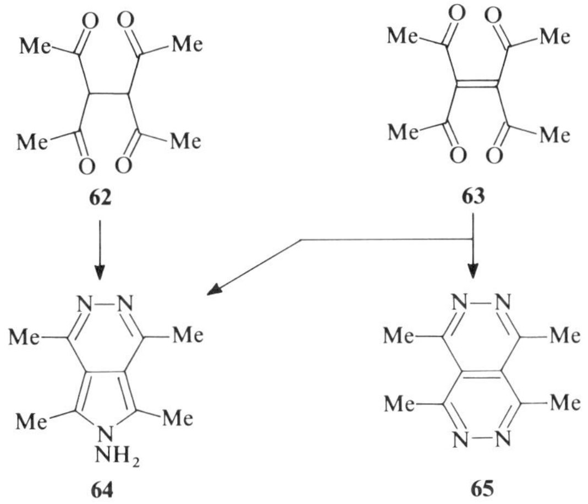

Scheme 20

Pyridazopyridazine **65** was also obtained, together with **64**, from the diketone **63**. A further complication in the reactions of hydrazines with these 1,4-dicarbonyl compounds arises from the fact that derivatives of type **66** can behave as a 1,3-diketone or 1,3-ketoester, giving rise to pyrazoles **67** or pyrazolones **68**, respectively, when they are reacted in ethanol with hydrazines, or to give the bis-derivatives **69** and **70** when reacted with an excess of hydrazine[88] (Scheme 21). The reaction of diacetylhexanedione **66** (R = R' = Me) with hydrazine in ethanol, for example, yields pyrazoles **67** (R = R' = Me) and **69** (R = R' = Me).[102] In the case of the reaction of the dibenzoylsuccinic ester **66** (R = OEt; R' = Ph) with phenylhydrazine, the pyrazolone **68**

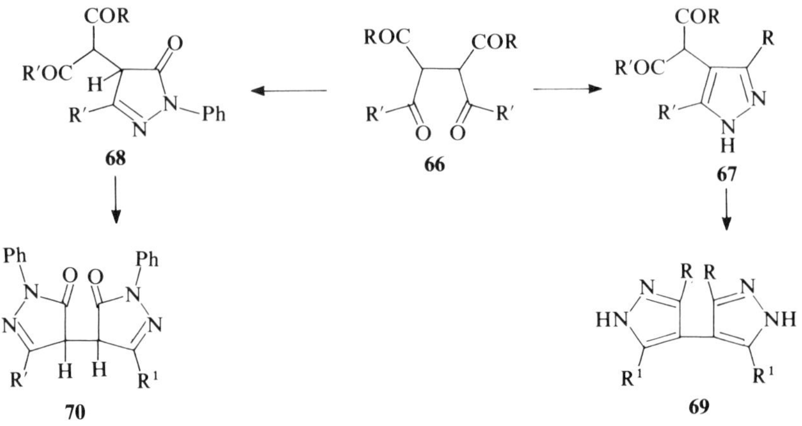

Scheme 21

(R = OEt; R' = Ph) and bis-pyrazolone **70** (R' = Ph) were isolated.[97] Similarly, ethyl acetonylbenzoylacetate and semicarbazide produces the monopyrazolone **68**.[69]

1-Aminopyrroles can also be obtained by the cyclization reaction of the open-chain intermediate mono- and bis-hydrazones/semicarbazones **50** and **51** (see Scheme 15), which are frequently thermally labile, so that even at room temperature they cyclize to the thermodynamically controlled reaction products.[62] In the case of the monosemicarbazone of phenacylacetone, a mixture of 1-aminopyrrole and pyridazine of type **33** (as main product) were formed at pH = 4.7.[69] Complete conversion into 1-aminopyrroles is generally achieved in good yield by simple heating[53,68] or under acidic conditions.[49,50,52,69,78,79,104] However, in some cases, cyclization to 1-aminopyrroles was not observed under any reaction condition; the hydrazones are recovered unchanged, or polimers are obtained.[55,61]

Problems associated with the synthesis of 1-aminopyrroles can be avoided by the use of *gem*-disubstituted hydrazines, such as *N*-aminophthalimide, to give 1-phthalimido-2,5-disubstituted pyrroles in good yield.[105] The free 1-aminopyrroles can be obtained by subsequent hydrazinolysis in ethanol. However, such an unambiguous synthesis of 1-aminopyrroles is not of general application. The reaction is viable with only a small number of 1,4-dicarbonyl compounds **54** (R = H, Me) and fails when bulky 1,4-diketones are employed (e.g., 1,2-dibenzoylethane, phenacylacetone).[67]

Unsaturated 1,4-diketones also give 1-aminopyrroles in their reaction with hydrazines, but the reaction is not general, and, depending on the pH of the medium and on the nature of the diketone, either 1-aminopyrroles and/or pyridazines are produced. Thus, **71a,b** react with hydrazine in strongly acidic media to give pyridazine derivatives **72a,b**, whereas in basic (ethanol or pyridine) or in weakly acidic (acetic acid) media 1-aminopyrroles **73a,b** (R = H) are mainly formed; [e.g., compound **73a** (R = H) is the only product in the case of

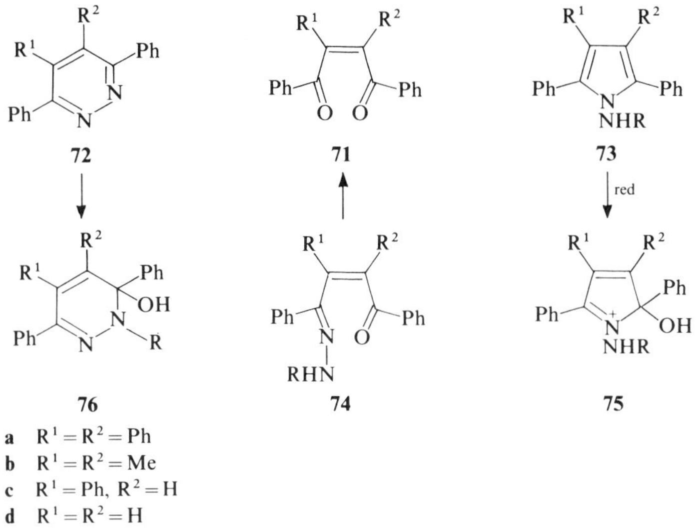

3.2. Synthesis

Scheme 22

a $R^1 = R^2 = Ph$
b $R^1 = R^2 = Me$
c $R^1 = Ph, R^2 = H$
d $R^1 = R^2 = H$

benzoylstilbene][106,107] (Scheme 22). The situation is reversed, however, in the case of **71c**; the pyrrole derivative **73c** (R = H) is obtained under acidic conditions, together with the pyridazine **72c**, which is the only product of the reaction under basic conditions.[106,107] It should be noted, however, that the erroneous structures **72** have been assigned[106] to the 1-aminopyrroles. It is noteworthy that the diketone **71d** fails to give 1-aminopyrrole under any standard experimental conditions.[107]

The proposed mechanism for these reactions involves the initial formation of the same monohydrazone intermediate **74**, and—depending on the pH of the medium and the nature of the substituents R^1 and R^2—it leads to either a pyridazine **72** or an 1-aminopyrrole derivative **73**.[107] Although the influence of pH is not easily understood, the favored formation of pyrrole derivatives can be rationalized in terms of the steric hindrance of the R^1 and R^2 groups. The final step leading to 1-aminopyrroles involves reduction of an intermediate 2H-hydroxypyrrole **75**, and the higher yields of 1-aminopyrrole, obtained when the diketone:hydrazine ratio is 1:1.5, are compatible with the oxidation of hydrazine to nitrogen, although the latter has never been detected.

Indirect proof for the mechanism is found in the products isolated from the reaction of **71a** and phenylhydrazine. Although Klingemann originally reported that the reaction in acetic acid gave the pyridazine derivative,[82] it has been shown that the product is 1-anilinotetraphenylpyrrole **73a** (R = Ph).[108] However, when the reaction was conducted in ethanol, the dihydropyridazine analog of **76**, which is capable of aromatization to **72**, was obtained.

The reaction of **71c** and phenylhydrazine in ethanol is reported to give 1-anilinopyrrole **73c** (R = Ph),[109,110] whereas diketone **71d** reacts with hydrazides in acetic acid to give aromatic pyridazine **72d**, as a result of hydrolysis of intermediates **76d** (R = $CONH_2$, $CSNH_2$) and subsequent aromatization. In all the other cases, only bis-hydrazones have been isolated.[61]

1-Arylamino-2-vinylpyrrole **78** has been obtained from the reaction of 2,4-dinitrophenylhydrazine and the intermediate ketoalkenes **77**, obtained from pivaloylcarbene and 2-alkoxy-1,3-butadienes[111] (Scheme 23).

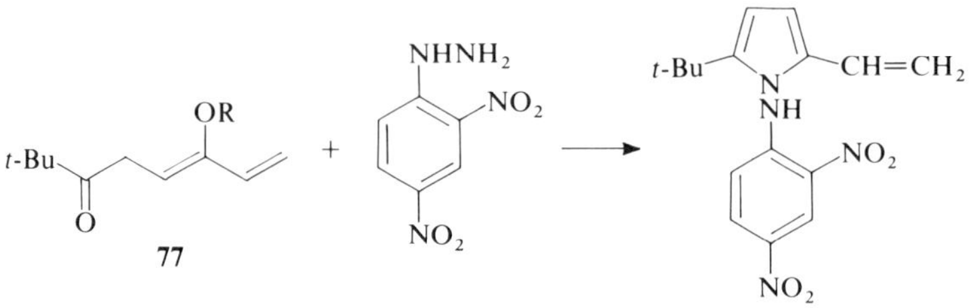

Scheme 23

3.2.2.1.2. From Ring Transformations

1-Aminopyrroles of type **81** have been synthesized by the condensation of 2,5-diethoxytetrahydrofuran **79**, which behaves as masked 1,4-dicarbonyl compound, and hydrazine,[102] arylhydrazines,[51,112] and arylhydrazides[51,67,113] under acidic conditions (acetic acid or ethanol saturated with hydrogen chloride) (Scheme 24). Such a condensation reaction presents the same difficulties as observed with corresponding reaction of 1,4-diketones and a similar reaction mechanism has been proposed.[113] The synthesis of 1-tosylaminopyrrole of type **81** is shown in Scheme 24. Pyrrole **81** is obtained in one step from **79** in hot acetic acid, but in a poor yield (8%), and better yields have been obtained by bubbling hydrogen chloride through a benzene solution of the intermediate **80**, which is isolated from the phosphoric acid–catalyzed reaction of **79** and tosylhydrazide in ethanol.

The same reactants in cold acetic acid lead to the tetrahydropyridazine **82** (R' = Ac), which, on treatment with a cold, dilute solution of hydrochloric acid

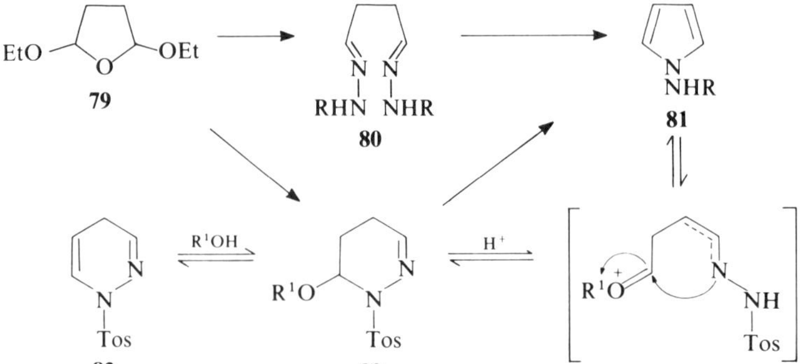

Scheme 24

in glyme, gives **81** in good yield. The dihydropyridazine **83**, obtained from rearrangement of **82** with cold boron trifluoride in glyme, produces *N*-tosylamidopyrrole **81** on reflux in acetic acid or in ethanol–hydrochloric acid. Such a rearrangement has been rationalized in terms of the acid-catalyzed addition of R′OH (R′ = Et, Ac) to the olefinic bond of **83**, followed by ring opening of **82** with fission of the sulfonamide nitrogen–carbon bond.

Suitable epoxyalkynes of type **84** undergo ring opening in the presence of hydrazines, affording 2,4-disubstituted 1-aminopyrroles **86**, through the intermediate addition product **85**, which has also been isolated[114] (Scheme 25).

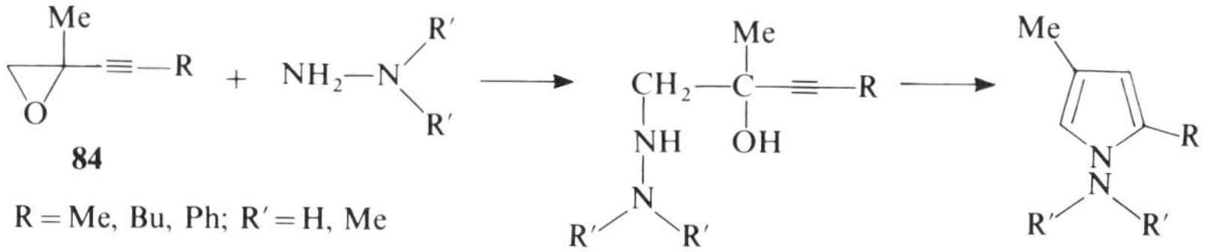

84

R = Me, Bu, Ph; R′ = H, Me

85

86

Scheme 25

3.2.2.1.3. From (2 + 2) Carbon Units

1-Ureidopyrroles **90** have been obtained in two steps from the reaction of the sodium salts of β-dicarbonyl compounds **88**, such as ethyl acetoacetate, ethyl benzoylacetate, acetylacetone, benzoylacetone, and dibenzoylmethane, with the semicarbazones of type **87** derived from ω-bromoacetophenone or from chloroacetone[115–118] (Scheme 26). It was originally believed that the reaction proceeds via the cyclization of the open-chain intermediate **89**, but more recent ^{13}C

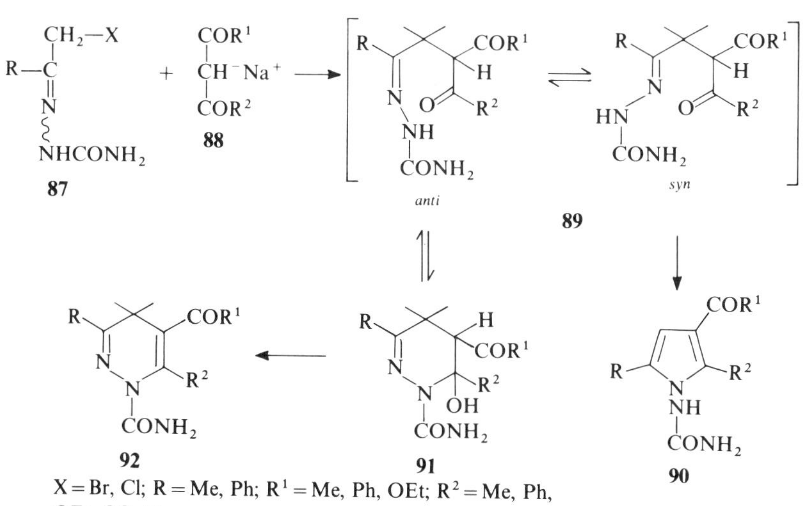

87 **88** **89** **92** **91** **90**

X = Br, Cl; R = Me, Ph; R¹ = Me, Ph, OEt; R² = Me, Ph, OEt, CO_2Et

Scheme 26

and ^{1}H NMR studies[119] demonstrated that the intermediates of these reactions are not the semicarbazones **89**, but the carbinolamines **91** that undergo a ring-opening reaction to give **89**, followed by *anti*- to *syn*-isomerization. The *syn*-isomers can produce the 1-ureidopyrroles, since the poor nucleophilicity of the sp^2 nitrogen atom is balanced by the enhanced reactivity of the carbonyl group in the acidic media. However, carbinolamines **91** can also dehydrate to give the dihydropyridazine **92**.[119,120]

The production of the intermediate **91** greatly depends on the relative stability of the final products. In some cases, especially when R = Me, acid catalysis is not necessary and the 1-ureidopyrroles are formed during any attempted crystallization of the intermediates.[115]

A synthetic method, which is particularly suited for regiospecific synthesis of alkyldimethylaminopyrroles of type **95** from ketones, involves the three-step reaction sequence outlined in Scheme 27.[121] *N,N*-Dimethylhydrazones **93a–d** were metallated with lithium diisopropylamide in THF/HMPT at 0°C and then alkylated at −78°C with 2-bromomethyl-1,3-dioxolane (BMD) or 2-iodomethyl-1,3-dioxolane (IMD). The ease of conversion depends on the nature of hydrazones. Thus, conversion of derivatives **94** into 1-aminopyrroles can be achieved by refluxing in anhydrous toluene with catalytic amount of *p*-toluenesulfonic acid, and the reaction is more effective from sterically noncrowded dimethylhydrazones, whereas a hindered dimethylhydrazone, such as that derived from camphor, fails to react. The reaction times for maximum yields also depend on the structure of amino acetals **94**.

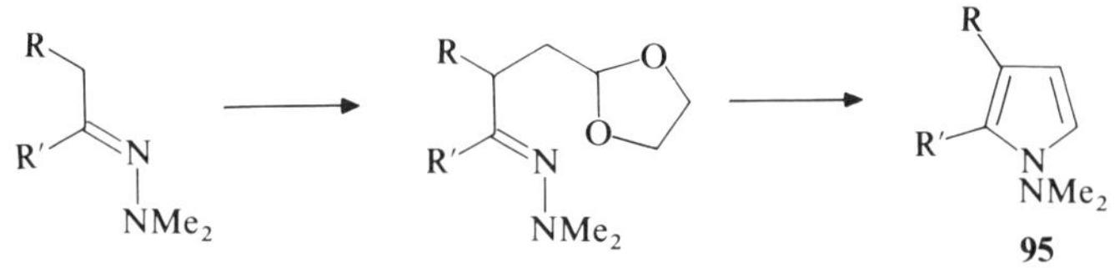

93
a R = H, R′ = C$_4$H$_9$
b R = H, R′ = Ph
c R = Me, R′ = Et
d R—R′ = —(CH$_2$)$_4$—

Scheme 27

1-Benzoylamino-2,3-diphenylpyrrole (**99**) has been obtained from the Wittig reaction of vinyltriphenylphosphonium bromide (**97**) with (*E*)-*N*-benzoylbenzylidene monohydrazone (**96**) in DMF/NaH[122] (Scheme 28). Only the *E*-isomer leads to 1-aminopyrrole; a similar reaction of the *Z*-isomer affords the corresponding dihydropyridazine **101** as expected from mechanistic studies of analogous reactions of oximes leading to 1-hydroxypyrroles.[122] Thus, the formation of **99** can be rationalized as occurring through the attack of nitrogen-1 on **97** to generate ylide **98**, whereas reaction of nitrogen-2 leads to the ylide **100**, which gives the pyridazine **101**.

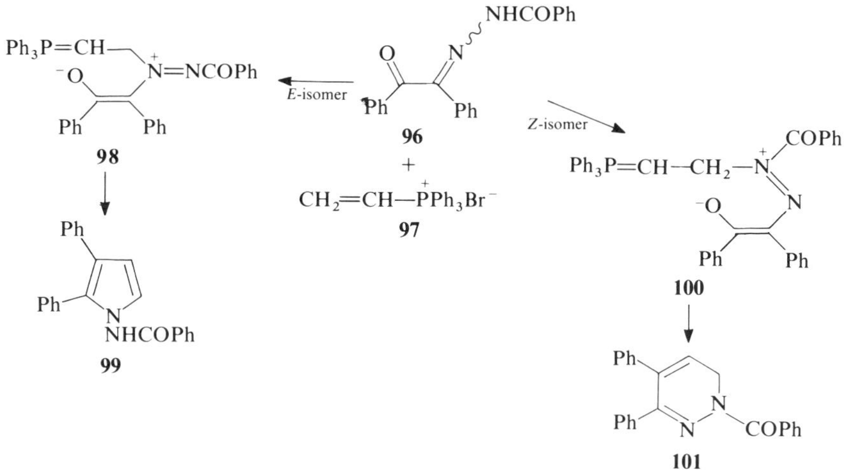

Scheme 28

1-Carbomethoxyaminopyrroles **104a,b** have been prepared in good yields from ethyl 3-carbomethoxyazo-2-butenoate **103** and enamines **102a,b**, obtained from the corresponding ketones and morpholine[123,124] (Scheme 29). The enamine, obtained from tetralone, reacts in an analogous fashion.[123,124]

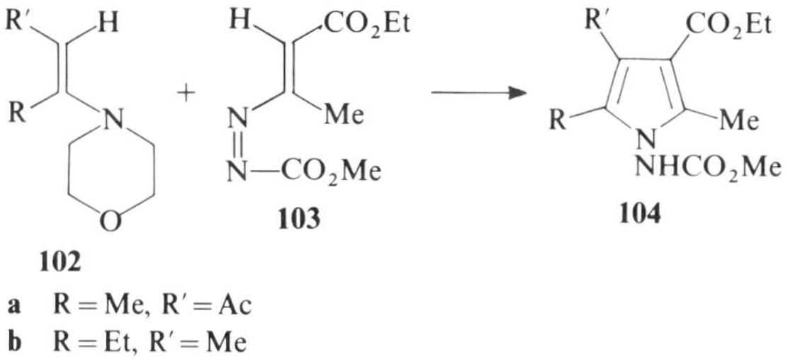

Scheme 29

The most useful direct synthesis of a large number of widely substituted 1-aminopyrroles is represented by the reaction between conjugated azoalkenes and compounds containing activated methylene groups. The reaction, which takes place under very mild conditions, at room temperature, in the absence of strong acids or bases, occurs by nucleophilic attack of activated methylene compound of type **105** on the heterodiene system of the conjugated azoalkene of type **106**. The resultant 1,4-adduct intermediate of type **107**[125,126] (Scheme 30) can theoretically undergo an intramolecular condensation that involves either of nitrogen atoms of the resulting hydrazone derivative to afford the 6-membered heterocycles **109** (pathway A) or 5-membered heterocycles **111** (pathway B), through the intermediacy of the corresponding dihydro derivatives **108** and **110**. Conceivably, the relatively unstable dihydropyridazine intermediate could also

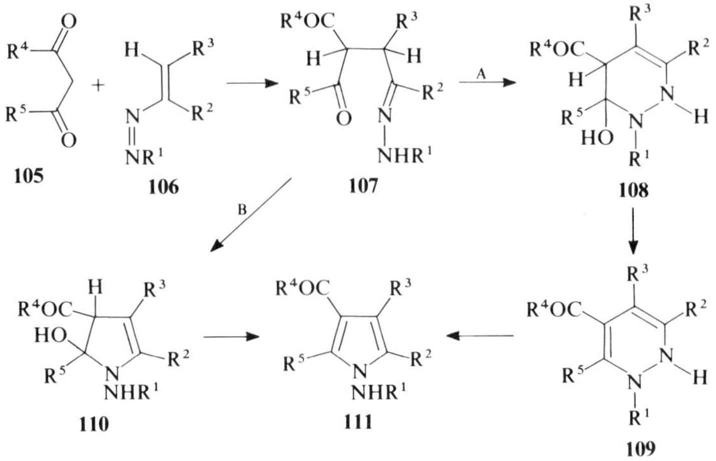

Scheme 30

rearrange to the more stable pyrrole systems (see Section 3.2.2.1.2). However, pathway A has been shown not to take place, although in the first report of this reaction, the isolated product was erroneously described as a dihydropyridazine.[127] Evidence in support of pathway B has been provided by the detection and/or isolation of two distinct intermediates **107**[128-131] and **110**,[126,132] as well as the interconversion of **107** into **110** and the subsequent aromatization of **110**.

Usually the reaction requires the presence of a copper(II) catalyst, and it was originally thought to involve an organometallic complex between the metal ions and the arylazoalkenes and/or β-dicarbonyl compounds.[133-135] However, more recently it has been demonstrated that neither the formation of **107**, nor its conversion to **110**, are affected by the presence of copper ions, and the role of the catalyst appears to be involved in the transformation of **110** into the 1-aminopyrrole **111**.[126] In the absence of the catalyst intermediates **107** or **110** have been isolated, depending on the nature of the reactants.[126,128,130,136] Thus, 2-hydroxy-4-pyrrolines are converted almost quantitatively into the corresponding 1-aminopyrroles by the addition of copper(II) chloride.[126]

This approach has been used for the synthesis of over 200 variously functionalized 1-aminopyrroles in 75–98% yields. Thus, 1-arylamino-3-carbonylpyrroles **111** (R^1 = Ph, 4-$NO_2C_6H_4$; R^4 = Me, Et, Ph, OMe) have been obtained from the reaction of β-diketones or β-ketoesters **105** (R^5 = Me, Ph) and arylazoalkenes **106**.[133-135] The reaction of **106** with 3-oxoalkanamides **105** (R^4 = NH_2, NEt_2, NHaryl; R^5 = Me) gives the corresponding 1-arylamino-3-carboxamidopyrroles.[137] In all of these reactions, the ring closure of the intermediate **107**, derived from ketoesters or ketoamides, occurs only with the ketonic group, and with the aliphatic ketonic group, in the case of asymmetric β-diketones.

3-Acyl-, 3-alkoxycarbonyl-, and 3-aminocarbonyl-1-ureidopyrroles **111** (R^1 = $CONH_2$, CONHaryl) have been prepared from the reaction of aminocarbonylazoalkenes **106** (R^2 = Me; R^3 = COOalkyl) and suitable β-diketones, β-ketoesters, and 3-oxoalkanamides, respectively.[129,138] β-Dicarbonyl compounds also react with alkoxycarbonylazoalkenes **106** (R^2 = Me; R^3 = COOalkyl) to give the corresponding N-substituted 1-aminopyrroles **111** (R^1 = COOalkyl) in the presence of copper(II) chloride, whereas, in the absence of the inorganic salt, the reactions proceed only as far as 1,4-adducts **107**.[136,139] 3-Acyl-, 3-amido-, and 3-alkoxycarbonyl-1-arylsulfonylaminopyrroles of type **111** can be synthesized by a similar route[128,130,140] in which the isolated intermediates **107** are slowly converted at room temperature into the corresponding pyrroles **111**. 3-Carbonyl- and 3-carboxy-1-heteroarylaminopyrroles **111** (R^1 = pyridinyl, pyrimidyl, benzothiazolyl) have been prepared by the analogous reaction from heterocyclic conjugated azoalkenes **106**.[141,142] However, in these reactions, the low reactivity of the azoalkenes requires the use of an excess of the dicarbonyl compound. Similarly, 1-aroylaminopyrroles **111** have been synthesized from the appropriate aroylazoalkenes.[126,143]

N-Phosphinic and N-phosphonic conjugated azoalkenes have been used in the preparation of 3-acyl- and 3-carboxy-N-phosphinic-1-aminopyrroles **111** (R^1 = POPh$_2$), 3-acyl-, 3-carboxy-, and 3-aminocarbonyl-N-phosphonic-1-aminopyrroles **111** [R^1 = PO(OEt)$_2$, PO(OPh)$_2$], respectively.[144,145] Although the role of the catalyst is not well understood, it has been noted that, in some cases, the catalytic role of copper(II) chloride was important, whereas in other reactions the presence of copper salt had no observable catalytic effect, and sometimes very complicated reaction mixtures were obtained and better yields were obtained in the absence of catalyst. Because of the low reactivity of this class of azoalkene, the yields are rarely high (45–65%) but are acceptable in view of the general inaccessibility of the N-phosphorous-aminopyrroles by other routes.

1-Alkoxycarbonyl-, 1-ureido-, and 1-aroylaminopyrroles **113** are generally obtained from β-ketosulfones **112** (R^4 = Me, aryl; R^5 = Me, Ph) and the appropriate azoalkenes[131,146,147] (Scheme 31) in the absence of a catalyst. In the reactions the rate of formation of the pyrroline intermediates **110** appears to be similar to the rate of their conversion into the related 1-aminopyrroles.

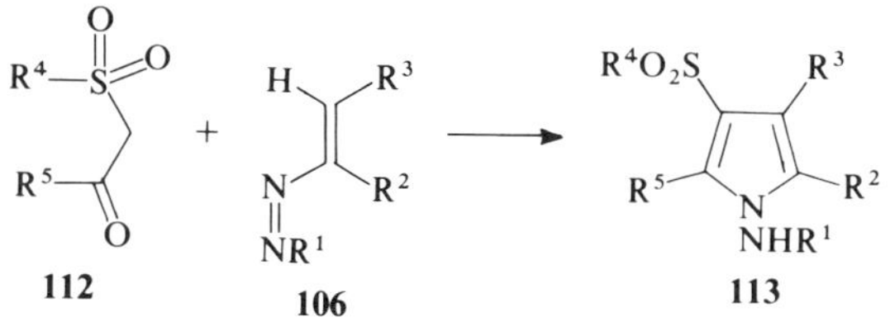

Scheme 31

Activated methyne systems (i.e., CH-substituted β-dicarbonyl derivatives) can also be used in this general synthetic approach, but the reaction with conjugated azoalkenes of type **114** stops at the stage of 1-amino-2-hydroxy-4-

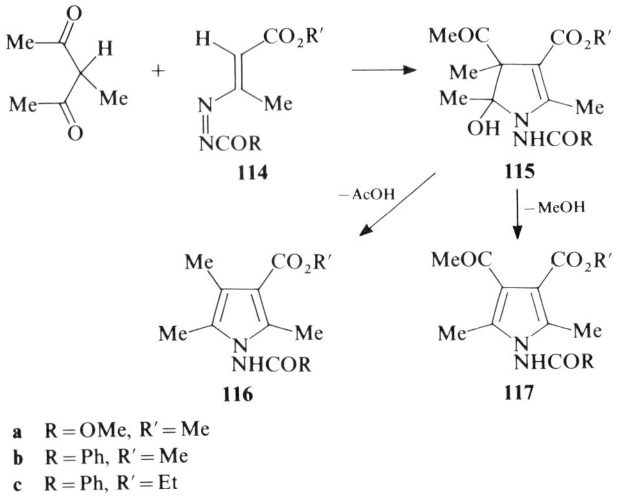

Scheme 32

a R = OMe, R' = Me
b R = Ph, R' = Me
c R = Ph, R' = Et

pyrrolines **115**[132] (Scheme 32). For example, the reaction of 3-methyl-2,4-pentanedione with **114a–c** can lead to different products depending on the nature of the azoalkene. With **114a**, the reaction produced a 1-amino-2-hydroxy-4-pyrroline **115**, which was isolated in high yield. In solution, and even in the solid state, this derivative **115a** readily eliminated acetic acid or methanol to afford **116** and **117**, respectively (58% and 27%). In contrast, **114b** directly forms a mixture of pyrroles **116** (65%) and **117** (10%), while in the reaction of **114c** the pyrrole **116** (50%), derived from the elimination of acetic acid, was formed.

In the case of the reaction of these azoalkenes with CH-substituted cyclic diketones, the elimination reaction involving the intermediate 1-amino-2-hydroxy-4-pyrroline analogs of **115** does not occur. This difference in behavior of cyclic and acyclic derivatives can be ascribed to the hindrance to aromatization in the case of the reaction involving cyclic β-dicarbonyl compounds.

3.2.2.2. 2-Aminopyrroles

Direct synthesis of 2-aminopyrroles can be conveniently achieved either by cyclization of suitable γ-aminonitriles or by condensation of α-aminoketones with methylene active nitriles.

3.2.2.2.1. From Four-Carbon Units

A versatile method for the preparation of 2-aminopyrroles of type **123** is represented by the cyclization reactions of the intermediates of type **122** under either acidic or basic conditions. The flexibility of this synthetic method is to be found in the different ways the key intermediate can be obtained (Scheme 33).

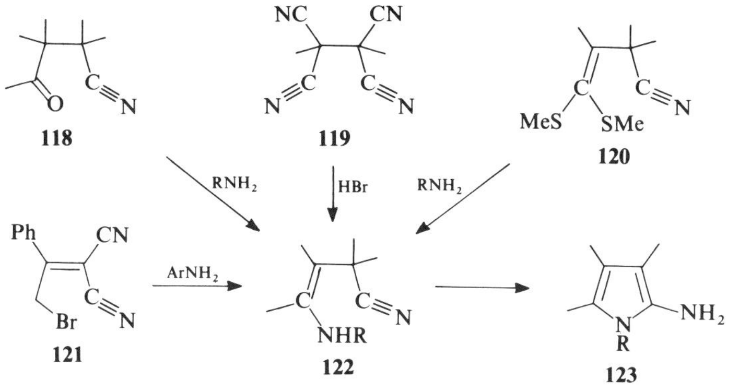

3.2. Synthesis

Scheme 33

Isolation of the enaminonitriles depends on the nature of the precursors and the experimental conditions. Thus, β-benzoyl-α-phenylpropionitrile, a ketonitrile of type **118**, when heated with ammonium formate or ammonia, led to 2-amino-3,5-diphenylpyrrole. This compound is relatively unstable (see Sections 3.3.11 and 3.4.3), whereas the more stable 1-formyl derivative can be obtained from the corresponding reaction with formamide.[148]

Hydroxymethylenesuccinonitrile, the enolic form of **118**, which can be obtained from succinonitrile and ethyl formate, reacts with alkyl-, aryl-, or aralkylamines to give the corresponding N-substituted aminomethylene succinonitriles **122** (R = alkyl, aryl, aralkyl).[149] Ring closure of the intermediates to the corresponding 1-substituted 2-amino-4-cyanopyrroles of type **123** is achieved with sodium ethoxide in ethanol. The yields are unaffected by the pK_a (range 1–10) of the amines, and only 2-nitroaniline ($pK_a = -0.3$) failed to react.[150] Pyrroles bearing ester groups adjacent to the amino substituent **123** [R = CH$_2$COOEt, CH(Me)COOEt] undergo spontaneous cyclization to give bicyclic lactones.[151]

2-Amino-5-bromo-3,4-dicyanopyrrole has been obtained from the reaction of tetracyanoethylene with hydrogen bromide in acetone[152] via the initial formation of tetracyanoethane **119**. The mechanism of the conversion of **119** into **123** is not clear, but probably involves the intermediacy of **122**. In an analogous fashion, the reaction of tetracyanoethane with aqueous sodium bisulfite gives 2-amino-3,4-dicyanopyrrole-2-sulfonic acid.

Reaction of S,S-diacetals of type **120** with primary amines leads to 1-substituted 5-methylthio-2-aminopyrroles **123** (R = alkyl, CH$_2$Ph),[153] while the analogous reaction between α-cyano-γ-bromocrotonitrile **121** and primary amines or ammonia gives the corresponding 2-amino-3-cyano-4-phenylpyrroles.[154] However, halonitriles of type **121**, which have several electrophilic centers, give different products depending on the site of reaction and on the basicity of the amine. Thus, amines with discrete basicity such as arylamines react by replacing the bromine to give the derivatives of type **122**, while strongly basic amines do not give definable products.

Ammonia, hydrazine, or phenylhydrazine react with **124** at the C=C double bond to give the adduct **125**, which fragments to malononitrile and the α-bromoketimine **126**. Nucleophilic displacement of the bromine atom by malononitrile gives the intermediate **127**, which undergoes intramolecular nucleophilic reaction to give the 2-aminopyrroles **128** (R = H, NH_2, NHPh). Reaction of **124** with primary aliphatic amines at one of the two cyano groups produces the intermediate **129**, which undergoes ring closure and subsequent elimination of hydrogen bromide to give 2-aminopyrroles **130** (R = Me, Et, CH_2Ph)[154] (Scheme 34). In an analogous fashion, 2-aminopyrroles of type **128** (R = H, aryl) or **130** (R = Me, CH_2Ph) are obtained from α-chlorocyclohexylidenemalononitrile.[154] In contrast, reaction of alkylidene malononitriles with diazocarbonyl compounds leads to 2-amino-3-cyanopyrroles analogous to **128** (R = N=CHCOR').[155]

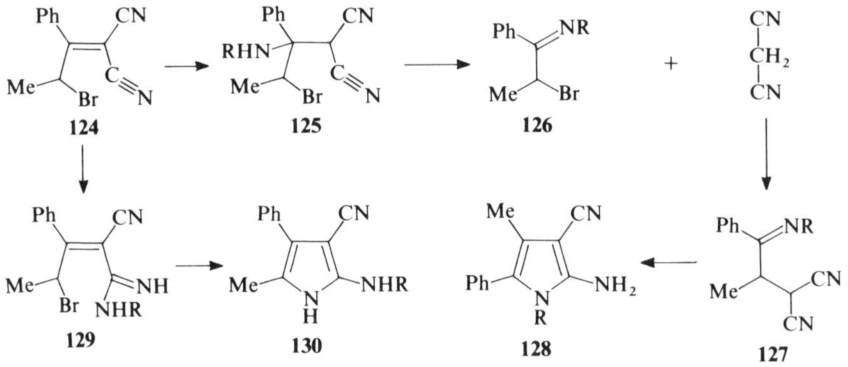

Scheme 34

2-Aminopyrroles **132** (R = CH_2CH_2OH, Ph) have been obtained from the butadienes **131** via the acid-catalyzed or thermally induced elimination of the thioalcohol.[152] Attempts to generalize this reaction have been unsuccessful; 1,4-bis-alkylthiobutadienes **131** (R = Me, Et) react in an alternative manner to give 2,5-bis(alkylthio)pyrroles **133** (Scheme 35).

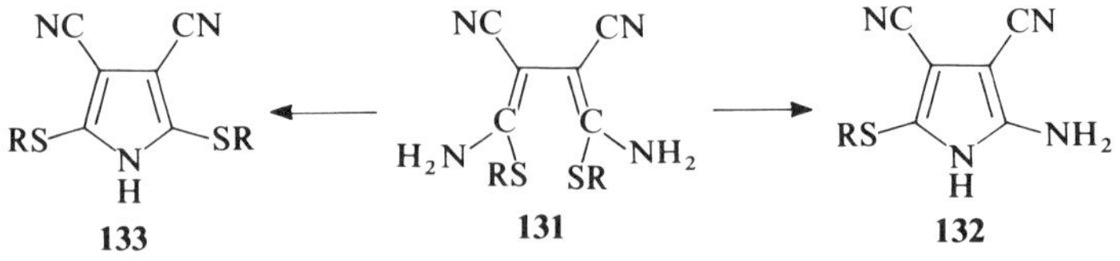

Scheme 35

3,3-Disubstituted 2-iminopyrrol-5-ones **136** (R and R' = alkyl or aryl) together with variable amounts of 3,3-disubstituted succinimides have been obtained by alkaline hydrolysis of α,α-disubstituted succinodinitriles **134**[156]

3.2. Synthesis

(Scheme 36). The cyclization proceeds via the initial preliminary preferential hydrolysis of the less hindered cyano group, followed by intramolecular addition at the second nitrile group, as shown in Scheme 36. The formation of 5-imino-3,3-disubstituted pyrrol-2-ones has also been observed.

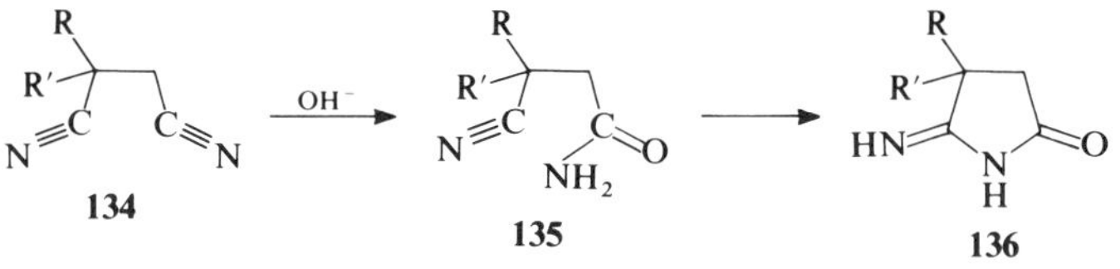

Scheme 36

The iminopyrrolones are not isolated from monosubstituted succinic dinitriles of type **134** (R = H), probably because of their instability under the reaction conditions. However, 5-imino-2-pyrrolidone **136** (R = R' = H) has been obtained from ethyl β-ethoxycarbonylpropionimidoate hydrochloride and ammonia.[157]

3-Substituted 2-imino-3-pyrrolidones **140** and **141** are produced in 37 and 16% yields, respectively, from the reaction of cyclohexanone and succinonitrile in the presence of sodium amide in diethyl ether.[158] The formation of these products has been rationalized in terms of a mechanism similar to that of the Stobbe condensation as shown in the Scheme 37. The initial aldol-type condensation of succinonitrile with cyclohexanone is followed by a reversible

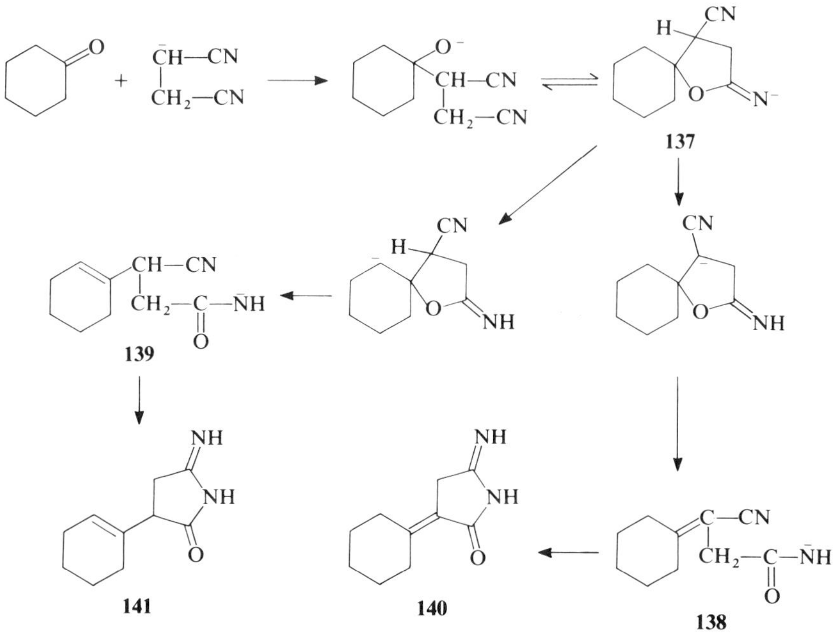

Scheme 37

cyclization to the lactim derivative **137**. Subsequent migration of the α- or γ-proton, accompanied by ring opening, leads to **138** and **139**, respectively, which cyclize to yield the iminopyrrolidones **140** and **141**.

3.2.2.2.2. From Ring Transformations

Reduction of the iminopyridazines **142** with zinc in acetic acid yields 2-aminopyrroles of type **144**[159,160] (Scheme 38). When the reduction is carried out with sodium borohydride, the dihydropyridazine **145a** can be isolated, which on treatment with zinc in acetic acid affords the 2-phenylaminopyrrole **147**, an isomer of **144a**. The difference in reactivity must be attributable to the intermediacy of the two different open-chain structures, **143** and **146**; in the former the nucleophilic attack is performed by the arylamino group, in the latter the nucleophilic attack is provided by the amino group.

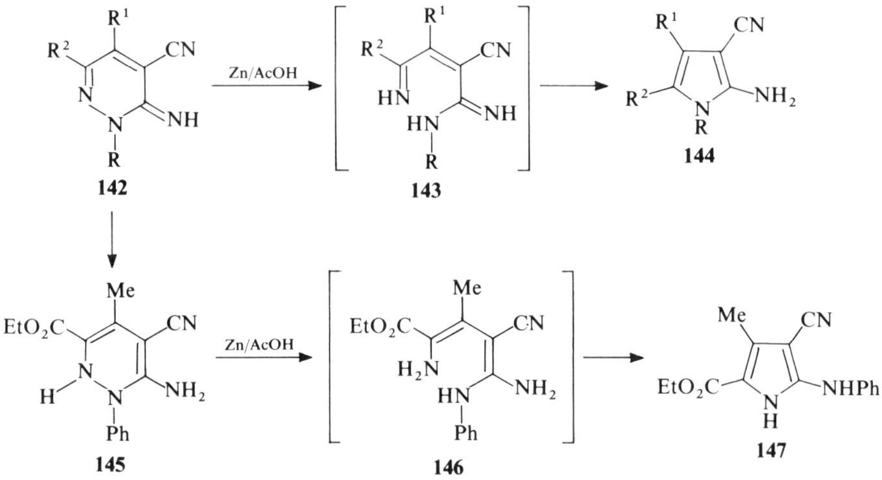

a $R = Ph, R^1 = Me, R^2 = CO_2Et$
b $R = Ph, R^1 = Me, R^2 = CONHPh$
c $R = 4\text{-MeC}_6H_4, R^1 = Ph, R^2 = CO_2Et$
d $R = 4\text{-Cl}—C_6H_4, R^1 = Ph, R^2 = CONHPh$

Scheme 38

Base-catalyzed rearrangement of 2,5-diamino-3,4-dicyanothiophene (**148**) produces 2-amino-3,4-dicyano-5-mercaptopyrrole (**150**) by an ANRORC mech-

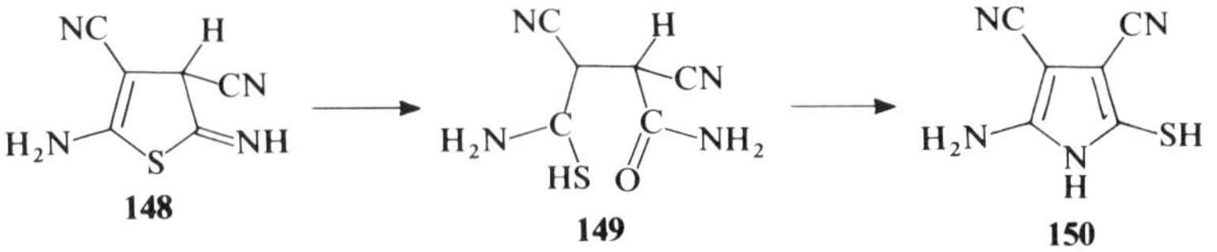

Scheme 39

3.2. Synthesis

anism with ring opening to **149**, followed by nucleophilic attack by the amino group on the carbonyl group and subsequent aromatization[152] (Scheme 39).

5,5-Disubstituted 2-aminopyrrol-4-ones of types **154** and **155** have been obtained from 4,4-disubstituted 5-oxazolones **151** by ring rearrangements, subsequent to nucleophilic attack by ynamines or methylene active compounds, respectively[161,162] (Scheme 40).

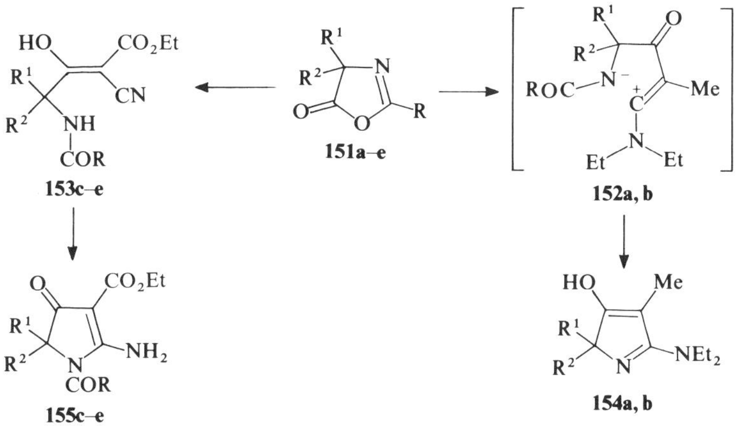

a R = CF$_3$, R^1 = R^2 = Me
b R = CF$_3$, R^1 = Me, R^2 = Ph
c R = Ph, R^1,R^2 = [=CHPh]
d R = Ph, R^1,R^2 = [=CH—CH=CHPh]
e R = Ph, R^1,R^2 = [=CH—C$_6$H$_{11}$]

Scheme 40

Ring enlargement of 2-dibenzoylmethylencyclobuten-1-one imine **156** on reaction with aromatic amines results in the formation of 2-amino-N-alkylpyrroles **157**[163] (Scheme 41). The corresponding reaction with aliphatic amines fails to give 2-aminopyrroles; the reaction occurs with retention of the 4-membered ring.

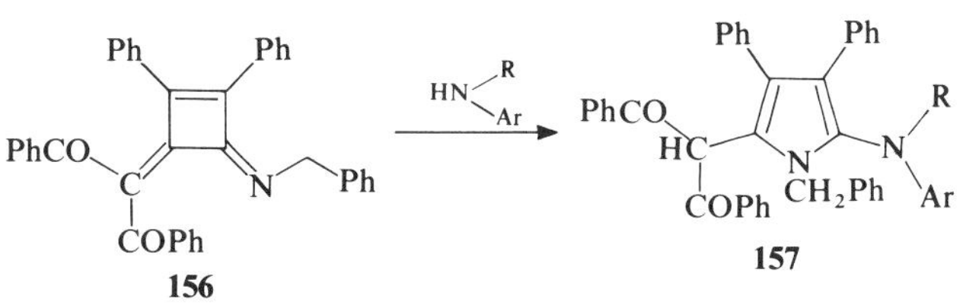

Scheme 41

The reaction between the 2H-azirine ester **158** and N,N-diethylpropynamine gives the 2-aminopyrrole derivatives **160**, via a bicyclic intermediate **159**[164] (Scheme 42). The yields are low, although azirine derivatives are reported to

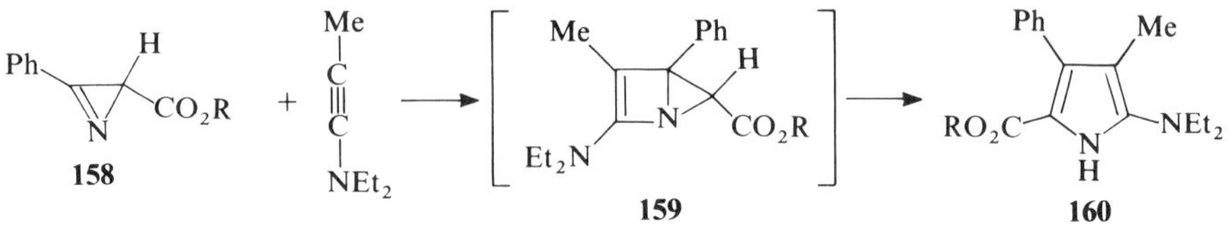

Scheme 42

react with enamines by a similar route to give 1H-pyrrole derivatives in good yields.

3.2.2.2.3. From 3 + 1 Carbon Units

Reaction of (2,2-dichloroalkylidene)imines 161 with an excess of potassium cyanide in DMSO at 120°C leads to 2-amino-5-cyanopyrroles 166 via the isolable 1,3-dicyano-2-alkylideneamines 164[165] (Scheme 43). The 2-aminopyrroles arise from 161 by dehydrochlorination to give 162, which undergoes Michael addition by cyanide ion to give the intermediate 163. Nucleophilic addition to the C=N double bond by a second cyanide ion, followed by elimination of a chloride ion, produces the intermediates 164, ring closure of which through an intramolecular nucleophilic attack of the amino group on the nitrile moiety and prototropic shift affords the 2-iminopyrroline 165. Subsequent tautomerism leads to the stable 2-aminopyrroles 166. Confirmation of this mechanism is provided by the reaction of N-1-(2,2-dichloro-3-methylbutylidene)tert-butylamine 161 (R = Me) with potassium cyanide in which both the 1,3-dicyanoenamine 164 and the iminopyrrole 165 have been isolated. In this case, tautomerism to the aromatic system is not possible due to the geminal methyl groups.

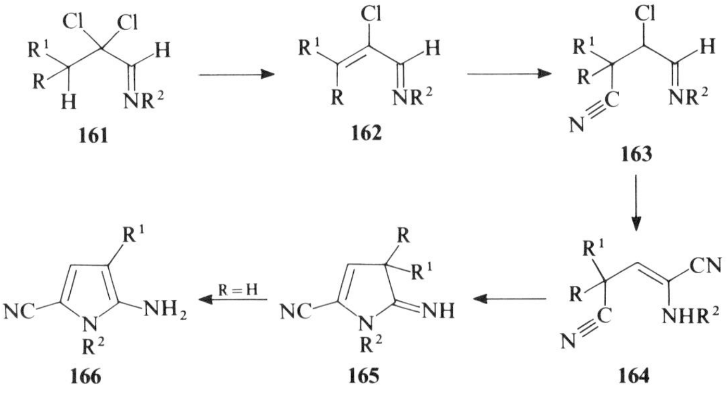

Scheme 43

2-Amino-4-benzoyl-3-cyano-5-ethoxypyrrole has been obtained from the reaction of phenacylmalononitrile with trichloroacetonitrile in refluxing ethanol.[166]

3.2. Synthesis

Tetracyanoethane (**119**) reacts readily with azomethynes to give the pyrrolines of type **167**, which spontaneously lose hydrogen cyanide to give the 2-aminopyrroles **168** (R = alkyl, aryl, heteroaryl; R' = aryl)[167,168] (Scheme 44). Compound **119** also produces alkylideniminopyrrolines **169**, which, on slight heating, are converted into the 2-aminopyrroles **170** or the pyrrolines **171**, depending on whether aldazines or ketazines have been employed.[169] In the case of azomethynes based on aromatic diamines, further reaction of the 1-substituted 2-aminopyrroles with **119** lead to the corresponding bis-derivatives.[168]

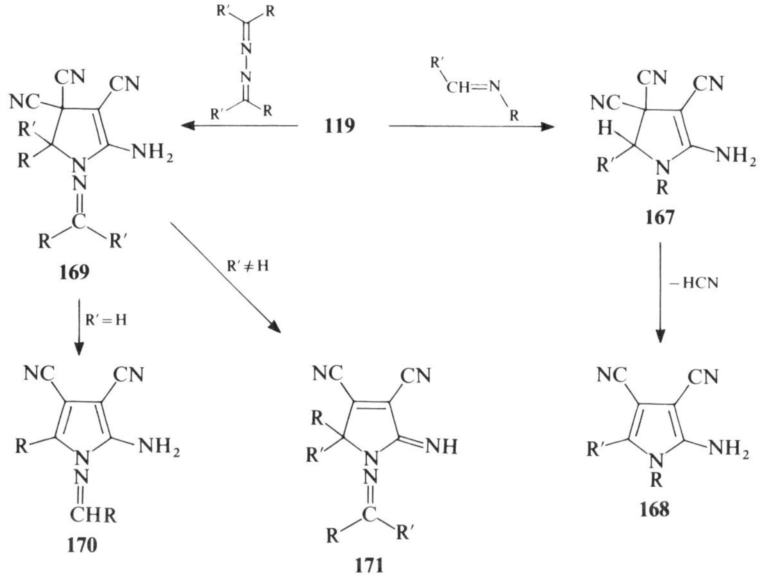

Scheme 44

3.2.2.2.4. From 2 + 2 Carbon Units

A typical example of this approach to the synthesis of 2-aminopyrroles **175** is represented by the base-catalyzed reaction of α-aminoketones **172** with methylene active nitriles **173**. Nucleophilic attack by the amino group on the nitrile gives the intermediates **174**, which cyclize, with the formation of the C(3)—C(4) bond to produce the 2-aminopyrroles (Scheme 45). Thus, 2-aminopyrroles **175** are obtained from the reaction of **172** (R^1 = H) with malononitrile or benzylcyanide.[170,171] However the use of the hydrochlorides of the α-aminoketones leads to the formation of substituted dihydropyrazines, as by-products, and modifications of this procedure involve the use of N-substituted α-aminoketones, either isolated or generated in situ from the suitable precursors. For example, reaction of desylaniline and malononitrile gives a high yield of 2-amino-3-cyano-1,4,5-triphenylpyrrole.[172]

The most useful and widely employed reaction for the preparation of 1-substituted 2-aminopyrroles in a one-pot synthesis using activated methylene

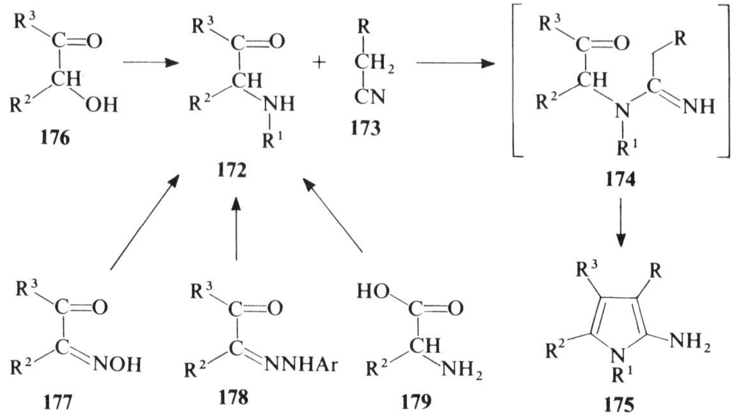

Scheme 45

nitriles, α-hydroxyketones **176** and primary amines without the isolation of the intermediate aminoketones **172**.

In a similar manner, 2-amino-3-cyanopyrroles **175** (R^1 = alkyl, aryl, aralkyl, heteroaryl) have been obtained from the reaction of malononitrile with acetoin, benzoin, or adipoin in the presence of a suitable amine.[172,173] A nonalkaline reaction medium is necessary to prevent polymerization of the malononitrile and eliminate the base-catalyzed reaction of the α-hydroxyketone and malononitrile, which would give a 2-aminofurane. The reactions are, therefore, carried out under the influence of an acidic catalyst. A modification of the reaction also employs aminocarbonyl compounds, such as o-cyano- and o-alkoxycarbonylanilines or β-aminoesters, in place of amines, and direct cyclization of the intermediates aminodicarbonyl compounds to the N-substituted 2-aminopyrroles occurs in the presence of malononitrile.[174] The reaction has been generalized by the use of α- and γ-aminoesters.[175] The formation of bicyclic and tricyclic ring systems has been observed when the substituents at the 1-position of the pyrrole ring bear groups that can react with the 2-amino group.

2-Amino-3-alkoxycarbonylpyrroles **175** (R^1 = alkyl, aryl, aralkyl) are obtained in a similar manner by the condensation reaction of acetoin with primary amines and cyanacetic esters,[176] while the acid-catalyzed condensation of acetoin, primary amines, and alkyl or arylsulfonylacetonitriles **173** (R = SO_2-alkyl, SO_2-aryl) produces 3-alkylsulfonyl- and 3-arylsulfonyl-2-aminopyrroles.[177]

Acetylaminoketones **172** represent another class of N-substituted intermediates that allow the synthesis of 1-unsubstituted 2-aminopyrroles. Catalytic reduction of monoximes of type **177**, obtained from symmetric 1,2-diketones, in acetic acid–acetic anhydride produces **172** (R^1 = Ac), which, with malononitrile, give the 2-amino-3-cyanopyrroles **175**.[178,179] Similarly, reduction of arylhydrazones **178** with zinc in acetic acid–acetic anhydride yields 2-amino-3-cyanopyrroles, via the intermediacy of **172**.[180] Compounds **172** (R^3 = Me) can

3.2. Synthesis

also be obtained from α-aminoacids **179**, but this route permits a variation of the substituent only at the 5-position of 2-amino-3-cyano-4-methylpyrroles.[181] α-Acetylaminoketones react with β-cyanoesters to give 2-amino-3-alkoxycarbonylpyrroles.[182]

2-Aminopyrroles unsubstituted at the 1-position are generally isolated from the N-acetylated intermediates **172** (R^1 = Ac), which undergo hydrolysis during the condensation step. Where the N-acetyl pyrroles can be isolated,[180,181,183] hydrolysis has been reported to proceed under mild basic conditions.

α-Halocarbonyl derivatives and α-haloamines have also been found to be useful in the synthesis of 1-unsubstituted 2-aminopyrroles. Thus, a versatile route to 2-amino-3-ethoxycarbonylpyrroles **184** utilizes the condensation reaction of ethoxycarbonylacetamidine **181** with chloroacetaldehyde or α-bromoketones in ethanolic sodium ethoxide[184-186] (Scheme 46). Cyclic, alkyl, and/or aryl substituents can be introduced at positions 4 and 5 of the pyrrole ring by a suitable choice of the starting α-bromoketone. When the substituents R^1 and R^2 are different, two isomers could, in principle, be obtained as a result of a competition between C- and N-alkylation of the ethoxycarbonylacetamidine in the first step of the reaction. However, the C-alkylation, leading to **182**, is favored, as demonstrated by mass spectroscopic studies[185] (see Section 3.3.7). A similar reaction of **180** with cyanoacetamidine **181** (R = CN) has been reported.[186]

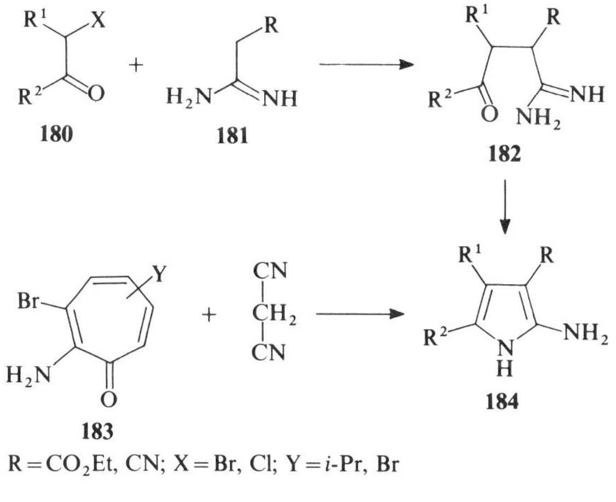

R = CO_2Et, CN; X = Br, Cl; Y = i-Pr, Br

Scheme 46

Substituted 2-amino-3-bromotropones **183** react with malononitrile under basic conditions in ethanol to give the corresponding 2-amino-3-cyanopyrroles **184**.[187] In this reaction, the formation of 2-amino-3-cyano-8-hydroxyquinoline, as by-products, is also observed.

Enamines **185** react with the diimmonium salts **186** to give 2-aminopyrroles **188** in good yield[188] (Scheme 47). The postulated mechanism involves a double electrophilic attack of the diimmonium salt **186** at the nitrogen and β-carbon

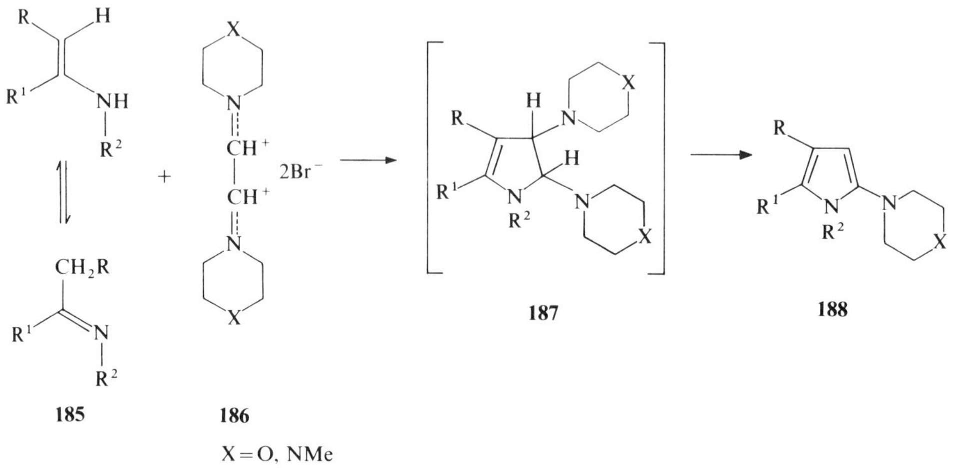

Scheme 47

atoms of the enamine form of **185** to produce the dihydropyrroles **187**, which aromatize by loss of a molecule of amine. The preparative utility of this reaction fails when R is hydrogen or methyl, probably because of the instability of the resulting 2-aminopyrroles that limits their isolation in a pure state.

2-Aminopyrroles **190** can also be synthesized by the reaction of cyanacetamide or ethyl cyanoacetate with sulfur in the presence of bases[189] (Scheme 48). The reaction proceeds through an initial dehydrodimerization, followed by addition of hydrogen sulfide to give intermediates **189**. Intramolecular nucleophilic cyclization of the amine on the cyano group leads to **190**. When R = NH_2, the 2-aminopyrrole, because of its instability, was isolated as a methyl derivatives when R = OEt, **190** undergoes aerial oxidation to **191**.

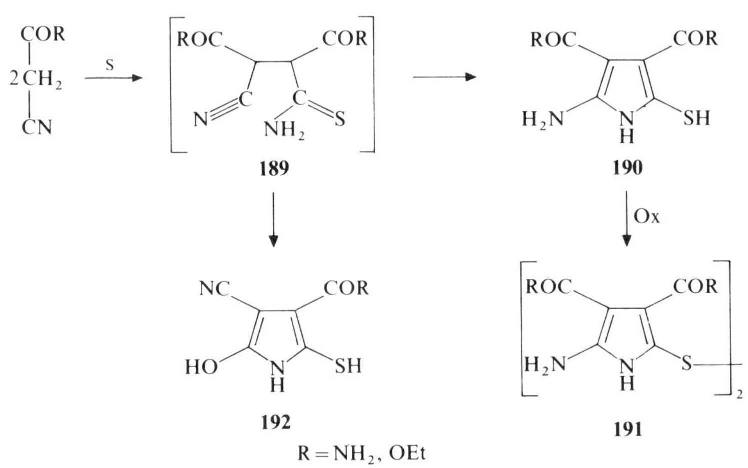

Scheme 48

However, in this latter case, the main product of the reaction is found to be **192**, as result of the nucleophilic attack on the ester function, during the cyclization of the intermediate **189**.

Primary or secondary α-aminoesters **193** add to ynamines **194** at room temperature to give the amidine derivatives **195b** or the ketene-N,N-acetals **195a**, respectively[190] (Scheme 49). Both types of intermediates cyclize by an intramolecular acylation that, in the case of secondary α-aminoesters, is almost spontaneous to give the N,N-disubstituted 2-amino-4-oxo-$2H$-pyrroles **196**. When the secondary α-aminoester is part of a cyclic system, as, for example, with methylpiperidine-2-carboxylate and 2-methyl-2-thiazolidine carboxylate, the nitrogen bridgehead derivatives, pyrrocoline and pyrrolo[2,1-b]thiazole, are respectively obtained. The oxoaminopyrroles **196** can also be obtained, although under more severe reaction conditions and in lower yields, from the reaction of the α-aminoesters **193** and ketene N,N-acetals **197** by a transamination reaction.

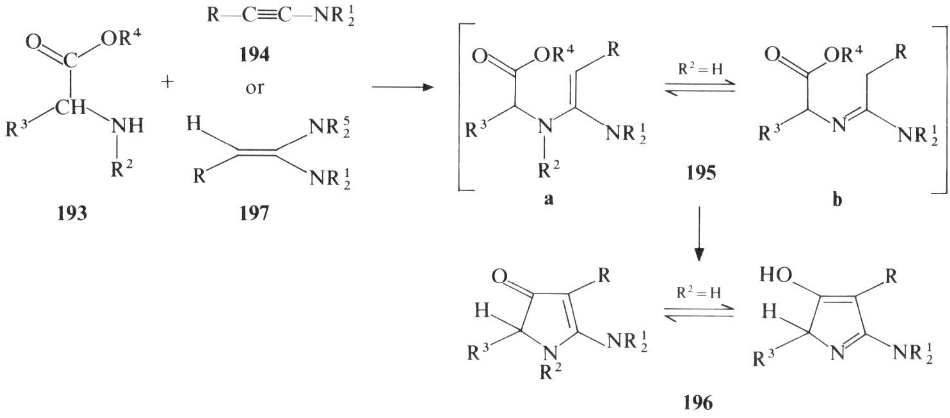

Scheme 49

3.2.2.2.5. From 2 + 1 + 1 Carbon Units

N-Substituted 2-butylaminopyrroles have been obtained by the reaction of acetylenes with *tert*-butylisocyanide in 1:3 molar ratio in presence of catalytic amount of nickel halides,[191,192] and palladium-catalyzed silylcyanation of silylacetylenes has been reported to give 4-substituted 2-[N,N-bis(trimethylsilyl)amino]-5-cyano-3-silylpyrroles with high regioselectivity.[193,194] Both these two reactions have been reviewed in Part 1, Section 2.9.

3.2.2.3. 3-Aminopyrroles

The most important method for the direct synthesis of 3-aminopyrroles is the condensation reaction of α-aminonitriles with β-dicarbonyl compounds, although the hydrogenolysis of isoxazole derivatives allows the preparation of a good variety of functionalized pyrroles.

3.2.2.3.1. From Four-Carbon Units

The synthesis of 3-aminopyrroles from four-carbon units is not of general application. Only two procedures are of any importance.

The reaction of *cis*-3-bromo-1,2-dibenzoylpropene **198** with primary amines gives 3-alkylamino-4-benzoyl-2-phenylpyrroles **201a,b**[195] (Scheme 50) via the isolable 2-(α-aminoacetophenonyl)acrylophenones **199**, which undergo either an amine addition reaction to give **200** (pathway A) or further substitution–rearrangement by amines to give **202** (pathway B). Intermediate **200** leads to **201**; intermediate **202**, which can be isolated, ring-closes to the 3-benzoylpyrrole **203**. The competition between pathways A and B is controlled by the space demand of the attacking amine and by the polarity of the solvent; for example, reaction of **198** with the bulky *tert*-butylamine fails to produce the 3-aminopyrrole, and for reactions carried out in chloroform or without solvent, the intermediate **199** gives **203a–c** almost exclusively, whereas in diethyl ether both **201** and **203** are produced.

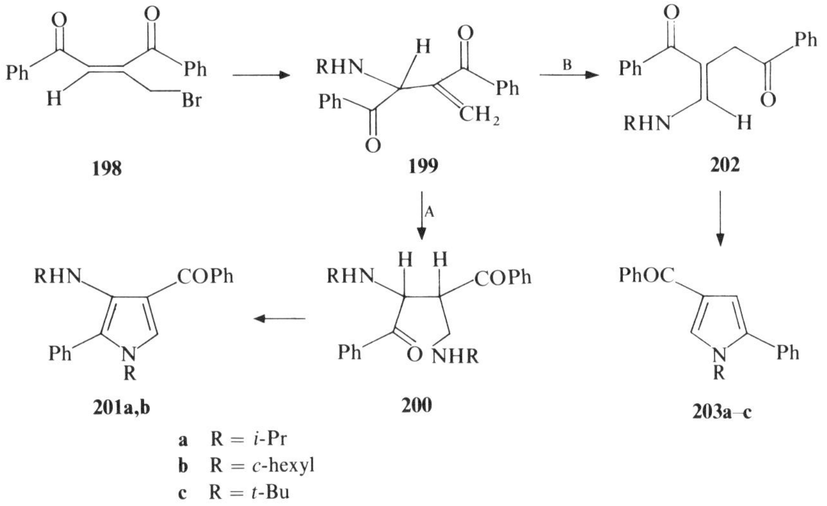

a R = *i*-Pr
b R = *c*-hexyl
c R = *t*-Bu

Scheme 50

In the second synthetic approach enaminonitriles **204**, obtained from the diethylacetal of *N*-cyanomethylpyrrolidone and activated methylene nitriles, are converted into 3-aminopyrroles **205**[196] by a Thorpe–Ziegler reaction (Scheme 51). When the diethylacetal of dimethylformamide is used as base, the 3-aminopyrroles are isolated together with the Schiff's bases.

3.2.2.3.2. From Ring Transformations

3-Aminopyrroles of type **209**, together with derivatives **211** as main products, have been obtained by reduction of 2-aryl-4-chloropyrimidines **206** with zinc

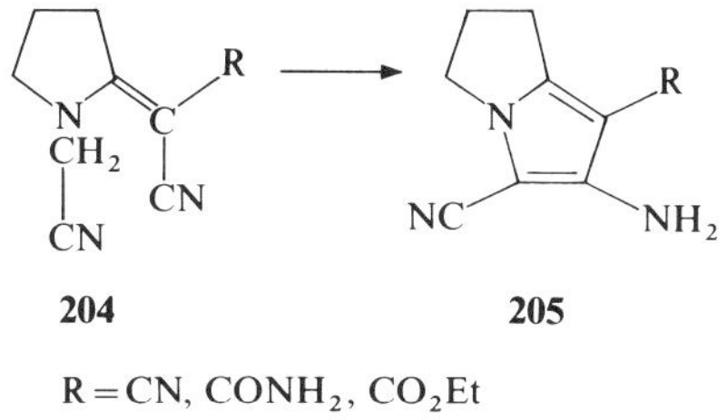

3.2. Synthesis

Scheme 51

R = CN, CONH$_2$, CO$_2$Et

and acetic acid[197] (Scheme 52). It was demonstrated that the first step of the reaction is the dechlorination of **206**, suggesting that the halogen substituent is not necessary for success of the reaction. The proposed mechanism involves the initial reduction of the 1,6 bond, and the key intermediate appears to have the bicyclic structure **207**. Cleavage of the aziridine ring takes place by either rupture **a** or rupture **b** to produce the intermediates **208** and **210**, respectively. The 3H-pyrrole intermediate **208** gives the 3-aminopyrroles **209** by a prototropic shift, whereas further reduction and elimination of ammonia of the intermediate **210** results in the formation of the pyrrole **211**. A different mechanism involving reductive cleavage via attack at the 2-position of the pyrimidine ring has also been proposed.

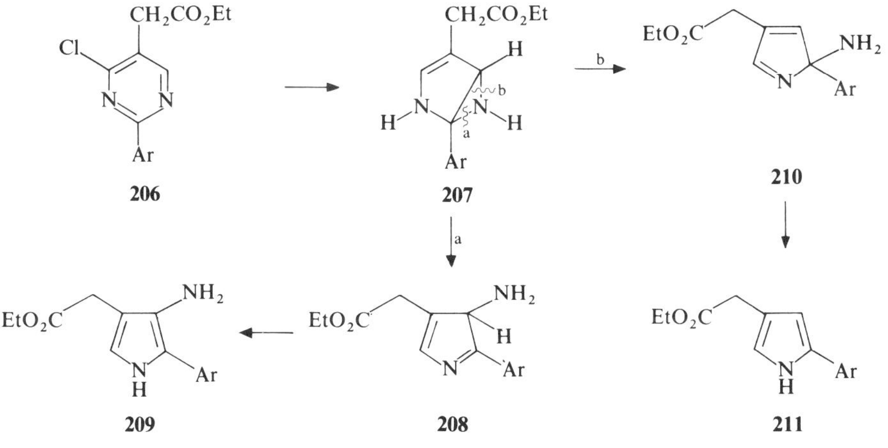

Scheme 52

3-Aminopyrroles **215** can be prepared from 3-isoxazolylketoximes **212** by a hydrogenolysis of the isoxazole nucleus.[198,199] The proposed mechanism involves the formation of the open-chain intermediate **213**, which by ring closure leads to the 3-imino-3H-pyrrole **214**. Further reduction of **214** affords the 3-amino-1H-pyrroles.

3-Aminopyrrole **217** has been obtained in good yield by the cycloaddition of dimethyl acetylendicarboxylate to the azirine **216**[200] (Scheme 54). This thermal

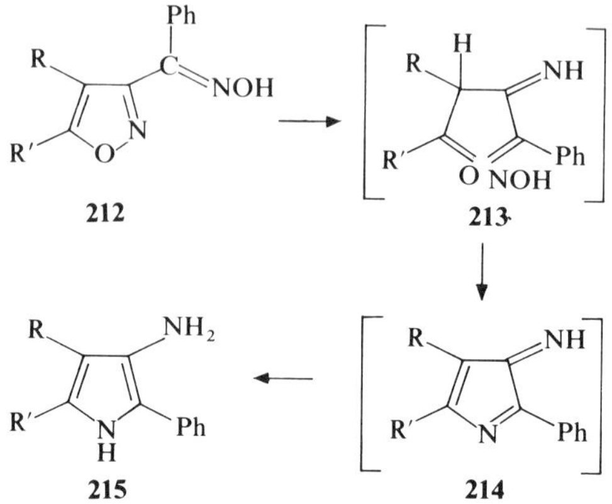

Scheme 53

R = H, Ph, COPh; R' = Ph; R–R' = (CH$_2$)$_4$, (CH$_2$)$_5$

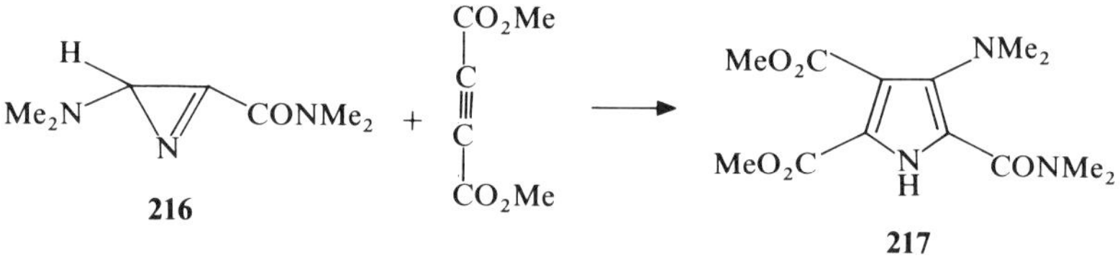

Scheme 54

addition takes place on the N(1)—C(2) bond of the azirine and contrasts with the photoaddition reaction, which takes place across the C(2)—C(3) bond.[201]

3.2.2.3.3. From 3 + 1 Carbon Units

3-Amino-2-ethoxycarbonylpyrroles **221** have been synthesized from the reaction of substituted formylacetonitriles **218** with glycine ethyl esters **219**[202, 203] (Scheme 55). The intermediate enaminonitriles **220**, on treatment with alkoxides, cyclize to give the pyrroles **221**. The yields are low when R' is hydrogen but are very high in all other cases. This suggests that the competitive ionization of the amino group of the intermediates **220** is a determining factor in inhibiting the cyclization to the corresponding pyrroles.

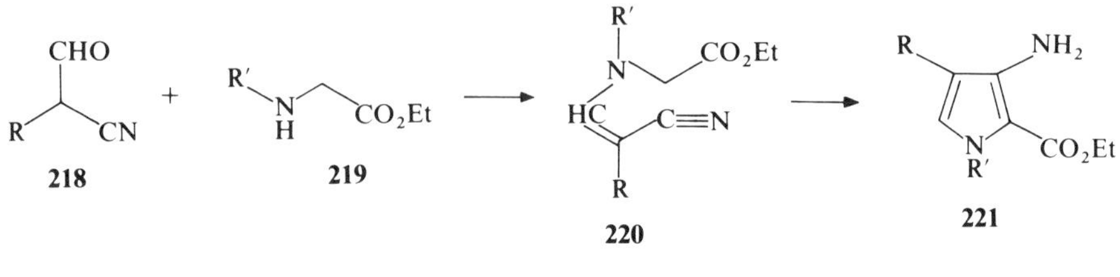

Scheme 55

3.2. Synthesis

The glycine esters **219** also reacts with thioacrylamides **222** to give 3-morpholinopyrrole-2-carboxylic esters **225**[204] (Scheme 56). The X group in compounds **222** is easily replaced by the amino group of the glycine ester, forming the morpholides of 3-[(alkoxycarbonylmethyl)amino]thioacrylic acids **223**, which are converted by dimethyl sulfate into the methylthio derivatives **224**, and subsequent reaction under basic conditions leads to the 3-morpholinopyrroles **225**. 1-Substituted analogs cannot be prepared by this procedure.

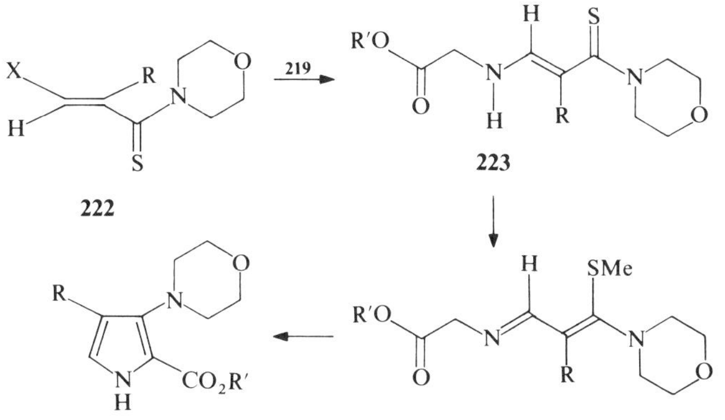

X = OH, NMe$_2$; R = Ph, aryl; R' = Me, Et

Scheme 56

3.2.2.3.4. From 2 + 2 Carbon Units

3-Aminopyrroles, unsubstituted at the 1-position (**228**), are prepared by the condensation of α-aminonitriles **226** with β-diketones, β-ketoesters, or β-ketoamides through the conjugated enaminocarbonyl intermediates **227**[205-207] (Scheme 57). This synthesis has already been discussed in Part 1 (Section 2.2.1.4). However, α-aminonitriles **226** also add to dimethyl acetylendicarboxylate to give the same intermediate iminonitriles **227**, which cyclize easily to the amines **228**.[205,206] A similar synthesis of 3-aminopyrroles from enamines, reported by Murata et al.,[208-210] has also been described in detail in Part 1 (Section 2.2.1.4). The difference between these two syntheses of 3-aminopyrroles

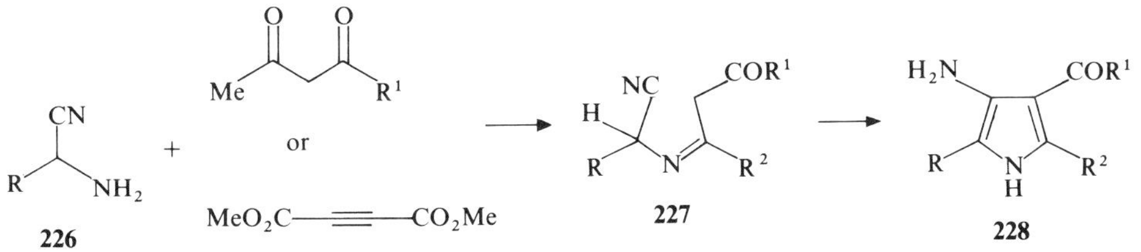

Scheme 57

relates to the protection of the reactive position α to the cyano group. In the former reaction, it is not necessary, and therefore makes the synthesis more flexible and of wider application, although all attempts to prepare 3-amino-2,5-dimethylpyrroles by this procedure have failed[211] (see Section 4.3).

3-Amino-2,4-diphenylpyrrole **232** has been obtained by self-condensation of ω-aminoacetophenone **229** in strongly alkaline media[212] (Scheme 58). The synthesis was first achieved by Gabriel,[213] and the pyrrole **232** was not the only product of the reaction. The open-chain intermediate **230** was also isolated, as was the dihydropyrazine **233** and its oxidation product **234**. Intermediate **231** was also isolated under acidic conditions and was easily dehydrated to **232**.[214] The corresponding reaction with *m*- and *p*-hydroxy- and *m*- and *p*-acetoxyphenacylamines is complicated by the lower reactivity of the carbonyl group, and the aldol condensation leading to **230** was observed only in the case of the *m*-acetoxyphenyl derivative.[212]

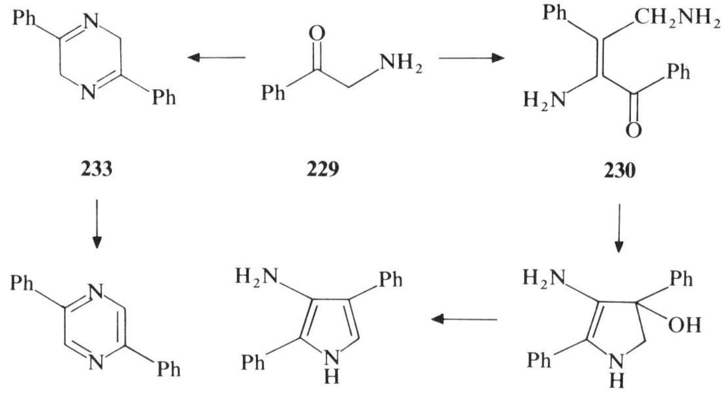

Scheme 58

In an analogous fashion, 2,4-disubstituted 3-acetylaminopyrroles are obtained from β-acetylaminocarbonyl compounds under mildly basic conditions.[215,216] In these cases, 1-acetylpyrroles are obtained initially, but are very easily hydrolyzed under the reaction conditions.

N,N-Disubstituted 3-aminopyrroles of type **239** have been obtained from the reaction of dialkylthioamides **235** and ethyl 2-bromopyruvate in acetonitrile[217] (Scheme 59). In the first step of the reaction, intermediate thioimmonium salt **236** is formed. Deprotonation and extrusion of sulfur converted **236** into **238**, which undergoes ring closure, followed by proton shift and dehydration, to yield the 3-aminopyrroles **239**.

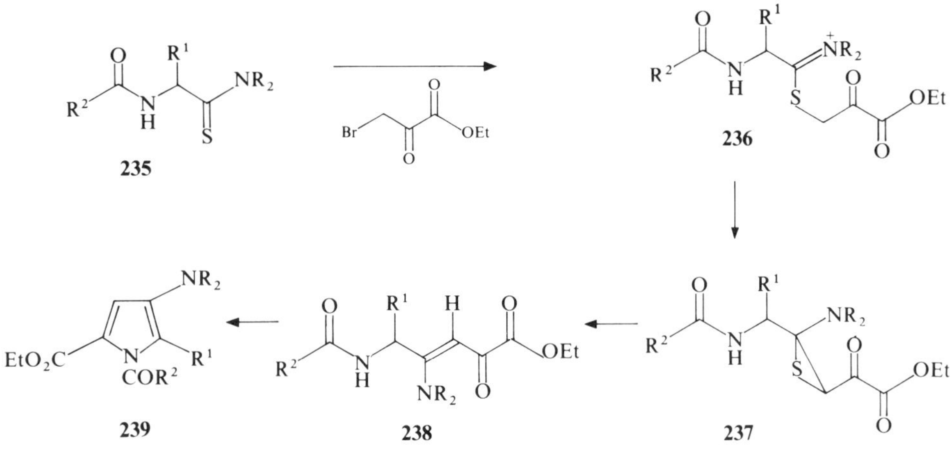

3.2. Synthesis   345

Scheme 59

3.2.2.4. Diaminopyrroles

3.2.2.4.1. 1,2-Diaminopyrroles

The reaction leading to the formation of 1,2-diaminopyrroles **128** (R = NH$_2$, NHPh) from α-cyano-γ-bromocrotonitrile and either hydrazine or phenylhydrazine is discussed in Section 3.2.2.2.1 (Scheme 34). 1,2-Diaminopyrrole **241** can also be obtained, although in low yield, from phenacyl hydrazine **240** and malononitrile in the presence of morpholine as the base; the analogous reaction of monohydrazones of 1,2-dicarbonyl compounds with malononitrile gives aminopyridazines[218] (Scheme 60).

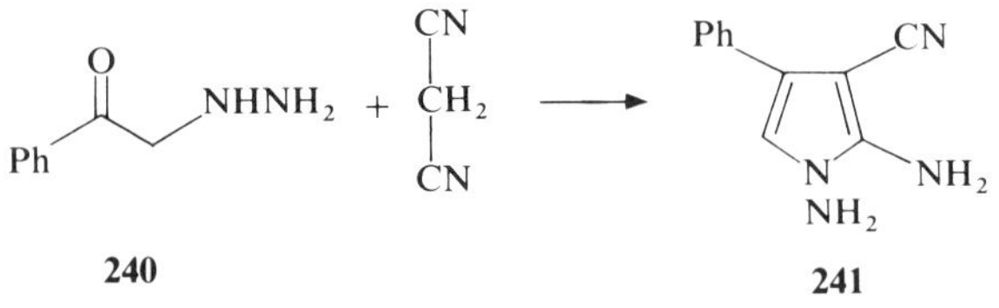

Scheme 60

1,2-Diamino-3-cyanopyrroles **244** have been prepared from malononitrile and alkoxycarbonyl or aminocarbonylazoalkanes **114**[219] (Scheme 61). High yields of the 1,2-diaminopyrroles are obtained when the molecular ratio of the two reactants is 5:1 but, in all the cases, large quantities of the tetrahydropyrrolo[2,3-b]pyrroles **246** are also obtained. The initial 1,4-addition of malononitrile to **114** can be followed either by ring closure to the intermediate

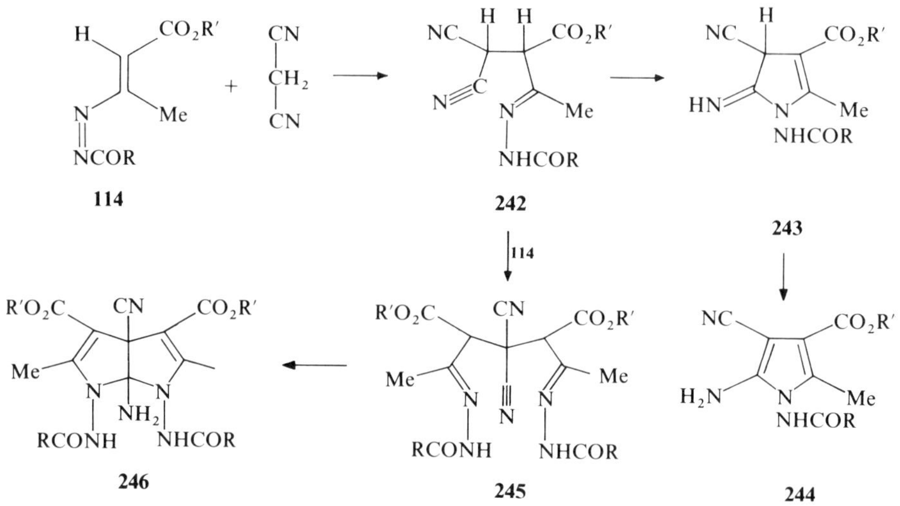

Scheme 61

R = OMe, OEt, O*t*-Bu, NH$_2$, NHPh; R′ = Me, Et

iminopyrroline **243**, or by further addition of **114** to give **245** and subsequent ring closure.

3.2.2.4.2. 2,4-Diaminopyrroles

Substituted 2,4-diaminopyrroles **252** can be prepared in good yields from the reaction of tris(imino)thietanes **247** with *trans*-β-dimethylaminostyrene[220]

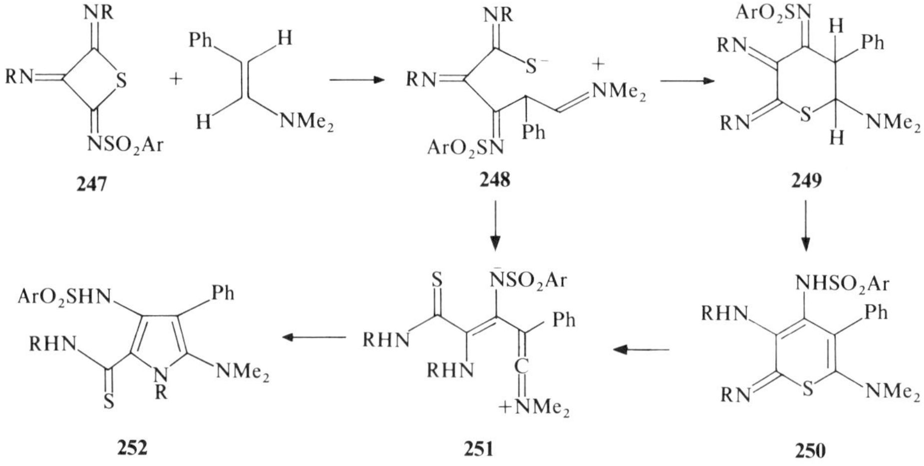

R = *t*-Bu; Ar = Ph, 4-Me—C$_6$H$_4$

Scheme 62

(Scheme 62). The formation of **252** was explained in terms of a stepwise reaction involving the nucleophilic attack of the enamine onto the electrophilic C(4) atom of the thietane, followed by ring closure to give **250**, which could not be isolated. Ring contraction via the open-chain intermediate **251** produces the pyrrole **252**. Involvement of the intermediate thiopyran was postulated, since the dehydro intermediate analogous to **249** was isolated from the reaction of **247** and ynamines. However, direct isomerization of the intermediate betaine **248** to **251** cannot be excluded.

3.2.2.4.3. 2,5-Diaminopyrroles

The synthesis of 2,5-diaminopyrroles is generally related to the formation of its tautomeric form, succinimidine **256** (R = H) and its symmetric *N*-derivatives (Scheme 63). Thus, succinamidine hydrochloride **253**, obtained from succinimido ether **254** and alcoholic ammonia, is converted into **256** (R = H) at ~100°C with loss of ammonia,[221,222] and **254** reacts with methylamine to give the corresponding dimethylamino derivative **256** (R = Me).[222]

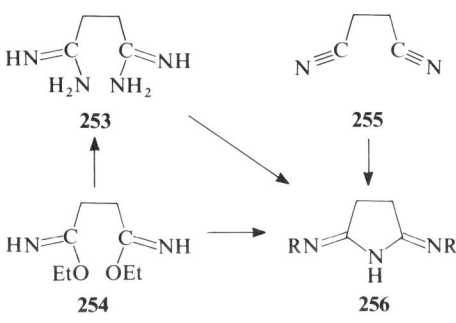

Scheme 63

Succinimidine has also been obtained from succinonitrile **255** and methanolic ammonia under pressure, or at ordinary temperature from succinonitrile and sodamide in formamide or DMF.[157,223] In an analogous fashion 2,5-diimino-3-phenyl-3-benzylpyrrolidine has been prepared from the corresponding dinitrile.[156]

2,5-Diphenyliminopyrrolidine **256** (R = Ph) is formed from **255** and aniline,[157] although in an early report[224] 2-amino-1-phenyl-5-phenylaminopyrrole was proposed as the structure for the product.

3.2.2.5. Triaminopyrroles

1,1,2,2-Tetracyanoethane **119** reacts in aqueous solution under mild conditions with hydrazine and methylhydrazine to give the corresponding 1,2,5-triaminopyrroles **257**[225,226] (Scheme 64).

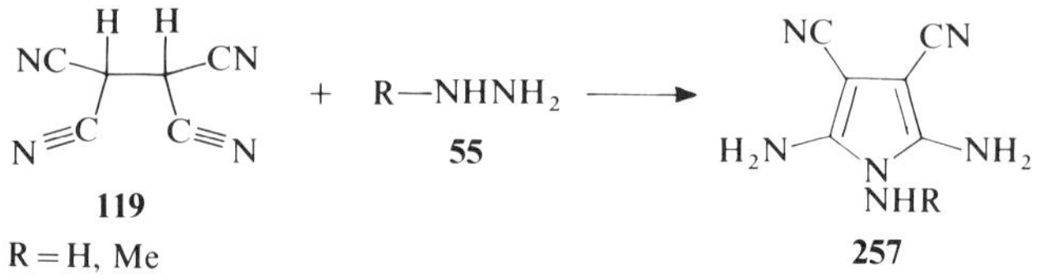

Scheme 64

3.2.3. Conversion of Functional Groups

The traditional methods used to prepare aromatic amino derivatives by the conversion of functional groups find only limited application in the synthesis of aminopyrroles on account of the unavailability of the suitable starting materials. The preparation of N-alkylamino- or N-carbonylaminopyrroles from the corresponding amino derivatives is reviewed in Section 3.4.2.

3.2.3.1. 1-Aminopyrroles

The only method to prepare 1-aminopyrroles by conversion of other functional groups is represented by the direct introduction of the amino function into the preformed ring. Thus, 1-amino-2,5-diphenylpyrrole and 1-amino-2,3,4,5-tetraphenylpyrrole are produced from the sodium salts of the corresponding 1H-pyrroles **258** on treatment with mesitoxyamine in DMF[66] (Scheme 65).

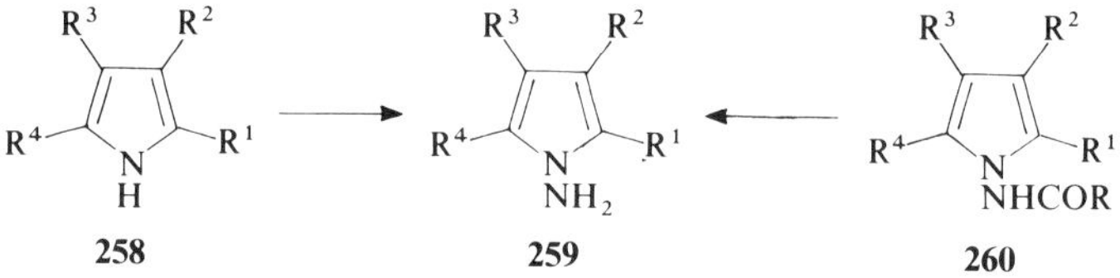

Scheme 65

The 1-aminopyrrole **259** (R^1 = H, R^2 = R^3 = Et, R^4 = Me) has been obtained by reaction of pyrrole **258** (R^1 = H, R^2 = Et, R^3 = Ac, R^4 = Me) with hydrazine.[227]

Other reactions leading to N-unsubstituted 1-aminopyrroles involve the cleavage, either reductive or hydrolytic, of the carboxamido group in derivatives of type **260**, e.g., **260** (R = CH_2Ph, R^1 = R^4 = Ph, R^2 = R^3 = H) → **259** (R^1 = R^4 = Ph, R^2 = R^3 = H),[54] while other N-acyl- and N-alkoxycarbonyl-aminopyrroles are hydrolyzed to the corresponding 1-aminopyrroles under either basic or acidic conditions.[52,66,123]

3.2. Synthesis 349

Although these reactions do not truly represent the conversion of functional groups, since the starting materials already bear the amino group, they are of relevance in the preparation of *N*-unsubstituted 1-aminopyrroles, which are unobtainable by direct ring syntheses.

3.2.3.2. 2-Aminopyrroles

2-Aminopyrroles can be synthesized from 2-pyrroloylazides by Curtius rearrangement. Thermal rearrangement of the azides **261** produces the isolable isocyanates **262**[228] (Scheme 66), which on hydrolysis in aqueous acetic acid leads to the 2-aminopyrroles **263**.[227] In some cases, 3-acetamidopyrroles,[229] or bis-derivatives of type **264**,[228] resulting from further reaction of the 2-aminopyrrole with the isocyanate, are also formed. When the Curtius rearrangement is carried out in presence of aniline, the 2-phenylureidopyrroles **265** are produced,[228,229] and, in the presence of alcohols, the corresponding urethanes **266** are obtained.[227–231] However, when the derivatives **266** bear only alkyl substituents on the pyrrole nucleus, the urethanes undergo aerial oxidation and 2-hydroxy-2*H*-pyrroles are isolated. These observations are in agreement with other reports of the autoxidation of (alkylpyrrolyl)urethanes[232] and of 2,5-dimethyl-3-aminopyrroles[211] (see Section 3.4.3).

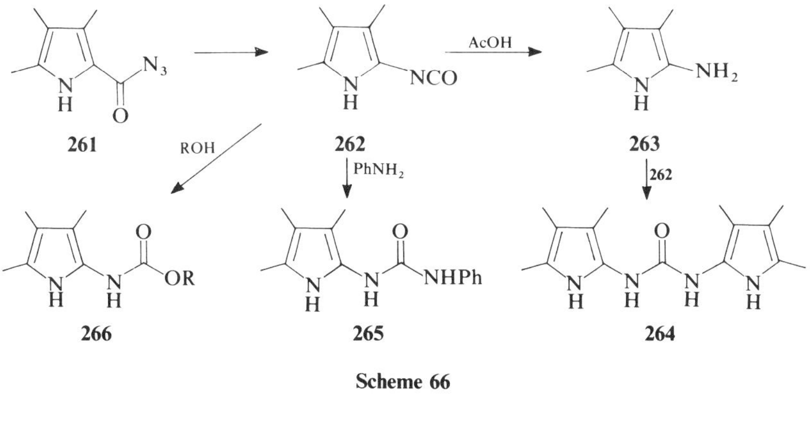

Scheme 66

Attempts to prepare 2-acylaminopyrroles by a Beckmann rearrangement of the oximes of 2-acylpyrroles have failed. Unexpectedly, only methyl group and hydrogen migration has been observed, and pyrrole 2-carboxamides are obtained.[233,234]

2-Aminopyrroles are also obtained by reduction of azo, nitroso, and nitro derivatives. Thus, 2-arylazopyrroles of type **267** are catalytically reduced over platinum in an alkaline medium, when $R' = SO_3H$,[235] and with tin(II) chloride in acetic acid, when $R' = NO_2$,[236] to give the corresponding 2-aminopyrroles **270** in good yields (Scheme 67). Catalytic reduction of the 2-nitroso-3,5-

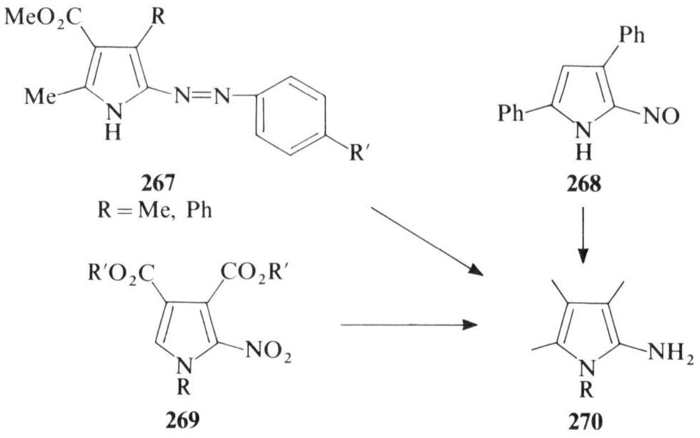

Scheme 67

diphenylpyrrole **268** in methanol over Adam's catalyst produces the 2-aminopyrrole,[237] and 2-amino-5-acetylpyrrole, the first 2-amino derivative to be isolated, is obtained by reduction of the corresponding 2-nitro compound with tin powder in hydrochloric acid.[238]

Several other 2-aminopyrroles have been obtained in excellent yield by catalytic reduction of the corresponding nitro derivatives, e.g., **269**[239] and pyrrolyl-2-urethanes.[228]

Nucleophilic substitution by secondary amines of polynitropyrroles has been reported to lead to 2-amino derivatives. Thus, 1-methyl-5-nitro-2-piperidinopyrrole (**272**) (73%) is obtained from 1-methyl-2,5-dinitropyrrole (**271**) by reaction with an excess of piperidine in DMSO at room temperature[240,241] (Scheme 68).

2-Piperidino- and 2-morpholino-4-nitro-1-methylpyrroles **272** can be obtained in excellent yield (86–90%) by cine nucleophilic substitution of

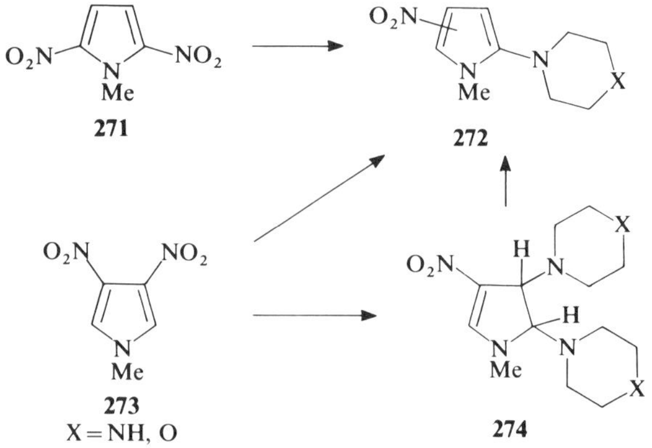

Scheme 68

1-methyl-3,4-dinitropyrrole 273.[242] The same 2-amino derivatives can also be formed, although in much lower yield, from the pyrroline 274, obtained from 273 at room temperature.[242] In an analogous fashion 1-*tert*-butyl-2-(N,N-dimethylamino)-4-nitropyrrole has been obtained from 1-*tert*-butyl-3,4-dinitropyrrole and dimethylamine.[243] A detailed discussion of nucleophilic substitution reactions is to be found in Part 1.

The 1,6-diazabicyclo[3.2.0]heptenone derivative 275 has been shown to be a good precursor for 2-benzamidopyrrol-5-ones 276 and 277[244] (Scheme 69). Thus, catalytic reduction of 275 leads to the 1-methyl derivative 276 in good yield, while the acid-catalyzed cleavage in water, alcohols, or acetic acid produces the derivatives 277 in excellent yield.

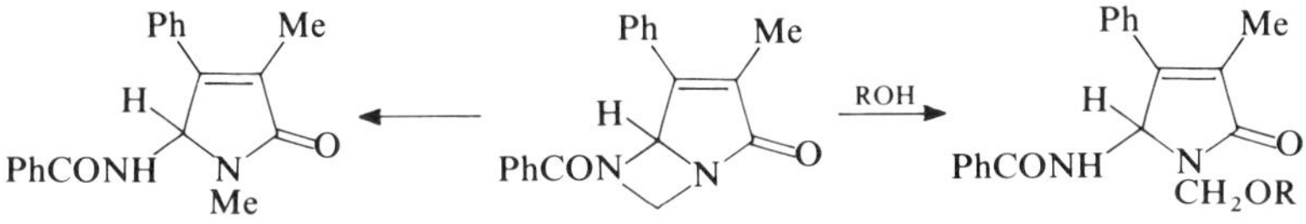

Scheme 69

3.2.3.3. 3-Aminopyrroles

In a manner analogous to that used for the preparation of 2-aminopyrroles the Curtius rearrangement of 3-pyrroloylazides can be used for the synthesis of the 3-amino derivatives. Thus, decomposition of 3-carbonylazidopyrrole 278 (R = H) in acetic acid gives the 3-acetylamino derivative 280[215] (Scheme 70). Similarly, the azide 278 (R = CPh$_3$) rearranges smoothly in anhydrous solvent to the isocyanate 279, which can be converted into the urethanes 280 (R' = COO—*tert*-butyl; COOCH$_2$Ph) by reaction with the appropriate alcohol,[245] and the aminopyrrole is obtained from the urethane by a catalytic hydride transfer using ammonium formate in the presence of Pd—C.

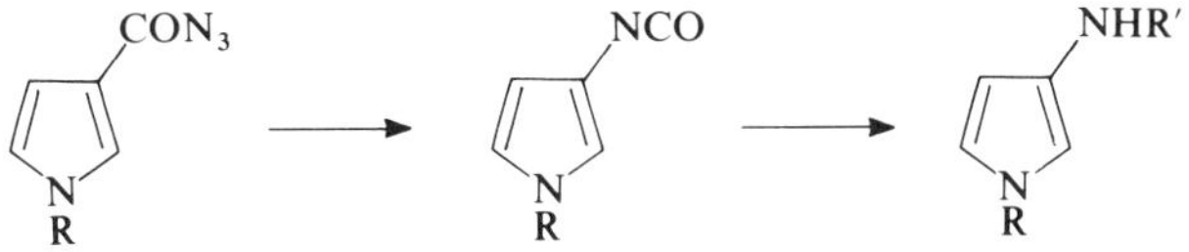

Scheme 70

Unlike the corresponding reaction for the 2-isomers, the oximes of the 3-acylpyrroles 281 undergo a Beckmann rearrangement in dilute hydrochloric acid to give the corresponding 3-acylamino derivatives 282[246] (Scheme 71).

3-Aminopyrroles are synthesized by reductive cleavage of the N=N bond in azo derivatives. Thus, arylazopyrroles of type 283 are reduced either catalytically in an alkaline medium[235] or by tin(II) chloride in acetic acid[211,247,248]

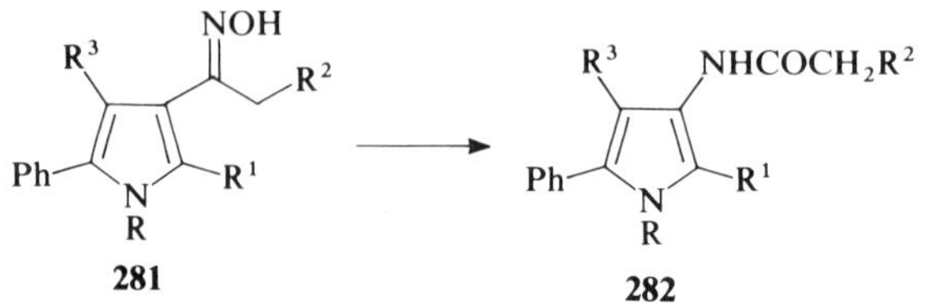

Scheme 71

to yield **284** (Scheme 72). The 3-aminopyrroles are generally obtained in high yields, although this method leads to complex mixtures in the case of 3-arylazo-1-methylpyrroles.[250]

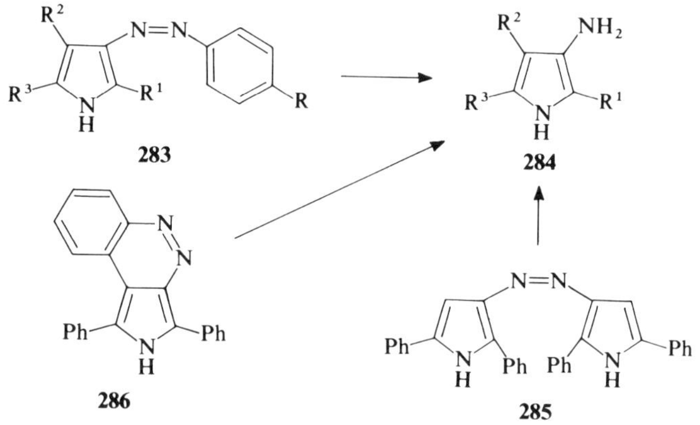

R = SO$_3$H, NO$_2$; R^1 = Me, Ph, 3-MeOC$_6$H$_4$;
R^2 = H, Me, COMe, CO$_2$Et, CN; R^3 = Me, Ph

Scheme 72

Reduction of the azo compounds by zinc in acetic acid–acetic anhydride is reported to yield the 3-acetylaminopyrrole.[250]

Symmetric azo bis-pyrrole, such as compound **285** is also catalytically reduced over palladium to the corresponding 3-amino derivatives, whereas reduction with Raney nickel affords a complex mixture and resinous material.[251] Hydrogenolysis is successful, even when the azo group is part of a polycyclic system; for example, **286** is reduced by zinc in acetic acid to the amine **284** (R^2 = 2-aminophenyl).[252]

Reduction of 3-nitroso or 3-nitro derivatives represents the most important method for the synthesis of 3-aminopyrroles, because of the easy availability of the starting materials.

The reduction of the 3-nitroso derivatives can be achieved under a variety of conditions. Thus, the nitroso compound **287** (R = Ph) has been reduced by an excess of hydroxylamine or with hydrazine in either presence or absence of alkali to the corresponding 3-aminopyrrole **288**, which was the first aminopyrrole

3.2. Synthesis 353

isolated as free base[253,254] (Scheme 73). However, a more general route to 3-aminopyrroles is achieved by reduction with zinc in acetic acid.[255,256]

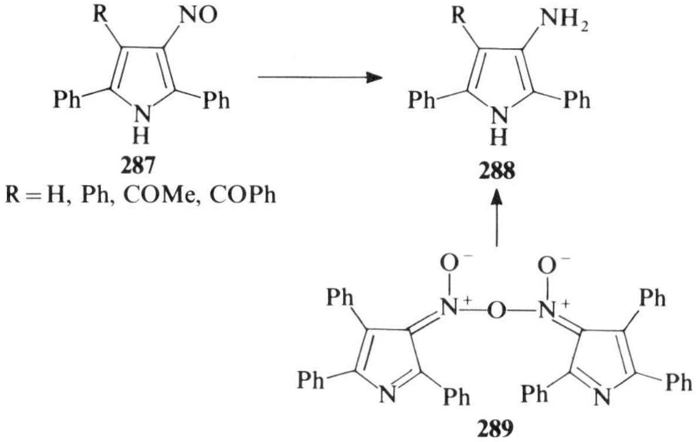

Scheme 73

3-Amino-2,4,5-triphenylpyrrole **288** (R = Ph) is also synthesized from compound **289**, by reduction either in acetic acid with zinc or under basic conditions with, for example, hydroxylamine, ammonium sulfite–ammonia, or sodium sulfite–sodium arsenite.[257] Reductions with aluminum amalgam[258,259] or zinc and ammonium chloride[260] are also effective. Less widely employed is the reduction with metal and acids either because of the lability of the pyrrole nucleus in the reaction conditions or because the intermediate 3-amino derivative undergoes further intramolecular reaction with other functional groups of the molecule.[256,260,261]

Catalytic reduction is also effective in the preparation of 3-aminopyrroles. Thus, with Raney nickel the 3-nitroso-2,5-diphenylpyrrole **287** (R = H) gives a 70% yield of the corresponding 3-aminopyrrole,[251] and an even better yield (93%) has been obtained by hydrogenation over 10% Pd—C.[262] In fact, the reduction of 3-nitropyrroles is more commonly achieved by hydrogenation over palladium or platinum. With this method, either 3-aminopyrroles[206,247,263] or their 1-substituted derivatives[8,249,264–266] are conveniently synthesized, and, generally, in high yields.

Nucleophilic displacement of a nitro group of 3,4-dinitropyrroles can lead to 3-alkylamino-4-nitropyrroles, although this reaction is of limited interest as a synthetic method. 3-Morpholino- (X = O) and 3-piperidinopyrroles (X = NH) of type **290** are produced, however, in good yield by action of sodium methoxide on the intermediate **274**, derived from **273**[242] (Scheme 74), while the action of primary amines (R = *tert*-butyl, *i*-propyl) on 1-methyl-3,4-dinitropyrrole **273** affords mixtures of the 3-alkylaminopyrroles **291**, **292**, and **293**, as a result of direct substitution reaction or of ANRORC sequence.[243]

3-Guanidinopyrrole **297** has been obtained in 25% yield from the amino acid viomycidine **294** on treatment with mercury(II) acetate in aqueous acetic acid[267] (Scheme 75). This reaction proceeds via the pyrroline **295**, resulting from an

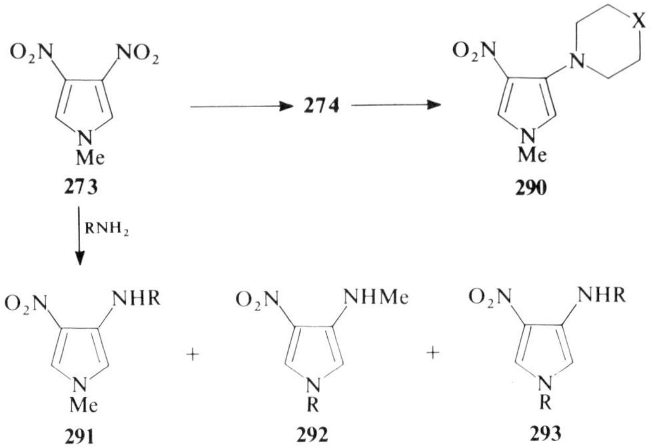

Scheme 74

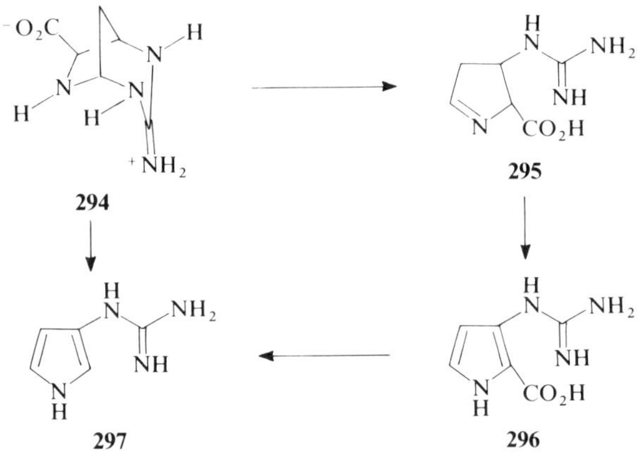

Scheme 75

acid-catalyzed ring opening. Oxidation of **295** to yield **296** and successive decarboxylation leads to pyrrole **297**.

3.2.3.4. Diaminopyrroles

Relatively few diaminopyrroles can be prepared by conversion of functional groups. 1,2-Diaminopyrrole **299** has been obtained from derivative **298** on *trans*-hydrazinolysis of the substituent at the 1-position with 2,4-dinitrophenylhydrazine in 30% perchloric acid[155] (Scheme 76).

2,3-Dimorpholino- and 2,3-dipiperidinopyrroles **302** (R = Me) can be obtained from **273** via the intermediate pyrroline **301**, which is dehydrogenated with chloranil[242] (Scheme 77). The 2,4-diamino derivative **302** (R = *tert*-butyl)

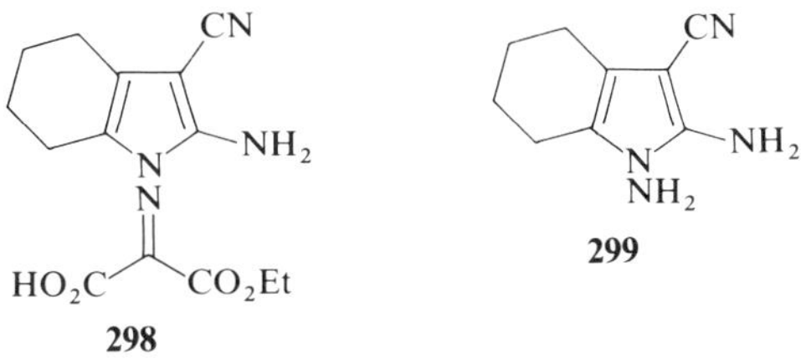

3.3. Structure and Physical Properties

Scheme 76

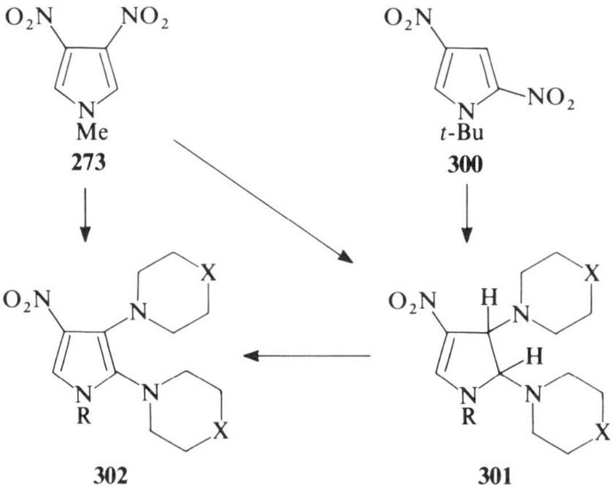

Scheme 77

has been obtained from 2,4-dinitro-1-*tert*-butylpyrrole **300** by an analogous route.[268]

Curtius rearrangement of the bis-azido derivative **303** in refluxing acetic acid produces 2,5-diaminopyrrole **304**[231] (Scheme 78).

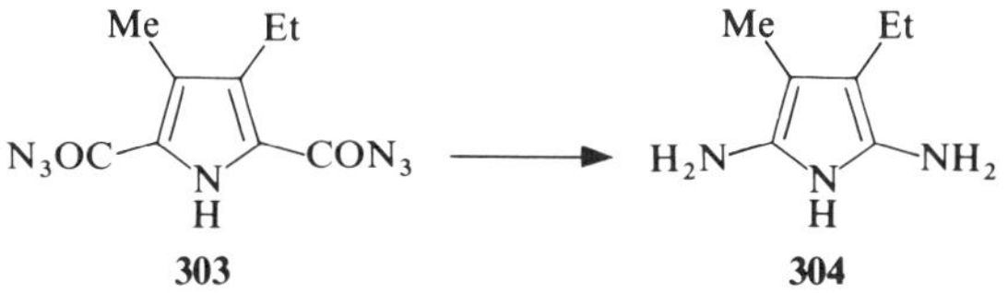

Scheme 78

3.3. STRUCTURE AND PHYSICAL PROPERTIES

All theoretical and spectroscopic studies on aminopyrroles have been directed to a clarification of their structure and have shown that they exist in the amino form, with the exception of the diamino derivatives, which adopt the diimino form.

The basicity of the aminopyrroles depends on the position of the amino group in the pyrrole ring. Thus N-aminopyrroles, in which interactions with the nucleus are virtually nonexistent, are less basic than the C-amino derivatives. Among the C-amino compounds, the 2-aminopyrroles are stronger bases. However, the derivatives undergo protonation on the pyrrole ring, in contrast with the 3-aminopyrroles, in which the amino group is the more basic site.

3.3.1. Theoretical Methods

Theoretical studies on the structure, tautomerism, and protonation of aminopyrroles have been reported.

The conformation, stabilities, and charge distributions in 2- and 3-aminopyrroles and 2- and 3-(N-fluoroamino)pyrroles have been studied using standard LCAO SCF (linear combination of atomic orbitals, self-consistent field) molecular orbital theory.[269] In each case the three distinct conformations, *cis*, *orthogonal*, and *trans*, together with several intermediate conformations, were considered. The calculated total energies for the preferred conformations and the relative conformational energies are shown in Table 3.1. The stabilizing

TABLE 3.1. CALCULATED TOTAL ENERGIES (HARTREES) AND STABILIZATION ENERGIES (ΔE_{SE}, kJ mol^{-1}) FOR PREFERRED CONFORMATIONS OF 2- AND 3-AMINOPYRROLES

Substituent	Total Energy		Stabilization Energy	
(-Z)	2-PyrZ	3-PyrZ	2-PyrZ	3-PyrZ
—NH$_2$ planar	− 260.53669	− 260.53314	− 13.00	− 22.32
—NH$_2$ optimized	− 260.54347	− 260.54034	− 6.59	− 14.81
—NH$_2$ pyramidal	− 260.54452	− 260.54028	− 2.42	− 13.54
—NHF planar	− 357.94247	− 357.94034	− 14.36	19.95
—NHF pyramidal	− 357.95950	− 357.95645	0.68	− 7.32

effects of substituents on the pyrrole ring, relative to their effect on the benzene ring, were analyzed in terms of stabilization energies, defined as energy changes (ΔE_{SE}) in the formal reaction shown in Scheme 79, as a measure of the stabilizing effect of the substituent Z on the pyrrole ring, compared with its effect

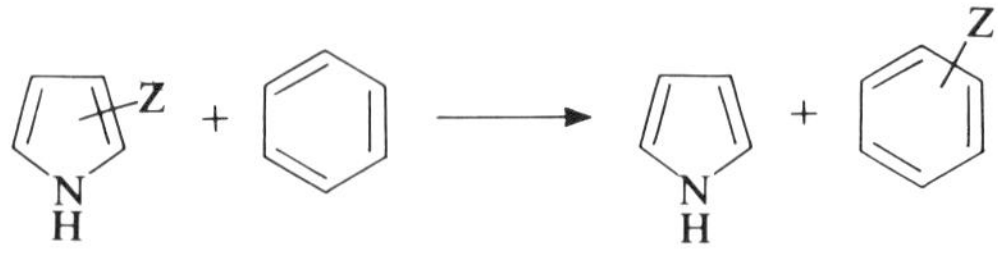

Scheme 79

TABLE 3.2. π-ELECTRON POPULATIONS IN 2- AND 3-AMINOPYRROLES

Substituent	2-Substituted Pyrrole Ring Atoms					3-Substituted Pyrrole Ring Atoms				
(-Z)	(1)	(2)	(3)	(4)	(5)	(1)	(2)	(3)	(4)	(5)
—H	1.647	1.087	1.090	1.090	1.087	1.647	1.087	1.090	1.090	1.087
—NH$_2$ planar	1.686	1.023	1.195	1.070	1.127	1.659	1.191	1.033	1.139	1.070
—NH$_2$ optimized	1.677	1.042	1.168	1.074	1.115	1.655	1.164	1.050	1.128	1.072

on benzene. Positive values for ΔE_{SE} imply that the substituent has a stabilizing effect in pyrrole relative to its effect in benzene and vice versa. The charge distributions (π-electron distributions and total σ and π charges, q_σ and q_π, donated to the ring by the substituent) and dipole moments were also determined. A negative value of q corresponds to donation of negative charge, i.e., electron donation, while a positive value of q corresponds to electron withdrawal by the substituent (Tables 3.2 and 3.3).

TABLE 3.3. DONATION OF σ- AND π-CHARGE BY AMINO GROUP FOR 2- AND 3-AMINOPYRROLES

Substituent	PhZ		2-PyrZ		3-PyrZ	
(-Z)	q_σ (B)	q_π (B)	q_σ (2-Pyr)	q_π (2-Pyr)	q_σ (3-Pyr)	q_π (3-Pyr)
—NH$_2$ planar	0.159	− 0.120	0.142	− 0.101	0.155	− 0.093
—NH$_2$ optimized	0.140	− 0.095	0.120	− 0.076	0.132	− 0.069

In the case of aminopyrroles, the amino substituent is taken as being planar trigonal, although experimental and theoretical data suggest that the amino group is nonplanar in the aniline. Consequently, pyramidal models, in which the bond angles at nitrogen are kept equal but varied from 120°, were also considered. The optimized value for out-of-plane angles (α) (cf. **305**) was 52.3° with an $\widehat{HNH}$ bond angle of 110.4°. The inversion barriers were calculated to be 17.8 and 18.9 kJ mol^{-1} for 2- and 3-aminopyrroles, respectively, which indicates a less effective interaction of the amino group lone pair with the pyrrole ring, compared with the benzene ring ($\alpha = 47.7°$ for the aniline), due the nonplanarity of the amino group (Scheme 80).

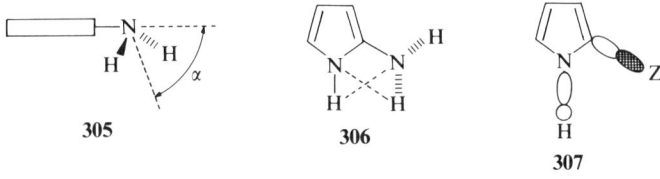

Scheme 80

Using the same pyramidal model (bond angles of 109.5° at nitrogen), the rotational potential function about the pyrrole–nitrogen bond was examined. Energy minima occur at $\widehat{NCN} = 60°$ for 2-aminopyrrole and at $\widehat{C(2)CN} = 93°$ for 3-aminopyrrole, suggesting that N—H····H—N electrostatic repulsion is probably important in 2-aminopyrrole.

The calculated stabilization energies for 2- and 3-aminopyrroles are negative, since the amino group is both a strong π-donor and a strong σ-acceptor. The value for the 2-aminopyrrole is less negative than that for 3-isomer, because compensating effects, such as intramolecular hydrogen bonding (**306**) and hyperconjugation (**307**), are important.

The π-electron distributions are compatible with the valence structures **308** and **309** (Scheme 81). Population analysis of both 2- and 3-aminopyrrole shows an increased π-electron density on the ring nitrogen, compared with the unsubstituted pyrrole. Thus, the electron-donating effect of the amino group appears to suppress the electron delocalization from the pyrrolyl nitrogen atom, and this effect is more pronounced at the 2-position.

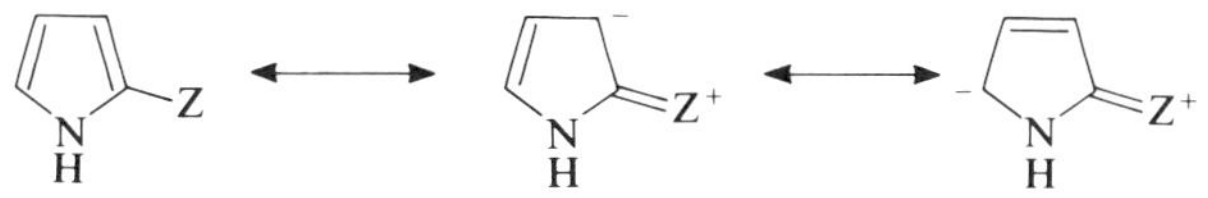

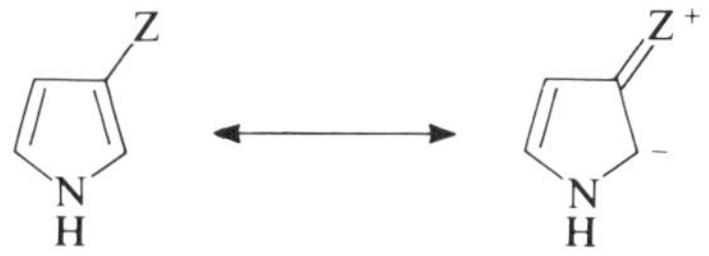

Scheme 81

N-Fluoroaminopyrroles were also studied with the standard and pyramidal ($\widehat{HNH}$ = 109.5°) geometries. Calculations using the planar model indicated *cis* conformations for both $\widehat{NCNF}$ and $\widehat{C(2)CNF}$. These findings are consistent with the dipolar interactions shown in **310** and **311**, but, for the 3-fluoroamino derivative, further stabilization can arise from 6π-electron cyclic conjugation (**312**) (Scheme 82). The pyramidal geometry was used to examine the rotation about the pyrrole–NHF bond. Energy minima were found at dihedral angles for $\widehat{NCNF}$ of 75° for the 2-(N-fluoroamino)pyrrole and 40° for the $\widehat{C(2)CNF}$ dihedral angle for the 3-isomer. These values are distorted from the expected orthogonal disposition of the nitrogen lone pair with respect to the ring.

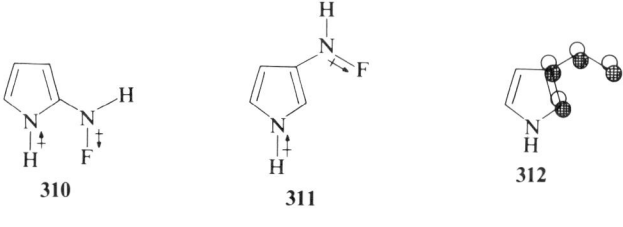

Scheme 82

The prototropic tautomerism in aminopyrroles has been studied[270] using a semiempirical SCF MO treatment for the conjugated molecule, using the

TABLE 3.4. HEATS OF ATOMIZATION OF TAUTOMERIC FORMS OF AMINOPYRROLE DERIVATIVES

Tautomer A	ΔH_a (A) eV	Tautomer B	ΔH_a (B) eV	Tautomer C	ΔH_a (C) eV	ΔH_a (A)–ΔH_a (B) or – ΔH_a (C) (eV)
313	52.854	313	52.372	313	51.900	0.482
						0.954
314	52.589	314	52.125			0.354
315	60.586	315	59.620			0.966
316	60.360	316	59.408			0.952

3.3. Structure and Physical Properties

Hückel approximation.[271] The calculated heats of atomization (ΔH_a) are reported in Table 3.4; positive quantities indicate that the amino tautomer (a) is more stable, while a negative value would imply that the imino form (b) is the more stable (Scheme 83). Data in the Table 3.4 show that the aminopyrroles are more stable than the corresponding tautomeric imines. The spectroscopic evidence supports these calculations, except in the case of 2,5-diaminopyrroles (see Chapter 3.3).

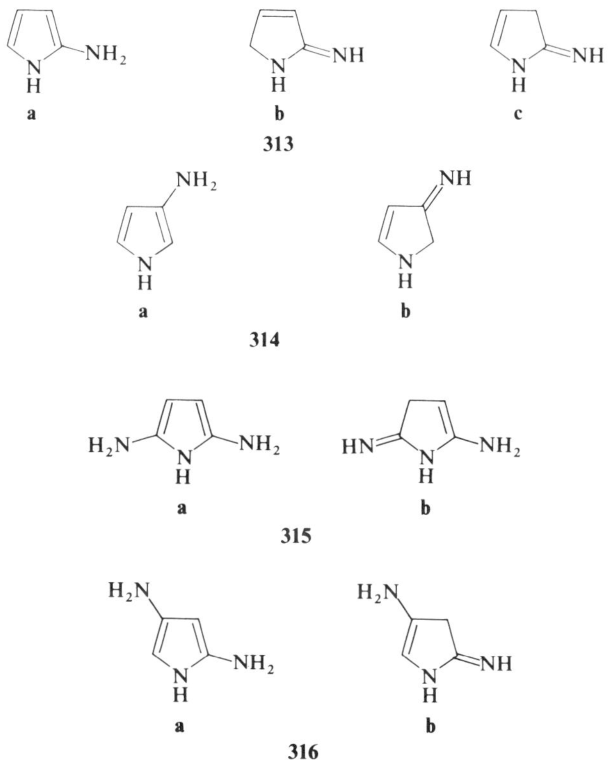

Scheme 83

Quantum-chemical studies of electrophilic attack by a proton on 2-amino and 3-aminopyrroles[272] considered protonation only at the exocyclic nitrogen atom, and proton affinity (PA) was calculated as the difference between the total energies of the protonated and unprotonated forms of the aminopyrroles. The total energies were calculated by nonempirical STO-3G methods; although this method gives high values for the PA, compared with experimental values, there is a good linear correlation between the two sets of values. In these calculations, the geometric parameters were optimized using the MNDO method, and it was assumed that the annular atoms and the nitrogen atom of the amino group are coplanar. As indicated earlier, the amino group is pyramidal in all cases and, in rotation of the amino group about the C—N bond, the hydrogen atoms can lie on either one side (ipsilateral) or on opposite sides (contralateral) of the pyrrole plane. The average lengths of the annular bonds in the protonated and unprotonated forms are shown in Scheme 84. The tendency for the bond lengths to become equal usually corresponds to an increase in the aromaticity of the heterocycle during protonation and can be rationalized by a better delocalization of the electron pair of the pyrrole nitrogen atom when the ring is attached to the σ-accepting ($^+NH_3$) group but not with the π-donating (NH_2) substituent. The increase in the lengths of the bonds α to the exocyclic nitrogen atom (for the C—NH bond, in particular) may result from an increase in the coulombic repulsion of like-charged atoms in the protonated forms.

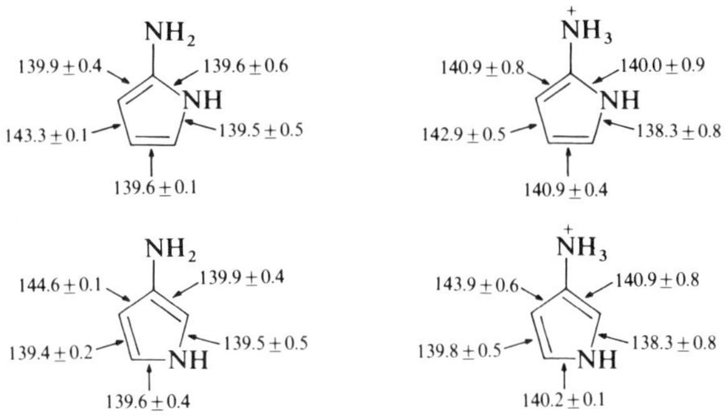

Scheme 84

The inductively induced positive charge on the annular atoms of the protonated aminopyrroles has the largest value at the atoms α to the $^+NH_3$ group, and the repulsion of these atoms leads to an increase in the bond angles at the carbon atom bearing the substituent while the other bond angles are decreased in size.

The proton affinities, calculated by MNDO and STO-3G methods, are -0.2428 AU (atomic unit) (MNDO) and -0.4170 AU (STO-3G) for 2-aminopyrrole, and -0.2674 AU (MNDO) and -0.4402 AU (STO-3G) for 3-aminopyrrole. Comparison of the values calculated by the two methods shows that there is a linear relation with a high correlation coefficient (Eq. 1):

$$PA_{STO-3G} = 1.195 PA_{MNDO} - 0.1200 \tag{1}$$

($r = 0.992$; $S = 0.00245$; $F = 937$). The position of the amino group has a significant effect on its proton affinity. An α-amino group has a lower proton affinity than when it is at the β-position, and the values are consistent with the σ constants for the α- and β-pyrrolyl substituent, where it has been confirmed by experimental data that the β-pyrrolyl group is a stronger electron donor than is the α-pyrrolyl group (see Section 3.3.10).

3.3. Structure and Physical Properties

3.3.2. X-Ray Diffraction

It is generally acknowledged that X-ray diffraction studies result in an unambiguous determination of structure, and they have had particular significance in the confirmation of the structure of 1-aminopyrroles (e.g., **317–320**),[126,133,145,273] especially where several authors had suggested the alternative pyridazine structure.

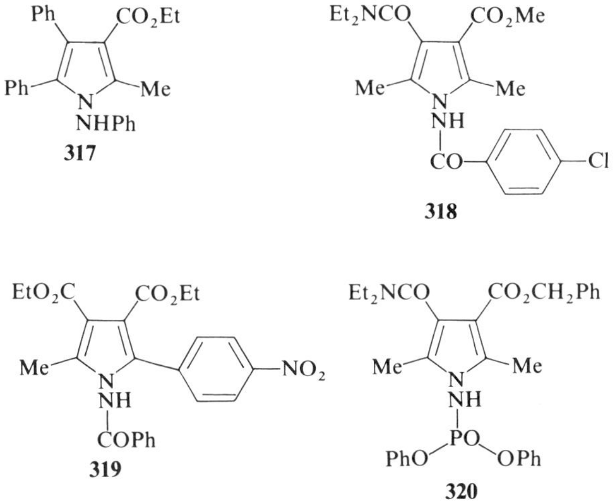

Scheme 85

Generally, the bond angles for the 1-aminopyrrole system indicate that the pyrrolyl nitrogen atom is sp^2-hybridized and that the ring is virtually a symmetric planar system. However, in the case of derivative **319**, the C(2)—C(3) and C(4)—C(5) bond distances of 1.32(2) Å and 1.36(2) Å, respectively, indicate a certain degree of double-bond fixation.[126]

The geometries of substituent groups are almost normal. In **317**, for example, the planes of the 3-phenyl group and the N-phenyl group are nearly orthogonal with the plane of the pyrrole ring, while the plane of the 2-phenyl group has a dihedral angle of 126°.[133] Additionally it has been found that the N-phenyl group is completely above the plane of the heterocycle.

In **318**, the geometries of the substituents at the 1- and 3-positions reveal their usual amidic character,[273] and the dihedral angle between the mean planes of the benzene ring and the pyrrole ring is 64.7(1)°. The crystal packing shows two intermolecular hydrogen bonds, involving the NH at the 1-position of one molecule with the oxygen of the 3-amido group of a second molecule leading to the formation of an unusual "eight-atom" pseudo–ring system.

The bond angles around the phosphorus atom of the phosphonoamino substituent of **320**, deviate little from the value typical for the tetrahedral arrangement expected of $d_\pi p_\pi$ bonds in analogous organophosphorus derivatives.[145] In the crystal packing, intramolecular hydrogen bonding analogous to that in **318** was observed.

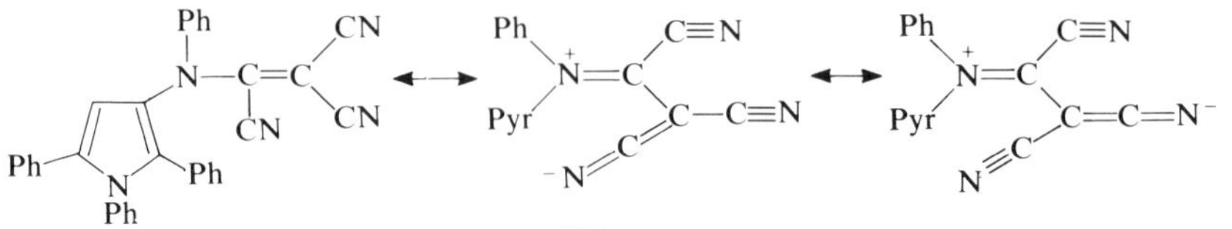

Scheme 86

3-(N-Phenyl-N-tricyanovinyl)amino-1,2,5-triphenylpyrrole (**321**) is the only C-aminopyrrole studied in detail by X-ray diffraction.[274] None of the pyrrole substituents are coplanar with the pyrrole ring, but there is near-coplanarity between the bond linking the ring with the 3-amino group and the tricyanovinyl moiety. Surprisingly, the pyrrole ring displays approximate C_{2v} symmetry, but there are deviations in the endocyclic angles about the β-carbon atoms, only one of which is substituted, and in the exocyclic angles about the α-carbon atoms. In the latter case, these deviations may reflect the different orientations of the respective phenyl substituents. With regard to the effect of substituents, the lack of any significant conjugation between pyrrole ring and the substituents is substantiated by the minor deviations from C_{2v} symmetry for the pyrrole ring, compared with large deviations reported for highly conjugated pyrrole derivatives. The N-phenyl substituent shows a net positive inductive effect, as can be deduced by the endocyclic angle [121.9(3)°], whereas for the other two phenyl groups the endocyclic angles value are 118.5(3)° and 117.8(3)°, compared with that for biphenyls (117.6°), indicating that there is little interaction between the pyrrole ring and the C-phenyl substituents. The electron-withdrawing β-substituent produces an increase in the internal angle at the 3-position, but has little effect on the other parameters of the ring. Of course, the geometry of the amino nitrogen is also sensitive to the substituent interactions. The conformation of the 3-substituent and the C—N and C—C distances reflect the importance of the dipolar canonical forms shown in Scheme 86.

Probably because of the difficulties in obtaining the compounds in a suitable crystalline form, X-ray diffraction data are available for only a few other C-aminopyrroles: **322**,[193] **323**,[211] and **324**.[220]

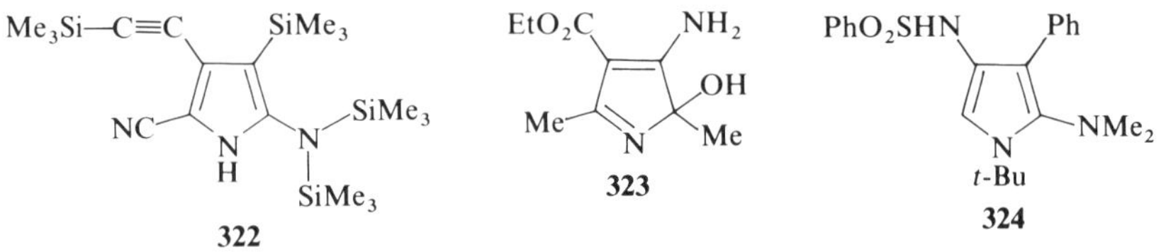

Scheme 87

3.3.3. ¹H NMR Spectra

The ¹H NMR chemical shift data available for 1-aminopyrroles are listed in Table 3.5.

The signal attributable to the amino protons of 1-aminopyrrole is to be found at 4.50 ppm,[67] and C-substituted 1-aminopyrroles have corresponding signals in the range 4.1–4.5 ppm, although, in the case of 1-amino-2,5-diphenylpyrrole, the signal is shifted downfield to 5.60 ppm as a result of the deshielding effect of the phenyl groups. Similarly, 1-phenylaminopyrrole derivatives have signals due to the amino proton in the range 5.9–6.9 ppm, and the introduction of a nitro group in the phenyl ring further deshields the signals. Thus, for 4-nitrophenyl derivatives, the signals appear at 6.6–8.0 ppm and are further deshielded to 10.1–10.2 ppm when the pyrrole ring is substituted at the 3-position by an amido group. The NH signal for the 2-nitrophenyl derivative is found at 9.95 ppm probably as a result of intramolecular hydrogen bonding. In contrast, 1-(4-chlorophenyl)aminopyrroles show NH signals in the range 6.7–7.0 ppm.

N-Heterocyclic derivatives, such as 2-pyridylamino- and 2-pyrimidylamino-pyrroles, have NH signals at 11.0–11.4 and 10.5–11.0 ppm, respectively, and the NH signals for the 2-benzothiazole derivatives overlap the aromatic protons in the range 6.7–8.0 ppm.

The introduction of an acyl function on the 1-amino group deshields the protons, and the signals appear in a range typical of the amido function; for example, the range for 1-benzylacylamino-, 1-benzoylamino-, and substituted 1-benzoylaminopyrroles is 9.30–12.4 ppm, with lower field values found for compounds in which electron-withdrawing groups are also present at the β-position of the pyrrole ring. Similarly, the signals for 1-alkoxycarbonylamino-pyrroles are to be found near 10.6–10.7 ppm, but move to 8.0–9.0 ppm when the pyrrole ring has a sulfonyl substituent in the 3-position.

1-Ureidopyrroles show two signals at 9.0–9.6 ppm and at 5.9–6.4 ppm assignable to the NH and NH_2 groups, respectively. 1-Phenylureidopyrroles generally show two distinct signals for the two NH protons in the ranges 9.2–9.6 ppm and 9.5–9.6 ppm, although, in the presence of sulfone substituents, the signals are broadened and overlap.

N-Phosphinic-1-amino- and N-phosphonic-1-aminopyrroles show the expected doublets for the NH protons in the range 9.5–9.7 ppm and 9.0–10.30 ppm, respectively, and 1-arylsulfonamidopyrroles generally exhibit signals between 11.6 and 11.7 ppm.

The ¹H NMR resonances for the amino protons of 2-aminopyrroles (Table 3.6) appear over a wide range (3.5–7.6 ppm) and are usually broad, demonstrating an increased interaction with the aromatic nucleus. The chemical shifts of the signal for the annular NH lie in the range 7.5–10.9 ppm.

In the case of 2-aminopyrrol-4-one derivatives, the range of the NH_2 signals is shifted downfield to 7.3–8.0 ppm.

The introduction of a substituent at the 1-position shifts the signals due to the amino proton upfield, narrowing the range (3.6–5.1 ppm). This effect, which is

TABLE 3.5. ^{1}H CHEMICAL SHIFTS OF THE 1-AMINOPYRROLES^a

R^2	R^3	R^4	R^5	ppm, mult., nH	Solvent^b	Ref.
				1-NH_2		
H	H	H	H	4.50,s,2; 5.89,t,2; 6.40,t,2	a	67
Me	H	H	Ph	2.25,d,3; 4.22,s,2; 5.82,dd,1; 6.1,d,1; 7.3,m,5	b	116
Ph	H	H	Ph	5.60,bs,2; 6.23,s,2; 7.50,m,10	c	66
Ph	Me	Me	Ph	1.97,s,3; 4.13,s,2; 7.1,m,5; 7.4,m,5;	d	107
Ph	Ph	H	Ph	4.46,s,2; 6.50,s,1; 7.1–7.7,m,15	d	107
Ph	Ph	Ph	Ph	4.4,s,2; 7.03,m,10; 7.10,m,10	d,e	107
Ph	Br	Br	Ph	4.18,s,2; 7.37,m,10;	f	66
				1-NHR [$R = C(Me)=CHCOOEt$]		
H	H	H	H	1.2,t,3; 1.66,s,3; 4.05,q,2; 4.63,s,1; 6.00,m,2; 6.52,m,2; 10.1,bs,1	d	292
				1-NHR ($R = Ph$)		
CH_2Ph	Ph	Ac	Me	1.95,s,3; 2.45,s,3; 6.1,s,1; 6.3–7.4,m,15; 7.60,s,2	d	133
CH_2Ph	Ph	COPh	Me	2.30,s,3; 3.80,s,2; 6.0,s,1; 6.10–7.85,m,20	d	134
CH_2Ph	Ph	$CONH_2$	Me	2.55,s,3; 3.6,s,2; 5.26,bs,2; 5.95–7.6,m,15; 6.0,s,1	d	137
CH_2Ph	Ph	CONHPh	Me	2.55,s,3; 3.67,s,2; 6.0–7.7,m,21; 6.05,s,1	d	137
CH_2Ph	Ph	4-MeOC$_6$H$_4$NHCO	Me	2.55,s,3; 3.73,s,5; 6.1–7.65,m,20; 6.15,s,1	d	137

CH$_2$Ph	Ph	CONEt$_2$	Me	0.5–1.3,m,6; 2.13,s,3; 2.5–3.7,m,4; 3.87,s,2; 6.1–7.6,m,15; 6.16,s,1	d	137
CH$_2$Ph	Ph	COOMe	Me	2.45,s,3; 3.60,s,5; 5.90,s,1; 6.25–7.45,m,15	d	134
CH$_2$Ph	Ph	COOEt	Me	1.05,t,3; 2.50,s,3; 3.65,s,2; 4.10,q,2; 5.95,s,1; 6.30–7.50,m,15	d	134
Ph	Ph	Ac	Me	1.95,s,3; 2.45,s,3; 6.35–7.35,m,15; 6.8,s,1	d	133
Ph	Ph	—CO(CH$_2$)$_3$—		1.95–2.95,m,6; 6.35; 7.45,m,15; 6.65,s,1	d	133
Ph	Ph	COPh	Me	2.35,s,3; 6.35–7.90,m,20; 6.85,s,1	d	134
Ph	Ph	COPh	Ph	6.30–8.15,m	d	134
Ph	Ph	CONH$_2$	Me	2.55,s,3; 5.26,bs,2; 6.3–7.4,m,15; 6.70,s,1	d	137
Ph	Ph	CONHPh	Me	2.55,s,3; 6.3–7.55,m,21; 6.7,s,1	d	137
Ph	Ph	4-MeOC$_6$H$_4$NHCO	Me	2.55,s,3; 3.75,s,3; 6.3–7.4,m,20; 6.73,s,1	d	137
Ph	Ph	4-ClC$_6$H$_4$NHCO	Me	2.55,s,3; 6.3–7.7,m,20; 6.76,s,1	d	137
Ph	Ph	CONEt$_2$	Me	0.5–1.4,m,6; 2.1,s,3; 2.5–4.0,m,4; 6.4–7.6,m,15; 6.87,s,1	d	137
Ph	Ph	COOMe	Me	2.50,s,3; 3.60,s,3; 6.25–7.35,m,15; 6.60,s,1	d	134
Ph	Ph	COOEt	Me	1,t,3; 2.5,s,3; 4,q,2; 6.3–7.3,m,15; 6.6,s,1	d	133

1-NHR (R = 2-NO$_2$C$_6$H$_4$)

H	H	H	H	6.25,m,2; 6.8,m,2; 7.2–8.4,m,4; 9.95,bs,1	d	112

1-NHR (R = 4-NO$_2$C$_6$H$_4$)

—(CH$_2$)$_4$—		COPh	Me	1.5–2.55,m,8; 2.0,s,3; 6.55,d,2; 7.25–8.10,m,5; 8.00,s,1; 8.20,d,2	d	134
—(CH$_2$)$_4$—		COPh	Ph	1.5–2.85,m,8; 6.50,d,2; 7.0–7.7,m,10; 7.40,s,1; 8.10,d,2	d	134
—(CH$_2$)$_4$—		CONHPh	Me	1.45–2.9,m,8; 2.3,s,3; 6.35–8.4,m,5; 6.57,d,2; 8.24,d,2; 9.3,s,1; 10.1,s,1	c	137
—CH$_2$)$_4$—		4-MeOC$_6$H$_4$NHCO	Me	1.4–3.0,m,8; 2.25,s,3; 3.78,s,3; 6.52,d,2; 6.88,d,2; 7.65,d,2; 8.18,d,2; 9.1,s,1; 10.1,s,1	c	137

TABLE 3.5. Continued

R^2	R^3	R^4	R^5	ppm, mult., nH	Solvent[b]	Ref.
—(CH$_2$)$_4$—		4-ClC$_6$H$_4$NHCO	Me	1.3–3.0,m,8; 2.27,s,3; 6.55,d,2; 7.35,d,2; 7.78,d,2; 8.18,d,2; 9.43,s,1; 10.1,s,1	c	137
—(CH$_2$)$_4$—		CONEt$_2$	Me	0.6–2.8,m,14; 1.65,s,3; 2.9–4.0,m,4; 6.5,d,2; 8.1,d,2; 10.2,s,1	d	137
—(CH$_2$)$_4$—		COOMe	Me	1.6–3.0,m,8; 2.40,s,3; 3.80,s,3; 6.50,d,2; 7.30,s,1; 8.20,d,2	d	134
—(CH$_2$)$_4$—		COOEt	Me	1.4,t,3; 1.6–3.0,m,8; 2.4,s,3; 4.3,q,2; 6.5,d,2; 7.9,s,1; 8.1,d,2	d	133

1-NHR (R = 4-ClC$_6$H$_4$)

R^2	R^3	R^4	R^5	ppm, mult., nH	Solvent[b]	Ref.
	Ph	CONH$_2$	Me	2.55,s,3; 5.26,bs,2; 6.3–7.3,m,14; 6.77,s,1	d	137
	Ph	CONHPh	Me	2.55,s,3; 6.3–7.8,m,19; 6.83,s,1; 6.84,s,1	d	137
	Ph	4-MeOC$_6$H$_4$NHCO	Me	2.55,s,3; 3.73,s,3; 6.3–7.5,m,19; 6.67,s,1	d	137
	Ph	4-ClC$_6$H$_4$NHCO	Me	2.55,s,3; 6.3–7.8,m,18; 6.73,s,1; 6.98,s,1	d	137

1-NHR (R = 2-Cl-4-OHC$_6$H$_3$)

R^2	R^3	R^4	R^5	ppm, mult., nH	Solvent[b]	Ref.
H	H	H	H	4.50,bs,1; 6.11,d,1; 6.15,t,2; 6.53,dd,1; 6.71,t,2; 6.79,d,1; 6.87,bs,1	d	51

1-NHR (R = 2-Cl-4-OAcC$_6$H$_3$)

R^2	R^3	R^4	R^5	ppm, mult., nH	Solvent[b]	Ref.
H	H	H	H	2.23,s,3; 6.10,d,1; 6.13,t,2; 6.65,t,2; 6.75,dd,1; 7.00,bs,1; 7.04,d,1	d	51

1-NHR (R = 3-NO$_2$-2-Pyridyl)

R^2	R^3	R^4	R^5	ppm, mult., nH	Solvent[b]	Ref.
Me	COOMe	Ac	Me	2.1,s,3; 2.24,s,3; 2.3,s,3; 3.75,s,3; 6.97–7.3,m,1; 8.37–8.77,m,2; 11.03,bs,1	c	141

Me	COOMe	COPh	Me	2.05,s,3; 2.3,s,3; 3.23,s,3; 7.0–7.33,m,1; 8.43–8.8,m,2; 7.33–7.9,m,5; 11.1,bs,1	c	141
Me	COOMe	COPh	Ph	2.38,s,3; 3.4,s,3; 6.83–7.93 and 8.3–8.63,m,13; 11.17,bs,1	c	141
Me	COOMe	CONHPh	Me	2.17,t,3; 2.30,s,3; 3.73,s,3; 7.00–8.00,m,6; 8.35–8.83,m,2; 10.43,s,1; 11.1,s,1	c	142
Me	COOMe	4-MeOC$_6$H$_4$NHCO	Me	2.17,t,3; 2.30,s,3; 3.75,s,6; 6.73–7.87,m,5; 8.35–8.80,m,2; 10.28,s,1; 11.07,s,1	c	142
Me	COOMe	COOMe	Me	2.13,s,6; 3.74,s,6; 6.95–7.3,m,1; 8.33–8.73,m,2; 11.03,bs,1	c	141
Me	COOMe	COOEt	Me	1.23,t,3; 2.15,s,6; 3.73,s,3; 4.17,q,2; 6.93–7.29,m,1; 8.35–8.78,m,2; 11.03,bs,1	c	141
Me	COOMe	COOEt	Ph	1.13,t,3; 2.28,s,3; 3.77,s,3; 4.1,q,2: 6.8–7.63 and 8.17–8.67,m,8; 11.13,bs,1	c	141
Me	COOCH$_2$Ph	CONHPh	Me	2.08,s,3; 2.25,s,3; 5.20,s,2; 6.93–7.93,11; 8.35–8.80,m,2; 10.36,s,1; 11.05,s,1	c	142
Me	COOCH$_2$Ph	4-ClC$_6$H$_4$NHCO	Me	2.1,s,3; 2.27,s,3; 5.24,s,2; 7.00–8.00,m,10; 8.40–8.85,m,2; 10.51,s,1; 11.12,s,1	c	142
4-NO$_2$C$_6$H$_4$	COOEt	COOMe	Me	1.15,t,3; 2.3,s,3; 3.8,s,3; 4.15,q,2: 6.9–7.27,m,1: 7.97,q,2: 8.37–8.77,m,2; 11.33,bs,1	c	141
4-NO$_2$C$_6$H$_4$	COOEt	COOEt	4-NO$_2$C$_6$H$_4$	1.17,t,6; 4.13,q,4; 6.78–7.1,m,1; 7.6–8.5,m,10; 11.43,b,1	c	141

1-NHR (R = 2-Pyrimidyl)

Me	COOMe	COOMe	Me	2.13,s,6; 3.74,s,6; 6.98,t,1; 8.51,d,2; 10.5,bs,1	c	141
Me	COOEt	CONHPh	Me	1.10,t,3; 2.13,s,3; 2.32,s,3; 4.13,q,2; 6.80–8.0,m,6; 8.52,d,2; 10.33,s,1; 10.55,s,1	c	142
Me	COOEt	CONHPh	Ph	1.32,t,3; 2.43,s,3; 4.35,q,2; 6.80,t,1; 7.30,s,10; 8.37,d,2; 9.97,s,1; 11.03,s,1	c	142
Me	COOEt	4-MeCO$_6$H$_4$NHCO	Me	1.10,t,3; 2.10,s,3; 2.28,s,3; 3.73,s,3; 4.13,q,2; 6.75–7.87,m,5; 8.48,d,2; 10.12,s,1; 10.45,s,1	c	142

TABLE 3.5. Continued

R^2	R^3	R^4	R^5	ppm, mult, nH	Solvent[b]	Ref.
Me	COOEt	4-ClC$_6$H$_4$NHCO	Me	1.08,t,3; 2.12,s,3; 2.30,s,3; 4.13,q,2; 6.98,t,1; 7.58,q,4; 8.50,d,2; 10.44,s,1; 10.55,s,1	c	142
Me	COOCH$_2$Ph	CONHPh	Me	2.10,s,3; 2.18,s,3; 5.20,s,2; 6.80–7.80,m,11; 8.50,d,2; 10.35,s,1; 10.57,s,1	c	142
Me	COOCH$_2$Ph	CONHPh	Ph	2.35,s,3; 5.20,s,2; 6.67–8.00,m,16; 8.60,d,2; 10.20,s,1; 10.58,s,1	c	142
Me	COOCH$_2$Ph	COOMe	Me	2.15,s,3; 2.2,s,3; 3.7,s,3; 5.23,s,2; 6.95,t,1; 7.45,s,5; 8.5,d,2; 10.52,bs,1	c	141
Me	COOCH$_2$Ph	COOCH$_2$Ph	Me	2.17,s,6; 5.17,s,4; 6.97,t,1; 7.42,s,10; 8.48,d,2; 10.52,bs,1	c	141
1-NHR (R = 2-Benzothiazolyl)						
Me	COOMe	CONHPh	Me	2.23,s,3; 2.40,s,3; 3.72,s,3; 7.00–8.00,m,10; 10.33,s,1	c	142
Me	COOMe	4-MeCO$_6$H$_4$NHCO	Me	2.22,s,3; 2.38,s,3; 3.75,s,6; 6.73–8.00,m,9; 10.20,s,1	c	142
Me	COOMe	COOMe	Me	2.3,s,6; 3.79,s,6; 6.93–7.97,m,5	c	141
1-NHR (R = Ac)						
H	H	H	H	1.83 and 2.02,s,3; 6.20,m,2; 6.67,m,2	d	67
1-NHR (R = COCH$_2$Ph)						
Me	COOMe	4-ClC$_6$H$_4$NHCO	Me	2.12,s,3; 2.39,s,3; 3.67,s,2; 3.71,s,3; 6.9–8.2,m,9; 10.28,bs,1; 11.74,bs,1	c	143
Me	COOMe	COOEt	4-NO$_2$C$_6$H$_4$	1.10,t,3; 2.32,s,3; 3.48,s,2; 3.80,s,3; 4.13,q,2; 7.27,s,5; 7.55,d,2; 8.19,d,2; 11.52,bs,1	c	126
Me	COOEt	CONHPh	Ph	1.17,t,3; 2.39,s,3; 3.67,s,2; 4.18,q,2; 6.9–8.0,m,15; 10.28,bs,1; 11.74,bs,1	c	143

1-NHR (R = COPh)

Me	COOMe	Me	1.90,s,3; 2.16,s,3; 2.26,s,3; 3.73,s,3; 7.30–8.03,m,5; 10.06,s,1	d	132
Me	COOMe	Ac	2.15,s,3; 2.28,s,3; 2.35,s,3; 3.77,s,3; 7.60–8.13,m,5; 11.82,bs,1	c	126
Me	COOMe	CONH$_2$	2.12,s,3; 2.39,s,3; 3.71,s,3; 7.12,bs,2; 7.3–8.3,m,5; 11.74,bs,1	c	143
Me	COOMe	CONHPh	2.12,s,3; 2.39,s,3; 3.71,s,3; 6.8–8.3,m,10; 10.28,bs,1; 11.74,bs,1	c	143
Me	COOMe	CONHPh	2.39,s,3; 3.71,s,3; 6.9–8.0,m,15; 10.28,bs,1; 11.74,bs,1	c	143
Me	COOMe	4-MeOC$_6$H$_4$NHCO	2.12,s,3; 2.39,s,3; 3.71,s,3; 3.78,s,3; 6.9–8.2,m,9; 10.28,bs,1; 11.74,bs,1	c	143
Me	COOMe	4-ClC$_6$H$_4$NHCO	2.12,s,3; 2.39,s,3; 3.71,s,3; 6.9–8.2,m,9; 10.28,bs,1; 11.74,bs,1	c	143
Me	COOMe	CONEt$_2$	0.7–1.4,m,6; 2.12,s,3; 2.39,s,3; 2.9–3.6,m,4; 3.71,s,3; 7.3–8.3,m,5; 11.74,bs,1	c	143
Me	COOMe	4-MeC$_6$H$_4$SO$_2$	2.23,s,3; 2.42,s,3; 2.52,s,3; 3.70,s3; 7.38–8.17,m,9; 12.38,bs,1	c	147
Me	COOEt	Me	1.26,t,3; 1.96,s,3; 2.10,s,3; 2.23,s,3; 3.86–4.30,q,2; 7.07–7.76,m,5; 9.30,s,1	d	132
Me	COOEt	COPh	0.82,t,3; 2.48,s,3; 3.89,q,2; 7.35,s,5; 7.47–8.03,m,10; 11.93,bs,1	c	126
Me	COOEt	CONH$_2$	1.17,t,3; 2.12,s,3; 2.39,s,3; 4.18,q,2; 7.12,bs,2; 7.3–8.3,m,5; 11.74,bs,1	c	143
Me	COOEt	CONHPh	1.17,t,3; 2.39,s,3; 4.18,q,2; 6.9–8.0,m,15; 10.28,bs,1; 11.74,bs,1	c	143
Me	COOEt	COOMe	1.25,t,3; 2.23,s,6; 3.73,s,3; 4.21,q,2; 7.58–8.13,m,5; 11.78,bs,1	c	126
Me	COOEt	COOEt	1.12,t,3; 1.27,t,3; 2.37,s,3; 3.93–4.42,m,4; 7.43,s,5; 7.50–7.93,m,5; 11.82,bs,1	c	126
Me	COOEt	4-NO$_2$C$_6$H$_4$	1.20,t,3; 1.33,t,3; 2.47,s,3; 4.10–4.57,m,4; 7.72–8.65,m,9; 12.37,bs,1	c	126
Ph	Ph	H	6.38,d,1; 6.65,d,1; 6.9–7.7,m,15; 9.31,s,1	d	122

TABLE 3.5. Continued

R^2	R^3	R^4	R^5	ppm. mult., nH	Solvent[b]	Ref.
\multicolumn{7}{c}{*1-NHR (R = 3-MeC$_6$H$_4$CO)*}		
Me	COOMe	COPh	Me	2.08,s,3; 2.35,s,3; 2.43,s,3; 3.22,s,3; 7.45–7.85,m,9; 11.77,bs,1	c	126
Me	COOMe	COPh	Ph	2.37,s,3; 2.47,s,3; 3.40,s,3; 7.37,s,5; 7.47–7.87,m,9; 11.93,bs,1	c	126
Me	COOMe	CONHPh	Ph	2.35,s,3; 2.39,s,3; 3.71,s,3; 6.9–8.0,m,14; 10.28,bs,1; 11.74,bs,1	c	143
Me	COOMe	CONEt$_2$	Me	0.7–1.4,m,6; 2.12,s,3; 2.35,s,3; 2.39,s,3; 2.9–3.6,m,4; 3.71,s,3; 7.4–8.2,m,4; 11.74,bs,1	c	143
Me	COOEt	COOEt	Me	1.27,t,6; 2.25,s,6; 2.43,s,3; 4.22,q,4; 7.47–7.98,m,4; 11.89,bs,1	c	126
Me	COOEt	4-MeC$_6$H$_4$SO$_2$	Me	1.13,t,3; 2.32,s,3; 2.42,s,3; 3.47,s,6; 3.87–4.61,m,2; 7.42–8.02,m,8; 10.93,bs,1	c	147
\multicolumn{7}{c}{*1-NHR (R = 3-ClC$_6$H$_4$CO)*}		
Me	COOMe	COPh	Me	2.10,s,3; 2.37,s,3; 3.23,s,3; 7.47–8.10,m,9; 11.98,bs,1	c	126
Me	COOMe	CONHPh	Ph	2.39,s,3; 3.71,s,3; 6.9–8.0,m,14; 10.28,bs,1; 11.74,bs,1	c	143
Me	COOMe	CONEt$_2$	Me	0.7–1.4,m,6; 2.12,s,3; 2.39,s,3; 2.9–3.6,m,4; 3.71,s,3; 7.4–8.2,m,4; 11.74,bs,1	c	143
Me	COOMe	SO$_2$Ph	Ph	2.28,s,3; 3.73,s,3; 7.45,s,5; 7.50–7.95,m,9; 12.01,bs,1	c	147
Me	COOMe	4-MeC$_6$H$_4$SO$_2$	Me	2.23,s,3; 2.42,s,3; 2.50,s,3; 3.68,s,3; 7.40–8.10,m,8; 12.10,bs,1	c	147
Me	COOEt	CONH$_2$	Me	1.17,t,3; 2.12,s,3; 2.39,s,3; 4.18,q,2; 7.12,bs,2; 7.4–8.2,m,4; 11.74, bs,1	c	143
Me	COOEt	CONHPh	Ph	1.17,t,3; 2.39,s,3; 4.18,q,2; 6.9–8.0,m,14; 10.28,bs,1; 11.74,bs,1	c	143
Me	COOEt	SO$_2$Ph	Ph	1.23,t,3; 2.37,s,3; 4.35,q,2; 7.67,s,5; 7.75–8.20,m,9; 12.33,bs,1	c	147
Me	COOEt	4-MeC$_6$H$_4$SO$_2$	Me	1.18,t,3; 2.25,s,3; 2.42,s,3; 2.50,s,3; 4.20,q,2; 7.38–8.12,m,8; 12.13,bs,1	c	147

1-NHR (R = CONH₂)

			1-NHR (R = CONH₂)		
Me	H	COPh	COOMe	2.20,d,3; 3.30,s,3; 6.15,q,1; 6.30,s,2; 7.4–7.8,m,5; 9.00,s,1	c	117
H	COOH	H	Me	2.02,d,3; 6.13,m,3; 7.16,d,1; 9.27,s,1; 11.65,s,1	c	118
Me	H	COOH	COOEt	1.22,t,3; 2.04,d,3; 4.20,q,2; 6.22,bs,3; 9.10,s,1	c	118
H	COOEt	H	Me	1.22,t,3; 2.01,d,3; 4.15,q,2; 6.15,m,3; 7.20,d,1; 9.31,s,1	c	118
Me	H	COOEt	COOH	1.26,t,3; 2.08,d,3; 4.25,q,2; 6.27,bs,3; 9.20,s,1	c	118
Me	H	COOEt	COOEt	1.20,t,6; 2.00,d,3; 4.15,q,4; 6.20,m,3; 9.10,s,1	c,g	117
Me	Ac	H	Me	2.02,d,3; 2.22,s,3; 2.25,s,3; 6.10,m,3; 9.12,s,1	b	120
Me	Ac	H	Ph	2.39,s,3; 2.42,s,3; 6.30,s,2; 6.78,s,1; 7.3–7.8,m,5; 9.47,s,1	b	120
Me	COPh	H	Me	2.10,d,3; 2.36,s,3; 6.05,q,1; 6.35,s,2; 7.5–8.0,m,5; 9.32,s,1	b	120
Me	COPh	H	Ph	2.45,s,3; 6.38,s,2; 6.49,s,1; 7.3–8.1,m,10; 9.56,s,1	b	120
Me	COOMe	Ac	Me	2.13,s,3; 2.27,s,3; 2.34,s,3; 3.73,s,3; 6.41,bs,2; 9.38,bs,1	c	138
Me	COOMe	COPh	Me	2.13,s,3; 2.34,s,3; 3.2,s,3; 6.41,bs,2; 7.3–7.5,m,5; 9.38,bs,1	c	138
Me	COOMe	CONH₂	Me	2.14,s,3; 2.35,s,3; 3.67,s,3; 6.28,bs,2; 6.96,bs,1; 7.78,bs,1; 9.30,bs,1	c	129
Me	COOMe	CONHPh	Me	2.14,s,3; 2.35,s,3; 3.67,s,3; 6.28,bs,2; 6.8–7.9,m,5; 9.30,bs,1; 10.26,bs,1	c	129
Me	COOMe	CONHPh	Ph	2.35,s,3; 3.67,s,3; 6.28,bs,2; 6.8–8.0,m,10; 9.30,bs,1; 10.26,bs,1	c	129
Me	COOMe	4-MeOC₆H₄NHCO	Me	2.14,s,3; 2.35,s,3; 3.67,s,3; 3.73,s,3; 6.28,bs,2; 6.9,d,2; 7.6,d,2; 9.30,bs,1; 10.26,bs,1	c	129
Me	COOMe	4-ClC₆H₄NHCO	Me	2.14,s,3; 2.35,s,3; 3.67,s,3; 6.28,bs,2; 7.4,d,2; 7.8,d,2; 9.30,bs,1; 10.26,bs,1	c	129
Me	COOMe	CONEt₂	Me	0.6–1.3,m,6; 2.14,s,3; 2.35,s,3; 2.7–3.5,m,4; 3.67,s,3; 6.28,bs,2; 9.30,bs,1	c	129
Me	COOMe	COOMe	Me	2.22,s,6; 3.73,s,6; 6.41,bs,2; 9.38,bs,1	c	138

TABLE 3.5. Continued

R²	R³	R⁴	R⁵	ppm. mult., nH	Solvent[b]	Ref.
Me	COOMe	COOMe	CH₂COOMe	2.2,s,3; 3.65,s,3; 3.72,s,3; 3.75,s,3; 3.8,s,2; 6.41,bs,2; 9.38,bs,1	c	138
Me	COOMe	COOEt	Me	1.24,t,3; 2.22,s,6; 3.73,s,3; 4.2,q,2; 6.41,bs,2; 9.38,bs,1	c	138
Me	COOMe	SO₂Ph	Ph	2.23,s,3; 3.73,s,3; 6.23,bs,2; 7.3–7.83,m,10; 9.2,bs,1	c	131
Me	COOMe	4-MeC₆H₄SO₂	Ph	2.2,s,3; 2.37,s,3; 3.7,s,3; 6.23,bs,2; 7.26–7.73,m,9; 9.2,bs,1	c	131
Me	COOEt	H	Me	1.24,t,3; 2.0,d,3; 2.30,s,3; 4.15,q,2; 6.12,q,1; 6.20,s,2; 9.15,s,1	b	120
Me	COOEt	H	Ph	1.25,t,3; 2.35,s,3; 4.21,q,2; 6.20,s,2; 6.52,s,1; 7.2–7.75,m,5; 9.37,s,1	b	120
Me	COOEt	Ac	Me	1.24,t,3; 2.13,s,3; 2.27,s,3; 2.34,s,3; 4.2,q,2; 6.41,bs,2; 9.38,bs,1	c	138
Me	COOEt	COPh	Me	0.7,t,3; 2.13,s,3; 2.34,s,3; 3.67,q,2; 7.3–7.8,m,5; 6.41,bs,2; 9.38,bs,1	c	138
Me	COOEt	CONH₂	Me	1.09,t,3; 2.14,s,3; 2.35,s,3; 4.11,q,2; 6.28,bs,2; 7.0,bs,1; 7.83,bs,1; 9.30,bs,1	c	129
Me	COOEt	CONHPh	Ph	1.09,t,3; 2.35,s,3; 4.11,q,2; 6.28,bs,2; 6.83–7.83,m,10; 9.30,bs,1; 10.26,bs,1	c	129
Me	COOET	4-ClC₆H₄NHCO	Me	1.09,t,3; 2.14,s,3; 2.35,s,3; 4.11,q,2; 6.28,bs,2; 7.37,d,2; 7.8,d,2; 9.30,bs,1; 10.26,bs,1	c	129
Me	COOEt	COOEt	Me	1.23,t,6; 2.22,s,6; 4.17,q,4; 6.41,bs,2; 9.38,bs,1	c	138
Me	COOEt	SO₂Ph	Ph	1.15,t,3; 2.23,s,3; 4.18,q,2; 6.3,bs,2; 7.27–7.93,m,10; 9.27,bs,1	c	131
Me	COOEt	4-MeC₆H₄SO₂	Ph	1.2,t,3; 2.23,s,3; 2.38,s,3; 4.22,q,2; 6.28,bs,2; 7.3–7.8,m,9; 9.27,bs,1	c	131
Ph	COPh	H	Ph	6.05,s,2; 6.68,s,1; 7.15–7.8,m,15; 9.52,s,1	b	120
Ph	COOEt	H	Me	1.05,t,3; 2.09,d,3; 4.01,q,2; 6.08,s,2; 6.29,q,1; 7.37,s,5; 9.10,s,1	b	120
Ph	COOEt	H	Ph	1.10,t,3; 4.12,q,2; 5.92,s,2; 6.72,s,1; 7.2–7.8,m,10; 9.40,s,1	b	120

1-NHR (R = CONHPh)

Me	COOMe	Ac	2.13,s,3; 2.27,s,3; 2.34,s,3; 3.73,s,3; 6.85–7.65,m,5; 9.38,bs,1; 9.59,bs,1	c	138
Me	COOMe	COPh	2.13,s,3; 2.34,s,3; 3.27,s,3; 6.85–7.97,m,10; 9.38,bs,1; 9.59,bs,1	c	138
Me	COOMe	CONH$_2$	2.14,s,3; 2.35,s,3; 3.67,s,3; 6.67–8.0,m,5; 7.77,bs,2; 9.30,bs,1; 9.53,bs,1	c	129
Me	COOMe	CONHPh	2.35,s,3; 3.67,s,3; 6.8–7.9,m,15; 9.30,bs,1; 9.53,bs,1; 10.26,bs,1	c	129
Me	COOMe	4-MeOC$_6$H$_4$NHCO	2.14,s,3; 2.35,s,3; 3.67,s,3; 3.78,s,3; 6.67–7.9,m,9; 9.30,bs,1; 9.53,bs,1; 10.26,bs,1	c	129
Me	COOMe	COOMe	2.22,s,6; 3.77,s,6; 6.85–7.65,m,5; 9.38,bs,1; 9.59,bs,1	c	138
Me	COOMe	4-MeC$_6$H$_4$SO$_2$	2.27,s,3; 2.43,s,3; 2.53,s,3; 3.7,s,3; 7.07–7.97,m,9; 9.63,bs,2	c	131
Me	COOMe	4-MeC$_6$H$_4$SO$_2$	2.27,s,3; 2.37,s,3; 3.73,s,3; 7.0–7.77,m,14; 9.23,bs,1; 9.5,bs,1	Ph	131
Me	COOEt	Ac	1.24,t,3; 2.13,s,3;2.27,s,3; 2.34,s,3; 4.2,q,2; 6.85–7.66,m,5; 9.38,bs,1; 9.59,bs,1	c	138
Me	COOEt	COPh	0.73,t,3; 2.13,s,3; 2.34,s,3; 3.7,q,2; 6.83–7.93,m,10; 9.38,bs,1; 9.59,bs,1	c	138
Me	COOEt	CONHPh	1.09,t,3; 2.14,s,3; 2.35,s,3; 4.11,q,2; 6.8–8.0,m,10; 9.30,bs,1; 9.53,bs,1; 10.26,bs,1	c	129
Me	COOEt	CONHPh	1.09,t,3; 2.35,s,3; 4.11,q,2; 6.8–7.8,m,15; 9.30,bs,1; 9.53,bs,1; 10.26,bs,1	Ph	129
Me	COOEt	4-MeOC$_6$H$_4$NHCO	1.09,t,3; 2.14, s,3; 2.35,s,3; 3.75,s,3; 4.11,q,2; 6.7–7.8,m,9; 9.30,bs,1; 9.53,bs,1; 10.26,bs,1	c	129
Me	COOEt	4-ClC$_6$H$_4$NHCO	1.09,t,3; 2.14,s,3; 2.35,s,3; 4.11,q,2; 6.6–7.7,m,9; 9.30,bs,1; 9.53,bs,1; 10.26,bs,1	c	129
Me	COOEt	CONEt$_2$	0.7–1.4,m,9; 2.14,s,3; 2.35,s,3; 2.83–3.73,m,4; 4.11,q,2; 6.8–7.7,m,5; 9.30,bs,1; 9.53,bs,1	c	129

TABLE 3.5. Continued

R^2	R^3	R^4	R^5	ppm. mult., nH	Solvent^b	Ref.
Me	COOEt	COOMe	Me	1.24,t,3; 2.22,s,6; 3.73, s,3; 4.2,q,2; 6.86–7.7,m,5; 9.38,bs,1; 9.59,bs,1	c	138
Me	COOEt	SO$_2$Me	Me	1.3,t,3; 2.3,s,3; 2.37,s,3; 3.38,s,3; 4.27,q,2; 7.03–7.67,m,5; 9.5,bs,1; 9.6,bs,1	c	131
Me	COOEt	SO$_2$Ph	Ph	1.17,t,3; 2.3,s,3; 4.22,q,2; 7.0–7.93,m,15; 9.23,bs,1; 9.47,bs,1	c	131
Me	COOEt	4-MeC$_6$H$_4$SO$_2$	Me	1.17,t,3; 2.23,s,3; 2.4,s,3; 2.48,s,3; 4.17,q,2: 7.0–7.97,m,9; 9.57,bs,2	c	131
Me	COOEt	4-MeC$_6$H$_4$SO$_2$	Ph	1.23,t,3; 2.3,s,3; 2.4,s,3; 4.25,q,2; 7.1–7.8,m,14; 9.2,bs,1; 9.47,bs,1	c	131
				1-NHR (R = COOMe)		
Me	COOMe	Me	Me	1.96,s,3; 2.07,s,3; 2.26,s,3; 3.66,s,6; 7.66,s,1	d	132
Me	COOMe	Ac	Me	1.86,s,3; 2.13,s,3; 2.30,s,3; 3.66,s,6; 8.67,s,1;	d	132
Me	COOMe	COPh	Me	2.12,s,3; 2.26,s,3; 2.32,s,3; 3.77,s,6; 10.66,bs,1	c	136
Me	COOMe	COOMe	Me	2.08,s,3; 2.34,s,3; 3.77,s,6; 7.3–8.0,m,5; 10.66,bs,1	c	136
Me	COOMe	CONH$_2$	Me	2.21,s,6; 3.73,s,6; 10.66,bs,1	c	136
Me	COOMe	CONHPh	Me	2.11,s,3; 2.31,s,3; 3.68,s,3; 3.77,s,3; 7.04,bs,1; 7.73,bs,1; 10.61,bs,1	c	139
Me	COOMe	CONHPh	Me	2.11,s,3; 2.31,s,3; 3.68,s,3; 3.77,s,3; 7.0–7.93,m,5; 10.20,bs,1; 10.61,bs,1	c	139
Me	COOMe	CONHPh	Ph	2.42,s,3; 3.7,s,6; 6.9–7.9,m,10; 10.20,bs,1; 10.61,bs,1	c	139
Me	COOMe	4-MeOC$_6$H$_4$NHCO	Me	2.11,s,3; 2.31,s,3; 3.68,s,3; 3.77,2,s,6; 6.9,d,2; 7.63,d,2; 10.20,bs,1; 10.61,bs,1	c	139
Me	COOMe	4-ClC$_6$H$_4$NHCO	Me	2.11,s,3; 2.31,s,3; 3.68,s,3; 3.77,s,3; 7.38,d,2; 7.77,d,2; 10.20,bs,1; 10.61,bs,1	c	139

Me	COOMe	CONEt$_2$	Me	0.67–1.31,m,6; 2.66–3.5,m,4; 2.11,s,3; 2.31,s,3; 3.68,s,3; 3.77,s,3; 10.61,bs,1	c	139
Me	COOMe	SO$_2$Me	Me	2.27,s,6; 3.32,s,3; 3.77,s,3; 3.83,s,3; 8.91,bs,1	d	146
Me	COOMe	SO$_2$Ph	Ph	2.23,s,3; 3.52,s,3; 3.67,s,3; 7.13–7.92,m,10; 8.28,bs,1	d	146
Me	COOMe	4-MeC$_6$H$_4$SO$_2$	Me	2.25,s,3; 2.4,s,3; 2.47,s,3; 3.66,s,3; 3.83,s,3; 7.52,q,4; 8.73,bs,1	d	146
Me	COOEt	Ac	Me	1.24,t,3; 2.12,s,3; 2.26,s,3; 2.32,s,3; 3.75,s,3; 4.18,q,2; 10.66,bs,1	c	136
Me	COOEt	CO-t-Bu	t-Bu	0.95–1.43,m,21: 2.23,s,3; 3.75,s,3; 4.18,q,2; 10.66,bs,1	c	136
Me	COOEt	COPh	Me	0.73,t,3; 2.08,s,3; 2.34,s,3; 3.5–4.0,m,5; 7.3–8.0,m,5; 10.66,bs,1	c	136
Me	COOEt	COOMe	Me	1.24,t,3; 2.21,s,6; 3.73,s,6; 4.18,q,2; 10.66,bs,1	c	136
Me	COOEt	SO$_2$Me	Me	1.37,t,3; 2.28,s,6; 3.33,s,3; 3.82,s,3; 4.32,q,2; 8.95,bs,1	d	146

1-NHR (R = COOEt)

Me	COOMe	Ac	Me	1.27,t,3; 2.12,s,3; 2.26,s,3; 2.32,s,3; 3.77,s,3; 4.22,q,2; 10.66,bs,1	c	136
Me	COOMe	COPh	Me	1.27,t,3; 2.08,s,3; 2.34,s,3; 3.3,s,3; 4.22,q,2; 7.3–8.0,m,5; 10.66,bs,1	c	136
Me	COOMe	CONH$_2$	Me	1.18,t,3; 2.11,s,3; 2.31,s,3; 3.68,s,3; 4.11,q,2; 7.04,bs,1; 7.73,bs,1; 10.61,bs,1	c	139
Me	COOMe	CONHPh	Ph	1.18,t,3; 2.42,s,3; 3.68,s,3; 4.11,q,2; 6.9–7.9,m,10; 10.20,bs,1; 10.61,bs,1	c	139
Me	COOMe	COOMe	Me	1.27,t,3; 2.21,s,6; 3.73,s,6; 4.22,q,2; 10.66,bs,1	c	136
Me	COOMe	4-MeC$_6$H$_4$SO$_2$	Me	1.3,t,3; 2.27,s,3; 2.37,s,3; 2.47,s,3; 3.67,s,3; 4.23,q,2; 7.5,q,4; 8.22,bs,1	d	146
Me	COOEt	COPh	Me	0.73,t,3; 1.27,t,3; 2.08,s,3; 2.34,s,3; 3.75,q,2; 4.22,q,2; 7.3–8.0,m,5; 10.66,bs,1	c	136
Me	COOEt	4-MeC$_6$H$_4$SO$_2$	Me	0.92–1.52,m,6; 2.27,s,3; 2.37,s,3; 2.47,s,3; 3.87–4.5,m,4; 7.5,q,4; 8.87,bs,1	d	146

TABLE 3.5. Continued

R^2	R^3	R^4	R^5	ppm, mult., nH	Solvent[b]	Ref.
				1-NHR (R = COO-t-Bu)		
Me	COOMe	Ac	Me	1.49,s,9; 2.12,s,3; 2.26,s,3; 2.32,s,3; 3.77,s,3; 10.66,bs,1	c	136
Me	COOMe	COPh	Me	1.49,s,9; 2.08,s,3; 2.34,s,3; 3.3,s,3; 7.3–8.0,m,5; 10.66,bs,1	c	136
Me	COOMe	CONHPh	Ph	1.44,s,9; 2.42,s,3; 3.68,s,3; 6.9–7.9,m,10; 10.20,bs,1; 10.61,bs,1	c	139
Me	COOMe	CONEt$_2$	Me	0.67–1.31,m,6; 1.44,s,9; 2.66–3.5,m,4; 2.11,s,3; 2.31,s,3; 3.68,s,3; 10.61,bs,1	c	139
Me	COOMe	COOMe	Me	1.49,s,9; 2.21.s.6; 3.73.s.6; 10.66,bs,1	c	136
Me	COOMe	SO$_2$Ph	Ph	1.3,s,9; 2.23,s,3; 3.65,s,3; 7.17–8.0,m,11	d	146
Me	COOMe	4-MeC$_6$H$_4$SO$_2$	Me	1.52,s,9; 2.25,s,3; 2.37,s,3; 2.45,s,3; 3.63.s.3; 7.47,q,4; 8.46,bs,1	d	146
Me	COOMe	4-MeC$_6$H$_4$SO$_2$	Ph	1.3,s,9; 2.27,s,3; 2.37,s,3; 3.68,s,3; 6.93–7.9,m,10	d	146
Me	COOEt	COPh	Me	0.73,t,3; 1.49,s,9; 2.08,s,3; 2.34,s,3; 3.75,q,2; 7.3–8.0,m,5; 10.66,bs,1	c	136
Me	COOEt	CONH$_2$	Me	1.14,t,3; 1.44,s,9; 2.11,s,3; 2.31,s,3; 4.15,q,2; 7.04,bs,1; 7.73,bs,1; 10.61,bs,1	c	139
Me	COOEt	CONHPh	Ph	1.14,t,3; 1.44,s,9; 2.42,s,3; 4.15,q,2; 6.9–7.9,m,10; 10.20,bs,1; 10.61,bs,1	c	139
Me	COOEt	COOMe	Me	1.24,t,3; 1.49,s,9; 2.21,s,6; 3.73,s,6; 4.18,q,2; 10.66,bs,1	c	136
Me	COOEt	SO$_2$Ph	Ph	1.13–1.5,s,12; 2.28,s,3; 4.22,q,2; 7.17–7.97,m,11	d	146
Ph	H	H	Ph	1.27,s,9; 6.31,s,2; 7.4,m,10	d	66
				1-NHR (R = PO(Ph)$_2$)		
Me	COOMe	COOEt	4-NO$_2$C$_6$H$_4$	1.05,t,3; 2.65,s,3; 3.75,s,3; 4.05,q,2; 7.87–8.33,m,14; 9.68,d,1	c	144
Me	COOEt	COPh	Me	0.70,t,3; 2.00,s,3; 2.42,s,3; 3.77,q,2; 7.37–8.13,m,15; 9.45,d,1	c	144

			1-NHR (R = PO(OEt)₂)		
Me	COOMe	COPh	Ph	0.70–1.30,m,6; 2.60,s,3; 3.37,s,3; 3.23–3.80,m,4; 7.13–7.80,m,10; 9.17,d,1	c	144
Me	COOMe	CONHPh	Me	1.01–1.47,m,6; 2.33,s,3; 2.50,s,3; 3.70,s,3; 3.83–4.63,m,4; 6.90–7.91,m,5; 9.00,d,1; 10.27,bs,1	c	145
Me	COOMe	CONHPh	Ph	0.87–1.50,m,6; 2.58,s,3; 3.67,s,3; 3.25–4.38,m,4; 6.95–7.92,m,10; 9.10,d,1; 10.15,bs,1	c	145
Me	COOMe	4-MeOC₆H₄NHCO	Me	1.10–1.53,m,6; 2.32,s,3; 2.50,s,3; 3.72,s,3; 3.78,s,3; 3.86–4.47,m,4; 6.95,d,2; 7.70,d,2; 9.08,d,1; 10.13,bs,1	c	145
Me	COOEt	4-MeOC₆H₄NHCO	Me	0.90–1.53,m,9; 2.30,s,3; 2.50,s,3; 3.77,s,3; 3.83–4.43,m,6; 6.93,d,2; 7.70,d,2; 8.97,d,1; 10.06,bs,1	c	145
Me	COOCH₂Ph	COPh	Me	1.00–1.53,m,6; 2.20,s,3; 2.52,s,3; 3.83–4.43,m,4; 4.78,s,2; 6.90–7.93,m,10; 9.09,d,1	c	144
Me	COOCH₂Ph	CONHPh	Me	1.10–1.48,m,6; 2.30,s,3; 2.50,s,3; 3.82–4.42,m,4; 5.20,s,2; 6.82–7.95,m,10; 9.03,d,1; 10.30,bs,1	c	145
Me	COOCH₂Ph	4-ClC₆H₄NHCO	Me	0.93–1.50,m,6; 2.30,s,3; 2.50,s,3; 3.80–4.43,m,4; 5.18,s,2; 7.00–7.93,m,9; 9.04,d,1; 10.38,bs,1	c	145
Me	COOCH₂Ph	CONEt₂	Me	0.75–1.25,m,6; 1.25–1.38,m,6; 2.13,s,3; 2.55,s,3; 3.00–3.65,m,4; 3.95–4.48,m,4; 5.32,s,2; 7.65,s,5; 9.26,d,1	c	145
Me	COOCH₂Ph	COOEt	4-NO₂C₆H₄	0.78–1.32,m,9; 2.53,s,3; 3.32–4.18,m,6; 5.28,s,2; 7.47,s,5; 7.77,d,2; 8.38,d,2; 9.34,d,1	c	144
			1-NHR (R = PO(OPh)₂)		
Me	COOMe	Ac	Me	2.18,s,3; 2.32,s,6; 3.77,s,3; 7.10–7.37,m,10; 10.04,d,1	c	144
Me	COOMe	CONH₂	Me	2.23,s,3; 2.33,s,3; 3.72,s,3; 6.85–7.82,m,12; 9.93,d,1	c	145
Me	COOMe	CONEt₂	Me	0.73–1.33,m,6; 2.00,s,3; 2.38,s,3; 2.83–3.53,m,4; 3.67,s,3; 7.07–7.80,m,10; 10.04,d,1	c	145

TABLE 3.5. Continued

R^2	R^3	R^4	R^5	ppm, mult., nH	Solvent[b]	Ref.
Me	COOMe	COOEt	Ph	1.07,t,3; 2.37,s,3; 3.75,s,3; 4.08,q,2; 6.67–7.63,m,15; 10.12,d,1	c	144
Me	COOEt	CONHPh	Ph	1.02,t,3; 2.47,s,3; 4.12,q,2; 6.68–7.92,m,20; 10.20,d,1; 10.28,bs,1	c	145
Me	COOEt	4-ClC$_6$H$_4$NHCO	Me	1.33,t,3; 2.50,s,3; 2.67,s,3; 4.37,q,2; 7.30–8.30,m,14; 10.28,d,1; 10.65,bs,1	c	145
Me	COOEt	COOMe	Me	1.23,t,3; 2.27,s,6; 3.73,s,3; 4.20,q,2; 7.13–7.73,m,10; 10.05,d,1	c	144
Me	COOCH$_2$Ph	Ac	Me	2.30,s,3; 2.40,s,3; 2.45,s,3; 5.43,s,2; 7.28–7.88,m,10; 10.20,d,1	c	144
Me	COOCH$_2$Ph	CONH$_2$	Me	2.23,s,3; 2.33,s,3; 5.26,s,2; 7.00–7.86,m,17; 10.00,d,1	c	145
Me	COOCH$_2$Ph	CONEt$_2$	Me	0.70–1.33,m,6; 1.97,s,3; 2.40,s,3; 2.80–3.67,m,4; 5.20,s,2; 7.07–7.73,m,15; 10.03,d,1	c	145

1-NHR ($R = SO_2Ph$)

Me	COOMe	CONH$_2$	Me	1.82,s,3; 2.14,s,3; 3.69,s,3; 7.1,bs,1; 7.3–8.0,m,6; 11.63,bs,1	c	130
Me	COOMe	CONHPh	Me	1.82,s,3; 2.14,s,3; 3.69,s,3; 7.0–8.1,m,10; 10.15,bs,1; 11.63,bs,1	c	130
Me	COOMe	CONHPh	Ph	2.14,s,3; 3.69,s,3; 6.8–7.8,m,15; 10.15,bs,1; 11.63,bs,1	c	130
Me	COOEt	CONHPh	Ph	1.03,t,3; 2.14,s,3; 4.12,q,2; 6.9–7.9,m,15; 10.15,bs,1; 11.63,bs,1	c	130
Me	COOEt	4-MeOC$_6$H$_4$NHCO	Me	1.03,t,3; 1.82,s,3; 2.14,s,3; 3.69,s,3; 4.12,q,2; 6.93,d,2; 7.67,d,2; 7.87,s,5; 10.15,bs,1; 11.63,bs,1	c	130
Me	COOEt	CONEt$_2$	Me	0.6–1.34,m,9; 1.82,s,3; 2.14,s,3; 2.77–3.83,m,4; 4.12,q,2; 7.83,s,5; 11.63,bs,1	c	130
Ph	H	H	Ph	6.26,s,2; 7.2,m,15; 8.1,bs,1	d	66

1-NHR ($R = 4\text{-}MeC_6H_4SO_2$)

H	H	H	2.38,s,3; 5.92,t,2; 6.31,t,2; 7.39,q,4	d	113
Me	COOMe	CONH$_2$	1.82,s,3; 2.14,s,3; 2.47,s,3; 3.69,s,3; 7.1,bs,1; 7.3–7.9,m,5; 11.63,bs,1	c	130
Me	COOMe	CONHPh	1.82,s,3; 2.14,s,3; 2.47,s,3; 3.69,s,3; 6.9–8.0,m,9; 10.15,bs,1; 11.63,bs,1	c	130
Me	COOMe	CONHPh	2.14,s,3; 2.47,s,3; 3.69,s,3; 6.8–7.9,m,14; 10.15,bs,1; 11.63,bs,1	c	130
Me	COOMe	4-MeOC$_6$H$_4$NHCO	1.82,s,3; 2.14,s,3; 2.47,s,3; 3.69,s,3; 3.78,s,3; 6.7–8.0,m,8; 10.15,bs,1; 11.63,bs,1	c	130
Me	COOMe	4-ClC$_6$H$_4$NHCO	1.82,s,3; 2.14,s,3; 2.47,s,3; 3.69,s,3; 7.2–8.0,m8; 10.15,bs,1; 11.63,bs,1	c	130
Me	COOMe	CONEt$_2$	0.6–1.34,m,6; 1.82,s,3; 2.14,s,3; 2.47,s,3; 2.7–3.95,m,4; 3.67,s,3; 7.63,q,4; 11.63,bs,1	c	130
Me	COOMe	COOMe	1.95,s,6; 2.45,s,3; 3.72,s,6; 7.61,q,4; 11.63,bs,1	c,h	140
Me	COOMe	COOCH$_2$Ph	1.95,s,6; 2.45,s,3; 3.68,s,3; 5.2,s,2; 7.23–7.9,m,9; 11.74,bs,1	c,h	140
Me	COOEt	CONHPh	1.03,t,3; 1.82,s,3; 2.14,s,3; 2.47,s,3; 4.12,q,2; 6.8–7.9,m,9; 10.15,bs,1; 11.63,bs,1	c	130
Me	COOEt	4-ClC$_6$H$_4$NHCO	1.03,t,3; 1.82,s,3; 2.14,s,3; 2.47,s,3; 4.12,q,2; 7.2–8.1,m,8; 10.15,bs,1; 11.63,bs,1	c	130
Me	COOEt	CONEt$_2$	0.6–1.37,m,9; 1.82,s,3; 2.14,s,3; 2.47,s,3; 2.8–3.8,m,4; 4.12,q,2; 7.62,q,4; 11.63,bs,1	c	130
Me	COOEt	COOEt	1.23,t,6; 1.95,s,6; 2.45,s,3; 4.16,q,4; 7.61,q,4; 11.74,bs,1	c,h	140
Me	COOEt	COO–t–Bu	1.18,t,3; 1.47,s,9; 1.95,s.6; 2.45,s,3; 4.1,q,4; 7.61,q,4; 11.74,bs,1	c,h	140
Ph	Ph	Ph	2.28,s,3; 7.52,s,1; 7.1,m,24	d	66

TABLE 3.5. Continued

R^2	R^3	R^4	R^5	ppm. mult., nH	Solvent[b]	Ref.
\multicolumn{7}{c}{1-NHR ($R = 4\text{-}MeC_6H_4SO_2$)}						
Me	COOMe	COOMe	Me	1.95,s,6; 3.72,s,6; 3.89,s,3; 7.48,q,4; 11.74,bs,1	c,h	140
Me	COOMe	COOEt	Ph	1.12,t,3; 2.42,s,3; 3.68,s,3; 3.89,s,3; 4.09,q,2; 6.6–7.5,m,9; 11.74,bs,1	c,h	140
Me	COOEt	COO—t-Bu	Me	1.18,t,3; 1.47,s,9; 1.95,s,6; 3.89,s,3; 4.1,q,2; 7.48,q,4; 11.74,bs,1	c,h	140
Me	COOEt	COOCH$_2$Ph	Me	1.18,t,3; 1.95,s,6; 3.89,s,3; 4.1,q,2; 5.2,s,2; 7.42,s,5; 7.48,q,4; 11.74,bs,1	c,h	140
\multicolumn{7}{c}{1-NHR ($R = 4\text{-}ClC_6H_4SO_2$)}						
Me	COOMe	COOMe	Me	1.95,s,6; 3.72,s,6; 3.68,s,3; 7.77,s,4; 11.74,bs,1	c,h	140
Me	COOMe	COOEt	Me	1.12,t,3; 1.95,s,6; 3.68,s,3; 4.09,q,2; 7.77,s,4; 11.74,bs,1	c,h	140
Me	COOMe	COOEt	Ph	1.12,t,3; 2.42,s,3; 3.68,s,3; 4.09,q,2; 7.17,s,5; 7.77,s,4; 11.74,bs,1	c,h	140
Me	COOMe	COO—t-Bu	Me	1.47,s,9; 1.95,s,6; 3.68,s,3; 7.77,s,4; 11.74,bs,1	c,h	140
Me	COOMe	COOCH$_2$Ph	Me	1.95,s,6; 3.68,s,3; 5.2,s,2; 7.42,s,5; 7.77,s,4; 11.74,bs,1	c,h	140
Me	COOEt	COOEt	Me	1.23,t,6; 1.95,s,6; 4.16,q,4; 7.77,s,4; 11.74,bs,1	c,h	140
Me	COOEt	COOCH$_2$Ph	Me	1.18,t,3; 1.95,s,6; 4.1,q,2; 5.2,s,2; 7.42,s,5; 7.77,s,4; 11.74,bs,1	c,h	140
\multicolumn{7}{c}{1-NHR ($R = 2,4,6\text{-}Me_3C_6H_2SO_2$)}						
Me	COOMe	CONH$_2$	Me	1.82,s,3; 2.14,s,3; 2.33,s,3; 2.47,s,6; 3.69,s,3; 7.18,s,3; 7.67,bs,1; 11.63,bs,1	c	130
Me	COOMe	4-MeOC$_6$H$_4$NHCO	Me	1.82,s,3; 2.14,s,3; 2.33,s,3; 2.47,s,6; 3.69,s,3; 3.78,s,3; 6.93,d,2; 7.2,s,2; 7.63,d,2; 10.15,bs,1; 11.63,bs,1	c	130

Me	COOMe	4-ClC$_6$H$_4$NHCO	Me	1.82,s,3; 2.14,s,3; 2.33,s,3; 2.47,s,6; 3.69,s,3; 7.2,s,2; 7.4,d,2; 7.78,d,2; 10.15,bs,1; 11.63,bs,1	c	130
1-NR$_2$ (R = Me)						
H	H	Ph		2.80,s,6; 6.23,d,2; 7.10,t,1; 7.26–7.83,m,5	f	121
4-MeC$_6$H$_4$	H	H		2.40,s,3; 2.80,s,6; 6.2–6.35,m,2; 7.1–7.65,m,5	d	104
1-NR$_2$[R =(CH$_2$)$_2$COPh]						
H	H	H		2.8–3.7,m,8; 6.15,t,2; 6.88,t,2; 7.4–8.0,m,10	d	294
1-NR$_2$ (R = Ac)						
Ph	Ph	Ph		2.12,s,6; 7.0–7.23,m,20	d,i	66
1-NRR' (R = Me; R' = Ph)						
4-MeC$_6$H$_4$	H	H		2.60,s,3; 3.25,s,3; 6.28–7.5,m,12	d	104
1-NRR' (R = Me, R' = 2-NH$_2$C$_6$H$_4$)						
H	H	H		3.10,s,3; 3.63,s,2; 6.0,m,2; 6.4–7.5,m,6	f	112
1-NRR' (R = Me, R' = 2-AcNHC$_6$H$_4$)						
H	H	H		2.05,s,3; 3.2,m,3; 6.0,m,2; 7.0–8.0,m,6; 9.1,s,1	c	112
1-NRR' (R = Me, R' = 2-NO$_2$C$_6$H$_4$)						
H	H	H		3.30,s,3; 6.2,m,2; 6.8,m,2; 6.9–7.8,m,4	d	112

[a] Uncompleted spectra were also reported.[49,120,121,294]

[b] Solvent: a, neat; b, not reported; c, dimethylsufoxide-d_6; d, deuterochloroform; e, other values reported in the same solvent are 4.37,s,2; 7.10,m,20;[66] f, carbon tetrachloride; g, other values reported in the same solvent are 1.20,t,3; 2.03,d,3; 4.13,q,2; 4.15,q,2; 6.15,bs,3; 9.10,s,1;[118] h, 3-(trimethylsilyl)propanesulfonic acid sodium salt as internal reference; i, other values reported in the same solvent are 2.12,s,6; 7.07,m,10; 7.27,m,10.[107]

TABLE 3.6. ^{1}H CHEMICAL SHIFTS OF THE 2-AMINOPYRROLES[a]

R^1	R^3	R^4	R^5	ppm, mult., nH	Solvent[b]	Ref.
				2-NH_2		
H	Ph	Ac	Me	1.84,s,3; 2.29,s,3; 4.06,bs,2; 7.15,m,3; 7.29,t,2; 10.56,bs,1	a	236
H	CN	Me	H	1.90,s,3; 5.35,s,2; 5.78,s,1; 9.6,b,1;	a	182
				2.55,s,3; 4.72,s,2	b	182
H	CN	Me	Me	1.89,s,3; 1.95,s,3; 5.05,s,2; 9.28,b,1;	a	279
				1.6,d,3; 2.5,s,3; 4.85,q,1; 7.85,d,2; 8.9,b,1	b	279
H	CN	Me	$(CH_2)_2SMe$	1.77,s,3; 1.96,s,3; 2.46,s,4; 5.18,s,2; 9.67,bs,1	a	181
H	CN	Me	CH_2Ph	2.00,s,3; 3.67,s,2; 3.80,bs,2; 7.10,s,5; 7.50,bs,1	c	181
H	CN	Me	$4\text{-}OHC_6H_4CH_2$	1.86,s,3; 3.45,s,2; 5.16,s,2; 6.66,q,4; 8.95,s,1; 9.56,s,1	a	181
H	CN	Me	3-indolylCH_2	1.90,s,3; 3.65,s,2; 5.1,s,2; 6.7–7.3,m,5; 9.4,bs,1; 10.6,bs,1	a	181
H	CN	Me	Ph	2.18,s,3; 5.16,bs,2; 7.28,m,5; 9.65,bs,1	c	181
H	CN	Ph	H	3.45,bs,2; 6.47,d,1; 7.15–7.50,m,5; 10.34,s,1	a	186
H	CN	COPh	OEt	1.5,t,3; 2.9,q,2; 5.1,bs,2; 7.4–7.8,m,5; 8.1,s,1	a	166
H	$CONH_2$	$CONH_2$	SCH_3	2.35,s,3; 3.65,s,4; 7.6,s,2; 8.8,s,1	a	189
H	COOEt	H	H	1.31,t,3; 4.24,q,2; 5.00,bs,2; 6.14,m,1; 6.29,m,1; 8.24,bs,1	c	184

				NMR		
H	COOEt	H	Me	1.31,t,3; 2.12,s,3; 4.29,q,2; 4.84,bs,2; 5.93,s,1; 7.96,bs,1	c	184
H	COOEt	H	Ph	1.31,t,t,3; 4.27,q,2; 5.22,s,2; 6.62,d,1; 7.1–7.4,m,5; 8.84,bs,1	c	185
H	COOEt	Me	Me	1.32,t,3; 2.02,s,3; 2.08,s,3; 4.24,q,2; 5.09,bs,2; 7.80,bs,1	c	185
H	COOEt	Me	Ph	1.25,t,3; 2.28,s,3; 4.13,q,2; 5.22,s,2; 7.20,m,5; 8.33,bs,1	c	185
H	COOEt	Ph	H	1.02,t,3; 3.98,q,2; 5.79,s,2; 6.14,d,1; 7.1–7.4,m,5; 10.2,bs,1	a	186
H	COOEt	Ph	Me	1.06,t,3; 2.02,s,3; 4.09,q,2; 5.07,bs,2; 7.31,m,5; 7.89,bs,1	c	185
H	COOEt	Ph	Ph	0.98,t,3; 4.02,q,2; 5.47,s,2; 7.0–7.3,m,10; 10.2,bs,1; 0.82,t,3; 3.81,q,2; 5.69,s,2; 6.90–7.20,m,10	c a	185 186
H	COOEt	Indeno[1,2-b]		1.37,t,3; 3.56,s,2; 4.29,q,2; 5.47,s,2; 7.0–7.4,m,4; 9.95,bs,1	c	185
H	COOEt	Naphthaleno[1,2-b]		1.47,t,3; 4.40,q,2; 5.95,bs,2; 7.32,ddd,1; 7.44,ddd,1; 7.55,d,1; 7.88,dd,1; 7.93,d,1; 8.04,dd,1; 10.84,bs,1	c	185
H	COOEt	=O	c-Hexyl	1.23,t,3; 1.4–1.8,m,6; 2.1–2.3,m,2; 2.9–3.2,m,2; 4.16,q,2; 7.3,bs,2; 9.23,s,1	a	162
H	COOEt	=O	=CHCH=CHPh	1.18,t,3; 4.13,q,2; 6.2–7.65,m,8; 7.95,bs,2; 10.9,s,1	a	162
H	COOEt	=O	=CHPh	1.4,t,3; 4.17,q,2; 6.4,s,1; 7.2–7.7,m,5; 7.85,bs,2; 9.76,s,1	a	162
H	COO—t-Bu	Me	H	1.43,s,9; 1.98,s,3; 5.41,s,2; 5.72,s,1; 9.5,bs,1; 1.67,s,9; 2.58,s,3; 4.56,s,2	a d	182 182
Me	CN	Me	Me	2.0,s,6; 3.25,s,3; 3.75,bs,2	c	300
Me	CN	Me	Et	1.03,t,3; 1.93,s,3; 2.58,q,2; 3.27,s,3; 3.83,bs,2	c	300
Me	CN	Me	i-Bu	0.88,d,6; 1.4–2.0,m,1; 1.97,s,3; 2.26,d,2; 3.24,s,3; 3.71,bs,2	c	300
Me	CN	Me	$(CH_2)_2SMe$	2.01,s,3; 2.09,s,3; 2.50–2.75,m,4; 3.30,s,3; 3.74,bs,2	c	300

TABLE 3.6. Continued

R^1	R^3	R^4	R^5	ppm, mult., nH	Solvent[b]	Ref.
Me	CN	Me	CH_2Ph	2.07,s,3; 3.06,s,3; 3.60,bs,2; 3.78,s,2; 7.18,m,5	c	300
Me	CN	Me	4-$OMeC_6H_4CH_2$	2.08,s,3; 3.08,s,3; 3.75,s,7; 6.86,m,4	c	300
Me	CN	Me	(1-Me-3-indolyl)CH_2	2.10,s,3; 3.11,s,3; 3.66,s,5; 3.87,s,2; 6.5,m,1; 7.2,m,4	c	300
Me	CN	Me	Ph	2.02,s,3; 3.23,s,3; 3.91,bs,2; 7.20–7.35,m,5	c	300
Me	COOEt	H	H	1.31,t,3; 3.36,s,3; 4.24,q,2; 4.87,bs,2; 6.07,d,1; 6.24,d,1	c	184
Me	COOEt	Me	Ph	1.34,t,3; 2.16,s,3; 3.22,s,3; 4.29,q,2; 5.09,bs,2; 7.2–7.5,m,5	c	185
n-Pr	COO—t-Bu	Me	Me	0.92,t,3; 1.4–1.8,m,2; 1.53,s,9; 2.0,s,3; 2.05,s,3; 3.52,t,2; 4.85,s,2	c	176
i-Pr	Me	H	CN	1.32,d,6; 1.97,s,3; 3.00,bs,2; 4.43,h,1; 6.75,s,1	a	165
i-Pr	Et	H	CN	1.22,t,3; 1.40,d,6; 2.51,q,2; 2.60,bs,2; 4.50,h,1; 6.90,s,1	a	165
n-Bu	COO—t-Bu	Me	Me·$HClO_4$	0.93,t,3; 1.1–1.7,m,4; 1.5,d,3; 1.6,s,9; 2.45,s,3; 3.50,m,2; 4.7,q,1; 7.9 and 8.15,2s,2	c	176
i-Bu	COO—t-Bu	Me	Me·$HClO_4$	0.85–1.10,m,6; 1.2–1.5,m,1; 1.5,d,3; 1.6,s,9; 2.5.s,3; 3.25–3.90,m,2; 4.65,q,1; 8.0 and 8.2,2s,2	c	176
t-Bu	Et	H	CN	1.16,t,3; 1.61,s,9; 2.36,q,2; 2.90,bs,2; 6.83,s,1	a	165
c-Pentyl	COO—t-Bu	Me	Me	1.57,s,9; 2.0,s,3; 2.10,s,3; 1.3–2.1,m,8; 4.4,m,1; 5.0,s,2	c	176
c-Hexyl	COO—t-Bu	Me	Me	1.2–2.0,m,10; 1.55,s,9; 2.10,s,6; 3.8,m,1; 5.0,s,2	c	176
1-Morpholyl-n-Pr	COO—t-Bu	Me	Me	1.4–1.7,m,2; 1.55,s,9; 2.05,s,3; 2.10,s,3; 2.3–2.6,m,6; 3.55–3.85,m,6; 5.8,s,2	c	176
C_2H_4Ph	COO—t-Bu	Me	Me	1.51,s,9; 1.92,s,3; 2.06,s,3; 2.80,t,2; 3.75,t,2; 4.50,s,2; 6.9–7.3,m,5	c	176

CH₂Ph	COOMe	Me	2.0,s,3; 2.15,s,3; 3.77,s,3; 4.8,s,2; 4.85,s,2; 6.85–7.35,m,5	c	176
CH₂Ph	COO—t-Bu	Me	1.53,s,9; 2.0,s,3; 2.1,s,3; 4.70,s,2; 4.83,s,2; 6.9–7.3,m,5	c	176
CH₂Ph	SO₂Me	Me	1.96,s,3; 2.10,s,3; 2.95,s,3; 4.45,bs,2; 4.83,s,2; 6.8–7.3,m,5	c	177
2-PyrrolylCH₂	COO—t-Bu	Me	1.53,s,9; 2.03,s,3; 2.10,s,3; 4.87,s,2; 5.45,s,2; 6.9–7.7,8.4–8.55,m,4	c	176
2-FurylCH₂	CN	Me	1.85,s,3; 2.0,s,3; 4.9,s,2; 5.45,s,2; 6.35,m,2; 7.5,m,1	a	173
2-ThienylCH₂	CN	Me	1.95,s,3; 2.05,s,3; 5.15,s,2; 6.75,s,2; 6.95,m,2; 7.4,m,1	a	173
3-PyridylCH₂	COO—t-Bu	Me	1.53,s,9; 1.96,s,3; 2.10,s,3; 4.83,s,2; 4.95,s,2; 7.15–7.25,m,2; 8.3–8.55,m,2	c	176
CH₂CH(OMe)₂	COO—t-Bu	Me	1.53,s,9; 2.03,s,3; 2.10,s,3; 3.4,s,6; 3.75,d,2; 4.35,t,1; 5.22,s,2	c	176
CH₂COOEt	CN	Me	1.23,t,3; 1.9,s,6; 4.15,q,2; 4.57,s,2; 5.58,s,2	a	175
CH₂OCH₂Ph	COOEt	H	1.29,t,3; 2.11,s,3; 4.26,q,2; 4.47,s,2; 5.08,s,2; 5.11,s,2; 6.02,s,1; 7.36,m,5	c	184
Ph	CN	Me	1.1,t,3; 2.4,s,3; 4.1,q,2; 4.3,s,2; 7.2–7.6,m,5	c	159
Ph	CONH₂	Ph	5.0,s,2; 5.5,s,2; 7.2,m,15	c	297
Ph	COOH	Ph	4.2,s,2; 5.8,s,1; 7.3,m,15	c	297
Ac	CN	Me	1.98,s,3; 2.15,s,1; 2.25,s,3; 2.55,s,3; 6.98,b,1	c	279
			1.97,s,3; 2.25,s,3; 2.55,s,3; 5.96,s,2	c	183
			1.8,d,3; 2.7,d,6; 5.50,s,1; 10.0,b,2	b	279
Ac	CN	3-IndolylCH₂	1.85,s,3; 2.46,s,3; 3.58,s,2; 5.09,s,2; 6.90–7.35,m,4; 7.95–8.15,bm,1; 9.45,bs,1	a	181
COPh	COOEt	=O	1.43,t,1; 4.4,q,2; 6.7–7.3,m,11; 8.1,s,1; 8.75,s,1	c	162
N=CHCOOEt	CN	—(CH₂)₄—	1.3,t,3; 1.65,m,8; 3.5,bs,2; 4.35,q,2; 7.95,s,1	a	155

387

TABLE 3.6. Continued

R^1	R^3	R^4	R^5	ppm, mult., nH	Solvent[b]	Ref.
				2-NHR (R = Me)		
Me	COOEt	Me	Ph	1.36,t,3; 2.18,s,3; 2.88,bd,3; 3.31,s,3; 4.29,q,2; 5.84,bs,1; 7.2–7.5,m,5	c	185
				2-NHR (R=Et)		
CH$_2$Ph	COO—t-Bu	Me	Me	1.03,t,3; 1.54,s,9; 1.87,s,3; 2.13,s,3; 2.78,q,2; 4.97,s,2	c	305
CH$_2$OCH$_2$Ph	COOEt	H	Me	1.14,t,3; 1.31,t,3; 2.21,s,3; 3.08,q,2; 4.23,q,2; 4.49,s,2; 5.13,bs,1; 5.20,s,2; 6.13,s,1; 7.38,m,5	c	184
				NHR (R = CH$_2$COPh)		
H	CN	Me	Me	1.87,s,3; 1.97,s,3; 4.75,d,2; 5.95,t,1; 7.50–8.05,m,5; 10.03,s,1	a	298
				NHR (R = 4-BrC$_6$H$_4$COCH$_2$)		
H	CN	Me	Me	1.84,s,3; 1.93,s,3; 4.69,d,2; 5.97,t,1; 7.82,q,2; 9.98,s,1	a	298
				NHR (R = Ph)		
CH$_2$Ph	Ph	Ph	CH(COPh)$_2$	4.51,s,2; 5.15,s,1; 6.3–7.5,m,30; 18.41,s,1	c	163
H	CN	Me	COOEt	1.28,t,3; 2.33,s,3; 4.32,q,2; 6.75–7.37,m,5; 8.6,s,1	a	159
				NHR (R = Ac)		
H	CN	Me	i-Bu	0.80,d,6; 1.2–1.7,m,1; 1.9,s,3; 2.0,s,3; 2.3,d,2; 10.1,bs,1; 10.9,bs,1	a	181
H	CONH$_2$	Me	H	2.13,s,3; 2.20,s,3; 6.30,s,1; 6.77,s,2; 10.93,s,1; 11.16,s,1	a	290
H	CONH$_2$	Me	i-Bu	0.82,d,6; 2.0,m,1; 2.18,s,6; 2.50,d,2; 6.93,s,2; 10.3,s,2	a	290

H	CONH$_2$	CH$_2$Ph	2.13,s,3; 2.20,s,3; 4.00,s,2; 6.8,s,2; 7.5,m,5; 11.13,s,2	a	290
H	CONH$_2$	H	2.17,s,3; 6.5,s,3; 7.1,m,5; 10.80,s,1	a	290
(2-Ac-Fur-5-yl)CH$_2$	CONH$_2$	Me	2.1,s,6; 2.2,s,3; 2.45,s,3; 5.05,s,3; 6.5,d,1; 6.7,bs,2; 7.4,d,1; 11.4,s,1	a	173
5-[2-Ac-thienyl]CH$_2$	CONH$_2$	Me	2.1,s,9; 2.35,s,3; 5.15,s,2; 6.7,bs,2; 7.0,d,1; 7.8,d,1; 11.5,s,1	a	173
Ph	CONH$_2$	Me	1.83,s,3; 1.97,s,3; 2.23,s,3; 6.90,s,2; 7.66,m,5; 9.60,s,1	a	290
Ph	CONH$_2$	—(CH$_2$)$_4$—	1.77,m,4; 2.3,m,5; 2.53,m,2; 6.66,s,2; 7.20–7.43,m,5; 9.4,s,1	a	290

2-NHR (R = COEt)

(2-COEt-fur-5-yl)CH$_2$	CONH$_2$	Me	1.1,t,6; 2.1,s,3; 2.2,s,3; 2.5,q,2; 2.8,q,2; 5.0,s,2; 6.4,m,2; 7.6,m,2; 9.5,s,1	a	173
5-(2-COEt-thienyl)CH$_2$	CONH$_2$	Me	1.05,t,6; 2.1,s,6; 2.35,q,2; 2.95,q,2; 5.2,s,2; 6.7,bs,2; 7.0,d,1; 7.8,d,1; 9.5,s,1	a	173

2-NHR [R = CO(CH$_2$)$_3$Cl]

H	CN	Me	2.05,s,3; 2.10,s,3; 2.20–2.80,m,4; 3.65,t,2; 9.75,s,1; 10.30,s,1	c	302
Ac	CN	Me	2.03,s,3; 2.29,s,3; 2.00–2.40,m,2; 2.57,s,3; 2.60,t,2; 3.62,t,2; 8.88,s,1	c	302

2-NHR [R = COCH(Me)NH$_2$]

i-Bu·HCl	CN	Me	0.82,d,6; 1.3–2.0,m,1; 1.50,d,3; 1.95,s,3; 2.31,d,2; 4.06,q,1; 9.01,bs,4; 11.32,bs,1	a	282

2-NHR [R = 1-piperidinyl-CH(Me)CO]

Me·HCl	CN	Me	1.56,d,3; 2.00,bm,4; 1.95,s,3; 2.05,s,3; 3.31,m,4; 4.25,q,1; 11.35,s,2; 11.55,s,1	a	282

TABLE 3.6. Continued

R^1	R^3	R^4	R^5	ppm, mult., nH	Solvent^b	Ref.
\multicolumn{7}{l}{2-NHR [R = COCH(Me)Br]}		

R^1	R^3	R^4	R^5	ppm, mult., nH	Solvent^b	Ref.
				2-NHR [R = COCH(Me)Br]		
H	CN	Me	Me	1.73,d,3; 1.94,s,3; 2.04,s,3; 4.72,q,1; 10.78,s,1; 11.25,s,1	a	282
				2-NHR (R = COCH$_2$NEt$_2$)		
H	CN	Me	Me·HCl	1.19,t,6; 1.84,s,3; 1.95,s,3; 3.10,q,4; 3.89,s,2; 10.10–10.70,bs,1; 11.18,bs,1; 11.30,bs,1	a	281
H	CN	Me	i-Bu·HCl	0.82,d,6; 0.8–1.2,m,1; 1.28,t,6; 1.95,s,3; 2.3,d,2; 3.28,q,4; 4.15,s,2; 10.5–11.0,bs,1; 11.47,s,2	a	281
H	CONH$_2$	Me	Me·HCl	1.19,t,6; 2.08,s,6; 3.14,q,4; 4.05,bs,2; 6.55,bs,2; 9.90–10.50,bs,1; 10.70,bs,1; 10.95,bs,1	a	301
				2-NHR (R = COCH$_2$Cl)		
H	CN	Me	Me	1.85,s,3; 1.95,s,3; 4.12,s,2; 10.41,bs,1; 10.95,bs,1	a	281
H	CONH$_2$	Me	Me	1.95,s,6; 4.20,s,2; 6.42,bs,2; 10.72,bs,1; 10.89,bs,1	a	301
				2-NHR (R = COCF$_3$)		
CH$_2$Ph	COO—t-Bu	Me	Me	1.53,s,9; 2.03,s,3; 2.15,s,3; 5.0,s,2; 6.75–7.3,m,5; 8.8–9.0,bs,1	c	305
Ph	CN	Me	Me	1.96,s,3; 2.13,s,3; 7.04–7.50,m,5; 8.78,s,1	c	290
				2-NHR (R = COPh)		
Me	H	COPh	SMe	2.05,s,3; 3.40,s,3; 7.12,s,1; 7.2–7.7,m,10	d	153
Me	H	4-MeOC$_6$H$_4$CO	SMe	1.98,s,3; 3.22,s,3; 3.56,s,3; 6.7–7.9,m,9; 8.67,s,1	c	153
Me	H	4-ClC$_6$H$_4$CO	SMe	2.23,s,3; 3.50,s,3; 6.93,s,1; 7.3–8.1,m,9; 8.44,s,1	c	153
Me	Ph	Me	OH	2.03,d,3; 2.88,s,3; 6.66,dd,1; 7.2–7.9,m,11	a	244

				NMR data		
Et	H	COPh	SMe	1.23,t,3; 2.20,s,3; 3.82,q,2; 6.90,s,3; 7.4,m,6; 7.8,m,4; 8.42,s,1	c	153
Et	H	4-MeOC$_6$H$_4$CO	SMe	1.10,t,3; 2.03,s,3; 3.65,s,3; 3.82,q,2; 6.85,s,1; 7.0–8.0,m,9	c	153
Et	H	4-ClC$_6$H$_4$CO	SMe	1.36,t,3; 2.23,s,3; 3.84,q,2; 6.97,s,1; 7.2–8.0,m,9; 8.30,s,1	c	153
CH$_2$Ph	H	COPh	SMe	2.34,s,3; 5.06,s,2; 7.00,s,1; 7.1–9.9,m,15; 7.90,s,1	c	153
CH$_2$Ph	H	4-MeOC$_6$H$_4$CO	SMe	2.38,s,3; 3.90,s,3; 5.12,s,2; 6.9–7.9,m,17	d	153
CH$_2$Ph	H	4-ClC$_6$H$_4$CO	SMe	2.04,s,3; 4.95,s,2; 6.6–7.7,m,16	d	153
CH$_2$OMe	Ph	Me	OH	1.98,s,3; 3.33,s,3; 4.84,d,2; 7.0–7.9,m,12	c	244
CH$_2$OAc	Ph	Me	OH	1.97,s,3; 2.12,s,3; 5.45,d,2; 6.91,bs,2; 7.3–7.75,m,10	e	244
Ph	CONH$_2$	Ph	Me	1.96,s,3; 2.20,s,3; 6.80,s,2; 7.2–7.9,m,10; 9.9,s,1	a	290
CHO	Ph	Ph	OH	2.16,s,3; 7.0–7.7,m,12; 9,10,bs,1	c	244

2-NHR (R = COOEt)

H	CN	Me	Me	1.21,t,3; 1.98,s,3; 2.07,s,3; 4.1,q,2; 9.5,bs,1; 11.05,bs,1	a	291
H	CN	Me	Et	1.05,t,3; 1.22,t,3; 1.95,s,3; 2.42,q,2; 4.11,q,2; 9.46,bs,1; 11.03,bs,1	a	291
H	CN	Me	i-Bu	0.82,d,6; 0.8–1.2,m,1; 1.20,t,3; 1.92,s,3; 2.27,d,2; 4.07,q,2; 9.38,bs,1; 11.03,bs,1	a	291
H	CN	Me	CH$_2$Ph	1.2,t,3; 1.97,s,3; 3.79,s,2; 4.1,q,2; 7.17,s,5; 9.52,bs,1; 11.24,bs,1	a	291
H	CN	Me	4-OHC$_6$H$_4$CH$_2$	1.2,t,3; 1.98,s,3; 3.58,s,2; 4.06,q,2; 6.78,m,4; 9.09,s,1; 9.45,bs,1; 11.13,bs,1	a	291
H	CN	Me	Ph	1.25,t,3; 2.24,s,3; 4.18,q,2; 7.4,s,5; 9.75,bs,1; 11.55,bs,1	a	291
H	CONH$_2$	Me	Me	1.22,t,3; 2.04,s,6; 4.24,q,2; 6.54,bs,2; 10.1,bs,1; 10.6,bs,1	a	291

TABLE 3.6. Continued

R^1	R^3	R^4	R^5	ppm, mult., nH	Solvent[b]	Ref.
H	CONH$_2$	Me	Et	1.03,t,3; 1.23,t,3; 2.07,s,3; 2.46,q,2; 4.11,q,2; 6.61,bs,2; 10.02,bs,1; 10.46,bs,1	a	291
H	CONH$_2$	Me	i-Bu	0.84,d,6; 0.8–1.2,m,1; 1.24,t,3; 2.07,s,3; 2.34,d,2; 4.14,q,2; 6.60,bs,2; 10.02,bs,1; 10.42,bs,1	a	291
H	CONH$_2$	Me	CH$_2$Ph	1.22,t,3; 2.14,s,3; 3.85,s,2; 4.12,q,2; 6.65,s,2; 7.15,s,5; 10.09,bs,1; 10.66,bs,1	a	291
H	CONH$_2$	Me	4-OHC$_6$H$_4$CH$_2$	1.2,t,3; 2.09,s,3; 3.65,s,2; 4.12,q,2; 6.61,s,2; 6.84,m,4; 9.03,s,1; 10.0,bs,1; 10.5,bs,1	a	291
H	CONH$_2$	Me	Ph	1.22,t,3; 2.26,s,3; 4.12,q,2; 6.75,bs,2; 7.32,s,5; 9.6,bs,1; 10.8,bs,1	a	291
2-NR_2 ($R = Me$)						
Me	Ph	OH	H	2.8,c6; 2.9,c3; 3.77,c2; 7.22,c5	c	190
Me	COOEt	Me	Ph	1.37,t,3; 2.13,s,3; 2.83,s,6; 3.30,s,3; 4.30,q,2; 7.27,m,5	c	185
t-Bu	H	NO$_2$	H	1.65,s,9; 2.61,s,6; 6.41,d,1; 7.19,d,1	e	243
2-NR_2 ($R = Et$)						
H	Me	Ph	COOMe	1.05,t,6; 1.81,s,3; 3.05,q,4; 3.56,s,3; 7.18,m,5; 8.45,br,1	c	164
H	Me	Ph	COOEt	1.06,t,9; 1.82,s,3; 3.04,q,4; 4.03,q,2; 7.19,m,5; 8.45,br,1	c	164
H	Me	=O	Me,Me	1.18,s,6; 1.23,t,6; 1.97,s,3; 3.54,q,4; 7.71,s,1	e	161
H	Me	=O	Me,Ph	1.25,t,6; 1.62,s,3; 1.85,s,3; 3.48,q,4; 4.75,s,1; 7.3,m,5	c	161
Et	Me	=O	Me,Ph	1.13,t,3; 1.15,t,6; 1.57,s,3; 1.89,s,3; 3.32,q,4; 3.88,q,2; 7.3,m,5	e	161

COCF$_3$	Me	=O	Me,Me	1.20,t,6; 1.41,s,6; 1.84,s,3; 3.36,q,4	e	161
COCF$_3$	Me	=O	Me,Ph	1.16,t,6; 1.77,s,6; 3.35,q,4; 7.20,s,5	e	161
2-NR$_2$ (R = Ac)						
2-Furyl-CH$_2$	CN	Me	Me	2.1,s,3; 2.2,s,6; 2.3,s,3; 5.0,s,2; 6.5,d,2; 7.65,m,1	a	173
2-Thienyl-CH$_2$	CN	Me	Me	2.1,s,3; 2.2,s,6; 2.3,s,3; 5.2,s,2; 7.0,d,2; 7.55,m,1	a	173
Ac	CN	Ph	H	2.32,s,6; 2.57,s,3; 7.3,s,1; 7.3–7.5,m,5	c	180
2-NR$_2$ (R = COEt)						
2-Furyl-CH$_2$	CN	Me	Me	1.0,t,6; 2.1,s,3; 2.3,s,3; 2.5,q,4; 5.0,s,2; 6.4,m,2; 7.6,m,1	a	173
2-Thienyl-CH$_2$	CN	Me	Me	1.0,t,6; 2.1,s,3; 2.3,s,3; 2.5,q,4; 5.2,s,2; 7.0,m,2; 7.55,m,1	a	173
2-NR$_2$ (R = SiMe$_3$)						
H	Ph	H		0.08,s,18; 6.98,d,1; 7.21,t,1; 7.34,t,2; 7.55,t,2; 8.71,bs,1	c	194
H	Ph	Ph		0.01,s,18; 7.09–7.27,m,10; 8.80,bs,1	c	194
H	SiMe$_3$	C≡CSiMe$_3$		0.1,s,18; 0.23,s,9; 0.32,s,9; 8.3–8.6,br,1	c	193
2-NRR [R–R = —(CH$_2$)$_4$—]						
H	CN	Ph		1.86,m,4; 3.44,m,4; 6.47,d,1; 7.10–7.50,m,5; 10.3,s,1	a	186
2-NRR [R–R = —(CH$_2$)$_5$—]						
H	CN	Me		1.52,m,6; 2.06,s,3; 3.35,m,4; 7.15–7.45,m,5; 10.58,s,1	a	186
Me	H	NO$_2$		1.63,m,6; 2.77–2.98,m,4; 3.68,s,3; 5.49,d,1; 6.88,d,1	e	240
2-NRR [R–R = —(CH$_2$)$_2$O(CH$_2$)$_2$—]						
Me	H	NO$_2$		2.8,m,4; 3.46,s,3; 3.7,m,4; 6.10,d,1; 7.19,d,1	c	242

TABLE 3.6. Continued

R^1	R^3	R^4	R^5	ppm, mult., nH	Solvent[b]	Ref.
			$2\text{-}NRR\ [R\text{-}R = -CO(CH_2)_3-]$			
H	CN	Me	Me	2.05,s,3; 2.15,s,3; 2.2–2.26,m,4; 3.90,t,2; 10.0,s,1	a	336
Me	CN	Me	Me	2.00,s,3; 2.10,s,3; 2.40,m,4; 3.10,s,3; 3.65,m,2	a	336
			$2\text{-}NRR'\ (R = Me;\ R' = Ph)$			
CH_2Ph	Ph	Ph	$CH(COPh)_2$	2.65,s,3; 4.16–4.54,d,2; 6.3–7.6,m,30; 18.37,s,1	c	163
			$2\text{-}NRR'\ (R = Me;\ R' = COOEt)$			
H	CN	Me	Me	1.19,t,3; 1.97,s,3; 2.06,s,3; 3.15,s,3; 4.11,q,2; 11.4,bs,1	a	291
H	$CONH_2$	Me	Me	1.14,t,3; 2.03,s,6; 3.08,s,3; 4.05,q,2; 6.45,bs,2; 10.8,bs,1	a	291
			$2\text{-}NRR'\ (R = Et;\ R' = C_2H_4COOEt)$			
CH_2OCH_2Ph	COOEt	H	Me	0.98,t,3; 1.22,t,3; 1.33,t,3; 2.22,s,3; 2.37,t,2; 3.16,q,2; 3.46,bm,2; 4.08,q,2; 4.27,q,2; 4.53,s,2; 5.27,bs,2; 6.29,s,1; 7.38,m,5	c	184
			$2\text{-}NRR'\ (R = Et;\ R' = COCF_3)$			
CH_2Ph	COO—t-Bu	Me	Me	0.93,t,3; 1.5,s,9; 1.97,s,3; 2.2,s,3; 3.5,m,2; 4.95,m,2; 6.8–7.4,m,5	c	305
			$NRR'\ (R = Ac;\ R' = 4\text{-}BrC_6H_4COCH_2)$			
H	CN	Me	Me	2.08,s,6; 2.13,s,3; 5.03,s,2; 7.71,q,4; 9.00–9.40,bs,1	c	298

[a] Uncompleted spectra were also reported.[180;188;190;191;239] In Ref. 150 the data are available on a supplementary publication.
[b] Solvent: a, dimethylsulfoxide-d_6; b, trifluoroacetic acid; c, deuterochloroform; d, deuterochloroform-trifluoroacetic acid; e, carbon tetrachloride.
[c] Not reported.

TABLE 3.7. ¹H CHEMICAL SHIFTS OF THE 3-AMINOPYRROLES[a]

3-NH_2

R^1	R^2	R^4	R^5	ppm, mult., nH	Solvent[b]	Ref.
H	Ph	H	Me	2.13,s,3; 3.91,bs,2; 5.40,d,1; 6.94,d,1; 7.26,t,2; 7.45,d,2;9.98,bs,1;	a	262
H	Ph	H	Ph	2.22,s,3; 5.93,d,1; 7.29,t,1; 7.45,t,2; 7.50,d,2; 9.81,bs,3; 11.28,bs,1	b	262
H	Ph	H	Ph	4.11,bs,2; 6.14,d,1; 7.08,t,1; 7.14,t,1; 7.35,m,4; 7.67,m,4; 10.40,bs,1	a	262
H	Ph	H	Ph	6.38,d,1; 6.98,t,1; 7.14,t,3; 7.25,t,2; 7.39,d,2; 7.48,t,2; 9.74,b,3; 11.43,bs,1	b	262
H	Ph	CN	Ph	4.51,s,2; 7.19–7.90,m,10; 11.44,s,1	a	247
H	Ph	Ac	Me	2.34,s,3; 2.50,s,3; 5.30,bs,2; 7.06,t,1; 7.34,t,2; 7.51,d,2; 10.94,bs,1	a	262
H	Ph	Ac	Me	2.48,s,3; 2.59,s,3; 7.40,t,1; 7.52,m,4; 10.20,bs,3; 12.03,bs,1	b	262
H	Ph	Ac	Ph	1.92,s,3; 5.36,s,2; 6.97–7.87,m,10; 11.33,s,1	a	263
H	Ph	3,5-Me₂isoxazol-4-ylCO	Ph	1.97,s,3; 2.22,s,3; 5.10,s,2; 7.10–7.80,m,10; 11.30,s,1	a	260
H	Ph	CONH₂	Me	2.50,s,3; 4.90,bs,2; 6.80,bs,2; 7.15–7.85,m,5; 10.70,s,1	a	207
H	Ph	CONHPh	Me	2.50,s,3; 4.40,s,2; 7.00–7.80,m,10; 10.30,s,1; 10.90,s,1	a	207

TABLE 3.7. Continued

R^1	R^2	R^4	R^5	ppm, mult., nH	Solvent[b]	Ref.
H	Ph	COOEt	Me	1.28,t,3; 2.41,s,3; 4.20,q,2; 4.84,bs,2; 7.05,t,1; 7.34,t,2; 7.52,d,2; 10.97,bs,1	a	262
				1.33,t,3; 2.47,s,3; 4.28,q,2; 7.40,t,1; 7.51,m,4; 9.80,b,3; 12.03,bs,1	b	211
H	Ph	COOEt	Ph	1.06,t,3; 4.07,q,2; 4.93,s,2; 7.09–7.72,m,10; 11.17,s,1	a	262
H	Ph	COOEt	COOEt	1.33,t,3; 1.38,t,3; 4.32,q,2; 4.40,q,2; 4.2–4.6,bs,2; 7.3–7.7,m,5; 9.75,s,1	c	247
H	2-$NH_2C_6H_4$	Ac	Me	2.27,s,3; 2.42,s,3; 4.77,s,2; 4.90,s,2; 6.50–7.17,m,4; 10.45,s,1	a	206
H	3-$MeOC_6H_4$	Ac	Me	2.33,s,3; 2.50,s,3; 3.80,s,3; 5.36,s,2; 6.70–7.50,m,4; 11.00,s,1	a	261
H	3-$MeOC_6H_4$	COOEt	Me	1.30,t,3; 2.50,s,3; 3.83,s,3; 4.30,q,2; 4.93,s,2; 6.70–7.45,m,4; 11.00,s,1	a	248
H	4-ClC_6H_4	CH_2COOEt	H	1.25,t,3; 3.36,s,2; 3.43,s,2; 4.17,q,2; 6.58,d,1; 7.35,s,4; 7.9–8.2,b,1	c	248
H	$CONH_2$	COOEt	Me	1.27,t,3; 2.36,s,3; 4.19,q,2; 5.66,bs,2; 6.61,bs,2; 10.9,b,1	a	197
H	CONHMe	COOEt	Me	1.30,t,3; 2.36,s,3; 2.74,d,3; 4.20,q,2; 5.60,s,2; 6.95,d,1; 10.77,bs,1	a	210
H	1-PiperimidylCO	COOEt	Me	1.27,t,3; 1.55,bs,6; 2.35,s,3; 3.41,bs,4; 4.20,q,2; 5.13,s,2; 10.91,b,1	a	210
H	COOEt	Me	H	1.35,t,3; 1.94,d,3; 4.31,q,2; 4.31,bs,2; 6.53,d,1; 8.02,bs,1	a	210
H	COO—t-Bu	—$CO(CH_2)_3$—		1.50,s,9; 1.98,bq,2; 2.28,bt,2; 2.65,t,2; 5.50,bs,2; 11.00,bs,1	c	202
					a	210

H	COO—t-Bu	Me	1.35,t,3; 1.60,s,9; 2.46,s,3; 4.29,q,2; 5.45,bs,2; 9.85,bs,1	c	210
H	COO—t-Bu	Me	1.57,s,9; 2.43,s,3; 5.2,bs,2; 7.26–7.81,m,5; 9.60,bs,1	c	210
Me	Ph	H	3.52,s,3; 5.22,s,2; 6.20,s,1; 6.62–7.49,m,10	a	249
Me	Ph	Ph	3.29,s,3; 7.16–7.50,m,17	a	249
Me	Ph	CN	3.38,s,3; 4.28,s,2; 7.44–7.57,m,10	a	249
Me	Ph	Ac	2.36,s,3; 2.46,s,3; 3.35,s,3; 4.85,bs,2; 7.32,s,5	a	249
Me	Ph	COOEt	1.29,t,3; 2.47,s,3; 3.36,s,3; 4.21,q,2; 4.41,bs,2; 7.28,m,3; 7.44,t,2;	a	262 249
			1.32,t,3; 2.53,s,3; 3.37,s,3; 4.28,q,2; 7.44,t,2; 7.53,m,3; 9.60,b,3	b	262
CH$_2$Ph	COOEt	Me	1.23,t,3; 1.94,s,3; 4.21,q,2; 4.49,bs,2; 5.31,s,2; 6.49,s,1; 7.39,m,5	c	202
CH$_2$Ph	COOEt	Ph	1.22,t,3; 4.21,q,2; 4.72,bs,2; 5.36,s,2; 6.77,s,1; 7.25,m,10	c	202
CH$_2$COOEt	COO—t-Bu	—CO(CH$_2$)$_3$—	1.26,t,3; 1.51,s,9; 2.10,q,2; 2.32,f,3; 2.54,bt,2; 4.16,q,2; 4.77,s,2; 5.66,s,2	c	210
CPh$_3$	H	H	2.87,b,2; 5.75,q,1; 6.11,t,1; 6.35,t,1; 7.17–7.27,m,15	c	245
		3-NHR (R = Me)			
t-Bu	H	NO$_2$	1.55,s,9; 2.81,s,3; 5.91,b,1; 5.94,d,1; 7.32,d,1	d	243
		3-NHR (R = i-Pr)			
i-Pr	Ph	COPh	0.92,d,6; 1.25,d,6; 2.95,h,1; 4.13,h,1; 5.50,s,1; 7.02,s,1; 7.40,s,5; 7.50,m,3; 7.80,m,2	c	195
		3-NHR (R = t-Bu)			
t-Bu	H	NO$_2$	1.32,s,9; 1.55,s,9; 5.5,b,1; 6.02,d,1; 7.34,d,1	d	243

TABLE 3.7. Continued

R¹	R²	R⁴	R⁵	ppm, mult., nH	Solvent[b]	Ref.
_____	_____	_____	3-NHR (R = c-Hexyl)			
c-Hexyl	Ph	COPh	H	1.25,b,22; 2.43,b,1; 3.53,b,1; 5.18,s,1; 6.97,s,1; 7.40,s,5; 7.47,m,3; 7.75,m,2	c	195
			3-NHR [R = C(NH)NH₂]			
H	H	H	H·AcOH	2.05,s,3; 6.12,t,1; 6.82,d,2	e	267
H	H	H	COOH·HCl	6.61,d,1; 6.89,d,1	d	267
H	H	H	COOMe·HCl	3.83,s,3; 6.87,d,1; 7.12,d,1	e	267
H	COOH	H	H·HCl	6.43,d,1; 6.72,d,1	f	267
H	COOMe	H	H·HCl	3.92,s,3; 6.38,d,1; 7.17,d,1	e	267
			3-NHR (R = CHO)			
CH₂COOEt	H	—CO(CH₂)₃—		1.27,t,3; 2.18,o,2; 2.46,q,2; 2.68,t,2; 4.23,q,2; 4.52,s,2; 7.26,s,1; 8.32,s,1; 9.22,bs,1	c	210
CH₂COOEt	COO—t-Bu	—CO(CH₂)₃—		1.33,t,3; 1.57,s,9; 2.1–2.8,m,6; 4.27,q,2; 5.00,s,2; 8.8,bs,1; 9.0,bs,1	c	210
			3-NHR (R = Ac)			
H	(CH₂)₂CN	—CO(CH₂)₃—		2.13,s,3; 8.56,bs,1; 9.47,bs,1	b	210
H	CH₂COOEt	—CO(CH₂)₃—		1.18,t,3; 1.94,s,3; 2.27,t,2; 2.70,t,2; 3.54,s,2; 4.05,q,2; 8.88,s,1; 11.09,s,1	a	210
H	Ph	Ac	Me	2.0,s,3; 2.28,s,3; 2.46,s,3; 7.25–7.70,m,5; 9.35,s,1; 11.62,s,1	a	317

H	Ph	COPh	Me	1.43,s,3; 2.32,s,3; 7.2–7.8,m,10; 8.85,s,1; 11.41,s,1	a	317
H	COO—*t*-Bu	—CO(CH_2)$_3$—		1.55,s,9; 2.1,g,2; 2.13,s,3; 2.4,g,2; 2.80,t,2; 8.18,s,1; 10.58,s,1	c	210
			3-NHR (R = $COCH_2Ac$)			
H	Ph	Ac	Me	2,18,s,3; 2.28,s,3; 2.45,s,3; 3.55,s,2; 7.2–7.7,m,5; 9.55,s,1; 11.6,s,1	a	317
			3-NHR (R = $COCH_2COOEt$)			
H	Ph	Ac	Me	1.15,t,3; 2.20,s,3; 2.40,s,3; 3.22,s,2; 4.05,q,2; 7.15–7.70,m,5; 9.40,s,1; 11.42,s,1	a	317
			3-NHR (R = COPh)			
H	Ph	COOMe	COOMe	3.69,s,3; 3,81,s,3; 7.2–8.0,m,10; 9.72,s,1; 12.35,s,1	a	206
			3-NHR (R = $CONH_2$)			
H	H	COPh	Me	1.78,s,3; 6.23,s,2; 7.10,d,1; 7.5,s,5; 8.78,s,1; 11.05,bs,1	a	317
H	Ph	Ac	Me	2.1,s,3; 2.8,s,3; 4.9,s,2; 6.8–8.5,m,6; 12.4,s,1	h	317
H	Ph	COOMe	COOMe	3.72,s,3; 3.78s,3; 5.68,s,2; 7.28–7.80,m,5; 12.30,bs,2	a	317
H	Ph	COOEt	Me	1.28,t,3; 2.43,s,3; 4.13,q,2; 3.8–4.4,bs,2; 5.60,s,1; 7.0–7.7,m,5; 11.35,s,1	a	317
			3-NHR (R = CONHMe)			
H	COOEt	Me	H	1.26,t,3; 1.88,s,3; 2.59,d,3; 4.19,q,2; 6.35,q,1; 6.67,d,1; 7.43,s,1; 11.15,bs,1	a	202
			3-NHR (R = CONHPh)			
H	Ph	COOEt	Me	1.28,t,3; 2.50,s,3; 2.76,s,2; 4.25,q,2; 7.0–8.0,m,5; 10.85,s,1	b	317

TABLE 3.7. Continued

R^1	R^2	R^4	R^5	ppm, mult., nH	Solvent[b]	Ref.
			3-NHR (R = CSNHMe)			
H	COOEt	Me	H	1.23,t,3; 1.85,s,3; 2.84,d,3; 4.14,q,2; 6.77,d,1; 7.14,bs,1; 8.64,s,1; 11.52,bs,1	a	202
			3-NHR (R = COOEt)			
H	COO—t-Bu	—CO(CH$_2$)$_3$—		1.27,t,3; 1.52,s,9; 2.14,bq,2; 2.44,bq,2; 2.80,t,2; 4.16,q,2; 7.50,s,1; 10.46,s,1	c	210
			3-NHR (R = COO-t-Bu)			
H	H	H	H	1.45,s,9; 5.98,m,1; 6.50,b,1; 6.57,m,1; 6.85,b,1; 8.45,b,1	i	245
H	COCF$_3$	H	H	1.50,s,9; 6.94,t,1; 7.10,t,1; 8.85,b,1; 9.12,b,1	i	245
CPh$_3$	H	H	H	1.41,s,9; 6.05,b,1; 6.18,b,1; 6.45,t,1; 6.64,b,2; 7.18–7.28,m,15	c	245
			3-NHR (R = COOCH$_2$Ph)			
CPh$_3$	H	H	H	5.11,s,2; 6.04,b,1; 6.37,b,1; 6.45,q,1; 6.75,b,1; 7.19–7.26,m,20	c	245
			3-NR$_2$ (R = Me)			
H	CONMe$_2$	COOMe	COOMe	2.71,s,6; 3.07,s,6; 3.80,s,3; 3.87,s,3;	c	200
t-Bu	H	NO$_2$	H	1.55,s,9; 2.66,s,6; 6.18,d,1; 7.48,d,1	d	243

		3-NR₂ (R = Ac)			
H	Ph	—(CH₂)₄—	1.50–2.70,m,14; 7.40,m,5; 11.10,s,1;	a	199
H	Ph	—(CH₂)₅—	1.56–2.80,m,14; 7.24,s,5; 8.22,bs,1	c	199
H	Ph	—(CH₂)₅—	1.50–2.00,m,6; 2.30–2.82,m,10; 7.10–7.80,m,5; 11.84,bs,1	h	199
		3-NRR [R—R = —(CH₂)₅—]			
Me	H	NO₂	1.7,m,6; 2.9,m,4; 3.59,s,3; 6.08,d,1; 7.36,d,1	c	242
Ac	CH₂Ph	H	1.32,t,3; 1.88,m,4; 2.03,s,3; 3.08,m,4; 4.17,s,2; 4.25,d,2; 6.78,s,1; 7.02,d,2; 7.24,d,2; 7.31,m,1	c	217
COO—t-Bu	C₂H₄SMe	H	1.28,t,3; 1.53,s,9; 1.84,m,4; 2.11,s,3; 2.68,m,2; 3.00,m,6; 4.21,q,2; 6.75,s,1	c	217
COO—t-Bu	CH₂Ph	H	1.24,s,9; 1.32,t,3; 1.86,m,4; 3.01,m,4; 4.23,s,2; 4.25,q,2; 6.75,s,1; 7.01,d,2; 7.15,d,1; 7.21,d,2	j	217
		3-NRR [R—R = —(CH₂)₂O(CH₂)₂—]			
H	COOMe	Ph	2.93,m,4; 3.61,m,4; 3.70,s,3; 6.68,d,1; 7.25,m,5; 9.32,b,1	c	204
Me	H	NO₂	3.0,m,4; 3.64,s,3; 3.9,m,4; 6.14,d,1; 7.46,d,1	c	242
COOEt	CH₂Ph	H	1.07,t,3; 1.32,t,3; 2.78,m,4; 3.73,m,4; 4.06,q,2; 4.18,s,2; 4.25,q,2; 6.88,s,1; 7.05,d,2; 7.19,m,3	c	217

[a] Other uncompleted spectra were also reported.[206,265]
[b] Solvent: a, dimethylsulfoxide-d_6; b, dimethylsulfoxide-d_6/trifluoroacetic acid; c, deuterochloroform; d, carbon tetrachloride; e, deuterium oxide, 3-(trimethylsilyl)propanesulfonic acid sodium salt as external standard; f, deuterium oxide–sodiumdeuteroxide, 3-(trimethylsilyl)propanesulfonic acid sodium salt as external standard; g, undefined; h, pyridine-d_5; i, dichloromethane-d_2; j, methanol-d_4.

TABLE 3.8. ^{1}H CHEMICAL SHIFTS OF THE 1,2-DIAMINOPYRROLES

R^3	R^4	R^5	ppm, mult., nH	Solvent[a]	Ref.
			1-NHR (R = CONH$_2$)		
CN	COOMe	Me	2.5,s,3; 4.0,s,3; 6.3,s,2; 6.6,s,2; 9.2,s,1	a	219
CN	COOEt	Me	1.2,t,3; 2.2,s,3; 3.9–4.3,q,2; 6.0,s,2; 6.3,s,2; 9.0,s,1	a	219
			1-NHR (R = CONHPh)		
CN	COOMe	Me	2.2,s,3; 3.7,s,3; 6.2, s,2; 6.9–7.6,m,5; 9.2,s,2	a	219
CN	COOEt	Me	1.3,t,3; 2.3,s,3; 4.0–4.4,q,2; 6.2,s,2; 7.0–7.6,m,5; 9.2,s,1; 9.3,s,1	a	219
			1-NHR (R = COOMe)		
CN	COOMe	Me	2.2,s,3; 3.7,s,3; 6.3,s,2; 10.3,s,1	a	219
CN	COOEt	Me	1.2,t,3; 2.2,s,3; 3.7,s,3; 4.0–4.3,q,2; 6.3,s,2; 10.5,s,1	a	219
			1-NHR (R = COOEt)		
CN	COOMe	Me	1.2,t,3; 2.2,s,3; 3.7,s,3; 3.9–4.3,q,2; 6.3,s,2; 10.3,s,1	a	219
			1-NHR (R = COO—t-Bu)		
CN	COOMe	Me	1.5,s,9; 2.2,s,3; 3.7,s,3; 6.2,s,2; 10.0,s,1	a	219
CN	COOEt	Me	1.4,t,3; 1.5,s,9; 2.2,s,3; 4.0–4.4,q,2; 6.2,s,2; 10.0,s,1	a	219

[a] solvent: a, dimethylsulfoxide-d_6.

3.3. Structure and Physical Properties

probably due to a reduced tautomerism of the amino group, becomes more evident with bulky 1-substituents as the planar imino form has a lower stability.

Alkylation of the amino group does not affect the chemical shifts, whereas 2-acylaminopyrroles show signals in the range 9.4–11.2 ppm; with benzoylamino derivatives a deshielding effect (2–3 ppm) is also observed.

The amino proton signals for 3-aminopyrroles (Table 3.7) lie in the range 3.4–5.7 ppm; the lower field resonances are observed for derivatives having electron-withdrawing groups on sites adjacent to the amino group. The electronic effect of the substituent is greater in the case of 2-substituted derivative as a result of bond fixation. The pyrrolyl NH signal is found in the range 8.0–11.4 ppm. Alkylation at the 1-position or on the amino group does not affect the chemical shifts to the amino protons. Acylamino derivatives, however, show resonances in the range 8.3–9.7 ppm, whereas for the 3-pyrrolylurethanes the amino proton signal is found at 6.1–8.9 ppm.

Tables 3.8 and 3.9 list ^{1}H NMR data for diaminopyrroles. The NMR spectra of 1,2-diaminopyrroles, unsubstituted at both the amino groups have not been recorded. However, the spectra of 2-amino-1-alkoxycarbonylamino- and 2-amino-1-ureidopyrroles show the signals due to the amino group at 6.0–6.3 ppm, whereas the signal for the urethane proton appears at 10.0–10.5 ppm and for the ureido proton, at 9.0–9.2 ppm.

TABLE 3.9. ^{1}H CHEMICAL SHIFTS OF THE 2,3-DIAMINOPYRROLES[a]

R^1	R^4	R^5	ppm, mult., nH	Solvent[b]	Ref.
			2,3 Di-NRR [R–R = —$(CH_2)_5$—]		
Me	NO_2	H	1.6,m,12; 3.1,m,8; 3.53,s,3; 7.50,s,1	a	242
t-Bu	NO_2	H	1.6,m,12; 2.3–2.7,m,8; 7.26,s,1	b	268
			2,3 Di-NRR [R–R = —$(CH_2)_2O(CH_2)_2$—]		
Me	NO_2	H	3.1,m,8; 3.49,s,3; 3.8,m,8; 7.29,s,1	c	242

[a] Other incomplete spectra in the series 2,4-diamino[220] and 2,5-diamino[284] were also reported.
[b] Solvent: a, acetone-d_6; b, carbon tetrachloride; c, deuterochloroform.

3.3.4. ^{13}C NMR Spectra

Only the ring carbon resonances of aminopyrroles are reported in Tables 3.10–3.12. The presence of an amino group at the 1-position of the ring does not affect the chemical shifts of the ring carbon atoms, which are found within the

TABLE 3.10. ^{13}C CHEMICAL SHIFTS OF THE 1-AMINOPYRROLES

R^2	R^3	R^4	R^5	ppm	Solvent^a	Ref.
\multicolumn{7}{c}{*1-NHR (R = Ph)*}						
Ph	CH$_2$Ph	CONEt$_2$	Me	C-2 126.7 C-3 114.3 C-4 119.9 C-5 130.9	a^b	357
Ph	Ph	CONHPh	Me	C-2 131.0 C-3 112.6 C-4 119.5 C-5 130.0	a^b	357
Ph	Ph	4-MeOC$_6$H$_4$NHCO	Me	C-2 130.9 C-3 112.9 C-4 119.4 C-5 130.1	a^b	357
Ph	Ph	4-ClC$_6$H$_4$NHCO	Me	C-2 128.2 C-3 112.5 C-4 119.6 C-5 129.9	a^b	357
Ph	Ph	CONEt$_2$	Me	C-2 131.0 C-3 115.7 C-4 119.0 C-5 128.8	a^b	357

1-NHR (R = 4-NO$_2$C$_6$H$_4$)

—(CH$_2$)$_4$—	CONHPh	Me	C-2 129.3 C-3 114.1 C-4 114.3 C-5 134.4	a[b]	357
—(CH$_2$)$_4$—	4-MeOC$_6$H$_4$NHCO	Me	C-2 126.8 C-3 113.6 C-4 114.2 C-5 132.2	a[b]	357
—(CH$_2$)$_4$—	4-ClC$_6$H$_4$NHCO	Me	C-2 127.2 C-3 113.6 C-4 114.3 C-5 131.3	a[b]	357
—(CH$_2$)$_4$—	CONEt$_2$	Me	C-2 124.6 C-3 112.9 C-4 114.1 C-5 126.3	a[b]	357

1-NHR (R = COPh)

Me	COOMe	Ac	Me	C-2 131.7 C-3 108.6 C-4 121.5 C-5 136.5	a	358
Me	COOMe	CONHPh	Me	C-2 139.6 C-3 107.3 C-4 115.3 C-5 132.5	a	358
Me	COOMe	4-MeOC$_6$H$_4$NHCO	Me	C-2 136.1 C-3 107.3 C-4 115.4 C-5 132.8	a	358

TABLE 3.10. Continued

R^2	R^3	R^4	R^5	ppm	Solvent[a]	Ref.
Me	COOMe	4-ClC$_6$H$_4$NHCO	Me	C-2 134.2 C-3 107.3 C-4 114.9 C-5 136.6	a	358
Me	COOEt	CONHPh	Ph	C-2 139.6 C-3 108.7 C-4 118.8 C-5 132.0	a	358
Me	COOEt	COOMe	Me	C-2 134.7 C-3 109.9 C-4 110.2 C-5 134.5	a[b]	358
	1-NHR (R = 3-MeC$_6$H$_4$CO)
Me	COOMe	Ac	Me	C-2 131.8 C-3 108.6 C-4 121.5 C-5 136.4	a	358
Me	COOEt	COOEt	Me	C-2 134.4 C-3 110.2 C-4 110.2 C-5 134.4	a	358
	1-NHR (R = CONH$_2$)
H	COOH	H	Me	C-2 126.73 C-3 112.81 C-4 105.46 C-5 130.79	b	118

406

H	COOEt	H	Me	C-2 126.66 C-3 112.04 C-4 105.08 C-5 130.94	b	118
Me	H	COOH	COOEt	C-2 134.88 C-3 106.64 C-4 118.35 C-5 123.67	b	118
Me	H	COOEt	COOH	C-2 135.54 C-3 106.12 C-4 115.67 C-5 124.58	b	118
Me	H	COOEt	COOEt	C-2 134.46 C-3 105.55 C-4 116.60 C-5 124.40	b	118
Me	COOMe	CONHPh	Ph	C-2 132.0 C-3 107.8 C-4 118.5 C-5 139.6	a	358
Me	COOMe	4-ClC$_6$H$_4$NHCO	Me	C-2 135.1 C-3 106.7 C-4 114.4 C-5 137.5	a	358
Me	COOMe	COOMe	Me	C-2 135.0 C-3 109.8 C-4 109.8 C-5 135.0	a	358
Me	COOEt	CONHPh	Ph	C-2 131.9 C-3 108.0 C-4 118.6 C-5 139.7	a	358

TABLE 3.10. Continued

R^2	R^3	R^4	R^5	ppm	Solvent[a]	Ref.
Me	COOEt	COOEt	Me	C-2 134.8 C-3 110.3 C-4 110.3 C-5 134.8	a	358
			1-NHR (R = CONHPh)			
Me	COOMe	COOMe	Me	C-2 135.3 C-3 110.1 C-4 110.1 C-5 135.3	a	358
Me	COOEt	4-MeOC$_6$H$_4$NHCO	Me	C-2 133.0 C-3 107.0 C-4 115.2 C-5 136.7	a	358
Me	COOEt	4-ClC$_6$H$_4$NHCO	Me	C-2 134.2 C-3 107.1 C-4 114.7 C-5 137.2	a	358
Me	COOEt	CONEt$_2$	Me	C-2 138.1 C-3 106.4 C-4 114.1 C-5 126.8	a	358
			1-NHR (R = COOMe)			
Me	COOMe	4-MeOC$_6$H$_4$NHCO	Me	C-2 137.1 C-3 107.0 C-4 114.5 C-5 135.2	a	358

Me	COOMe	4-ClC$_6$H$_4$NHCO	Me	C-2 137.6 C-3 106.9 C-4 113.8 C-5 137.5	a[c]	358
Me	COOMe	COOMe	Me	C-2 134.7 C-3 110.0 C-4 110.0 C-5 134.7	a	358
Me	COOEt	COOMe	Me	C-2 134.6 C-3 110.0 C-4 110.3 C-5 134.4	a	358

1-NHR (R = COOEt)

Me	COOMe	Ac	Me	C-2 132.2 C-3 108.6 C-4 121.2 C-5 136.5	a	358
Me	COOMe	COPh	Me	C-2 132.3 C-3 109.6 C-4 118.3 C-5 136.8	a	358
Me	COOMe	CONHPh	Ph	C-2 139.2 C-3 108.2 C-4 118.1 C-5 134.0	a	358

1-NHR (R = COO—t-Bu)

Me	COOMe	CONEt$_2$	Me	C-2 137.1 C-3 105.6 C-4 114.0 C-5 126.5	a	358

TABLE 3.10. Continued

R^2	R^3	R^4	R^5	ppm	Solvent[a]	Ref.
Me	COOMe	COOMe	Me	C-2 134.8 C-3 110.2 C-4 110.2 C-5 134.8	a	358
Me	COOMe	SO$_2$Me	Me	C-2 138.4 C-3 108.4 C-4 115.5 C-5 137.8	a	358
Me	COOMe	SO$_2$Ph	Ph	C-2 139.3 C-3 110.4 C-4 118.8 C-5 137.6	a	358
Me	COOMe	4-MeC$_6$H$_4$SO$_2$	Me	C-2 138.3 C-3 109.0 C-4 115.2 C-5 138.2	a	358
Me	COOMe	4-MeC$_6$H$_4$SO$_2$	Ph	C-2 138.9 C-3 110.4 C-4 119.3 C-5 137.3	a	358
Me	COOEt	CONH$_2$	Me	C-2 137.4 C-3 107.3 C-4[d] C-5 136.9	a	358

1-NHR (R = SO$_2$Ph)						
Me	COOMe	CONHPh	Ph	C-2 138.3 C-3 108.8 C-4 119.3 C-5 131.9	a	358
Me	COOEt	4-MeOC$_6$H$_4$NHCO	Me	C-2 137.2 C-3 108.0 C-4 115.8 C-5 133.6	a	358
Me	COOEt	CONEt$_2$	Me	C-2 138.2 C-3 106.6 C-4 114.3 C-5 127.2	a	358
1-NHR (R = 4-MeC$_6$H$_4$SO$_2$)						
Me	COOMe	CONH$_2$	Me	C-2 136.9 C-3 107.9 C-4 114.8 C-5 133.9	a	358
Me	COOMe	4-ClC$_6$H$_4$NHCO	Me	C-2 137.1 C-3 107.9 C-4 115.7 C-5 133.5	a	358
Me	COOEt	CONEt$_2$	Me	C-2 137.3 C-3 107.3 C-4 115.9 C-5 126.7	a	358
Me	COOEt	COOEt	Me	C-2 134.9 C-3 110.9 C-4 110.9 C-5 134.9	a	358

TABLE 3.10. Continued

R²	R³	R⁴	R⁵	ppm	Solvent[a]	Ref.
			1-NHR (R = 4-MeOC₆H₄SO₂)			
Me	COOMe	COOMe	Me	C-2 135.3 C-3 110.7 C-4 110.7 C-5 135.3	a	358
			1-NHR (R = 4-ClC₆H₄SO₂)			
Me	COOEt	COOEt	Me	C-2 134.6 C-3 111.0 C-4 111.0 C-5 134.6	a	358
			1-NHR (R = 2,4,6-Me₄C₆H₂)			
Me	COOMe	CONH₂	Me	C-2 137.3 C-3 107.8 C-4 114.4 C-5 133.3	a	358
			1-NHR [R = PO(OEt)₂]			
Me	COOMe	CONHPh	Me	C-2 139.1 C-3 106.7 C-4 114.1 C-5 137.3	b	145
Me	COOMe	CONHPh	Ph	C-2 139.1 C-3 108.0 C-4 117.5 C-5 134.9	b	145

Me	COOMe	4-MeOC$_6$H$_4$NHCO	Me	C-2 138.8 C-3 106.5 C-4^e C-5 136.7	b	145
Me	COOEt	4-MeOC$_6$H$_4$NHCO	Me	C-2 138.7 C-3 106.9 C-4^e C-5 136.8	b	145
Me	COOCH$_2$Ph	CONHPh	Me	C-2 139.0 C-3 106.6 C-4 114.3 C-5 135.9	b	145
Me	COOCH$_2$Ph	4-ClC$_6$H$_4$NHCO	Me	C-2 139.1 C-3 106.6 C-4 113.9 C-5 135.7	b	145
Me	COOCH$_2$Ph	CONEt$_2$	Me	C-2 138.5 C-3 105.4 C-4 113.9 C-5 127.6	b	145

1-NHR [R = PO(OPh)$_2$]

Me	COOMe	CONH$_2$	Me	C-2 137.8 C-3 107.3 C-4 113.7 C-5 135.9	b	145
Me	COOMe	CONEt$_2$	Me	C-2 138.1 C-3 105.9 C-4 114.4 C-5 127.5	b	145

TABLE 3.10. Continued

R²	R³	R⁴	R⁵	ppm	Solvent[a]	Ref.
Me	COOEt	CONHPh	Ph	C-2 139.2 C-3 108.5 C-4 117.2 C-5 134.9	b	145
Me	COOEt	4-ClC₆H₄NHCO	Me	C-2 138.8 C-3 107.1 C-4 114.2 C-5 136.8	b	145
Me	COOCH₂Ph	CONH₂	Me	C-2 139.1 C-3 107.1 C-4[e] C-5 135.9	b	145
Me	COOCH₂Ph	CONEt₂	Me	C-2 138.5 C-3 105.7 C-4 114.2 C-5 127.6	b	145

[a] Solvent: a, deuterochloroform; b, dimethylsulfoxide-d_6.
[b] The assignments of C-2 and C-5 as well as those of C-3 and C-4 can be reversed.
[c] The assignment of C-2 and C-5 can be reversed.
[d] Resonance not observed.
[e] Not assigned.

TABLE 3.11. ^{13}C CHEMICAL SHIFTS OF THE 2-AMINOPYRROLES[a]

$$\begin{array}{c} R^4 \diagdown\!\!\diagup R^3 \\ R^5 \diagdown\!\!\!\diagdown\!\!\!\diagup\!\!\!\diagup Z \\ \underset{R^1}{N} \end{array}$$

R^1	R^3	R^4	R^5	ppm	Solvent[c]	Ref.
				2-NH_2		
H	Ph	Ac	Me	C-2 133.58 C-3 103.38 C-4 119.45 C-5 127.54	a	236
H	COOEt	Ph	Me	C-2 146.95 C-3 90.82[b] C-4 117.71 C-5 117.31	a	185
2-Furyl-CH$_2$	CN	Me	Me	C-2 146.3 C-3 71.3 C-4 111.6 C-5 117.3	a	173
Ph	CN	CN	Ph	C-2 149.47 C-3 71.26 C-4 91.29 C-5 136.56	a	167
Ph	CN	CN	2-OHC$_6$H$_4$	C-2 148.77 C-3 70.50 C-4 91.83 C-5 134.58	a	167

TABLE 3.11. Continued

R^1	R^3	R^4	R^5	ppm		Solvent[c]	Ref.
4-OHC$_6$H$_4$	CN	CN	Ph	C-2	150.04	a	167
				C-3	73.01		
				C-4	92.29		
				C-5	137.69		
4-ClC$_6$H$_4$	CN	CN	Ph	C-2	150.23	a	167
				C-3	72.36		
				C-4	92.59		
				C-5	137.44		
2-Pyridyl	CN	CN	Ph	C-2	149.85	a	167
				C-3	71.63		
				C-4	92.21		
				C-5	135.77		
Ac	CN	Me	Me	C-2	149.59	b	183
				C-3	74.70		
				C-4	115.64		
				C-5	118.33		

N=CHPh	CN	Ph	C-2 147.34 C-3 71.88 C-4 92.69 C-5 134.45	a	169
2-FurylCH=N	CN	2-Furyl	C-2 147.72 C-3 70.67 C-4 87.98 C-5 133.12	a	169
		2-NRR [R–R = –$[CH_2]_2O(CH_2)_2$–]			
Me	H	COOEt	Me	C-2 142.0 C-3 96.1 C-4 109.6 C-5 132.3	a	188

[a] Other uncompleted or unassigned spectra were also reported.[193,194] In Ref. 150 the data are available on a supplementary publication.
[b] In the original paper the assignment of the C-2 and C-3 resonances was erroneously reversed.
[c] Solvent: a, dimethylsulfoxide-d_6; b, deuterochloroform.

TABLE 3.12. ^{13}C CHEMICAL SHIFTS OF THE 3-AMINOPYRROLES

R^1	R^2	R^4	R^5	ppm	Solvent[a]	Ref.
				3-NH_2		
H	Ph	H	Me	C-2 112.35 C-3 131.29 C-4 100.99 C-5 126.63	a	262
				C-2 122.29 C-3 110.82 C-4 103.24 C-5 127.77	b	262
H	Ph	H	Ph	C-2 116.22 C-3 132.52 C-4 100.17 C-5 129.92	a	262
				C-2 125.59 C-3 112.67 C-4 102.84 C-5 131.18	b	262
H	Ph	Ac	Me	C-2 110.11 C-3 133.57 C-4 112.74 C-5 133.57	a	262

H	Ph	COOEt	Me	C-2 123.57 C-3 112.41 C-4 114.95 C-5 135.76	b	262
	Ph	COOEt	Me	C-2 110.86 C-3 132.76 C-4 101.79 C-5 133.54	a	262
				C-2 123.77 C-3 111.53 C-4 105.60 C-5 135.37	b	262
Me	Ph	COOEt	Me	C-2 114.61 C-3 131.74 C-4 100.64 C-5 133.31	a	262
				C-2 125.81 C-3 113.78 C-4 104.42 C-5 135.52	b	262

3-NHR (R = COOCH$_2$Ph)

CPh$_3$	H	H	H	C-2 113.0 C-3 122.9 C-4 100.9 C-5 121.9	c	245

3-NRR [R–R = —(CH$_2$)$_5$—]

Ac	CH$_2$Ph	H	COOEt	C-2 127.4 C-3 136.9 C-4 111.3 C-5 120.7	d[b]	217

TABLE 3.12. Continued

R¹	R²	R⁴	R⁵	ppm	Solvent[a]	Ref.
COO—t-Bu	C_2H_4SMe	H	COOEt	C-2 126.5 C-3 135.5 C-4 111.0 C-5 121.8	d[b]	217
COO—t-Bu	CH_2Ph	H	COOEt	C-2 125.8 C-3 136.1 C-4 110.5 C-5 121.9	d[b]	217
3-NRR [R–R = $—(CH_2)_2O(CH_2)_2—$]						
COOEt	CH_2Ph	H	COOEt	C-2 128.2 C-3 137.8 C-4 112.6 C-5 122.7	d[b]	217

[a] Solvent: a, dimethylsulfoxide-d_6; b, dimethylsulfoxide-d_6/trifluoroacetic acid; c, dichloromethane-d_2; d, deuterochloroform.
[b] Assignments made by the reviewers.

TABLE 3.13. ULTRAVIOLET ABSORPTION MAXIMA OF THE 1-AMINOPYRROLES

R^2	R^3	R^4	R^5	nm	$\varepsilon \cdot 10^{-3}$	Solvent[a]	Ref.
				1-NH$_2$			
Me	H	H	Me	216	6.45	b	61
Me	H	H	Ph	204, 289	4.17, 4.15	a	116
Me	COOH	H	Me	238, 267	3.63, 3.63	a	74
Me	COOEt	H	Me	238, 267	3.63, 3.63	a	74
Me				370	40[b]	b	275
Me				375	50[b]	c	275
Me	COOEt	Me	Me	360	80[b]	b	275
Me	COOEt	COOEt	Me	370	40[b]	b	275
Me				400	60[b]	c	275
Ph	H	H	Ph	212, 310	15.70, 23.50	b	61
Ph	Me	Me	Ph	314	14.00	b	107
Ph	Ph	H	Ph	262, 304	12.50, 14.50	b	107
Ph	Ph	Ph	Ph	260, 300	24.00, 18.00	b	107
				1-NHR (R = Me)			
=CHPH	CN	CN	=CHPH	421	14.0	d	226
				1-NHR (R = Ph)			
Ph	Ph	Ph	Ph	250, 290	30.00, 17.60	e	108

TABLE 3.13. Continued

R²	R³	R⁴	R⁵	nm	$\varepsilon \cdot 10^{-3}$	Solvent[a]	Ref.
			1-NHR (R = 4-NO₂C₆H₄)				
Me	COOEt	COOEt	Me	335	14.21	b	49
			1-NHR (R = 2,4-diNO₂C₆H₃)				
Me	H	H	Me	220, 328	13.36, 11.31	b	49
Me	COOEt	H	Me	258, 323	13.74, 12.97	b	49
Me	COOEt	COOEt	Me	235, 258	22.22, 17.85	b	49
				320	15.33		
			1-NHR [R = C(=NH)NH₂]				
Me	H	H	Me	214	8.65	b	61
			1-NHR (= CHO)				
Me	H	H	Me	214–15	8.05	b	61
			1-NHR (R = COPh)				
Me	Ac	Ac	Me	230, 271	4.36; 3.83	f	102
			1-NHR (R=CONH₂)				
Me	H	H	Me	216	6.05[c]	b	61
Me	H	H	Ph	206, 286	13.80, 14.00	b	69
Me	H	COOEt	Ph	207, 285	13.70, 4.50	b	69
Me	Me	Me	H	218	3.78	b	70

Me	Ac	H	Me	246, 288	3.80, 4.96	b	120
Me	Ac	H	Ph	242, 278	4.10, 4.24	b	120
Me	COPh	H	Me	242, 314	3.74, 4.11	b	120
Me	COPh	H	Ph	258, 320	4.25, 3.64	b	120
Me	COOEt	H	Me	229, 264	6.20, 6.30[d]	b	69
Me	COOEt	H	Ph	270	4.22	b	120
Me	COOEt	Me	Me	380	40[b]	a	72
Me	COOEt	COOEt	Me	372	40[b]	b	275
Ph	COPh	H	Me	252, 326	4.19, 3.65	b	120
Ph	COPh	H	Ph	264	4.50	b	120
Ph	COOEt	H	Me	286	3.87	b	120
Ph	COOEt	H	Ph	280	4.17	b	120

1-NHR (R = CSNH$_2$)

Me	H	H	Me	210, 246	12.75, 16.60	b	61
Ph	H	H	Ph	208, 249 305	25.80, 17.10 22.90	b	61

1-NHR [R = (2,5-Me$_2$pyrrole-1-yl)CS]

Me	H	H	Me·HBr	211, 245	18.75, 16.00	b	61

1-NHR (R = 4-MeC$_6$H$_4$SO$_2$)

H	H	H	H	222	15.80	b	113

[a] Solvent: a, not reported; b, ethanol; c, ethanol–hydrogen chloride; d, dioxane; e, chloroform; f, methanol.
[b] The intensity of the maximum was measured in millimeters.
[c] Other value reported for the same solvent is 216 nm ($\varepsilon = 6710$).[69]
[d] Values reported for the same solvent are 370 nm (60 mm^b)[275] and 262 nm ($\varepsilon = 3890$).[120]

usual range observed for polysubstituted pyrroles. Thus the chemical shifts for the α-carbon atoms are found to be in the range 123–140 ppm, and the β-carbon signals are at 105–122 ppm. In a series of 1-ureidopyrroles, the CO signals lie in the range 156.9–157.4 ppm with a $^2J_{NH,CO}$ of 4.4–5.5 Hz.[118]

The C(2) atom signal for 2-aminopyrroles experiences a downfield shift relative to pyrrole and is found at 146.3–150.2 ppm. The chemical shift for the C(3) atom is in the range 70.5–90.8 ppm, as a result of the shielding effect of the amino group, which is additive to that of the cyano group generally present in all the derivatives studied.

Analogous considerations can be extended to the 3-aminopyrroles. The signal for the C(3) atom is found in the range 131.3–137.8 ppm, whereas those for the two C(2) and C(4) atoms occur in the ranges 110.1–116.2 ppm and 100.2–112.7 ppm, respectively. In these derivatives, substitution on the amino group (morpholino and piperidino derivatives) does not affect the position of the C(3) atom signal, but it has a deshielding effect on the adjacent C(2) and C(4) atoms of ~10 ppm.

3.3.5. UV Spectra

Ultraviolet spectral studies have proved to be critical in the identification of 1-aminopyrroles and distinguish them from the isomeric dihydropyridazines (see Section 3.2.2.1). Consequently, the majority of systematic studies have been conducted on 1-aminopyrroles (Table 3.13).

Comparison of the UV spectra of series of 1-amino **325** and 1-ureidopyrroles **326** with those of the corresponding 1-unsubstituted and 1-methyl derivatives shows that the presence of the auxochromic amino group in the 1-position produces a bathochromic shift with an increase in the intensity of the absorption band relative to the N-unsubstituted compound such that the spectra are similar to those of the N-methyl derivatives[275] (Scheme 88). In contrast, the long wavelength absorption bands of N-ureido derivatives are shifted by 5–10 nm toward lower wavelengths, and its intensity is greater than that of the other 1-amino compounds.

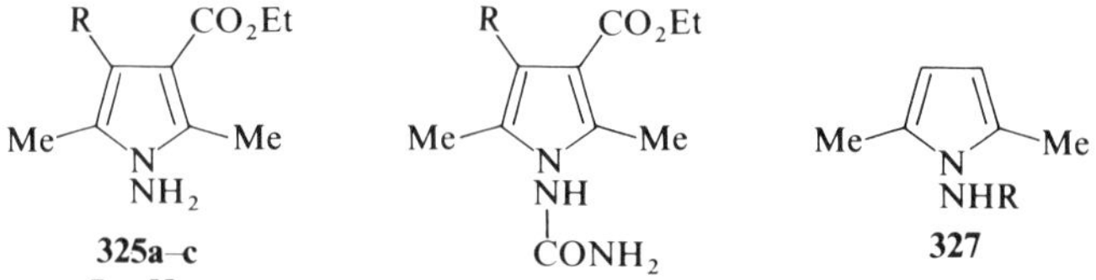

Scheme 88

The introduction of a methyl group at the 3-position shifts the absorption maxima to higher wavelengths with a simultaneous decrease in intensity of the

3.3. Structure and Physical Properties

first maximum and an increased intensity of the long wavelength band, so that the two maxima could merge into a single band.[72]

A comparison of the spectra of compound **325c** and the corresponding diacetyl derivative showed a bathochromic shift.[102] These same effects are observed in other series of 1H-pyrrole derivatives.

The UV absorption data for a series of 1-amino-2,5-dimethylpyrroles **327**, although not included in Table 3.13, have been reported.[50] The cross-conjugative effects of the two methyl groups and the syngergic effect result in the UV spectra being similar to those of the corresponding N-substituted bis-hydrazones of acetonylacetone, although the absorption below 230 nm is less significant for the 1-aminopyrrole derivatives.

The intensities of the absorption bands for 1-aminopyrroles **325a,c**, recorded in ethanol saturated with hydrochloric acid[275] are lower than those of the free bases and are accompanied by a hypsochromic effect. It is noteworthy that spectra of **325c**, measured as the free base and as the protonated form, are very similar. Similar observations were recorded with 1-ureidopyrroles **326a,c**, and the spectra of the anions, recorded in aqueous sodium hydroxide, showed a bathochromic effect.

The UV absorption maxima of 2-amino- and 3-aminopyrroles are listed in Tables 3.14 and 3.15.

In a homologous series of 2-amino-3-cyanopyrroles the absorption maxima were found in the range 264–298 nm.[170] The introduction of a methyl substituent on the amino group shifted the maxima 12 nm to longer wavelengths.[154]

Of the 3-aminopyrrole derivatives, only the UV spectra of **328** have been recorded[195] (Scheme 89). The main absorption at ~290 nm in the UV spectra of **328** can be attributed to a very strong vinylamide delocalization, due to the presence of the β-amino group. The spectrum of its hydrochloride salt does not

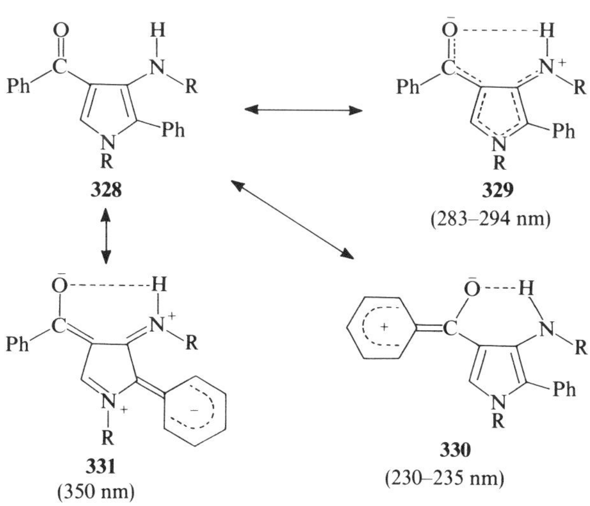

Scheme 89

TABLE 3.14. ULTRAVIOLET ABSORPTION MAXIMA OF THE 2-AMINOPYRROLES

$$\begin{array}{c} R^4 \quad R^3 \\ \diagdown \diagup \\ R^5 \diagup N \diagdown Z \\ R^1 \end{array}$$

2-NH_2

R^1	R^3	R^4	R^5	nm	$\varepsilon \cdot 10^{-3}$	Solvent[a]	Ref.
H	H	H	CH(CN)COOEt	310	38.60	a	283
H	H	H	CH(CN)COO—t-Bu	310	34.30	a	283
H	H	H	OH	227	22.00[b]	b	157
H	c-Hex-1-enyl	H	OH	226	4.30	b	158
H	CN	Me	H	264	3.85	c	170
				262	3.77	c	182
H	CN	Me	Me	269	3.68	c	170
				268	3.83	a	180
H	CN	Me	Et	266	3.93	a	180
H	CN	Me	Pr	268	3.83	a	180
H	CN	Me	Ph	286, 309	4.05, 4.20	a	154
H	CN	Et	Me	268	3.85	a	180
H	CN	Pr	Et	268	2.92	a	180
H	CN	—(CH$_2$)	4—	269	3.80	c	154
H	CN	Ph	H	292	3.72	c	170
				268	3.85	a	180
H	CN	Ph	Me	298	3.89	c	170
H	CN	CN	SEt	224, 256	14.4, 6.2	a	307
				300	10.5		
H	CONH$_2$	CONH$_2$	SCH$_3$	314	3.87	c	189
				324	—[c]	d	189
H	COOEt	COOEt	H	223, 293	4.20, 4.17	a	189

H	COOEt	=O	c-Hexyl	246, 273	—c	c	162
H	COOEt	=O	=CHPh	238, 308	—c	c	162
H	COOEt	=O	=CH—CH=CHPh	240, 355	—c	c	162
H	COO—t-Bu	Me	H	213, 285	3.95, 3.63	c	182
[H	COOEt	S-]bis		235, 287	3.90, 3.87	c	189
				387	3.77		
$CH_2CH(OEt)_2$	H	H	H	258	3.63	a	149
N=CHPh	CN	CN	N=CHPh	434	12.8	a	226
Ph	CN	Me	COOEt	303	4.58	a	159
Ph	CN	Ph	H	247, 293	4.59, 3.96	a	154
Ph	CN	Ph	4-EtOCOC$_6$H$_4$N=N	273, 455	4.27, 4.54	e	154
Ph	CN	Ph	6-MeO-2-benzothiazolyl N=N	299, 351	4.13, 3.74	a	154
MeOC$_6$H$_4$	CN	—(CH$_2$)$_4$—		270	3.81	c	154
MeOC$_6$H$_4$	CN	Ph	H	241, 289	4.67, 4.07	a	154
MeOC$_6$H$_4$	CN	Ph	4-MeOC$_6$H$_4$N=N	285, 431	4.21, 4.39	e	154
4-BrC$_6$H$_4$	CN	Ph	H	242, 289	4.61, 3.95	a	154
Ac	CN	Me	Me	232	4.08	a	180
COPh	COOEt	=O	=CHPh	235, 285	—c	c	162
PhCOCH=N	CN	—(CH$_2$)$_4$—		266, 311	4.15, 4.18	f	155
4-NO$_2$C$_6$H$_4$COCH=N	CN	—(CH$_2$)$_4$—		266, 327	4.35, 4.26	g	155
4-NO$_2$C$_6$H$_4$COCH=N	CN	Ph	H	273, 338	—c	f	155
				430			
N=CHCOOEt	CN	—(CH$_2$)$_4$—		261, 297	4.20, 4.19	c	155
N=C(COOH)COOEt	CN	—(CH$_2$)$_4$—		243, 269	4.33, 4.37	a	155
				387	3.84		
		2-NHR (R = Me)					
H	CN	—(CH$_2$)$_4$—		2.75	3.77	c	154
H	CN	Ph	Me	290, 314	4.00, 4.08	c	154
		2-NHR (R = Et)					
H	CN	Ph	Me	291, 314	4.18, 417,	c	154

TABLE 3.14. Continued

R^1	R^3	R^4	R^5	nm	$\varepsilon \cdot 10^{-3}$	Solvent[a]	Ref.
			2-NHR (R = CH$_2$Ph)				
H	CN	Ph	Me	291, 312	4.12, 4.20	c	154
			2-NHR (R = Ph)				
H	H	H	OH	227, 265	5.6, 10.6	c	157
H	CN	Me	COOEt	304	4.33	a	159
CH$_2$Ph	Ph	Ph	CH(COPh)$_2$	243, 332	4.59, 4.11	g	163
				431	3.30		
			2-NHR (R = Ac)				
H	CN	Me	Me	284	3.93	a	180
H	CN	Me	Et	284	3.89	a	180
H	CN	Me	Pr	287	3.97	a	180
H	CN	Et	Me	284	4.25	a	180
H	CN	Pr	Et	286	4.01	a	180
H	CN	Ph	H	225, 286	4.62, 4.48	a	180
[H	COOEt	COOEt	S-]bis	279	4.32	e	189
			2-NMR [R = CO(CH$_2$)$_3$Cl]				
H	CN	Me	Me	210, 284	9.99, 8.19	c	302
Ac	CN	Me	Me	209, 284	13.3, 9.8	c	302
			2-NHR (R = COPh)				
H	Ph	Me	OH	224, 233	20, 21	a	244
				266	11		
Me	Ph	Me	OH	225, 234	20, 20	a	244
				271	8.2		

CH₂OH	Ph	Me	OH	224, 234 270	18, 18 9.8	a	244
CH₂OMe	Ph	Me	OH	224, 234 270	19, 19 11	a	244
CH₂OEt	Ph	Me	OH	223, 234 271	20, 21 12	a	244
CH₂OAc	Ph	Me	OH	224, 233 273	19, 19 11	a	244
CHO	Ph	Me	OH	225, 283	19, 14	a	244

2-NR₂ (R = Me)

H	Ph	OH	H	222, 287 241	12.0, 19.8 16.2	c	190
H	Ph	OH	Me	222, 287 240	12.0, 17.8 17.0	c	190
H	Ph	OH	i-Pr	223, 287 242	13.6, 19.5 18.0	c	190
H	Ph	OH	i-Bu	225, 288	12.0, 17.5	c	190
H	Ph	OH	(CH₂)₂COOEt	224, 287	15.0, 21.0	c	190
H	Ph	OH	(CH₂)₂COOCH₂Ph	225, 287	12.0, 16.9	c	190
H	Ph	OH	CH₂Ph	225, 290	12.0, 15.3	c	190
Me	Ph	OH	H	243, 263 302	9.5, 9.5 12.6	c	190
t-Bu	H	NO₂	H	240–260	14.1		190
Ph	Ph	OH	H	280, 336	5.0, 7.0	c	243
			H	261, 306	14.4, 18.4	c	190

2-NH₂ (R = Et)

H	Me	OH	Me	228, 289	9.9, 19.4	c	190
H	Me	OH	Pr	228, 289	10.0, 19.4	c	190
H	Me	=O	Me, Me	226, 293	10.0, 22.2	a	161
H	Me	=O	Me, Ph	220, 293	12.2, 16.7	a	161
				242	16.0	h	161

TABLE 3.14. Continued

R^1	R^3	R^4	R^5	nm	$\varepsilon \cdot 10^{-3}$	Solvent[a]	Ref.
COCF$_3$	Me	=O	Me, Me	266, 305	11.5, 11.8	i	161
COCF$_3$	Me	=O	Me, Ph	274, 311	10.7, 10.1	i	161
			2-NR_2 ($R = Ac$)				
Ac	CN	Ph	H	242	4.38	a	180
			2-$NRR[R\!-\!R = -CO(CH_2)_3-]$				
H	CN	Me	Me	271	—[c]	c	336
			2-$NRR[R\!-\!R = -(CH_2)_5-]$				
Me	H	H	NO$_2$	399	16.3	e	241
Me	H	NO$_2$	H	281, 358	3.9, 3.6	c	242
			2-$NRR[R\!-\!R = -(CH_2)_2O(CH_2)_2-]$				
Me	H	NO$_2$	H	280, 348	7.2, 3.8	c	242
			2-NRR' ($R = Me; R' = Ph$)				
CH$_2$Ph	Ph	Ph	CH(COPh)$_2$	245, 323 428	4.61, 4.06 3.30	e	163

[a] Solvent: a, ethanol; b, water; c, methanol; d, methanol–hydrogen chloride; e, acetonitrile; f, dichloromethane; g, chloroform; h, ethanol–hydrogen chloride; i, hexane.
[b] Other value reported in the same solvent is 226 nm.[223]
[c] Not reported.

TABLE 3.15. ULTRAVIOLET ABSORPTION MAXIMA OF THE 3-AMINOPYRROLES

3-NH_2

R¹	R²	R⁴	R⁵	nm	$\varepsilon \cdot 10^{-3}$	Solvent[a]	Ref.
H	H	Ac	Me	263, 294	13.82, 6.01	a	206
H	H	COPh	Me	246, 280	6.11, 8.26	a	206
H	H	COOEt	Me	241, 294	12.14, 6.22	a	206
H	Et	Ac	Me	262, 306	12.24, 4.28	a	206
H	Et	COOMe	COOMe	265, 341	6.68; 8.95	a	206
H	Ph	Ph	H	260, 301	17, 19.6	b	214
H	Ph	Ac	Me	263	23.17	a	206
H	Ph	—CO(CH₂)₃—		262, 292	26.07, 14.98	a	206
H	Ph	COPh	H	232, 285	10.19, 32.63	a	206
H	Ph	COPh	Me	288	28.0	a	206
H	Ph	COPh	COOEt	258, 296	15.02, 19.96	a	206
H	Ph	CONH₂	Me	235, 295	15.76, 14.60	a	206
H	Ph	CONHMe	Me	232, 297	20.76, 13.05	a	206
H	Ph	COOMe	CH₂COOMe	238, 293	19.85, 16.86	a	206
H	Ph	COOMe	COOMe	225, 264 359	11.61, 9.65 8.55	a	206
H	Ph	COOEt	H	238, 289	20.49, 14.93	a	206
H	Ph	COOEt	Me	265	26.25	a	206
H	Ph	Ph, OH	H, H	244	14.4	b	214
H	2-MeC₆H₄	Ac	Me	264	19.68	a	206
H	3-MeC₆H₄	Ac	Me	266	18.47	a	206
H	4-MeC₆H₄	Ac	Me	265	25.80	a	206

TABLE 3.15. Continued

R^1	R^2	R^4	R^5	nm	$\varepsilon \cdot 10^{-3}$	Solvent[a]	Ref.
H	4-MeC$_6$H$_4$	COOMe	COOMe	264	12.90	a	206
H	4-MeC$_6$H$_4$	COOEt	Me	240, 293	21.90, 18.54	a	206
H	4-MeOC$_6$H$_4$	Ac	Me	268	23.11	a	206
H	4-MeOC$_6$H$_4$	COOMe	COOMe	263, 354	15.17, 12.68	a	206
H	4-MeOC$_6$H$_4$	COOEt	Me	243, 288	19.80, 17.13	a	206
H	4-FC$_6$H$_4$	Ac	Me	265, 304	22.49, 14.95	a	206
H	4-FC$_6$H$_4$	COPh	Me	284	22.25	a	206
H	4-FC$_6$H$_4$	COOEt	Me	242, 288	19.54, 14.73	a	206
H	4-ClC$_6$H$_4$	Ac	Me	265, 304	22.49, 14.95	a	206
H	4-ClC$_6$H$_4$	COOMe	COOMe	228, 266 350	14.12, 9.80 9.38	a	206
H	CONH$_2$	COOEt	Me	236.5, 281	—[b]	b	210
H	CONHMe	COOEt	Me	235.5, 280	—[b]	b	210
H	COOEt	Me	H	270	16.96	c	202
H	COO—t-Bu	COOEt	Me	236.5, 280	34.3, 13.6	b	210
H	COO—t-Bu	—CO(CH$_2$)$_3$—		254.5, 283	29.9, 11.5	b	210
CH$_2$COOEt	COO—t-Bu	—CO(CH$_2$)$_3$—		254, 286	32.0, 11.5	b	210

3-NHR (R = i-Pr)

i-Pr	Ph	COPh	H	235, 288	15.3, 20.1	d	195

3-NHR (R = c-Hexyl)

c-Hexyl	Ph	COPh	H	233, 289	15.6, 20.1	d	195

3-NHR [R = c(NH)NH$_2$]

H	H	H	H·HCl	—[c]		c	267
H	COOMe	H	H·HCl	266	13.5	—[c]	267

			3-NHR (R = Ac)			
H	(CH$_2$)$_2$CN		246	8.2	d	210
H	CH$_2$COOEt		245	9.8	b	210
H	—CO(CH$_2$)$_3$—	Ph	205, 232	4.38, 4.36	b	250
	—CO(CH$_2$)$_3$—		325	4.40		
Me	H	Ph	209, 305	4.32, 4.22	b	250
			3-NHR (R = CONHMe)			
H	COOEt	Me	275	12.66	c	202
			3-NHR (R = CSNHMe)			
H	COOEt	Me	236, 275	19.97, 20.87	c	202
			3-NR$_2$ (R = Me)			
t-Bu	H	NO$_2$	310	9.3	d	243
			3-NR$_2$ (R = Ac)			
H	Ph	—(CH$_2$)$_4$—	296	4.1	b	199
H	Ph	—(CH$_2$)$_5$—	296	4.1	b	199
			3-NRR[R-R = (CH$_2$)$_5$]			
Me	H	NO$_2$	310	3.9	d	242
			3-NRR[R-R = (CH$_2$)$_2$O(CH$_2$)$_2$]			
Me	H	NO$_2$	308	3.9	d	242

[a] Solvent; a, buffer pH 7.38, b, ethanol; c, pH 1; d, methanol; c, pH 7.
[b] Not reported.
[c] End absorption only.

exhibit vinylamide absorption, indicating that the β-amino group alone is responsible for nearly all of the delocalization (**329**).[195] Less dominant absorptions at ~230 nm, due to benzoyl absorption (**330**), and at ~350 nm, due to a four-charge-center resonance interaction as in structure **331**, were observed. Similar behavior was observed for 5-substituted derivatives analogs of **328**.[206]

UV data of diaminopyrroles are listed in Tables 3.16–3.18.

TABLE 3.16. ULTRAVIOLET ABSORPTION MAXIMA OF THE 1,2-DIAMINOPYRROLES

R^3	R^4	R^5	nm	$\varepsilon \cdot 10^{-3}$	Solvent[a]	Ref.
			1-NH_2			
CN	Me	Ph	279, 301	3.98, 4.02	a	154
CN	—(CH$_2$)—		253	3.73	a	155
CN	Ph	H	299	3.93	b	218
			1-NHR (R=Ph)			
CN	Me	Ph	231, 283 305	4.33, 4.12 4.06	a	154

[a] Solvent: a, methanol; b, ethanol.

TABLE 3.17. ULTRAVIOLET ABSORPTION MAXIMA OF THE 2,3-DIAMINOPYRROLES

R^1	R^4	R^5	nm	$\varepsilon \cdot 10^{-3}$	Solvent[a]	Ref.
			2,3 di-NRR [R–R = —(CH$_2$)$_5$—]			
Me	NO$_2$	H	225, 294	—[b]	a	242
t-Bu	NO$_2$	H	293	—[b]	b	268
			2,3 di-NRR [R–R = —(CH$_2$)$_2$O(CH$_2$)$_2$—]			
Me	NO$_2$	H	224, 290	—[b]	a	242

[a] Solvent: a, acetonitrile; b, methanol.
[b] Not reported.

3.3. Structure and Physical Properties

TABLE 3.18. ULTRAVIOLET ABSORPTION MAXIMA OF THE 2,5-DIAMINOPYRROLES

$$\text{Z} \underset{\underset{R^1}{N}}{\overset{R^4 \quad R^3}{\diagdown\!\diagup}} \text{Z}$$

R^1	R^3	R^4	nm	$\varepsilon \cdot 10^{-3}$	Solvent[a]	Ref.
			2,5-diNH_2			
H	H	H	237	20.0	a	157
			235	—[b]	b	223
			2-NH_2; 5-NHR (R = Ph)			
H	H	H·H_2O	240, 251	19.6, 13.3	a	157
			280	12.6		
			2,5-di-NHR (R = Ph)			
H	H	H	227, 280	13.0, 16.7	a	157
			294	17.4		
H	H	H·HCl·½H_2O	227, 251	15.0, 7.7	a	157
			258, 265	8.5, 8.5		
			312	23.0		

[a] Solvent: a, methanol; b, ethanol.
[b] Not reported.

3.3.6. IR Spectra

The v_{NH} absorption bands for 1-aminopyrroles are reported in Table 3.19. The parent compound shows the expected two bands at 3300 and 3200 cm^{-1}. The introduction of a methyl or phenyl group on the pyrrole nucleus shifts the bands to higher frequencies by 50–80 cm^{-1}, while the single v_{NH} band resulting from the introduction of an aryl or heteroaryl substituent on the amino function is not substantially affected by the substituent. Predictably, v_{NH} for the acylamino derivatives is to be found at lower frequencies (~ 50 cm^{-1}).

1-Ureidopyrroles show bands in the ranges 3480–3420 and 3320–3280 cm^{-1}, assignable to the amino stretching vibrations and in the range 3250–3200 cm^{-1} for the NH vibration. The two v_{NH} absorption bands for the 1-phenylureidopyrroles are found in the ranges 3340–3280 and 3260–3200 cm^{-1}. However, ureidopyrroles bearing an alkanesulfonyl or arenesulfonyl substituent at the 4-position have remarkably high v_{NH} frequencies (3630 cm^{-1}).

1-Pyrrolylurethanes absorb at 3255–3145 cm^{-1}; in this series of compounds, sulfonyl derivatives show absorption at higher frequencies (3300–3260 cm^{-1}) and the v_{NH} band of 1-arenesulfonylaminopyrroles generally lies between 3200 and 3100 cm^{-1}, but acylamino groups on the pyrrole ring produce a shift of

TABLE 3.19. NH$_2$ STRETCHING BANDS OF THE 1-AMINOPYRROLES

R^2	R^3	R^4	R^5	cm^{-1}	Solvent[a]	Ref.
			1-NH$_2$			
H	H	H	H	3300, 3200	a	67
Me	H	H	Me	3350, 3250	a	67
Me	H	H	Ph	3360, 3280	a	116
Ph	Ph	H	Ph	3360	b	107
Ph	Ph	Ph	Ph	3350	b	107
			1-NHR [R=(CH$_2$)$_2$COPh]			
H	H	H	H	3300	c	294
			1-NHR (R=Ph)			
CH$_2$Ph	Ph	Ac	Me	3220	d	133
CH$_2$Ph	Ph	COPh	Me	3250	d	134
CH$_2$Ph	Ph	CONH$_2$	Me	3485, 3240, 3160	d	137
CH$_2$Ph	Ph	CONHPh	Me	3420, 3220	d	137
CH$_2$Ph	Ph	4-MeOC$_6$H$_4$NHCO	Me	3420, 3225	d	137
CH$_2$Ph	Ph	CONEt$_2$	Me	3200	d	137
CH$_2$Ph	Ph	COOMe	Me	3250	d	134
CH$_2$Ph	Ph	COOEt	Me	3305	d	134
Ph	Ph	Ph	Ph	3360	c	108
Ph	Ph	Ac	Me	3240	d	133
Ph	Ph	—CO(CH$_2$)$_3$—		3230	d	133

Ph		COPh	Me	3240	d	134
Ph		COPh	Ph	3320	d	134
Ph		CONH$_2$	Me	3485, 3255, 3140	d	137
Ph		CONHPh	Me	3400, 3210	d	137
Ph		4-MeOC$_6$H$_4$NHCO	Me	3400, 3240	d	137
Ph		4-ClC$_6$H$_4$NHCO	Me	3390, 3295	d	137
Ph		CONEt$_2$	Me	3210	d	137
Ph		COOMe	Me	3300	d	134
Ph		COOEt	Me	3300	d	133
	1-NHR (R=2-NH$_2$C$_6$H$_4$)					
H		H	H	3400, 3300, 3250	d	112
	1-NHR (R=2-AcNHC$_6$H$_4$)					
H		H	H	3300–3250	d	112
	1-NHR (R=2-NO$_2$C$_6$H$_4$)					
H		H	H	3350	d	112
	1-NHR (R=4-NO$_2$C$_6$H$_4$)					
—(CH$_2$)$_4$—		COPh	Me	3210	d	134
—(CH$_2$)$_4$—		COPh	Ph	3315	d	134
—(CH$_2$)$_4$—		CONHPh	Me	3360, 3220	d	137
—(CH$_2$)$_4$—		4-MeOC$_6$H$_4$NHCO	Me	3340, 3200	d	137
—(CH$_2$)$_4$—		4-ClC$_6$H$_4$NHCO	Me	3455, 3265	d	137
—(CH$_2$)$_4$—		CONEt$_2$	Me	3190	d	137
—(CH$_2$)$_4$—		COOMe	Me	3270	d	134
—(CH$_2$)$_4$—		COOEt	Me	3240	d	133
	1-NHR (R=4-ClC$_6$H$_4$)					
Ph		CONH$_2$	Me	3490, 3245, 3160	d	137
Ph		CONHPh	Me	3395, 3345	d	137

TABLE 3.19. Continued

R^2	R^3	R^4	R^5	cm^{-1}	Solvent^a	Ref.
Ph	Ph	4-MeOC$_6$H$_4$NHCO	Me	3425, 3250	d	137
Ph	Ph	4-ClC$_6$H$_4$NHCO	Me	3425, 3265	d	137
		1-NHR (R=2,4-DiNO$_2$C$_6$H$_3$)				
Me	COOEt	H	Me	3310	c	49
Me	COOEt	COOEt	Me	3270	c	49
		1-NHR (R=3-NO$_2$-2-Pyridyl)				
Me	COOMe	Ac	Me	3340	d	141
Me	COOMe	COPh	Me	3340	d	141
Me	COOMe	COPh	Ph	3290	d	141
Me	COOMe	CONHPh	Me	3330	d	142
Me	COOMe	4-MeOC$_6$H$_4$NHCO	Me	3280	d	142
Me	COOMe	COOMe	Me	3325	d	141
Me	COOMe	COOEt	Me	3370	d	141
Me	COOMe	COOEt	Ph	3330	d	141
Me	COOCH$_2$Ph	CONHPh	Me	3280	d	142
Me	COOCH$_2$Ph	4-ClC$_6$H$_4$NHCO	Me	3280	d	142
4-NO$_2$C$_6$H$_4$	COOEt	COOMe	Me	3320	d	141
4-NO$_2$C$_6$H$_4$	COOEt	COOMe	4-NO$_2$C$_6$H$_4$	3330	d	141
		1-NHR (R=2-Pyrimidyl)				
Me	COOMe	COOMe	Me	3320	d	141
Me	COOEt	CONHPh	Me	3380	d	142
Me	COOEt	CONHPh	Ph	3340, 3180	d	142
Me	COOEt	4-MeOC$_6$H$_4$NHCO	Me	3360, 3320	d	142
Me	COOEt	4-ClC$_6$H$_4$NHCO	Me	3200	d	142

Me	COOCH$_2$Ph	CONHPh	Me	3290, 3190	d	142
Me	COOCH$_2$Ph	CONHPh	Ph	3340, 3210	d	142
Me	COOCH$_2$Ph	COOMe	Me	3230	d	141
Me	COOCH$_2$Ph	COOCH$_2$Ph	Me	3220	d	141

1-NHR (R = 2-Benzothiazolyl)

Me	COOMe	CONHPh	Me	3490	d	142
Me	COOMe	4-MeOC$_6$H$_4$NHCO	Me	3340	d	142
Me	COOMe	COOMe	Me	3330	d	141

1-NHR (R = Ac)

H	H	H	H	3080	a	67

1-NHR (R=COCH$_2$Ac)

H	H	H	H	3250	c	294

1-NHR (R=COCH$_2$Ph)

Me	COOMe	4-ClC$_6$H$_4$NHCO	Me	3310, 3170	d	143
Me	COOMe	COOEt	4-NO$_2$C$_6$H$_4$	3190	d	126
Me	COOEt	CONHPh	Ph	3290, 3190	d	143

1-NHR (R=COPh)

H	H	H	H	3280	a	67
Me	COOMe	Me	Me	3230	d	132
Me	COOMe	Ac	Me	3200[b]	d	126
Me	COOMe	CONH$_2$	Me	3450, 3340, 3200	d	143
Me	COOMe	CONHPh	Me	3300, 3140	d	143
Me	COOMe	CONHPh	Ph	3310, 3200	d	143
Me	COOMe	4-MeOC$_6$H$_4$NHCO	Me	3315, 3170	d	143
Me	COOMe	4-ClC$_6$H$_4$NHCO	Me	3315, 3190	d	143
Me	COOMe	CONEt$_2$	Me	3150	d	143

TABLE 3.19. Continued

R^2	R^3	R^4	R^5	cm^{-1}	Solvent[a]	Ref.
Me	COOMe	4-MeC$_6$H$_4$SO$_2$	Me	3210	d	147
Me	COOEt	Me	Me	3240	d	132
Me	COOEt	COPh	Ph	3260	d	126
Me	COOEt	CONH$_2$	Me	3370, 3190	d	143
Me	COOEt	CONHPh	Ph	3300, 3200	d	143
Me	COOEt	COOMe	Me	3225	d	126
Me	COOEt	COOEt	Ph	3265	d	126
Me	COOEt	COOEt	4-NO$_2$C$_6$H$_4$	3160	d	126
Ph	Ph	H	H	3250	d	122

1-NHR (R=3-MeC$_6$H$_4$CO)

Me	COOMe	COPh	Me	3175	d	126
Me	COOMe	COPh	Ph	3245	d	126
Me	COOMe	CONHPh	Ph	3230	d	143
Me	COOMe	CONEt$_2$	Me	3170	d	143
Me	COOEt	COOEt	Me	3250	d	126
Me	COOEt	4-MeC$_6$H$_4$SO$_2$	Me	3200	d	147

1-NHR (R=3-ClC$_6$H$_4$CO)

Me	COOMe	COPh	Me	3170	d	126
Me	COOMe	CONHPh	Ph	3310, 3210	d	143
Me	COOMe	CONEt$_2$	Me	3170	d	143
Me	COOMe	SO$_2$Ph	Ph	3200	d	147
Me	COOMe	4-MeC$_6$H$_4$SO$_2$	Me	3260	d	147
Me	COOEt	CONH$_2$	Me	3430, 3360, 3280	d	143
Me	COOEt	CONHPh	Ph	3310, 3210	d	143
Me	COOEt	SO$_2$Ph	Ph	3190	d	147
Me	COOEt	4-MeC$_6$H$_4$SO$_2$	Me	3260	d	147

1-NHR (R=CONH$_2$)

H	COOH		H	Me	3400, 3300, 3200	d	118
H	COOEt		H	Me	3400, 3280	d	118
Me	H		H	Me	3480, 3320, 3200	d	69
Me	H		COPh	COOMe	3400, 3200	d	117
Me	H		COOH	COOEt	3400, 3250	d	118
Me	H		COOEt	COOH	3450, 3280	d	118
Me	H		COOEt	COOEt	3250	d	117
Me	COOMe		Ac	Me	3410, 3200	d	138
Me	COOMe		COPh	Me	3450, 3290	d	138
Me	COOMe		COOMe	Me	3420, 3260	d	138
Me	COOMe		COOMe	CH$_2$COOMe	3430, 3260, 3210	d	138
Me	COOMe		CONH$_2$	Me	3380, 3265, 3200	d	129
Me	COOMe		CONHPh	Me	3395, 3250, 3195	d	129
Me	COOMe		CONHPh	Ph	3435, 3275, 3220	d	129
Me	COOMe		4-MeOC$_6$H$_4$NHCO	Me	3395, 3250, 3190	d	129
Me	COOMe		4-ClC$_6$H$_4$NHCO	Me	3425, 3275, 3210	d	129
Me	COOMe		CONEt$_2$	Me	3405, 3260, 3210	d	129
Me	COOMe		COOEt	Me	3410, 3260, 3200	d	138
Me	COOMe		SO$_2$Ph	Ph	3460, 3330, 3250	d	131
Me	COOMe		4-MeC$_6$H$_4$SO$_2$	Ph	3430, 3340, 3200	d	131
Me	COOEt		Ac	Me	3420, 3270, 3210	d	138
Me	COOEt		COPh	Me	3410, 3260	d	138
Me	COOEt		CONH$_2$	Me	3385, 3250, 3195	d	129
Me	COOEt		CONHPh	Ph	3430, 3295	d	129
Me	COOEt		4-ClC$_6$H$_4$NHCO	Me	3415, 3300, 3190	d	129
Me	COOEt		COOEt	Me	3420, 3260	d	138
Me	COOEt		SO$_2$Ph	Ph	3470, 3310, 3235	d	131
Me	COOEt		4-MeC$_6$H$_4$SO$_2$	Ph	3630, 3330, 3210	d	131

1-NHR (R=CONHPh)

Me	COOMe		Ac	Me	3330, 3260	d	138
Me	COOMe		COPh	Me	3330, 3260, 3210	d	138

441

TABLE 3.19. Continued

R^2	R^3	R^4	R^5	cm^{-1}	Solvent[a]	Ref.
Me	COOMe	CONH$_2$	Me	3445, 3400, 3305	d	129
Me	COOMe	CONHPh	Ph	3380, 3270, 3200	d	129
Me	COOMe	4-MeOC$_6$H$_4$NHCO	Me	3305, 3220	d	129
Me	COOMe	COOMe	Me	3340, 3240	d	138
Me	COOMe	4-MeC$_6$H$_4$SO$_2$	Me	3280	d	131
Me	COOMe	4-MeC$_6$H$_4$SO$_2$	Ph	3350, 3270	d	131
Me	COOEt	Ac	Me	3280, 3210	d	138
Me	COOEt	COPh	Me	3330, 3260, 3210	d	138
Me	COOEt	CONHPh	Me	3300,3200	d	129
Me	COOEt	CONHPh	Ph	3380, 3270, 3240	d	129
Me	COOEt	4-MeOC$_6$H$_4$NHCO	Me	3320, 3210	d	129
Me	COOEt	4-ClC$_6$H$_4$NHCO	Me	3300, 3250, 3200	d	129
Me	COOEt	CONEt$_2$	Me	3365, 3330, 3200	d	129
Me	COOEt	COOMe	Me	3290, 3200	d	138
Me	COOEt	SO$_2$Me	Me	3590, 3280	d	131
Me	COOEt	SO$_2$Ph	Ph	3350, 3260	d	131
Me	COOEt	4-MeC$_6$H$_4$SO$_2$	Me	3290	d	131
Me	COOEt	4-MeC$_6$H$_4$SO$_2$	Ph	3610, 3320	d	131
		1-NHR (R=COOMe)				
Me	COOMe	Me	Me	3210	d	132
Me	COOMe	Ac	Me	3255[c]	d	136
Me	COOMe	COPh	Me	3255	d	136
Me	COOMe	CONH$_2$	Me	3420, 3160	d	139
Me	COOMe	CONHPh	Me	3345, 3150	d	139
Me	COOMe	CONHPh	Ph	3285	d	139
Me	COOMe	4-MeOC$_6$H$_4$NHCO	Me	3320, 3150	d	139
Me	COOMe	4-ClC$_6$H$_4$NHCO	Me	3315, 3175	d	139

442

Me	COOMe	CONEt$_2$	Me	3145	d	139
Me	COOMe	COOMe	Me	3255	d	136
Me	COOMe	SO$_2$Me	Me	3275	d	146
Me	COOMe	SO$_2$Ph	Ph	3275	d	146
Me	COOMe	4-MeC$_6$H$_4$SO$_2$	Me	3270	d	146
Me	COOEt	Ac	Me	3255	d	136
Me	COOEt	CO—t-Bu	t-Bu	3255	d	136
Me	COOEt	COPh	Me	3255	d	136
Me	COOEt	COOMe	Me	3255	d	136
Me	COOEt	SO$_2$Me	Me	3260	d	146

1-NHR (R=COOEt)

Me	COOMe	Ac	Me	3255	d	136
Me	COOMe	COPh	Me	3255	d	136
Me	COOMe	CONH$_2$	Me	3375, 3170	d	139
Me	COOMe	CONHPh	Ph	3260	d	139
Me	COOMe	COOMe	Me	3255	d	136
Me	COOMe	4-MeC$_6$H$_4$SO$_2$	Me	3280	d	146
Me	COOEt	COPh	Me	3255	d	136
Me	COOEt	4-MeC$_6$H$_4$SO$_2$	Me	3290	d	146

1-NHR (R=COO—t-Bu)

Me	COOMe	Ac	Me	3255	d	136
Me	COOMe	COPh	Me	3255	d	136
Me	COOMe	CONHPh	Ph	3290	d	139
Me	COOMe	CONEt$_2$	Me	3155	d	139
Me	COOMe	COOMe	Me	3255	d	136
Me	COOMe	SO$_2$Ph	Ph	3295	d	146
Me	COOMe	4-MeC$_6$H$_4$SO$_2$	Me	3330	d	146
Me	COOMe	4-MeC$_6$H$_4$SO$_2$	Ph	3290	d	146
Me	COOEt	COPh	Me	3255	d	136
Me	COOEt	CONH$_2$	Me	3385, 3180	d	139
Me	COOEt	CONHPh	Ph	3300	d	139

TABLE 3.19. Continued

R^2	R^3	R^4	R^5	cm^{-1}	Solvent[a]	Ref.
Me	COOEt	COOMe	Me	3255	d	136
Me	COOEt	SO_2Ph	Ph	3280	d	146
		1-NHR (R=COOCH$_2$Ph)				
H	H	H	H	3330	a	67
Ph	H	H	Ph	3220	a	67
		1-NHR [R=PO(Ph)$_2$]				
Me	COOMe	COOEt	$4\text{-}NO_2C_6H_4$	3430	d	144
Me	COOEt	COPh	Me	3425	d	144
		1-NHR [R=PO(OEt)$_2$]				
Me	COOMe	COPh	Ph	3390	d	144
Me	COOMe	CONHPh	Me	3290, 3150	d	145
Me	COOMe	CONHPh	Ph	3280, 3130	d	145
Me	COOMe	$4\text{-}MeOC_6H_4NHCO$	Me	3320, 3110	d	145
Me	COOEt	$4\text{-}MeOC_6H_4NHCO$	Me	3330, 3110	d	145
Me	$COOCH_2Ph$	COPh	Me	3380	d	144
Me	$COOCH_2Ph$	CONHPh	Me	3300, 3100	d	145
Me	$COOCH_2Ph$	$4\text{-}ClC_6H_4NHCO$	Me	3300, 3100	d	145
Me	$COOCH_2Ph$	$CONEt_2$	Me	3385, 3070	d	145
Me	$COOCH_2Ph$	COOEt	$4\text{-}NO_2C_6H_4$	3390	d	144
		1-NHR [R=PO(OPh)$_2$]				
Me	COOMe	Ac	Me	3430	d	144
Me	COOMe	$CONH_2$	Me	3340, 3230	d	145
Me	COOMe	$CONEt_2$	Me	3390, 3100	d	145

Me	COOMe	COOEt	Ph	3405	d	144
Me	COOEt	CONHPh	Ph	3200, 3080	d	145
Me	COOEt	4-ClC$_6$H$_4$NHCO	Me	3180, 3100	d	145
Me	COOEt	COOMe	Me	3400	d	144
Me	COOCH$_2$Ph	Ac	Me	3380	d	144
Me	COOCH$_2$Ph	CONH$_2$	Me	3340, 3200	d	145
Me	COOCH$_2$Ph	CONEt$_2$	Me	3375, 3070	d	145

1-NHR (R = SO$_2$Ph)

Me	COOMe	CONH$_2$	Me	3430, 3325, 3260	d	130
Me	COOMe	CONHPh	Me	3300, 3215	d	130
Me	COOMe	CONHPh	Ph	3360	d	130
Me	COOEt	CONHPh	Ph	3360	d	130
Me	COOEt	4-MeOC$_6$H$_4$NHCO	Me	3300, 3210	d	130
Me	COOEt	CONEt$_2$	Me	3400	d	130

1-NHR (R = 4-MeC$_6$H$_4$SO$_2$)

H	H	H	H	3280	b	113
Me	COOMe	CONH$_2$	Me	3290, 3160	d	130
Me	COOMe	CONHPh	Me	3310, 3210	d	130
Me	COOMe	CONHPh	Ph	3350	d	130
Me	COOMe	4-MeOC$_6$H$_4$NHCO	Me	3340, 3100	d	130
Me	COOMe	4-ClC$_6$H$_4$NHCO	Me	3315, 3215	d	130
Me	COOMe	CONEt$_2$	Me	3400	d	130
Me	COOMe	COOMe	Me	3110	d	140
Me	COOMe	COOCH$_2$Ph	Me	3110	d	140
Me	COOMe	CONHPh	Me	3350, 3130	d	130
Me	COOEt	4-ClC$_6$H$_4$NHCO	Me	3305, 3210	d	130
Me	COOEt	CONEt$_2$	Me	3400	d	130
Me	COOEt	COOEt	Me	3110	d	140
Me	COOEt	COO—t-Bu	Me	3110	d	140

TABLE 3.19. Continued

R²	R³	R⁴	R⁵	cm⁻¹	Solvent[a]	Ref.
			1-NHR (R = 4-MeOC₆H₄SO₂)			
Me	COOMe	COOMe	Me	3110	d	140
Me	COOMe	COOEt	Ph	3110	d	140
Me	COOEt	COO—t-Bu	Me	3110	d	140
Me	COOEt	COOCH₂Ph	Me	3110	d	140
			1-NHR (R = 4-ClC₆H₄SO₂)			
Me	COOMe	COOMe	Me	3110	d	140
Me	COOMe	COOEt	Me	3110	d	140
Me	COOMe	COOEt	Ph	3110	d	140
Me	COOMe	COO—t-Bu	Me	3110	d	140
Me	COOMe	COOCH₂Ph	Me	3110	d	140
Me	COOEt	COOEt	Me	3110	d	140
Me	COOEt	COOCH₂Ph	Me	3110	d	140
			1-NHR (R = 2,4,6-TriMeC₆H₂SO₂)			
Me	COOMe	CONH₂	Me	3315, 3180	d	130
Me	COOMe	4-MeOC₆H₄NHCO	Me	3290, 3220	d	130
Me	COOMe	4-ClC₆H₄NHCO	Me	3320, 3210	d	130
			1-NRR' (R = Me, R' = 2-NH₂C₆H₄)			
H	H	H	H	3450, 3330	d	112
			1-NRR' (R = Me, R' = 2-AcNHC₆H₄)			
H	H	H	H	3400	d	112

[a] Solvent: a, not reported; b, chloroform; c, potassium bromide disk; d, nujol mull.
[b] Other value reported is 3290 cm⁻¹.[1,132]
[c] Other value reported is 3220 cm⁻¹.[1,132]

TABLE 3.20. NH$_2$ STRETCHING BANDS OF THE 2-AMINOPYRROLES[a]

R^1	R^3	R^4	R^5	cm^{-1}	Solvent[b]	Ref.
			2-NH$_2$			
H	Ph	Ph	H	3420, 3220	a	171
H	Ph	Ac	Me	3420, 3340	a	236
H	CN	Me	H	3390, 3290	b	182
H	CN	Me	Me	3360, 3220	a	180
H	CN	Me	Et	3370, 3200	a	180
H	CN	Me	Pr	3370, 3140	a	180
H	CN	Me	(CH$_2$)$_2$SMe	3420, 3340	a	181
H	CN	Me	CH$_2$Ph	3440, 3340	b	181
H	CN	Me	4-OHC$_6$H$_4$CH$_2$	3380, 3275	b	181
H	CN	Me	3-IndolylCH$_2$	3420, 3360	b	181
H	CN	Me	Ph	3330, 3275	b	181
H	CN	Et	Me	3350, 3230	a	180
H	CN	Pr	Et	3360, 3240	a	180
H	CN	Ph	H	3380, 3250[c]	a	171
						186
H	CN	COPh	OEt	3400–3150	c	166
H	CONH$_2$	Ph	H	3250, 3100	a	239
H	CONH$_2$	CONH$_2$	SCH$_3$	3480, 3450	b	189
				3360, 3200		
H	COOMe	COOMe	H	3450, 3275	a	239
H	COOEt	H	H	3450, 3400	a	184

447

TABLE 3.20. Continued

R^1	R^3	R^4	R^5	cm^{-1}	Solvent[b]	Ref.
H	COOEt	H	Me	3500, 3350	a	184
H	COOEt	H	Ph	3500, 3300	a	185
H	COOEt	Me	Me	3450, 3310	a	185
H	COOEt	Me	Ph	3500, 3450	a	185
H	COOEt	Ph	H	3500, 3370	a	186
H	COOEt	Ph	Me	3500, 3330	a	185
H	COOEt	Ph	Ph	3500, 3350	a	185
H	COOEt	Indeno[1,2-b]		3490, 3360	a	186
H	COOEt	Naphthalenol[1,2-b]		3450, 3350	a	185
H	COOEt	COOEt	H	3500, 3400	a	185
H	COOEt	COOEt	H	3490, 3350	b	189
H	COOEt	=O	c-Hexyl	3450, 3300	a	239
H	COOEt	=O	=CHPh	3400, 3340	b	162
H	COOEt	=O	=CHCH=CHPh	3280, 3000	b	162
H	COO—t-Bu	Me	H	3350	b	162
[H	COOEt	COOEt	S-]bis	3380, 3310	b	182
Me	CN	Me	Me	3490, 3380	b	189
Me	CN	Me	Et	3350	b	300
Me	CN	Me	i-Bu	3370, 3330	b	300
Me	CN	Me	(CH$_2$)$_2$SMe	3360, 3320	b	300
Me	CN	Me	CH$_2$Ph	3380, 3330	b	300
Me	CN	Me	4-MeOC$_6$H$_4$CH$_2$	3370, 3320	b	300
Me	CN	Me	(1-Me-3-indolyl)CH$_2$	3370, 3320	b	300
Me	CN	Me	Ph	3360, 3320	b	300
Me	COOMe	COOMe	H	3320, 3220	b	300
Me	COOMe	COOMe	H	3450, 3350	a	239

Me	COOEt	H		3450, 3400	a	184
n-Pr	COO—t-Bu	Me		3430, 3370	b	176
i-Pr	Me	CN		3320	d	165
i-Pr	Et	CN		3200	d	165
n-Bu	COO—t-Bu	Me		3400, 3280	b	176
i-Bu	COO—t-Bu	Me		3380, 3260	b	176
t-Bu	Et	CN		3320	d	165
c-Pentyl	COO—t-Bu	Me		3450, 3350	b	176
c-Hexyl	COO—t-Bu	Me		3450, 3340	b	176
(CH$_2$)$_3$Morph	COO—t-Bu	Me		3410, 3290	b	176
(CH$_2$)$_2$Ph	COO—t-Bu	Me		3400, 3320	b	176
CH$_2$CH(OMe)$_2$	COO—t-Bu	Me		3430, 3340	b	176
CH$_2$CH(OEt)$_2$	H	CN		3450, 3350	a	149
CH$_2$Ph	COOMe	Me		3440, 3330	b	176
CH$_2$Ph	COO—t-Bu	Me		3420, 3240	b	176
CH$_2$Ph	SO$_2$Me	Me		3450, 3350	b	177
2-PyrrolylCH$_2$	COO—t-Bu	Me		3380, 3270	b	176
3-PyridylCH$_2$	COO—t-Bu	Me		3360, 3240	b	176
CH$_2$COOEt	CN	Me		3400	b	175
CH$_2$OCH$_2$Ph	COOEt	H		3550, 3400	a	184
Ph	CN	Me	COOEt	3430, 3330	b	159
Ph	CN	Ph	COOEt	3420, 3320	b	160
4-MeC$_6$H$_4$	CN	Me	CONHPh	3460, 3360	b	160
4-MeC$_6$H$_4$	CN	Ph	CONHPh	3400, 3360	b	160
Ac	CN	Me	Me	3420, 3300	a	180
Ac	CN	Me	3-IndolylCH$_2$	3460, 3340	b	181
COPh	COOEt	=O	=CHPh	3430, 3280	b	162
N=CHPh	CN	CN	N=CHPh	3390, 3280	c	226
N=CHCOPh	CN	—(CH$_2$)$_4$—		3437, 3355	b	155
4-NO$_2$C$_6$H$_4$COCH=N	CN	—(CH$_2$)$_4$—		3470, 3335	b	155
4-NO$_2$C$_6$H$_4$COCH=N	CN	Ph	H	3482, 3362	b	155
N=CHCOOEt	CN	—(CH$_2$)$_4$—		3455, 3300	b	155
N=C(COOH)COOEt	CN	—(CH$_2$)$_4$—		3500, 3385	b	155
N=CHSO$_2$Ph	CN	—(CH$_2$)$_4$—		3500, 3350	e	155

TABLE 3.20. Continued

R^1	R^3	R^4	R^5	cm^{-1}	Solvent[b]	Ref.
			2-NHR(R = Me)			
H	CN	Ph	Me	3440	b	154
			2-NHR(R = Et)			
H	CN	Ph	Me	3420	b	154
CH$_2$Ph	COO—t-Bu	Me	Me	3320	f	305
CH$_2$OCH$_2$Ph	COOEt	H	Me	3400	f	184
			2-NHR(R = CH$_2$Ph)			
H	CN	Ph	Me	3410	b	154
			2-NHR(R = CH$_2$COPh)			
H	CN	Me	Me	3385	b	298
			2-NHR(R = 4-ClC$_6$H$_4$COCH$_2$)			
H	CN	Me	Me	3385	b	298
			2-NHR(R = 4-BrC$_6$H$_4$COCH$_2$)			
H	CN	Me	Me	3385	b	298
			2-NHR(R = Ph)			
H	CN	Me	COOEt	3270	b	159
CH$_2$Ph	Ph	Ph	CH(COPh)$_2$	3410	b	163

			2-NHR(R = Ac)		
H	CN	Me	3300	a	180
H	CN	Et	3310	a	180
H	CN	Pr	3320	a	180
H	CN	i-Bu	3280	b	181
H	CN	Me	3310	a	180
H	CN	Et	3310	a	180
H	CN	Pr	3310	a	180
H	CN	Ph	3300	a	239
					180
H	CONH$_2$	Me	3330	b	290
H	CONH$_2$	i-Bu	3300	b	290
H	CONH$_2$	CH$_2$Ph	3320	b	290
H	CONH$_2$	Ph	3350	a	239
			3340	b	290
H	COOEt	CONH$_2$	3400	a	239
H	COOEt	COOEt	3400	a	239
[H	COOEt	COOEt	3350	b	189
CH$_2$CH(OEt)$_2$	H	CN	3260	a	149
Ph	CONH$_2$	Me	3280	b	290
Ph	CONH$_2$	—(CH$_2$)$_4$—	3260	b	290
			2-NHR[R = CO(CH$_2$)$_3$Cl]		
H	CN	Me	3250	b	302
Ac	CN	Me	3200	b	302
			2-NHR[R = COCH(Me)NH$_2$]		
H	CN	i-Bu·HCl	3380	b	282
			2-NHR[R = 1-piperidinylCH(Me)CO]		
H	CN	Me·HCl	3300–2800	b	282

TABLE 3.20. Continued

R^1	R^3	R^4	R^5	cm^{-1}	Solvent[b]	Ref.
2-NHR[R = COCH(Me)Br]						
H	CN	Me	Me	3360	b	282
2-NHR(R = COCH$_2$NEt$_2$)						
H	CN	Me	Me·HCl	3300–2500	b	281
H	CN	Me	i-Bu·HCl	3400, 2800–2300	b	281
H	CONH$_2$	Me	Me·HCl	3340, 3200–2600	b	301
2-NHR(R = COCH$_2$Cl)						
H	CN	Me	Me	3345	b	281
H	CONH$_2$	Me	Me	3350	b	301
2-NHR(R = COPh)						
H	Ph	Me	OH	3240	b	244
H	CN	Ph	H	3340	a	180
Me	H	COPh	SMe	3330	a	153
Me	H	4-MeOC$_6$H$_4$CO	SMe	3307	a	153
Me	H	4-ClC$_6$H$_4$CO	SMe	3305	a	153
Me	Ph	Me	OH	3340	b	244
Et	H	COPh	SMe	3300	a	153
Et	H	4-MeOC$_6$H$_4$CO	SMe	3305	a	153
Et	H	4-ClC$_6$H$_4$CO	SMe	3210	a	153

452

CH₂Ph	H	COPh	SMe	3295	a	153
CH₂Ph	H	4-MeOC₆H₄CO	SMe	3260	a	153
CH₂Ph	H	4-ClC₆H₄CO	SMe	3300	a	153
CH₂OH	Ph	Me	OH	3360	b	244
CH₂OMe	Ph	Me	OH	3440	e	244
				3360	b	244
CH₂OEt	Ph	Me	OH	3360	b	244
CH₂OAc	Ph	Me	OH	3360	b	244
Ph	CONH₂	Me	Me	3260	b	290
CHO	Ph	Me	OH	3330	b	244

2-NHR(R = COCF₃)

CH₂Ph	COO-t-Bu	Me		3250	b	305
Ph	CN	Me		3180	b	290

2-NHR(R = CONHPh)

H	Ph	Ph		3300	a	171
H	CN	Ph		3400ᵈ	a	171

2-NHR(R = 4-NO₂C₆H₄NHCO)

H	Ph	Ph		3400, 3300	a	171

2-NHR(R = CONHAc)

H	Ph	Ph		3300	a	171

2-NHR(R = COOEt)

H	CN	Me	Me	3400	b	291
H	CN	Me	Et	3380	b	291
H	CN	Me	i-Bu	3380	b	291
H	CN	Me	CH₂Ph	3570	b	291
H	CN	Me	4-OHC₆H₄CH₂	3570	b	291

TABLE 3.20. Continued

R^1	R^3	R^4	R^5	cm^{-1}	Solvent[b]	Ref.
H	CN	Me	Ph	3370	b	291
H	CONH$_2$	Me	Me	3530	b	291
H	CONH$_2$	Me	Et	3460	b	291
H	CONH$_2$	Me	i-Bu	3415	b	291
H	CONH$_2$	Me	CH$_2$Ph	3480	b	291
H	CONH$_2$	Me	4-OHC$_6$H$_4$CH$_2$	3480	b	291
H	CONH$_2$	Me	Ph	3420	b	291
			2-NHR[R = COO(CH$_2$)$_2$Cl]			
H	CN	Me	CH$_2$Ph	3560	b	304
H	CONH$_2$	Me	CH$_2$Ph	3500	b	304
			2-NHR(R = COOCH$_2$CCl$_3$)			
H	CONH$_2$	Me	Me	3525	b	304
			2-NHR(R = Tosyl)			
H	CN	Ph	H	3300	a	180

[a] In Ref. 150 the data are available on a supplementary publication.
[b] Solvent: a, nujol mull; b, potassium bromide disk; c, not reported; d, sodium chloride; e, chloroform; f, neat.
[c] Other values reported in the same solvent are 3390, 3250 cm^{-1}.[180]
[d] The same author reported for the same compound different NH stretchings at 3380 and 3375 cm^{-1},[239] probably one of the two compounds is the 1-anilido derivative.

TABLE 3.21. NH$_2$ STRETCHING BANDS OF THE 3-AMINOPYRROLES

R^1	R^2	R^4	R^5	cm^{-1}	Solvent[a]	Ref.
			3-NH$_2$			
H	H	Ac	Me	3400, 3300	a	206
H	H	Ac	Me·HCl	2800, 2780, 2700	a	206
H	H	COPh	Me	3450, 3300	a	206
H	H	COPh	Me·HCl	2750, 2700, 2670	a	206
H	H	COOEt	Me·HCl	2800, 2600	a	206
H	Et	Ac	Me·HCl	2610	a	206
H	Et	COOMe	COOMe	3450, 3300	a	206
H	Et	COOMe	COOMe·HCl	2800, 2750, 2650	a	206
H	Ph	CN	Ph	3400, 3330	a	247
H	Ph	Ac	Me	3400, 3320	a	206
H	Ph	Ac	Me·HCl	2680, 2580	a	206
H	Ph	Ac	Ph	3440, 3340	a	263
H	Ph	—CO(CH$_2$)$_3$—	HCl	2750, 2690, 2550	a	206
H	Ph	COPh	H	3400, 3350	a	206
H	Ph	COPh	H·HCl	2590	a	206
H	Ph	COPh	Me	3350, 3300	a	206
H	Ph	COPh	Me·HCl	2800, 2710, 2700	a	206
H	Ph	COPh	Ph	3455, 3360	a	198
H	Ph	COPh	COOEt·HCl	2750, 2700, 2600	a	206
H	Ph	(3,5-Me$_2$ isoxazol-4-yl)CO	Ph	3470, 3370	a	260
H	Ph	CONH$_2$	Me	3390, 3310	b	207
H	Ph	CONH$_2$		3400, 3300	a	206
H	Ph	CONH$_2$	Me·HCl	2750, 2700, 2650	a	206

TABLE 3.21. Continued

R^1	R^2	R^4	R^5	cm^{-1}	Solvent^a	Ref.
H	Ph	CONHMe	Me	3400, 3300	a	206
H	Ph	CONHMe	Me·HCl	2680	a	206
H	Ph	CONHPh	Me	3440, 3360	b	207
H	Ph	COOMe	CH$_2$COOMe	3500, 3400	a	206
H	Ph	COOMe	CH$_2$COOMe·HCl	2750, 2550	a	206
H	Ph	COOMe	COOMe	3450, 3300	a	206
H	Ph	COOMe	COOMe·HCl	2600	a	206
H	Ph	COOEt	H·HCl	2750, 2550	a	206
H	Ph	COOEt	Me	3450, 3350	a	206
H	Ph	COOEt	Me·HCl	2700, 2650, 2550	a	206
H	Ph	COOEt	Ph	3420, 3340	a	247
H	Ph	COOEt	COOEt	3500, 3400	a	206
H	Ph	COOEt	COOEt·HCl	2700, 2550	a	206
H	2-MeC$_6$H$_4$	Ac	Me	3400, 3300	a	206
H	3-MeC$_6$H$_4$	Ac	Me	3400, 3300	a	206
H	4-MeC$_6$H$_4$	Ac	Me	3500, 3400	a	206
H	4-MeC$_6$H$_4$	Ac	Me·HCl	2680, 2580	a	206
H	4-MeC$_6$H$_4$	COOMe	COOMe·HCl	2700, 2600	a	206
H	4-MeC$_6$H$_4$	COOEt	Me	3450, 3350	a	206
H	4-MeC$_6$H$_4$	COOEt	Me·HCl	2750, 2680, 2580	a	206
H	2-NH$_2$C$_6$H$_4$	Ac	Me	3400–3060	a	261
H	3-MeOC$_6$H$_4$	Ac	Me	3440, 3340	a	248
H	3-MeOC$_6$H$_4$	Ac	Me	3440, 3350	a	248
H	4-MeOC$_6$H$_4$	Ac	Me	3400, 3300	a	206
H	4-MeOC$_6$H$_4$	Ac	Me·HCl	2700, 2600	a	206
H	4-MeOC$_6$H$_4$	COOMe	COOMe	3450, 3300	a	206
H	4-MeOC$_6$H$_4$	COOMe	COOMe·HCl	2700, 2580	a	206
H	4-MeOC$_6$H$_4$	COOEt	Me	3400, 3300	a	206
H	4-FC$_6$H$_4$	Ac	Me	3400, 3300	a	206
H	4-FC$_6$H$_4$	COPh	Me	3450, 3250	a	206
H	4-FC$_6$H$_4$	COOEt	Me	3400, 3300	a	206
H	4-FC$_6$H$_4$	COOEt	Me·HCl	2850, 2550	a	206
H	4-ClC$_6$H$_4$	CH$_2$COOEt	H	3490, 3390	c	197

H	4-ClC_6H_4	Ac	Me	3400, 3300	a	206
H	4-ClC_6H_4	COOMe	COOMe	3400	a	206
H	4-ClC_6H_4	COOMe	COOMe·HCl	2750, 2650, 2550	a	206
H	COO—*t*-Bu	COOEt	Me	3490, 3370	a	210
H	COO—*t*-Bu	—$CO(CH_2)_3$—		3515, 3390	d	210
H	COO—*t*-Bu	$COOCH_2Ph$	Me	3550, 3430	d	210
Me	Ph	H	Ph	3420, 3380	b	249
Me	Ph	Ph	Ph	3400, 3330	b	249
Me	Ph	Ac	Me	3325, 3315	b	249
Me	Ph	CN	Ph	3440, 3360	b	249
Me	Ph	COOEt	Me	3330, 3300	b	249
			3-NHR (R = i-Pr)			
i-Pr	Ph	COPh	H	3330	e	195
			3-NHR (R = c-Hexyl)			
c-Hexyl	Ph	COPh	H	2335	e	195
			3-NHR [R = C(NH)NH₂]			
H	H	H	H·AcOH	3450	d	267
H	H	H	COOH·HCl	3325	d	267
			3-NHR (R = CHO)			
CH_2COOEt	H	—$CO(CH_2)_3$—		3350	a	210
CH_2COOEt	COO—*t*-Bu	—$CO(CH_2)_3$—		3220	a	210
			3-NHR (R = Ac)			
H	CH_2COOEt	—$CO(CH_2)_3$—		3350–2700	d	210
H	Ph	H	Ph	3440, 3220	a	250
H	Ph	Ac	Me	3400	a	317
H	Ph	COPh	Me	3380	a	317
H	CONHMe	COOEt	Me	3310	a	210
H	COO—*t*-Bu	—$CO(CH_2)_3$—		3415	f	210

TABLE 3.21. Continued

R^1	R^2	R^4	R^5	cm^{-1}	Solvent[a]	Ref.
Me	Ph	H	Ph	3230	a	250
Ac	H	H	H	3300	g	215
3-NHR (R = COCH$_2$Ac)						
H	Ph	Ac	Me	3200	a	317
3-NHR (R = COCH$_2$COOEt)						
H	Ph	Ac	Me	3300	a	317
3-NHR (R = COPh)						
H	Ph	COOMe	COOMe	3250	a	206
3-NHR (R = CONH$_2$)						
H	H	COPh	Me	3500	a	317
H	Ph	Ac	Me	3400	a	317
H	Ph	COOMe	COOMe	3400	a	317
H	Ph	COOEt	Me	3300	a	317
3-NHR (R = CONHPh)						
H	Ph	COOEt	Me	3300	a	317
3-NHR (R = COO—t-Bu)						
H	COCF$_3$	H	H	3468	f	245
CPh$_3$	H	H	H	3435	f	245
3-NHR (R = COOCH$_2$Ph)						
CPh$_3$	H	H	H	3450	e	245

[a] Solvent: a, nujol mull; b, bromoform; c, not reported; d, potassium bromide disk; e, carbon tetrachlcride; f, chloroform; g, potassium chloride disk.

3.3. Structure and Physical Properties

TABLE 3.22. NH$_2$ STRETCHING BANDS OF THE 1,2-DIAMINOPYRROLES

R^3	R^4	R^5	cm^{-1}	Phase[a]	Ref.
			1-NH$_2$		
CN	—(CH$_2$)$_4$—		3421, 3373 3335, 3225	a	155
CN	Ph	H	3435, 3315	a	218
			1-NHR (R = CONH$_2$)		
CN	COOMe	Me	3430, 3340, 3200	b	219
CN	COOEt	Me	3460, 3360, 3310 3250, 3200	b	219
			1-NHR (R = CONHPh)		
CN	COOMe	Me	3500, 3300, 3190	b	219
CN	COOEt	Me	3450, 3300, 3260	b	219
			1-NHR (R = COOMe)		
CN	COOMe	Me	3420, 3330, 3250	b	219
CN	COOEt	Me	3440, 3310, 3185	b	219
			1-NHR (R = COOEt)		
CN	COOMe	Me	3440, 3330, 3250	b	219
			1-NHR (R = COO—t-Bu)		
CN	COOMe	Me	3390, 3325 3280, 3210	b	219
CN	COOEt	Me	3380, 3320, 3210	b	219

[a] Phase: a, potassium bromide disk, b, nujol mull.

~100 cm^{-1} to higher frequencies. *N*-Phosphinic- and *N*-phosphonicaminopyrroles have absorption bands at 3430–3280 cm^{-1}.

2-Aminopyrroles show v_{NH_2} absorption bands in the ranges expected for aromatic amines at 3500–3370 and 3340–3140 cm^{-1} (Table 3.20, pages 447–454). The corresponding bands for 3-alkoxycarbonyl-2-aminopyrroles occur at ~50 cm^{-1} higher than those for the corresponding 3-cyano derivatives.

The single v_{NH} absorption band of 2-acylaminopyrroles is to be found in the range 3400–3260 cm^{-1}.

The amino stretching bands for 3-aminopyrroles (Table 3.21, pages 455–458) are found in the ranges 3500–3350 and 3400–3300 cm^{-1}, showing the aromatic character of these compounds. The corresponding protonated compounds show the expected broad absorption in the region 2850–2550 cm^{-1}. The amino v_{NH}

absorption band for 3-acylaminopyrroles lies in the same range as that for the corresponding 2-acylamino derivatives.

The pyrrolyl NH stretching vibration for both 2-amino and 3-aminopyrroles is found in the range 3300–3150 cm^{-1}.

Amino stretching bands for 1,2- and 2,4-diaminopyrroles, listed in Tables 3.22 and 3.23, respectively, do not differ greatly from those of the monoaminopyrroles.

The spectrum of 2,5-diaminopyrrole shows a broad absorption band centered at 3310 cm^{-1}, due to the symmetric diimino tautomer[223] (see Section 3.3.10).

TABLE 3.23. NH$_2$ STRETCHING BANDS OF THE 2,4-DIAMINOPYRROLES

R^1	R^3	R^5	cm^{-1}	Phase[a]	Ref.
		2-NR$_2$—4-NHR' (R = Me; R' = Ph)			
t-Bu	Ph	H	3280	a	220
t-Bu	Ph	CSNH—t-Bu	3280, 3240	a	220
		2-NR$_2$—4-NHR' (R = Me; R' = 4-MeC$_6$H$_4$))			
t-Bu	Ph	H	3280	a	220
t-Bu	Ph	CSNH—t-Bu	3280, 3240	a	220

[a] Phase: a, potassium bromide disk.

3.3.7. Mass Spectra

The mass spectra of 1-amino-, 2-amino-, and 3-amino-pyrroles are reported in Tables 3.24–3.26.

Fragmentation of 1-aminopyrroles has not been studied in detail, but it can be used, in some instances, to distinguish between 1-aminopyrrole and the corresponding dihydropyridazine. The mass spectra of 1-arylaminopyrroles always show a sequential loss of the aryl residue and the arylamino fragment, followed by the usual fragmentation pattern of the substituted pyrrole nucleus.[108,128,133]

Detailed studies on the electron impact fragmentation on 2-aminopyrroles have been conducted on only a few derivatives. The fragmentation pattern of compounds **332** and **333**, shown in Scheme 90, was used to establish the synthetic pathway of the reaction reported in Scheme 46 (see Section 3.2.2.2.4).[185] In particular, the presence of the two fragments with $m/z = 105$

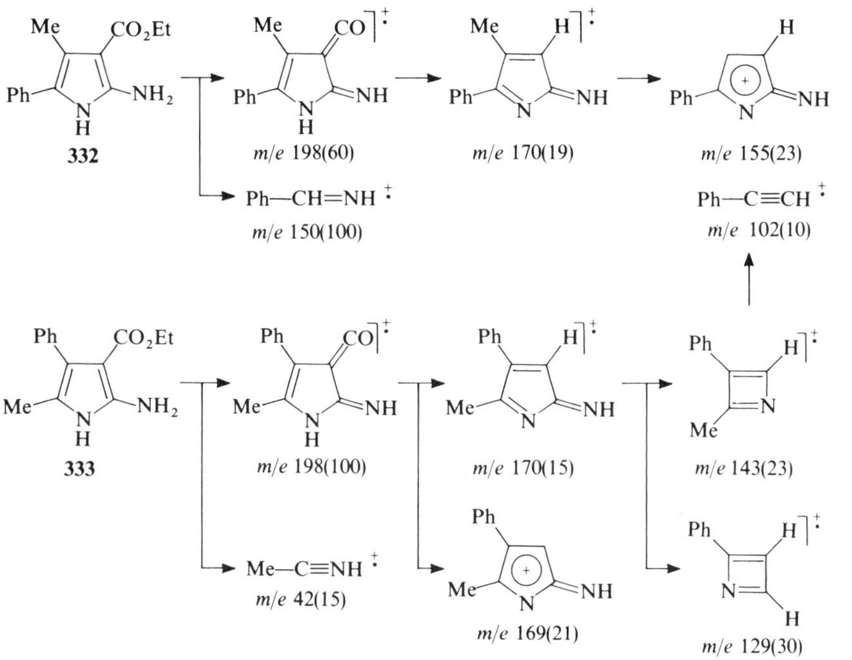

3.3. Structure and Physical Properties

Scheme 90

and 42 demonstrated the correct structures of the two isomers **332** and **333**, respectively.

The mass-spectrometric fragmentation behavior of 5-substituted 1-alkylideneimino-2-aminopyrroles is shown in Scheme 91.[276] The main fragmentation of the dicyanopyrroles **334** is characterized by "benzylic" α-cleavage of the C—C bond of the side chain to give the stable ions **335**, which lose the

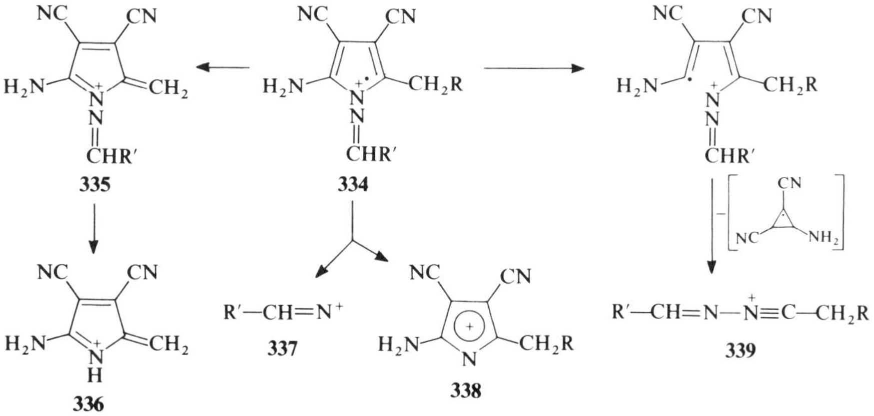

Scheme 91

alkyl nitrile to give **336**. Another important process is the cleavage of the N—N bond leading to ions **337** and **338**. Finally, the characteristic ring opening of the pyrrole nucleus produces the ions **339**, which have a high relative abundance in the aryl- and heteroaryl-substituted pyrroles, whereas ions **335** and **336** are not produced by these derivatives. However, the overall abundance of the molecular ions and the fragment ions **335–339** is generally greater than 40% of the total ion current. This indicates a high selectivity of the fragmentation.

The electron impact fragmentation of several 3-aminopyrroles have been studied.[277] The molecular ions and the fragmentation depicted in Scheme 92 have high relative abundance, which has been interpreted as being due to the stabilizing effect that the amino group has on the positively charged ions. The primary decomposition of the molecular ion **340** giving rise to the base peak (M–MeOH) follows the standard fragmentation of pyrrole carboxylic esters. The ions **341** and **341'** lose a further molecule of methanol to give **342**, which leads to **343**. Ions **341** and **341'** can produce ion **346** by two different decomposition paths (Scheme 92).

TABLE 3.24. MASS SPECTRA OF THE 1-AMINOPYRROLES

R^2	R^3	R^4	R^5	m/z (%)	Ref.
		1-NHR (R = Ph)			
Ph	CH$_2$Ph	Ac	Me	380(100); 365(16); 338(24); 337(24); 289(54); 288(92); 274(19); 246(81); 244(43); 230(19); 220(14); 202(14); 168(16); 128(19); 127(18); 127(18); 115(22); 106(16); 103(15); 93(57); 91(43)	133
Ph	Ph	—(CH$_2$)$_3$—		378(84); 323(12); 301(10); 287(44); 286(100); 274(10); 258(17); 244(17); 230(25); 127(26); 93(22)	133
Ph	Ph	Ac	Me	366(58); 351(6); 324(9); 323(10); 275(23); 274(100); 259(5); 258(5); 231(6); 230(17); 189(6); 128(6); 127(5)	133

3.3. Structure and Physical Properties 463

TABLE 3.24. Continued

R²	R³	R⁴	R⁵	m/z (%)	Ref.
Ph	Ph	COOEt	Me	343(100); 328(10); 315(12); 314(52); 298(15); 282(10); 270(28); 207(10); 206(25); 205(10); 178(35); 177(10); 162(10); 161(10); 160(26); 158(14); 149(15); 134(36); 133(20); 132(32); 130(13); 117(13); 91(17)	133
		1-NHR (R = 4-NO₂C₆H₄)			
—(CH₂)₄—		COOEt	Me	396(58); 381(7); 367(7); 351(4); 323(8); 305(5); 304(20); 277(5); 276(21) 260(4); 259(24); 258(100); 230(10); 220(4); 189(4); 128(7); 127(16); 93(8); 92(5)	133
		1-NHR (R = tosyl)			
—(CH₂)₄—		—CH₂C(Me₂)CH₂—		386;[a] 231; 216	128
—(CH₂)₄—		Ac	Me	346;[a] 191; 176	128
—(CH₂)₄—		COOEt	Me	376;[a] 221; 206	128
—(CH₂)₄—		COOEt	Ph	438;[a] 283; 268	128
		1-NRR' (R = tosyl; R' = Ac)			
—(CH₂)₄—		Ac	Me	388;[a] 233; 176	128
—(CH₂)₄—		COOEt	Me	418;[a] 263; 206	128
—(CH₂)₄—		COOEt	Ph	480;[a] 325; 268	128

[a] Intensities not reported.

3.3.8. Other Physical Measurements

The densities and the refractive indices of 1-aminopyrroles have been reported[105,114] (Table 3.27).

3.3.9. Thermodynamic Aspects

N-Aminopyrroles have physical properties different from those of C-aminopyrroles. They are generally thermally very stable and resistant to common hydrolytic agents,[44,71,121] and are also stable to oxidizing agents.[66] Nevertheless, some derivatives are photosensitive or are susceptible to X-rays.[64,107,133]

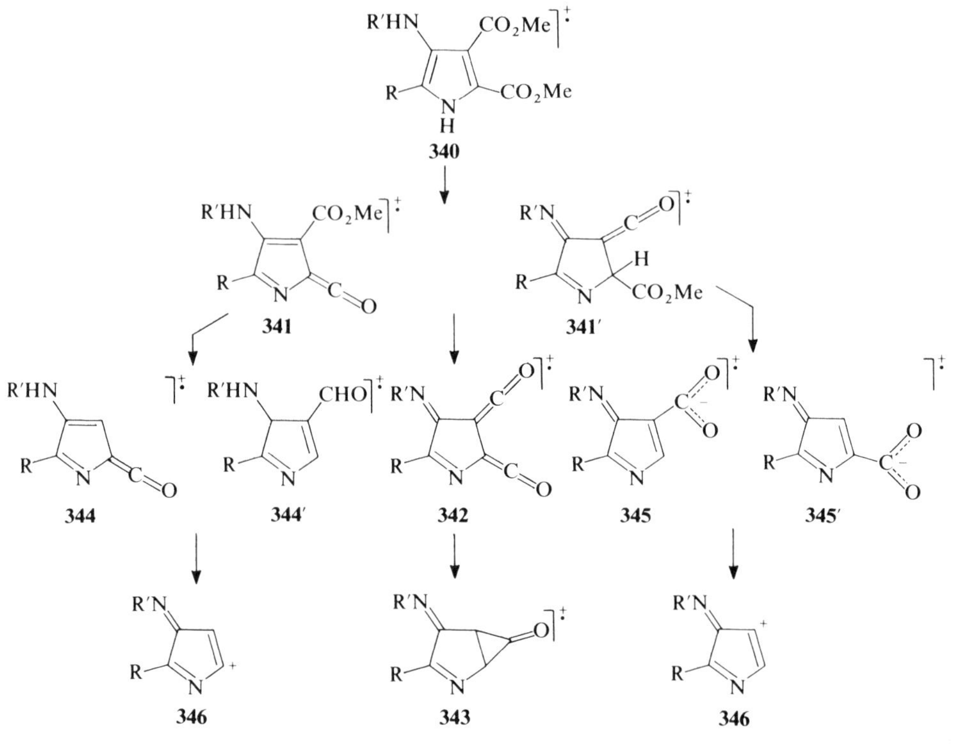

Scheme 92

1-Aryl- and 1-aroylamino-2,5-dimethylpyrroles are generally unstable on exposure to air, with formation of red or black products. The instability increases in the order benzamido < anilino < phenyl, although nitro substitution of the benzene ring increases the stability to some degree.[50]

1-Amino-2,5-diphenylpyrrole is water-soluble,[52] but the introduction of substituents on either the carbon atoms or the nitrogen atom lowers the solubility in water and increases their solubility in the most common polar organic solvents.

The heat of combustion of diethyl 1-amino-2,5-dimethylpyrrole-4,5-dicarboxylate has been measured in an attempt to differentiate between this class of compounds and the isomeric dihydropyridazines. However, the experimental error was such that no unequivocal conclusion could be reached.[278]

The physical properties of the C-aminopyrroles are controlled by the conjugative interaction of the amino group with the pyrrole ring. The electron-rich 2-aminopyrroles are considerably more unstable than the corresponding 3-amino isomers and are therefore difficult to handle and sometimes difficult to isolate as pure samples. They are generally hygroscopic, are easily oxidized, and decompose during the chromatographic purification on silica gel.[152,153,185,189,237]

TABLE 3.25. MASS SPECTRA OF THE 2-AMINOPYRROLES[a]

R^1	R^3	R^4	R^5	m/z (%)	Ref.
			2-NH_2		
H	COOEt	H	Me	168(80); 122(100); 94(90); 53(42); 42(49)	184
i-Pr	Me	H	CN	163(38); 121(95); 120(100); 119(8); 93(19); 66(15); 57(7); 43(42); 42(10); 41(30)	165
i-Pr	Et	H	CN	177(17); 162(5); 150(15); 149(12); 148(60); 135(17); 121(66); 109(9); 107(13); 106(12); 95(12); 81(100); 68(24); 54(24); 43(26); 42(32); 41(49); 39(20)	165
t-Bu	Et	H	CN	191(6); 136(5); 135(67); 134(8); 121(9); 120(100); 93(6); 66(5); 57(91); 42(5); 41(44)	165
Ph	CN	CN	Ph	284(100); 283(14); 256(6); 229(3); 207(7); 180(17); 154(8); 104(12); 78(8); 77(26); 51(10)	167

TABLE 3.25. Continued

R^1	R^3	R^4	R^5	m/z (%)	Ref.
4-OHC$_6$H$_4$	CN	CN	Ph	300(100); 272(5); 223(6); 207(5); 196(13); 180(9); 170(6); 120(10); 100(9); 94(9); 77(10)	167
N=CHMe	CN	CN	Me	187(100); 161(18); 146(45); 145(36); 118(82); 104(13); 91(15); 77(32); 76(14); 43(37); 41(15)	276
N=CH—n-Pr	CN	CN	n-Pr	243(34); 214(25); 173(100); 159 (36); 158(11); 145(74); 133(15); 132(11); 131(40); 77(12); 70(15)	276
N=CH—i-Pr	CN	CN	i-Pr	243(27); 228(13); 173(100); 159(39); 158(13); 157(8); 133(17); 132(10); 131(6); 105(7); 77(6)	276
N=CHPh	CN	CN	Ph	311(45); 208(43); 207 (100); 180(17); 153(6); 129(5); 104(32); 103(9); 79(7); 77(21); 52(5)	276
2-Furyl-CH=N	CN	CN	2-Furyl	231(40); 197(100); 169(9); 142(8); 115(4); 95(12); 94(21); 77(4); 69(5); 66(5); 39(14)	276

3-Pyridyl-CH=N	CN	CN	3-Pyridyl	313(50); 208(100); 181(20); 155(5); 154(10); 1458(8); 106(100); 104(50); 103(5); 54(8); 52(5)	276
			2-NHR (R = Et)		
Me	COOEt	H	H	196(60.8); 167(12.6); 150(21.0); 149(32.2); 139(17.2); 135(100.0); 122(22.8); 121(24.2); 108(40.6)	184
			NHR (R = Ph)		
CH₂Ph	Ph	Ph	CH(COPh)₂	622(17); 621(28); 531(10); 530(32); 517(11); 105(40); 91(14); 77(15); 58(30); 44(14); 42(100)	163
			2-NHR (R = Ac)		
H	CONH₂	Me	H	181(68); 122(100)	290
H	CONH₂	Me	i-Bu	237(40); 194(100)	290
H	CONH₂	Me	CH₂Ph	271(100); 212(35)	290
H	CONH₂	Ph	H	243(72); 184(100)	290
Ph	CONH₂	Me	Me	271(42); 212(100)	290
Ph	CONH₂	—(CH₂)₄—		297(34); 238(100)	290
			2-NHR (R = COCF₃)		
Ph	CN	Me	Me	307(62); 210(100)	290
			2-NHR (R = COPh)		
Ph	CONH₂	Me	Me	333(27); 316(100); 212(24); 211(68)	290

TABLE 3.25. MASS SPECTRA OF THE 2-AMINOPYRROLES[a]

R^1	R^3	R^4	R^5	m/z (%)	Ref.
			2-NRR [R–R = CO(CH$_2$)$_3$]		
Me	CN	Me	Me	218; 189; 175; 162[b]	336
			2-NRR' (R = Me; R' = Ph)		
CH$_2$Ph	Ph	Ph	CH(COPh)$_2$	636(26); 635(55); 531(26); 335(16); 105(100); 91(29); 77(34); 43(17); 40(43)	163

[a] In Ref. 150 the data are available on a supplementary publication.
[b] Intensities not reported.

TABLE 3.26. MASS SPECTRA OF THE 3-AMINOPYRROLES

$$R^5 \underset{R^1}{\overset{R^4}{\underset{N}{\bigvee}}} R^2$$

R^1	R^2	R^4	R^5	m/z (%)	Ref.
				3-NH$_2$	
H	Et	COOMe	COOMe	226 (45); 194 (100); 162 (7); 151(8); 136(14); 134(15); 107(7)	277
H	Ph	COOMe	COOMe	274(58); 242(100); 210(21); 199(12); 188(23); 184(14); 155(31)	277
H	4-MeC$_6$H$_4$	COOMe	COOMe	288(76); 256(100); 224(7); 213(12); 198(7); 196(17); 169(16)	277
H	4-OMeC$_6$H$_4$	COOMe	COOMe	304(50); 272(100); 240(4); 229(34); 214(11); 212(23); 185(23)	277
H	4-ClC$_6$H$_4$	COOMe	COOMe	308(55); 276(100); 244(12); 233(7); 218(13); 216(17); 189(14)	277
H	CONH$_2$	COOEt	Me	211;[a] 194; 165; 148(100); 120; 92; 67	210
H	COO—*t*-Bu	COOEt	Me	268;[a] 223; 212(100); 195; 194; 167; 166; 122; 121	210
H	COO—*t*-Bu	—CO(CH$_2$)$_3$—		250;[a] 195; 194(100); 177; 176; 148	210
				3-NHR (R = Ac)	
H	CH$_2$COOEt	—CO(CH$_2$)$_3$—		278;[a] 235; 190; 163(100)	210
H	Ph	H	Ph	276(100); 234(80); 233(38); 206(7); 130(10); 127(10); 105(9); 103(11); 77(12); 44(16); 43(12)	250
				3-NHR (R = COO—t-Bu)	
CPh$_3$	H	H	H	424(1); 243(35); 165(53); 91(19); 57(100)	245
H	COCF$_3$	H	H	278(5); 149(54); 91(42); 59(66); 58(100); 57(70)	245

TABLE 3.26. Continued

R^1	R^2	R^4	R^5	m/z (%)	Ref.
				3-NHR (R = COOCH$_2$Ph)	
CPH$_3$	H	H	H	454(3); 244(30); 243(100); 155(56); 91(74); 74(36)	245
				3-NHR (R = SO$_2$Me)	
H	Ph	COOMe	COOMe	352(27); 320(24); 288(3); 260(1)	277
				3-NR$_2$ (R = Ac)	
H	Ph	—(CH$_2$)$_4$—		296;[a] 254; 236; 226; 212; 195; 184; 169; 156; 145; 130; 115	199
H	Ph	—(CH$_2$)$_5$—		310;[a] 268; 250; 239; 225; 209; 197; 182; 169; 156; 145; 132; 115	199
				3-NRR [R–R = —(CH$_2$)$_2$O(CH$_2$)$_2$—]	
H	COOMe	Ph	H	286; 271; 255; 254; 242; 227; 225; 223; 197; 196; 195(100); 169; 168; 167; 140; 128; 115	204

[a] Intensities not reported.

TABLE 3.27. OTHER PHYSICAL MEASUREMENTS OF THE 1-AMINOPYRROLES

R^2	R^3	R^4	R^5	Density	Refractive Index	Ref.
				1-NH_2		
H	H	H	H	1.049° at 23°C	1.5350 at 21°C	105
Me	H	Bu	H	0.9492 at 20°C	1.4990 at 20°C	114
				1-NR_2 (R = Me)		
Me	H	Bu	H	0.8829 at 20°C	1.4811 at 20°C	114

Consequently, 2-aminopyrroles are better isolated as salts or as the more stable acetyl or benzoyl derivatives.

However, when the pyrrole ring is substituted by phenyl groups or by strong electron-withdrawing groups, the stability of the system is increased by the greater delocalization of the π-electrons, and these 2-aminopyrroles are, as a result, less prone to autooxidation.

In contrast with the 2-aminopyrroles, only a few 3-aminopyrroles appear to be unstable. For example, 3-amino-5-methyl-2-phenylpyrrole polymerizes rapidly on exposure on air, but most 3-aminopyrroles can be stored at low temperatures or under nitrogen. In some instances 3-aminopyrroles are better isolated as salts or as the more stable acyl derivatives.[265,266]

However, there are examples where the 3-aminopyrroles cannot be isolated, as they undergo oxidation or condensation reactions.[211,212,266]

The stability of polyaminopyrroles has not been studied, although 2,5-diaminopyrrole has been reported to be remarkably sensitive to moisture and is best kept in a sealed vial at 0°C. It is not very soluble in organic solvents, but dissolves easily in ammonia and water.[157,222] However, the aqueous solution, of pH > 10.5, rapidly decomposes to give 5-imino-2-pyrrolidone with the liberation of ammonia.

3.3.10. Tautomeric and Basicity Studies

3.3.10.1. 1-Aminopyrroles

The tautomerism of simple 1-aminopyrroles is not possible, and, although prototropic tautomerism could occur with 1-ureido and 1-guanidinopyrroles, they have not been examined.

1-Aminopyrroles are far less basic than primary C-aminopyrroles or N,N-disubstituted hydrazines.[86] Quaternarization of N,N-dimethylaminopyrroles fails both with methyl iodide under reflux and with trimethyloxonium tetrafluoroborate in dichloromethane,[121] and 1-amino-2,5-diphenylpyrrole does not react with sulfonyl chlorides in the presence of a range of bases and solvents,[66] indicating the low nucleophilic character of the 1-amino group.

In contrast, 1-ureido- and 1-arylsulfonylaminopyrroles are weakly acidic, and are soluble in alkali, from which they can be reprecipitated by the addition of carbon dioxide.[98,99,113]

Not unexpectedly, 1-guanidinopyrroles are also more basic than the 1-aminopyrroles.[61]

3.3.10.2. 2-Aminopyrroles

In contrast with the stability of the pyrrol-2-one system, amino–imino tautomeric equilibrium of 2-aminopyrroles is shifted toward the amino form. Evidence supporting the predominance of the amino form comes from spectroscopic data as well as theoretical studies (see Section 3.3.1). The IR spectra show the normal absorption of an aromatic amine, and the ^{1}H NMR spectra are consistent with the amino form. Earlier work suggested that the ^{1}H NMR data for the 2-aminopyrrole **347** were compatible with the imino structure,[279] but a more recent examination of ^{1}H and ^{13}C NMR spectra indicates that **347** exists exclusively in the amino form.[183] The two ^{1}H resonance signals for the amino protons, described in Ref. 279, might be rationalized in terms of an intramolecular hydrogen bond, as shown in Scheme 93. A similar effect was observed in the ^{1}H NMR spectrum of 1-benzoyl-2-aminopyrrole.[162]

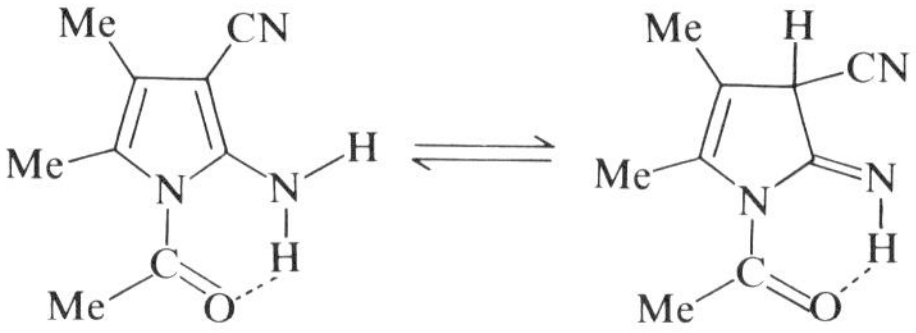

347

Scheme 93

The more complex tautomeric equilibria of 2-amino-4 hydroxypyrrole derivatives **348** have been studied in detail.[190] These compounds can exist in four potentially tautomeric forms **A–D** (see Scheme 94). The preferred structure for **348** has been identified as **B** on the basis of IR and NMR data. In particular, the absence of absorption in the CO region and a broad absorption band at 3225–3030 cm^{-1} excludes structures **C** and **D**, whereas the presence of typical ^{1}H NMR spectral pattern for geminal hydrogen atoms (when R^2 = H) and for vicinal hydrogen atoms eliminates structure **A**. The corresponding N-substitu-

3.3. Structure and Physical Properties

ted derivatives **349** can exist in the tautomeric forms **A'** and **D'**, and it has been demonstrated that structure **D'** predominates since the IR spectra show a v_{CO} absorption band at 1695–1640 cm^{-1}, and a CH$_2$ peak at 3.77–4.10 ppm is found in the ^{1}H NMR spectra.

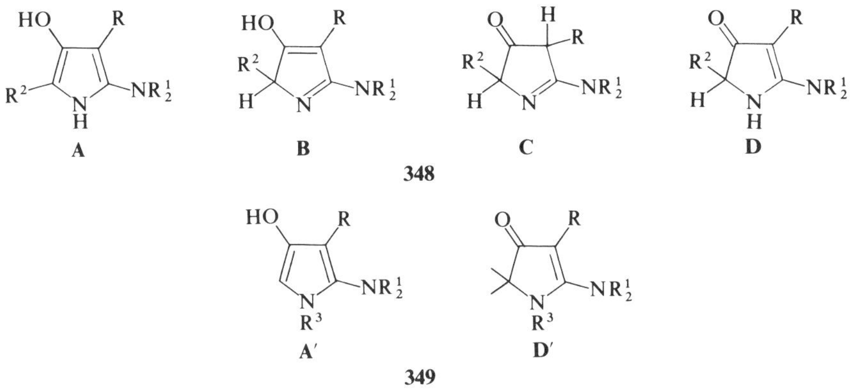

Scheme 94

The complex system of tautomeric forms for the 2-amino-5-hydroxypyrroles of type **350** are shown in Scheme 95. The compounds are amphoteric, as shown by their solubilities; they give salts with acids, but are also soluble in strong bases.[156,158] Spectral evidence has been obtained in support of the predominance of tautomeric form **D**.

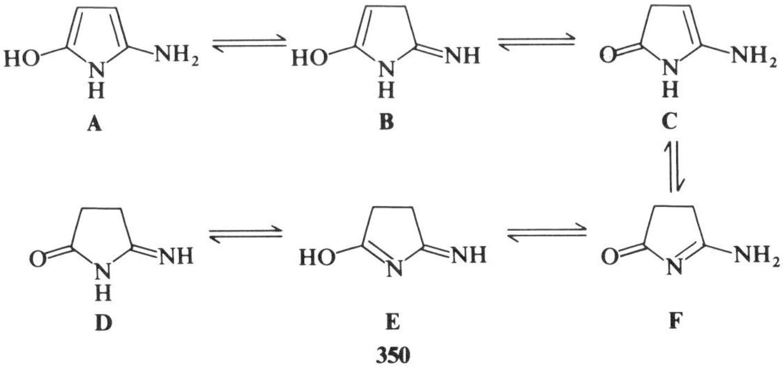

Scheme 95

The pK_a of the conjugate acids 2-aminopyrroles has been measured only for 2-amino-3,4-dicyano-5-(hydroxyethylthio)pyrrole (pK_a = 9.4),[152] but there is qualitative evidence that 2-aminopyrroles are strong bases. Strongly electron-withdrawing groups in the pyrrole ring enhance the acidity of the pyrrole NH. Thus, for example, 2-amino-3,4-dicyanopyrroles dissolve in aqueous sodium bicarbonate,[152] and 2-amino-3-cyanopyrrole cannot be titrated potentiometrically as a base with perchloric acid in nitrosobenzene and acetic acid as solvents, but it can be titrated as acid with tetrabutylammonium hydroxide in pyridine.[280]

Protonation of 2-aminopyrroles has been studied using ^{1}H NMR spectroscopy. A comparison of the spectra recorded in DMSO and in the presence of TFA showed that 2-aminopyrroles **351** are not protonated like typical aromatic amines to give the ammonium salt **352**, but their behavior is characteristic of the zwitterionic form **353** or **354**[279] to give the *C*-protonated salts **355** or **356**, respectively (Scheme 96). The splitting pattern of the 5-methyl group signal in the ^{1}H NMR spectrum indicates that the protonation occurs predominantly at the 5-position. Protonation also occurs predominantly at the 5-position in the case of 2-amino-3-carboxamido and 2-amino-3-alkoxycarbonyl derivatives.[182,239]

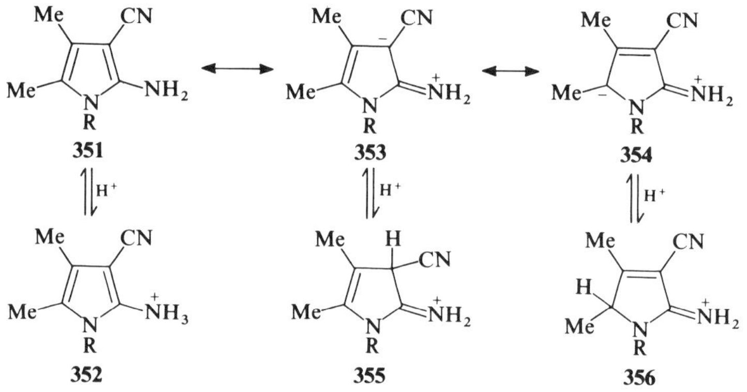

Scheme 96

Hydrogen NMR spectra of the isolated perchlorate salts of 2-aminopyrroles are also compatible with protonation at the 5-position, rather than at the primary amino group.[176]

In specific 2-amino derivatives, the protonation occurs on other exocyclic basic centers.[190,281,282]

3.3.10.3. 3-Aminopyrroles

3-Aminopyrroles are weaker bases than the corresponding 2-amino isomers with pK_a values for the conjugated acids ranging from 2.50 to 5.85 (Table 3.28). For this class of compounds the predominant tautomer is again the amino form, as confirmed by spectroscopic data.[262] However, anomalous behavior has been observed in the case of 3-amino-1-tritylpyrrole **357**, which, in deuterochloroform, exists exclusively as the imino form, as shown by its ^{1}H NMR spectrum[245]

TABLE 3.28. pK_a VALUES OF THE 3-AMINOPYRROLES

3-NH_2

R^1	R^2	R^4	R^5	pK_a	Method[a]	Ref.
H	H	Ac	Me	5.70	a	206
H	H	COPh	Me	5.30	a	206
H	H	COOEt	Me	5.70	a	206
H	Et	Ac	Me	5.85	a	206
H	Et	COOMe	COOMe	3.40	a	206
H	Ph	Ac	Me	3.45	a	206
H	Ph	—CO(CH$_2$)$_3$—		2.95	a	206
H	Ph	COPh	H	3.10	a	206
H	Ph	COPh	Me	3.35	a	206
H	Ph	COPh	COOEt	2.70	a	206
H	Ph	CONH$_2$	Me	4.35	a	206
H	Ph	CONHMe	Me	3.00	a	206
H	Ph	COOMe	CH$_2$COOMe	3.20	a	206
H	Ph	COOMe	COOMe	2.55	a	206
H	Ph	COOEt	H	3.05	a	206
H	Ph	COOEt	Me	3.45	a	206
H	2-MeC$_6$H$_4$	Ac	Me	4.97	b	206
H	3-MeC$_6$H$_4$	Ac	Me	3.74	b	206
H	4-MeC$_6$H$_4$	Ac	Me	3.95	b	206
H	4-MeC$_6$H$_4$	COOMe	COOMe	2.50	a	206

TABLE 3.28. Continued

R^1	R^2	R^4	R^5	pK_a	Method[a]	Ref.
H	4-MeC$_6$H$_4$	COOEt	Me	3.65	a	206
H	4-MeOC$_6$H$_4$	Ac	Me	4.05	a	206
H	4-MeOC$_6$H$_4$	COOMe	COOMe	2.55	a	206
H	4-MeOC$_6$H$_4$	COOEt	Me	3.85	a	206
H	4-FC$_6$H$_4$	Ac	Me	4.50	b	206
H	4-FC$_6$H$_4$	COPh	Me	4.73	b	206
H	4-FC$_6$H$_4$	COOEt	Me	3.35	a	206
H	4-ClC$_6$H$_4$	Ac	Me	4.51	b	206
H	4-ClC$_6$H$_4$	COOMe	COOMe	2.50	a	206

[a] Method: a, titration of the hydrochloride with 0.1 N sodium hydroxide in methylcellosolve-water; b, spectrometric determination in absolute ethanol.

3.3. Structure and Physical Properties

in which the 4-H and 5-H resonances appear as two 3-line multiplets at 5.75 and 6.35 ppm, respectively, the α-methylene signal at 2.87 ppm is a broad unresolved multiplet, and the single imino proton appears at 6.10 ppm. The two latter signals disappear on treatment with D_2O, leaving the 4-H and 5-H signals as doublets. These experimental data are at variance with MO calculations (see Section 3.3.1), which predict that the amino tautomer of 3-aminopyrroles is the more stable form. The discrepancy is difficult to rationalize, but the effect of the steric trityl group in **357** (Scheme 97) on the position of the tautomeric equilibrium, which cannot be incorporated into the theoretical analysis, may be significant.

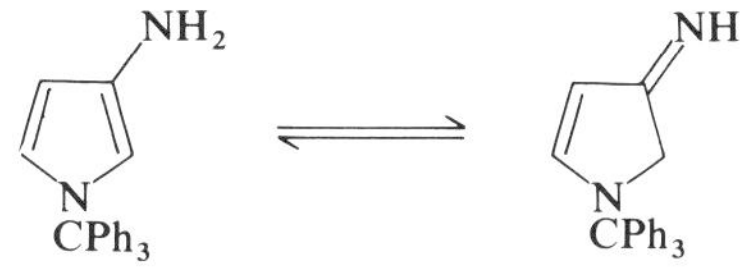

357

Scheme 97

Protonation of 3-aminopyrroles has been studied by ^1H and ^{13}C NMR spectroscopy.[262] The ^1H spectral data for the amines **358** and their protonated species **359** indicate that protonation occurs on the amino group. The ^1H resonance signals of the protonated species show downfield shifts relative to those of the amine, with no changes in their multiplicity, and, significantly, there is no evidence of upfield shifts expected for ring protonation. The broad signal, due to the amino group experiences a downfield shift of ~ 5 ppm, and integration increases from two to three protons, while the pyrrolic NH signal exhibits a downfield shift of ~ 1.2 ppm. The ^{13}C data for the protonated species show a marked upfield shift for the C(3) resonance and a smaller, but distinct, downfield shift in the C(2) resonance signal, clearly indicating protonation of the amino group.

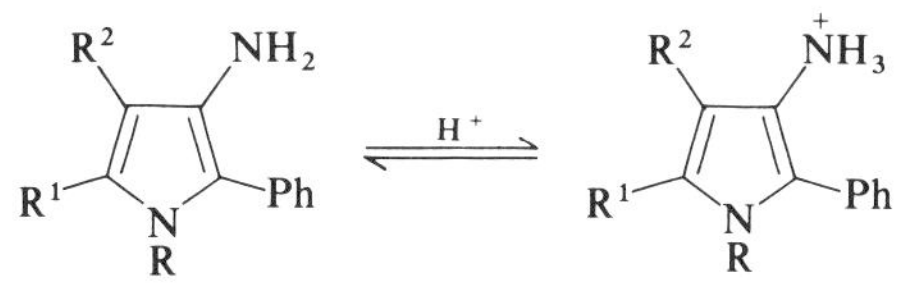

358 **359**

R = H, Me; R^1 = Me, Ph; R^2 = H, Ac, CO_2Et

Scheme 98

3.3.10.4. Diaminopyrroles

The only study of the tautomeric equilibrium of 2,4-diaminopyrroles relates to **252** (R = *tert*-butyl) (Scheme 99); the ^1H NMR spectrum,[220] measured in

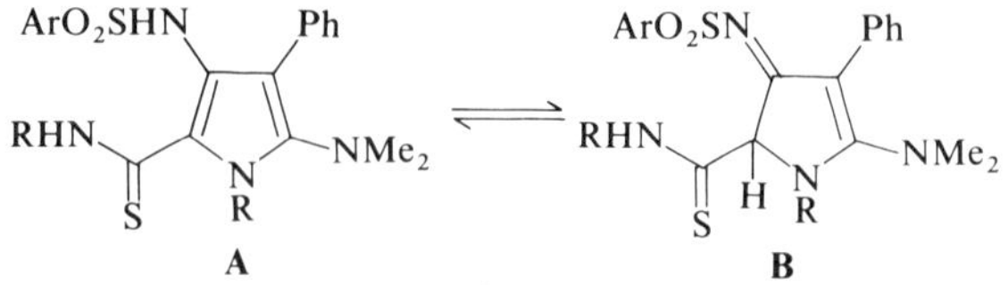

478   Aminopyrroles

Scheme 99

252
R = t-Bu; Ar = Ph, 4-MeC₆H₄

deuterochloroform, indicates the presence of a 10% of the 5H-tautomer B, based on the signals at 1.38 and 1.58 (s, *tert*-butyl), 2.87 (s, NMe) and 6.0 ppm (s, 5-CH).

Extensive spectroscopic studies of the 2,5-diaminopyrrole **256** (R = H) and its symmetric diphenyl derivative (R = Ph) have established that it exists predominantly in the tautomeric form **A**. Although the data are not entirely unambiguous, the UV spectra suggested that an imino form is important in the tautomeric equilibrium.[157,223,283] ^{1}H NMR spectroscopy provides more definitive conclusions; the high symmetry of the spectra, measured in pyridine and in DMSO, with CH$_2$ signals in the range 2.66–2.67 ppm, demonstrates that **256** (Scheme 100) exists as the diimino form **A**. This observation is in good agreement with the chemical shifts of the methylene signals of succinimide (2.74 ppm in chloroform) but totally at variance with the typical chemical shifts expected for the other symmetric tautomer **C**.[284] These observations contrast with findings that other analogous heterocyclic compounds exist predominantly in the diamino form.[285]

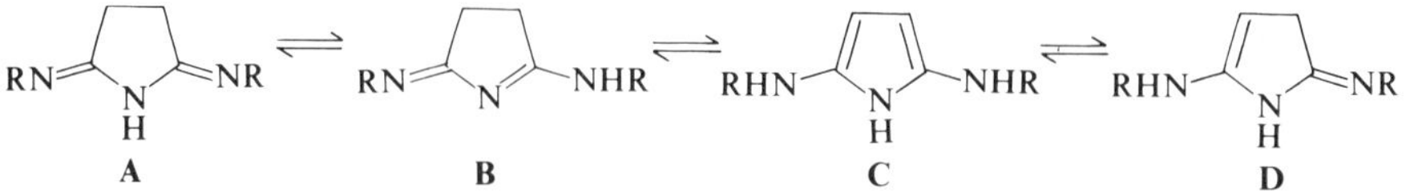

Scheme 100

3.4. REACTIVITY

Predictably, the reactivity of aminopyrroles depends on the position of the amino group. Thus N-aminopyrroles, which show weak basic character, generally behave in a fashion analogous to N,N-disubstituted hydrazines, whereas C-aminopyrroles generally react as aromatic amines.

3.4.1. Thermolysis and Photolysis Reactions

3.4.1.1. 1-Aminopyrroles

Thermolysis of 1-aminopyrroles has not been extensively studied, but they have a remarkable thermal stability, compared with that of the 2- and 3-isomers.

3.4. Reactivity 479

Thus, the sodium salt of *N*-tosylaminopyrrole **360** can be recovered unchanged after heating for 1 h at 225°C in tetraglyme, but it decomposes after 8 h at 275°C to give pyrrole **364**, together with a complex mixture of unidentified products.[286] The mechanism of this decomposition (Scheme 101), proposed by analogy with the behavior of the corresponding carbazole derivative, involves the intermediate diazene **361** and tetrazene **362**, which extrudes nitrogen to form the pyrrolyl radical **363**. The alternative pathway, involving direct fission of N—N bond in the starting compound, cannot be rigorously excluded.

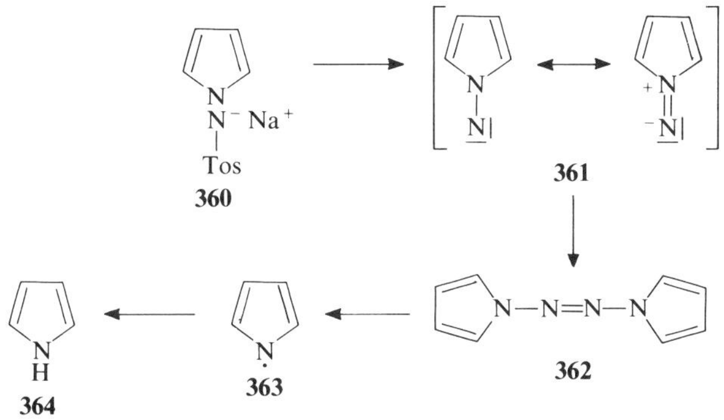

Scheme 101

Thermolysis of substituted 2-phenyl-1-ureidopyrroles **365** results in the production of a complex mixture from which pyrrolophthalazinones **366** can be

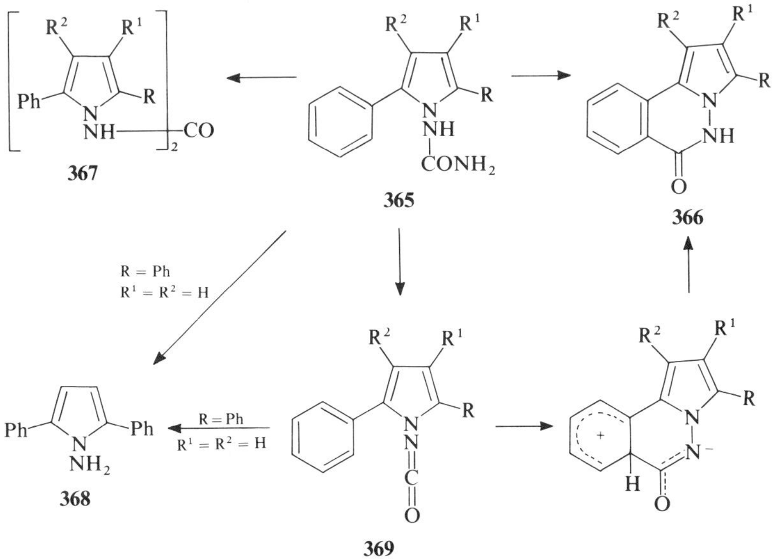

R = Me, Ph; R^1 = H, CO_2H, CO_2Et; R^2 = H, CO_2H, CO_2Et

Scheme 102

isolated as the major product, together with the N,N'-bis(1-pyrrolyl)urea **367** and 1-amino-2,5-diphenylpyrrole (**368**) (from 2,5-diphenyl-1-ureidopyrrole)[287-289] (Scheme 102). The mechanism proposed for the formation of **366** requires the initial loss of ammonia to give the isocyanate derivative **369**, which interacts with the phenyl group to yield the cyclic product. The presence of the intermediate isocyanate can also account for the formation of the 1-aminopyrrole on hydrolysis.

Other thermal reactions of 1-aminopyrroles, not involving the amino moiety, have been reported. Such reactions essentially involve the normal decarboxylation of acid groups on the pyrrole ring.[83,97,100,116]

In the only report on photolysis of 1-aminopyrroles, the N—N bond of 1-aminotetraphenylpyrrole **370** (in benzene, ether, chloroform, propanone, or acetic acid) cleaves to give tetraphenylpyrrole **371**[107] (Scheme 103). Although peroxide intermediates were not detected in sensitized photolysis at low temperature ($-60°C$), the main product of the reaction in chloroform or ether was the 1-benzoyltriphenylpyrazole **376** (45–40%) with a lesser amount of tetraphenylpyridazine (15%), cis-dibenzoylstilbene (15–10%), and 3,4,5-triphenylpyrazole **377** (5–2%). Tars are also produced in ether. The postulated mechanism for the formation of the pyrazoles involves the intermediate hydrazidochalcone **374**, obtained by cleavage of the peroxides **372** and **373**.

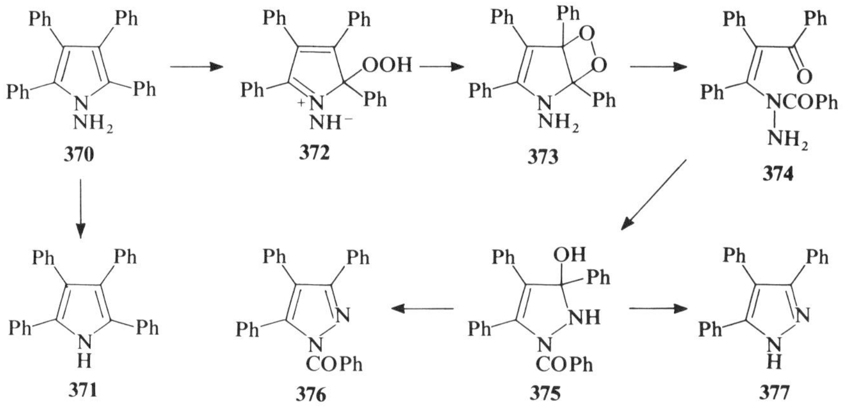

Scheme 103

Photolysis in propanone produces, as the main product, the stilbene derivative (30%), but the pyrazole **377** (15%) and tetraphenylpyridazine (15%) are also isolated, while photolysis in chloroform or propanone in the presence of strong acids leads to the formation of pyridazine (75–70%) and stilbene derivatives (20–10%).

At low temperature, in the presence of a reducing agent, the major product of the photolysis in chloroform is the stilbene (50%), together with the pyridazine derivative (20%). Photooxidation does not occur at room temperature in chloroform or at $-60°C$ in propanone.

3.4.1.2. 2-Aminopyrroles

Thermolysis of substituted 2-aminopyrroles **378** leads, by an intramolecular cyclization, to pyrrolo[1,2-b]-as-triazines **379**[155] (Scheme 104). The same cyclization reaction has been achieved on treatment of **378** with acids or bases.

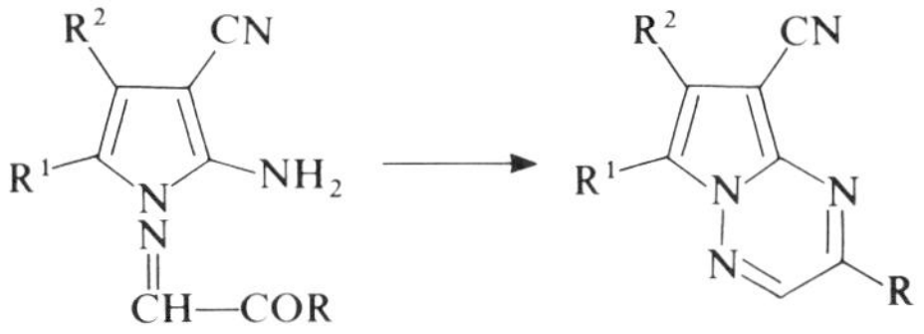

R = Ph, 4-NO$_2$C$_6$H$_4$, OH, OEt; R^1 = H;
R^2 = Ph; R^1—R^2 = —(CH$_2$)$_4$—

Scheme 104

Thermolysis of 2-carbonylamino-3-amidopyrroles **380** (R = Me, Ph), with or without solvent, has been reported to yield the corresponding deazahypoxanthines **381**,[290] whereas 2-carbethoxyamino-3-amidopyrroles **380** (R = OEt) gave pyrrolo[2,3-d]pyrimidine-2,4-diones **382**[291] (Scheme 105).

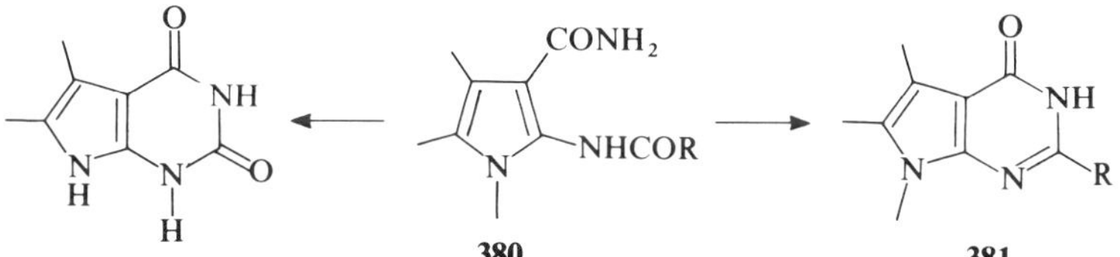

Scheme 105

Thermolysis of the urethane **383** produces the β,δ-diimidoetioporphyrin **384**,[231] while the isomeric urethane **385**, on heating to 200°C, yields the α,γ-diimidoetioporphyrin **386** (Scheme 106).

2-Iminopyrrol-5-one in boiling water gave succinimide, with evolution of ammonia.[157]

Pyrolysis of the 2-aminopyrrole **277** (R = H) yields **387** and traces of the 2-amidopyrrol-5-one **388**[224] (Scheme 107).

3.4.1.3. Diaminopyrroles

Thermolysis of 2,4-diaminopyrroles **252** in chloroform or dichloromethane results in the formation of the corresponding 5-unsubstituted pyrroles with elimination of *tert*-butylisothiocyanate.[220] Such an elimination probably pro-

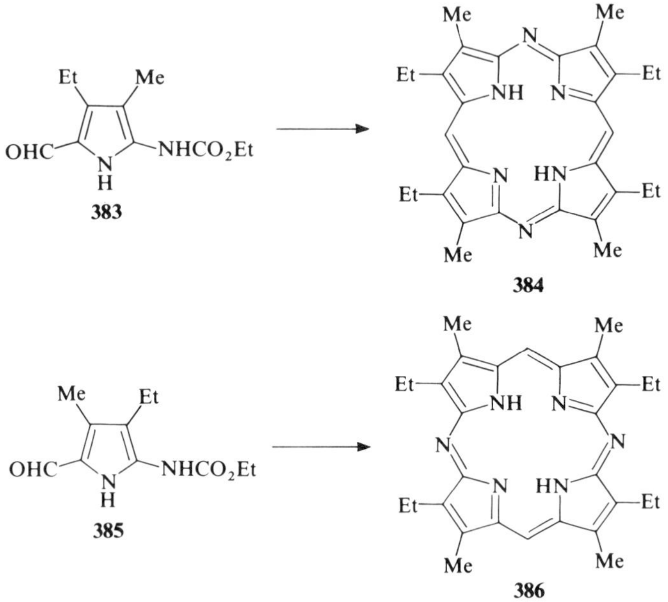

Scheme 106

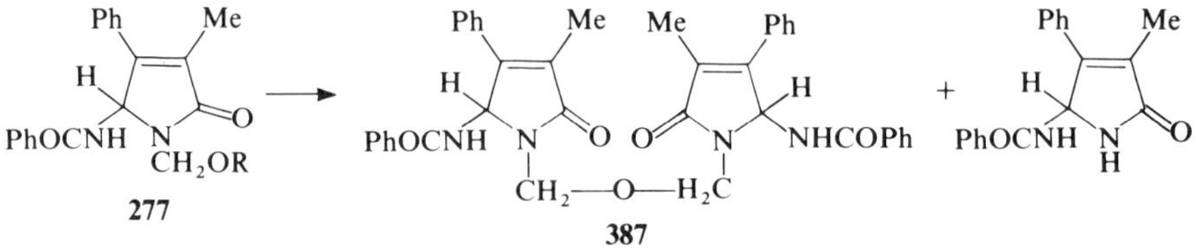

Scheme 107

ceeds via the tautomeric 2H-pyrrole form **B** (Scheme 99), since the corresponding 4-N-methylated derivative is stable under similar conditions.

Thermolysis of 2,5-diphenyliminopyrrolidine in aqueous methanol at 70°C results in nucleophilic ring opening to give the succindianilide, whereas under more vigorous conditions 1-phenylsuccinimide is formed.[157]

3.4.2. Reactions with Electrophiles

3.4.2.1. 1-Aminopyrroles

1-Acylamino-2,5-dimethylpyrroles decompose on treatment with mineral acids at high temperature, giving rise to hydrazine and the carboxylic acid.[52,83] Under less vigorous conditions, 1-acylaminopyrroles give rise to hydrazines and dicarbonyl compounds that react further to produce stable, isolable compounds.[292] 1-Amino- and 1-ureidopyrroles, bearing two ethoxycarbonyl groups at the 2,3- or 3,4-position, undergo normal acid-catalyzed hydrolysis and/or decarboxylation depending on the strength of acids used.[86,118]

N-Substituted 1-aminopyrroles are N-methylated by dimethyl sulfate[85,86,98] or with methyl iodide,[112] while 1-thioureido-2,5-diphenylpyrrole **389** reacts with α-haloketones under reflux to yield the 1-thiazolylamino derivative **390**[61] (Scheme 108).

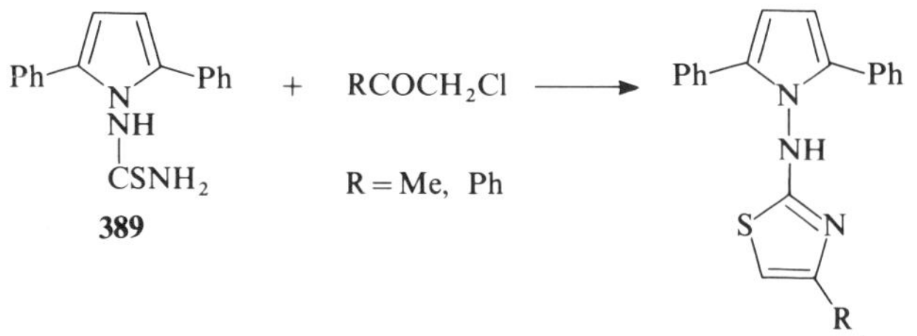

Scheme 108

N-Arylation of 1-amino-3,4-ethoxycarbonylpyrrole has been achieved by nucleophilic aromatic substitution on treatment with 1-fluoro-2,4-dinitrobenzene in the absence of a solvent at 100°C.[49]

Acylation reactions of 1-aminopyrroles have been widely employed as a diagnostic test, although in some instances the reactions fail, either because of the reduced nucleophilicity of the amino group or because of steric hindrance. Thus 1-aminopyrroles are acetylated with acetic anhydride in pyridine to give the corresponding mono and/or bis-acetyl compounds.[61,66,107,116] 1-Arylamino and 1-tosylamidopyrroles similarly give the corresponding acetyl derivatives,[49,128] and 1-acylaminopyrroles have also been obtained by reaction with formic or acetic acids.[67,86]

Benzoylation of 1-aminopyrroles with benzoyl chloride in pyridine or under Schotten–Baumann conditions is effective for the preparation of the mono- or bis-benzoyl derivatives.[54,83,105] In the case of 1-ureido- or 1-arylthioureidopyrroles, 1-benzoylamino derivatives have been isolated with the simultaneous loss of the amido or thioamido function.[85,98]

In the case of several 1-arylaminopyrroles, bearing other nucleophilic sites on the aryl ring, the acylation reaction can also occur on the aryl substituent,[51,112] and, in some cases, cyclic products have been isolated.[112]

In the reaction of 1-aminopyrroles with oxalyl chloride, acylation occurs both on the amino nitrogen and at the 2-position of the pyrrole ring, giving rise to cyclic compounds.[51]

1-Aminopyrroles react with isocyanic acid and isothiocyanates to give the corresponding 1-ureidopyrroles and N-substituted thiocarbamidopyrroles, respectively.[52,61,85,86]

Variously functionalized 1-aminopyrroles react with aromatic aldehydes, with acetone and acetophenone to give the corresponding Schiff's bases in generally good yields.[61,64,74,86,207,226] The imines **392** and **393** can be obtained from 1-aminopyrrole (**391**) and, respectively, 4-nitrosophenol or substituted quinones under acidic conditions[51] (Scheme 109). Reaction of 1-aminopyrroles with 1,3- or 1,4-dicarbonyl compounds produces nitrogen bridgehead polycyclic heterocycles, bis-pyrrolyl derivatives, and open-chain compounds. Thus, 5-azaindolizines **394** are obtained in good yields from 1-aminopyrrole (**391**) and 1,3-dicarbonyl compounds.[293–295] Asymmetric dicarbonyl compounds could lead to two isomers, but only the corresponding enamines **395** are isolated from the reaction with β-ketoesters, such as ethyl acetoacetate, ethyl cyclopentan-2-one-carboxylate, or ethyl cyclohexan-2-one-carboxylate.[292]

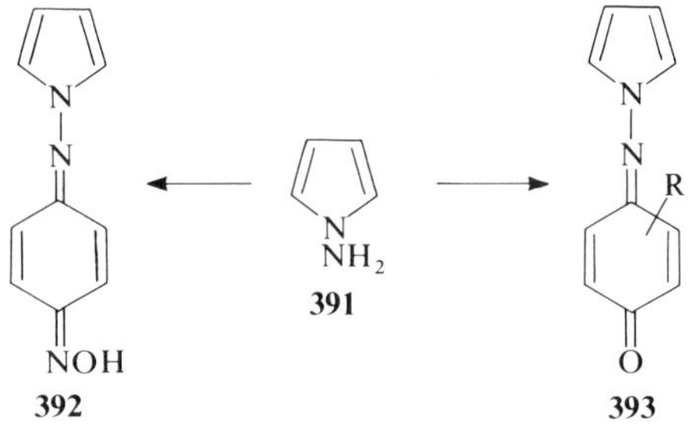

Scheme 109

1-Aminopyrrole reacts with diketene, in a fashion analogous to primary amines, to give the acetoacetamide **396** and the enamine **397** in a 1:2 ratio;[294] the acetoacetamide **396** cyclizes to **394** in the presence of an acid (Scheme 110).

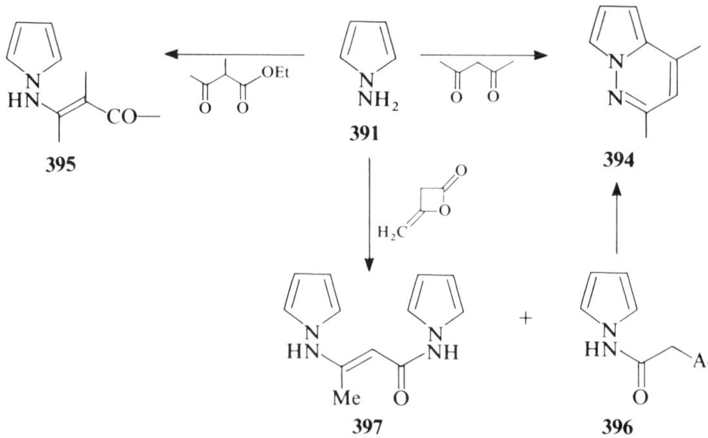

Scheme 110

1-Aminopyrroles **259** react with 1,4-dicarbonyl compounds to give the bisimines **398** (Scheme 111) in absence of acids,[67] but, in the presence of acids

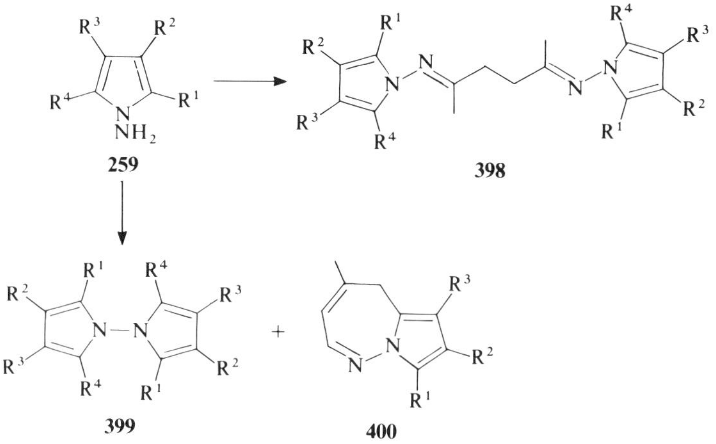

3.4. Reactivity 485

Scheme 111

(glacial acetic acid or hydrochloric acid), the 1,1'-bipyrroles **399** (58–7%) have been obtained, together with 6H-diaza-azulenes **400** (5–40%) when R^4 = H.[67,84] The bipyrroles **399** ($R^1 = R^4$ = Me; R^3 = H; R^4 = COOR) have the optical characteristic typical of biphenyl and are separable into two stereoisomers.[73]

N,N-Linked heterocycles **403** (Scheme 112), analogous to bipyrrole **399**, were also obtained from 1-aminopyrroles of type **401** and masked 1,5-diketones of type **402**.[296] The reaction of 1-aminopyrrole with an excess of acetonylacetone results in a further cyclocondensation at the unsubstituted 2,3-positions of the pyrrole ring, and tricyclic derivatives have been isolated,[67] while 1-thioureido-2,5-dimethylpyrrole reacts with acetonylacetone in hydrobromic acid to give the C,N-bis(2,5-dimethyl-1-pyrrolyl)thioformamide hydrobromide.[61]

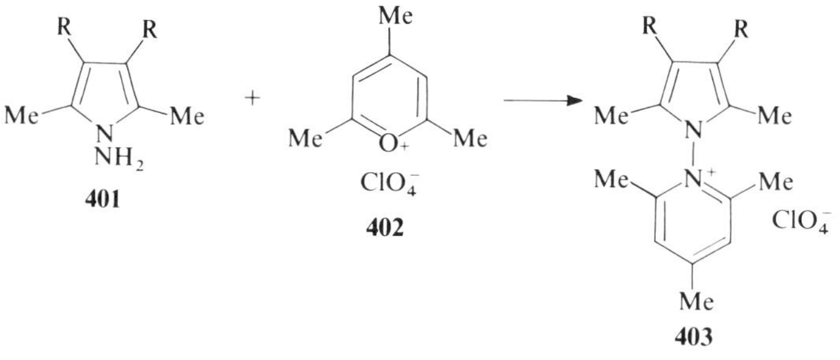

Scheme 112

Diazotization of 1-aminopyrrole derivatives leads to the corresponding deaminated compounds.[83,86,116] Evidently the unstable diazonium salts lack resonance interaction with the ring and lose nitrogen as soon they are formed.

Diazotization with amyl nitrite in acids results in oxidation of the 1*H*-pyrrole ring.[107]

3.4.2.2. 2-Aminopyrroles

2-Aminopyrroles, when treated with mineral acids, decompose,[148,152] and, in the case of iminopyrroline, lead to pyrrol-2-ones as a result of the hydrolysis of the imino function.[158,283] In contrast, treatment of 2-amino-3-cyanopyrroles with phosphoric acid results in hydrolysis of the cyano group to the corresponding acid or amide,[239,291,297] which, in the case of 2-acylamino derivatives, spontaneously cyclizes to produce pyrimidines.[290]

The isolation of 1-unsubstituted derivative **388** from the benzamido derivative **277** (R = Me) without hydrolysis of the amide function occurs under acidic conditions.[244]

The amino group of 2-amino-1*H*-pyrroles generally is not very reactive toward alkylating agents, since the annular nitrogen atom reacts preferentially. Thus, 2-alkylamino derivatives of type **405** are obtained only with very reactive alkyl halide (such as ω-bromoacetophenone)[298] or by converting the amino compound into an alkylimidate derivative of type **404**, which can be reduced to **405** (R = Me)[299] (Scheme 113). In all the other cases, derivatives analogous to **405** can be obtained only when the nuclear nitrogen has been protected before alkylation of the amino group.[184] However, in the case of the reaction with benzyl chloride, both nitrogen atoms are alkylated.[152] Methylation under basic conditions with methyl iodide[152,300] or with dimethyl sulfate under PTC conditions[185] results in the exclusive formation of the 1-methylpyrroles.

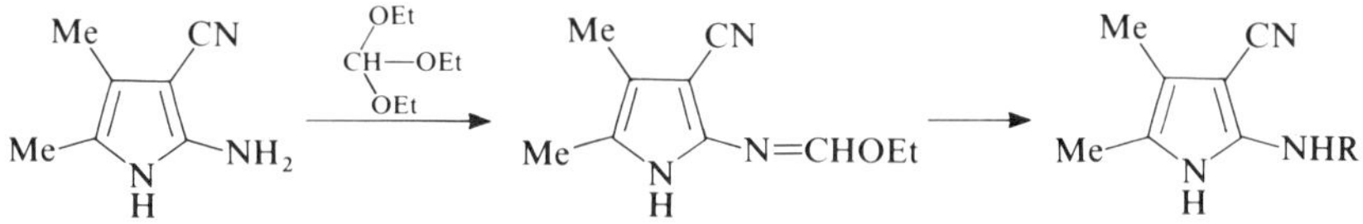

351 **404** **405**

Scheme 113

Alkylation of reactive pyrrole ring positions[180] or at other nucleophilic centers[152,189] has also been observed.

The acylation of 2-aminopyrroles has been employed to characterize this class of compounds,[193] and the conversion of unstable 2-aminopyrroles into the acylamino derivative is frequently used for their isolation.[153,181] Acetylation of 2-aminopyrroles is easily achieved with acetic anhydride or acetyl chloride in presence of bases[149,151,152,159,239,281,301] and, in some cases, di- or triacetyl derivatives have been isolated.[148,180,189] Similarly, reaction with mixed anhydrides produces mono- and diacyl derivatives, whose initial identification was complicated by the possibility that these compounds exist in more than one rotational isomeric form.[148]

2-Benzamido derivatives are obtained in good yields using benzoyl chloride in dry solvents,[149,151,180] and other acylation reactions, even by using less common reactants, do not present any particular difficulty.[180,282,291,302–305]

2-Aminopyrroles **406** react with isocyanates, in dry pyridine under reflux condition, to give the pyrrolotriazines **408**[171] (Scheme 114). It has been proposed that an initial attack of the amino derivative on the isocyanate gives the intermediate pyrrolylureas **407**, which react further with a second molecule of the isocyanate to yield the pyrrolotriazine system. The basic catalysis of the pyridine is necessary for the ring closure, as it has been shown that the reaction of **406** with isocyanates in benzene stops at the stage of pyrrolylureas **407**. The presence of an electron-withdrawing group on the 3-position of **407** imparts an enhanced acidity on the imidic group, which under basic conditions leads to **408**, whereas a less acidic pyrrolylurea (**407**, $R^1 = R^2 = $ Ph) does not cyclize. It has been suggested that **409–411** are precursors in the formation of **408**. However, the formation of **407**, followed by addition of isocyanate at the 1-position, which gives **409** with a final transamidation to **408**, seems the more likely pathway.

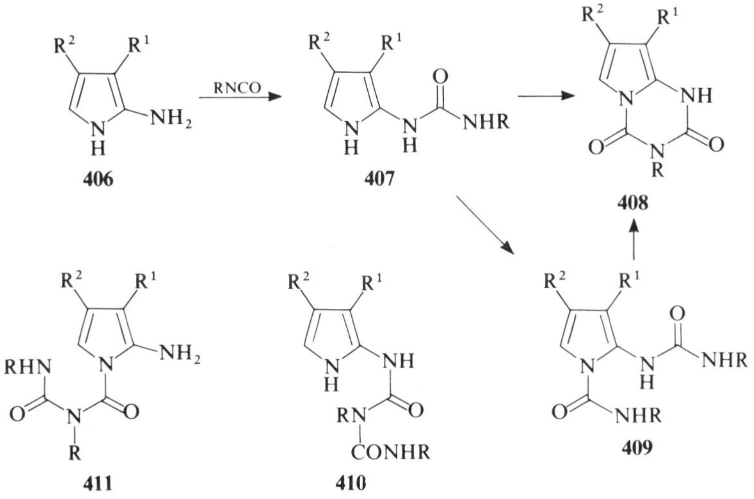

Scheme 114

The reaction of 2-aminopyrroles with aromatic aldehydes results in the formation of the corresponding Schiff bases.[152,155] In the case of 5-unsubstituted 2-aminopyrroles, a positive Ehrlich reaction, leading to 1-azafulvenium salts, has been observed.[154,189]

2-Aminopyrroles yields the corresponding imidate on reaction with trimethylorthoformate,[306,307] while the reaction of the 2-amino-3-cyanopyrroles with formamidine or with formamide in formic acid leads directly to the pyrrolopyrimidines **412** ($R = NH_2$)[308,309] (Scheme 115).

2-Amino-3-cyanopyrroles react with carbon disulfide (2 h under reflux conditions) to give the pyrrolothiazine **413**, while prolonged reflux results in the formation of **412** ($R = SH$), probably via rearrangement of **413**.[239]

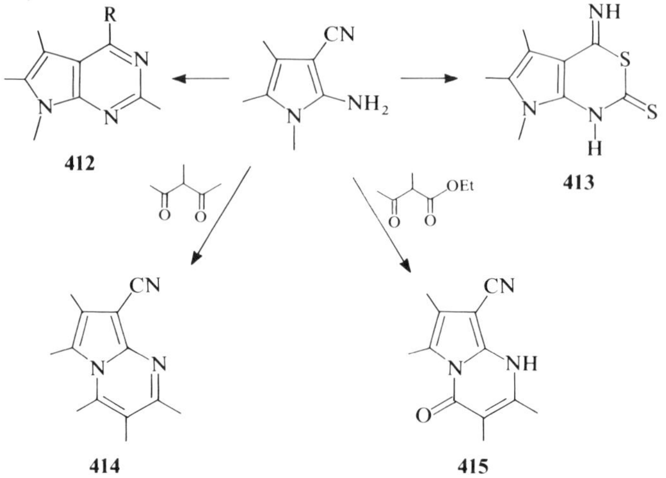

Scheme 115

The reaction of 2-aminopyrroles with 1,3-dicarbonyl compounds generally leads to cyclic derivatives. Thus, 2-amino-3-cyano-1H-pyrroles react with 1,3-diketones in either boiling pyridine or acetic acid, or without solvent at 150–160°C, to give directly the pyrrolo[1,2-a]pyrimidines **414** in 41–74% yield.[310,364] When acetoacetic esters are used, the 4-oxo derivatives **415** are produced. Conversely, 2-aminopyrroles **416**, bearing no substituent at the 3-position, undergo condensation with 1,3-dicarbonyl compounds and their acetals in presence of a catalytic amount of hydrochloric acid to give, generally in one step, the pyrrolo[2,3-b]pyridines **417**.[150] With unsymmetric β-diketones, isomeric products are sometimes isolated, depending on the nature of the

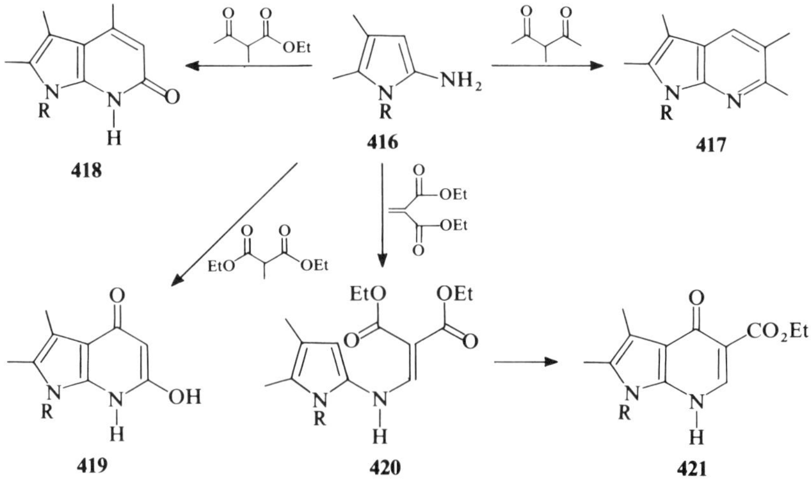

Scheme 116

diketone and the substituents at the 1-position of the ring. 1,3-Ketoesters react with **416** to give the 6,7-dihydro-6-oxo derivatives **418**, and with 1,3-diesters the pyrrolopyridine **419** are obtained. With ethoxymethylenemalonates, **416** undergoes a two-stage reaction to yield the oxo derivatives **421** via the thermally induced cyclization of the intermediates **420**.

In contrast with **416**, the 1-substituted 2-amino-3-cyanopyrroles react with 1,3-ketoesters to give the intermediate Schiff bases **422**, which cyclize to 4-aminopyrrolo[2,3-b]pyridines **423**.[311]

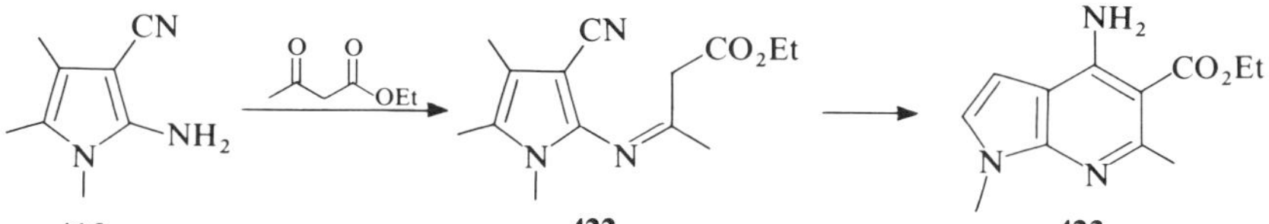

Scheme 117

Diazotization of 2-amino-3-carboxamidopyrrole **424** produces the intermediate diazonium salt **425**, which cyclizes to the pyrrolotriazine **426** in good yield.[297]

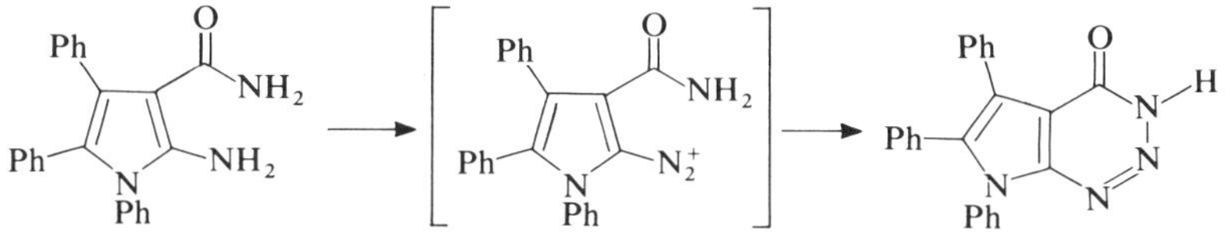

Scheme 118

The diazotization of 2-aminopyrroles with a half equivalent (0.5 equiv) of sodium nitrite in acetic acid yields products of self-coupling.[154] Azocoupling of 2-aminopyrroles has also been observed in their reactions with aromatic diazonium salts.[154,189]

The products of the reaction of 2-amino-3-carboxamidopyrrole **424** with thionyl chloride depends on the reaction conditions. Thus, in DMF at 0°C dehydration of the amido group results in the formation of the aminopyrrole **427**,[297,312] while in the same solvent at 30°C **424** reacts with the solvent to give the azomethine **428**, and at higher temperatures, the bicyclic derivative **429** is obtained. In THF and in dioxane, **424** reacts with thionyl chloride to give the pyrrolothiadiazine **430**, and, in acetone, reaction with the solvent leads to the pyrrolodiazinone **431** (Scheme 119). Several other pyrrolothiadiazines **430**, functionalized on the pyrrole ring, have also been obtained from the suitable 2-amino-3-carboxamidopyrroles.[313]

3.4.2.3. 3-Aminopyrroles

3-Aminopyrroles have been shown to be more reactive toward alkylating agents than the corresponding 2-amino derivatives. Reaction with dimethyl

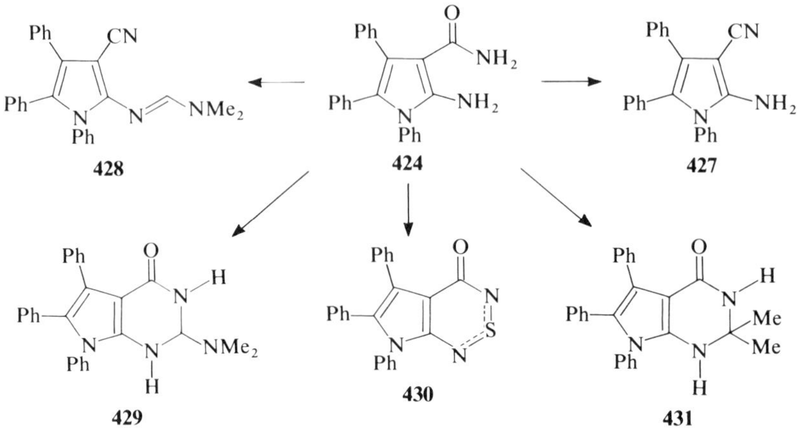

490   Aminopyrroles

Scheme 119

sulfate or methyl iodide always lead to the *N*,*N*-dimethyl derivatives, whereas methylation on the ring nitrogen atom generally does not take place under these reaction conditions.[205,243,258]

The ring nitrogen atom can be alkylated by reaction with alkyl or benzyl halide in the presence of a strong base, such as an alkali metal or alkali metal hydride, in an inert solvent, such as DMF, at room temperature. In such conditions the amino hydrogen can also be alkylated. Thus, in order to secure selective alkylation on the ring nitrogen atoms, it is necessary to protect the amino group by reaction with a carbonyl compound. Subsequent to *N*-alkylation of the pyrrole ring, the amino group can be regenerated by acidic hydrolytic cleavage.

Monomethylation of the 3-aminopyrroles can be achieved by initial sulfonation of the amino group and subsequent methylation of the sulfonamide by the usual procedure, followed by cleavage of the sulfonamido group.[205]

Nucleophilic reaction by the 3-aminopyrroles on 2-bromobenzoic acid results in arylation of the amino group.[205]

3-Aminopyrroles are readily converted into their stable acyl derivatives by reaction with aliphatic anhydrides or acyl halides.[198,199,205,210,212,252,259,265,314] Formylation can be achieved by action of formic acid,[210] and the 3-aminopyrroles are benzoylated with benzoyl chloride in pyridine or under Schotten–Baumann conditions.[253,259,315] In the case of 5-unsubstituted 3-aminopyrroles, *C*-benzoylation of the pyrrole ring has also been observed.[213]

Variously substituted 3-aroylamino and 3-sulfonylaminopyrroles can be prepared from the reaction with the appropriate acyl and sulfonyl halides.[205,316] However, 3-acylaminopyrroles of type **432**, bearing a carbonyl function at the 4-position, can lead to bicyclic compounds. Thus, **432** (R = Me) reacts with ammonium acetate to give the pyrrolopyrimidine **433** (R = Me);[317,318] the derivatives **432** (R = CH$_2$NH$_2$, R' = aryl) afford the pyrrolodiazepin-2-one ring system **434** under acidic conditions.³¹⁹ Several patents report the analogous synthesis of derivatives of **434** (Scheme 120) functionalized at the 1-, 3-, 4-, 5-, 7-, or 8-position to improve their biological activity.³²⁰⁻³²²

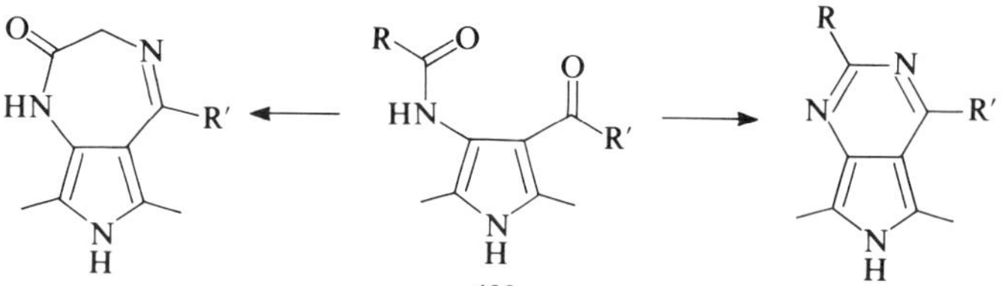

Scheme 120

The 2-aminopyrrolopyrimidine derivative of type **433** (R = NH_2; R' = Me) is isolated from the fusion of 4-acetyl-3-aminopyrrole with cyanamide.³¹⁷,³¹⁸

3-Aminopyrroles react with isocyanates and isothiocyanates, producing the corresponding 3-ureido and thioureido derivatives of type **435**.²⁰⁵,²⁵³ These derivatives can be cyclized when they bear a carbonyl group at either the 2- or 4-position. Thus, 2-ethoxycarbonyl derivatives of type **435** (X = O, S) are easily converted into pyrrolo[3,2-*d*]pyrimidines **436**,²⁰² while 3-ureidopyrroles **435** (X = O) bearing an ester or ketone group in the 4-position yield pyrrolo[3,4-*d*]pyrimidin-2,4-dione **437** or pyrrolo[3,4-*d*]pyrimidin-2-one **438** (Scheme 121), respectively, under basic reaction conditions.²¹⁰,³¹⁷,³²³,³²⁴

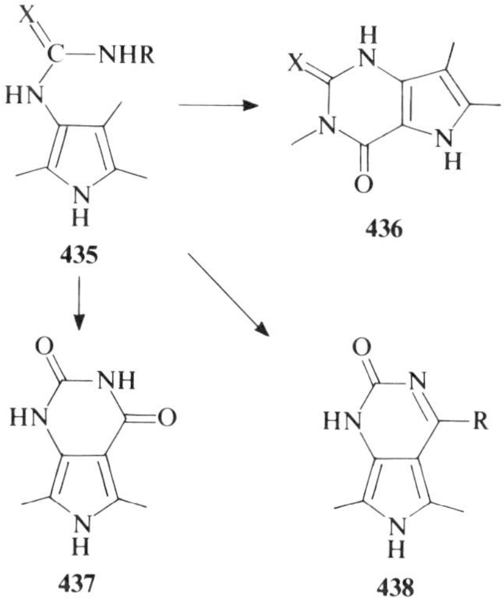

Scheme 121

Schiff's bases are formed from the reaction of 3-aminopyrroles with aldehydes or ketones.²⁰⁵,²¹³ In the case of 3-aminopyrroles **232**, the reaction with benzaldehyde under oxidizing conditions produces an intensely blue compound, which was initially thought to be **439**,[213] but was later identified as having structure **440**[315] (Scheme 122).

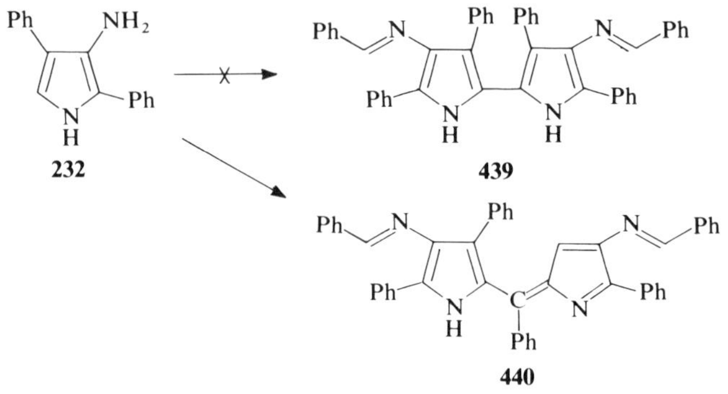

Scheme 122

3-Aminopyrroles, with carboxamido or ester groups in the adjacent positions, are useful intermediates in the synthesis of various condensed heterocycles, on reaction with acylating agents. Thus, 3-amino-4-carboxamidopyrroles react with formamide or ethyliminobenzoate at 140–160°C directly to give the pyrrolopyrimidin-4-ones **441**,[317,318] whereas reaction with phosgene leads to **442**.[325] Reaction of 3-amino-2-ethoxycarbonylpyrroles with dimethylformamide diethyl acetal leads to pyrrolopyrimidin-4-one **443**,[202,203] whereas, from the reaction with ethyl chloroformate or potassium xantogenate, the dioxo and thioxo derivatives **436** are isolated.[209,210] 3-Aminopyrroles react with 1,3-dicarbonyl compounds to give generally bicyclic systems through open-chain intermediates that, in some cases, can be isolated. Thus, reaction with β-diketones or β-ketoesters leads to intermediates of type **444**, which cyclize, when the 2-position of the pyrrole ring is unsubstituted, to give compounds **445** (when R = OEt) and the aromatic derivative **446** (when R = alkyl).[266] When the 2-position of the pyrrole ring is substituted and an acetyl group is present at the 4-position, the intermediate **444** leads to the pyrrolopyridine **447** (Scheme 123) by Dieckmann-type condensation with the acetyl group.[317,326] The same 4-acetyl-3-aminopyrroles in their reaction with diketene or with ethylmalonyl chloride give intermediates of type **432** (R = Me; R' = CH$_2$COMe, CH$_2$COOEt), that cyclize, on treatment with bases to yield pyrrolopyridin-2-ones **448**.[317,326]

Pyrrolopyrimidin-4-ones **443** are isolated from 3-amino-2-carboxamidopyrroles with 2,4-pentanedione.[209,210]

The diazotization reaction of 3-aminopyrroles produces the corresponding 3-diazonium salts, which are stable enough for their isolation.[251,264] However, depending on the reaction conditions and the nature of the 3-aminopyrrole substrates, the intermediate diazonium salts can give rise to the corresponding 3-unsubstituted pyrroles on loss of nitrogen,[210] or lead to the diazo derivatives **450**, on neutralization of the reaction mixture.[247,248,251,255,263,314]

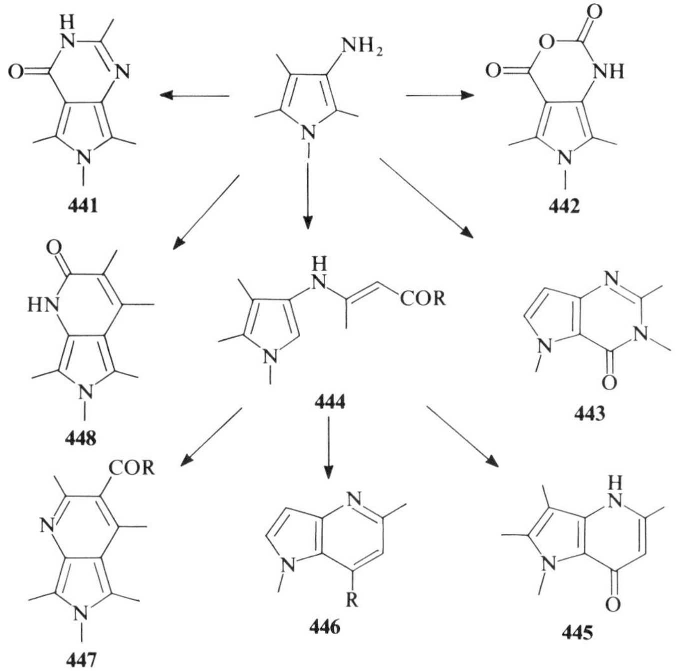

3.4. Reactivity 493

Scheme 123

When the 3-aminopyrrole derivatives bear a carboxamido function at the 4-position, the diazotization reaction directly leads to the pyrrolo[3,4-*d*][1,2,3]triazines **451** through an immediate intramolecular coupling of the diazonium group with the carboxamido substituent.[207] Similarly, diazotization of 3-amino-2-carboxamidopyrroles leads to the pyrrolo[3,2-*d*][1,2,3]triazines **452** by analogous ring closure onto the 2-substituent.[209, 210] The 3-pyrrolyl diazonium salts can also couple with aromatic substrates to give the azo derivatives of type **453**[251, 258] (Scheme 124).

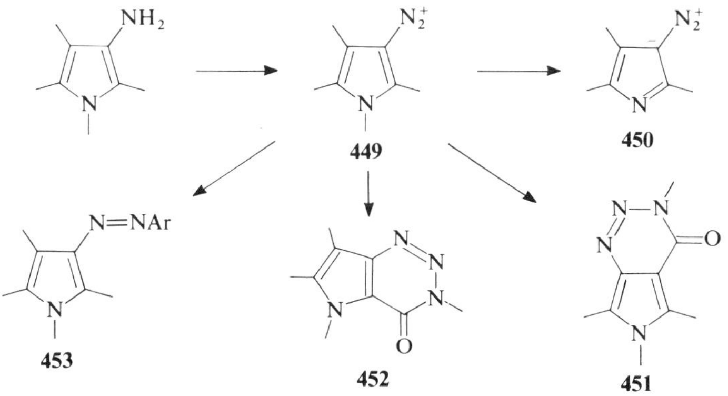

Scheme 124

3.4.2.4. Diaminopyrroles

Reaction of diaminopyrroles with electrophiles is less well studied. Methylation of the 2,4-diamino derivative **252** with diazomethane has been reported.[220]

2,5-Diphenyliminopyrrolidine **256** (R = Ph) yields *N*-phenylsuccinimide under acidic reaction conditions,[157] also obtained in thermolytic reactions (Section 3.4.13). The same compound is reported to react with sodium nitrite in hydrochloric acid to give the diazonium salt, as identified by its coupling product with β-naphthol.[157]

3.4.2.5. Triaminopyrroles

1,2,5-Triamino-3,4-dicyanopyrrole (**454**) reacts with DMF in the presence of 3 mol of *p*-toluensulfonyl chloride to give the trisamidine **457** in a reaction that is typical of weakly basic amines[226] (Scheme 125). With 1 mol of *p*-toluensulfonyl chloride, the monoamidine at the 2-position is formed.

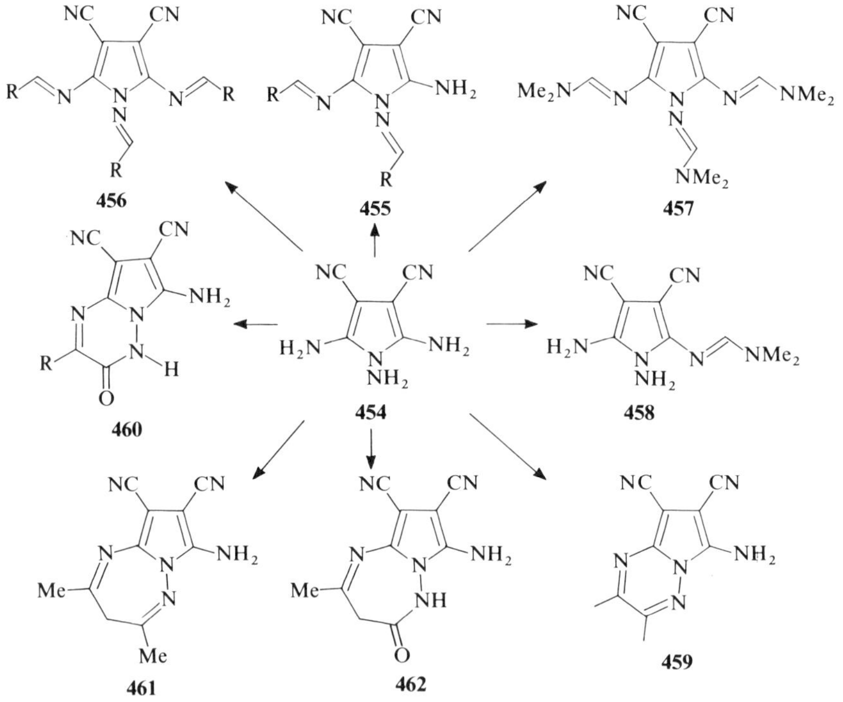

Scheme 125

The triamine **454** reacts with aromatic aldehydes to give either a dianil, or the trianil **456**, depending on the aldehydes and the reaction conditions. With benzaldehyde, for example, it has been found impossible to isolate a monoimine, and structure **455** has been suggested for the bisimine.

Reactions of **454** with 1,2- and 1,3-dicarbonyl compounds have been reported. The 6-aminopyrrolotriazines **459** were obtained from biacetyl and benzil, and only one of the two possible isomers is isolated from the reaction with ethyl pyruvate and ethyl oxomalonate; the structure **460** has been proposed on the basis of the more reactive amino group condensing with the keto group. In the corresponding condensation reactions with acetylacetone and ethyl acetoacetate, the 7-aminopyrrolo[b]-1,2,4-triazepines **461** and **462**, respectively, are obtained.

3.4.3. Reactions with Oxidizing Agents

3.4.3.1. 1-Aminopyrroles

1-Aminopyrroles are totally decomposed on exposure to strong oxidizing agents, such as potassium permanganate in potassium hydroxide, or nitric acid.[96,115,116] However, partial oxidation by manganese dioxide in organic solvents has been reported.[66]

The reaction of 1-aminopyrroles with lead tetraacetate has been studied in detail. When the reaction is carried out in benzene, nitrogen is evolved and a dark tar is formed, from which either the deaminated pyrrole[327] or fragments containing the four-carbon-atom skeleton of the original ring have been isolated.[328] It has been assumed that a nitrene intermediate is formed in these reactions. Evidence for the nitrene is provided by the oxidation of 1-aminopyrroles **463** with lead tetraacetate in the presence of excess alkenes. The pyrrolyl-diazene **464** or the aziridine **465** are isolated, depending on the nature of the substrate and the solvent[329,330] (Scheme 126). The formation of the aziridine **465** can be rationalized in terms of the intermediacy of a nitrene species, while the formation of **464** can be understood in terms of electrophilic attack of a nitrene intermediate on the 1-amino group of a second molecule of **463**.

Treatment of 1-amino-2,5-diphenylpyrrole with NBS results in oxidation of the 1-amino group and simultaneous bromination of the pyrrole ring to yield

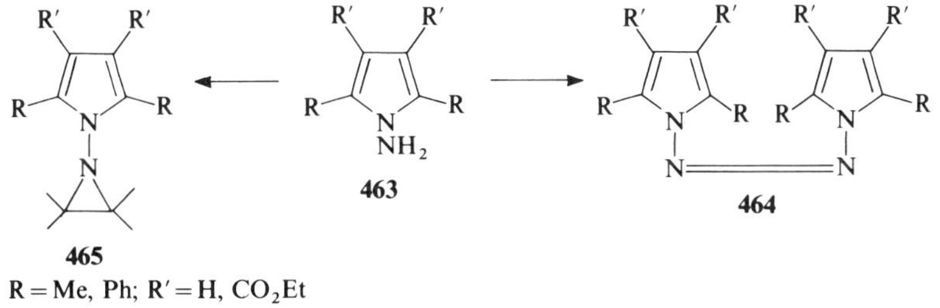

Scheme 126

464 (R = Ph, R' = Br), which exhibit reversible phototropy.[66] The same tetrabromo compound has been obtained from 1-amino-2,5-diphenyl-3,4-dibromopyrrole upon treatment with NCS or NBS.

1-(4-Hydroxyphenylamino)pyrroles are oxidized by mercury(II) oxide or by silver oxide to the corresponding 4-(1H-pyrrol-1-ylimino)-2,5-cyclohexadienes **393**, which can be reduced back to the 1-arylaminopyrroles with sodium hydrosulfite.[51]

3.4.3.2. 2-Aminopyrroles

2-Amino-3,5-diphenylpyrrole is readily oxidized by air, giving, among other products, the tetraphenylazadipyrromethine **466**, which can be reduced by glucose under alkaline conditions to **467**[148,237] (Scheme 127). The oxidation is catalyzed by formic, acetic, or phosphoric acid and even, although with a much slower reaction, by a protic solvent, such as ethanol, and can be observed during the preparation of the 2-aminopyrrole. Although the mechanism for its formation is somewhat obscure, it is reasonable to assume an initial protonation of the 5-position of the ring to give **468**, followed by nucleophilic attack by the second molecule of amine and subsequent elimination of ammonia to give **467**, which is immediately oxidized. When the preparation of the aminopyrrole is carried out in a sealed tube in absence of air, the leuco-derivative **467** is formed exclusively but, on exposure to air, is immediately oxidized to the azamethine **466**.

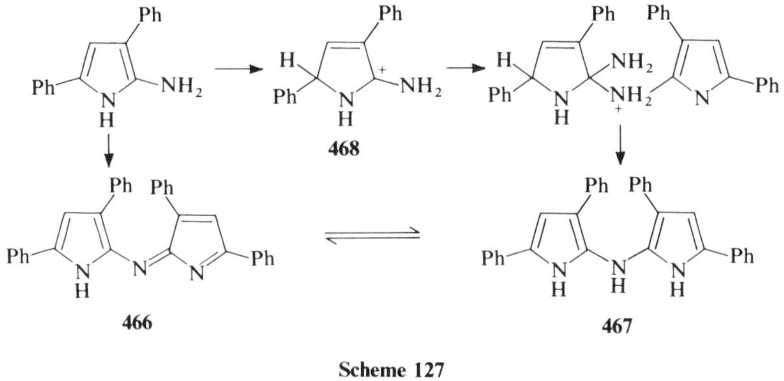

Scheme 127

Mild oxidation by chromic acid of 2-aminopyrrole **277** (R = H) leads to the N-formyl derivative **469** (R' = CHO), which, in acidic methanol, gives the deformylated product **469** (R' = H).[244] The same product has been obtained directly in useful yields by chromic acid oxidation of **277** (R = H, Me). Vigorous oxidation of 2-aminopyrroles **277** (R = H, Me) and **469** (R' = CHO) leads to the 3-methyl-4-phenylmaleimide **470** (R' = H) (Scheme 128). 2-Aminopyrrole **277** (R = Me) in methanolic alkali undergoes aerial oxidation to **471** or to **470** (R' = CH_2OMe), depending on the concentration of the base.

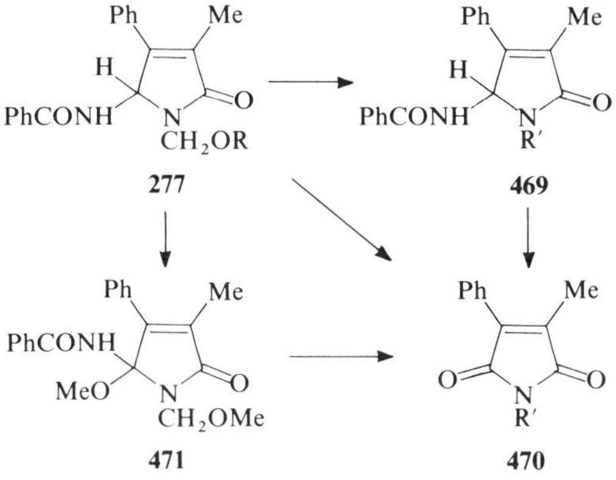

3.4. Reactivity   497

Scheme 128

Oxidation of 2-amino-5-mercaptopyrrole **150** with iodide in bicarbonate solution does not affect the amino group, but results in the formation of a sulfur-bridged dimer.[152]

3.4.3.3. 3-Aminopyrroles

The oxidation of the 3-amino-2,5-diphenylpyrrole with reagents such as H_2O_2, Fe(III) salts, CrO_3 gives a single product to which structure **472** (Scheme 129) has been assigned.[331] The azoxy derivative can be reduced quantitatively back to the amine.

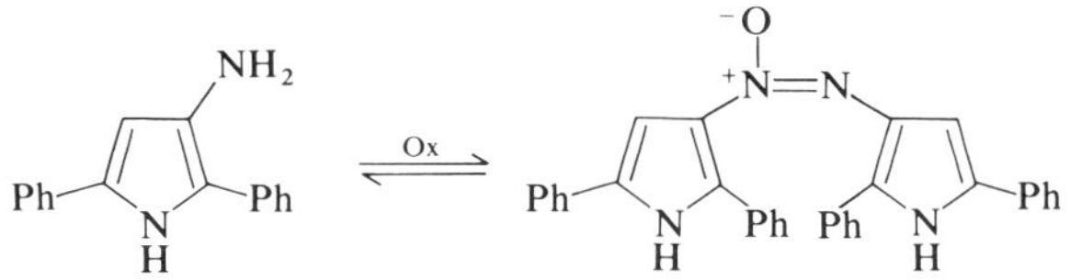

Scheme 129

Detailed studies of the oxidation of 3-amino-2,4,5-triphenylpyrrole have been carried out. It has been found to be remarkably stable, but oxidation in acidic, neutral, or basic media can lead to a quite complex mixture of products.[323,333] In neutral (PbO_2) or basic media [$K_3Fe(CN)_6$], the isolated product, which melts at 170°C, was assigned the structure **473**, by analogy with the behavior of 3-amino-2-phenylindole, which by oxidation gave the corresponding iminophenylindole.[334] In an acidic medium ($FeCl_3$/AcOH; H_2O_2/AcOH; CrO_3/AcOH), two main products have been isolated with melting points of 168 and 290°C and are identical to **474** and **475**, which are isolated from the reaction of the nitrosotriphenylpyrrole with hydroxylamine or hydrazine.[254,335] Moreover, **473**, which on reduction is reconverted into the aminotriphenylpyrrole, undergoes an acid-catalyzed oxidation at room temperature to give the two derivatives **474** and **475**. Prolonged reaction times result in a decrease of the ratio of **474:475**, suggesting that **474** is a precursor of **475**. When the oxidation is conducted in hydrochloric acid or sulfuric acid, the oxo derivative **476** can be isolated. Under alkaline conditions **476** is hydrated to yield **475**. The release of ammonia has been detected in these reactions, but not hydroxylamine, thereby suggesting the structure **474'** for the product with m.p. 168°C, and a mechanistic route of **473** to **475** via **474'** and not via **474** (Scheme 130).

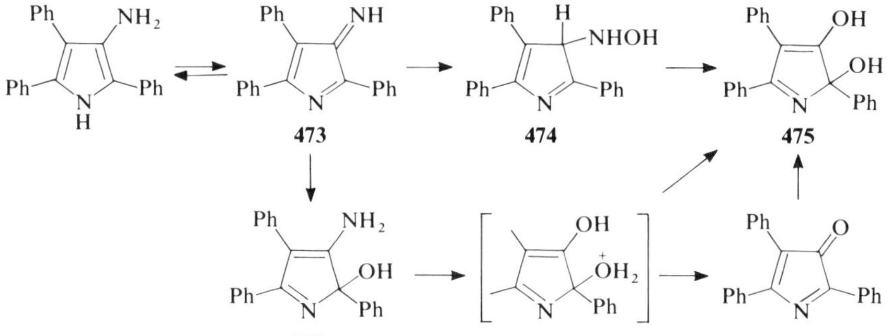

Scheme 130

Aerial oxidation of 3-aminopyrroles has been observed during the attempted preparations of 4-substituted 3-amino-2,5-dimethylpyrroles **478** (R = Me) from the reaction of 2-aminopropionitrile with ethyl 3-oxobutanoate or with pentan-2,4-dione.[211] The 3-amino-2-hydroxy-2H-pyrroles **479** are produced via oxidation of the initially formed 3-aminopyrroles, and evidence for this postulate comes from the observation that the corresponding 3-amino-2-phenyl-5-methylpyrroles can be isolated and have been observed to oxidize to give 2-hydroxy-2H-pyrroles of type **479** (R = Ph) (Scheme 131). A solution of **478** (R = Ph, R' = COOEt) in dichloromethane is completely transformed into the corresponding oxidized derivative **479** in 5 days.

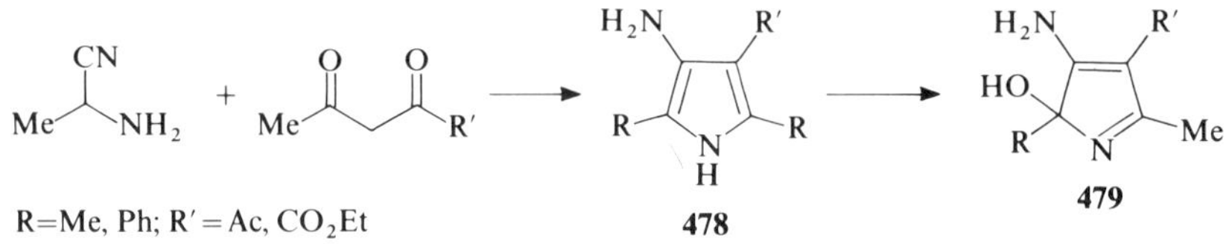

Scheme 131

3.4.4. Reactions with Nucleophiles

The low electrophilicity of the aminopyrrole system makes it unreactive toward nucleophiles, and very few reactions with nucleophiles have been studied.

3.4.4.1. 1-Aminopyrroles

Various alkoxycarbonyl-substituted 1-aminopyrroles have been reported to undergo base-catalyzed hydrolysis of the ester function and/or decarboxylation.[83,97,98,100,118] Under more vigorous conditions, 1-acylamino derivatives cleave with the formation of either the 1-aminopyrroles or 1H-pyrroles.[52,61,115,116,123] Similarly, although 1-phosphinylamino derivatives are remarkably stable toward bases, they can also be hydrolyzed to the corresponding 1-aminopyrroles.[62]

In contrast, the 1-ureidopyrroles **480**, when refluxed with sodium hydroxide in water–methanol, cyclize by an intramolecular nucleophilic reaction to produce the pyrrolotriazines **481**[117,118] (Scheme 132).

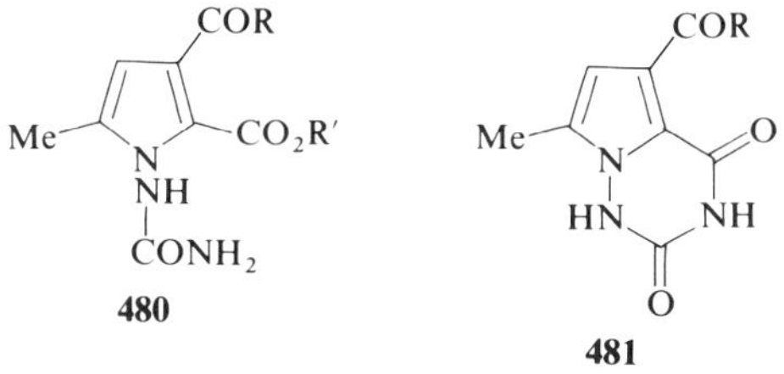

Scheme 132

3.4.4.2. 2-Aminopyrroles

2-Aminopyrroles possessing cyano or ester groups undergo basic hydrolysis to the carboxylic acids,[152,239] but, in contrast, 2-iminopyrrol-5-ones are hydrolyzed to the corresponding succinimido derivatives.[156] The corresponding reaction with hydroxylamine transforms the imino group into an oxime.[157]

Intramolecular nucleophilic cyclization of **482** in the presence of potassium tert-butoxide was initially reported to form pyrrolo[1,2-a][1,3]diazepine **483**,[302] but following an unambiguous synthesis of **483**,[175] the structure of the product was revised to **484** (Scheme 133) and confirmed by an alternative synthetic route.[336]

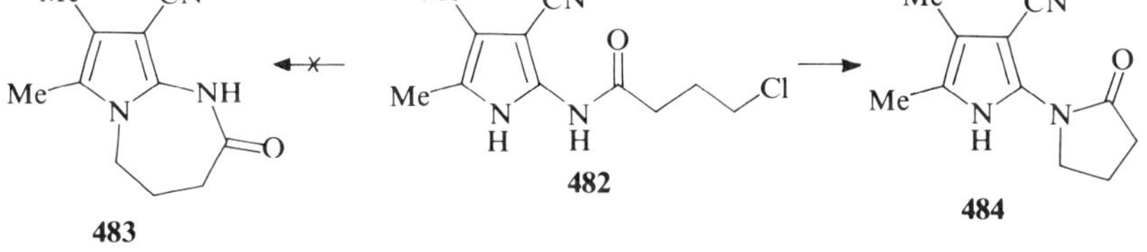

Scheme 133

Cyclization of the pyrrolyl-2-urethane **485** (Scheme 134) to give the pyrrolo[2,3-d]pyrimidindione **486** has been achieved by the action of potassium

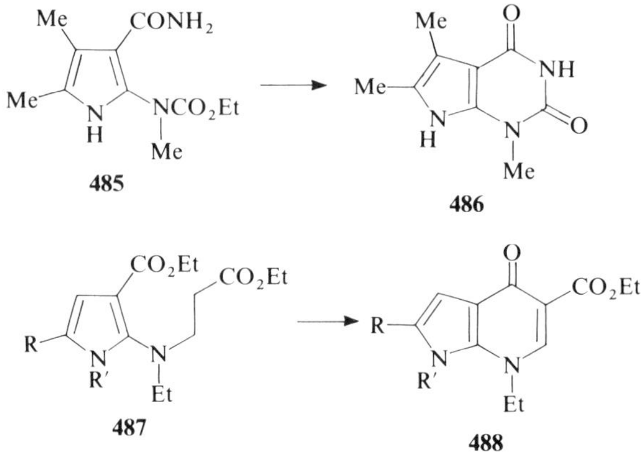

Scheme 134

tert-butoxide in anhydrous THF. Attempts to produce **486** by a thermal reaction in the same fashion in which **382** was obtained (see Section 3.4.1.2) were unsuccessful.[291]

The 2-aminopyrroles **487** undergo the base-catalyzed Dieckmann cyclization, to give the pyrrolopyridines **488**,[184] and pyrrolo[1,2-*a*]pyrimidines **490** are obtained from 2-(phenacylamino)pyrroles **489**, on treatment with potassium *tert*-butoxide or potassium hydroxide in absolute ethanol.[298] The proposed mechanism, shown in Scheme 135, involves nucleophilic attack by pyrrolyl anion of **489** at the carbonyl group of a second molecule of **489**, followed by an intramolecular cyclization of **492** and aerial oxidation of **493** to give the final red product.

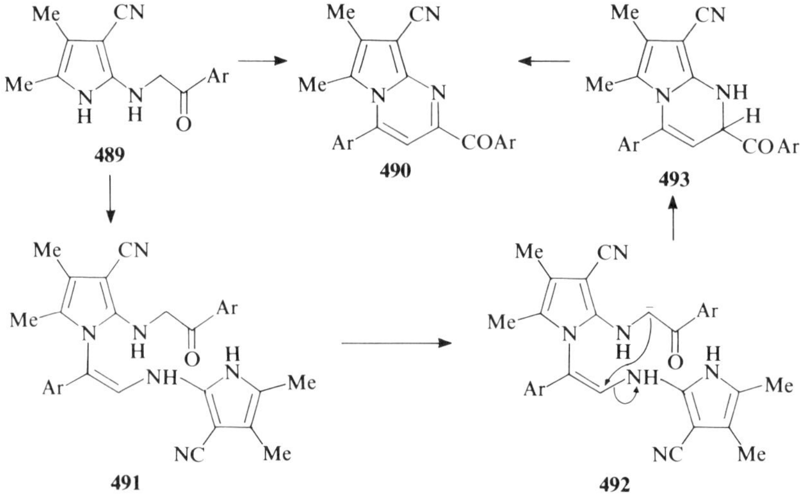

Scheme 135

3.4. Reactivity

The (5-formyl-2-pyrrolyl)urethanes **494** (R = Et, Br) undergo the expected reaction with kryptopyrrole (see Chapter 1, above) to give the ethylurethanedipyrromethene **495**,[227,231] which can be converted into γ-monoimidoetioporphyrin **496**. Alternatively, **494** (R = Et), can be converted into the ethylurethanedipyrromethene **497** by reaction with haemopyrrole, and subsequently into β-monoiminoetioporphyrin **498**.

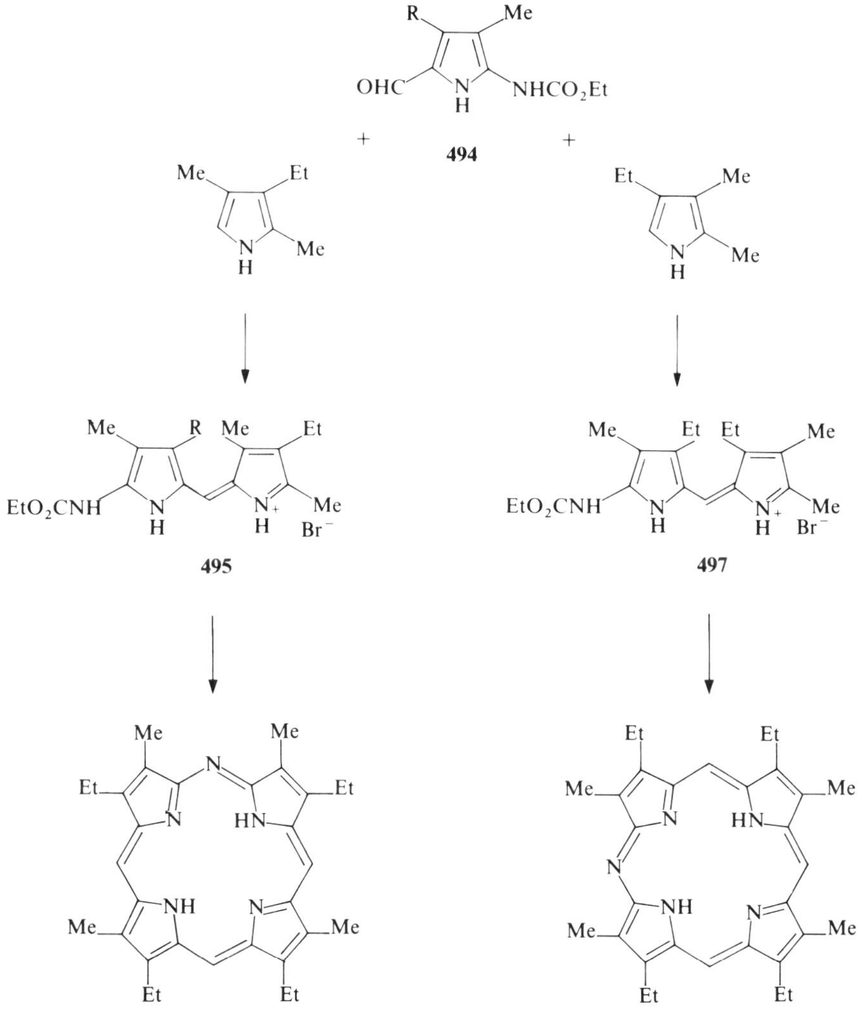

Scheme 136

The dipyrromethene **499**, when heated in presence of sodium carbonate or phenylhydrazine at 160–70°C, produces β,δ-diimidoetioporphyrin **384**.[230]

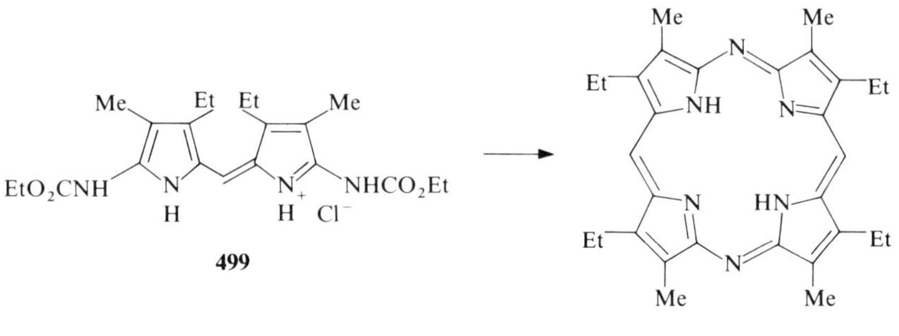

3.4.4.3. 3-Aminopyrroles

3-Aminopyrrolyl carboxylic esters undergo base-catalyzed hydrolysis and/or decarboxylation of the ester function in a manner analogous to that observed for 2-amino series.[314]

The 3-aminopyrroles **500**, formed by the reduction of the corresponding dinitro compounds by iron in acetic acid, cyclize under acidic conditions to the pyrrolo[3,2-*b*]indole ring system **501**.[261] The most probable mechanism involves initial protonation at the 2-position of the pyrrole ring followed by nucleophilic attack at the 3-position and extrusion of ammonia from **503** (Scheme 138). Evidence for this mechanism is based on the extrusion of ^{15}N isotopically labeled ammonia and the behavior of other leaving groups (NO_2, Br, Cl).[337]

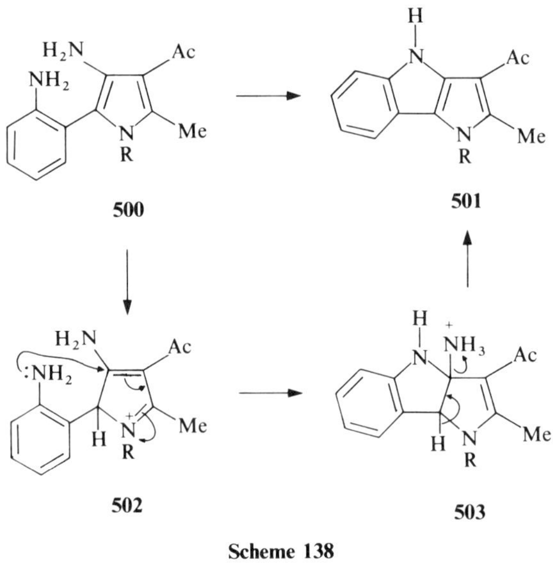

Scheme 138

3.4.4.4. Diaminopyrroles

Substituted succinimidines, on treatment at room temperature with alkali and sometimes even with water, are easily hydrolyzed to the corresponding 2-imino-5-one derivatives; at higher temperature the pyrrol-2,5-diones are generally obtained, although ring opening to succinic acid has also been observed.[156,157,222-224]

Succinimidine **504** reacts with hydroxylamine in boiling ethanol to give the dioxime **256** (R = OH),[157] and reaction with aniline initially produces the monophenylimine **505** (R = Ph),[157] which reacts further to yield the 2,5-diphenylimino derivative **256** (R = Ph) (Scheme 139), the structure of which has been chemically confirmed by degradation to succindianilide. 2-Phenylimino-5-iminopyrrole **505** (R = Ph) reacts with β-naphthylamine to give the corresponding 5-(β-naphthylimino)pyrrole.

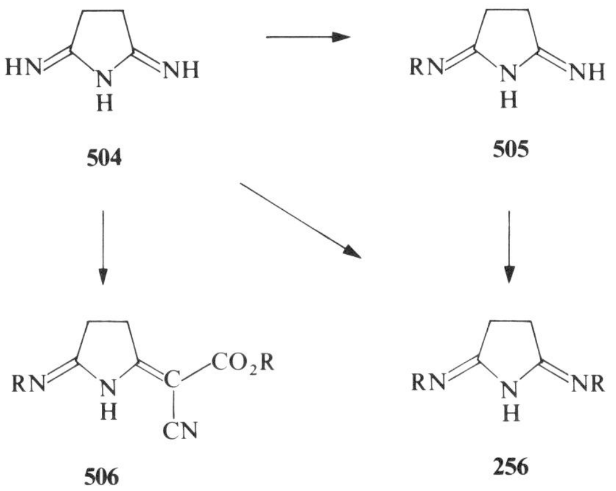

Scheme 139

In contrast with the reaction of diiminoisoindoline, which produces a bis-condensation product, succinimidine **504** reacts with active methylene compounds to yield the monocondensation products, e.g., **506**.[283]

3.4.5. Reactions with Reducing Agents

3.4.5.1. 1-Aminopyrroles

1-Amino-2,5-dimethylpyrrole is catalytically hydrogenated over rhodium on alumina in acetic acid to *cis*-1-amino-2,5-dimethylpyrrolidine.[54]

The reduction of *N*-substituted 1-aminopyrroles either catalytically (Pd, Raney nickel), or by sodium in alcohol, generally results in cleavage of the N—N bond with the isolation of the corresponding 1*H*-pyrroles.[61,104,110,120,128]

However, in the presence of bulky substituents on the 1-amino group, there is increased resistance to the N—N bond cleavage.[121] In the case of 1-ureido-2,5-dimethylpyrrole, reduction with zinc in hydrochloric acid results in cleavage of the N—N bond and reduction of the ring.[52]

3.4.5.2. 2-Aminopyrroles

Although 2-pyrrolylurethanes are generally reduced catalytically to the 2-aminopyrrole,[228] many reductions generally occur on other functional groups. Thus, cyano groups are reduced to the aldehyde,[338] and reduction of the disulfide **191** (see Scheme 48) results in elimination of the sulfur atom.[189]

Hydrogenation of **507** (R = H) over Raney nickel leaves the starting material unchanged; with a palladium catalyst an unseparable mixture is obtained. However, hydrogenation over platinum gives the corresponding cyclohexylpyrrolidinones **508**[244] (Scheme 140).

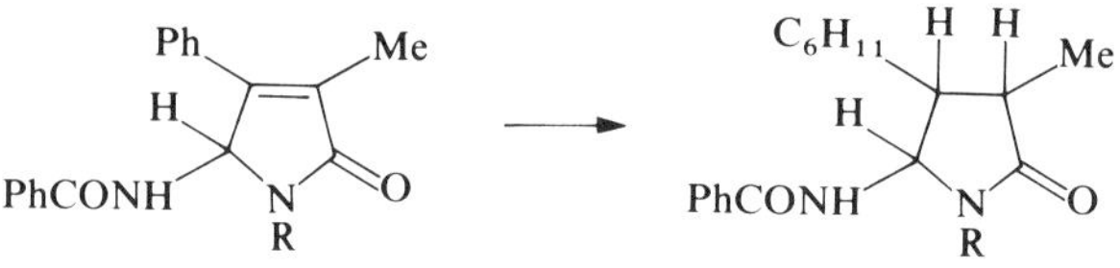

R = H, CH$_2$OH, CH$_2$OMe

Scheme 140

3.4.5.3. 3-Aminopyrroles

Generally, metal–acid reduction of 3-aminopyrroles leaves the pyrrole ring intact. Thus, reduction of the 3-aminopyrrole **509** with iron in acetic acid leads directly to the pyrrolo[3,4-*b*]pyridin-4-ones **511**[260] (Scheme 141). The reaction probably goes through the intermediate formation of **510**, followed by a nucleophilic ring closure.[261]

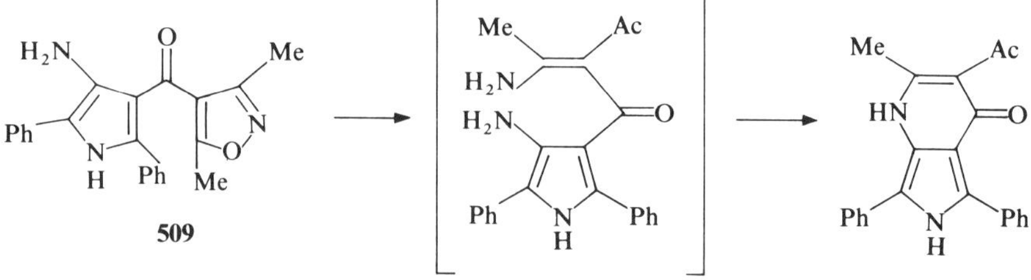

Scheme 141

3.4. Reactivity

The dipyrromethane **512** undergoes the expected reaction in formic acid with iron to produce tetraacetylaminoporphyrin **513**[314] (Scheme 142). The mechanism of this reaction is not clear and could involve a decarboxylation–formylation process, followed by the normal condensation to yield the macrocycle.

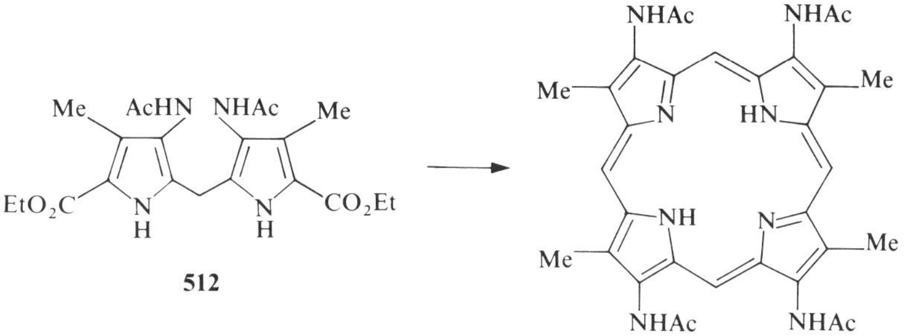

Scheme 142

3.4.6. Cycloaddition Reactions

Although in general pyrroles are reluctant to participate in Diels–Alder reactions and prefer instead to undergo Michael addition with electron-deficient electrophiles, suitable substitution on the pyrrole nitrogen atom can reverse this behaviour (see Section Part 1, 3.4.2.1.3). Thus, N-acylaminopyrroles—or, better, N-aminopyrroles—can lead to cycloaddition reactions.[339]

1-Aminopyrroles undergo efficient Diels–Alder reaction with electron-deficient alkynes, providing a remarkably simple route to substituted benzenes. Thus, reaction of N-carbomethoxyaminopyrroles **104** with an excess of dimethyl acetylenedicarboxylate (DMAD) leads to the arenes **514** ($R^2 = R^3 = $ COOMe) in good yield[123] (Scheme 143). However, the cycloaddition is feasible only when one electron-withdrawing group is present at C(3) and fails when two such groups are present at both the β-positions.

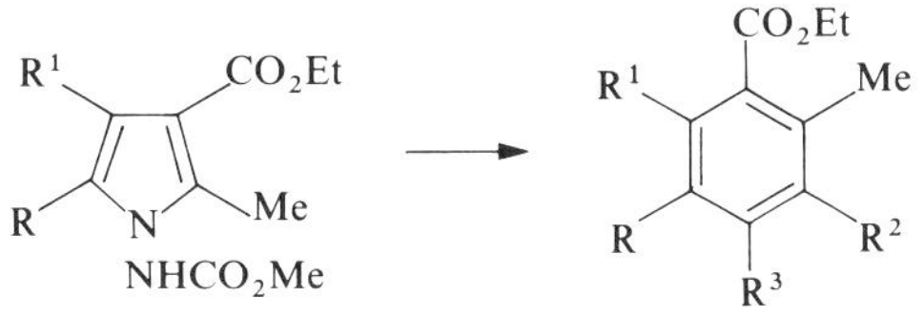

Scheme 143

With asymmetric dienophiles, such as ethyl propiolate, the reaction proceeds with little regiochemical control and mixtures of both isomers are isolated. The major product generally results in a *para* orientation between the alkyl substituent (R^1) on the pyrrole (diene component) and the electron-withdrawing substituent on the acetylenic dienophile.

With labile systems, such as the annelated 1-aminopyrroles **515** (Scheme 144), the Diels–Alder reaction with, for example, DMAD requires milder conditions (chloroform at room temperature), but the yields of **516** are generally low (13–50%).[123] However, in the case of unsubstituted 1-aminopyrroles **515** (R = H) the resulting yields are high.

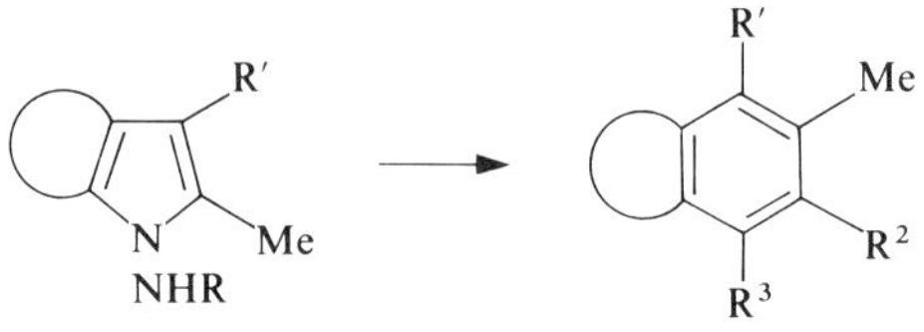

Scheme 144

Cycloaddition reaction of 1-aminopyrroles **517** with DMAD results in the formation of **518** and eventually **519**. The rate of cycloaddition to **517** increases in the order **a** > **b** > **c**, while decomposition of the adduct to **519** increases in the order **c** > **b** > **a**. Adduct **520** from the reaction of **517b** and N-phenylmaleimide (NPMI) has been detected by ^{1}H NMR studies, but it was unstable, and, when DMAD was added, the benzene derivative **519** was formed in 70% yield, confirming the reversibility of the Diels–Alder addition reaction of 1-aminopyrroles. The heteroatom bridge in the adduct **518** is extruded as the relatively stable nitrene **521a–c** with the driving force of aromatization of the cyclic system.

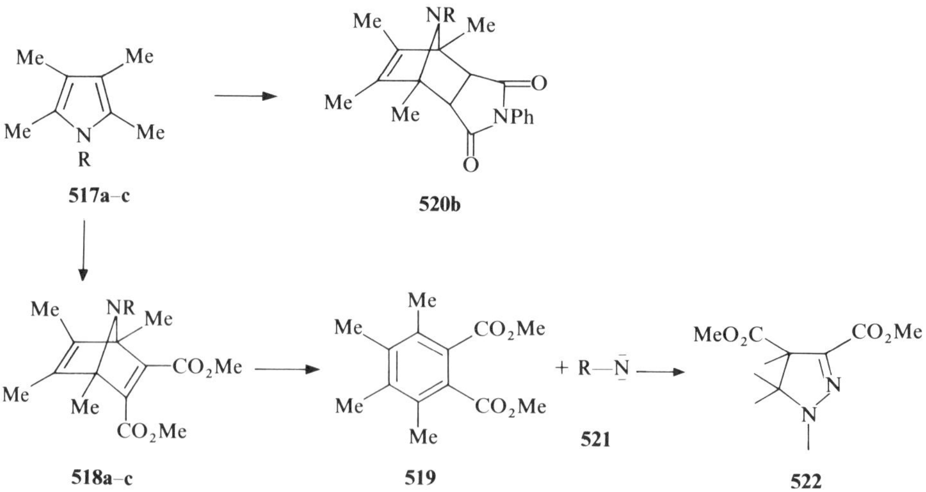

a R = NH_2
b R = NHMe
c R = NMe_2

Scheme 145

In these Diels–Alder addition reactions the general production of dimethyl maleate has also been observed. A reasonable mechanism for its formation involves rearrangement of the aminonitrene **521** to a diimide, which reduces DMAD to dimethyl maleate.[340] When dialkylaminonitrenes (e.g. **521c**) are formed as intermediates, the pyrazolines **522** (Scheme 145) have been isolated (30–43%) together with the benzene compounds (87–80%).

With less reactive pyrroles, such as 1-(*N*,*N*-dimethylamino)-2,5-dimethylpyrrole or 2,5-dimethyl-1-morpholinopyrrole, the reaction with an excess of DMAD tends to produce the corresponding aromatic pyrazoles (72–57%) instead of **522**.[341]

The 1-methoxycarbonylaminopyrrole **523** reacts at room temperature in chloroform with ethyl *β*-benzenesulfonylpropiolate to give the isolable adduct **524** in 75% yield.[342] On heating, the intermediate **524** extrudes the *N*-aminonitrene to produce ethyl 2,5-dimethyl-6-benzenesulfonylbenzoate **525** (Scheme 146) in 42% yield.

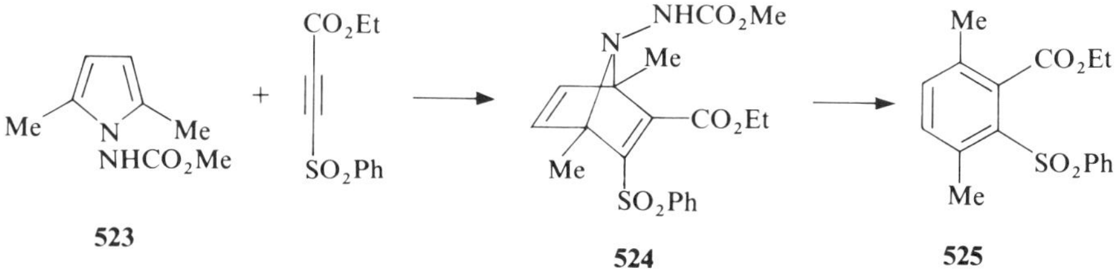

CO2Et
Me
N
Me
NHCO2Me
SO2Ph
NHCO2Me
N
Me
CO2Et
Me
SO2Ph
Me
CO2Et
SO2Ph
Me
523
524
525

Scheme 146

In a similar manner, 1-methoxycarbonylaminopyrrole **526** might be expected to react smoothly with electron-deficient alkynes with a high degree of regioselectivity to give derivatives of the type **527**, rather than **528**. However, the

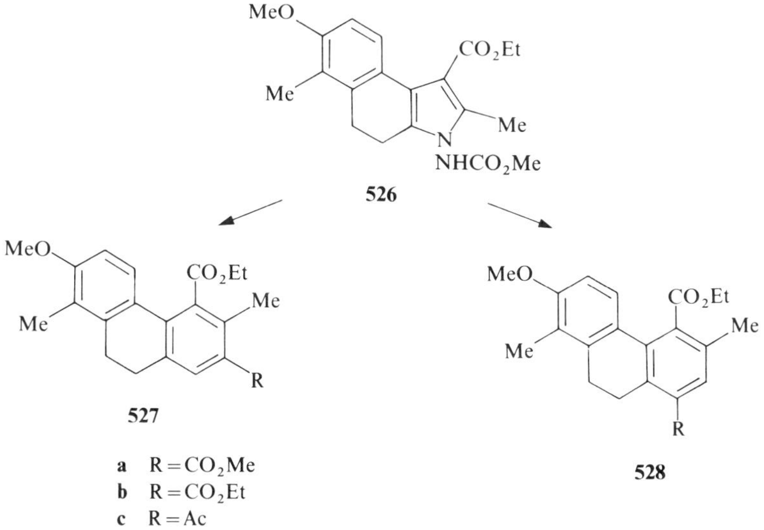

MeO
Me
CO2Et
Me
N
NHCO2Me
526
MeO
Me
CO2Et
Me
R
527
MeO
Me
CO2Et
Me
R
528
a R = CO2Me
b R = CO2Et
c R = Ac

Scheme 147

reaction with propiolaldehyde fails, and vigorous conditions (in dry xylene under nitrogen at reflux temperature for 24 h) are required for the cycloaddition with methyl propiolate, ethyl propiolate, and butyn-2-one to give both **527** and **528** (Scheme 147) in a ratio of 3–2:1.[124] The cycloadduct **527c** is a useful intermediate for the total synthesis of the highly active antineoplastic agent juncusol.[124]

3.5. APPLICATIONS

Aminopyrroles have been shown to have a wide spectrum of biological activities, and they are also useful as synthons for the preparation of pharmaceutically interesting derivatives.

Among the aminopyrroles with interesting biological applications, an important role is played by the naturally occurring oligopeptide antibiotics. These derivatives, although active on different biological systems, qualitatively show similar spectra of action.

3.5.1. Biological and Medical Uses

1-Acylamino-2,5-dimethylpyrroles of type **260** have been screened for hypoglycemic activity, using tolbutamide and the analog 1-benzoyl-3,5-dimethylpyrazole as reference.[57] Several derivatives were more active than tolbutamide, but none exhibited hypoglycemic activity similar to the pyrazole derivative. Only in the case of 1-(isonicotinoylamino)-2,5-dimethylpyrrole was the activity close to that of the standard. However, derivatives with R = alkyl, are more active than the compounds with R = aryl, and the valeric derivative is the most active. In the aryl series, the activity is increased with the introduction of a hydroxyl group and lowered in the presence of methyl or *ortho* and *para* nitro groups. Among the pyridyl derivatives, maximum activity was found with the isonicotinic compound.

Derivatives of type **260** (R = heteroaryl) are antituberculosis agents[46] and, in particular, 1-(isonicotinoylamino)-2,5-dimethylpyrrole has an LD_{50} in mice of 555 mg/kg wt. Its in vitro tuberculostatic activity (against strain H37RV on Dubos medium) is 0.62 μg/ml.[55, 343]

Several 1-(*N*,*N*-substituted aminoacetamido)-2,5-dimethylpyrroles, that are structurally related to xylocaine (with anaesthetic and bronchidilatator properties), show analgesic and local anaesthetic properties.[56] The most active as anaesthetics are the derivatives with cyclohexyl, ethyl, and *tert*-butyl substituents. Also these derivatives partially inhibit the action of barbiturates and improve spontaneous ventilation. This latter effect is more important when the substituent is an unbranched alkyl group (ethyl, *n*-propyl, *n*-butyl). However, all compounds show a high degree of toxicity, which is more evident with branched alkyl groups (*i*-propyl > *i*-butyl > *n*-propyl > cyclohexyl).

Several 1-arylaminopyrroles and 4-(1H-pyrrol-1-ylimino)-2,5-cyclohexadienes of type **393** show antibacterial activity (MIC ≤ 31.3 µg/ml) against *Staphylococcus aureus* and *Proteus mirabilis*.[51] Antibacterial activity against *Escherichia coli, Klebsiella pneumoniae,* and *Pseudomonas aeruginosa* has been observed only at MIC ≥ 62.5 µg/ml. Compounds of type **393** and 1-(4-hydroxyphenylamino)pyrroles inhibited *S. aureus* at concentrations ≤ 7.8 µg/ml. All the compounds showing good inhibitory activity are components of hydroquinone–quinone pairs. Closely related derivatives that are not capable of undergoing the redox interconversion are not inhibitory. A similar lack of inhibitory activity has been observed against *P. mirabilis*. Metabolites of these compounds, present in urine, show similar antibacterial activity against the same strains. This selective activity is identical to that observed in the in vitro assay. It is noteworthy that the corresponding oxygen-substituted phenylamino derivatives, that were inactive in vitro, are metabolically activated to show antibacterial activity similar to the hydroquinone–quinone. This activation suggests that the metabolism of these compounds involves formation of metabolites that can oxidatively–reductively interconvert. Some of these derivatives have a curative effect on mice infected with *Mycobacterium tuberculosis* H37RV. This antitubercular activity appears to be unrelated to the activity of these compounds against *S. aureus* and *P. mirabilis*, but can be related to oxidative–reductive interconversion.

N-Substituted 2-butylaminopyrroles are useful as plant protecting agents.[192]

Variously 5-substituted 2-aminopyrroles analogous to **351** and their corresponding acylamino derivatives show antiarrhythmic and anaesthetic activities, and, in some cases, the activity was comparable or even greater than that of the structurally correlated lidocaine.[281,301]

Selected data on the pharmacological activities of some of the most active derivatives of this series show that they are potent agents with respect to blood pressure, while having minimal toxicity and appearing to be free of other central and peripheral side effects typical of lidocaine.[344] The most potent antiarrhythmic agent, which does not show the marked ataxia of lidocaine, was found to be the 2-(diethylaminoacetamido)-5-benzylpyrrole derivative. Moreover, this compound has local anaesthetic activity and an ability to protect against $CaCl_2$ induced arrhythmias.

Some of these derivatives have also been tested for antiarrhythmic activity and acute CNS toxicity.[282] They have potency similar to that of lidocaine and, in general, N-methyl substituted pyrroles appear to be more potent and more toxic than the corresponding N-unsubstituted derivatives. All these compounds show acute CNS side effects at doses that are not significantly different from those required for efficacy.

The oxamate derivative of **351** shows antiallergy activity and lowers serum cholesterol–lipoprotein level in rats.[303]

3-Cyano- and 3-carboxamido-substituted derivatives of type **266** have been tested for anticonvulsant activity in mice.[304] In some cases they showed anticonvulsant activity, which is higher than that of trimethadione, against pentylenetetrazol-induced convulsions. Generally, the nitrile derivatives are more active compared with the carbamyl analogs, due to higher lipid solubility.

2-Acetylamino-3-cyano-4-phenylpyrrole shows herbicide activity in post emergent tests against *Solanum lycopersicum, Sinapis alba, Stellaria media*, and *Phaseolus vulgaris* at 4 kg/ha (kilograms per hectare). The same compound also shows plant regulatory activity.[290]

The 3-aminopyrroles **228** show remarkable antiinflammatory activity, evinced by the carrageenin-induced edema test in rats.[205] A 40–60% decrease of the induced edema was noted, when administered at dose levels between 2 and 20% of their LD_{50} values. Under the same experimental conditions, the decrease of the induced edema by phenylbutazone, a widely employed antiinflammatory drug, is about 45%, but only if administered at a dose level > 25% its LD_{50}. The toxicities of the aminopyrroles, with few exceptions, are very low, and their LD_{50} values are always higher than 1000 mg/kg p.o. (orally administered) in mice, whereas the LD_{50} of phenylbutazone is $\sim$ 390 mg/kg p.o.

3-Aminopyrroles also possess other biological activities. They have outstanding antipyretic and analgesic activities, which are respectively $\sim$ 10 and 4 times those displayed by acetylsalicylic acid, and also show a very low ulcerogenic activity, which is $\sim$ 5–10 times less than those observed for acetylsalicylic acid and phenylbutazone.

3-Aminopyrroles, particularly derivatives of type **228** (R = Ph, 4-OH—C_6H_4, $R^1 = R^2$ = Me), exhibit activity as CNS sedatives and myorelaxants, which is a valuable characteristic of drugs for patients affected by severe inflammatory diseases. Several other compounds of this series also show ansiolytic properties.

3-Aminopyrroles of type **228** ($R^1 = NH_2$), have also been shown to be active as prostaglandin synthetase inhibitors and possess a valuable degree of activity on the water balance of warm-blooded animals.[325]

Various 3-sulfonamidopyrroles and their 1-methyl derivatives have been tested for antibacterial activity.[316]

Netropsin has been tested in vitro and shows activity against several bacteria, and, in particular, the MIC against *S. aureus, S. albus, B. subtilis, E. coli*, and *A. aerogenes* was reported to be 3–7 µg/ml.[1] Netropsin also shows activity against clothes moth larvae and the black carpet beetle.

Compounds of type **15** have been patented as antitubercular agents, virucides, and, more specifically, as trypanocides, e.g., against *T. congolense* in cattle.[12] Other arene analogs of congocidine show trypanocidal activity and a low activity against the vaccinia virus,[15,16] while thiophen and pyridine analogs of congocidine show no useful trypanocidal or antiviral activity.[14]

The analog of congocidine containing one more pyrrole unit does not show improved antiparasitic activity, although it increases the therapeutic index.[345]

Distamycin A (DA) exhibits a wide spectrum of interesting biological activities. Partially purified samples inhibit to various degrees experimental tumors in mice and interfere with the process of cell division in vitro.[17] This latter activity

3.5. Applications

remarkably diminishes or disappears in samples repeatedly recrystallized.[17,18] DA protects the K_{12} strain of *E. coli* from T1 and T2 phages, inhibiting the phage adsorption by the bacterial cells and blocking the bacteria lysis already infected with consequent accumulation of intracellular mature phage particles. DA inhibits the actinomycete lysis by actinophage and the formation of inducible enzymes in *E. coli*.[18]

The oligopeptide antibiotics distamycin and netropsin are very effective inhibitors of the DNA and RNA synthesis in vitro and in vivo owing to their extremely strong binding affinity to the DNA template. Both form highly stable and highly ordered complexes with duplex DNA in which the *N*-methylpyrroles of the antibiotic are aligned with the double helix. A property of these complexes in the remarkable A-T specificity.[23,346] The interaction of netropsin with A-T cluster is accompained by a pronounced elongation of the contour length of the DNA double helix without intercalation; the changes of DNA contour length and flexibility have been quantitatively determined by viscosimetry measurements.[347]

The DNA–distamycin complex is partly dissociated by enzymatic hydrolysis, and the antibiotic can be quantitatively removed from the complex by extraction with a two-phase phenol–water system.[348]

Complex formation between distamycin and netropsin oligopeptides and DNA occurs at low and high ionic strength. UV spectrophotometry and circular dichroism analysis has been used to study the influence of organic solvents and salts on the stability of these complexes.[349–351] Thus, urea in 2 M LiCl is effective in the complete disorganization of the DNA–netropsin complex while, in the case of distamycin, complex formation is still observable. The complexes are effectively disorganized by glycol, DMSO/water, or by hydrophobic bond-breaking agents, such as 7 M LiCl and 7.2 M $NaClO_4$, although the dissociation is not totally complete. On this basis, it has been considered that, besides electrostatic interactions, the factors contributing to the stability of the complexes are hydrogen bonding, hydrophobic interaction, and dispersion forces.

Although these naturally occurring antibiotics have been widely investigated and several analogs have been synthesized, only limited information is available on the structural requirements for the antiviral activity of DA. Replacement of the formamido side-chain by other groups, such as nitro, amino, acetamido, cyclopentylpropionamido, or *N*-formylglycinamido, leads to a reduction of the activity against vaccinia virus.[23] When DA was converted into the analogous tetrapeptide or pentapeptide, the activity against some viruses increased. At the same time the cytotoxicity was reduced, indicating that these two activities are separable. Replacement of the β-propionamidine side-chain by a γ-butyramidine or a *p*-aminobenzamidine results in a considerable reduction of the activity against phage T2, whereas the activity against vaccinia virus and the cytotoxicity is still present.

The peptide bond at the 2- and 4-positions of the pyrrole ring is critical for the antiviral activity in the distamycin and congocidine series. In derivatives of type 35 the biological activity is severely reduced and an analog of 33, having only one pyrrole unit, was the most active of this series and was less toxic, but it had almost the same antiviral activity of DA.[24]

The permethyl analogs of distamycin have been tested as antimalarials and anticancer agents, but showed no activity.[28]

The antibiotic TAN-868 A (**38c**) shows a broad spectrum of antibacterial activity and is especially active against *Micrococcus flavus* IFO 3242, *Micrococcus lutens* IFO 12708, and *Acinetobacter calcoaceticus* IFO 13006.[37] It is also more active than kikumycin A (**38a**) and DA against bacteria. TAN-868 A has cytotoxic activity, as it is active against murine tumor cells, while kikumycin A is active not only against tumor cells but also against L-929 fibroblasts. Thermal denaturation studies suggest that interaction of kikumycin A and TAN-868 A with DNA have similar profiles and bind with an A-T-rich region of double-strained DNA.[37] Such an interaction appears to be sensitive to the absolute configuration of the oligopeptides, since the (4R)-(−)dihydrokikumycin B (**39b**) binds more strongly to DNA than its respective enantiomer (4S)-(+).[39]

Anthelvencin A (**45**) effectively controls nematode infection in mice and swine by oral administration and also shows a moderate antibacterial activity, but its unfavorable potency–toxicity ratio has precluded further interest.

A slight activity has been detected for **45** against vaccinia virus in tissue culture, but it has no activity against mouse tumors.[40] DNase I footprinting studies have shown that both enantiomeric forms of anthelvencin A bind to identical DNA sites and that there is only weak stereoselectivity. A slight preference for binding of the natural (4S)-(+) isomer with DNA, compared with laevo isomer, has been also detected by ethidium displacement and corroborated by ^{1}H NMR studies. These results have also been confirmed by thermodynamic data, which revealed that the two enantiomers exhibit nearly identical binding free energies. Thus, the stereochemical difference between the two enantiomers does not significantly alter the binding affinity ($\Delta G°$) for the poly[d(AT)] poly[d(AT)] duplex, and the binding constants of both enantiomers exhibit almost identical salt dependence reflecting equal electrostatic contributions to the DNA binding of the two optical isomers.[352]

3.5.2. As Synthons for Drugs

N-(2-Aroylaminophenyl)-N-methyl-1-aminopyrroles are precursors of several 5H-pyrrolo[1,2-b][1,2,5]benzotriazepines and dihydro-5H-pyrrolo[1,2-b][1,2,5]benzotriazepines, with psychotropic activity.[112,353]

5-Mercapto-3,4-dicyano-2-aminopyrrole (**132**) has been used in the synthesis of the aglycone of toyocamycin and tubercidin derivatives that have shown antibiotic and antitumor activities.[306,307,354,355]

The 5-bromo-2-aminopyrrole analog of **132** has been used as synthon in the total synthesis of the toyocamycin, sangivamycin, and tubercidin, a series of pyrrolo[2,3-d]pyrimidine nucleoside antibiotics.[308]

2-Aminopyrroles **351** (R = H, Ac) and the corresponding 2-acylamino derivatives are synthons for the synthesis of the pyrrolo[1,2-a][1,3]diazepine ring system, which shows acute hypotensive action in rats.[302] The unsubstituted amines have also been utilized as the starting materials for the synthesis of Schiff bases, which shows antiviral activity.[356]

Pyrrolo[3,4-e][1,4]diazepine-2-(1H)-ones **434**, which show activity as CNS depressants, anticonvulsants, antiinflammatories, and inhibitors of the enzymes that bring about the synthesis of prostaglandins, have been synthesized from 3-amino-4-benzoylpyrroles.[319] More recently, other derivatives of the same ring system, obtained by the same synthetic route, showed remarkable anticonvulsant and antianxiety activities together with a very low toxicity that in one case was > 20,000 mg/kg.[320-322] 3-Amino-4-acylpyrroles have also been used as starting materials for the synthesis of pyrrolo[3,4-b]pyridines **448** and pyrrolo[3,4-d]pyrimidines **441**, which exhibit antiinflammatory activity causing a 35–55% decrease of the "carrageenin-induced edema" at dose levels ranging from 20 to 100 mg/kg p.o. for the pyrrolopyridines[326] and 100 to 200 mg/kg p.o. for pyrrolopyrimidines.[318,324] Pyrrolo[3,4-b]pyridines and pyrrolo[3,4-d]pyrimidines also show activity as prostaglandin synthetase inhibitors and CNS depressants and have a valuable degree of activity on the water balance of warm-blooded animals.[318,324,326] All these compounds show low toxicity with their LD_{50} always being higher than 500 mg/kg p.o.[324,326] and, for some pyrrolo[3,4-d]pyrimidin-4-ones, higher than 1000 mg/kg p.o.[318]

3.6. APPENDIX

This appendix reports briefly on the latest papers dealing with aminopyrroles.

Pseudopeptide analogues of netropsin and distamycin, containing the aminoacridine chromophore, have been synthesized.[359] Their DNA binding properties are consistent with a model in which the acridine nucleus occupies an intercalation site and the netropsin or distamycin residue binds to the DNA minor groove. These compounds also show cytostatic and cytotoxic activities against murine cells.

Sequence-specific cleavage of G4 phage DNA fragment is also observed with a distamycin analog containing the metal binding site of bleomycin (PYML-6).[360]

2-Amino-3-cyano-5-phenylpyrroles of type **123** (R = aryl) have been prepared from phenacylmalononitrile and arylamines under acidic conditions.[361] The intermediacy of enaminonitriles of type **122** (see Scheme 33) has been postulated. Attempts to prepare the corresponding 1-alkyl derivatives resulted in the formation of 2-amino-3-cyano-5-phenylfuran.

3-Aminopyrroles **531** have been synthesized in high yields from aspartic acid diamide β-esters **529** through a ring transformation of oxazole acetates **530**

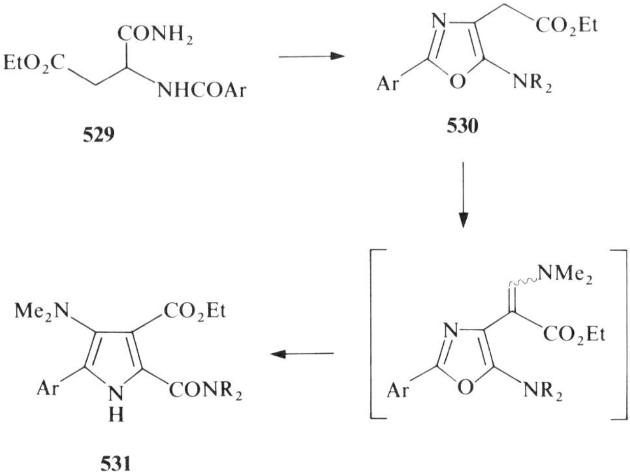

Scheme 148

under Vilsmeier–Haack reaction conditions[362] (Scheme 148). The multifunctional 3-aminopyrroles **531** exhibit high hypolipidemic activities.

1-Nucleoside 2-aminopyrroles of type **190** and their 3,4-dicyano analogs have been cyclized to pyrrolo[2,3-*d*]pirimidines **412**, several of which have shown activity against HIV virus in in vitro screens.[363]

Spectroscopic data of the aminopyrroles mentioned in this section are not included in the tables.

3.7. ACKNOWLEDGMENTS

The authors wish to express appreciation to Dr. Patrizia Diana for the helpful collaboration during the preparation of the manuscript.

3.8. REFERENCES

1. A. C. Finlay, F. A. Hochstein, B. A. Sobin, and F. X. Murphy, *J. Am. Chem. Soc.*, **73**, 341 (1951).
2. C. Cosar, L. Ninet, S. Pinnert-Sindico, and J. Preud'homme, *Compt. rend.*, **234**, 1498 (1952).
3. K. Watanabe, *J. Antibiotics*, **A9**, 102 (1956).
4. C. W. Waller, C. F. Wolf, W. J. Stein, and B. L. Hutchings, *J. Am. Chem. Soc.*, **79**, 1265 (1957).
5. E. E. van Tamelen, D. M. White, I. C. Kogon, and A. D. G. Powell, *J. Am. Chem. Soc.*, **78**, 2157 (1956).
6. E. E. van Tamelen and A. D. G. Powell, *Chem. Ind.* (*London*), **1957**, 365.
7. M. Julia and N. Preau-Joseph, *Compt. rend.*, **243**, 961 (1956).
8. M. J. Weiss, J. S. Webb, and J. M. Smith, Jr., *J. Am. Chem. Soc.*, **79**, 1266 (1957).
9. M. Julia and N. Preau-Joseph, *Compt. rend.*, **257**, 1115 (1963).
10. M. Julia and N. Preau-Joseph, *Bull. Soc. Chim. Fr.*, **1967**, 4348.

3.8. References

11. J. W. Lown and K. Krowicki, *J. Org. Chem.*, **50**, 3774 (1985).
12. Rhone-Poulenc S.A., Br. Pat. 1004974 (1965); *Chem. Abstr.*, **67**, 32589 (1967).
13. M. Bialer, B. Yagen, and R. Mechoulam, *J. Med. Chem.*, **23**, 1144 (1980).
14. D. H. Jones and K. R. H. Wooldridge, *J. Chem. Soc.* (C), **1968**, 550.
15. M. Julia and R. Gombert, *Bull. Soc. Chim. Fr.*, **1968**, 369.
16. M. Julia and R. Gombert, *Bull. Soc. Chim. Fr.*, **1968**, 376.
17. F. Arcamone, S. Penco, P. Orezzi, V. Nicolella, and A. Pirelli, *Nature*, **203**, 1064 (1964).
18. F. Arcamone, P. G. Orezzi, W. Barbieri, V. Nicolella, and S. Penco, *Gazz. Chim. Ital.*, **97**, 1097 (1967).
19. S. Penco, S. Redaelli, and F. Arcamone, *Gazz. Chim. Ital.*, **97**, 1110 (1967).
20. F. Arcamone, S. Penco, and F. Delle Monache, *Gazz. Chim. Ital.*, **99**, 620 (1969).
21. F. Arcamone, V. Nicolella, S. Penco, and S. Redaelli, *Gazz. Chim. Ital.*, **99**, 632 (1969).
22. M. Bialer, B. Yagen, and R. Mechoulam, *Tetrahedron*, **34**, 2389 (1978).
23. M. Bialer, B. Yagen, and R. Mechoulam, *J. Med. Chem.*, **22**, 1296 (1979).
24. M. Bialer, B. Yagen, R. Mechoulam, and Y. Becker, *J. Pharm. Sci.*, **69**, 1334 (1980).
25. L. Grehen and U. Ragnarsson, *J. Org. Chem.*, **46**, 3492 (1981).
26. L. Grehen, U. Ragnarsson, B. Eriksson, and B. Oberg, *J. Med. Chem.*, **26**, 1042 (1983).
27. L. Grehn, U. Ragnarsson, and R. Datema, *Acta Chem. Scand.*, B, **40**, 145 (1986).
28. P. L. Gendler and H. Rapoport, *J. Med. Chem.*, **24**, 33 (1981).
29. A. A. Khorlin, S. L. Grokhovskii, A. L. Zhuze, and B. P. Gottikh, *Bioorg. Khim.*, **8**, 1063 (1982); *Chem. Abstr.*, **97**, 216683 (1982).
30. S. L. Grokhovskii, A. L. Zhuze, and B. P. Gottikh, *Bioorg. Khim.*, **8**, 1070 (1982); *Chem. Abstr.*, **97**, 215877 (1982).
31. N. G. Plekhanova, E. N. Glibin, B. V. Tsukerman, and O. F. Ginzburg, *Zhur. Org. Khim.*, **19**, 1533 (1983); *Chem. Abstr.*, **99**, 176269 (1983).
32. F. Arcamone, N. Mongelli, and S. Penco, Ger. Offen. DE 3623880 (1987); *Chem. Abstr.*, **106**, 156157 (1987).
33. E. Lazzari, F. Arcamone, S. Penco, M. A. Verini, and N. Mongelli, Eur. Pat. EP 246868 (1987); *Chem. Abstr.*, **109**, 54578 (1988).
34. M. Kikuchi, K. Kumagai, N. Ishida, Y. Ito, T. Yamaguchi, T. Furumai, and T. Okuda, *J. Antibiotics*, **A18**, 243 (1965).
35. T. Takaishi, Y. Sugawara, and M. Suzuki, *Tetrahedron Lett.*, **1972**, 1873.
36. T. Takaishi, M. Suzuki, and A. Tatematsu, *Org. Mass Spectr.*, **9**, 635 (1974).
37. M. Takizawa, S. Tsubotani, S. Tanida, and T. Hasegawa, *J. Antibiotics*, **40**, 1220 (1987).
38. T. Hasegawa and S. Harada, Jpn. Kokai Tokkio Hoko JP 62190161 (1987); *Chem. Abstr.*, **109**, 5279 (1988).
39. M. Lee and J. W. Lown, *J. Org. Chem.*, **52**, 5717 (1987).
40. G. W. Probst, M. M. Hoehn, and L. Woods, *Antimicrobial Agents and Chemotherapy*, **1965**, 789.
41. M. Lee, D. M. Coulter, and J. W. Lown, *J. Org. Chem.*, **53**, 1855 (1988).
42. T. L. Jacobs, in *Heterocyclic Compounds*, Vol. 6, Wiley, New York, 1957, p. 101.
43. L. Knorr, *Chem. Ber.*, **18**, 299 (1885).
44. J. Thiele and E. Dralle, *Liebigs Ann. Chem.*, **302**, 275 (1898).
45. H. L. Yale, K. Losee, J. Martins, M. Holsing, F. M. Perry, and J. Bernstein, *J. Am. Chem. Soc.*, **75**, 1933 (1953).
46. H. L. Yale and J. Bernstein, U.S. Pat 2,727,896 (1955); *Chem. Abstr.*, **50**, 12115 (1956).
47. L. Knorr, *Chem. Ber.*, **22**, 168 (1889).

48. F. Nerdel, E. Henkel, R. Kayser, and G. Kannelbley, *J. Prakt. Chem.*, **3**, 153 (1956).
49. T. D. Binns and R. Brettle, *J. Chem. Soc. (C)*, **1966**, 341.
50. P. Grammaticakis, *Compt. rend.*, **272C**, 1574 (1971).
51. R. E. Johnson, A. E. Soria, J. R. O'Connor, and R. A. Dobson, *J. Med. Chem.*, **24**, 1314 (1981).
52. E. E. Blaise, *Compt. rend.*, **172**, 221 (1921).
53. H. V. Euler and B. Hagglund, *Arkiv. Kemi. Mineral. Geol.*, **A19**, 29 (1945); *Chem. Abstr.*, **41**, 1660 (1947).
54. C. G. Overberger, L. C. Palmer, B. S. Marks, and N. R. Byrd, *J. Am. Chem. Soc.*, **77**, 4100 (1955).
55. S. Fatutta, *Gazz. Chim. Ital.*, **90**, 1645 (1960).
56. R. Rips, A. Magnin, F. Barale, and P. Magnin, *Chimie Therapeutique*, **1968**, 445.
57. B. Lotti and O. Vezzosi, *Farmaco Ed. Sci.*, **27**, 317 (1972).
58. N. P. Buu-Hoi and N. D. Xuong, *J. Org. Chem.*, **20**, 850 (1955).
59. R. Rips, C. Derappe, and N. P. Buu-Hoi, *J. Org. Chem.*, **25**, 390 (1960).
60. V. S. Misra, V. K. Saxena, and R. Srivastava, *Indian Drugs*, **19**, 55 (1981); *Chem. Abstr.*, **96**, 122678 (1982).
61. H. Beyer, T. Pyl, and C. E. Volcker, *Liebigs Ann. Chem.*, **638**, 150 (1960).
62. H. J. Jahns and M. Baeker, *Z. Chem.*, **1972**, 417.
63. E. E. Blaise, *Compt. rend.*, **171**, 34 (1920).
64. R. Kuhn and K. Dury, *Liebigs Ann. Chem.*, **571**, 44 (1951).
65. G. Korschun and C. Roll, *Bull. Soc. Chim. Fr.*, **39**, 1223 (1926).
66. L. A. Carpino, *J. Org. Chem.*, **30**, 736 (1965).
67. W. Flitsch, U. Kramer, and H. Zimmermann, *Chem. Ber.*, **102**, 3268 (1969).
68. O. Wallach, *Liebigs Ann. Chem.*, **362**, 261 (1908).
69. T. Kubota, T. Matsuura, and Y. Kakuno, *Bull. Chem. Soc. Jpn.*, **38**, 1191 (1965).
70. W. Keller-Schierlein, M. L. Mihailovic, and V. Prelog, *Helv. Chim. Acta*, **41**, 220 (1958).
71. G. Korschun, *Chem. Ber.*, **37**, 2183 (1904).
72. G. Korschun and C. Roll, *Bull. Soc. Chim. Fr.*, **33**, 55 (1923).
73. C. Chang and R. Adams, *J. Am. Chem. Soc.*, **53**, 2353 (1931).
74. N. M. Timoshevskaya, *Trudy Khar'kov. Politekh. Inst.*, **4**, 73 (1954); *Chem. Abstr.*, **52**, 7279 (1958).
75. N. M. Timoshevskaya, *Trudy Khar'kov Politekh. Inst.*, **4**, 73 (1954); *Chem. Abstr.*, **52**, 7279 (1958).
76. W. Borsche and M. Spannagel, *Liebigs Ann. Chem.*, **331**, 298 (1904).
77. W. Borsche and A. Klein, *Liebigs Ann. Chem.*, **548**, 74 (1941).
78. A. Smith, *Liebigs Ann. Chem.*, **289**, 310 (1896).
79. A. Smith and H. N. McCoy, *Chem. Ber.*, **35**, 2169 (1902).
80. S. Capuano, *Gazz. Chim. Ital.*, **68**, 521 (1938).
81. A. Smith, *J. Chem. Soc.*, **57**, 643 (1890).
82. F. Klingemann, *Leibigs Ann. Chim.*, **269**, 104 (1892).
83. C. Bulow, *Chem. Ber.*, **35**, 4311 (1902).
84. C. Bulow and C. Sautermeister, *Chem. Ber.*, **37**, 2697 (1904).
85. C. Bulow and C. Sautermeister, *Chem. Ber.*, **39**, 647 (1906).
86. C. Bulow and E. Klemann, *Chem. Ber.*, **40**, 4749 (1907).
87. C. Paal and J. Ubber, *Chem. Ber.*, **36**, 497 (1903).
88. C. Bulow, *Chem. Ber.*, **37**, 91 (1904).
89. L. Knorr, *Liebigs Ann. Chem.*, **236**, 290 (1886).

90. C. Bulow, *Chem. Ber.*, **51**, 399 (1918).
91. S. Kawai and S. Tanaka, *Bull. Chem. Soc. Jpn.*, **33**, 674 (1960).
92. C. Bulow, *Chem. Ber.*, **37**, 2424 (1904).
93. C. Bulow and R. Weidlich, *Chem. Ber.*, **40**, 4326 (1907).
94. C. Bulow, *Chem. Ber.*, **42**, 3311 (1909).
95. C. Bulow and R. Huss, *Chem. Ber.*, **51**, 24 (1918).
96. C. Bulow and R. Engler, *Chem. Ber.*, **52**, 632 (1919).
97. J. Schimdt, *Liebigs Ann. Chem.*, **293**, 107 (1896).
98. C. Bulow, *Chem. Ber.*, **38**, 2366 (1905).
99. G. Korschun and C. Roll, *Gazz. Chim. Ital.*, **41**, I, 186 (1911).
100. C. Bulow, *Chem. Ber.*, **38**, 3914 (1905).
101. C. Bulow and R. Weidlich, *Chem. Ber.*, **39**, 3372 (1906).
102. W. L. Mosby, *J. Chem. Soc.*, **1957**, 3997.
103. G. Adembri, F. De Sio, R. Nesi, and M. Scotton, *J. Chem. Soc. (C)*, **1970**, 1536.
104. T. Seerin and H. Poehlmann, *Chem. Ber.*, **110**, 491 (1977).
105. R. Epton, *Chem. Ind. (London)*, **1965**, 425.
106. F. R. Japp and J. Wood, *J. Chem. Soc.*, **87**, 707 (1905).
107. G. Rio and A. Lecas-Nawrocka, *Bull. Soc. Chim. Fr.*, **1971**, 1723.
108. W. Ried and R. Lantzsch, *Liebigs Ann. Chim.*, **750**, 97 (1971).
109. F. R. Japp and G. N. Huntly, *J. Chem. Soc.*, **55**, 184 (1888).
110. F. R. Japp and F. Klingemann, *J. Chem. Soc.*, **57**, 662 (1890).
111. Y. I. Kheruze and A. A. Petrov, *Zhur. Org. Khim.*, **10**, 1412 (1974); *Chem. Abstr.*, **81**, 120043 (1974).
112. G. Stefancich, M. Artico, F. Corelli, and S. Massa, *Synthesis*, **1983**, 757.
113. D. M. Lemal and T. W. Rave, *Tetrahedron*, **19**, 1119 (1963).
114. F. Y. Perveev and V. Ershova, *Zhur. Obshehei Khim.*, **30**, 3554 (1960); *Chem. Abstr.*, **55**, 19896 (1961).
115. V. Sprio and P. Madonia, *Ann. Chim. (Rome)*, **50**, 1791 (1960).
116. V. Sprio, *Ann. Chim. (Rome)*, **55**, 301 (1965).
117. O. Migliara, S. Petruso, and V. Sprio, *J. Heterocycl. Chem.*, **16**, 833 (1979).
118. L. Lamartina, O. Migliara, and V. Sprio, *J. Heterocycl. Chem.*, **19**, 1381 (1982).
119. L. Ceraulo, L. Lamartina, O. Migliara, S. Petruso, and V. Sprio, *Heterocycles*, **20**, 551 (1983).
120. V. Sprio, S. Plescia, and J. Fabra, *Atti Acad. Sci. Lett., Arti Palermo, Parte 1*, **31**, 179 (1972); *Chem. Abstr.*, **79**, 105026 (1972).
121. G. Chelucci and M. Marchetti, *J. Heterocycl. Chem.*, **25**, 1135 (1988).
122. E. E. Schweizer and C. M. Kopay, *J. Org. Chem.*, **37**, 1561 (1972).
123. A. G. Schultz and M. Shen, *Tetrahedron Lett.*, **1979**, 2969.
124. A. G. Schultz and M. Shen, *Tetrahedron Lett.*, **22**, 1775 (1981).
125. O. Attanasi and L. Caglioti, *Org. Prep. Proced. Int.*, **18**, 299 (1986).
126. O. Attanasi, M. Grossi, F. Serra-Zanetti, and E. Foresti, *Tetrahedron*, **43**, 4249 (1987).
127. S. Brodka and H. Simon, *Liebigs Ann. Chem.*, **745**, 193 (1971).
128. L. Bernardi, P. Masi, and G. Rosini, *Ann. Chim. (Rome)*, **63**, 601 (1973).
129. O. Attanasi, P. Filippone, A. Mei, and S. Santeusanio, *Synthesis*, **1984**, 671.
130. O. Attanasi and F. R. Perrulli, *Synthesis*, **1984**, 874.
131. O. Attanasi, P. Filippone, S. Santeusanio, and F. Serra-Zanetti, *Synthesis*, **1987**, 381.
132. O. Attanasi, S. Santeusanio, and F. Serra-Zanetti, *Gazz. Chim. Ital.*, **119**, 631 (1989).

133. O. Attanasi, P. Bonifazi, E. Foresti, and G. Pradella, *J. Org. Chem.*, **47**, 684 (1982).
134. O. Attanasi, P. Bonifazi, and F. Buiani, *J. Heterocycl. Chem.*, **20**, 1077 (1983).
135. O. Attanasi, *Chim. Ind. (Milan)*, **66**, 19 (1984).
136. O. Attanasi, P. Filippone, A. Mei, and F. Serra-Zanetti, *Synth. Commun.*, **16**, 343 (1986).
137. O. Attanasi and S. Santeusanio, *Synthesis*, **1983**, 742.
138. O. Attanasi, P. Filippone, A. Mei, S. Santeusanio, and F. Serra-Zanetti, *Synthesis*, **1985**, 157.
139. O. Attanasi, P. Filippone, A. Mei, and S. Santeusanio, *Synthesis*, **1984**, 873.
140. O. Attanasi, F. R. Perrulli, and F. Serra-Zanetti, *Heterocycles*, **23**, 867 (1985).
141. O. Attanasi, P. Filippone, A. Mei, and F. Serra-Zanetti, *J. Heterocycl. Chem.*, **23**, 25 (1986).
142. O. Attanasi, P. Filippone, A. Mei, F. R. Perrulli, and F. Serra-Zanetti, *Synth. Commun.*, **16**, 1411 (1986).
143. O. Attanasi, M. Grossi, and F. Serra-Zanetti, *Org. Prep. Proced. Int.*, **18**, 1 (1986).
144. O. Attanasi, P. Filippone, P. Guerra, and F. Serra-Zanetti, *Heterocycles*, **27**, 149 (1988).
145. O. Attanasi, P. Filippone, P. Guerra, F. Serra-Zanetti, E. Foresti, and V. Tugnoli, *Gazz. Chim. Ital.*, **118**, 533 (1988).
146. O. Attanasi, P. Filippone, A. Mei, S. Santeusanio, and F. Serra-Zanetti, *Bull. Chem. Soc. Jpn.*, **59**, 3332 (1986).
147. O. Attanasi, M. Grossi, and F. Serra-Zanetti, *J. Heterocycl. Chem.*, **25**, 1263 (1988).
148. W. H. Davies and M. A. T. Rogers, *J. Chem. Soc.*, **1944**, 126.
149. C. A. Grob and H. Utzinger, *Helv. Chim. Acta*, **37**, 1256 (1954).
150. A. Brodrick and D. G. Wibberley, *J. Chem. Soc. Perkin Trans. 1*, **1975**, 1910.
151. C. A. Grob and P. Ankli, *Helv. Chim. Acta*, **33**, 273 (1950).
152. W. J. Middleton, V. A. Engelhardt, and B. S. Fisher, *J. Am. Chem. Soc.*, **80**, 2822 (1958).
153. S. Apparao, H. Ila and H. JunJappa, *Synthesis*, **1981**, 65.
154. K. Gewald and M. Hentschel, *J. Prakt. Chem.*, **318**, 663 (1976).
155. Von E. Fanghanel, K. Gewald, K. Putsch, and K. Wagner, *J. Prakt. Chem.*, **311**, 388 (1969).
156. A. Foucaud, *Bull. Soc. Chim. Fr.*, **1964**, 123.
157. J. A. Elvidge and R. P. Linstead, *J. Chem. Soc.*, **1954**, 442.
158. P. E. Fanta and S. Smith, *J. Am. Chem. Soc.*, **76**, 2915 (1954).
159. K. Gewald and U. Hain, *Synthesis*, **1984**, 62.
160. A. O. Abdelhamid and N. M. Abed, *Heterocycles*, **24**, 101 (1986).
161. W. Steglich, G. Hofle, W. Konig, and F. Weygand, *Chem. Ber.*, **101**, 308 (1968).
162. H. D. Stachel, K. K. Harigel, H. Poschenrieder, and H. Burghard, *J. Heterocycl. Chem.*, **17**, 1195 (1980).
163. T. Eicher and U. Stapperfenne, *Synthesis*, **1987**, 619.
164. K. W. Law, T. F. Lai, M. P. Sammes, A. R. Katritzky, and T. C. W. Mak, *J. Chem. Soc. Perkin Trans. 1*, **1984**, 111.
165. R. Verhe, N. De Kimpe, L. De Buyck, M. Tilley, and N. Schamp, *Tetrahedron*, **36**, 131 (1980).
166. F. M. Abdelrazek, A. W. Erian, and A. M. Torgoman, *Chem. Ind. (London)*, **1988**, 30.
167. O. E. Nasakin, V. V. Alekseev, P. B. Terent'ev, A. Kh. Bulai, and M. Yu. Zablotskaya, *Khim. Geterotsikl. Soedin.*, **1983**, 1062; *Chem. Abstr.*, **100**, 68112 (1984).
168. O. E. Nasakin, M. P. Vorob'eva, I. A. Abramov, L. S. Shevnitsyn, A. Y. Chernikhov, M. N. Yakovlev, A. V. Sukhobokov, and L. A. Alekseeva, USSR Pat. 1178746 (1985); *Chem. Abstr.*, **105**, 190912 (1986).
169. O. E. Nasakin, V. V. Alekseev, P. B. Terent'ev, A. Kh. Bulai, and M. Yu. Zablotskaya, *Khim. Geterotsikl. Soedin.*, **1983**, 1067; *Chem. Abstr.*, **100**, 103102 (1984).

3.8. References

170. K. Gewald, *Z. Chem.*, **1**, 349, (1961).
171. J. R. Traynor and D. G. Wibberley, *J. Chem. Soc. Perkin Trans. 1*, **1974**, 1786.
172. H. J. Roth and K. Eger, *Arch. Pharm.*, **308**, 179 (1975).
173. H. Pichler, G. Folkers, H. J. Roth, and K. Eger, *Liebigs Ann. Chem.*, **1986**, 1485.
174. W. Zimmermann and K. Eger, *Arch. Pharm.*, **312**, 552 (1979).
175. J. M. Sinambela, W. Zimmermann, H. J. Roth, and K. Eger, *J. Heterocycl. Chem.*, **23**, 393 (1986).
176. J. A. S. Laks, J. R. Ross, S. M. Bayomi, and J. W. Sowell, Sr., *Synthesis*, **1985**, 291.
177. R. J. Mattson, L. C. Wang, and J. W. Sowell, Sr., *J. Heterocycl. Chem.*, **17**, 1793 (1980).
178. S. Sunder, M. S. thesis, Auburn University, Auburn, Alabama, 1968, p. 52.
179. J. W. Sowell, Sr., Ph.D. dissertation, University of Georgia, Athens, Georgia, 1972, p. 54.
180. V. I. Shvedov, M. V. Mezentseva, and A. N. Grinev, *Khim. Geterotsikl. Soedin.*, **1975**, 1217; *Chem. Abstr.*, **84**, 59299 (1976).
181. R. W. Johnson, R. J. Mattson, and J. W. Sowell, Sr., *J. Heterocycl. Chem.*, **14**, 383 (1977).
182. H. Wamhoff and B. Wehling, *Synthesis*, **1976**, 51.
183. G. Cirrincione, A. M. Almerico, G. Dattolo, and E. Aiello, unpublished results 1989.
184. E. Toja, G. Tarzia, P. Ferrari, and G. Tuan, *J. Heterocycl. Chem.*, **23**, 1555 (1986).
185. E. Toja, A. De Paoli, G. Tuan, and J. Kettenring, *Synthesis*, **1987**, 272.
186. M. T. Cocco, C. Congiu, A. Maccioni, A. Plumitallo, M. L. Schivo, and G. Palmieri, *Farmaco, Ed. Sc.*, **43**, 103 (1988).
187. G. Sunagawa and N. Soma, Jpn. Pat. 8543 (1964); *Chem. Abstr.*, **61**, 11973 (1964).
188. S. Baroni and R. Stradi, *J. Heterocycl. Chem.*, **17**, 1221 (1980).
189. K. Gewald, M. Kleinert, B. Thiele, and M. Hentschel, *J. Prakt. Chem.*, **314**, 303 (1972).
190. R. Fuks and H. G. Viehe, *Tetrahedron*, **25**, 5721 (1969).
191. M. Jautelat and K. Ley, *Synthesis*, **1970**, 593.
192. M. Jautelat and K. Ley, Ger. Offen. 1,951,965 (1971); *Chem. Abstr.*, **75**, 35719 (1971).
193. T. Kusumoto, T. Hiyama, and K. Ogata, *Tetrahedron Lett.*, **27**, 4197 (1986).
194. N. Chatani and T. Hanafusa, *Tetrahedron Lett.*, **27**, 4201 (1986).
195. D. E. Weiss and N. H. Cromwell, *J. Heterocycl. Chem.*, **11**, 905 (1974).
196. A. V. Kadushkin, T. V. Stezhko, and V. G. Granik, *Khim. Geterotsikl. Soedin.*, **1986**, 564; *Chem. Abstr.*, **106**, 49937 (1987).
197. J. L. Longridge and T. W. Thompson, *J. Chem. Soc. (C)*, **1970**, 1658.
198. V. Sprio and E. Aiello, *Ann. Chim. (Rome)*, **56**, 858 (1966).
199. V. Sprio, S. Plescia, and O. Migliara, *Ann. Chim. (Rome)*, **61**, 271 (1971).
200. G. J. de Voghel, T. L. Eggerichs, B. Clamot, and H. G. Viehe, *Chimia*, **30**, 191 (1976).
201. D. J. Anderson and A. Hassner, *Synthesis*, **1975**, 483.
202. M. I. Lim, R. S. Klein, and J. J. Fox, *J. Org. Chem.*, **44**, 3826 (1979).
203. M. I. Lim, W. Y. Ren, B. A. Otter, and R. S. Klein, *J. Org. Chem.*, **48**, 780 (1983).
204. A. P. Knoll and J. Liebscher, *Khim. Geterotsikl. Soedin.*, **1985**, 628; *Chem. Abstr.*, **103**, 123423 (1985).
205. G. Tarzia and G. Panzone, Br. Pat. 1427945 (1974); *Chem. Abstr.*, **83**, 79069 (1975).
206. G. Tarzia and G. Panzone, *Ann. Chim. (Rome)*, **64**, 807 (1974).
207. G. Dattolo, G. Cirrincione, A. M. Almerico, and E. Aiello, *J. Heterocycl. Chem.*, **26**, 1747 (1989).
208. T. Murata, T. Sugawara, and K. Ukawa, *Chem. Pharm. Bull.*, **21**, 2571 (1973).
209. T. Murata and K. Ukawa, *Chem. Pharm. Bull.*, **22**, 240 (1974).

210. T. Murata, T. Sugawara, and K. Ukawa, *Chem. Pharm. Bull.*, **26**, 3080 (1978).
211. G. Cirrincione, G. Dattolo, A. M. Almerico, E. Aiello, R. A. Jones, H. M. Dawes, and M. B. Hursthouse, *J. Chem. Soc. Perkin Trans. 1*, **1987**, 1959.
212. MM. G. Muller, G. Amiard, and J. Mathieu, *Bull. Soc. Chim. Fr.*, **1949**, 533.
213. S. Gabriel, *Chem. Ber.*, **41**, 1127 (1908).
214. MM. G. Nomine, L. Penasse, and V. Delaroff, *Annales Pharm. Fr.*, **16**, 436 (1958).
215. J. W. Cornforth, *J. Chem. Soc.*, **1958**, 1174.
216. S. I. Zav'yalov, N. I. Aronova, and I. F. Mustafaeva, *Izv. Akad. Nauk SSSR, Ser. Chim.*, **1973**, 1906; *Chem. Abstr.*, **80**, 36940 (1974).
217. T. S. Mansour and G. Sauve, *Heterocycles*, **27**, 315 (1988).
218. K. Gewald and J. Oelsner, *J. Prakt. Chem.*, **321**, 71 (1971).
219. O. Attanasi, S. Santeusanio, F. Serra-Zanetti, E. Foresti, and A. McKillop, *J. Chem. Soc. Perkin Trans. 1*, **1990**, 1669.
220. G. L'Abbe, L. Huybrechts, S. Toppet, J. P. Declercq, G. Germain, and M. Van Meerssche, *Bull. Soc. Chim. Belg.*, **87**, 893 (1978).
221. A. Pinner, *Chem. Ber.*, **16**, 352 (1883).
222. A. Pinner, *Chem. Ber.*, **16**, 1655 (1883).
223. J. Schurz, A. Ullrich, and H. Bayzer, *Monatsh. Chem.*, **90**, 29 (1959).
224. R. Blochmann, *Chem. Ber.*, **20**, 1856 (1887).
225. W. J. Middleton, U.S. Pat. 2,961,447; (1960); *Chem. Abstr.*, **55**, 8874 (1961).
226. C. L. Dickinson, W. J. Middleton, and V. A. Engelhardt, *J. Org. Chem.*, **27**, 2470 (1962).
227. H. Fischer, H. Guggemos, and A. Schafer, *Liebigs Ann. Chem.*, **540**, 30 (1939).
228. H. Fischer and A. Waibel, *Liebigs Ann. Chem.*, **512**, 195 (1934).
229. H. Fischer, O. Sus, and F. G. Weilguny, *Liebigs Ann. Chem.*, **481**, 159 (1930).
230. W. Metzger and H. Fischer, *Liebigs Ann. Chem.*, **527**, 1 (1936).
231. F. Endermann and H. Fischer, *Liebigs Ann. Chem.*, **538**, 172 (1939).
232. A. Treibs and D. Grimm, *Liebigs Ann. Chem.*, **752**, 44 (1971).
233. A. P. Terent'ev and A. N. Makarova, *Vestnik Moskow. Univ.*, **4**, 101 (1947); *Chem. Abstr.*, **42**, 1590 (1948).
234. A. P. Terent'ev and A. N. Makarova, *Zhur. Obshchei Khim.*, **21**, 270 (1951); *Chem. Abstr.*, **45**, 7105 (1951).
235. H. Fischer and F. Rothweller, *Chem. Ber.*, **56B**, 512 (1923).
236. G. Cirrincione, A. M. Almerico, G. Dattolo, and E. Aiello, unpublished results, 1990.
237. M. A. T. Rogers, *J. Chem. Soc.*, **1943**, 590.
238. G. Ciamician and P. Silber, *Gazz. Chim. Ital.*, **15**, 315 (1885).
239. T. D. Duffy and D. G. Wibberley, *J. Chem. Soc. Perkin Trans. 1*, **1974**, 1921.
240. G. Doddi, P. Mencarelli, and F. Stegel, *J. Chem. Soc. Chem. Commun.*, **1975**, 273.
241. G. Doddi, G. Illuminati, P. Mencarelli, and F. Stegel, *J. Org. Chem.*, **41**, 2824 (1976).
242. G. Devincenzis, P. Mencarelli, and F. Stegel, *J. Org. Chem.*, **48**, 162 (1983).
243. P. Mencarelli and F. Stegel, *J. Chem. Soc. Chem. Commun.*, **1980**, 123.
244. J. M. Eby and J. A. Moore, *J. Org. Chem.*, **32**, 1346 (1967).
245. D. J. Chadwick and S. T. Hodgson, *J. Chem. Soc. Perkin Trans. 1*, **1983**, 93.
246. V. Sprio, P. Madonia, and R. Caronia, *Ann. Chim. (Rome)*, **49**, 169 (1959).
247. G. Dattolo, G. Cirrincione, A. M. Almerico, G. Presti, and E. Aiello, *Heterocycles*, **20**, 829 (1983).
248. G. Dattolo, G. Cirrincione, A. M. Almerico, G. Presti, and E. Aiello, *Heterocycles*, **22**, 2269 (1984).

3.8. References

249. A. M. Almerico, G. Cirrincione, E. Aiello, and G. Dattolo, *J. Heterocycl. Chem.*, **26**, 1631 (1989).
250. A. N. Grinev, M. V. Mezentseva, E. F. Kuleshova, and L. M. Alekseeva, *Khim. Geterotsikl. Soedin.*, **1986**, 612; *Chem. Abstr.*, **106**, 119593 (1987).
251. A. Kreutzberger and P. A. Kalter, *J. Org. Chem.*, **26**, 3790 (1961).
252. D. E. Ames, H. R. Ansari, and A. W. Ellis, *J. Chem. Soc. (C)*, **1969**, 1795.
253. F. Angelico, *Rend. Accad. Lincei*, **14**(I), 699 (1905).
254. T. Ajello and S. Gianferrara, *Gazz. Chim. Ital.*, **66**, 228 (1936).
255. F. Angelico, *Rend. Accad. Lincei*, **14**(II), 167 (1905).
256. V. Sprio and J. Fabra, *Ric. Sci.*, **34**, 581 (1964).
257. F. Angelico and C. Labisi, *Gazz. Chim. Ital.*, **40**, 417 (1910).
258. H. Fischer and A. Stern, *Liebigs Ann. Chem.*, **446**, 229 (1926).
259. S. Giambrone and J. Fabra, *Ann. Chim. (Rome)*, **50**, 237 (1960).
260. G. Dattolo, E. Aiello, S. Plescia, G. Cirrincione, and G. Daidone, *J. Heterocycl. Chem.*, **14**, 1021 (1977).
261. E. Aiello, G. Dattolo, and G. Cirrincione, *J. Chem. Soc. Perkin Trans. 1*, **1981**, 1.
262. G. Cirrincione, G. Dattolo, A. M. Almerico, E. Aiello, R. A. Jones, and W. Hinz, *Tetrahedron*, **43**, 5225 (1987).
263. G. Dattolo, G. Cirrincione, A. M. Almerico, and E. Aiello, *Heterocycles*, **20**, 255 (1983).
264. T. Ajello and S. Giambrone, *Ric. Sci.*, **24**, 49 (1954).
265. H. J. Anderson and S. J. Griffiths, *Can. J. Chem.*, **45**, 2227 (1967).
266. A. Z. Britten and G. W. G. Griffiths, *Chem. Ind. (London)*, **1973**, 278.
267. G. Buchi and J. A. Raleigh, *J. Org. Chem.*, **36**, 873 (1971).
268. P. Mencarelli and F. Stegel, *J. Chem. Res., (S)*, **1984**, 18.
269. J. Kao, A. L. Hinde, and L. Radom, *Nouv. J. Chim.*, **3**, 473 (1979).
270. N. Bodor, M. J. S. Dewar, and A. J. Harget, *J. Am. Chem. Soc.*, **92**, 2929 (1970).
271. M. J. S. Denar and T. Morita, *J. Am. Chem. Soc.*, **91**, 796 (1969).
272. V. G. Andrianov, M. A. Shokhen, A. V. Eremeev, and S. V. Barmina, *Khim. Geterosikl. Soedin.*, **1987**, 54; *Chem. Abstr.*, **108**, 36868 (1988).
273. G. Giuseppetti, C. Tadini, O. Attanasi, M. Grossi, and F. Serra-Zanetti, *Acta Cryst.*, **C41**, 450 (1985).
274. J. J. Stezowski, *Acta Cryst.*, **B33**, 2472 (1977).
275. G. Korschun and C. Roll, *Bull. Soc. Chim. Fr.*, **37**, 130 (1925).
276. P. B. Terent'ev, O. E. Nasakin, V. V. Alekseev, A. G. Kalandarishvili, and M. Sh. Kaviladze, *Khim. Geterotsikl. Soedin.*, **1983**, 1071; *Chem. Abstr.*, **99**, 194243 (1983).
277. G. Tuan, L. F. Zerilli, G. Panzone, and G. Tarzia, *Ann. Chim. (Rome)*, **67**, 593 (1977).
278. A. Gounder and C. Roll, *Bull. Chem. Soc. Fr.*, **39**, 1222 (1926).
279. C. T. Wie, S. Sunder, and C. DeWitt Blanton, Jr., *Tetrahedron Lett.*, **1968**, 4605.
280. T. Gunduz, E. Kilic, V. Ertuzun, and G. Cetinel, *Analyst.*, **111**, 1439 (1986).
281. R. W. Johnson, T. H. Keenan, J. W. Kosh, and J. W. Sowell, Sr., *J. Pharm. Sci.*, **68**, 317 (1979).
282. J. W. Sowell, Sr., A. J. Block, M. E. Derrick, J. J. Freeman, J. W. Kosh, R. J. Mattson, P. F. Mubarak, and P. A. Tenthorey, *J. Pharm. Sci.*, **70**, 135 (1981).
283. J. A. Elvidge, J. S. Fitt, and R. P. Linstead, *J. Chem. Soc.*, **1956**, 235.
284. J. Armand, S. Deswarte, J. Pinson, and H. Zamarlik, *Bull. Soc. Chim. Fr.*, **1971**, 671.
285. A. R. Katritzky and J. M. Lagowski, *Adv. Heterocycl. Chem.*, **2**, 20 (1963).
286. D. M. Lemal, T. W. Rave, and S. D. McGregor, *J. Am. Chem. Soc.*, **85**, 1944 (1963).
287. V. Sprio, *Ric. Sci. Rend. A*, **8**, 197 (1965); *Chem. Abstr.*, **63**, 13188 (1965).

288. S. Plescia, E. Aiello, and V. Sprio, *Atti Accad. Sci. Lett., Arti Palermo, I*, **33**, 305 (1974); *Chem. Abstr.*, **83**, 97181 (1975).
289. S. Plescia, E. Aiello, and V. Sprio, *J. Heterocycl. Chem.*, **12**, 375 (1975).
290. N. S. Girgis, A. Jorgensen, and E. B. Pedersen, *Synthesis*, **1985**, 101.
291. S. R. Etson, R. J. Mattson, and J. W. Sowell, Sr., *J. Heterocycl. Chem.*, **16**, 929 (1979).
292. M. Zupan, B. Stanovnik, and M. Tisler, *J. Heterocycl. Chem.*, **8**, 1 (1971).
293. W. Flitsch and U. Kramer, *Tetrahedron Lett.*, **1968**, 1479.
294. W. Flitsch and U. Kramer, *Liebigs Ann. Chim.*, **735**, 35 (1970).
295. G. M. Coppola, G. E. Hardtmann, and B. S. Huegi, *J. Heterocycl. Chem.*, **11**, 51 (1974).
296. A. R. Katritzky and J. W. Suwinski, *Tetrahedron Lett.*, **1974**, 4123.
297. H. J. Roth and K. Eger, *Arch. Pharm.*, **308**, 186 (1975).
298. J. W. Sowell, Sr., and C. DeWitt Blanton, Jr., *J. Heterocycl. Chem.*, **10**, 287 (1973).
299. R. A. Crochet, Jr., and C. DeWitt Blanton, Jr., *Synthesis*, **1974**, 55.
300. R. J. Mattson and J. W. Sowell, Sr., *Synthesis*, **1979**, 217.
301. R. W. Johnson, T. H. Keenan, J. W. Kosh, and J. W. Sowell, Sr., *J. Pharm. Sci.*, **68**, 955 (1979).
302. J. W. Sowell, Sr. and C. DeWitt Blanton, Jr., *J. Pharm. Sci.*, **65**, 908 (1976).
303. F. H. Briggs, W. T. Pelletier, and C. DeWitt Blanton, Jr., *J. Pharm. Sci.*, **67**, 735 (1978).
304. D. L. Powers, J. W. Sowell, Sr., J. J. Freeman, and J. W. Kosh, *J. Pharm. Sci.*, **69**, 473 (1980).
305. J. R. Ross, J. A. S. Laks, D. Li-Chang Wang, and J. W. Sowell, Sr., *Synthesis*, **1985**, 796.
306. E. C. Taylor and R. W. Hendess, *J. Am. Chem. Soc.*, **86**, 951 (1964).
307. E. C. Taylor and R. W. Hendess, *J. Am. Chem. Soc.*, **87**, 1995 (1965).
308. R. L. Tolman, R. K. Robins, and L. B. Townsend, *J. Am. Chem. Soc.*, **90**, 524 (1968).
309. H. J. Roth and K. Eger, *Arch. Pharm.*, **308**, 252 (1975).
310. V. I. Shvedov, I. A. Kharizomenova, L. B. Altukhova, and A. N. Grinvev, *Khim. Geterotsikl. Soedin.*, **6**, 428 (1970): *Chem. Abstr.*, **73**, 25403 (1970).
311. W. Zimmermann, K. Eger, and H. J. Roth, *Arch. Pharm.*, **309**, 597 (1976).
312. M. Kim-Su, K. Eger, and H. J. Roth, *Arch. Pharm.*, **309**, 721 (1976).
313. W. Offermann, K. Eger, and H. J. Roth, *Arch. Pharm.*, **314**, 168 (1981).
314. H. Fischer and K. Zeile, *Liebigs Ann. Chem.*, **483**, 251 (1930).
315. M. A. T. Rogers, *J. Chem. Soc.*, **1943**, 598.
316. G. Cirrincione, A. M. Almerico, G. Dattolo, E. Aiello, and J. K. Seydel, unpublished results, 1990.
317. G. Tarzia, G. Panzone, P. Carminati, P. Schiatti, and D. Selva, *Farmaco Ed. Sci.*, **31**, 81 (1976).
318. G. Tarzia and G. Panzone, Br. Pat. 1446153 (1975): *Chem. Abstr.*, **84**, 5003 (1976).
319. L. Fontanella, L. Mariani, and G. Tarzia, Br. Pat. 1433103 (1975); *Chem. Abstr.*, **84**, 5013 (1976).
320. L. Mariani and G. Tarzia, U. S. Pat. 4402970 (1983); *Chem. Abstr.*, **99**, 126168 (1983).
321. L. Mariani and G. Tarzia, Br. Pat. 22177 (1981); *Chem. Abstr.*, **98**, 107326 (1983).
322. L. Mariani and G. Tarzia, Eur. Pat. EP 102602 (1984); *Chem. Abstr.*, **101**, 90990 (1984).
323. T. Murata and K. Ukawa, *Chem. Pharm. Bull.*, **22**, 1212 (1974).
324. G. Tarzia and G. Panzone, Br. Pat. 1447426 (1975); *Chem. Abstr.*, **84**, 5004 (1976).
325. G. Tarzia and G. Panzone, Br. Pat. 1492663 (1976); *Chem. Abstr.*, **86**, 139842 (1977).
326. G. Tarzia and G. Panzone, Br. Pat. 1449364 (1975); *Chem. Abstr.*, **84**, 4933 (1976).
327. K. Sakai and J. P. Anselme, *Tetrahedron Lett.*, **1970**, 3851.
328. K. Sakai and J. P. Anselme, *Bull. Chem. Soc. Jpn.*, **45**, 307 (1972).
329. K. K. Mayer, F. Schroppel, and J. Sauer, *Tetrahedron Lett.*, **1972**, 2899.

3.8. References

330. M. Scotton and L. R. Lampariello, *Chim. Ind. (Milan)*, **62**, 843 (1980).
331. T. Ajello, *Gazz. Chim. Ital.*, **69**, 453 (1939).
332. T. Ajello, *Gazz. Chim. Ital.*, **68**, 602 (1938).
333. T. Ajello and G. Sigillo, *Gazz. Chim. Ital.*, **68**, 681 (1938).
334. L. Kalbe and J. Bayer, *Chem. Ber.*, **45**, 2150 (1912).
335. T. Ajello, *Gazz. Chim. Ital.*, **66**, 608 (1936).
336. W. Zimmermann, J. M. Sinambela, H. J. Roth, and K. Eger, *J. Heterocycl. Chem.*, **23**, 397 (1986).
337. E. Aiello, G. Dattolo, G. Cirrincione, and A. M. Almerico, *J. Heterocycl. Chem.*, **21**, 721 (1984).
338. K. Eger, J. G. Pfahl, G. Folkers, and H. J. Roth, *J. Heterocycl. Chem.*, **24**, 425 (1987).
339. M. Petrzilka and J. I. Grayson, *Synthesis*, **1981**, 753.
340. A. G. Schultz, M. Shen, and R. Ravichandran, *Tetrahedron Lett.*, **22**, 1767 (1981).
341. A. G. Shultz and R. Ravichandran, *Tetrahedron Lett.*, **22**, 1771 (1981).
342. M. Shen and A. G. Schultz, *Tetrahedron Lett.*, **22**, 3347 (1981).
343. J. M. Gazave, N. P. Buu-Hoi, N. D. Xuong, J. Mallet, J. Pillot, J. Savel, and G. Dufraisse, *Therapie*, **12**, 486 (1957); *Chem. Abstr.*, **53**, 8438 (1959).
344. T. H. Keenan, J. J. Freeman, J. W. Sowell, and J. W. Kosh, *Pharmacology*, **19**, 36 (1979).
345. M. Bialer, J. El-On, B. Yagen, and R. Mechoulam, *J. Pharm. Sci.*, **70**, 822 (1981).
346. C. Zimmer, K. E. Reinert, G. Luck, U. Wahnert, G. Lober, and H. Thrum, *J. Mol. Biol.*, **58**, 329 (1971).
347. K. E. Reinert, *J. Mol. Biol.*, **72**, 593 (1972).
348. A. K. Krey, R. G. Allison, and F. E. Hahn, *FEBS Lett.*, **29**, 58 (1973).
349. F. E. Hahn and A. K. Krey, *Proc. Int. Congr. Chem. (7th) Adv. Antimicrobial Antineoplastic Chem.*, **1**, (Part 2), 825 (1971).
350. G. Luck and C. Zimmer, *Studia Biophysica Berlin*, **40**, 9 (1973).
351. C. Zimmer and G. Luck, *Biochim. Biophys. Acta*, **287**, 376 (1972).
352. M. Lee, R. G. Shea, J. A. Hartley, K. Kissinger, R. T. Pon, G. Vesnaver, K. J. Breslaver, J. C. Dabrowiak, and J. W. Lown, *J. Am. Chem. Soc.*, **111**, 345 (1989).
353. G. Stefancich, M. Artico, F. Corelli, S. Massa, G. C. Pantaleoni, G. Palumbo, D. Fanini, and R. Giorgi, *Farmaco Ed. Sci.*, **40**, 930 (1985).
354. K. Anzai, G. Nakamura, and S. Suzuki, *J. Antibiotics*, **A10**, 201 (1957).
355. G. Acs, E. Reich, and M. Mori, *Proc. Natl. Acad. Sci. (USA)*, **52**, 493 (1964).
356. M. V. Mezentseva, I. S. Nikolaeva, A. N. Fomina, and M. I. Akimova, *Khim-Farm. Zhur.*, **21**, 1206 (1987); *Chem. Abstr.*, **109**, 6360 (1988).
357. O. Attanasi, S. Santeusanio, G. Barbarella, and V. Tugnoli, *Org. Mag. Res. Chem.*, **23**, 383 (1985).
358. O. Attanasi, M. Grossi, F. Perrulli, S. Santeusanio, F. Serra-Zanetti, A. Bongini, and V. Tugnoli, *Mag. Res. Chem.*, **26**, 714 (1988).
359. C. Bailly, N. Pommery, R. Houssin, and J. Henichart, *J. Pharm. Sci.*, **78**, 910 (1989).
360. M Otsuka, T. Masuda, A. Haupt, M. Ohno, T. Shiraki, Y. Sugiura, and K. Maeda, *J. Am. Chem. Soc.*, **112**, 838 (1990).
361. K. M. H. Hilmy and E. B. Pedersen, *Liebigs Ann. Chem.*, **1989**, 1145.
362. H. Kubota, T. Moriya, and K. Matsumoto, *Chem. Pharm. Bull.*, **38**, 570 (1990).
363. S. M. Bennett, N. Nguyen-Ba, and K. K. Ogilvie, *J. Med. Chem.*, **33**, 2162 (1990).
364. S. M. Bayomi, H. A. Al-Khamees, A. K. M. Ismail, H. M. Eissa, and M. El-Kerdawy, *J. Chin. Chem. Soc.*, **36**, 159 (1989); *Chem. Abstr.*, **111**, 194499 (1990).

CHAPTER 4

3-Hydroxypyrroles

Hamish McNab and Lilian C. Monahan

*Department of Chemistry,
University of Edinburgh,
West Mains Road, Edinburgh EH9 3JJ, U.K.*

4.1. Introduction	527
4.2. Synthesis of 3-Hydroxypyrroles and 1H-Pyrrol-3(2H)-ones	527
4.2.1. 3-Hydroxypyrrole [1H-Pyrrol-3(2H)-one] and Its Simple (Alkyl or Aryl) Derivatives	528
4.2.2. 3-Hydroxypyrroles [1H-Pyrrol-3(2H)-ones] Containing Electron-Withdrawing Groups	532
4.2.3. 3-Hydroxypyrroles [1H-Pyrrol-3(2H)-ones] Containing Electron-Donating Groups	536
4.2.4. The Cyclopropenone Route to Polysubstituted 1H-Pyrrol-3(2H)-ones	540
4.3. Physical Properties of 3-Hydroxypyrroles and 1H-Pyrrol-3(2H)-ones	543
4.3.1. Tautomerism of 3-Hydroxypyrroles	543
4.3.1.1. Keto-enol Tautomerism	543
4.3.1.2. Ring–Chain Tautomerism of 2-Hydroxy-1H-pyrrol-3(2H)-ones	545
4.3.2. Structure of 3-Hydroxypyrroles and 1H-Pyrrol-3(2H)-ones	546
4.3.3. IR Spectra of 3-Hydroxypyrroles and 1H-Pyrrol-3(2H)-ones	547
4.3.4. UV Spectra of 3-Hydroxypyrroles and 1H-Pyrrol-3(2H)-ones	548
4.3.4.1. 3-Hydroxypyrroles	548
4.3.4.2. 1H-Pyrrol-3(2H)-ones	548
4.3.5. NMR Spectra of 3-Hydroxypyrroles and 1H-Pyrrol-3(2H)-ones	550
4.3.5.1. 3-Hydroxypyrroles	550
4.3.5.2. 1H-Pyrrol-3(2H)-ones	552
4.3.6. Mass Spectra of 3-Hydroxypyrroles and 1H-Pyrrol-3(2H)-ones	556
4.3.7. ESR Spectra of 3-Hydroxypyrrole Radicals	556
4.4. Chemical Properties of 3-Hydroxypyrroles and 1H-Pyrrol-3(2H)-ones	557
4.4.1. Alkylation of 3-Hydroxypyrroles and 1H-Pyrrol-3(2H)-ones	557
4.4.2. Acylation of 3-Hydroxypyrroles and 1H-Pyrrol-3(2H)-ones	560
4.4.3. Reactions of 3-Hydroxypyrroles and 1H-Pyrrol-3(2H)-ones with Electrophiles, and Properties of the Products	562
4.4.3.1. General Considerations	562
4.4.3.2. Reactions with Acids	563

Pyrroles Part Two: The Synthesis, Reactivity, and Physical Properties of Substituted Pyrroles, Edited by R. Alan Jones.
ISBN 0-471-51306-7 © 1992 John Wiley & Sons, Inc.

	4.4.3.3. Halogenation, and Reactions of the Products	566
	4.4.3.4. Nitrosation, and Reactions of the Products.	567
	4.4.3.5. Diazo-Coupling Reactions and Reactions of the Products	569
	4.4.3.6. Friedel–Crafts Acylation and Vilsmeier and Related Reactions	571
	4.4.3.7. Reactions with Carbonyl Compounds.	572
	4.4.3.8. Mannich Reaction	575
	4.4.3.9. Reactions with Dimethyl Acetylenedicarboxylate	575
	4.4.3.10. Michael and Conjugate Addition Reactions.	577
4.4.4.	Reactions with Bases.	578
4.4.5.	Oxidation of 3-Hydroxypyrroles and $1H$-Pyrrol-3($2H$)-ones.	578
	4.4.5.1. Aerial Oxidation	579
	4.4.5.2. "One-Electron" Oxidizing Agents	581
	4.4.5.3. Ozonolysis.	582
	4.4.5.4. Miscellaneous Oxidation.	583
4.4.6.	Reduction of 3-Hydroxypyrroles and $1H$-Pyrrol-3($2H$)-ones.	583
4.4.7.	Hydrolysis and Decarboxylation Reactions of 3-Hydroxypyrrole Carboxylic Esters.	584
4.4.8.	Photolysis and Photochemical Reactions of 3-Hydroxypyrroles and $1H$-Pyrrol-3($2H$)-ones.	585
4.5. 3-Alkoxypyrroles.	588
4.5.1.	Preparation of 3-Alkoxypyrroles.	588
4.5.2.	Physical Properties of 3-Alkoxypyrroles	591
	4.5.2.1. Structure of 3-Alkoxypyrroles.	591
	4.5.2.2. IR Spectra of 3-Alkoxypyrroles.	592
	4.5.2.3. UV Spectra of 3-Alkoxypyrroles.	592
	4.5.2.4. NMR Spectra of 3-Alkoxypyrroles.	593
	4.5.2.5. Mass Spectra of 3-Alkoxypyrroles	593
4.5.3.	Chemical Properties of 3-Alkoxypyrroles.	595
	4.5.3.1. Alkylation Reactions.	595
	4.5.3.2. O-Dealkylation	595
	4.5.3.3. Reactions with Electrophiles	595
	4.5.3.3.1. General Considerations	595
	4.5.3.3.2. Reactions with Acids	595
	4.5.3.3.3. Halogenation, and Reactions of the Products	596
	4.5.3.3.4. Diazo-Coupling Reactions.	597
	4.5.3.3.5. Vilsmeier and Related Reactions	597
	4.5.3.3.6. Reactions with Carbonyl Compounds	598
	4.5.3.3.7. Reactions with Dimethyl Acetylenedicarboxylate and Other Potential Dienophiles.	598
	4.5.3.3.8. Sulfenylation.	599
	4.5.3.4. Coupling Reactions of 2-Formyl-3-alkoxypyrroles with Pyrrole Derivatives.	599
	4.5.3.5. Oxidation Reactions	600
	4.5.3.6. Reduction Reactions	601
	4.5.3.7. Hydrolysis and Decarboxylation Reactions of 3-Alkoxypyrrole Carboxylic Esters.	602
	4.5.3.8. Claisen Reaction	602
4.6. 3-Acyloxypyrroles.	603
4.6.1.	Preparation of 3-Acyloxypyrroles	603
4.6.2.	Physical Properties of 3-Acyloxypyrroles.	604
4.6.3.	Chemical Properties of 3-Acyloxypyrroles.	605
	4.6.3.1. Deacylation Reactions.	605
	4.6.3.2. Reactions with Electrophiles	605
	4.6.3.3. Oxidation and Reduction	606

4.7. 3-Alkoxypyrrole Natural Products	606
4.8. Acknowledgment	610
4.9. References	610

4.1. INTRODUCTION

In this chapter, we cover the synthesis and physical and chemical properties of 3-hydroxypyrroles **1a** from the report of the first example in 1913[1] until mid-1990. This particular topic has not been previously reviewed, although there is much useful information in the appropriate chapter of Jones and Bean's monograph.[2]

3-Hydroxypyrroles **1a** are highly electron-rich compounds, and so the system often compensates by tautomerism to the 1H-pyrrol-3(2H)-one form **1b** (also called "2-pyrrolin-4-ones" or "1,2-dihydro-3H-pyrrol-3-ones," etc). Throughout this chapter, the term "pyrrolone" refers exclusively to these isomers. Many of the 3-hydroxypyrroles that have been made are substituted with electron-withdrawing groups (often carboxylic esters) in order to increase their stability. Both tautomeric forms are considered in the following sections, including those compounds that are "locked" by 2,2-disubstitution in the pyrrolone form. 3-Alkoxy- and 3-acyloxy derivatives appear in separate sections, followed by a brief discussion of 3-alkoxypyrrole natural products. We specifically *exclude* a treatment of 2,3- and 2,4-dione derivatives, including tetramic acids and of reduced (pyrrolidin-3-one) and oxidized (azacyclopentadienone) systems. Generally, we consider 3-hydroxypyrrole and its simple (alkyl or aryl) derivatives before the more highly functionalized examples; within these sections, we deal first with N-unsubstituted systems, followed by N-substituted, N- and C-substituted, and finally N-2,2-substituted derivatives. In order to avoid confusion, the above nomenclature and the numbering scheme shown in **1** is maintained throughout, even when the ring may be substituted by a functional group that should take precedence.

4.2. SYNTHESIS OF 3-HYDROXYPYRROLES AND 1H-PYRROL-3(2H)-ONES

Although general routes to the pyrrole ring system were discussed in Part 1,[4] many 3-hydroxypyrroles have been prepared by rather specific methods that can be placed in context here. A large number of 3-hydroxypyrroles have been obtained by treatment of reducing sugars with amino compounds[5] (Maillard reaction), but yields are low, and the method seldom has any synthetic importance.

4.2.1. 3-Hydroxypyrrole [1H-Pyrrol-3-(2H)-one] and Its Simple (Alkyl or Aryl) Derivatives

3-Hydroxypyrrole **2** (R^1–R^5 = H) is highly unstable, and has been made only once;[6,7] it was released under mild conditions by methanolysis of the trimethylsilyl derivative **3**, which was itself obtained by in situ silylation of **2** (R^1–R^5 = H) generated by hydrogenolysis-decarboxylation of the corresponding benzyl 2-carboxylate.

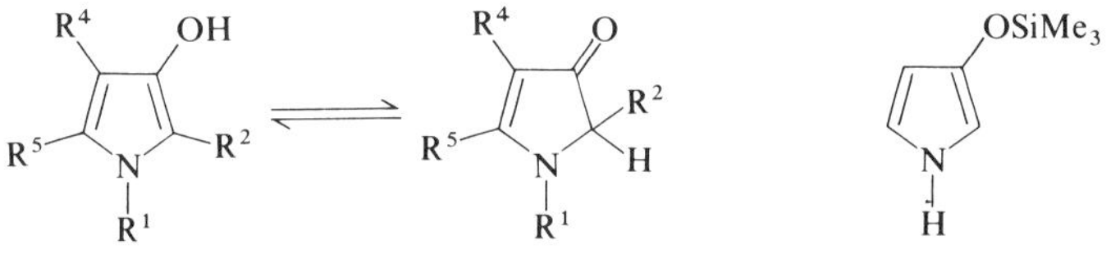

Degradation of such 2-carboxylic esters is perhaps the most general route to other simple 3-hydroxypyrroles (see Section 4.4.7), although yields and conditions are variable. Thus, Bauer's extensive work on 4,5-dialkyl derivatives, e.g., **2** (R^1 = R^2 = H, R^4 = R^5 = Me) relies on such a method[8,9] and excellent yields have been obtained,[9] whereas the corresponding synthesis of the 5-methyl derivative **2** (R^1–R^4 = H, R^5 = Me) is apparently less satisfactory.[10] An alternative route to this compound, via an intramolecular Wittig reaction followed by N-deacylation (Scheme 1), has been reported.[11,12]

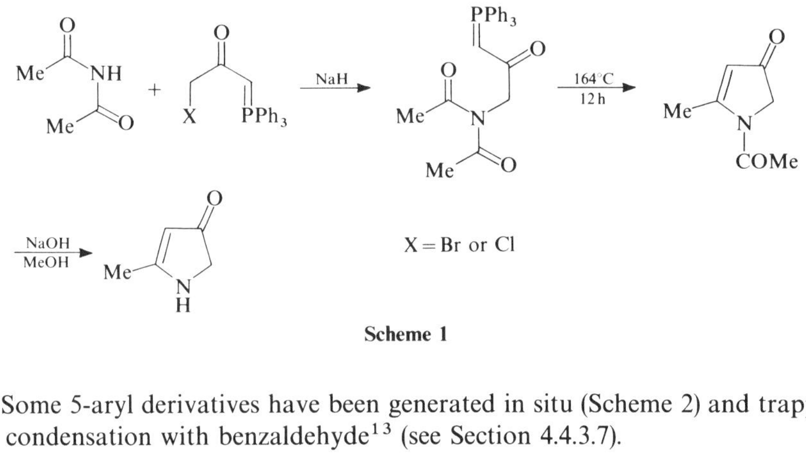

Scheme 1

Some 5-aryl derivatives have been generated in situ (Scheme 2) and trapped by condensation with benzaldehyde[13] (see Section 4.4.3.7).

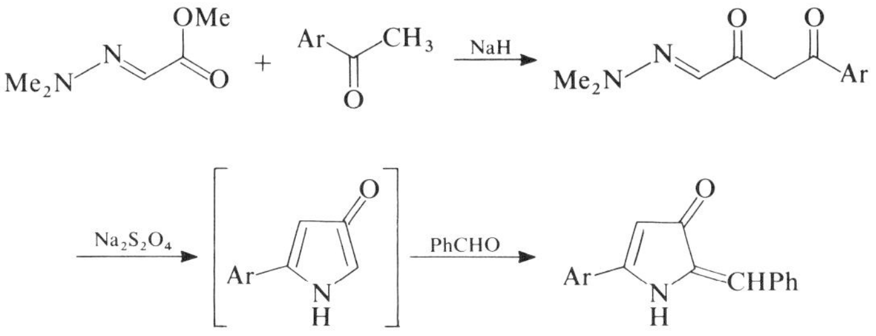

4.2. Synthesis of 3-Hydroxypyrroles and 1H-Pyrrol-3(2H)-ones

Scheme 2

A number of isolated routes to N-unsubstituted 2,2-dialkyl (or diaryl)-1H-pyrrol-3(2H)-ones (**4**) have been reported. The simplest such derivative **4** ($R^2 = R^{2'} = Me$, $R^4 = R^5 = H$) has been obtained by two routes,[14] either a (slow) photolytic oxidative dealkylation of the corresponding N-isopropyl compound, or an alternative 7-step route from methylalanine. The latter route is preferred owing to the slow rate of photoconversion.[14] Cyclization of the acetylenic ketone **5** with aqueous dimethylamine is reported[15] to give the 2,2,5-trimethyl-pyrrolone **4** ($R^2 = R^{2'} = R^5 = Me$, $R^4 = H$). 2,2,5-Triaryl derivatives can be obtained by butyl lithium–induced ring contraction of the oxazines **6** (3 examples; 40–73%),[16] or by cycloaddition of α-azidostyrenes to diphenyl-ketene[17] (2 examples; 45–51% yield). One 2-pyrrolyl-2,5-diphenylpyrrolone has been obtained in low yield by an oxidative method.[18]

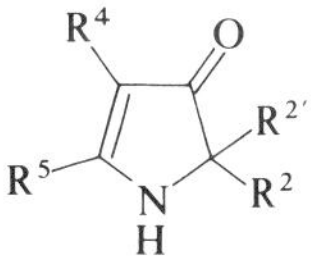

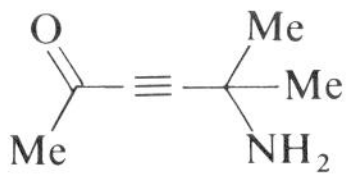

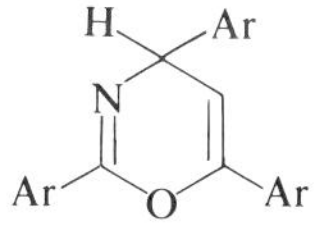

The most general route to N-substituted 3-hydroxypyrroles [1H-pyrrol-3(2H)-ones] involves flash-vacuum pyrolysis (FVP) of dialkylaminomethylene Meldrum's acid derivatives **7** or **8**[19–29] (Scheme 3), which are readily obtained from methoxymethylene Meldrum's acid or bis(methylthio)methylene Meldrum's acid and lead to 5-unsubstituted or 5-substituted hydroxypyrroles, respectively.

This method is readily applicable to N-alkyl or N-aryl-C-unsubstituted,[20–23] 2-substituted,[23] 5-substituted,[26,28,29] 2,2-disubstituted,[19,21–25] or 2,5-disubstituted[26,29] derivatives. Yields are generally in the range 60–80%. The mechanism of the pyrolysis has been studied[21,22,24,27] and involves the generation of a methyleneketene intermediate, followed by a hydrogen shift and cyclization. When $R^1 = H$, an alternative process takes place, which does not lead to cyclic products.[19,30,31] When R^1 is an alkyl group containing a CH in a position α to the nitrogen atom, then little selectivity is observed between ring

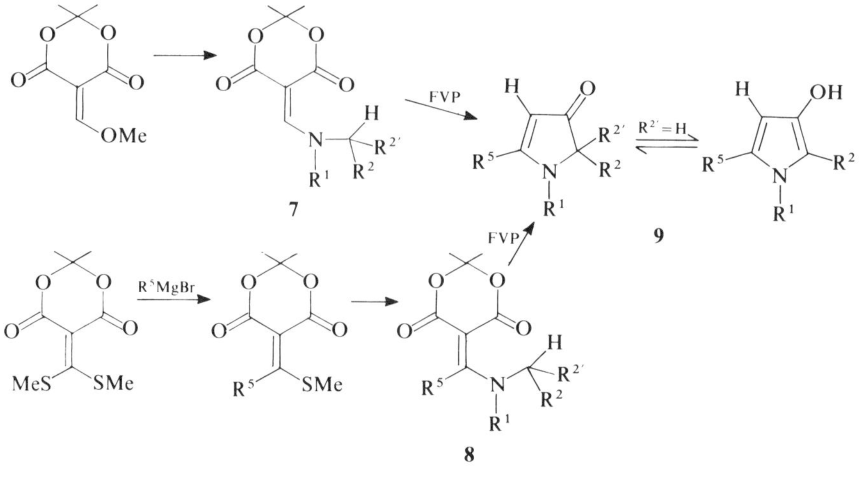

Scheme 3

formation at the two possible sites, although, when R^2 = benzyl, cyclization occurs regiospecifically to give a good route to *N*-substituted 3-hydroxy-2-phenylpyrroles.[23]

The FVP method is particularly convenient for the preparation of 1,2-disubstituted derivatives, which are highly sensitive toward oxygen (Section 4.4.5), although some such derivatives have been prepared by traditional "wet" chemistry. Thus, Davoll synthesized the 2,5-dimethyl-1-phenyl derivative **9** (R^1 = Ph, R^2 = R^5 = Me, $R^{2'}$ = H), in 66% yield by condensation of aniline with 3-hydroxyhexane-2,5-dione, or by a decarboxylation method.[32] A similar condensation has been used to make 3-hydroxy-2-methyl-5-phenylpyrrole.[33] Some highly hindered 1,2,5-trialkyl derivatives have been obtained (~ 50%) by treatment of α-lactams or α-bromoamides with alkynyl lithium reagents,[34] e.g., **9** (R^1 = R^2 = *t*-Bu, $R^{2'}$ = H; R^5 = Me). The highly substituted pyrrolones **10** (R^1 = Me or Ph) have been made by a onepot displacement–condensation process (Scheme 4) (2 examples; 60–70% yield).[35,36]

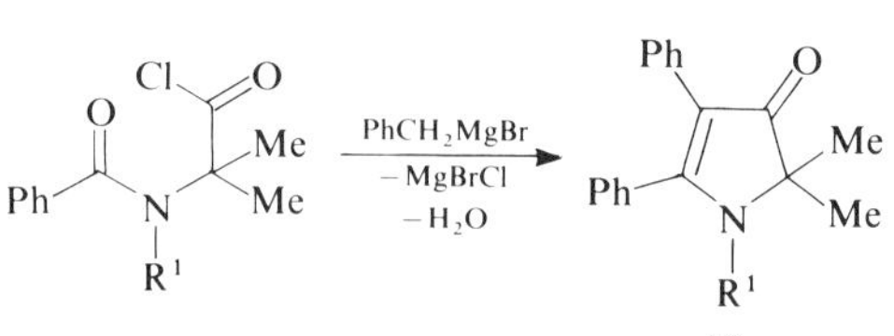

Scheme 4

1-Aryl-3-hydroxy-2,4-diphenylpyrroles are thought to be formed by the action of aniline on the β-isomer of "diphenacyl chloride";[37] this precursor was later shown to be the *E*-epoxide **11** (Scheme 5),[38,39] and various intermediates

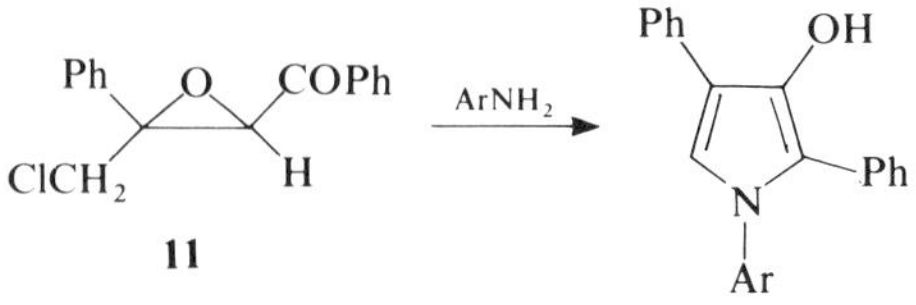

4.2. Synthesis of 3-Hydroxypyrroles and 1H-Pyrrol-3(2H)-ones

Scheme 5

isolated from both the α-[39] and β-[38] series also yield the hydroxypyrrole on subsequent treatment.

Thermolysis of alkynylenaminones **12** in refluxing xylene[40] gives a direct route to compounds that were later[41] shown to be 1-aryl-2-benzylidene-5-phenylpyrrol-3-ones (**13**) (Scheme 6).

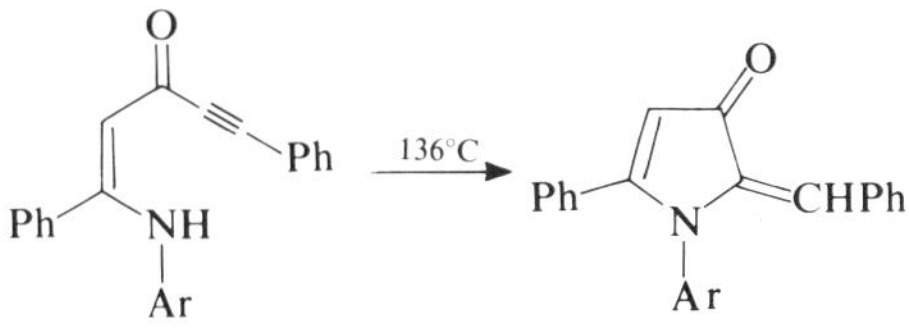

Scheme 6

One example of a 2-methylenepyrrolone has been obtained in $\sim$ 40% yield by a condensation method (Scheme 7).[42]

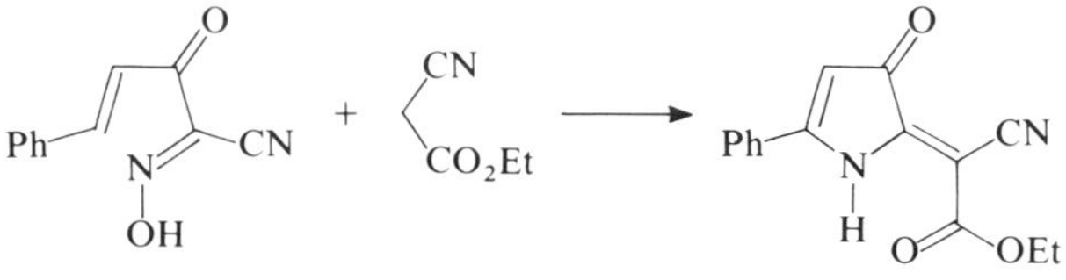

Scheme 7

A series of unusual 1,1-disubstituted pyrrolium salts **14** has been obtained by cyclisation of the enolates **15** with toluene-p-sulfonyl chloride (Scheme 8).[43] Examples include the 1,1,2-trimethyl **14** ($R^1 = R^2 = Me$, $R^2 = Me$, $R^4 = H$) (73%) and the 1,1-dimethyl derivatives **14** ($R^1 = Me$, $R^2 = R^4 = H$) (41%).

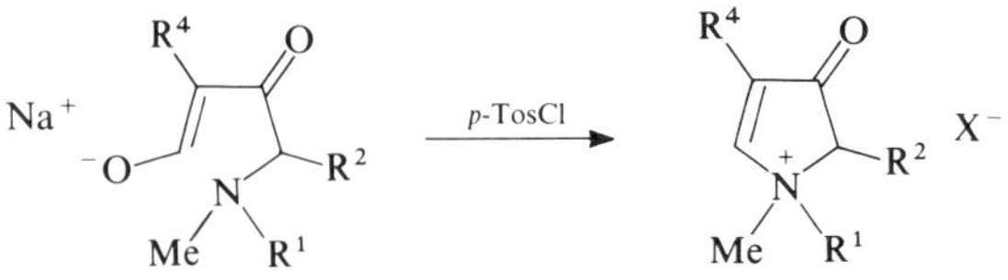

Scheme 8

4.2.2. 3-Hydroxypyrroles [1*H*-Pyrrol-3(2*H*)-ones] Containing Electron-Withdrawing Groups

A great deal is known about the synthesis of 3-hydroxypyrrole carboxylic esters, much of which has been covered in Part 1.[4] Only a brief summary is given here, with emphasis on the scope and limitations of the routes to 2-, 4-, and 5-substituted esters, and to hydroxypyrroles containing two electron-withdrawing groups. Routes to 3-hydroxypyrroles containing other electron-withdrawing groups are also included.

The most direct synthesis of 3-hydroxypyrrole-2-carboxylic esters involves the Dieckmann cyclization of enaminoesters **16** (Scheme 9), derived from amino esters and β-ketoesters under basic conditions. The method is particularly useful for the preparation of 4-substituted[44] or 4,5-disubstituted products[8,9,44–46] e.g., **17** (R^1 = H, R^2 = R^4 = Et, R^5 = Me; 50%[45] or R^1 = H, R^2 = Et, R^4 = R^5 = Me; 72%,[44] 60%[8]), since, when R^4 = H, alternative cyclization of the enamine to the CO_2R^2 ester group can also take place to give mixtures of products.[45]

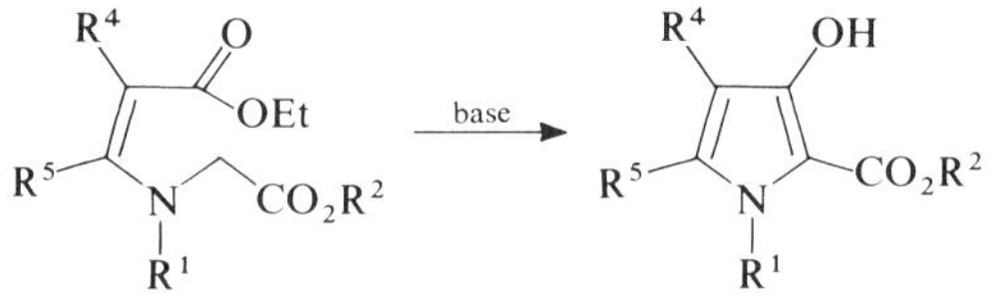

Scheme 9

Few *N*-substituted hydroxypyrroles have been made by this route, although this is not in principle a serious limitation.[46] In contrast to the above esters, 4-unsubstituted 5-arylhydroxypyrrole-2-carboxamides **18** (R^1, R^2 = H or Me) are readily obtained regiospecifically by the Dieckmann approach.[47] 1,5-Diaryl-3-hydroxypyrrole-2-carboxylic esters **19** (R^1 = aryl, R^2 = Me) are available in excellent yield by a completely different process, which involves silica gel–induced ring contraction of oxazines, obtained by cycloaddition of nitrosobenzenes with 2-acetoxy-4-arylbutadiene-1-carboxylates.[48] Some 3-hydroxypyrrole-2-carboxylic esters with unusual 5-substituents have been isolated in low yields by solvolysis of thienamycin[49,50] and related β-lactams,[51,52] and some 4,5-unsubstituted 2-carboxylic esters and 2-acyl derivatives have been obtained from azetidinones derived from clavulanic acid.[53]

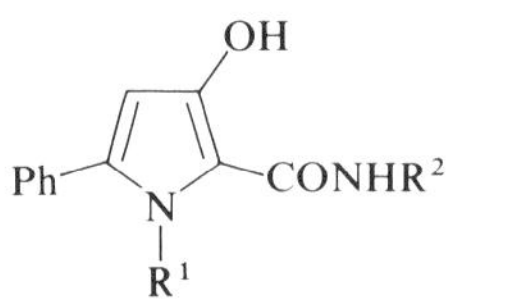

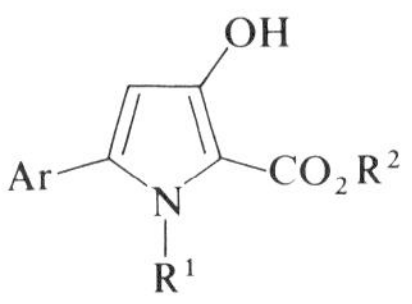

4.2. Synthesis of 3-Hydroxypyrroles and 1H-Pyrrol-3(2H)-ones

1,4,5-Unsubstituted 3-hydroxypyrrole-2-carboxylic esters are probably best made by a decarboxylation procedure (Section 4.4.7) (see Ref. 54), although N-substituted derivatives can be obtained directly by the Meldrum's acid method[26] (Scheme 3). Alternatively, condensation of a vinyl dioxo ester[55-57] with a primary amine, followed by dehydration, has given a wide variety of such derivatives **20** in 50–80% yield (Scheme 10).[58] Wasserman has recently adopted this approach extensively in the synthesis of alkaloids,[59] related model compounds,[58,60] and other natural products.[61]

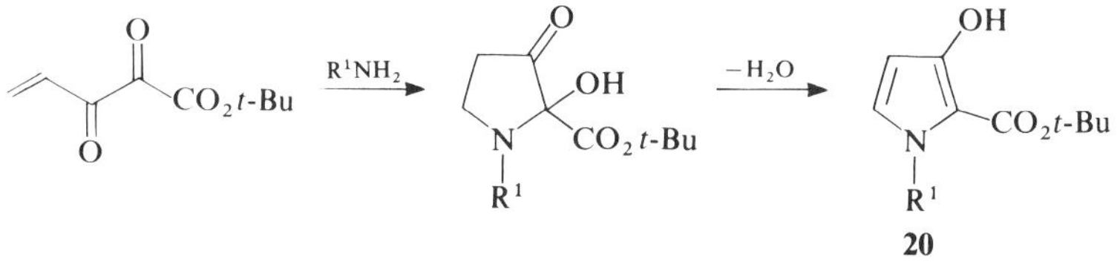

Scheme 10

Two general routes to 3-hydroxypyrrole-4-carboxylic esters (Scheme 11) have been employed to give a wide range of substituted derivatives. Benary's original[1] base-catalyzed cyclizations of α-chloroacetyl-β-aminocrotonic esters **21** (or the related amides[62]) (route a) give 1-substituted,[62,63] 5-substituted,[1] or 1,5-disubstituted[63] esters **22** (or the related amides[62]) in 50–80% yield. The starting materials can be readily made from ethyl 4-chloroacetoacetate.[52] Alternatively, the Dieckmann cyclization route (route b) using **16** ($R^4 = H$) proceeds regiospecifically under well-defined conditions to give 5-substituted or 2,5-disubstituted products **22** (Scheme 11) in 40–80% yield[45] (c.f. Scheme 9).

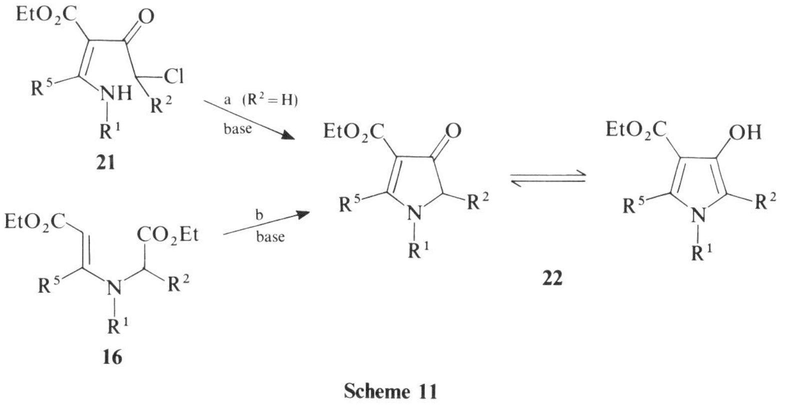

Scheme 11

A variation of this method employs the enaminoester derived from the free amino acid (c.f. **16**) under trifluoroacetic anhydride-induced cyclization conditions.[64] The yields of 5-methyl-2-phenyl derivatives are moderate and, in some cases, O-trifluoroacetyl derivatives are obtained.[64] A related cyclization,[65] in which a 2-carboxylic acid is reported to be formed, is overdue for reinvestigation, since there is some doubt[66] about the constitution of products derived therefrom.[32] In a more recent procedure, such 3-hydroxy-5-methyl-2-phenylpyrrole-4-carboxylic esters are readily available by reaction of phenylglyoxal with β-aminocrotonic esters[67] (Scheme 12) under neutral conditions. Yields are 50–60% for this one-pot method.

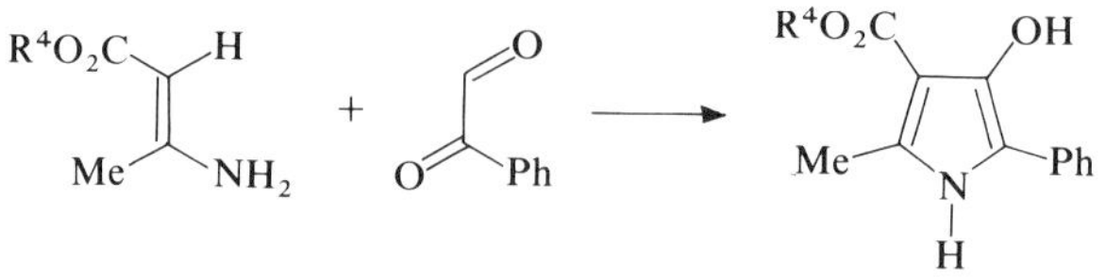

Scheme 12

In a second variation of method b (Scheme 11), in situ generation of **16** (R^1 = Me, R^5 = H) from the triester **23** in the presence of base gives **22** (R^1 = Me, R^5 = H) in up to 45% yield.[68]

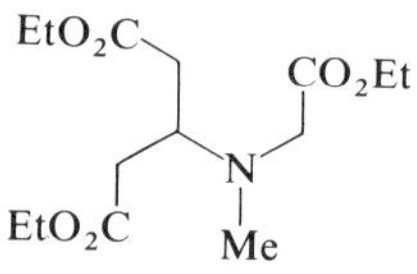

2-Aryl-2-substituted derivatives are available from the corresponding 2-substituted-2-hydroxy compound (Section 4.4.3.2). The 4-carboxylic acids can be obtained from the esters by hydrolysis (Section 4.4.7).

The procedures, shown in Scheme 11[69] (route a) and Scheme 12,[67] have been applied to give 4-cyano-3-hydroxypyrroles; 1,5-disubstituted[69] and 2,5-disubstituted[67] derivatives are available when the appropriate nitrile is used. A similar route, dating from 1914, may also give 4-cyano derivatives, but the properties of the pyrrolones so obtained (formation of oxime, hydrazone, etc) require reinvestigation.[70] Other 4-cyano derivatives can be obtained from the allenes **24** (R^1, R^2 = alkyl) and N-methylglycine esters.[71] Yields are good, but the procedure is limited by the availability of the starting materials, and the inherent formation of α-branched side-chains at the 5-position (Scheme 13).[71]

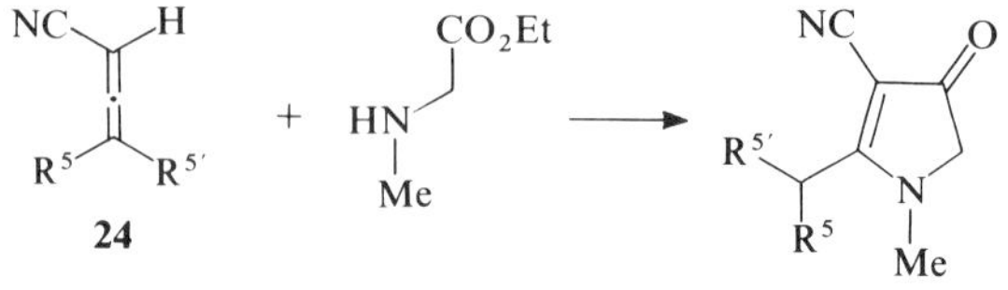

Scheme 13

4.2. Synthesis of 3-Hydroxypyrroles and 1H-Pyrrol-3(2H)-ones

In contrast to the well-developed synthetic routes to 3-hydroxypyrrole-2- and -4-carboxylic acid derivatives, very little is known of the corresponding 5-substituted products. 3-Hydroxypyrrole-5-carboxylic acid was first prepared in 1956 by a multistep procedure,[72] but it was not fully characterized. A highly substituted 5-carboxylic ester was obtained (as a foam) by degradation of a β-lactam derivative.[73] A potentially more useful route (Scheme 14) involves the Wittig reaction and cyclization of ethyl N-alkoxycarbonyl oxamates 25, but only one example has been published, and no experimental details are currently available.[74]

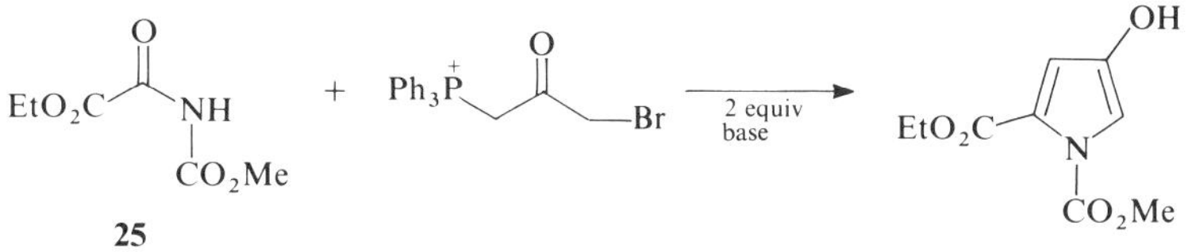

Scheme 14

3-Hydroxypyrrole-2,4-dicarboxylic esters are readily prepared by an extension of Scheme 9 in which the Dieckmann cyclization is performed on an aminomethylenemalonate derivative[75–78] 16 ($R^4 = CO_2Et$). Different ester groups (R^2, R^4) retain their integrity through this process and N-alkyl and N-aryl pyrrolones are obtained in 60–90% yield.[76] The N-unsubstituted compound can be obtained either directly[76] (24%), or by using the N-ethoxycarbonyl derivative of the precursor[76,77] 16 ($R^1 = R^4 = CO_2Et$; $R^5 = H$) (55%[76]): the protecting group is lost in situ. The use of an aminomethyleneacetoacetate 15 ($R^4 = COMe$) gives the expected 4-acetyl-3-hydroxypyrrole 2-carboxylic ester.[78]

1-Alkyl-5-aryl-3-hydroxypyrrole-2,4-dicarboxylic esters 26 have been obtained by a two-step route via the pyrrolidinones 27 (Scheme 15).[66,79,80]

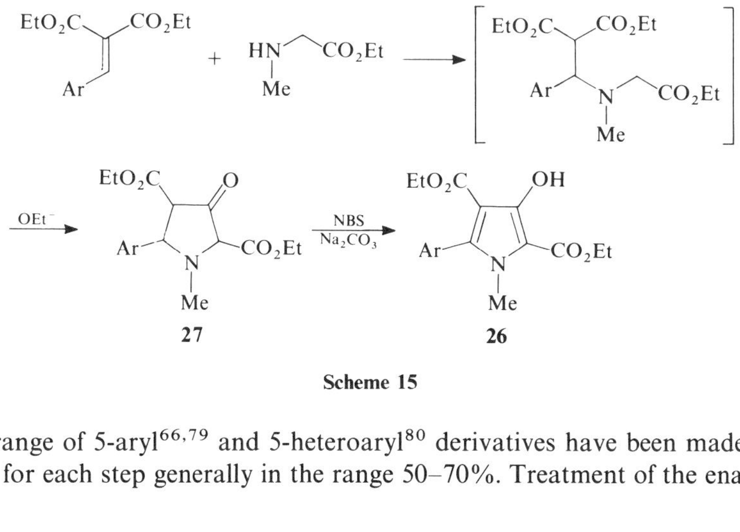

Scheme 15

A range of 5-aryl[66,79] and 5-heteroaryl[80] derivatives have been made, with yields for each step generally in the range 50–70%. Treatment of the enaminoester **28** with dimsyl sodium gives the 2-cyano compound **29**, whereas, in the presence of Michael acceptors, the 2,2-disubstituted compounds **30** are obtained in 50–70% yield (Scheme 16).[81,82]

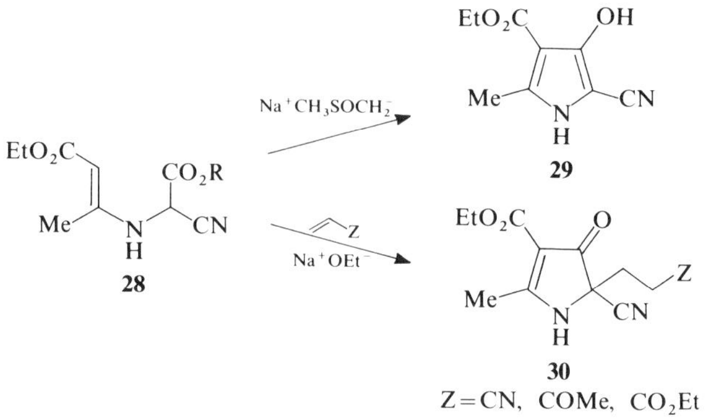

Scheme 16

The standard Dieckmann route (Scheme 9) is also applicable to 3-hydroxypyrrole-2,5-dicarboxylic esters, although yields are generally low. The product **17** (R^1 = H, R^2 = Et, R^4 = Me, R^5 = CO_2Et) is available by cyclization of the condensation product of ethyl glycinate and ethyl 3-methyl-2-oxosuccinate,[44] and the *N*-ethyl compound **17** (R^1 = Et, R^2 = Me, R^4 = H, R^5 = CO_2Me) is obtained in 20% yield by cyclization of the Michael addition product of methyl ethylaminoacetate with dimethyl acetylenedicarboxylate.[83] Similarly, *N*-phenacylaniline derivatives give the 2-benzoyl compounds **17** (R^1 = Ar, R^2 = PhCO, R^4 = H, R^5 = CO_2Me) in 30–60% yields.[84]

4.2.3. 3-Hydroxypyrroles [1*H*-Pyrrol-3(2*H*)-ones] Containing Electron-Donating Groups

A series of 1-amino derivatives **31** has been obtained by condensation of substituted acetaldehydes with glyoxal dimethylhydrazone, followed by hydrolysis of the resulting 3-amino compound[85] (Scheme 17).

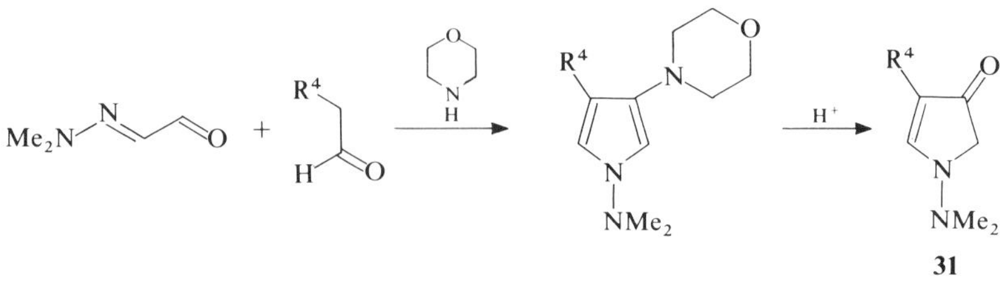

Scheme 17

4.2. Synthesis of 3-Hydroxypyrroles and 1H-Pyrrol-3(2H)-ones

The route is applicable to 4-alkyl or 4-aryl derivatives, and the yield for the two steps is ~ 30%.[85] A complex 1-hydroxy-1H-pyrrol-3(2H)-one has been isolated as a component of the mixture obtained by treatment of 2,4,5-triphenyl-3H-pyrrol-3-one-1-oxide with dimethyl acetylenedicarboxylate.[86]

Although the only example of a 2-monosubstituted 3-hydroxypyrrole containing a 2-(electron-donating) substituent is the 2-amino compound **32**, obtained in 28% yield by photolytic ring contraction of a dihydropyridazine derivative,[87] a great deal is known about 2-substituted-2-(electron-donating)-1H-pyrrol-3(2H)ones. 2-Alkyl or 2-aryl-2-hydroxypyrrolones can be obtained—often unexpectedly—by aerial oxidation of the corresponding pyrrolone (see Section 4.4.5), and a number of other examples have been made directly by the cyclopropenone route (Section 4.2.4).

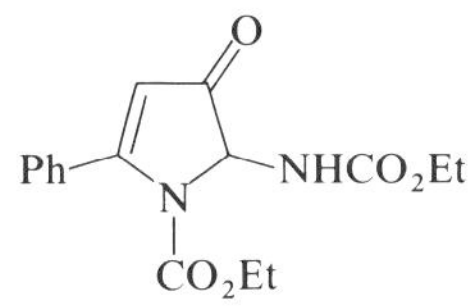

A range of 2,5-disubstituted and 2,4,5-trisubstituted 2-hydroxypyrrol-3(2H)-ones **33** has been obtained in low–moderate (10–40%) yield by treatment of isoxazoline derivatives with o-phenylenediamine and triethylamine (Scheme 18),[88] although the major reaction path gives an α-diketone that is trapped by the diamine to give a quinoxaline derivative.[88] Catalytic hydrogenation of certain oxazole derivatives also leads to 2-hydroxypyrrolones **33**.[88] 2,5-Dimethyl-2-methoxy-1H-pyrrol-3(2H)-one has been isolated in 13% yield by photooxidation of 2,3,5-trimethylpyrrole.[89]

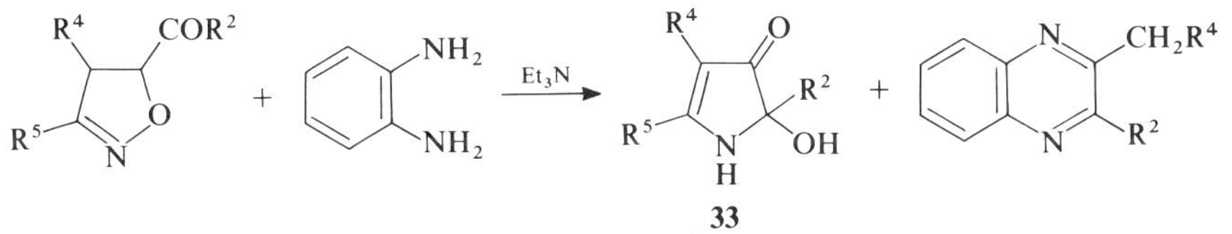

Scheme 18

An extensive series of 1,2-disubstituted-2-hydroxypyrrol-3(2H)-ones has been isolated in yields of ~ 80% from the reaction of 2-methoxy-3(2H)-furanones **34**, or their butenedione precursors **35**, with primary amines (Scheme 19).[90–92] The products **36** are highly fluorescent, and the reaction has found widespread use as an assay for primary amines, particularly in biological studies.

Indeed, the reaction of phenylalanine with phenyl acetaldehyde and ninhydrin was long used for clinical diagnosis before the major products were found to be the pyrrolones **37**.[90] In the light of these findings,[90] the improved reagent

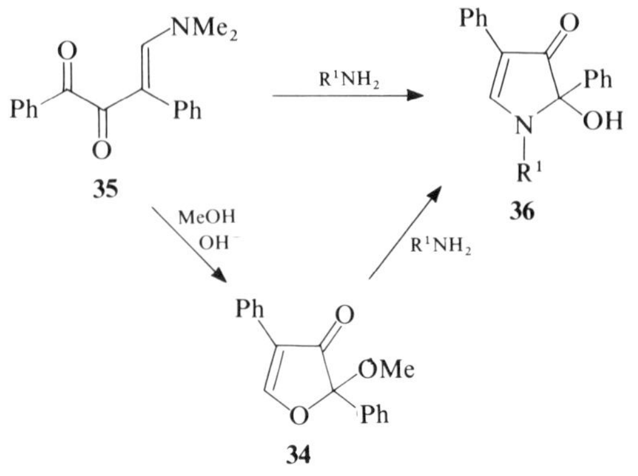

Scheme 19

"fluorescamine" (**38**) was designed.[91,92] Leading references to extensive applications in amino acid and peptide chemistry are given in Ref. 93. The reaction of either **34**[94-96] or **38**[97] with chiral amino acids gives products that show a characteristic Cotton effect enabling the absolute configuration of the amino acid to be determined.

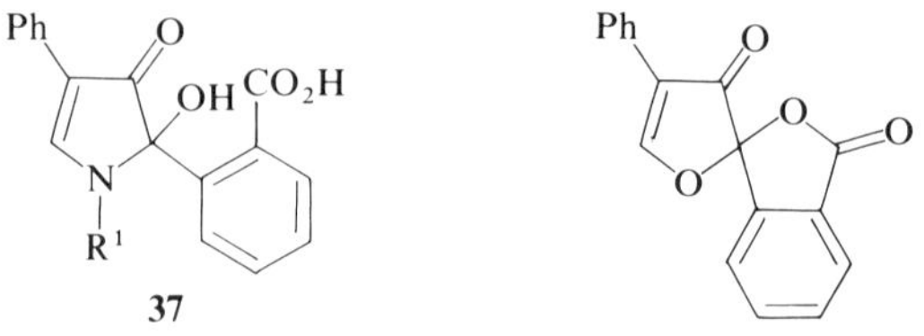

1-Substituted-2-hydroxypyrrol-3(2H)-one-2-carboxylic esters have been obtained in 30–70% yield by cyclisation of acetylenic tricarbonyls with primary amines.[98] (Scheme 20; cf. scheme 10); yields are low for N-aryl derivatives.

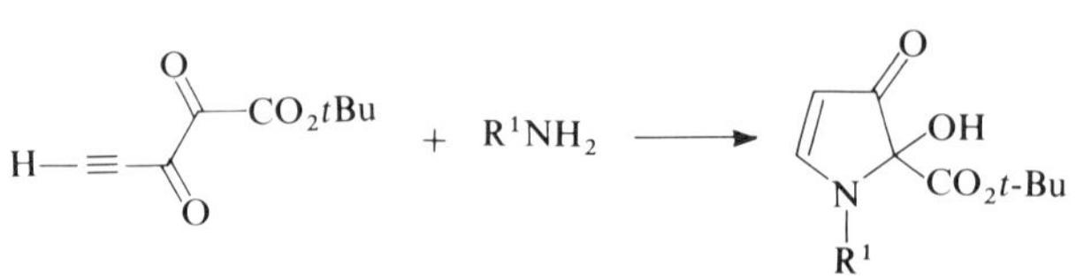

Scheme 20

Gelin has prepared a vast number of 2-substituted 2-hydroxypyrrolone-4-carboxylic esters **39**, either by adaptation of the furanone route (Scheme 19) to give **39** (R^1, R^5 = alkyl or aryl; R^2 = Me) in 70–95% yield,[99] or by the use of 2-arylidene-3(2H)-furanones to give **39** (R^1 = aryl, R^2 = alkyl, R^5 = Me) in similar yield.[100]

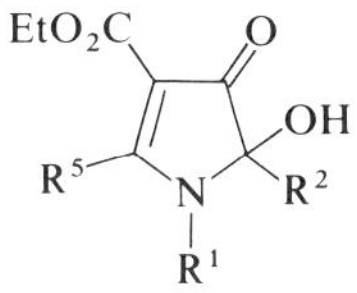

4.2. Synthesis of 3-Hydroxypyrroles and 1H-Pyrrol-3(2H)-ones

One 3-hydroxypyrrole series containing an electron-donating group at the 4-position is known. The 3,4-dihydroxypyrroles **40** (R^1 = Ph or CH_2CO_2Et) are obtained in ~ 70% yield by the surprisingly simple route shown in Scheme 21.[101] However, the 4-aminopyrrolone (**41**) has been made by reductive cleavage of the corresponding 4-azo derivative[102] (Section 4.4.3.5).

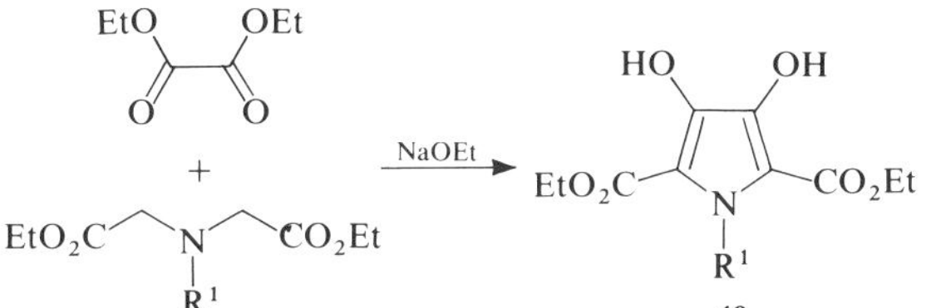

Scheme 21

1H-Pyrrol-3(2H)-one derivatives with electron-donating substituents in the 5-position are restricted to a range of amino compounds **42**. Application of the Meldrum's acid route (Scheme 3) to the bis(dimethylamino) derivative **43** gives **42** ($R^1 = R^5 = R^{5'}$ = Me, $R^2 = R^{2'} = R^4$ = H) in good yield.[103]

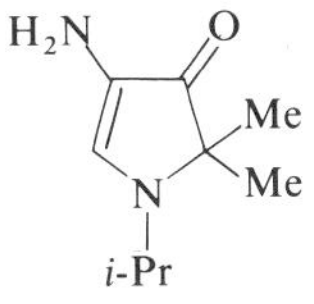

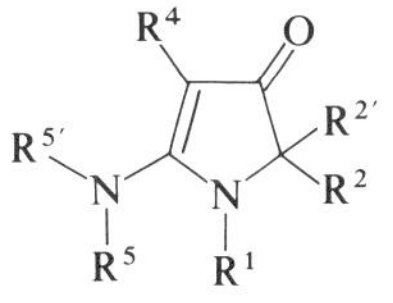

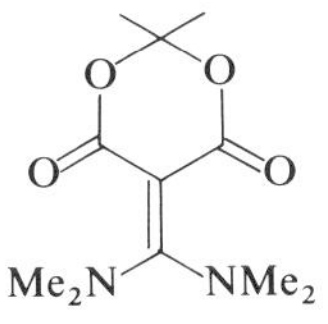

Most other routes lead to polysubstituted products. 4-Substituted compounds are available from α-aminoesters by three methods (Scheme 22) involving either reaction with ynamines[104,105] [to give **42** (R^1 = alkyl, R^2, $R^{2'}$ = H, alkyl, aryl, R^4 = alkyl, aryl, R^5, $R^{5'}$ = alkyl)] or ketene aminals,[104] cyclization of an iminoester[106,107] [to give **42** ($R^1 = R^2 = R^{2'}$ = H, R^4 = Ph; $NR^5R^{5'}$ = morpholino)], or condensation with a benzylic nitrile.[108]

Other methods of limited applicability include reaction of ynamines with certain fluorinated oxazolones[105] or oxazolidinones.[109] Stachel and coworkers have prepared a series of 5-amino-2-ylidene derivatives (e.g., **44**) by treatment of the corresponding azlactone with active methylene compounds; the products all

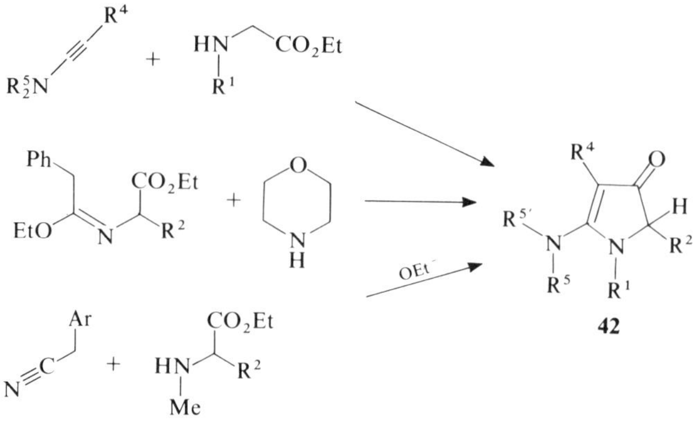

Scheme 22

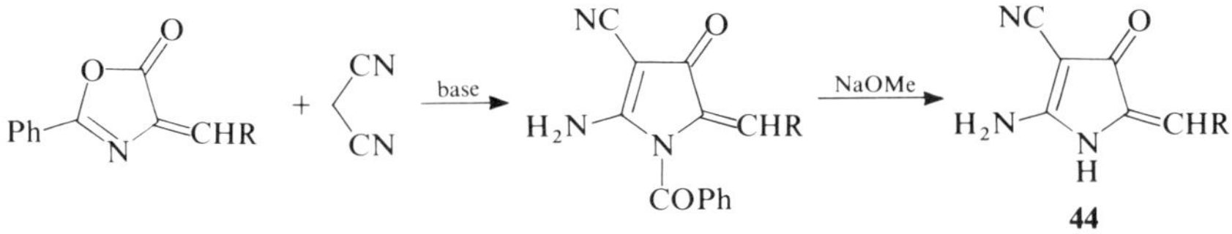

Scheme 23

have electron-withdrawing groups in the 4-position[110,111] (Scheme 23). The palladated pyrrolone **45** was made by treatment of a metal imine complex with dimethyl acetylenedicarboxylate.[112]

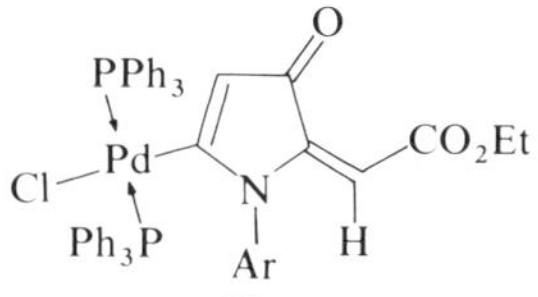

Treatment of α-cyano-γ-chloroacetoacetic ester with aromatic amines gives good (generally 75–85%) yields of N-aryl-5-aminopyrrolone-4-carboxylic esters.[113]

4.2.4. The Cyclopropenone route to Polysubstituted 1H-Pyrrol-3(2H)ones

The reaction of a wide variety of imine compounds with cyclopropenones has been the source of a number of (usually) 1,2,2,4,5-pentasubstituted 1H-pyrrol-3(2H)-ones. The atoms C(3), C(4), and C(5) of the products arise from the three ring carbon atoms of the cyclopropenones; the lack of good general routes to these precursors is the chief limitation of the procedure, and only the 4,5-diaryl

4.2. Synthesis of 3-Hydroxypyrroles and 1H-Pyrrol-3(2H)-ones 541

products are readily obtained. Thus treatment of diphenylcyclopropenone with N-phenyl- or N-methylketenimines **46** (R^1 = Me or Ph, R^2, $R^{2'}$ = alkyl and/or aryl) under reflux in dimethoxyethane gives the pyrrolones **47** in excellent (70–90%) yield[35,36] (Scheme 24).

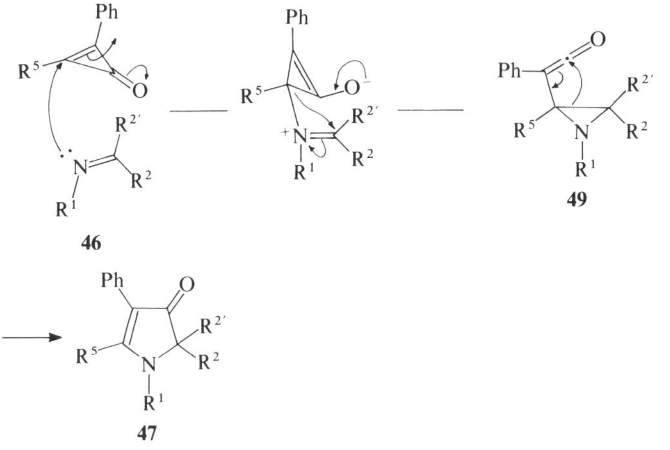

Scheme 24

The use of methylphenylcyclopropenone gives the 5-methyl derivatives **47** (R^5 = Me) regiospecifically.[36] Aldimines **46** (R^1 = alkyl, R^2 = aryl, $R^{2'}$ = H) similarly give pyrrolones **47** ($R^{2'}$ = H) in high yield,[36,114] but these compounds are very readily oxidized (see Section 4.4.5), and, unless special precautions are taken to exclude air, a range of secondary products is usually isolated.[36,114] Dimeric[115] secondary oxidation products **48** are also obtained from diphenylcyclopropenone and either benzaldimine[115] (generated in situ from benzaldehyde and ammonia) or aromatic azines,[116,117] via the pyrrolone **47** (R^1 = R^2 = H, $R^{2'}$ = Ar, R^5 = Ph) (Section 4.4.5), which, in turn, is formed from the azines by cleavage of ArCN from the initial cyclization products.[116,117] The monomeric form **47** of these derivatives **48** can be obtained independently by lithium aluminum hydride reduction of a pyrrolone-N-oxide.[115]

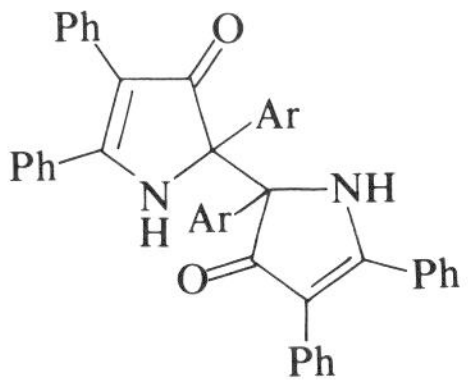

48

The mechanism of the cyclopropenone reaction may involve Michael-type addition of the imine followed by consolidation of the resulting adduct. There is evidence for an intermediate ketene (e.g., **49**) in some cases.[36]

Other synthetic applications include the use of the cyclic enaminone **50**, which reacts via its imine tautomer to give the pyrrolone **51** in 43% yield;[118] cyclic imines give access to a range of fused systems containing a pyrrolone unit.[119] However, if diphenylcyclopropenthione is used in place of the cyclopropenone, the reaction is diverted to bicyclic products and the corresponding pyrrolethiones are not obtained.[120,121] More complex "imines" have also been employed. Thus, benzamidines **46** (R^1 = H, R^2 = Ph, $R^{2'}$ = NR_2) give 2-aryl-2-dialkylaminopyrrolones **47** (R^1 = H, R^2 = R^5 = Ph, $R^{2'}$ = NR_2),[122] and some fused systems may be obtained from certain cyclic amidines.[123] N,N,N',N'-Tetraalkylguanidines **46** (R^1 = H, R^2 = $R^{2'}$ = NR_2) give azacyclopentadienone derivatives **52** after loss of a mole of amine from the initial adduct **47** (R^1 = H, R^2 = $R^{2'}$ = NR_2),[122,124] but other guanidine derivatives yield 6-membered ring products.[124] Imidates, thioimidates, and trisubstituted amidines all cyclize normally to give 60–90% yields of the 2-functionalised pyrrolones **47** (R^2 = OR, SMe, or NR_2).[125] The use of 1,4-diaryl-1,4-diazabutadienes gives the imines **53** (presumably as their hydroxypyrrole tautomers) in 20–80% yield, together with a small amount of 1,2,3-triarylpyrroles as coproducts.[126] Two papers record the application of dipolar imine derivatives in these reactions.[127,128] Both **54**[127] and **55**[128] give the expected 2-substituted pyrrolone derivatives by reaction at the formal imine function. Aniline reacts with diphenylcyclopropenone in a complex dehydration-dehydrogenation sequence to give the 2-methylene derivative **56**,[129] the structure of which was determined by X-ray crystallography.

4.3. PHYSICAL PROPERTIES OF 3-HYDROXYPYRROLES AND 1H-PYRROL-3(2H)-ONES

4.3.1. Tautomerism of 3-Hydroxypyrroles

4.3.1.1. Keto-Enol Tautomerism

The keto-enol tautomerism of the 3-hydroxypyrrole system was briefly mentioned in the introduction (Section 4.1). Early chemical work on the equilibrium produced equivocal results,[130] and IR and UV studies[32,44] were unable to resolve cases in which genuine mixtures of tautomers were present. Theoretical predictions were also inconsistent.[131,132] NMR spectroscopy is clearly the method of choice for the resolution of such problems (Table 4.1), and the positions of the equilibria as well as the factors that influence them are now well understood for the majority of cases.

3-Hydroxypyrrole itself and its N-methyl derivative have been studied as part of a comprehensive investigation of tautomerism in 5-membered hydroxy-heterocycles.[67] It is clear (Table 4.1) that the equilibrium is strongly solvent-dependent, and that the hydroxy form is favored in hydrogen bond acceptor solvents (e.g., DMSO), whereas the pyrrolone form is favored in hydrogen bond donor solvents (H_2O, MeOH).[7] Surprisingly, in view of these results, the 4,5-dimethyl compound **2** ($R^1 = R^2 = H$, $R^4 = R^5 = Me$) exists exclusively in the pyrrolone tautomer, even in DMSO.[8] The acid-catalyzed rearrangement of 3-hydroxypyrrole to 1H-pyrrol-3(2H)-one is very much faster than the corresponding indole, thiophene, or furan reaction,[6,7] presumably because of more efficient electron-donation from the ring heteroatom in the pyrrole case.

The general rule that the pyrrolone tautomer is favored in nonpolar solvents (e.g., $CHCl_3$), and the hydroxypyrrole is favored in polar hydrogen bond acceptor solvents (e.g., DMSO) is followed by a range of alkyl- and aryl-N-substituted derivatives[133] (Table 4.1), which also show the pyrrolone form in the solid state (Section 4.3.2). The amount of the hydroxy form is increased when a 2-phenyl group is present, e.g., **2** ($R^1 = Me$, $R^2 = Ph$, $R^4 = R^5 = H$), since delocalization (structure **57**) is possible, but this is reduced when a 1-*tert*-butyl substituent destroys the planarity needed for conjugation, e.g., **2** ($R^1 = t$-Bu, $R^2 = Ph$, $R^4 = R^5 = H$);[133] 1- or 5-aryl substituents have comparatively little effect.[26,133]

57   **58**

TABLE 4.1. NMR STUDIES OF KETO-ENOL TAUTOMERISM OF PYRROLONES IN A RANGE OF SOLVENTS

Derivatives					K_{enol}^a/Solvent					
R^1	R^2	R^4	R^5		$CDCl_3$	d_6-DMSO	d_4-MeOH	H_2O	d_6-Acetone	Ref.
H	H	H	H		—	> 50	0.406	0.133		7
Me	H	H	H		0.21	> 50	0.458	0.184		7, 10
t-Bu	H	H	H		0.11	19	0.35		3.0	133
Ph	H	H	H		< 0.01	19				133
Et	Me	H	H		0.14	7.3				133
Me	Ph	H	H		2.3	> 99				133
Me	H	H	Ph		0.20					26
Ph	Me	H	H		< 0.01	2.3				133
t-Bu	Ph	H	H		< 0.01	0.28				133
Me	CO_2Me	H	H		> 99					26
Ph	H	CO_2Et	H		13.2	7.3	0.81			62
H	Me	CN	Me			~ 1.0				67
Me	H	H	NMe_2		< 0.1	≪ 0.1				103

$^a K_{enol}$ = [enol]/[keto].

4.3. Physical Properties of 3-Hydroxypyrroles and 1H-Pyrrol-3(2H)-ones

2-Carboxylic esters stabilize the hydroxypyrrole form both by hydrogen-bonding and by conjugation (structure **58**), and for these no trace of the pyrrolone form has ever been detected.[26,44,76]

On the other hand, N-substituted 4-carboxylic esters exist as genuine mixtures[62] (Table 4.1), although the hydroxypyrrole form predominates,[62,68] particularly when the 5-position is unsubstituted.[10] This suggests that the conjugative interactions shown above are more important than the hydrogen bonding. Clearly the solvent–solute interactions of these compounds differ substantially from the simple alkyl and aryl derivatives since the pyrrolone form is relatively favored in methanol vis-à-vis chloroform (see Table 4.1).[62] Early work on N-unsubstituted 5-methyl 4-carboxylic esters, e.g., **2** ($R^1 = R^2 = H$, $R^4 = CO_2R$, $R^5 = Me$), has shown that the keto form is present exclusively in chloroform.[134] This surprising effect of NH groups has already been noted for other 4,5-disubstituted derivatives[8] (see above). The hydroxy form is relatively favored in certain mixed solvents.[135] The 4-cyano compound **2** ($R^1 = H$, $R^2 = R^5 = Me$, $R^4 = CN$) exists as a 50 : 50 mixture of tautomers in DMSO,[67] whereas **2** ($R^1 = Me$, $R^2 = H$, $R^4 = CN$, $R^5 = i\text{-Pr}$) is entirely in the keto form in $CDCl_3$,[71] but no systematic study of the effect of substituents and solvent has been carried out in this series.

N-Substituted 5-amino derivatives exist exclusively in the keto form;[103,104] there is no evidence for the enol form of **2** ($R^1 = Me$, $R^2 = R^4 = H$, $R^5 = NMe_2$) even in DMSO solution. It seems likely that N-unsubstituted derivatives will also adopt this tautomer,[106,136] to allow conjugative interaction through the exocyclic enaminone unit **59** though Fuks and Viehe favor the unusual 2H-tautomer **60** in the light of some IR spectroscopic evidence.[104]

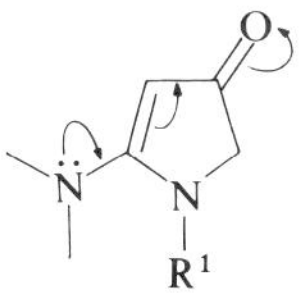

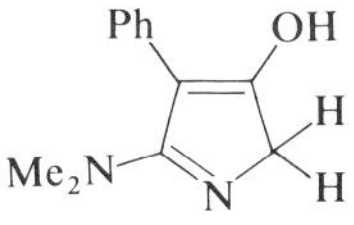

4.3.1.2. Ring-Chain Tautomerism of 2-Hydroxy-1H-pyrrol-3(2H)-ones

2-Hydroxy-1H-pyrrol-3(2H)-ones **61** are cyclic hemiaminals, and may be expected to exist in equilibrium with the ring-opened (α-diketone) tautomer **62**. The ring-closed form is favored in the solid state[88] and in DMSO solution,[88,137] in which hydrogen bonding to the solvent can take place, but the open-chain tautomer is favored in chloroform solution where intramolecular hydrogen bonding is possible.[137] Mixtures can be obtained in other solvents.[88] The Z-configuration of the enamine, and the s-trans configuration of the C—N bond in **61** ($R^4 = R^5 = H$) can be characterized by the ^1H NMR vicinal coupling constants.[137] The equilibria are apparently established quite rapidly[88] (even when R^2 = aryl), and so the extreme conditions used in one case to induce ring closure[36,114] are particularly surprising. In reactions with o-phenylenediamine the pyrrolones **61** can behave as α-diketones to give quinoxaline derivatives.[32]

4.3.2. Structure of 3-Hydroxypyrroles and 1H-Pyrrol-3(2H)-ones

No crystal structure of a 3-hydroxypyrrole has been reported, although results for a 3-methoxypyrrole[138] (Section 4.5.2.1) suggest that the substituent should only slightly perturb standard pyrrole geometry. Theoretical predictions of conformation and charge distribution have been reported.[139] 1-Phenyl-1H-pyrrol-3(2H)-one **63** (Fig. 4.1) adopts this tautomeric form rather than the hydroxypyrrole form in the solid state, and selected structural parameters are shown in Fig. 4.1.[133] The 5-membered ring is essentially planar, and the dihedral angle between the two rings is only 12.6°. N-Alkyl rather than N-aryl substitution causes a slight reduction in the length of the N(1)—C(5) bond,[21] but there is otherwise little effect.

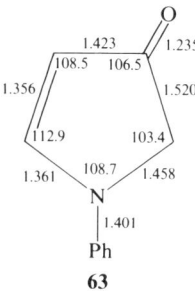

Figure 4.1. Selected bond lengths (Å) and angles (degrees) of 1-phenyl-1H-pyrrol-3(2H)-one (**63**)

More complex derivatives whose structures have been determined include the dimer **64**[140] the 2-hydroxy derivative **65**[90] and the 2-alkylidene compounds **45**,[112] **56**,[129] and **66**.[42] The main effect of the alkylidene group is to increase the length of the N(1)—C(5) bond and reduce the length of the N(1)—C(2) bond relative to the 1-phenyl derivative **63**.

4.3. Physical Properties of 3-Hydroxypyrroles and 1H-Pyrrol-3(2H)-ones 547

64, **65**, **66**

4.3.3. IR Spectra of 3-Hydroxypyrroles and 1H-Pyrrol-3(2H)-ones

A considerable number of IR spectra of 3-hydroxypyrroles and 1H-pyrrol-3(2H)-ones have been quoted in the literature. The most important bands are associated with the NH (where present), the OH (of the hydroxypyrrole form), and the enaminone conjugated system (of the pyrrolone form). These can be highly dependent on hydrogen bonding and may occur at quite different positions in solution and in the solid state.[8,9] Thus, the v_{NH} absorptions of **67** appear at 3150 and 3050 cm^{-1} when measured in the solid state as a potassium bromide disk, whereas a single band at 3445 cm^{-1} is observed for a solution in chloroform.[8] Two highly coupled[36] vibrations at 1620–1670 and 1525–1580 cm^{-1} are due to the enaminone system of N-unsubstituted,[8,17] N-substituted,[10] or C-substituted[32,36] pyrrolones: these are characteristic of vinylogous amides,[141,142] and their positions are relatively insensitive to the substitution pattern of the ring.

67, **68**

3-Hydroxypyrrole-2-carboxylic esters **68** and the corresponding ketones [both of which exist exclusively in the enol form (Section 4.3.1)] generally show OH absorptions at 3400–3500 cm^{-1},[44,48,76,84] although hydrogen bonding may again be important.[76] There is clear evidence of intramolecular hydrogen bonding in the low stretching frequency of the 2-carbonyl group.[44,48,76,84] The enaminone system of 2-substituted-2- or-4-carboxylic esters apparently absorbs in a similar range to that of the simple derivatives,[100,143] and the v_{OH} vibration of the enol form of the 4-carboxylic esters also appears in the expected position.[62,68]

4.3.4. UV Spectra of 3-Hydroxypyrroles and 1*H*-Pyrrol-3(2*H*)-ones

4.3.4.1. 3-Hydroxypyrroles

Despite the fact that UV spectroscopy was used extensively in the early literature to study tautomerism,[32] it appears that the spectrum of a simple (alkyl or aryl) 3-hydroxypyrrole in the enol form has never been recorded, even though choice of an appropriate solvent system may be all that is required (see Section 4.3.1.1). However, the UV spectra of 3-hydroxypyrrole-2-carboxylic esters and *N*-substituted 4-carboxylic esters, all of which predominate as the enol tautomer, have often been reported (Table 4.2). In general, the 2-carboxylic esters show an intense maximum (ε 11000–20000) at 260–270 nm, and this is effectively independent of substitution at the 1,4, or 5-position, even by other ester groups (Table 4.2). Other subsidiary maxima may be seen in some cases, e.g., **2a** (R^1 = Et, R^2 = R^5 = CO_2Me), λ_{max} 300 nm (ε 9750). The 4-carboxylic ester derivatives absorb at rather shorter wavelength (Table 4.2), but again this is almost independent of the *N*-substituent.

TABLE 4.2. UV SPECTRA OF 3-HYDROXYPYRROLES (2a)

R^1	R^2	R^4	R^5	λ_{max}/nm	ε	Solvent	Ref.
H	CO_2Et	H	H	263	17,000	EtOH	76
H	CO_2Et	Me	H	266	18,100		44
Me	CO_2Et	H	H	264	16,700	EtOH	76
Ph	CO_2Et	H	H	262	12,500	EtOH	76
Me	H	CO_2Et	H	243	12,600	EtOH	68
Ph	H	CO_2Et	H	246, 255	18,000	EtOH	32
Ph	CO_2Et	CO_2Et	H	262	11,800	EtOH	78
Et	CO_2Me	H	CO_2Me	274	19,800	MeOH	83

4.3.4.2. 1H-Pyrrol-3(2H)-ones

In contrast to the situation for 3-hydroxypyrroles, there is a considerable amount of UV spectral data for the 1*H*-pyrrol-3(2*H*)-one system **2b**. The enaminone chromophore of these compounds produces an intense maximum (ε 7000–16,000) in the range 300–360 nm (Table 4.3) clearly shifted to much

TABLE 4.3. UV SPECTRA OF 1*H*-PYRROL-3(2*H*)-ONES

R^1	R^2	$R^{2'}$	R^4	R^5	λ_{max}/nm	ε	Solvent	Ref.
H	Me	Me	H	H	301	7,900	MeCN	14
H	H	H	H	Me	305		EtOH	10
H	Me	Me	H	Me	308	12,100		15
H	Ph	Ph	H	Ph	337	7,900		16
H	H	H	Me	Me	315	7,600	MeOH	8
Me	H	H	H	H	328		EtOH	10
Ph	H	H	H	H	354		EtOH	10
i-Pr	Me	Me	H	H	322	9,300	Benzene	25
Ph	Me	Me	H	H	330		CHCl$_3$	133
Ph	Me	H	H	Me	325	16,200		32
H	CO$_2$Et	Me	H	H	308	10,700	MeCN	140
H	OH	*p*-Tolyl	H	Me	333	7,200	EtOH	88
H	OH	Me	H	Ph	364	8,300	EtOH	88
H	H	H	CO$_2$Et	Me	294	9,800	EtOH	32
Me	H	H	Ph	NMe$_2$	302	12,600	MeOH	104

549

longer wavelength than the hydroxypyrroles. The parent 1H-pyrrol-3(2H)-one probably absorbs toward the low end of the range. 2,2-Dimethyl- and 5-methyl substituents have relatively little effect on the position of the maximum, although 1-alkyl, 4-alkyl, 5-aryl, and, particularly, 1-aryl groups cause bathochromic shifts of up to 50 nm (Table 4.3). When an N-phenyl group is flanked by two methyl groups (in either the 2,2-[133] or the 2,5-[32] positions), this bathochromic shift is substantially reduced owing to the reduction in coplanarity of the two rings; one 2-methyl group is not enough to cause this effect.[133] The UV maxima of 2-substituted-2-carboxylic ester derivatives are hardly shifted from those of nonpolar analogs[143] (Table 4.3), but related 2-hydroxy substituents apparently cause a substantial bathochromic shift.[88] Those 4-carboxylic esters that exist in the keto form[32] show a slight hypsochromic shift in the maximum (Table 4.3) to below 300 nm, and a similar effect is apparently shown by 5-dimethylamino substituents.[104]

4.3.5. NMR Spectra of 3-Hydroxypyrroles and 1H-Pyrrol-3-(2H)-ones

4.3.5.1. 3-Hydroxypyrroles

Most simple 3-hydroxypyrroles [1H-pyrrol-3(2H)-ones] adopt the enol form in DMSO (Section 4.3.1), and their NMR spectra have been most often recorded in this solvent (see Tables 4.4, 4.5). The ^{1}H spectra are characterized by typical pyrrole multiplets[144] with the β-proton (4-H) signal appearing at lower frequency (δ_H 5.3–5.9) than either of those of the α-protons (2-H, δ_H 5.9–6.3: 5-H, δ_H 6.1–6.6).[7,145] Conjugative electron donation from the hydroxy group causes shielding of 2-H relative to 5-H. An N-phenyl group causes deshielding of the two α-protons due to its anisotropic effect, but only in the absence of adjacent substituents that can destroy the necessary coplanarity (Table 4.4). Three- and four-bond couplings ($^3J_{4,5}$, $^4J_{2,4}$ and $^4J_{2,5}$) are also typical of simple pyrroles[144] (2.0–3.0 Hz) (Table 4.4): the NH can also interact, although in the case of the parent compound **2a** (R^1–R^5 = H), these couplings were not analyzed.[7]

Neither 2- nor 4-carboxylic ester substituents apparently cause any major effect on the ^{1}H NMR parameters[76,135] (Table 4.4). However, these data allow identification of an unknown minor component of the reaction between ethyl methylaminoacetate and ethyl propiolate[146] as **69**.[68] In certain cases, three- and four-bond couplings to NH (2.0–3.0 Hz) were resolved.[76,135]

TABLE 4.4. ^{1}H NMR PARAMETERS OF TYPICAL 3-HYDROXYPYRROLES (**2a**)

$$\underset{\textbf{2a}}{\underset{R^5}{\overset{R^4}{\diagdown}}\underset{\underset{R^1}{|}}{\overset{OH}{\diagup}}R^2}$$

				δ_H						
R^1	R^2	R^4	R^5	2-H	4-H	5-H	$^3J_{4,5}$	$^4J_{2,4}$	$^4J_{2,5}$	Ref.
H	H	H	H	6.22	5.60	6.43	2.6	2.0	2.6	7
Me	H	H	H	5.94	5.34	6.10	2.9	2.0	2.6	7
t-Bu	H	H	H	6.26	5.50	6.49	3.1	1.9	2.6	145
Ph	H	H	H	6.74	5.85	7.06	3.0			145
Et	Me	H	H		5.52	6.30	3.0			145
Me	Ph	H	H		5.72	6.51	2.8			145
Ph	Me	H	H		5.78	6.56	3.0			145
H	CO$_2$Et	H	H		5.86	6.71	2.8^a			76
Me	CO$_2$Et	H	H		5.58	6.42	3.0			76
H	H	CO$_2$Et	Me	6.02^b						135
Me	H	CO$_2$Et	H	6.86		6.15			3.0	68

^{a} Also $^3J_{1,5} = {}^4J_{1,4} = 2.8$ Hz.
^{b} Also $^3J_{1,2} = 2.0$ Hz.

TABLE 4.5. ^{13}C NMR PARAMETERS OF TYPICAL 3-HYDROXYPYRROLES (2a)

R^1	R^2	R^4	R^5	δ_C				Ref.
				C-2	C-3	C-4	C-5	
t-Bu	H	H	H	101.16	143.37	97.60	114.18	145
Ph	H	H	H	101.54	146.07	102.18	115.96	145
Me	Ph	H	H	116.50	141.62	98.11	120.29	145
Ph	Me	H	H	110.47	141.40	100.45	116.41	145
H	COR^a	H	H	117.2	157.35	98.63	125.74	53
Me	CO$_2$Me	H	H	106.11	155.24	95.53	128.02	26

^{a} R = COCH$_2$CH$_2$OCO$_2$CH$_2$Ph.

Typical ^{13}C NMR parameters for 3-hydroxypyrroles are given in Table 4.5. The order of chemical shifts is the same as in the ^{1}H spectra, viz. δ_C(C-4) < δ_C(C-2) < δ_C(C-5) [< δ_C(C-3)]. N-Aryl groups cause a slight deshielding at C-4.[145] 2-Phenyl groups cause a substantial deshielding at C-5,[145] presumably due to conjugative electron withdrawal from the ring 70. This effect is particularly noticeable with 2-carboxylic ester and 2-acyl derivatives[26,53] (Table 4.5), where the 3-position is also deshielded.

One-bond coupling constants $^1J_{CH}$ show the expected[147] differences between the pyrrolyl α- and β-carbon atoms[145] (Fig. 4.2). Long-range couplings show a rich pattern of two- and three-bond interactions;[145] a typical example is shown in Fig. 4.2.

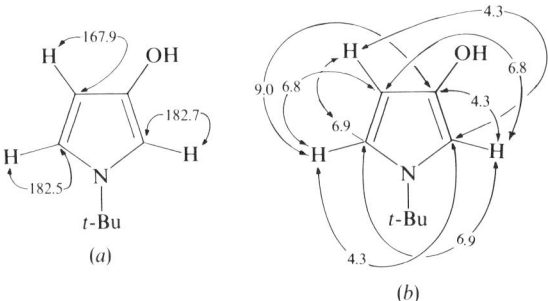

Figure 4.2. (a) One-bond couplings ($^1J_{CH}$) and (b) long-range couplings ($^nJ_{CH}$) of 1-tert-butyl-3-hydroxypyrrole.

4.3.5.2. 1H-Pyrrol-3(2H)-ones

The ^{1}H NMR spectra of the pyrrolone tautomers **2b** (which predominate in chloroform solution) are characterized by a widely spaced pair of doublets at δ_H 4.9–5.2 and 7.7–8.0, due to 4-H and 5-H, respectively[148] (Table 4.6). This pattern is typical of the enaminone conjugated system,[148] and is due to the

TABLE 4.6. ^{1}H NMR PARAMETERS OF TYPICAL 1H-PYRROL-3(2H)-ONES

R^1	R^2	$R^{2'}$	R^4	R^5	2-H	4-H	5-H	$^3J_{4,5}$	Ref.
H	H	H	H	Me	3.90	5.05			12
Me	H	H	H	H	3.72	5.07	7.75	3.0	10
t-Bu	H	H	H	H	3.70	5.09	7.93	3.4[a]	148
Ph	H	H	H	H	4.10	5.46	8.40	3.6	148
Et	Me	H	H	H	3.44	4.92	7.71	3.3	148
Me	Ph	H	H	H	4.34	5.08	7.86	3.2	148
Ph	Me	H	H	H	4.15	5.42	8.30	3.6	148
H	Me	Me	H	H		5.16[b]	7.95[c]	3.4	14
i-Pr	Me	Me	H	H		5.06	7.88	3.4	148
Ph	Me	Me	H	H		5.37	8.10	3.6	148
Me	H	H	Me	Ph	3.83				26
H	CO_2Et	Allyl	H	Me	3.58	5.16			8
H	H	H	CO_2Et	Me	3.80	5.64	8.07	3.5	143
H	H	H	H	H[d]	4.11	5.79			8
CHO	H	H	H	H	4.11	5.79	8.24	3.97	12
Me	H	H	H	NMe_2	3.63	4.62			103

[a] Also $^4J_{2,5}$ 0.9 Hz.
[b] Also $^4J_{1,4}$ 1.2 Hz.
[c] Also $^3J_{1,5}$ 3.4 Hz.
[d] Major rotamer.

electron-rich nature of the 4-position, compared with the electron-deficient nature of the 5-position of **71**. The signals for the methylene protons appear at δ_H 3.7–3.9 (Table 4.6). N-Aryl groups cause deshielding at all three ring positions, due either to their ring current effect on the α-protons or to delocalization which reduces electron density at the 4-position.[148] Similar effects are found with N-formyl or N-acetyl substituents[12] (Table 4.6). The three-bond coupling $^3J_{4,5}$ (3.5 ± 0.5 Hz) is also characteristic, and a minor coupling ($^4J_{2,5} \sim 0.9$ Hz) is often present. Few examples of coupling to the NH have been reported; in particular, Bauer consistently reports the 2-H protons of 4,5-dialkyl derivatives as "singlets."[8] However, **2a** ($R^1 = R^4 = R^5 = H$, $R^2 = R^{2'} = Me$) shows $^3J_{1,5}$ (3.4 Hz) and $^4J_{1,4}$ (1.2 Hz),[14] and these values are probably typical.

2-Substituted 2-carboxylic esters show significant high-frequency shifts of the 4-H signal, although the effect at 5-H is much smaller:[143] electron-donating groups at the 5-position cause the anticipated low-frequency shifts of the signal for the adjacent proton (4-H).[103]

The charge distribution in the enaminone system is also reflected by the ^{13}C NMR chemical shifts of C-4 (δ_C 94–104) and C-5 (δ_C 158–163) (Table 4.7).[46,148] The carbonyl (C-3) signal occurs at δ_C 198–207 and that for the methylene (C-2) group, at $\delta_C \sim 55.0$. Additional substituents have their expected effects. For example, alkyl substitution causes significant high-frequency shifts at the point of attachment (Table 4.7). Phenyl groups at the 1- or 5-position cause slight deshielding at C-4, whereas a 5-dimethylamino substituent causes a large shift of C-4 to low frequency.[103] For those pyrrolones with only nonpolar substituents, the chemical shift of the carbonyl (C-3) shows a regular deshielding with increased substitution at the 2-position.[148]

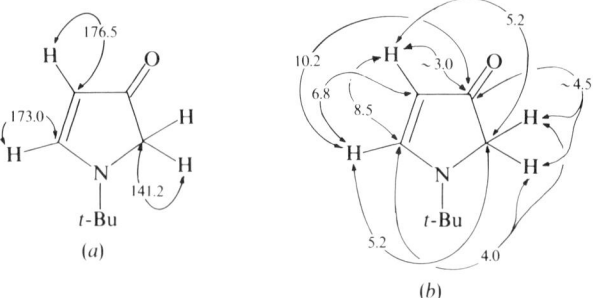

Figure 4.3. (a) One-bond couplings ($^1J_{CH}$) and (b) long-range couplings ($^nJ_{CH}$) of 1-tert-butyl-1H-pyrrol-3(2H)-one.

TABLE 4.7. ^{13}C NMR PARAMETERS OF TYPICAL 1H-PYRROL-3-(2H)-ONES

R^1	R^2	R$^{2'}$	R^4	R^5	C-2	C-3	δ_C C-4	C-5	Ref.
t-Bu	H	H	H	H	54.19	199.76	99.13	162.58	148
Ph	H	H	H	H	55.65	198.71	103.81	158.31	148
Et	Me	H	H	H	61.85	203.07	96.46	163.93	148
Ph	Me	H	H	H	61.13	202.92	101.36	158.73	148
H	Me	Me	H	H	63.7	206.7	96.7	162.5	14
i-Pr	Me	Me	H	H	72.69	203.50	94.34	160.73	148
Me	H	H	H	Ph	63.16	197.90	101.65	179.19	148
H	H	H	Bu	Me	54.2	200.0	112.5	176.0	26
H	Me	H	CN	Me	60.83	197.13	82.99	178.84	46
Me	H	H	H	NMe$_2$	63.32	193.76	87.36	179.87	103

555

In contrast to the hydroxypyrroles (Section 4.3.5.1), the one-bond coupling constants show little distinction between the 4- and 5-positions, although, as expected, both are much larger than the sp^3-type coupling at C-2 (Fig. 4.3).[148] Typical values of long-range couplings for a 1-substituted derivative are also shown in Fig. 4.3.[148]

4.3.6. Mass Spectra of 3-Hydroxypyrroles and 1H-Pyrrol-3(2H)-ones

Mass spectroscopy of 3-hydroxypyrroles has been routinely employed for characterization purposes. Substantial molecular ions are usually obtained under electron-impact conditions, but no systematic study of the fragmentation has been carried out. For simple alkyl or aryl derivatives the most characteristic breakdown of the ring apparently involves loss of CO (or HCO),[12,23,36] and this can give rise to the base peak in certain cases.[17] Cleavage of N- and/or C-alkyl groups is also a common fragmentation,[9,23,36] which is observed even with 3-hydroxy-5-methylpyrrole.[12] 2,2-Disubstituted-1H-pyrrol-3(2H)-ones can undergo both of these processes sequentially,[21,36] and further breakdown is thought to involve cleavage of the remaining ring atoms into two 2-atom fragments (Scheme 25).[36]

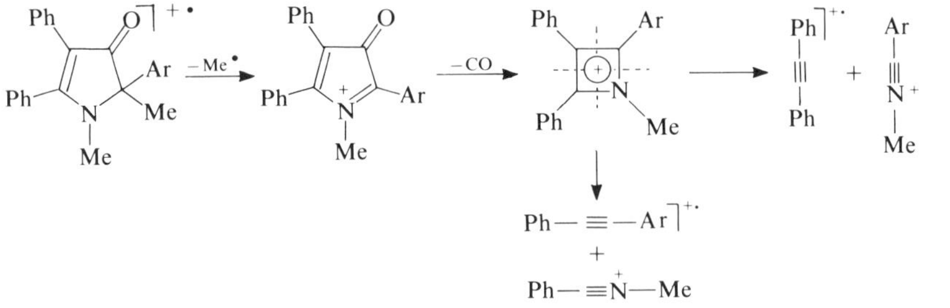

Scheme 25

N-Unsubstituted 3-hydroxypyrrole-2-carboxylic esters lose the alcohol component of the ester function, followed by two molecules of carbon monoxide; the NH rather than the OH group is thought to participate preferentially in the initial cleavage,[149] although the same fragmentation is also found in the corresponding N-substituted derivatives[26] and in 3-hydroxypyrrole-4-carboxylic esters.[64]

4.3.7. ESR Spectra of 3-Hydroxypyrrole Radicals

Attempts to observe a pyrrol-3-oxyl radical by hydrogen abstraction from a 3-hydroxypyrrole using photochemically generated *tert*-butoxyl radicals were unsuccessful. However, prolonged irradiation of these solutions gave a strong spectrum of persistent dimeric radicals (e.g. **72**) whose hyperfine splittings are given in Fig. 4.4.[137]

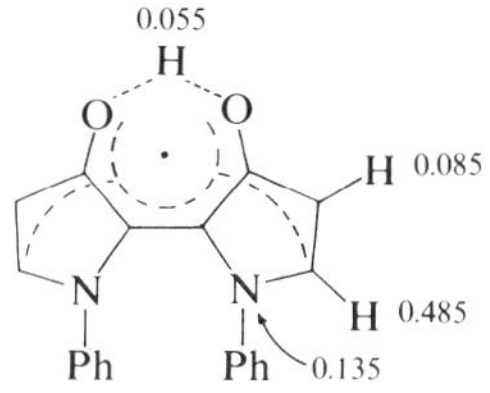

[structure **72** with hyperfine splittings: 0.055 (H), 0.085 (H), 0.485 (H), 0.135]

Figure 4.4. Hyperfine splittings (mT) for the radical (**72**) at 250 K.

4.4. CHEMICAL PROPERTIES OF 3-HYDROXYPYRROLES AND 1H-PYRROL-3(2H)-ONES

4.4.1. Alkylation of 3-Hydroxypyrroles and 1H-Pyrrol-3(2H)-ones

Alkylation under basic conditions via the enolate anion **73** can, in principle, give rise to C- and/or O-alkylated products (Scheme 26) and, when $R^1 = H$, N-alkylation is also possible. Each of these modes of reaction has been observed in specific cases. Under neutral conditions, O-alkylation of hydroxypyrroles by diazomethane, and O-alkylation of 1H-pyrrol-3(2H)-ones by Meerwein's reagent have both been studied, but these reactions are useful only in certain cases. Preparatively, alkylation of 3-hydroxypyrroles is most often used as a route to the corresponding alkoxy compound (see Section 4.5).

The only known reaction of 3-hydroxypyrrole itself is in situ O-trimethylsilylation using chlorotrimethylsilane and triethylamine in dry tetrahydrofuran.[7] Other simple N-unsubstituted 3-hydroxypyrroles do not appear to have been alkylated, although the appropriate alkoxy derivatives are readily prepared via a decarboxylation procedure (Section 4.5.3.7). In contrast, O- and C-alkylation of simple N-substituted hydroxypyrroles has been studied in detail,[138] and the influence of base, solvent, alkylating agent, and substrate on product distribution has been analyzed (see Scheme 26). The results may be explained by application of HSAB (hard–soft, acid–base) theory.[150] Using excess sodium hydride as the base, the C-alkylated product **74** was favored with nonpolar solvents (e.g., THF) and "soft" alkylating agents (e.g. MeI).[20] The alkoxy compound **76** was formed exclusively using sodium hydride and the "hard" alkylating agent methyl p-toluenesulfonate in dimethylimidazolidinone (DMI), a dipolar aprotic solvent.[138,151] This regiospecific O-alkylation takes place in excellent preparative yield with a range of alkyl p-toluenesulfonates (65–90%). It is applicable to a wide range of 1-substituted, 1,2-disubstituted, and

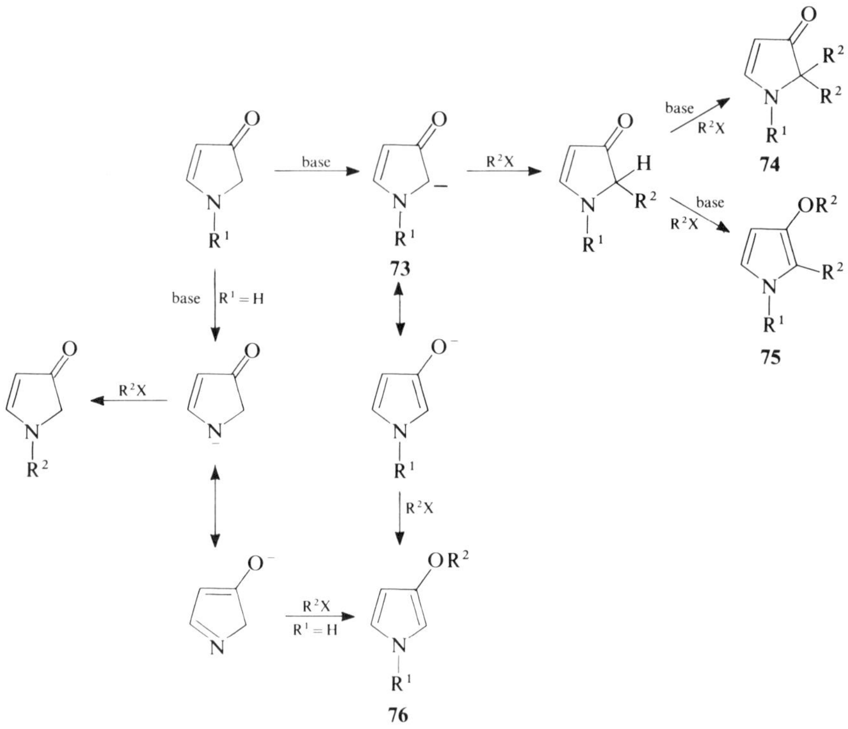

Scheme 26

1,5-disubstituted substrates, and is probably the best general route to *N*-substituted 3-alkoxypyrroles. The use of other alkylating agents (e.g., bromoalkanes) or other solvents (e.g., DMSO) usually results in a mixture of *O*-alkyl, *O*,*C*-dialkyl, and *C*,*C*-dialkylated products **76**, **75**, and **74** (Scheme 26).

2,5-Dimethyl-1-phenyl-1*H*-pyrrol-3(2*H*)-one does not give a methoxy derivative with diazomethane.[32]

3-Hydroxypyrrole-2-carboxylic esters are presumably more acidic than simple 3-hydroxypyrroles, and successful *O*-alkylation has been accomplished under relatively mild conditions. Dimethyl sulfate in dilute sodium hydroxide,[130] or in acetone in the presence of potassium carbonate,[44] specifically *O*-methylate *N*-unsubstituted compounds, and either set of conditions may be the method of choice for specific cases.[44] The reaction has been carried out with 4-substituted,[44,130] with 4,5-disubstituted,[44,130] and, using NaH/Me$_2$SO$_4$/THF, with 5-substituted[151] 2-carboxylic esters, with *N*-substituted precursors,[66] and with 2-carboxamides.[47] The yield is often >80%. However, a mixture of *O*-alkylated and *N*,*O*-dialkylated products is obtained in 3 : 2 ratio when ethyl 3-hydroxypyrrole-2-carboxylate is treated with benzyl chloride in acetone/K$_2$CO$_3$.[10] As with *C*/*O*-alkylation regioselectivity, the choice of alkylating agent may therefore be important in controlling the *N*/*O*-regiochemistry.

4.4. Chemical Properties of 3-Hydroxypyrroles and 1H-Pyrrol-3(2H)-ones 559

Although diazomethane generally fails to give products with these substrates,[44,130] the methylation of an N-unsubstituted 2-acyl derivative is more successful with this reagent than under standard basic conditions.[53]

C-Alkylated products are isolated in 65–85% yield from N-unsubstituted 2-carboxylic ester substrates when the reagents employed are NaH/MeI (or allyl bromide[143]) in degassed benzene,[140] although in one case an N,O-dialkylated product is formed under similar conditions.[140] If required, this product can be obtained in other cases by using a tenfold excess of the alkylating agent.[140] Similar results are obtained with N-substituted substrates,[140,143] although, in one instance, allylation of ethyl 3-hydroxy-1-methylpyrrole-2-carboxylate was reported to give a 2 : 1 mixture of C- and O-allyl products.[153] Treatment of some 4-substituted–N-unsubstituted 3-hydroxypyrrole-2-carboxylic esters with 2,3-dihydropyran in the presence of acid gives either exclusively O-alkylation or exclusively C-alkylation: the regiochemistry is somewhat surprisingly controlled by the presence (C-alkylation) or absence (O-alkylation) of a substituent in the 5-position.[44] Examples of intramolecular C-alkylation lead to functionalised pyrrolizidine or indolizidine ring systems[60] (Scheme 27).

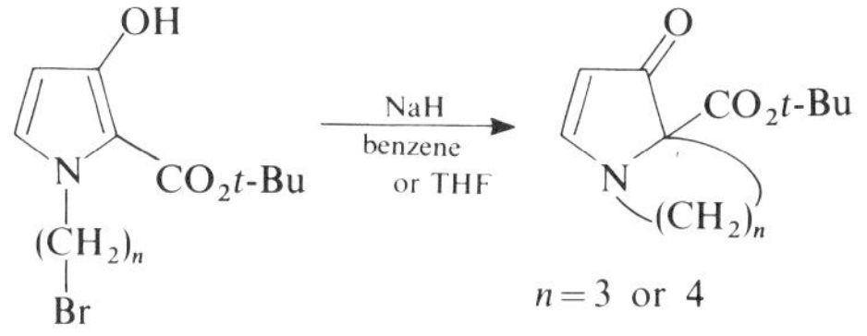

Scheme 27

In the 3-hydroxypyrrole-4-carboxylic ester series, regiospecific O-alkylation is possible using Meerwein's reagent,[135] whereas N-alkylation using either dimethyl sulfate in acetone/K_2CO_3, or methyl fluorosulfonate, or diazomethane is recorded.[135] N-Alkyl[68] or N-aryl[63] derivatives are readily O-alkylated with dimethyl sulfate in aqueous base. These conditions have also been used to O-alkylate an N-substituted 5-formyl-3-hydroxypyrrole.[154]

1-Alkyl-3-hydroxypyrrole-2,4-dicarboxylic esters are readily O-methylated[66] (Me_2SO_4/K_2CO_3/acetone), but benzylation (sodium salt, $PhCH_2Br$/MeOH) gives a 2 : 1 mixture of O- and C-alkylation products.[79] One example of O-alkylation of an N-unsubstituted 2-acetyl-3-hydroxypyrrole-4-carboxylic ester (Me_2SO_4/dilute NaOH) is recorded.[130] A 2,5-dicarboxylic ester gives the N,O-dialkylated product under standard conditions.[44]

Little is known of the alkylation of electron-rich 1H-pyrrol-3(2H)-ones. The 1-methyl-5-dimethylamino derivative has been O-alkylated using NaH/methyl p-toluenesulfonate/DMI[138] to give a highly reactive 3-methoxy-5-dimethylaminopyrrole.[103] The ethylenedioxy derivative **77** is obtained from the corresponding dihydroxy compound using dibromoethane and K_2CO_3 in DMF;[101] bromochloromethane gives the dimer **78** in >70% yield.[101]

Alkylation of 2,2-disubstituted 1H-pyrrol-3(2H)-ones has been less well studied, but one example of N-methylation of an N-unsubstituted derivative under basic conditions has been reported.[118] Meerwein's reagent transforms the corresponding N-substituted pyrrolone into the ethoxypyrrolium salt **79**.[155] One example of N-alkylation of an N-unsubstituted pyrrolone using this reagent has been published, but the spectroscopic data are equally compatible with the O-alkyl product.[105] (cf Ref. 135) A 2-methoxypyrrolone has been obtained from its 2-hydroxy analogue by methylation using dimethyl sulphate in hexamethylphosphoramide in the presence of potassium *tert*-butoxide.[90]

4.4.2. Acylation of 3-Hydroxypyrroles and 1H-Pyrrol-3(2H)-ones

O-Acylation of 3-hydroxypyrroles under basic conditions is a common and facile process. Simple N-alkyl or N-aryl derivatives have been acylated to give **80** with benzoyl chloride in pyridine,[32] or by using an excess of an acid chloride and triethylamine in THF.[156] Similar conditions, using ethyl chloroformate, give the appropriate carbonates **81**.[156] However, the majority of these reactions have been carried out on substrates bearing electron-withdrawing substituents. In those cases where the ring nitrogen atom is unsubstituted it is *usually* only the O-acyl derivative that is isolated, although in some cases this may be due to facile hydrolysis (e.g., Refs. 12 and 105) of the N-acyl compound on workup. Examples include acetylation (neat acetic anhydride) of 2-carboxamides to give **80** ($R^1 = R^4 = H$, $R^2 = CONHMe$, $R^3 = Me$, $R^5 = Ph$),[47] and of 2-carboxylic esters (using acetic anhydride/sodium acetate) to give **80** ($R^1 = H$, $R^2 = CO_2Et$, $R^3 = R^4 = R^5 = Me$).[44] 4-Carboxylic acid derivatives give **80** ($R^1 = R^2 = H$, alkyl

4.4. Chemical Properties of 3-Hydroxypyrroles and 1H-Pyrrol-3(2H)-ones 561

or aryl, $R^4 = CO_2Et$ or CO_2H, $R^5 = H$, alkyl), using sodium acetate in acetic anhydride,[63] acetic anhydride in pyridine,[67] or neat acetic[68] or trifluoroacetic[64] anhydrides. Carbamates **82** (R = H or Me) can be obtained from the hydroxypyrrole by using either methyl isocyanate and triethylamine in benzene, or dimethyl carbamoyl chloride and sodium hydride in the same solvent.[71] O-Acetyl derivatives have also been isolated when the hydroxypyrrole contains an electron-withdrawing group in the 5-position,[157,158] or even an electron-donating substituent at the 4-position.[101] Yields for all these acylations are generally good to excellent. Interpretation of the acetylation reactions of **83** caused problems for nearly half a century.[159] The N,O-diacetate **84**[159] is formed using acetic anhydride–sodium acetate,[130] and this can be readily N-deacylated to give **85**[159] or further C-acylated using acetic anhydride/perchloric acid to give the N,O,C-triacetyl derivative **86**[159] (Scheme 28).

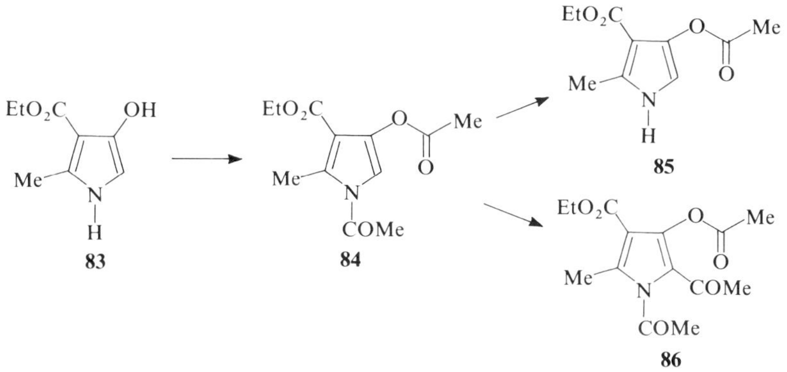

Scheme 28

In one example of the acetylation of an N-unsubstituted 2,2-disubstituted 1H-pyrrol-3(2H)-one, the 2H-pyrrole structure **87** is claimed for the product.[82] 1,2-Disubstituted-2-hydroxy derivatives are directly O-acylated at the 2-hydroxy group using acetic anhydride in pyridine.[99] Similar products **88** (R^2 = OCOMe) are obtained from **88** (R^2 = NMe_2 or OMe) and acetic anhydride, although these reactions probably take place via an intermediate azacyclopentadienone.[122] In one case, a ring-opened diacetate is probably formed from a 2-hydroxy derivative.[41]

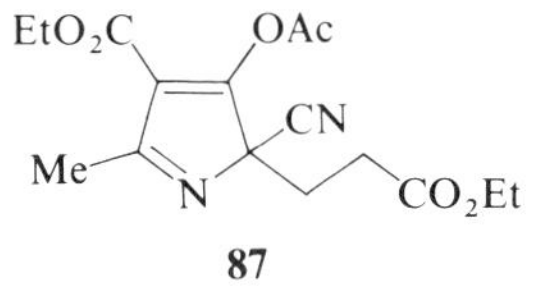

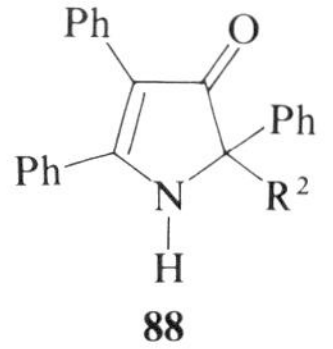

4.4.3. Reactions of 3-Hydroxypyrroles and 1*H*-Pyrrol-3(2*H*)-ones with Electrophiles, and Properties of the Products

4.4.3.1. General Considerations

Pyrroles in general are active towards electrophiles,[2,4] and the conjugative electron-donating effect of the 3-hydroxy substituent increases this reactivity. Simple considerations of resonance structures would predict that the 2-position should be most reactive and the 4-position least reactive (Scheme 29). However,

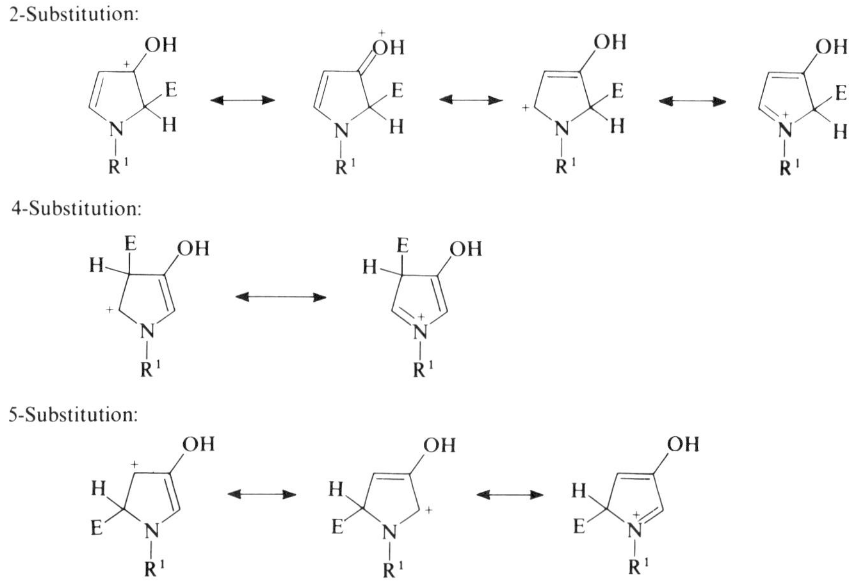

Scheme 29

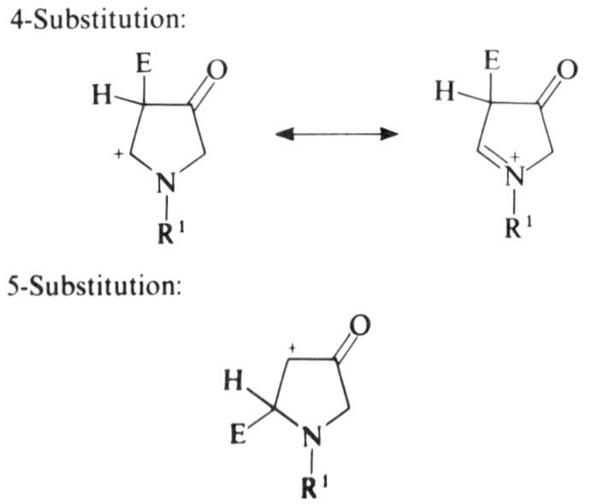

Scheme 30

4.4. Chemical Properties of 3-Hydroxypyrroles and 1H-Pyrrol-3(2H)-ones

for hydroxypyrroles that can tautomerise to 1H-pyrrol-3(2H)-ones the situation is more complicated, since the only viable site for reaction of these tautomers is the 4-position (Scheme 30), i.e., the electron-rich site of the enaminone system.

In practice, the following generalizations may be made:

1. 2,2-Disubstituted 1H-pyrrol-3(2H)-ones always react at the 4-position.
2. 2-Unsubstituted 3-hydroxypyrroles will usually react at the 2-position, although in some cases (bulky electrophile and bulky N-substituent) 4-substituted products are obtained.
3. 2-Substituted 3-hydroxypyrroles will normally react at the 5-position unless bulky N-substituents divert the reaction to the 4-position.

Strong electron-withdrawing substituents can also modify this reactivity; for instance, the 5-position of a hydroxypyrrole is substantially deactivated by a 2-carboxylic ester group.

4.4.3.2. Reactions with Acids

Simple N-substituted 1H-pyrrol-3(2H)-ones are smoothly O-protonated by trifluoroacetic acid to give solutions that are stable at room temperature indefinitely.[155] The ^{1}H and ^{13}C NMR spectra of these solutions show changes, when compared with the neutral species, that are typical of acyclic O-protonated enaminones.[155] The pyrrolones **89–91** were also sufficiently basic to form monomeric crystalline salts with picric acid, and the results of an X-ray crystal structure determination of the *tert*-butyl derivative **89** are shown in Fig. 4.5.[155] The shortening of the N(1)—C(5) and C(3)—C(4) bond lengths and lengthening of the C(4)—C(5) and C(3)—O(3) bond lengths relative to those of a typical neutral species (cf. Fig. 4.1) due to the increased delocalisation, is particularly marked.

Deuterium exchange at the 2- and 4-positions of **89** ($t_{1/2} \sim 6$ and <1 min, respectively) takes place in [^{2}H]trifluoroacetic acid. Control experiments

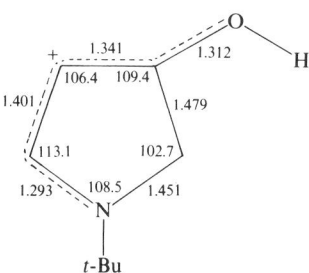

Figure 4.5. Selected bond lengths (Å) and angles (degrees) of the 1-*tert*-butyl-3-hydroxy-1,2-dihydropyrrolium cation.

suggest that the exchange at the 4-position takes place via attack on the *O*-unprotonated pyrrole form, whereas exchange at the 2-position probably occurs via the hydroxypyrrole tautomer[155] (Scheme 31). The rate constant for the ketonisation of 3-hydroxypyrrole in D_2O/dioxan is $\sim 10^6$ times greater than that for deuterium exchange of pyrrole itself.[7] *O*-Protonation of the electron-rich[104] and electron-deficient[134] pyrrolones **92** and **93** has also been reported, together with the effect of protonation on the UV spectra.[134] Under more vigorous conditions, the ester **93** ($R^4 = Et$) undergoes a self-condensation to give a dimer,[1,130] which is thought to have the pyrrolone structure **94**.[159]

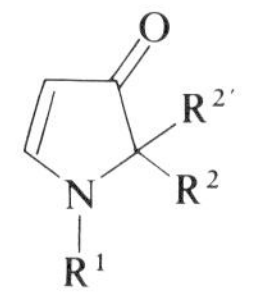

89 $R^1 = t\text{-Bu}; R^2 = R^{2'} = H$
90 $R^1 = Ph; R^2 = R^{2'} = H$
91 $R^1 = i\text{-Pr}; R^2 = R^{2'} = Me$

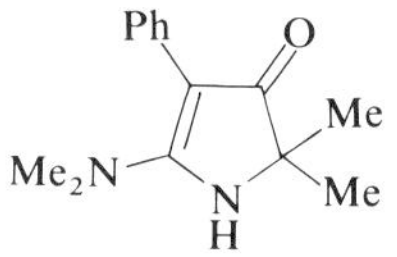

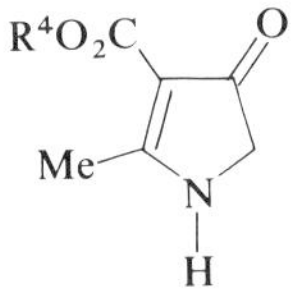

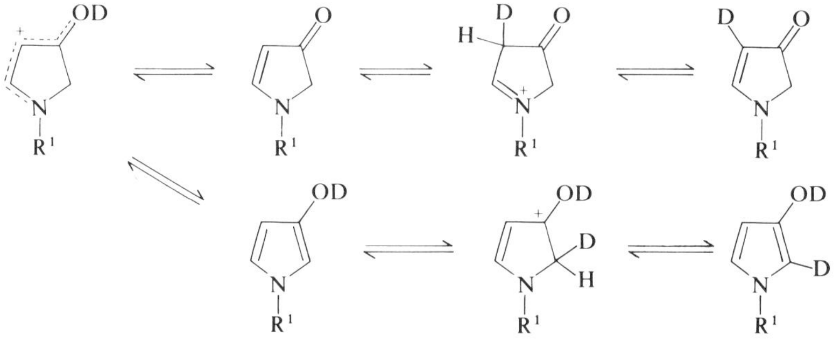

Scheme 31

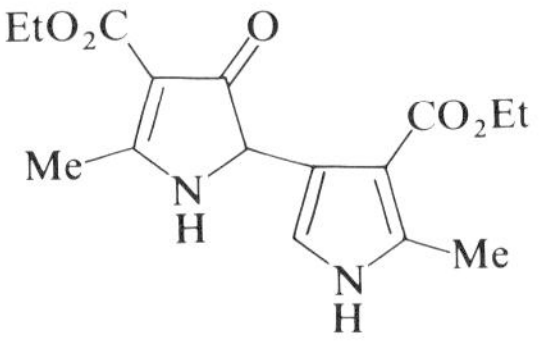

A number of processes have been shown to occur when 2-substituted 2-hydroxy (or related) derivatives are treated with acids. In the absence of other reagents, rearrangement to a tetramic acid derivative can take place, and the constitution of the product **95**, obtained from the corresponding 2-aryl-2-hydroxypyrrol-3(2*H*)-one, was established by an X-ray crystal structure.[90]

4.4. Chemical Properties of 3-Hydroxypyrroles and 1H-Pyrrol-3(2H)-ones 565

Drastic hydrolysis to 2-hydroxyfuran-3-ones has also been observed.[125,128] Treatment of the 2-hydroxypyrrolones **96** (R^1 = H, alkyl; R^2, R^5 = alkyl, aryl) with acid (H_2SO_4/H_3PO_4) in the presence of active aromatics (phenols, anisoles, etc.) gives the 2-aryl derivatives **97** in excellent yield, although often as a mixture of the *ortho*- and *para*-isomers[160] (Scheme 32).

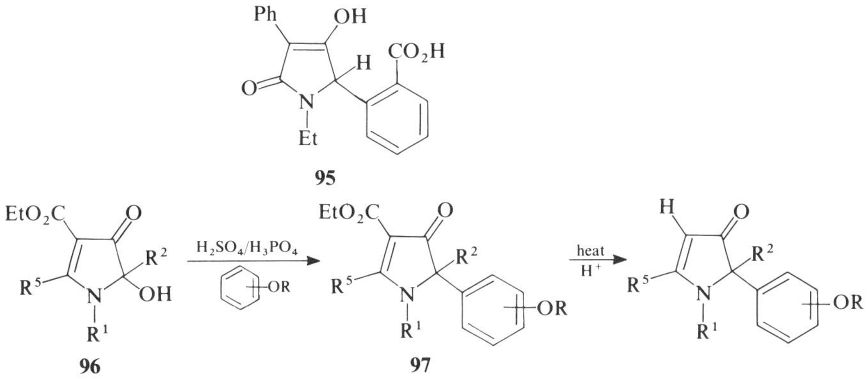

Scheme 32

Under milder conditions (H^+ or Lewis acid catalysis), the corresponding intramolecular reaction can take place (Scheme 33). Although this process is more efficient with activated aromatics as the pendant group, a 17% yield of the product **98** (X = H) could be obtained.[161]

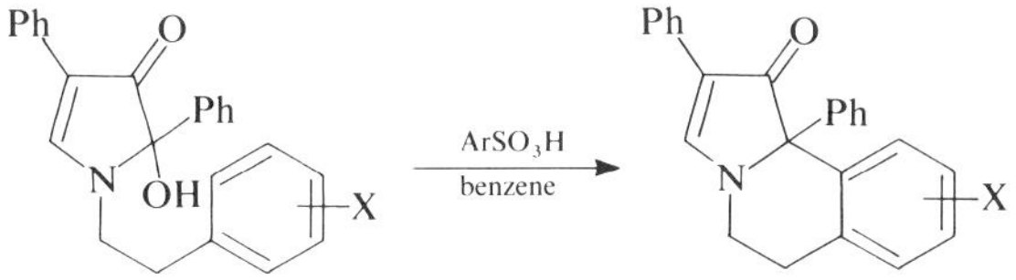

Scheme 33

The direction of cyclization of the tryptamine derivatives **99** and **100** is apparently strongly dependent on the substituent pattern. The diphenyl derivative **99** cyclizes to give to the "expected" product **101**,[161] whereas the ester **100** undergoes Michael-type addition of the tryptamine at the 5-position of the pyrrolone, followed by consolidation to **102** by loss of H_2O^{98} (Scheme 34). This type of process had been observed earlier when the pendant group contained amino or mercaptan functions.[161]

The sulfoximines **103** (R^1 = alkyl, aryl) are decomposed by acids to give dark brown solutions, which may contain the cations **104** (Scheme 35). Treatment of these solutions with nucleophiles (alcohols, thiols, thiocyanate, or acetylacetone) gives the trapped product **105** in 60–95% yield.[162]

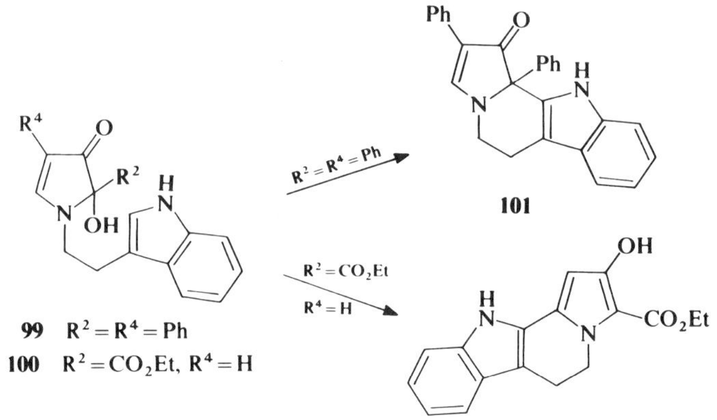

99 $R^2 = R^4 = Ph$

100 $R^2 = CO_2Et$, $R^4 = H$

Scheme 34

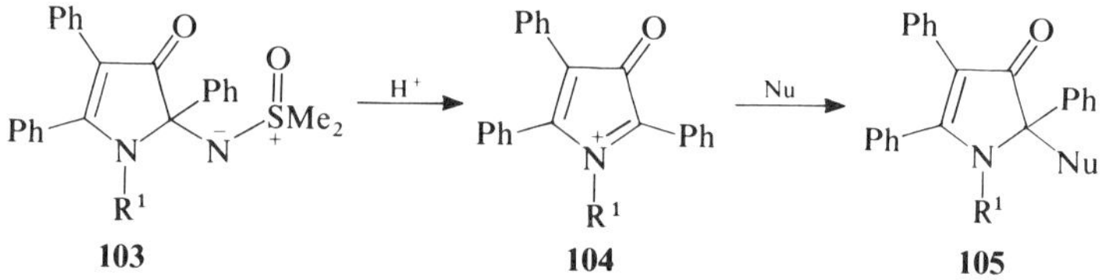

Nu = OR, SR, SCN, Ac_2CH

Scheme 35

4.4.3.3. Halogenation, and Reactions of the Products

No halogenation reactions of simple alkyl or aryl 3-hydroxypyrroles have been reported. Indeed, the 1-phenyl derivative undergoes rapid decomposition on treatment with bromine or NBS. However, the 2,2-disubstituted pyrrolone **106a** is smoothly brominated by either of these reagents to give the 4-bromo compound **106b** in >80% yield,[102] and the appropriate halogeno succinimide gives access to the corresponding chloro **106c** (91%) and iodo **106d** (76%) derivatives.[102] These are stable compounds. Protodehalogenation under acidic

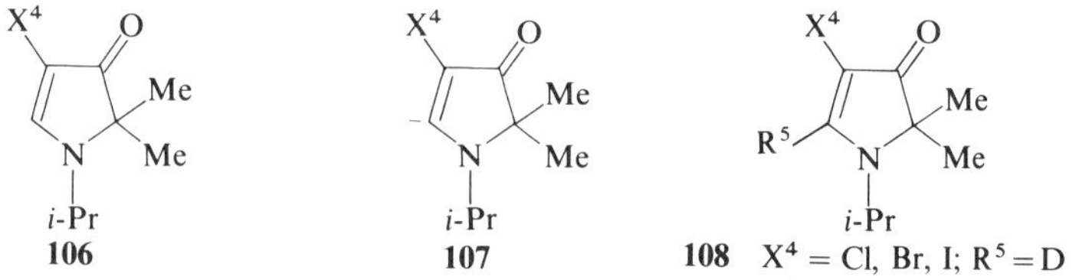

106

a $X^4 = H$
b $X^4 = Br$
c $X^4 = Cl$
d $X^4 = I$

107 $X^4 = Cl, Br, I$

108 $X^4 = Cl, Br, I; R^5 = D$
109 $X^4 = Cl, R^5 = Et$

4.4. Chemical Properties of 3-Hydroxypyrroles and 1H-Pyrrol-3(2H)-ones

conditions is slow, although it occurs readily in the presence of soft nucleophiles such as triphenylphosphine.[102] Reductive dehalogenation, for example, with zinc in methanol–water, has also been reported.[163] Surprisingly, bases can abstract the 5-proton to give a vinyl anion derivative **107** that may be quenched by D_2O to give the 5-deuterio derivatives **108**, or by addition of excess iodomethane to give a low yield of the 5-ethyl compound **109**, formed by further reaction of the initial 5-methylated product.[102]

Chlorination of N-unsubstituted 1H-pyrrol-3(2H)-ones with elemental chlorine or with PCl_5 can lead to a 4,4-dichloro derivative, e.g., **110**, from **111**.[17] If the 4-position is blocked, monochlorination takes place as expected to give the analogous product.[17] The hydroxypyrrole-2-carboxylic ester **112** reacts with 1 mol of bromine to give the 4-substituted product **113** and with 2 mol to give the 4,5-disubstituted compound **114**.[153]

112 $X^4 = X^5 = H$
113 $X^4 = Br, X^5 = H$
114 $X^4 = X^5 = Br$
115 $X^4 = H, X^5 = Br$

Similarly, the diester **116a** is smoothly brominated in high yield at the 5-position. Selective hydrolysis and decarboxylation at the 4-position leads to the 5-bromo compound **115** in low yield after chromatography. As expected, this can be brominated under standard conditions to give **114**.[153] There is one example of chlorinative deoxygenation of a pyrrolone. Thus, treatment of **117** with neat phosphorus oxychloride for 1 h at 40°C gave the 3-chloro derivative **118** in >60% yield.[135]

116a $X^5 = H$
116b $X^5 = Br$

4.4.3.4. Nitrosation, and Reactions of the Products

Nitrosation with dilute nitrous acid to give 2-oximinopyrrolones **119** has been known since the early days of hydroxypyrrole chemistry.[1,63,69] In the

4-carboxylic ester series, the reaction proceeds well (∼ 70% yield) with N-unsubstituted, N-substituted, 5-unsubstituted, or 5-substituted substrates,[1,63] and the corresponding product is also obtained from 4-cyano derivatives.[69] Nitrimine derivatives may be formed in the presence of excess nitrite.[1,63,69] A careful study of the reaction of 4,5-dimethyl-1H-pyrrol-3(2H)-one with nitrous acid has shown that the initial product is the N-nitroso compound **120**, which can be isolated in >**70%** yield when the reaction is carried out at 0°C.[164] This is transformed into the oximino compound **119** (R^1 = H, R^4 = R^5 = Me) on heating.[164]

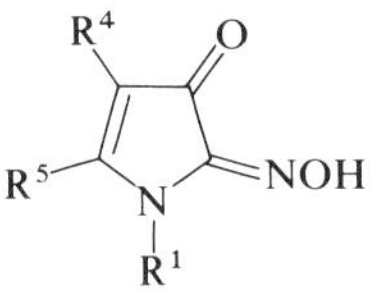

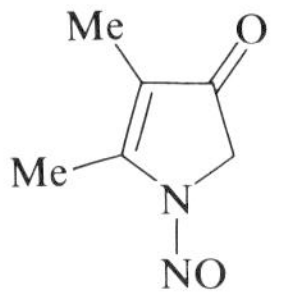

Attempted nitrosation of the ester **121** using sodium nitrite in acetic acid produces a curious rearrangement;[165] ring expansion to the pyrimidinone **122** takes place. The structure of the product was proved by hydrolysis to products including 1-benzyluracil **123** (30%), and hydrogenation to the ester **124** (84%). A suggested mechanism for the ring expansion is given in Scheme 36[165] (cf Ref. 36).

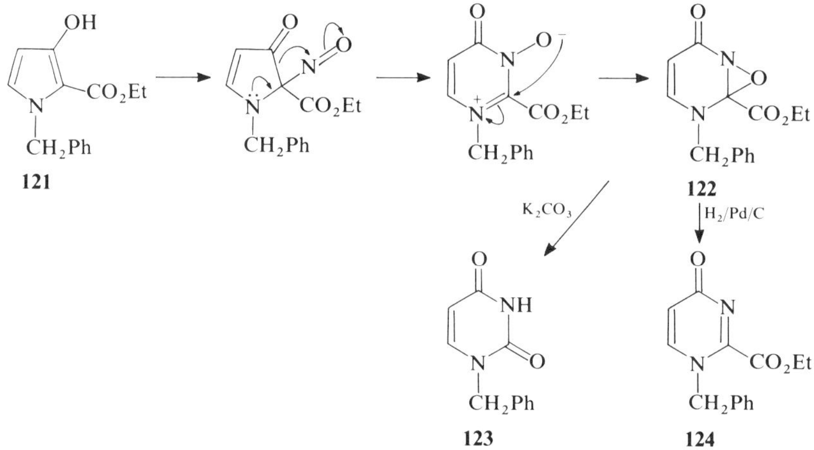

Scheme 36

A related ring expansion has been observed with a 2-hydrazono derivative (Section 4.4.3.5) and also on thermolysis of the betaines **125** (Scheme 37).[127]

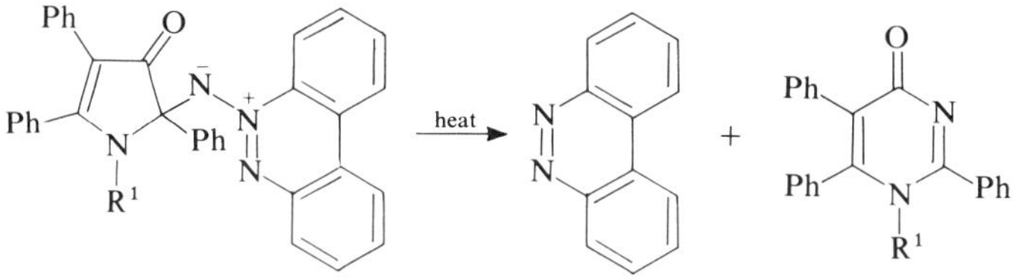

4.4. Chemical Properties of 3-Hydroxypyrroles and 1H-Pyrrol-3(2H)-ones

Scheme 37

125
R¹ = H or Ar

4.4.3.5. Diazo-Coupling Reactions and Reactions of the Products

The situation regarding azo-coupling reactions of 3-hydroxypyrroles is very similar to that of nitrosation (Section 4.4.3.4). Treatment of 4,5-dimethyl-1H-pyrrol-3(2H)-one with benzene diazonium chloride in methanol–water gives the 2-coupled product **126** (R¹ = H, R⁴ = R⁵ = Me; 62%) as a yellow precipitate.[164] Comparison of the UV spectrum of this product with an O-benzyl analog suggested that the product adopts the keto-hydrazone tautomer **126 K** in solution.[164] Similar reactions with 1-phenyl- or 1-*tert*-butyl-1H-pyrrol-3(2H)-one generate the products **126** (R¹ = Ph, R⁴ = R⁵ = H and R¹ = *t*-Bu, R⁴ = R⁵ = H respectively), which were also shown to adopt the keto-hydrazone tautomer in solution, by detailed analysis of their ¹³C NMR spectra.[166] An X-ray crystal structure of **127** has confirmed that the same tautomer is the stable form in the solid state.[166]

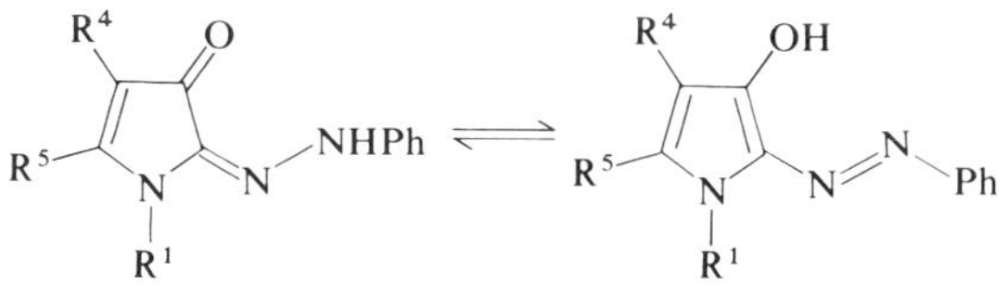

126 (keto form)   126 (enol form)

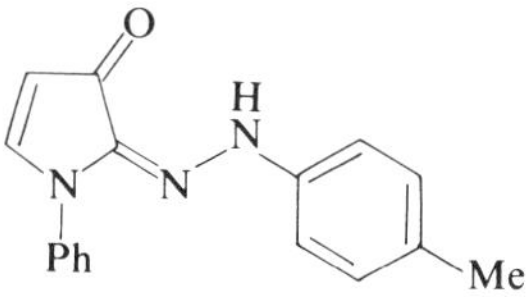

127

A range of 1-substituted 4,5-diaryl-2-hydrazonopyrrolones, (**126**) has been made in 30–90% yield by condensation of 2-imino-3(2H)-furanones with hydrazine derivatives, rather than by a coupling process.[167]

The 2,2-disubstituted pyrrolone **128**, in contrast, gives true azo compounds **129** (60–80%) by reaction at the 4-position.[102] Reductive cleavage of the azocarboxylic acid **129** (Ar = —C$_6$H$_4$CO$_2$H) with tin(II) chloride gave a good route to the amino compound **130** (72%) since the coformed *p*-aminobenzoic acid could be readily separated on workup. The amine **130** showed typical properties. It formed a picrate salt, could be acetylated, and could be diazotised to give a stable diazonium salt which coupled with β-naphthol[102] (Scheme 38). 5-Aminopyrrolones give amidines with DMF acetals.[110]

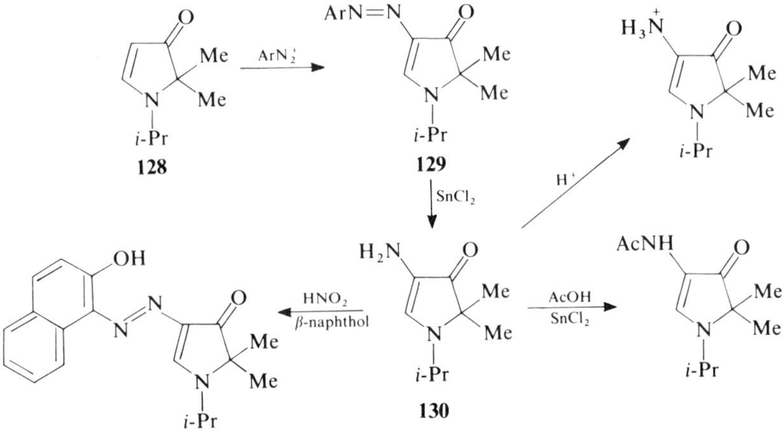

Scheme 38

Even hydroxypyrrole-4- or -5-carboxylic acid derivatives are sufficiently reactive to couple with arene diazonium salts,[1,63,72,130] although certain tetra-substituted derivatives are inactive.[130] In most of these cases, the reaction is assumed to occur at the 2-position. Indeed, reaction of the 2-carboxylic esters **131** still takes place at this site to give either the straightforward coupled product **132**, via a Jaap–Klingeman reaction at high pH, or the ring-expanded product **133** (67%) at neutral pH[165] (Scheme 39). A similar ring expansion takes place in the corresponding nitrosation (Section 4.4.3.4).

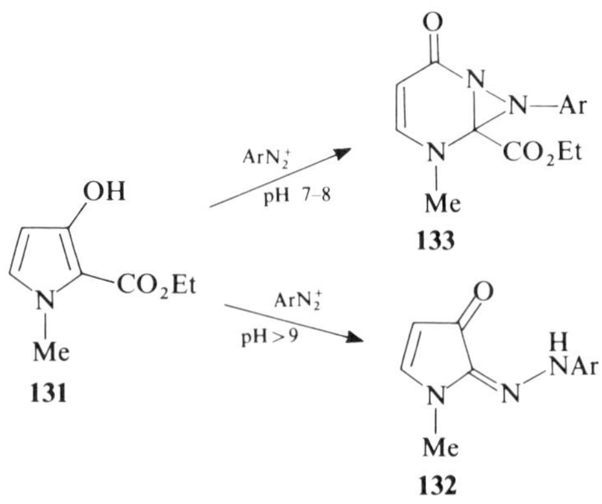

Scheme 39

4.4.3.6. Friedel–Crafts Acylation and Vilsmeier and Related Reactions

Only isolated examples of these reactions are known (cf Scheme 28), though the Vilsmeier reaction in particular has been thoroughly studied in the 3-methoxypyrrole series (Section 4.5.3.3.5).

The only genuine Friedel–Crafts acylation involves the formation of the 2-benzoylated product **135** (80% yield) by reaction of **134** with phthalic anhydride in acetic acid.[130] The related compounds **136**, **137**, and **138** are similarly available using acetonitrile/HCl, ethyl cyanoacetate, and phenylisocyanate, respectively.[130] Vilsmeier reaction of **134** with $DMF/POCl_3$ gives the expected aldehyde **139** in an extremely poor yield ($< 3\%$), together with a small amount of the chloro compound **140**.[168]

134

135 $R^2 = 2\text{-}C_6H_4CO_2H$
136 $R^2 = COMe$
137 $R^2 = CH_2CO_2Et$
138 $R^2 = NHPh$
139 $R^2 = H$

140

In contrast to these poorly defined reactions, methoxymethylene Meldrum's acid **141** acts as a mild C-electrophile and generally gives well-characterized

141

products in good yield on treatment with N-substituted 3-hydroxypyrroles, 3-methoxypyrroles (Section 4.5.3.3.5), or 1H-pyrrol-3(2H)-ones,[26,102,169] at room temperature in acetonitrile. 2-Unsubstituted compounds react preferentially in the 2-position,[169] unless the N-substituent is bulky (e.g., t-Bu), whereupon the reaction is diverted (via the pyrrolone form) to the 4-position[26] (Scheme 40). In either case, pyrolysis of the product leads to fused pyrone derivatives[26,169] (Scheme 40).

If the 2-position is blocked by monosubstitution, then reaction with **141** occurs at the 5-position, although again bulky N-substituents can give rise to the 4-substituted product.[26] 2,2-Disubstituted 1H-pyrrol-3(2H)-ones give rise to the

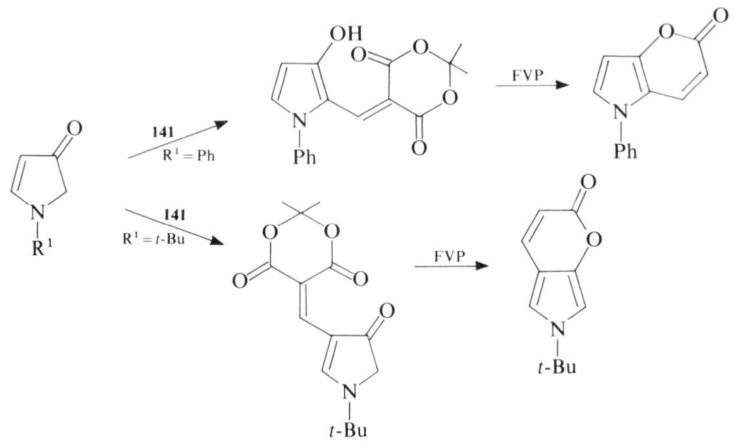

572   3-Hydroxypyrroles

Scheme 40

4-substituted products e.g., **142** (73%).[102] The 6-membered ring of this compound is readily cleaved with alkoxide to give the half-ester **143** (100%), which is easily transformed into malonate **144** or acrylate **145** derivatives (Scheme 41).[102]

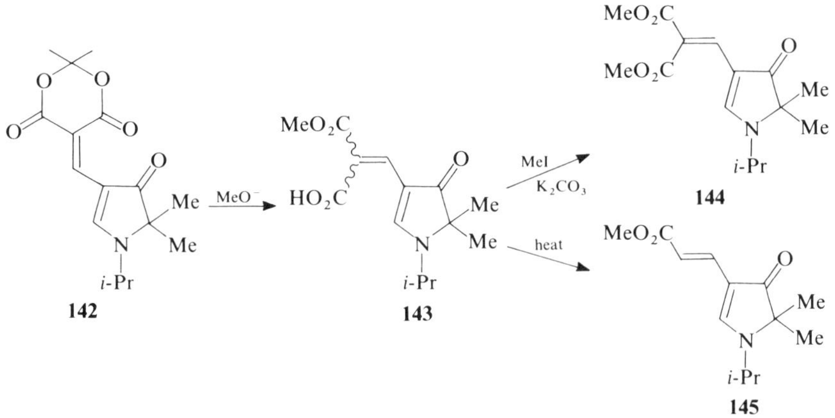

Scheme 41

4.4.3.7. Reactions with Carbonyl Compounds

The reaction of 3-hydroxypyrroles with carbonyl compounds, in particular with aldehydes, is well known to lead to 2-ylidene products by a formal aldol reaction. Surprisingly, the reaction is most often reported for *N*-unsubstituted derivatives, and, indeed, attempted reactions of simple *N*-substituted hydroxypyrroles with benzaldehyde derivatives gave materials that could not be characterized[169] (see, however, Ref. 46).

4.4. Chemical Properties of 3-Hydroxypyrroles and 1H-Pyrrol-3(2H)-ones 573

4,5-Dialkyl derivatives, e.g., **146** ($R^4 = R^5 = Me$), have been well studied (Scheme 42), and give the benzylidine derivatives **147** on treatment with the appropriate aldehyde in methanol at reflux.[8,46,164,170] Condensation with oxalacetic ester has also been reported.[171] Yields are in the range 40–80%. An N-substituted example gives a mixture of E- and Z-isomers of the product.[46] A formylpyrrole has also been used.[172] Reaction with glyoxal gives the expected 2 : 1 condensation products **148**[46,164] although generally in low yield, whereas trimethyl orthoformate in the presence of boron trifluoride etherate gives the product **149** almost quantitatively[8] (see also Ref. 62).

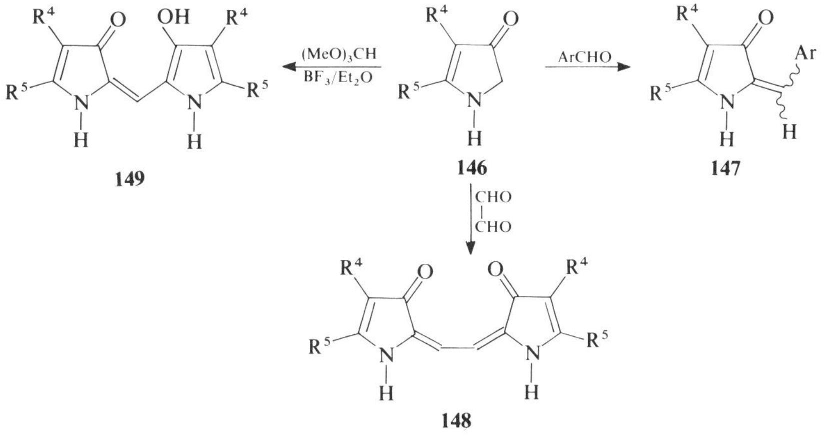

Scheme 42

Although aliphatic ketones have not been employed (see Section 4.4.8), the isomeric condensation products **150** and **151** were obtained by reaction with isatin and 2-methoxyindol-3-one, respectively.[38]

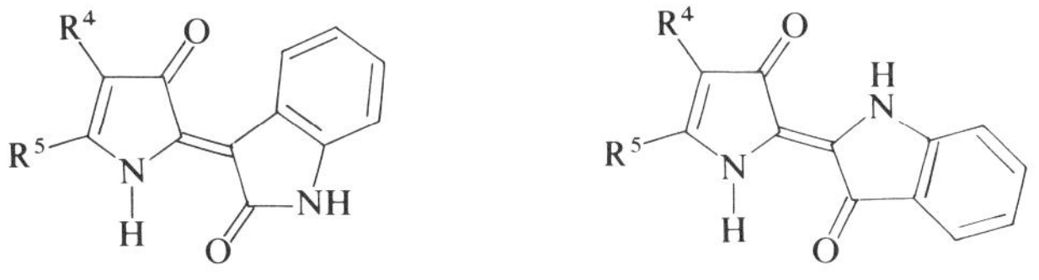

Severin has reported the formation of 2-benzylidine derivatives from 5-aryl-1H-pyrrol-3(2H)-ones.[13] Flitsch has condensed the N-acyl derivative **152** with the pentamethinium salt **153** (which acts as an aldehyde equivalent) in the presence of base.[173,174] The product **154** was deacylated and pyrolysed in refluxing trichlorobenzene to give the unstable azaazulenone **155** in 28% yield[173,174] (Scheme 43). 3-Hydroxypyrrole-5-carboxylic acid condenses with isatin in good yield.[175]

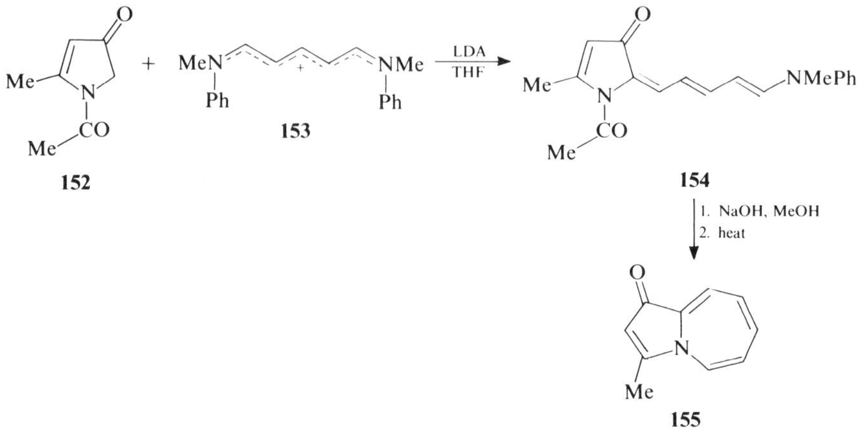

Scheme 43

2-Ylidene derivatives have also been obtained from the 4-carboxylic ester **156**. Condensation with benzaldehyde,[100] substituted benzaldehydes,[176] heterocyclic aldehydes,[130,135] and isatin[1] have all been reported. Treibs and Ohorodnik[130] have obtained a number of interesting products by acid-catalyzed reaction with β-dicarbonyl compounds. Although acetylacetone gives the 2 : 1 product **157**, 3-oxobutyraldehyde acetal gives a 1 : 1 product **158**, which can

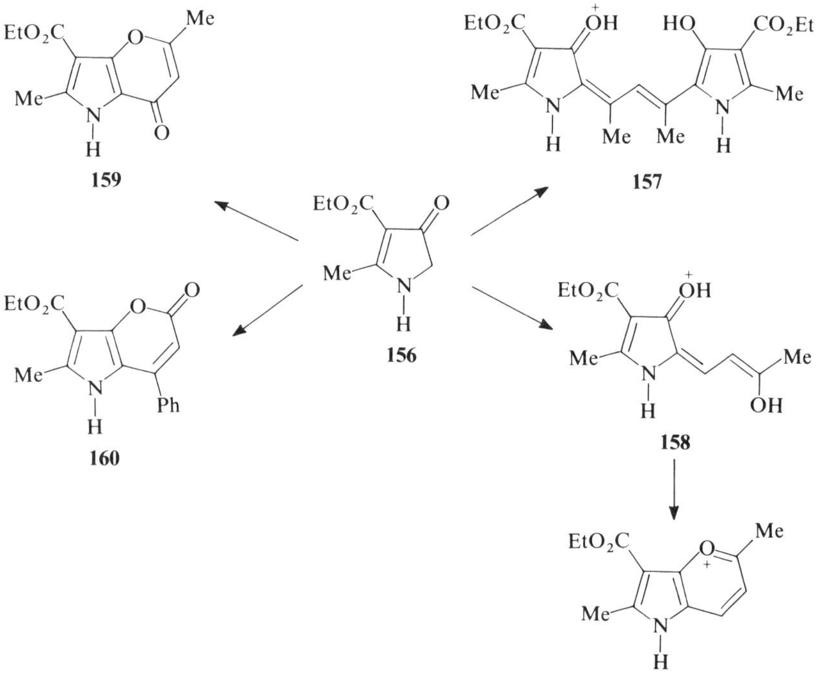

Scheme 44

4.4. Chemical Properties of 3-Hydroxypyrroles and 1H-Pyrrol-3(2H)-ones

subsequently cyclize. Pyrones **159** and **160** are obtained from ethyl acetoacetate and ethyl benzoylacetate, respectively,[130] but the evidence, and reasons, for the change in regiochemistry are unclear (Scheme 44).

A series of 2-substituted 3-hydroxypyrrole-4-carboxylic esters and 4-cyano derivatives has been shown to condense with pyruvic aldehyde to give the aldol products, e.g., **161**.[67]

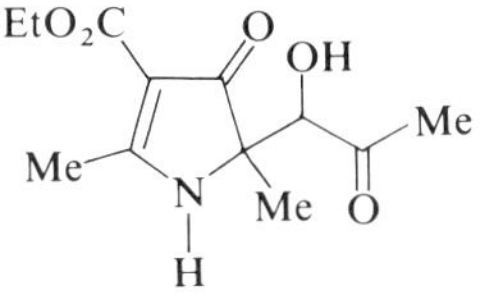

161

4.4.3.8. Mannich Reaction

The 3-hydroxypyrrole **162** (R^4 = H) reacts with formaldehyde and dimethylamine (1 h, at 20°C) to give a quantitative yield of the Mannich base **163** (Scheme 45).[153] The fact that reaction takes place at the 4-position confirms the deactivating effect that 2-carboxylic ester functions have toward electrophilic attack at the 5-position.

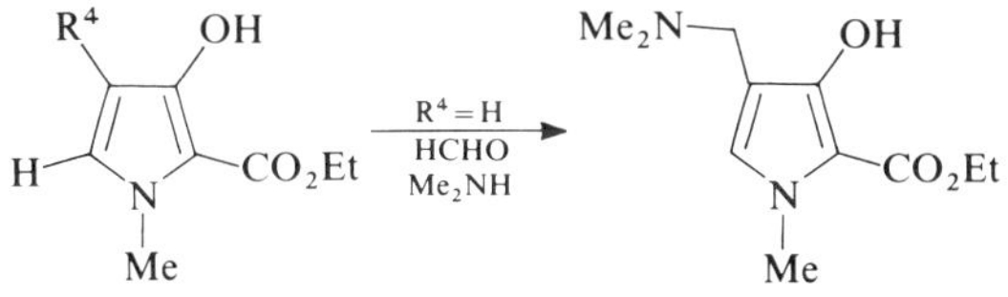

Scheme 45

Indeed, attempted reaction of the bromo compound **162** (R^4 = Br) under similar conditions gave a very low yield of the expected product despite long reaction times.[153]

No other applications have been reported.

4.4.3.9. Reactions with Dimethyl Acetylenedicarboxylate

Reaction of the 4,5-dimethylpyrrolone **164** with dimethyl acetylenedicarboxylate (DMAD) in methanol gives an almost quantitative yield of a 1 : 1 adduct assigned as **165**, formed by an initial Michael-type addition to the 2-position of the pyrrolone, followed by tautomerism of the adduct (Scheme 46).[171] Cyclization to the pyrone **166** takes place in good yield on treatment with base.[171]

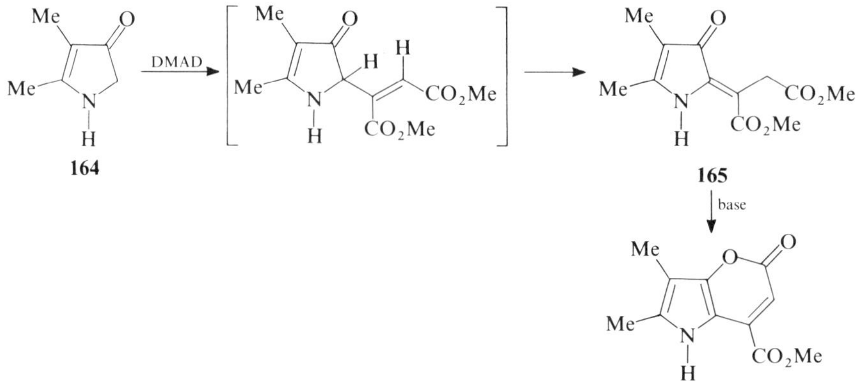

Scheme 46

The reaction of DMAD with a range of *N*-substituted hydroxypyrroles has been carried out using DMSO as solvent in order to maximize the amount of the enol tautomer of the pyrrole.[26] Depending on the substitution pattern of the pyrrole, a number of products have been isolated, including Michael adducts **167** (pyrrolone and/or hydroxypyrrole forms; *E*- and/or *Z*-isomers) and formal cycloadducts **168**. The factors that control these reactions are incompletely understood, but it appears that the presence of a 2-substituent favors the formal cycloaddition. Similar reactions have been carried out for corresponding methoxypyrroles (Section 4.5.3.3.7). These are rare examples of uncatalyzed Diels–Alder-type reactions of pyrrole derivatives, although a dipolar stepwise mechanism cannot be excluded. The adducts **168** eliminate ketene on flash-vacuum pyrolysis (FVP) at 600°C (10^{-3} torr) to give pyrrole 3,4-dicarboxylic ester derivatives.[26]

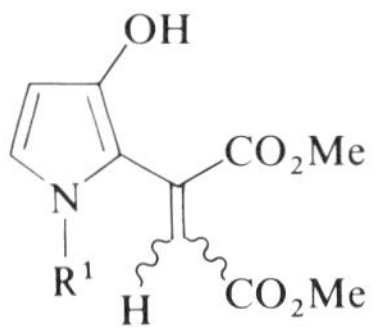

The azacyclopentadienone **169**, generated in situ from either the acetate[122] **170** or the dimer[86,115] **171**, undergoes cycloaddition with DMAD to give the pyridine derivative **172** after loss of carbon monoxide (see Scheme 47).

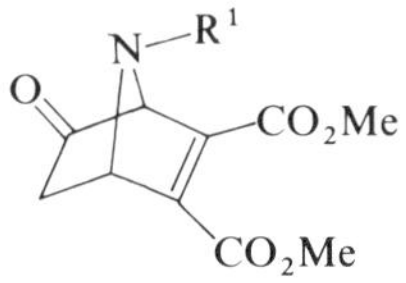

The dicarboxylic ester derivative **173** is sufficiently deactivated that on reaction with DMAD in the presence of base it behaves as an *O*-nucleophile and gives the product **174** of conjugate addition.[83]

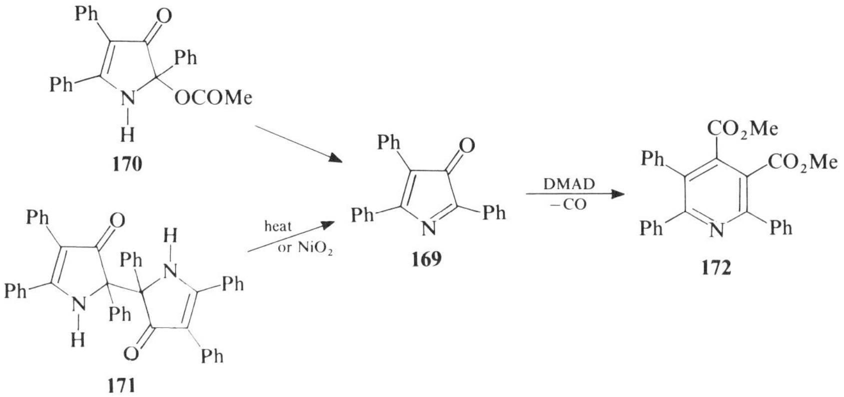

4.4. Chemical Properties of 3-Hydroxypyrroles and 1H-Pyrrol-3(2H)-ones

Scheme 47

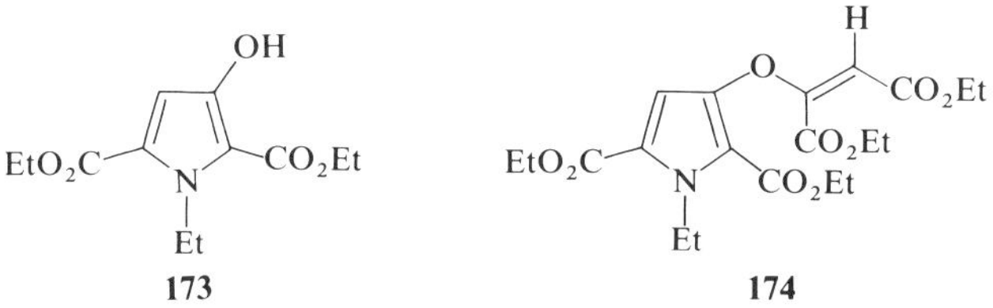

4.4.3.10. Michael and Conjugate Addition Reactions

Michael addition of DMAD has been presented in Section 4.4.3.9. Treatment of the 3-hydroxypyrrole-2-carboxylic ester **175** with ethyl acrylate in the presence of base gives the "expected" Michael adduct **176** (29%), together with a small amount of the bis-adduct **177** (7%) after cleavage of the 2-carboxylic ester under the reaction conditions[75,153] (Scheme 48). The corresponding N-unsubstituted precursor gives a mixture of the expected C(2)-adduct and the bis-adduct formed by sequential reaction at the 2- and 1-positions.[75,153]

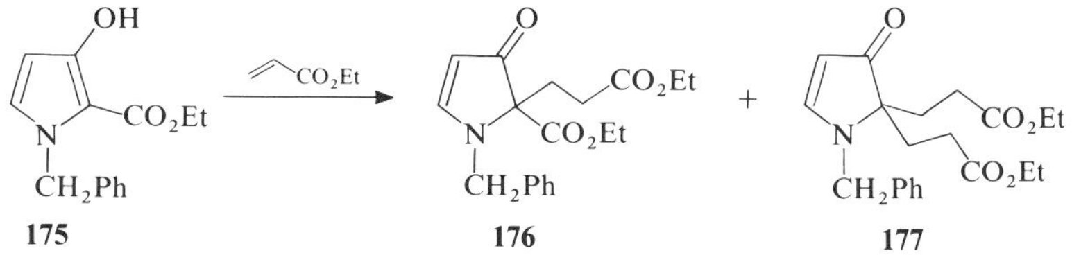

Scheme 48

An intramolecular conjugate addition via **178**, obtained on degradation of an acetoxyazetidinone, has been proposed as a rationalization for the formation of the bicyclic product **179**.[53]

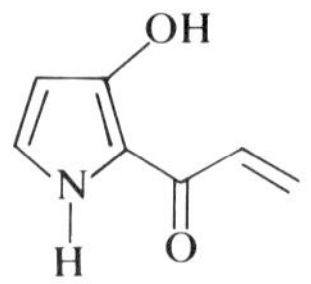

4.4.4. Reactions with Bases

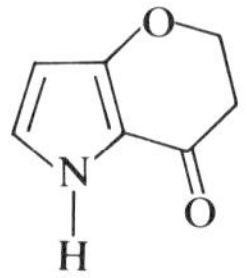

Few reactions come into this category. Although alkylation reactions take place smoothly under basic conditions (Section 4.4.1), simple pyrrolyloxy salts have never been isolated and no pK_a data have been recorded. A (rather small) effect of mild base on the UV spectrum of a 1H-pyrrol-3(2H)-one-4-carboxylic ester has been reported.[134] It is thought that simple 3-hydroxypyrroles are rather less acidic than phenols.[138]

Provided R^2 has a high migratory aptitude (benzyl, furfuryl, etc), acyloin rearrangement of the 2-hydroxy compounds **180** (R^1 = H, alkyl, R^5 = alkyl) to the 3-hydroxy derivatives **181** takes place in 60–80% yield[177] (Scheme 49).

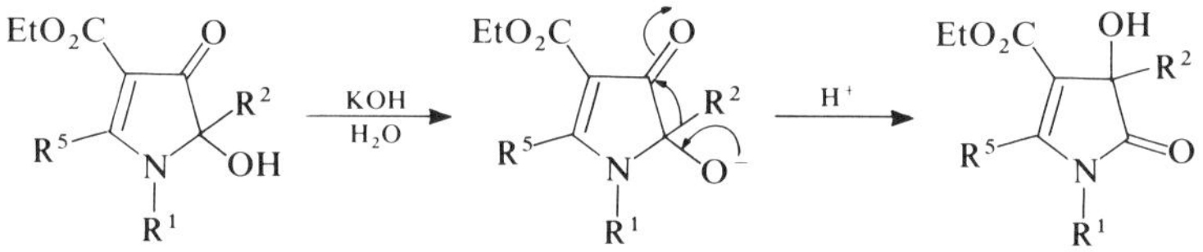

Scheme 49

Certain N-unsubstituted substrates fail to react, perhaps owing to the formation of the anion **182**.[177]

The salts **183** are somewhat acidic, and show deuterium exchange at position 2 in D_2O at room temperature ($t_{1/2} < 1$ h).[43]

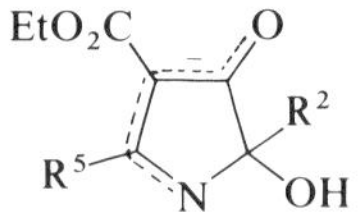

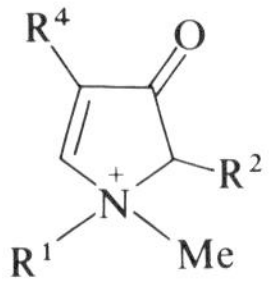

4.4.5. Oxidation of 3-Hydroxypyrroles and 1H-Pyrrol-3(2H)-ones

Oxidation processes are among the more important reactions of 3-hydroxypyrroles. Simple 2-substituted 3-hydroxypyrroles are exceedingly prone to aerial

4.4. Chemical Properties of 3-Hydroxypyrroles and 1H-Pyrrol-3(2H)-ones 579

oxidation to give 2-hydroxy-1H-pyrrol-3(2H)-ones **184**, whereas 2-unsubstituted derivatives undergo dimerization to give indigo-like structures **185**. Both of these processes have been rationalized by a single unifying mechanism[137] in which the captodative-stabilized[178] radical **186** is the key intermediate (Scheme 50). When the 2-position is substituted, dimerization of this species is reversible, and dedimerization can initiate an autoxidation chain that normally leads to the 2-hydroxy compound **184**. When the 2-position is unsubstituted, tautomerization and further hydrogen abstraction from the dimer **187** leads to the persistent radical **188** (Section 4.3.7) and hence to the indigoid product **185**.[137]

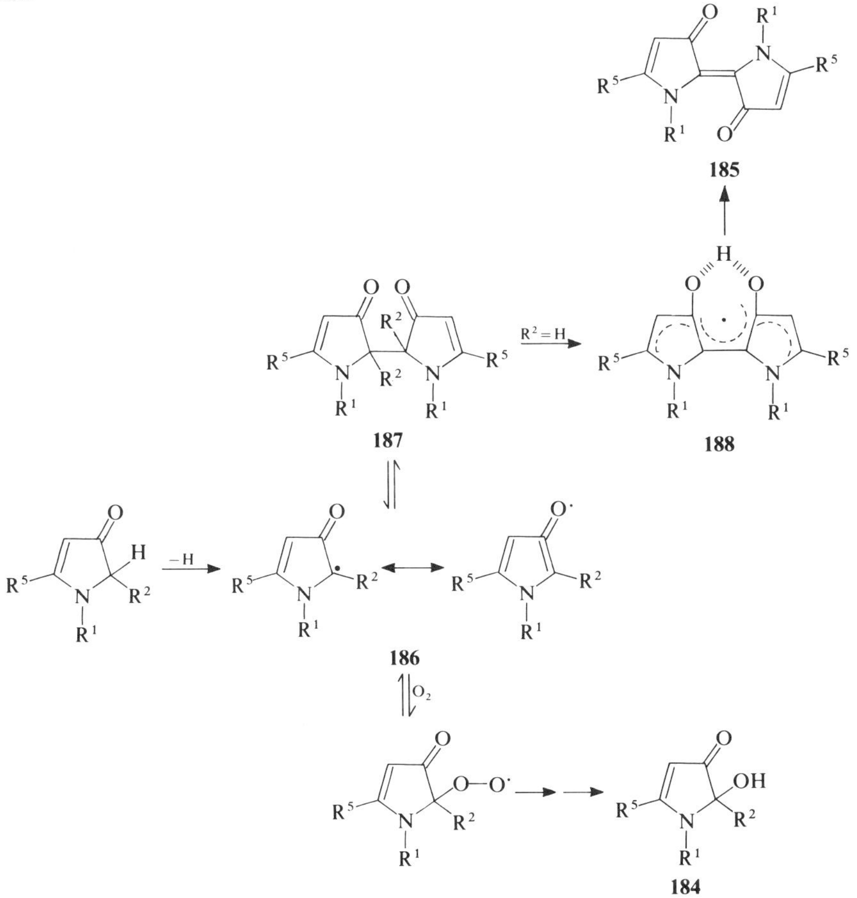

Scheme 50

4.4.5.1. Aerial Oxidation

2-Alkyl- and 2-aryl-3-hydroxypyrroles, in particular, are frequently very unstable toward oxidation and absorb oxygen on exposure to air. The 2-

hydroxypyrrolones **184** are generally isolated from 2-alkyl derivatives, but yields are usually low (10–20%) and seldom greater than 30%. The reaction takes place with N-unsubstituted[67,99,100] or N-substituted[32,41,126,137] derivatives, and is also compatible with the presence of electron-withdrawing groups in the 4-position.[67,99,100,179]

The reactions of 2-aryl-3-hydroxypyrroles with oxygen are more complex. The 2,4,5-triphenyl derivative is unusual in that the dimeric form **187** ($R^1 = H$, $R^2 = R^4 = R^5 = Ph$) is relatively stable,[115] possibly owing to intramolecular hydrogen bonding. Indeed, early reports of the preparation of 2,4,5-triaryl-3-hydroxypyrroles, either by reduction of a pyrrolone-N-oxide[180] or by the cyclopropenone route,[117] were later shown to be in error, and it was established that the dimeric form **187** was in fact obtained.[115] However, continued exposure of such dimers[105] to oxygen in pyridine[117] gives rise to oxazinone derivatives, e.g., **189**, presumably by ring expansion via the initial hydroperoxide (Scheme 51). Other oxidative procedures also give these products in variable (14–85%) yield.[117] The corresponding 1-substituted-2,4,5-trisubstituted compounds cannot undergo the dehydration step and three products are formed by fortuitous oxidation,[36] including the 2-hydroxy compound **190** (26%) [as its open-chain tautomer (Section 4.3.1.2)], the indanone **191** (38%) and the acrylamide **192** (28%). The latter products are thought to arise via an oxazinium

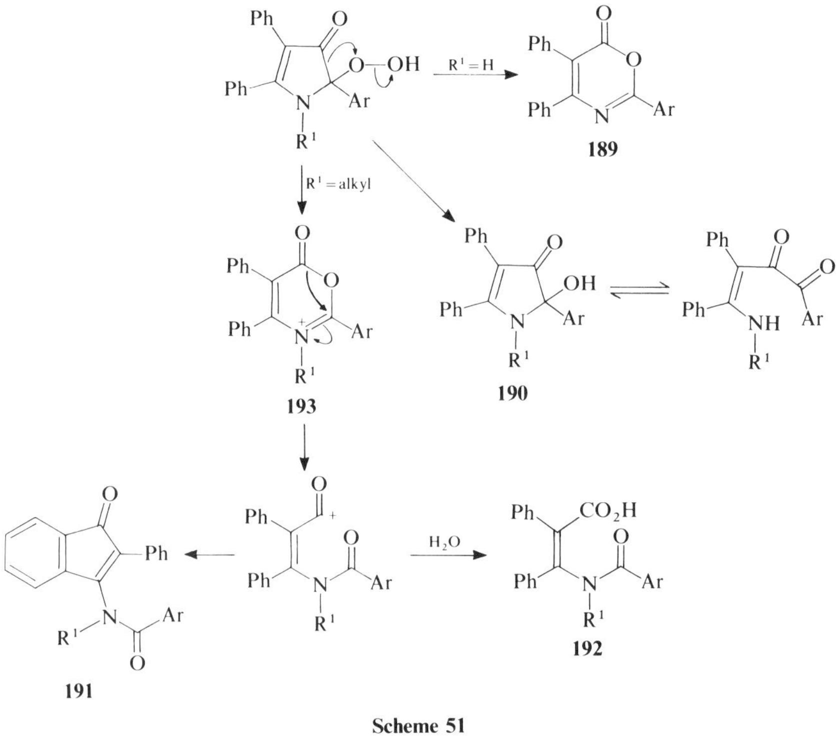

Scheme 51

4.4. Chemical Properties of 3-Hydroxypyrroles and 1H-Pyrrol-3(2H)-ones 581

species **193** (Scheme 51). An electron-donating substituent at the 5-position apparently confers some stability on 2-aryl-1H-pyrrol-3(2H)-ones.[103]

Although 3-hydroxypyrrole-2-carboxylic esters are generally stable, autoxidations of N-substituted[79] and N-unsubstituted[140] derivatives have been reported in the presence of base or by dye-sensitised photooxygenation in the presence of diphenylsulfide.[98] Attempted alkylation of ethyl 4,5-dimethyl-3-hydroxypyrrole-2-carboxylate with iodomethane under basic conditions gave an oxidation product that contained only C-alkyl groups, and to which the structure **194** was assigned.[140] This is the only example of oxidative functionalization taking place at the 4-position. In contrast, the dicarboxylic ester **195** gives two pyrrol-2-ones **197** and **198** in a total yield of 27% after overnight treatment with oxygen and dilute base[79] (Scheme 52). In this case, the base is probably needed to hydrolyze the 2-carboxylic ester at key steps in the sequence. The "normal" hydroxy compound **196** was isolated, together with **197**, when the reaction was performed in nonaqueous base, and was shown to be converted to **198** on treatment with aqueous sodium hydroxide.[79] Other base-induced oxidations have been reported.[115,117]

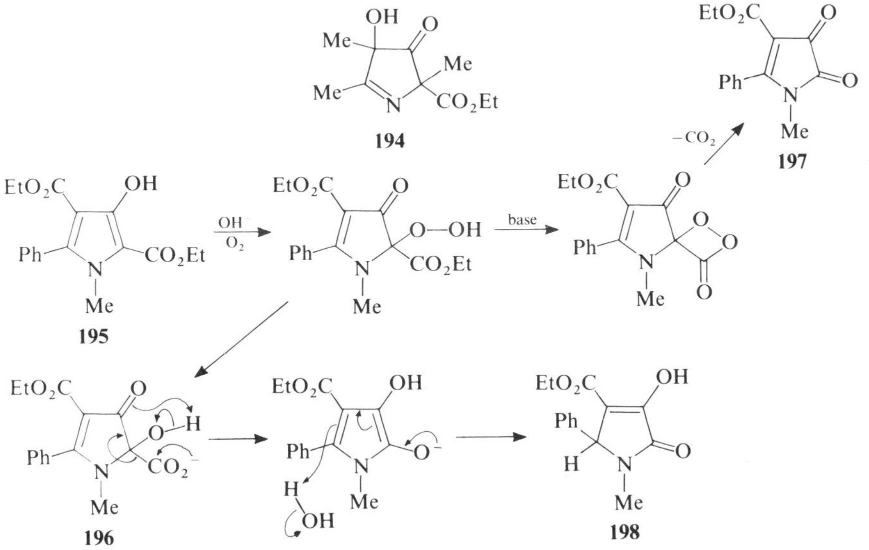

Scheme 52

4.4.5.2. "One-Electron" Oxidizing Agents

Treatment of 2-substituted 3-hydroxypyrroles with potassium ferricyanide is reported to give an increased yield (40–50%) of the 2-hydroxy derivatives obtained by aerial oxidation.[32,137] (Section 4.4.5.1). A related example has been reported for the reaction of a 2-unsubstituted 3-hydroxypyrrole with this reagent under mild oxidation conditions. Thus, the pyrrol-2-one **199** (49%)

probably arises by "dehydration" of the enol form of the hydroperoxide **200**[164] (Scheme 53). Under more vigorous conditions (70–80°C) an oxidative coupling reaction takes place to give, in 47% yield, the highly colored indigoid product **201**.[8,170]

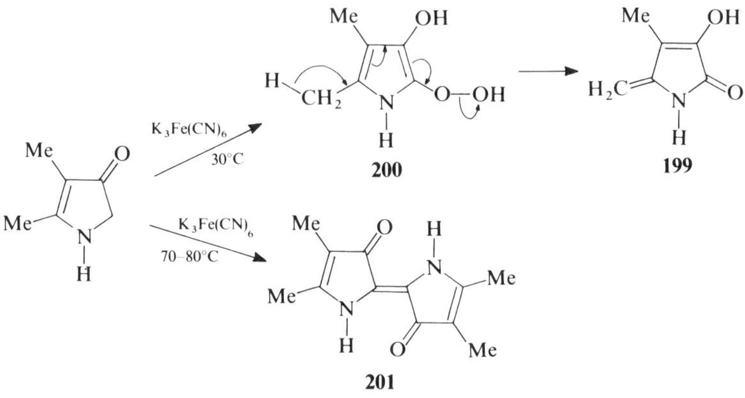

Scheme 53

This type of reaction is more typical of these 2-unsubstituted compounds,[46,481] and a number of blue indigoid-type products have been obtained under a variety of oxidation conditions. For example, the diester **202** has been isolated, in varying yields, from the coupling reaction using ferric chloride[1] (3 equivs), potassium ferricyanide, Fremy's salt, lead tetraacetate, iodine–potassium iodide, silver nitrate, or selenium dioxide.[159] However, such dimerizations are apparently unsuccessful with *N*-substituted substrates.[159] In addition, dyes of unknown constitution have been obtained in a few cases;[72,130,175] the formation of a colored complex between a metal and a 3-hydroxypyrrole-2- or-4-carboxylic acid derivative is a possible complication.[130]

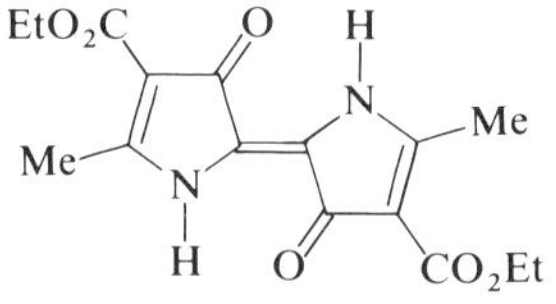

4.4.5.3. Ozonolysis

In one example, ozonolysis of the 4,5-double bond of a 1*H*-pyrrol-3(2*H*)-one, followed by reductive ($Na_2S_2O_5$) workup, gave the expected α-diketone derivative[35,36] (Scheme 54).

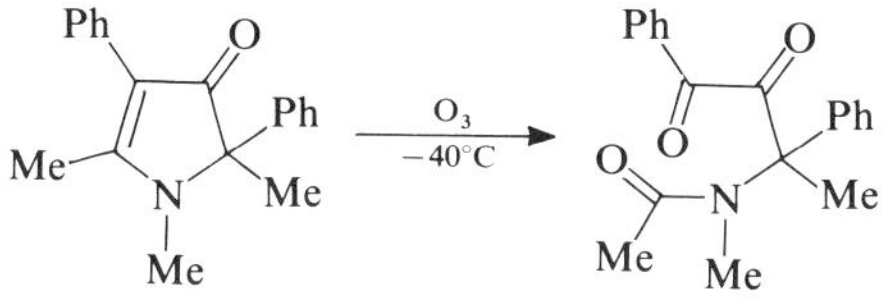

4.4. Chemical Properties of 3-Hydroxypyrroles and 1H-Pyrrol-3(2H)-ones 583

Scheme 54

4.4.5.4. Miscellaneous Oxidations

The oxidation of thiourea with iodine–potassium iodide in the presence of the pyrrole **203** gave a good yield of the isothiouronium salt **204**[182] (Scheme 55). The reaction is a general one[182] but has not been further applied in the hydroxypyrrole series.

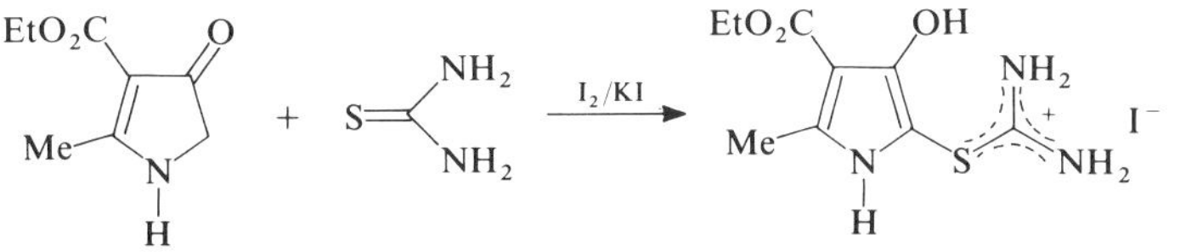

Scheme 55

4.4.6. Reduction of 3-Hydroxypyrroles and 1H-Pyrrol-3(2H)-ones

The 3-hydroxypyrrole [1H-pyrrol-3(2H)-one] nucleus is generally resistant to reduction. 1-Unsubstituted[17] and 2-unsubstituted[169] (and, presumably, 2-monosubstituted) derivatives are inert to lithium aluminum hydride, probably because an anion is formed that is resistant to further attack.[17] In contrast, 1,2,2-trisubstituted derivatives are reduced smoothly, regiospecifically, and in high yield by this reagent to give pyrrolidin-3-one derivatives[35,36] **205**. The Z-configuration of one product **205** ($R^4 = R^5 = Ph$) was assigned on the basis of the magnitude of the vicinal coupling $^3J_{4,5}$.[35,36] The 4,5-double bond of 1,2-dialkylpyrrolone-2-carboxylic esters can be reduced regiospecifically at $-78°C$ by "superhydride" in the presence of boron trifluoride etherate.[60] Pyrrolone derivatives, which were shown [115] to have the dimeric structure **206** can be cleaved and deoxygenated to the triarylpyrroles **207** in 25–70% yield by phosphorus pentasulphide in refluxing benzene;[116,117] some fused pyrrolones have been transformed to the thione by this reagent.[119]

Facile reduction of 2-methylene groups by catalytic hydrogenation[41,100,126] or with sodium borohydride[41] gives rise to 2-monosubstituted derivatives. These products are particularly sensitive to aerial oxidation (Section 4.4.5.1), and so it is the 2-hydroxy compounds **208** which are normally isolated.[41,100,126]

Because of this, the yields are at best only moderate. Reduction of the indigoid compound **209** with zinc in acetic anhydride leads to the corresponding 3,3′-bisacetoxy compound in 45% yield.[157]

Reductive cyclization of a 2-aryl-2-hydroxypyrrolone using Pd/C in acetic acid has been reported.[90]

4.4.7. Hydrolysis and Decarboxylation Reactions of 3-Hydroxypyrrole Carboxylic Esters

Because of the number and variety of 3-hydroxypyrrole carboxylic esters available by direct synthesis (Section 4.2.2), hydrolysis–decarboxylation strategies provide excellent preparative routes to simpler members of the series. Similar reactions of alkoxypyrroles are also important (Section 4.5.3.7).

In many cases, the precise choice of conditions is critical. Thus, Chong and Clezy obtained "intractable tars" from acid hydrolysis, and recovered starting material from attempted base-catalyzed hydrolysis of the 2-carboxylic ester **210a** (R^1 = H, R^4 = R^5 = Me).[44] However, Bauer and Pfeiffer obtained the corresponding decarboxylated product **211** (R^1 = H, R^4 = R^5 = Me) directly in 92% yield by hydrolysis of the ester in boiling 6 M hydrochloric acid (3 h) under an atmosphere of nitrogen.[9] Other 4,5-disubstituted derivatives **211** have been obtained by this method,[9,46] but the route could not be applied to 4,5-unsubstituted compounds.[10] However, hydrogenolysis of the benzyl esters **210b** (R^1 = alkyl, R^4 = R^5 = H) (obtained from the ethyl esters by ester exchange) using 5% Pd/C at atmospheric pressure gave the decarboxylated products **211** (R^1 = alkyl, R^4 = R^5 = H), albeit in unstated yield,[10,183] and acid cleavage of the *tert*-butyl ester **210c** (R^1 = Ph, R^4 = R^5 = H) with boron trifluoride etherate in acetic acid/ether (48 h) gave the *N*-phenyl compound **211** (R^1 = Ph, R^4 = R^5 = H).[10,183] [This compound is, however, better prepared by the pyrolysis route (Section 4.2.1)]. It is of interest that no 3-hydroxypyrrole-2-carboxylic acid has yet been reported; decarboxylation apparently occurs spontaneously under the conditions of their formation.

4.4. Chemical Properties of 3-Hydroxypyrroles and 1H-Pyrrol-3(2H)-ones 585

$$
\begin{array}{cc}
\textbf{210} & \textbf{211} \\
\end{array}
$$

a $R^2 = Et$
b $R^2 = CH_2Ph$
c $R^2 = t\text{-}Bu$

In contrast, 3-hydroxypyrrole-4-carboxylic acids are well characterized, and are available by standard basic hydrolysis of the esters.[63] Decarboxylation is more of a problem, and its success is apparently dependent on the remaining substitution pattern of the ring. Thus, Davoll was able to obtain **212** ($R^2 = R^5$ = Me) in 74% yield by decarboxylation of the 4-carboxylic acid at its melting point, yet attempted formation of **212** ($R^2 = R^5 = H$) by a similar reaction was unsuccessful.[32] One-pot hydrolysis–decarboxylation of some 2,2-disubstituted 4-carboxylic esters in H_2SO_4/H_3PO_4 at 120°C in 50–60% yield has been observed.[160] One example of the decarboxylation of a 3-hydroxypyrrole-5-carboxylic acid derivative in low yield has been reported.[175] Selective hydrolysis of 2,4-dicarboxylic esters is well known in pyrrole chemistry.[184] Treatment of diesters **213** ($R^2 = R^4 = Et$) with either concentrated sulfuric acid,[66] or an excess of base in aqueous[75,76] or DMSO[66] solution, gives only the 4-carboxylic acid **213** ($R^2 = Et$, $R^4 = H$); the reaction is compatible with both the presence or absence of substituents R^1 and R^5. Decarboxylation to the monoesters **214** is also a general process, with both acidic[66,76] and basic[75,76] conditions being advocated. Alternatively, the use of a mixed 2-*tert*-butyl 4-ethyl diester **213** ($R^2 = t\text{-}Bu$, $R^4 = Et$) allows selective hydrolysis and decarboxylation of the 2-ester substituent (using the boron trifluoride etherate reagent as above), which gives a concise route to 2,5-unsubstituted 4-carboxylic esters **215**.[10,183]

$$
\begin{array}{cccc}
\textbf{212} & \textbf{213} & \textbf{214} & \textbf{215} \\
\end{array}
$$

4-Cyanopyrrolones may be hydrolyzed with base to the corresponding amide.[110]

4.4.8. Photolysis and Photochemical Reactions of 3-Hydroxypyrroles and 1H-Pyrrol-3(2H)-ones

Margaretha and his coworkers (Anklam et al.) have made a detailed study of the photochemical reactivity of 1H-pyrrol-3(2H)-ones (and 3-hydroxypyrroles) as part of a general investigation of heteroenone systems. Simple derivatives, without electron-withdrawing groups, undergo slow, poorly defined photochemical reactions, and do not show cyclodimerisation or the [2 + 2] cycloadditions to olefins that are typical of oxa- or thia-enones.[25] Self-quenching of the excited state is a possible reason for this.[25] In the presence of ketone sensitisers, however, two possible reactions can take place. 2-Unsubstituted derivatives undergo electron transfer from the excited ketone followed by proton transfer, coupling, and loss of water to give the aldol-type products **216** (Scheme 56, route a).[185] This method is applicable to 1-substituted or 1-unsubstituted derivatives and gives good (50–70%) yields of the products **216** (R = alkyl) that have not been prepared by simple condensation reactions.[185] If the 2-position is substituted, an exceedingly slow N-dealkylation can take place[14] (Scheme 56, route b), but this is not a recommended preparative route.[14]

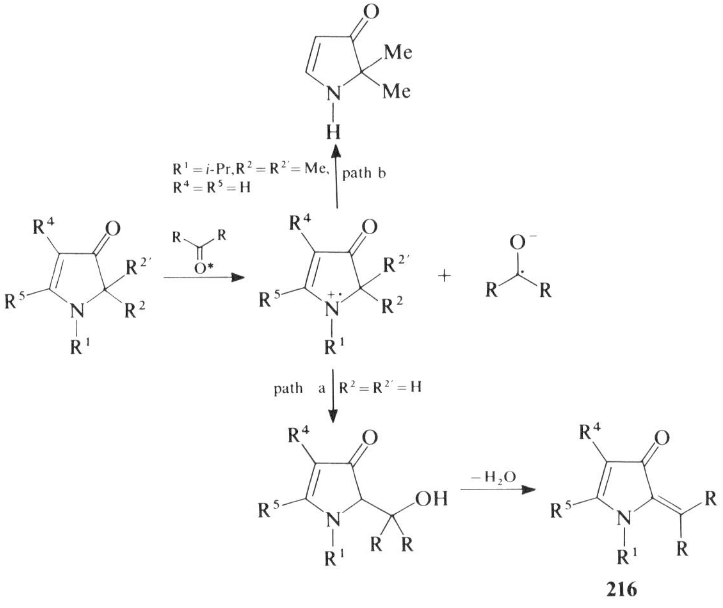

Scheme 56

When the electron donation of the nitrogen atom is moderated by an electron-withdrawing group (CO_2Et), the $1H$-pyrrol-3($2H$)-one system behaves in a manner similar to that of the corresponding oxa- or thia-enones.[163] Thus, photolysis of **217** (Scheme 57) in acetonitrile gives a single photodimer **218** (73%), and a range of cyclobuta[b]pyrrole derivatives **219** were obtained in 80–90% yield in the presence of alkenes.[163] Only one regioisomer, **219**, was obtained from 2-methylpropene, and similar diastereoisomeric mixtures resulted when the photolysis was carried out in the presence of dimethyl maleate or dimethyl fumarate.[163] These results are explained by spin inversion of the pyrrolone excited state to give a triplet, which reacts with the alkenes to give the formal [2 + 2] cycloaddition products.[163]

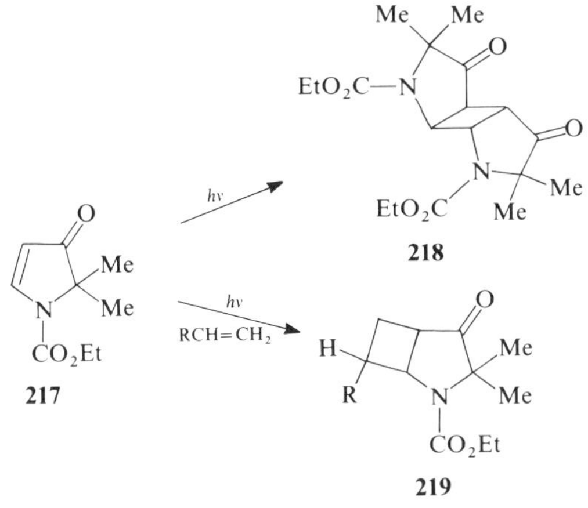

4.4. Chemical Properties of 3-Hydroxypyrroles and 1H-Pyrrol-3(2H)-ones 587

Scheme 57

N-Unsubstituted 3-hydroxypyrrole-2-carboxylic esters undergo a more unusual dimerisation on photolysis[140,186] in acetonitrile solution. The dehydrodimers[140] **220** are formed in reasonable yields by a process that involves loss of the NH proton from the initial cation radical **221** (Scheme 58); the corresponding N-alkyl compounds cannot undergo this step, and the reaction is diverted slowly to other ill-defined products. The constitution of the products **220** was confirmed by X-ray crystallography.[140]

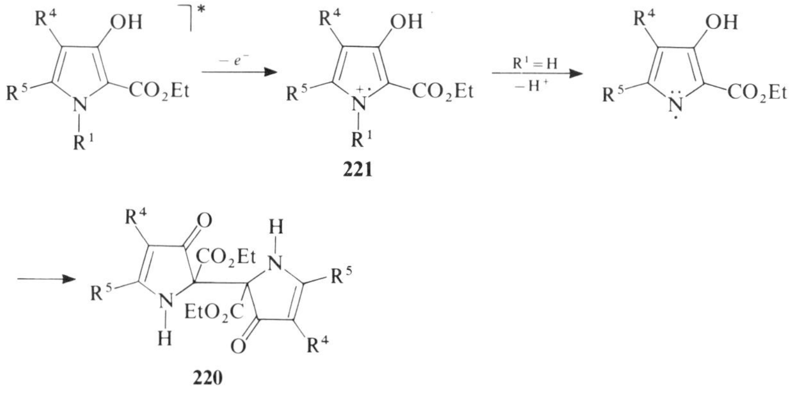

Scheme 58

The corresponding 2-allyl 2-carboxylic esters **222** undergo allyl migration to the 4-position (40–50% yield) on sensitized photolysis in acetone[143] (Scheme 59). The reaction appears to be general, provided the 4-position in the starting material is unsubstituted.[143]

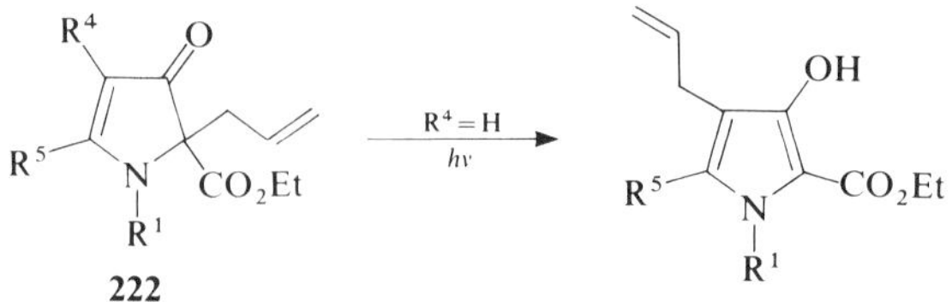

Scheme 59

4.5. 3-ALKOXYPYRROLES

4.5.1. Preparation of 3-Alkoxypyrroles

Most alkoxypyrroles are obtained by alkylation of the corresponding 3-hydroxypyrrole (Section 4.4.1). Optimum conditions for O-alkylation of simple N-substituted 3-hydroxypyrroles involve an alkyl p-toluenesulfonate and sodium hydride in DMI,[138] although, in general, dimethyl sulfate is the reagent of choice for methylation when the pyrrole contains an electron-withdrawing group. However, a number of direct routes to alkoxypyrroles, not involving hydroxypyrroles, have been developed, and these are discussed here. Many have only been carried out for specific examples, and may be of limited applicability.

Thus, the aryl derivatives **223** and **224** have been isolated in 55 and 48% yields, respectively, by condensation of the diketone **225** with ammonium acetate,[33] and by reductive cyclization of the ketohydrazone **226**[187] (Scheme 60).

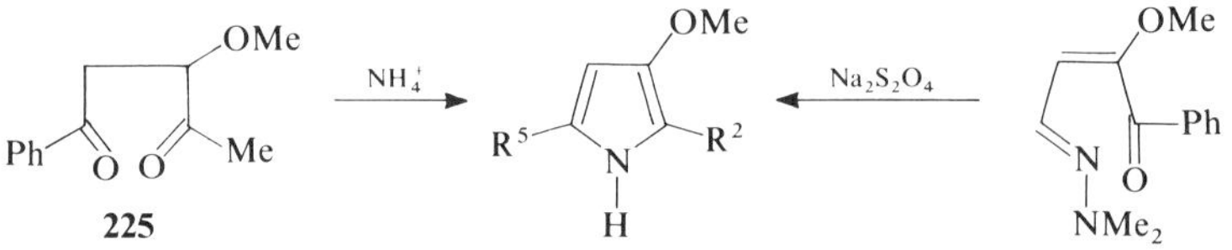

Scheme 60

Much lower yields are obtained by the latter route for alkyl-substituted examples.[13] Photolysis of 3-diazo-2,5-diphenylpyrrole in methanol gives a low yield (12%) of the 3-methoxy compound.[188] Two routes via chromium carbonyl complexes have been reported[189,190] (Scheme 61).

The imino complex **227** cyclizes with an alkoxyacetylene to give **228** in 94% yield,[189] whereas the better-known[191] alkenyl complex **229** gives the N-cyclohexyl derivative **230** (56%).[190] The only general direct route to simple 3-alkoxypyrroles available at present is due to Kochhar and Pinnick[192] (Scheme 62), although N-unsubstituted examples cannot be made by this method. The route involves Dibal reduction of 4-alkoxypyrrol-2(5H)-ones **231**, which are

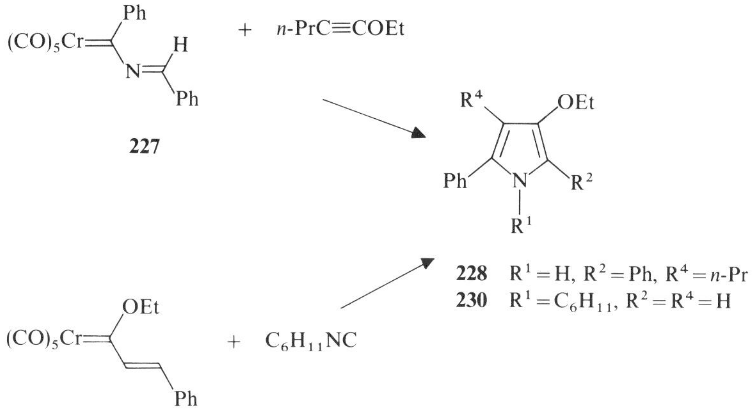

4.5. 3-Alkoxypyrroles

Scheme 61

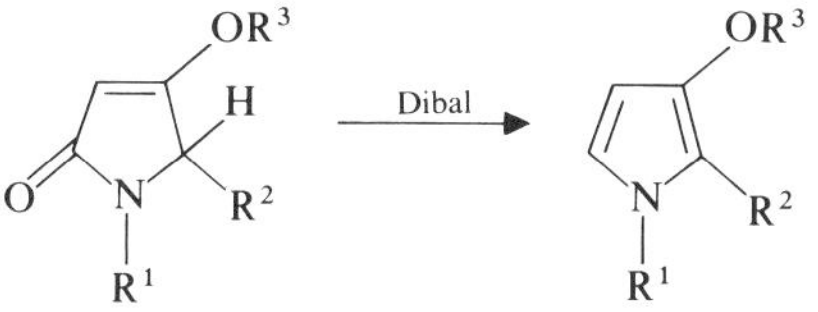

228 $R^1 = H$, $R^2 = Ph$, $R^4 = n\text{-Pr}$
230 $R^1 = C_6H_{11}$, $R^2 = R^4 = H$

Scheme 62

themselves readily available in three steps and high overall yield from β-ketoesters. The final reduction step takes place in 40–80% yield, and has been applied to a series of eight 1-alkyl and 1,2-dialkyl 3-alkoxypyrroles.[192]

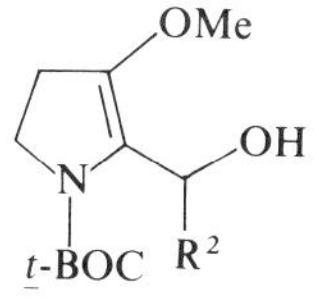

Some *N*-protected 2-substituted 3-alkoxypyrroles have been made via a dehydration–rearrangement sequence in an attempted silylation of the dihydropyrroles **232** (R^2 = Ph, *i*-Pr, etc.)[193] Two 2-acetyl-3-methoxypyrroles have been isolated by degradation of a macrolide antibiotic,[194] but most examples containing an electron-withdrawing group in the 2-position have been obtained either by alkylation or by a decarboxylation method (Section 4.5.3.7). In a unique photolytic ring contraction, 3-alkoxy-2-formylpyrroles are obtained by irradiation of an aqueous solution of 4-alkoxypyridine-*N*-oxides.[195] The yield is substantially improved (to ~40%) in the presence of copper sulfate solution,[196,197] and preparatively useful quantities of the products can be made in this way. Photolytic ring contraction of 1-benzoyl-5-methoxy-1,4-diazepine leads to 2-benzoyl-3-methoxypyrrole, but only in 15% yield.[198]

A number of N-unsubstituted,[199] N-alkyl and N-aryl 3-alkoxypyrrole-4-carboxylic esters and 4-carbonitriles have been obtained by dehydrogenation (DDQ or Pd/C at 235°C[199]) of the corresponding 2,5-dihydropyrrole,[68,200,201] and this route has also been used to obtain a 4,5-dicarboxylic ester[202] and a 2,5-dicarboxylic ester.[203] The reduced pyrroles can be prepared by reaction of a stabilized phosphorane with the azomethine ylide resulting from ring opening of an alkoxycarbonylaziridine[200,201] (Scheme 63). A 3-methoxy-4-nitropyrrole has been made from a 3,4-dinitro compound by a two-step formal displacement of one of the nitro groups.[204]

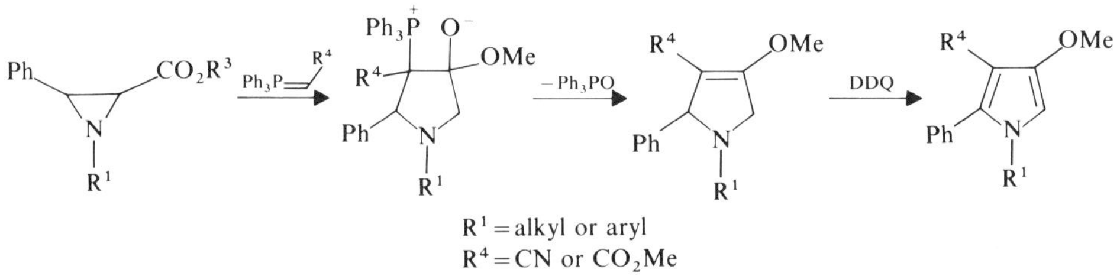

Scheme 63

Dehydrogenation methods have been widely used to make 3-alkoxypyrrole-5-carboxylic esters.[72,199] Kuhn and Osswald's classic route involves treatment of the enol ether **233** with NBS and triethylamine, followed by deprotection with base (Scheme 64).[72]

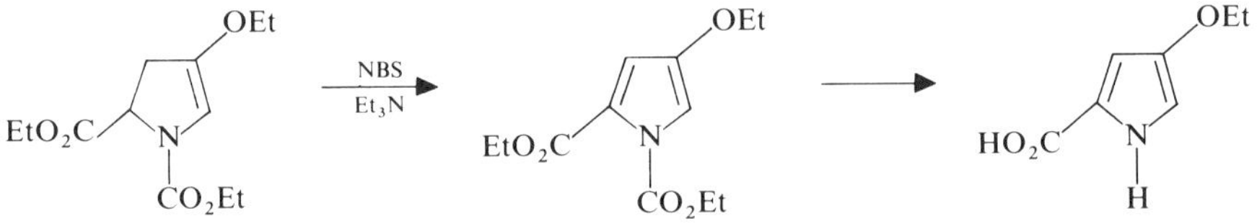

Scheme 64

However, the best route to the corresponding methoxy compound involves dehydrogenation of the protected acetal **234**[199] (Scheme 65).

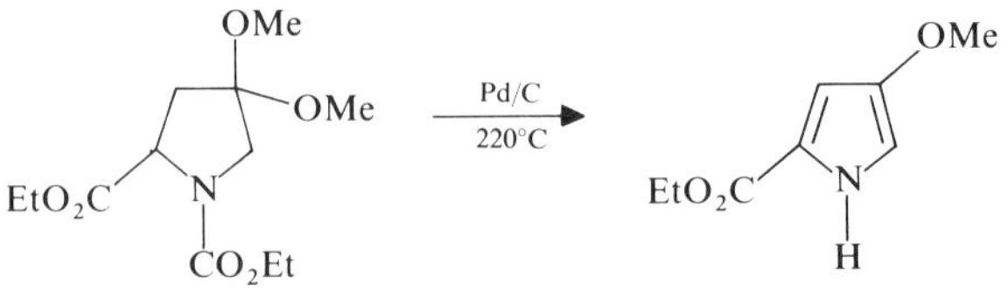

Scheme 65

4.5. 3-Alkoxypyrroles

A sequence similar to that of Kuhn and Osswald has been employed to produce *N*-protected 5-aryl-3-methoxypyrrole-2,4-dicarboxylates.[205] The final oxidation step was accomplished with NBS/Et$_3$N (5-phenyl derivative) or by treatment of the sodium enolate of the dihydropyrrole with NBS (5-thienyl derivative).[205] A range of alkoxypyrrole-2,5-dicarboxylic esters has been obtained in 20–40% overall yield by reverse electron-demand Diels–Alder reaction of a 1,2,4,5-tetrazine with electron-rich alkenes,[206,207] followed by reduction of the resulting pyridazine with zinc in acetic acid. An example is shown in Scheme 66.[206]

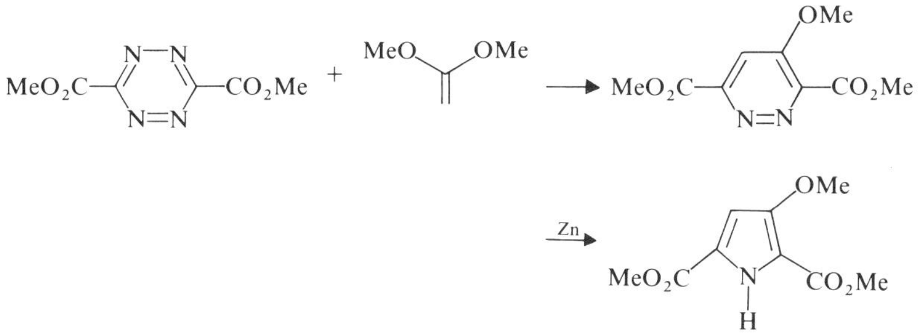

Scheme 66

4.5.2. Physical Properties of 3-Alkoxypyrroles

4.5.2.1. Structure of 3-Alkoxypyrroles

Although most simple 3-alkoxypyrroles are volatile oils, the X-ray crystal structure of one derivative **235** has been determined, and the results are shown in Fig. 4.6.[138] The 5-membered ring is essentially planar, and the nitrogen atom adopts a trigonal configuration. The planes of the 5- and 6-membered rings are

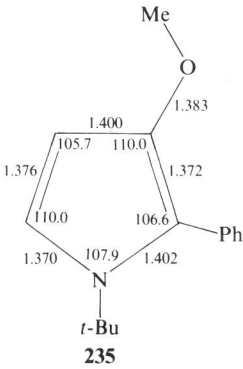

Figure 4.6. Selected bond lengths (Å) and angles (degrees) of the alkoxypyrrole (**235**).

almost orthogonal, and so there is little conjugative interaction between them, but there is probably some steric repulsion between the phenyl and *tert*-butyl groups, which results in the N(1)—C(2) bond being longer than the N(1)—C(5) bond. It is thought that the methoxy group has little influence on the geometry of the 5-membered ring, although the C(3)—O bond is comparable in length to that of anisole, which indicates that electron density from the oxygen atom is indeed being delocalized into the ring.[245]

4.5.2.2. IR spectra of 3-Alkoxypyrroles

Unlike 3-hydroxypyrroles and 1*H*-pyrrol-3(2*H*)-ones, there is no readily identifiable IR-active group in 3-alkoxypyrroles. However, the NH absorptions of a wide variety of *N*-unsubstituted derivatives appear in the range 3200–3500 cm^{-1}; the position is relatively independent of whether the compound contains only alkyl,[8] only aryl,[187] or alkyl and aryl substituents[33] or whether it contains electron-withdrawing groups in the 2-[8,10,53,172,196] or 4-[135] positions. Kochhar and Pinnick report that simple *N*-substituted 3-alkoxypyrroles show characteristic IR peaks at 1570, 1325, and 1040 cm^{-1}.[192]

4.5.2.3. UV Spectra of 3-Alkoxypyrroles

A considerable amount of UV spectroscopic data have been reported and some representative examples are listed in Table 4.8. Simple alkoxypyrroles absorb at around 220 nm,[8] although in the presence of traces of acid a much

TABLE 4.8. UV SPECTRA OF 3-ALKOXYPYRROLES (236)

R^1	R^2	R^3	R^4	R^5	λ_{max}/nm	ε	Solvent	Ref.
H	H	Me	Me	H	222	4300	MeOH	8
H	H	Me	Me	Me	221		MeOH	8
Ph	H	Me	H	H	272	10800	EtOH	32
H	Me	Me	H	Ph	312	15100	MeOH	33
H	CO$_2$Et	Me	H	H	264		MeOH	77
H	H	Me	CO$_2$Et	H	234	12500	MeOH	77
H	H	Me	H	CO$_2$Me	286	10020	MeOH	77

longer wavelength absorption (at 292 nm) due to the protonated form may be detected.[8] It is possible that the abnormally large bathochromic shifts observed for 1-[32] and, particularly, 5-[33] aryl-substituted derivatives may be due to this effect. A comparison between the spectrum of 2,5-dimethyl-1-phenylpyrrole and 3-methoxy-2,5-dimethyl-1-phenylpyrrole shows no change in λ_{max} (208 nm) and less than 10% difference in extinction coefficient ($\varepsilon \sim 15,500$).[32] A variety of 2-carboxylic ester derivatives all show maxima in the range 264–277 nm,[8,10,153,199] and again this appears to be independent of the methoxy substituent.[199] 4-Carboxylic ester derivatives absorb at a rather shorter wavelength (234–245 nm)[32,135,199] and 5-carboxylic ester derivatives, at a longer wavelength (~ 286 nm).[199] In this case, the methoxy group has a bathochromic effect of 22 nm relative to the 3-unsubstituted compound.[199]

4.5.2.4. NMR *Spectra of 3-Alkoxypyrroles*

The ^{1}H NMR spectra of 3-alkoxypyrroles bear close resemblance to the corresponding hydroxypyrrole (Section 4.3.5.1), with δ(5-H) > δ(2-H) > δ(4-H).[145] Some typical values from the extensive data in the literature are given in Table 4.9; further tabulations of data for N-substituted examples are given in Refs. 145 and 192. Three- and four-bond couplings ($^3J_{4,5}$, $^4J_{2,4}$, and $^4J_{2,5}$) are again in the range 2.0–3.0 Hz,[145] and, where measured,[33,187] couplings to the NH are apparently of similar magnitude. A five-bond coupling of ~ 1.0 Hz has been observed between the proton at the 5-position and the proton of the 2-formyl group.[208]

Electron-withdrawing groups at the 2- and 4-positions both cause a deshielding effect on the proton at the 5-position, but have little effect on the 4- and 2-protons, respectively (Table 4.9).

Although ^{13}C NMR data for the 3-alkoxypyrrole series are scarce, a compilation of N-alkyl and N-aryl examples has recently appeared.[145] Again, these data (Table 4.10) are found to be very similar to those of hydroxypyrroles (Section 4.3.5.1) [δ(C-3) $\gg$ δ(C-5) > δ(C-2) > δ(C-4)], although a slight deshielding at C-3 (4–5ppm) is observed in the alkoxy series relative to the corresponding hydroxy compound. In the ^{1}H-coupled spectra the C-3 signal is complicated by further coupling to the alkoxy group, but otherwise the parameters are similar to those shown in Fig. 4.2. The dramatic deshielding of C-5 in the presence of an electron-withdrawing substituent at the 2-position[196,209] (Table 4.10) has already been noted in the 3-hydroxy series (Section 4.3.5.1). An early study of the ^{13}C NMR spectra of some complex methoxypyrroles related to prodigiosin has been published[109] (Section 4.7).

4.5.2.5. Mass *Spectra of 3-Alkoxypyrroles*

The mass spectra of 3-alkoxypyrroles generally show intense molecular ions, and the major breakdown is usually the loss of the O-substituent.[26,192] Ex-

TABLE 4.9. ¹H NMR PARAMETERS OF TYPICAL 3-ALKOXYPYRROLES (236)

$$\underset{236}{\text{[pyrrole with R}^1\text{ on N, R}^2\text{ at 2, OR}^3\text{ at 3, R}^4\text{ at 4, R}^5\text{ at 5]}}$$

						δ_H					
R^1	R^2	R^3	R^4	R^5	2-H	4-H	5-H	$^3J_{4,5}$	$^4J_{2,4}$	$^4J_{2,5}$	Ref.
H	H	CH₂Ph	H	H	6.35	5.98	6.55				10
H	H	Me	Me	H	6.21		6.34				8
H	Ph	Me	H	H		6.10	6.65				187
Me	H	Me	H	H	6.18	5.80	6.35	3.0^a			192
Ph	H	Me	H	H	6.73	6.14	6.95	3.1	2.0	2.5	145
H	CHO	Me	H	H		5.80	6.85	2.7^b			196
H	COMe	Me	H	H		5.88	6.88	3.1			194
Me	H	Me	CO₂Et	H	6.11		7.02			2.5	68

^{a} Also $^3J_{1,5}$ and $^4J_{1,4}$ 3.0 Hz.
^{b} Also $^3J_{1,5}$ and $^4J_{1,4}$ 2.6 Hz.

TABLE 4.10. ¹³C NMR PARAMETERS OF TYPICAL 3-ALKOXYPYRROLES (236)

					δ_C			
R^1	R^2	R^3	R^4	R^5	C-2	C-3	C-4	C-5	Ref.
t-Bu	H	Me	H	H	99.76	148.36	95.80	115.04	145
Ph	H	Me	H	H	100.86	150.83	100.32	116.99	145
H	CHO	Me	H	H	118.95	159.14	95.40	127.23	196

ceptions include certain benzyloxypyrroles,[192] 1-*tert*-butyl compounds (which show preferential loss of isobutene[26]), and certain,[172] but not all,[154] formyl derivatives, where standard aldehyde cleavage is more important.

4.5.3. Chemical Properties of 3-Alkoxypyrroles

4.5.3.1. Alkylation Reactions

Section 4.4.1. covered *N*- and *O*-alkylation of hydroxypyrroles. Triethyl phosphate has been suggested as a useful reagent for *N*-ethylation of 2-formyl-3-methoxypyrrole.[210] *N*-Methylation has been accomplished using iodomethane and thallium ethylate.[13] In one case of *C*-alkylation, the reaction of a 1,2-unsubstituted 3-methoxypyrrole-5-carboxylic ester with a 2-bromomethylpyrrole in ethanol at 40°C gave a 40% yield of a dipyrrolylmethane derivative.[211]

4.5.3.2. O-Dealkylation

The formation of a 3-hydroxypyrrole from a 3-alkoxypyrrole using aluminum tribromide in benzene was first reported by Kuhn and Osswald.[72] Boron tribromide has also been employed for this reaction,[175] and, in the bipyrrole series, a good (67%) yield of a subsequent oxidation product was achieved (Scheme 67).[157] There is an indication that lithium iodide in refluxing pyridine may be effective for demethylating 3-methoxypyrrole-2-carboxylic esters.[66]

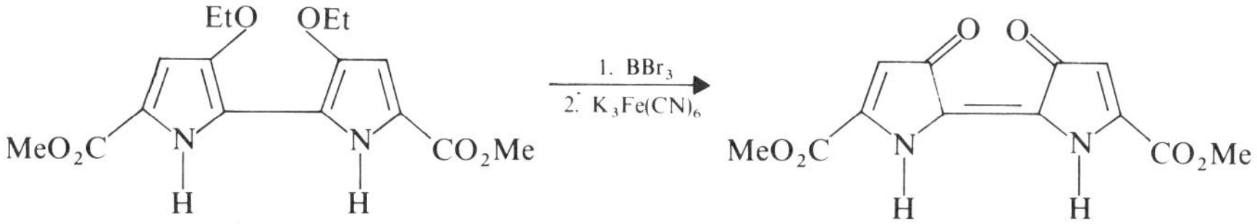

Scheme 67

4.5.3.3. Reactions with Electrophiles

4.5.3.3.1. General Considerations

The situation for 3-hydroxypyrroles (Section 4.4.3.1 and Scheme 29) obtains, with reactivity toward electrophiles being greatest at the 2-position and least at the 4-position in the absence of complicating steric or electronic effects.

4.5.3.3.2. Reactions with Acids

Simple *N*-substituted 3-alkoxypyrroles are quantitatively protonated at the 2-position by trifluoroacetic acid[26] (TFA) to give resonance-stabilized cations

237. Treatment with [^{2}H]TFA gives relatively slow exchange at this site ($t_{1/2}$ ca. several hours) followed by even slower exchange at the 4-position ($t_{1/2} \sim$ 1–2 days) and the 5-position ($t_{1/2} \sim$ 1 month). Taken together with the hydroxypyrrole results (Section 4.4.3.2), this suggests that the exchange process at the 4- and 5-positions take place via the free-base form of the alkoxypyrrole rather than the 2-protonated species **237**, although the relative reactivity of the 4- and 5-positions is perhaps surprising.

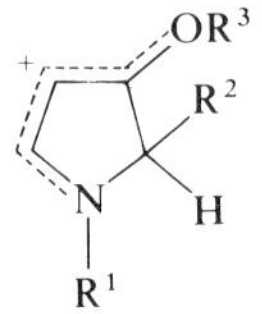

237

2-Hydroxymethyl-3-methoxypyrroles condense in the presence of a trace of acid with loss of formaldehyde to give symmetric dipyrrylmethanes in moderate (40–50%) yield[212] (Scheme 68).

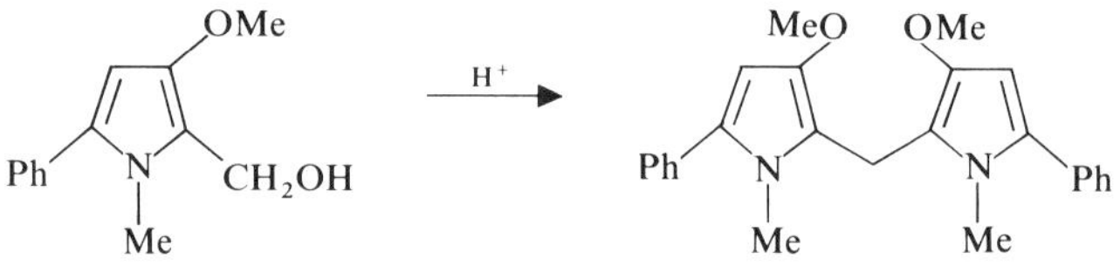

Scheme 68

4.5.3.3.3. Halogenation and Reactions of the Products

These reactions are confined to the formation of iodo compounds. Thus, treatment of the ester[157] **238a** or acid[175] **238b** with I_2/KI gives good yields of the 2-iodo derivatives **238c** and **238d**, respectively. Under rather similar conditions, the acids **239a**[207] and **240a**[44] undergo iodinative decarboxylation to give the iodo compounds **239b**[207] and **240b**[44], respectively. Oxidative coupling reactions (Section 4.5.3.5) can also take place with this reagent[8]. Ullmann coupling of **238c**

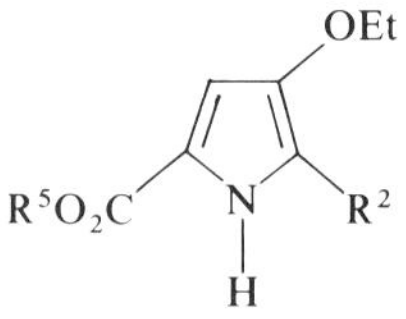

238a $R^2=H$, $R^5=Me$	**239a** $R^4=H$, $R^5=CO_2H$
238b $R^2=R^5=H$	**239b** $R^4=R^5=I$
238c $R^2=I$, $R^5=Me$	**239c** $R^4=R^5=H$
238d $R^2=I$, $R^5=H$	

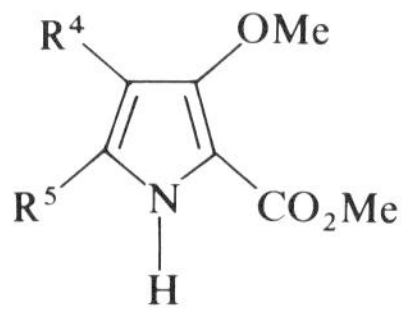

4.5. 3-Alkoxypyrroles

240a $R^5 = CO_2H$
240b $R^5 = I$
240c $R^5 = H$

241

with copper-bronze in toluene gives a low (< 20%) yield of the bipyrrole **241**, together with some of the hydrogenolysed product and an unidentified terpyrrole.[157] Intentional hydrogenolysis of the iodo derivatives **239b** and **240b** (Pd/C/H_2, 25°C) gives the parent methoxypyrrole carboxylic esters **239c**[207] and **240c**[44] in excellent yield (e.g., 96%[207]).

4.5.3.3.4. Diazo-Coupling Reactions

Yields of 60–70% of the azo-coupled products **242**,[8] **243**,[164] and **244**[66] have been obtained from either the 2-unsubstituted compound or the 2-carboxylic acid[164] under standard diazotization conditions in mixed organic–aqueous solvents. This reaction has also been used to characterise certain unstable 3-alkoxypyrroles, including hydroxymethyl derivatives.[212] Preliminary experiments with 4,5-unsubstituted derivatives also suggest that reaction takes place at the 2-position.[26] The simple azo compounds **242** and **243** show intense (ε 25,000) UV absorption at 397–398 nm,[8,164] whereas in the hydrochloride salt of **243**, this is shifted to 430 nm.

242 $R^1 = H$, $Ar = Ph$, R^3–$R^5 = Me$
243 $R^1 = H$, $Ar = Ph$, $R^3 = CH_2Ph$, $R^4 = R^5 = Me$
244 $R^1 = Me$, $Ar = 4$-$C_6H_4NO_2$, $R^3 = Me$, $R^4 = CO_2Et$, $R^5 = Ph$

4.5.3.3.5. Vilsmeier and Related Reactions

Vilsmeier formylation has been shown to take place at the 2-position for a range of 5-aryl-,[205] N-substituted 5-heteroaryl,[151] and 4,5-dialkyl[172] 3-alkoxypyrroles; yields are invariably > 70%. The reaction also takes place efficiently with a 3-alkoxypyrrole-4-carboxylic ester.[135] These are key intermediates for the synthesis of prodigiosins (Section 4.7) and their analogs (see also Section 4.5.3.6). If the 2-position is blocked, reaction takes place instead at the 5-position.[13]

Reaction of simple 3-alkoxypyrroles with methoxymethylene Meldrum's acid is also facile (see Section 4.4.3.6). This normally gives 2-substituted products, e.g., **245**, unless the *N*-substituent is bulky (whereupon a mixture of 2- and 4-substituted products is obtained) or the 2-position is blocked (whereupon the 5-substituted derivative is obtained).[26]

245

4.5.3.3.6. Reactions with Carbonyl Compounds

The dipyrrolylmethane **246** is obtained by treatment of 4,5-dimethyl-3-methoxypyrrole with benzaldehyde in boiling methanol,[8] although more complex products may be analogously obtained.[213] However, reaction of various 2-formylpyrroles to give dipyrrolylmethenes **247** is a more common process en route to prodigiosin[152,205] (Section 4.7) or corrin[172] analogs. Thus, the reaction has been carried out (in the presence of HBr) with 4,5-dialkyl-,[172] 5-aryl-,[205] and 5-heteroaryl-[152,205] 3-alkoxypyrroles to give the hydrobromide salts of **247**, from which the free bases may be obtained if desired. Yields can be as high as 80–90%.[152,205]

246   **247**

The unsubstituted 2-position may be, if necessary, generated in situ from the 2-carboxylic acid.[172] Use of a diformyldipyrrolylmethane gives a biladiene derivative, which can be cyclized to an alkoxy-substituted corrin ring system.[172] In a related reaction, 3-methoxy-4-methylpyrrole was treated with formaldehyde to give a 52% yield of a tetramethoxyporphyrin derivative.[8]

The converse reaction (viz., condensation of an alkoxyformylpyrrole with a pyrrole derivative) has also been extensively employed as a route to dipyrrolylmethenes (Section 4.5.3.4).

4.5.3.3.7. Reactions with Dimethyl Acetylenedicarboxylate and Other Potential Dienophiles

Reaction of *N*-substituted 3-alkoxypyrroles with dimethyl acetylenedicarboxylate in chloroform solution generally gives a mixture of *E* and *Z* Michael

4.5. 3-Alkoxypyrroles

adducts, e.g., **248** and **249**, and the formal cycloadduct, e.g., **250**.[26] When the 2-position is unsubstituted, the reaction takes place at that site, even with bulky N-substituents. When the 2-position is blocked by substitution, then the Michael addition proceeds at the 5-position. The cycloadduct **250** adopts the configuration shown, as demonstrated by X-ray crystallography;[26] the enol ether function is readily hydrolyzed to the corresponding ketone.[26]

Other potential dienophiles such as maleic anhydride,[66,213] or tetracyanoethylene[26] invariably give the Michael adduct by reaction at the 2-position.

4.5.3.3.8. Sulfenylation

The 2-phenylthio compound **251** has been obtained in 50% yield by sulfenylation of 3-methoxy-1-methyl-5-phenylpyrrole with N-phenylthiophthalimide **252**.[213]

4.5.3.4. Coupling Reactions of 2-Formyl-3-alkoxypyrroles with Pyrrole Derivatives

The formation of dipyrrolylmethenes **253** is most commonly accomplished by condensation of an 2-formyl-3-alkoxypyrrole with an α-unsubstituted pyrrole derivative (see, however, Section 4.5.3.3.6).

The reaction is catalyzed by acid (or phosphoryl chloride[151]), and takes place with 2,3-,[61,207] 3,4-,[172] or 2,4-disubstituted[77,135,213] or 2,3,4-trisubstituted[151] pyrroles; yields are usually 50–70%. In one case, the isolated product undergoes in situ O-dealkylation.[213] The coupling process is tolerant to a range of substituents in the alkoxypyrrole, with 5-substituted,[61,77,135,207] 1,5-disubstituted,[151,213] or 4,5-disubstituted[172] derivatives all being employed. These products are of interest because of their similarity to the prodigiosin system (Section 4.7).

The use of 2,2'-dipyrrolylmethane-5,5'-dicarboxylic acid substrates in this process has given access to biladienes that have been cyclized to alkoxyporphyrin derivatives.[149]

4.5.3.5. Oxidation Reactions

Generally, most 3-alkoxypyrroles are remarkably stable to aerial oxidation. Indeed, samples of 3-methoxy-1-phenylpyrrole and its 2-methyl derivative still contain some unchanged starting material after many months at room temperature. However, the presence of an α-phenyl group dramatically reduces this stability[66,214] (cf. ref. 190). 2-Phenyl derivatives[214] are transformed after some days in air to the pyrrol-2-one (e.g., **254**), whereas the isomeric structure e.g., **255** is obtained (in just 15 min) from 5-phenyl derivatives.[66] This compound, rather than the carboxylic acid, is also obtained by oxidation of the corresponding 2-hydroxymethyl derivative using silver oxide in pyridine.[66]

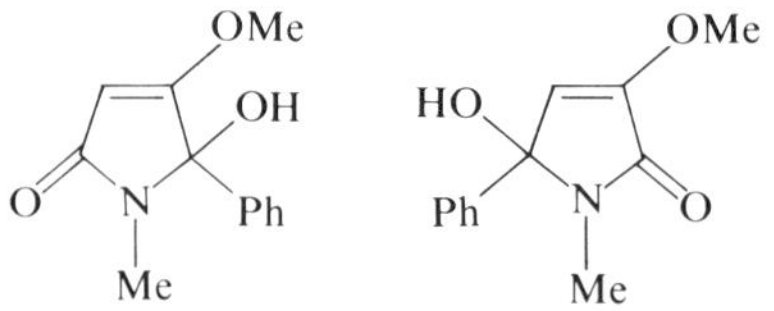

It seems likely that these products in general are formed by a free-radical autoxidation mechanism rather than by fortuitous photoxidation.[214]

Treatment of the N-unsubstituted alkoxypyrrole **256** with an excess of manganese dioxide or an equivalent of lead tetraacetate gives 60–70% yields of the 2-functionalized derivatives **257** and **258**, respectively (Scheme 69).[33]

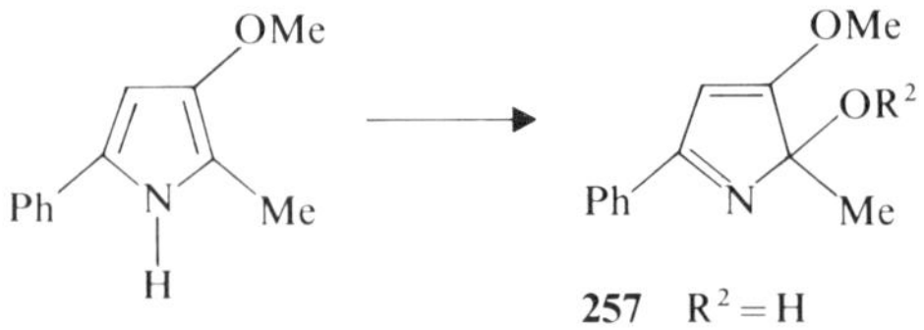

257 R² = H
258 R² = COMe

Scheme 69

No reaction at the 2-methyl group was detected, in contrast to the behavior of the corresponding 3-acetoxy compound (see Section 4.6.3.3). Lead tetraacetate oxidation of the 2-unsubstituted compound **259a**, however, gives the oxidative coupling product **260a**, but in <10% yield.[175] The corresponding bipyrrole carboxylic acid **260b** can itself be oxidized, with O_2, ferricyanide, or, better, with I_2/KI, to the tetramer **261**, which is also obtained in 47% yield directly from the carboxylic acid **259b** and ferric chloride[175] (Scheme 70).

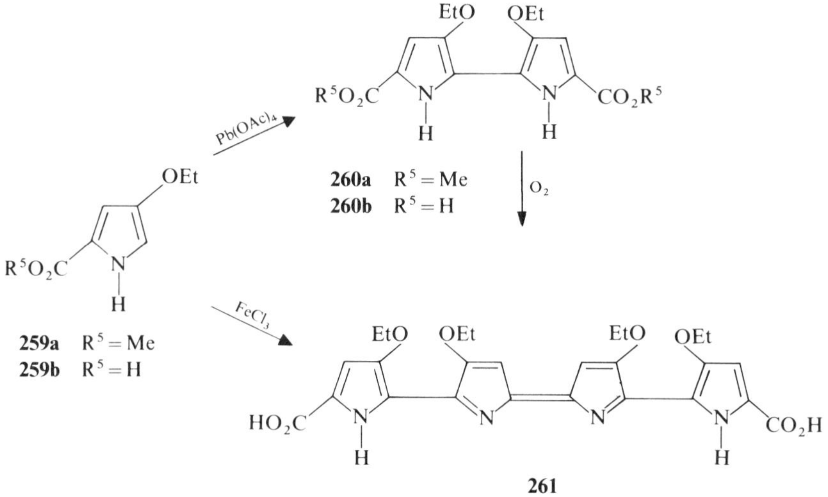

Scheme 70

Formal oxidation of a 3-methoxy-4-methylpyrrole-2-carboxylic ester to the 4-carboxylic acid is possible by treatment with sulfuryl chloride, followed by hydrolysis.[44]

4.5.3.6. Reduction Reactions

Although the alkoxypyrrole ring is stable to most reducing agents, these reactions have been used extensively in substituent functional group interconversions. Thus lithium aluminum hydride smoothly reduces N-substituted 2-[66], 4-[202,212] and 5-[202] carboxylic ester functions to the corresponding primary alcohol, whereas diborane[212] or lithium aluminum hydride–aluminum chloride[212] gave the corresponding methyl derivative. Although these esters are inert, or at least poorly reactive, to sodium borohydride,[212] the α-diketone **262** is transformed into the aldehyde **263** on treatment with this reagent followed by in situ oxidative cleavage with periodate.[154]

The formation of 2-carboxaldehydes from 2-carboxylic esters is best achieved in moderate overall yield by the MacFadyen–Stevens reaction. Although 2-formyl-3-methoxy-4,5-dimethylpyrrole has been made in this way,[44] the reac-

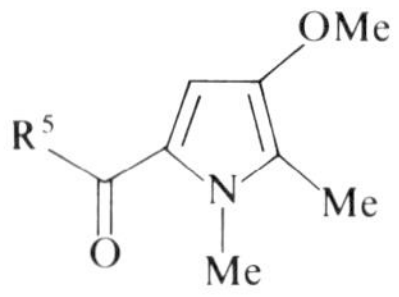

262 R⁵ = COMe
263 R⁵ = H

tion has been most often employed with 5-pyrrolyl derivatives en route to prodigiosin (Section 4.7).[61,77,207] A related 5-formyl-2-pyrrolyl derivative has also been synthesized.[77]

4.5.3.7. Hydrolysis and Decarboxylation Reactions of 3-Alkoxypyrrole Carboxylic Esters

Hydrolysis of 3-alkoxypyrrole-2-, -4-, or -5-carboxylic esters takes place under reflux in dilute base.[8,10,66,72,205] Under these conditions, 2,4-dicarboxylic esters can be selectively hydrolyzed at the 2-position,[66,77,205] whereas the 4-ester is selectively hydrolyzed in concentrated sulfuric acid;[77] both esters are hydrolyzed using an excess of potassium hydroxide.[66] Decarboxylation of the 2-carboxylic acids can take place simply by heating[10,66,77] (even in the presence of a 4-carboxylic acid[66]), or by treatment with strong acid;[8,172,205] the 4-carboxylic acid function is normally removed by heating.[32,63,77,205] All of these reactions can take place in the presence, or absence, of further substituents. A 3,3′-diethoxy-2,2′-bipyrrole-5,5′-dicarboxylic ester is readily hydrolyzed with base, and the resulting diacid can be decarboxylated at 230°C in the presence of copper bronze to give the parent 3,3′-diethoxybipyrrole in 79% yield.[157]

4.5.3.8. Claisen Reaction

The Claisen rearrangement of the O-allyl derivative **264** to the C-allyl compound **265** takes place at 180–200°C (Scheme 71). Corresponding thermolysis of a crotyl derivative confirms that the standard sigmatropic shift mechanism is operating.[75,153]

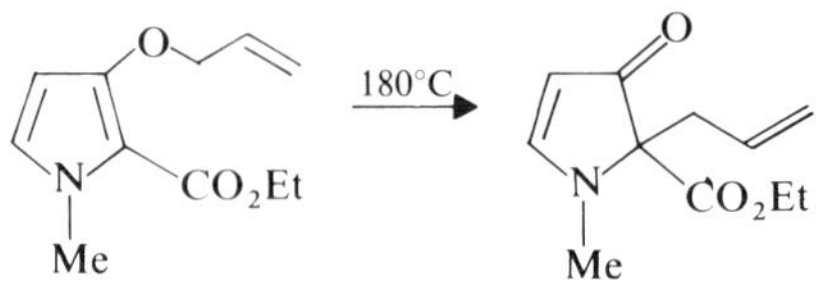

Scheme 71

4.6. 3-ACYLOXYPYRROLES

4.6.1. Preparation of 3-Acyloxypyrroles

The best general route to 3-acyloxypyrroles is by acylation of the corresponding 3-hydroxypyrrole (Section 4.4.2), but a number of derivatives have been made either by direct introduction of the substituent into the pyrrole ring system, or by direct ring synthesis.

Treatment of 2,5-diphenyl-, 1,2,5-triphenyl- or 2,3,5-triphenyl-pyrrole with benzoyl peroxide at 25°C gives the corresponding β-benzoyloxypyrrole in 30–80% yield.[215] The reaction is unsuccessful with pyrrole itself or with alkylpyrroles, and probably takes place by attack of molecular dibenzoyl peroxide rather than the usual free-radical process.[215] Lead tetraacetate oxidation of 2,5-disubstituted pyrroles is tolerant of at least one alkyl group in the ring, and the acetoxy compounds **266** (R^4 = H or CO_2Et) were obtained, apparently regiospecifically, in 40–50% yield directly from the 3-unsubstituted pyrrole after chromatography.[33] Dehydrogenation of a 2,5-dihydropyrrole using DDQ leads to the diester **267** in 66% yield.[68]

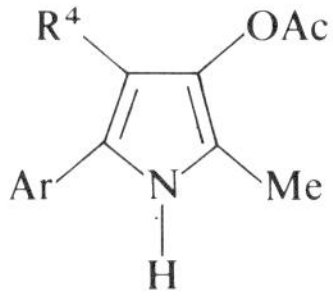

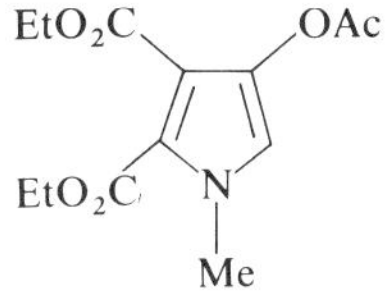

Two methods involving diazo compounds have been used: photolysis[188] or thermolysis[216] of **268** (R^4 = H, CN, CO_2Et) in glacial acetic acid gives a 20–30% yield of the corresponding products **269** (Scheme 72).

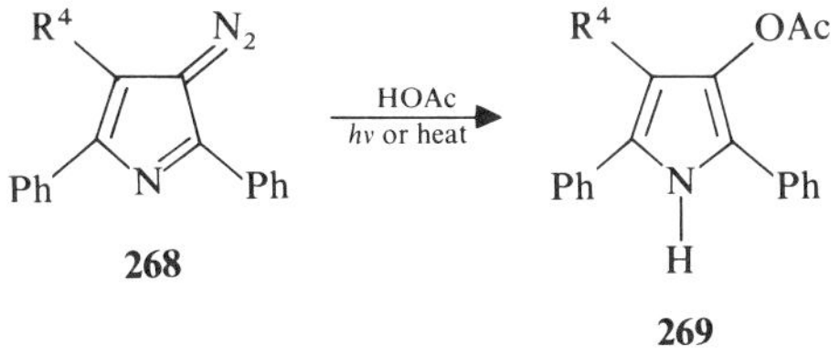

Scheme 72

Although yields are not high, the cycloaddition of acetoxybutadienes with the azodicarboxylic ester **270**, followed by in situ reduction with zinc (Scheme 73), has been used as a route to 3-acetoxypyrroles **271**, and the products have been further elaborated to prodigiosin analogs[152] (Section 4.7).

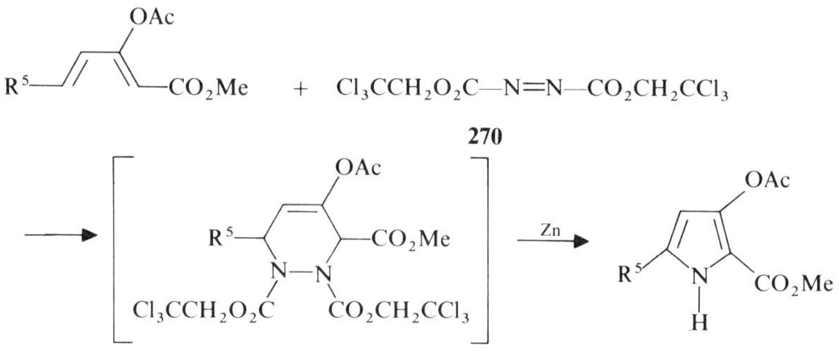

Scheme 73

3-Acyloxypyrroles are formed as by-products (5–15% yield) from treatment of oxazolium oxides **272** with anhydrides;[217] a mechanism for this process has been suggested.[217]

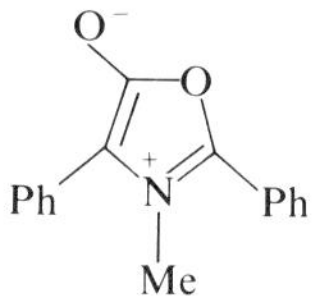

272

4.6.2. Physical Properties of 3-Acyloxypyrroles

No crystal structure data have been reported. The IR spectra of 3-acetoxy-pyrroles[33,188,216,217] show carbonyl absorptions at ~ 1740–1760 cm^{-1}, and benzoyloxypyrroles,[215] at ~ 1720 cm^{-1}. Although their UV spectra are often quoted (Refs. 32, 159, 217, etc.), there are insufficient data to assess the effect on the spectra of acyloxy groups relative to alkoxy or hydroxy groups. In one case, the maximum at 298 nm is shifted to lower wavelength by 14 nm relative to the corresponding methoxy compound, with a slight increase in extinction coefficient.[33]

In the ^{1}H NMR spectra of 3-acyloxypyrroles, the 4-H and 5-H resonances ($\delta_H \sim 6.0$, 6.8) are little shifted relative to the corresponding methoxy compound, but the signal for 2-H is deshielded by ~ 0.5 ppm ($\delta_H \sim 7.0$).[169] Similarly, C-4 and C-5 (at δ_C 100 and 115, respectively) are similar in acetoxy and methoxy derivatives, whereas C-2 is deshielded by ~ 8 ppm and C-3 is shielded by ~ 10 ppm (δ_C 108 and 137, respectively) relative to the methoxy compounds.[169] One-bond couplings ($^1J_{CH}$) are slightly larger in the acetoxy series.[169]

The mass spectra of the O-acetyl derivatives show initial loss of ketene from the molecular ion,[149] and the benzoyl compounds show cleavage of PhCO.[215]

4.6.3. Chemical Properties of 3-Acyloxypyrroles

4.6.3.1. Deacylation Reactions

3-Acetoxypyrroles are smoothly deacylated in high yield by treatment with sodium hydroxide,[33] potassium carbonate,[33] or sodium methoxide at room temperature,[156] or under reflux,[152] to give the corresponding 3-hydroxy compound. Such deprotection is also possible using lithium aluminum hydride,[156] or, for the special case of *tert*-butyl carbonate derivatives, by flash-vacuum pyrolysis (FVP) at 650°C.[156] Treatment of an acetoxypyrrole with *p*-toluidine duly gave the *N*-arylacetamide, but the pyrrole could not be recovered.[217]

4.6.3.2. Reactions with Electrophiles

3-Acetoxypyrroles are much less reactive than their 3-hydroxy- or 3-alkoxy analogs. For example, no reaction takes place with methoxymethylene Meldrum's acid (see Sections 4.4.3.6 and 4.5.3.3.5), even after extended reaction times.[156] More active electrophiles, such as oxalyl chloride, give the 2-substitution product.[156] Iodination of the *N*,*O*-diacetate **273** (R^1 = COMe, R^2 = H) gives the iodo compound **273** (R^1 = H, R^2 = I) in 76% yield on treatment with iodine and sodium acetate in ethanol.[159] Ullmann coupling of this iodo compound has been achieved.[159] Acylation at the 2-position with acetic anhydride (catalyzed by perchloric acid) also takes place,[159] and Fries rearrangement of the monoacetate **273** (R^1 = R^2 = H) to the 2-acyl derivative **274** occurs either thermally or under the influence of aluminum trichloride.[159]

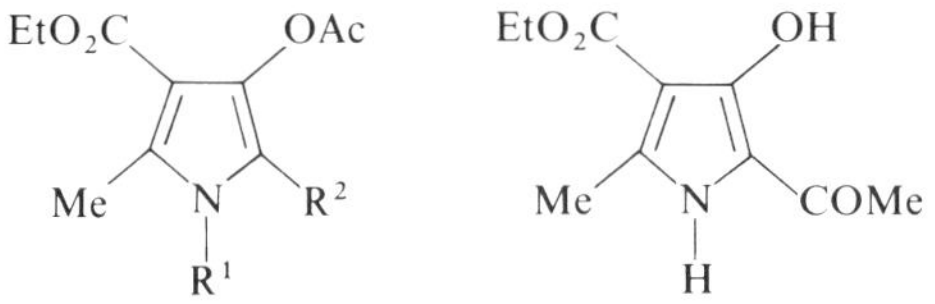

Treatment of the acetoxy compound **275** with the bromomethylpyrrole **276** in benzene gave, after 1 h at reflux, the coupled product **277**[211] (Scheme 74).

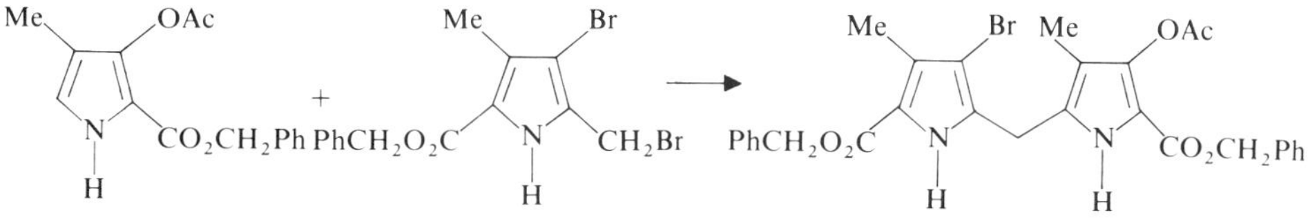

Scheme 74

Hydrogenolysis and further condensation gave a porphyrin-like structure that could not be fully characterized.[211] A related dipyrrolylmethene, **278**, has been made by coupling of an acetoxypyrrole carboxaldehyde with a pyrrolophane derivative.[33]

278

4.6.3.3. Oxidation and Reduction

Although the acetoxy group is cleaved with lithium aluminum hydride,[156] it is stable to a variety of other oxidation and reduction conditions. Thus, a 3-acetoxy-2-methylpyrrole was oxidized to the 2-carboxaldehyde in >80% yield using manganese dioxide in dioxane,[33] and hydrogenolysis of the iodo compound **273** (R^1 = H, R^2 = I) to **273** (R^1 = R^2 = H) over Adam's catalyst took place, leaving the acetoxy substituent intact.[159]

4.7. 3-ALKOXYPYRROLE NATURAL PRODUCTS

There is such a vast chemical and microbiological literature on the prodigiosin series of alkoxypyrrole natural products that it cannot be covered here in detail. Prodigiosin (**279**) is a bright-red tripyrrole pigment produced by the saphrophytic bacterium *Serratia marcescens*, and its obvious effects on foodstuffs have been noted since Greek times. It was first isolated in 1929,[218] but the correct structure was not proved by total synthesis until 1962.[77] An extensive family of prodigiosin pigments has been isolated from a group of eubacteria and actinomycetes.[219] These pigments generally differ only in the substitution pattern in ring C, although some have a link between the 5-positions of rings A and C. All have a 5-pyrrolyl-substituted 3-methoxypyrrole ring, with a modified aldehyde function at the 2-position. Some examples, e.g., **279**–**282**, are shown. The chemistry and microbiology of prodigiosin pigments was reviewed in 1975;[220] leading up-to-date references are given by Boger and Patel[207] and by Brown et al.[221] Prodigiosins possess antimicrobial and cytotoxic properties, and it is known that the methoxy group is essential for biological activity.[221] Metabolites isolated from mutant strains of *S. marcescens* include "norprodigiosin" (**283**)[222] (the only natural product in the series that does not contain a

4.7. 3-Alkoxypyrrole Natural Products

279 R = Me
283 R = H

methoxy group), the aldehyde **284**[223] (which is a biosynthetic precursor of prodigiosin), and the tetrapyrrole **285**.[224] This has also been isolated from two marine sources, viz., the Japanese bryozoan *Bugula dentata*[225] and an Australian ascidian,[226] and possesses antimicrobial properties.[225]

Prodigiosins themselves have been isolated from both aerobic[227] and anaerobic[228] marine bacteria.

The unique dipyrroles **286a–286d** occur in certain Californian nembrothid nudibranches,[229] although the metabolites have been traced to a bryozoan *Sessibugula translucens*, which forms part of the diet of the nudibranches.[229]

	X	Y	R²
286a	H	H	H
286b	Br	H	H
286c	H	H	n-Bu
286d	H	Br	n-Bu

The biosynthesis of prodigiosin metabolites is well understood,[219,228,230,231] and its elucidation by labeling studies required an analysis of the ^{13}C NMR[135,209,219] spectra of the compounds. Ring A is derived from proline; ring B, from the carboxylic acid group of the proline, plus acetate and serine with the methoxy group derived from methionine.[230] Ring C is constructed from alanine and a polyacetate,[230] and the condensation of the ring C pyrrole with the formylbipyrrole **284** takes place at a late stage in the biosynthesis[223] (Scheme 75).

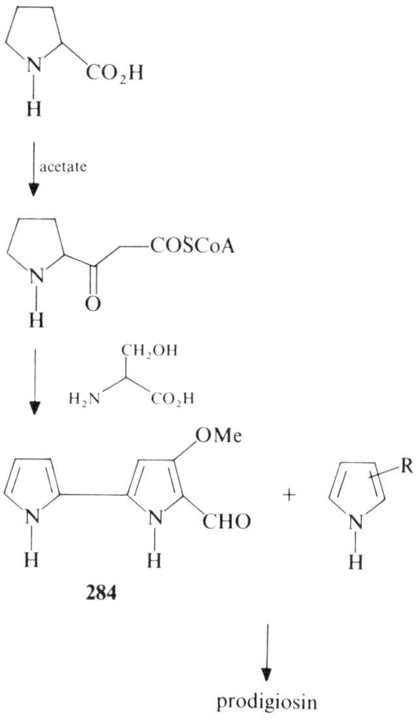

Scheme 75

A number of total syntheses of prodigiosins and prodigiosin analogs have been carried out. Since the condensation of the ring C pyrrole with the formylbipyrrole **284** is facile,[232–235] the main interest, from an alkoxypyrrole point of view, centers on the preparation of the aldehyde **284**. Rapoport's original method[77] employed the malonate route (Section 4.2.2) to the hydroxypyrroledicarboxylic ester **287**, followed by O-methylation (diazomethane), and selective hydrolysis and decarboxylation to the monoester **288**. The key introduction of the 5-pyrrolyl substituent was effected in 10% yield by condensation with pyrroline and dehydrogenation. MacFadyen–Stevens degradation of the ester gave the required aldehyde **284** (Scheme 76).

More recently, Boger and Patel[207] employed a reverse electron-demand cycloaddition route (Section 4.5.1) to the ester **289** and improved the introduc-

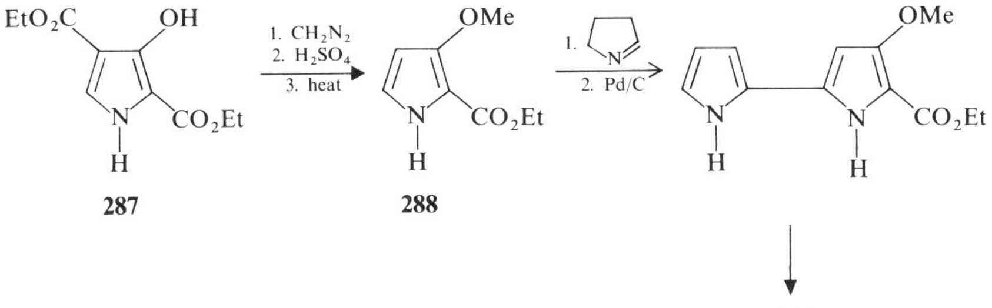

4.7. 3-Alkoxypyrrole Natural Products

Scheme 76

tion of the 5-pyrrolyl substituent by employing an intramolecular Pd(II)-promoted coupling reaction (Scheme 77). The overall yield from **289** to **290** was 54%. The MacFadyen–Stevens route to the aldehyde **284** was again used.

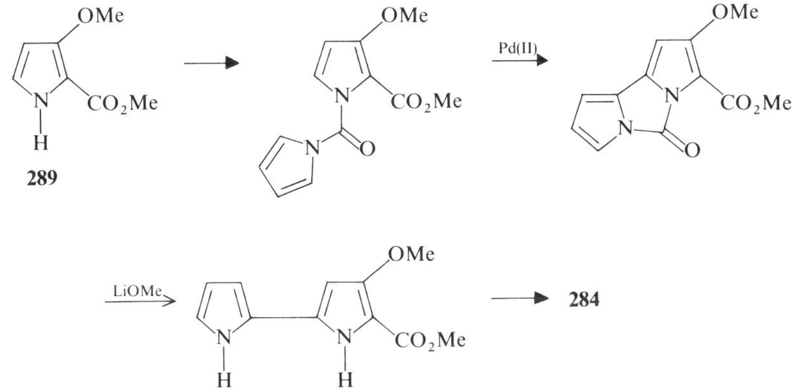

Scheme 77

In a third synthesis, Wasserman[61] used his tricarbonyl methodology (Section 4.2.3) to build ring B onto a preformed ring A, and hence avoided the difficulties of the coupling reaction (Scheme 78). However, the yield of this key

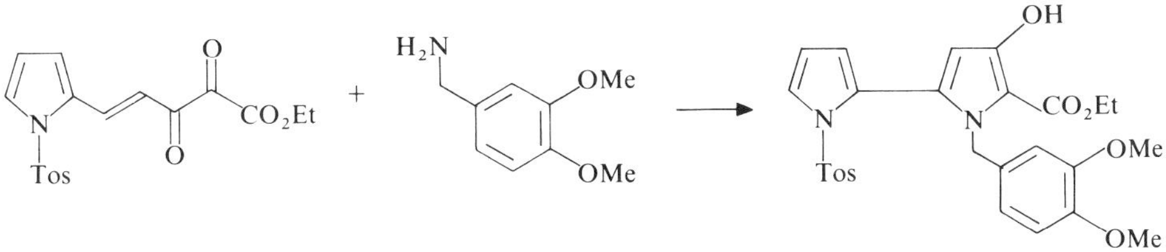

Scheme 78

step was just 23%. The aldehyde **284** was obtained from **291** by successive *O*-methylation, deprotection, and MacFadyen–Stevens degradation.

Structural analogs of prodigiosins that have been synthesized include those derived from 5-aryl-3-methoxypyrroles,[33,205] and from a 3-methoxy-5-methylpyrrole-4-carboxylic ester.[135] 5-Thienyl analogs have also been made[151,152,205] including an *N*-alkyl derivative whose crystal structure (Fig. 4.7) shows clear evidence of hydrogen bonding between the N*H* of ring C and the 3-methoxy substituent.[151] Since prodigiosin itself is sufficiently basic (pK_a 8.25) to exist as the cation at neutral pH,[236] it is possible that this is the stable configuration of the natural product.

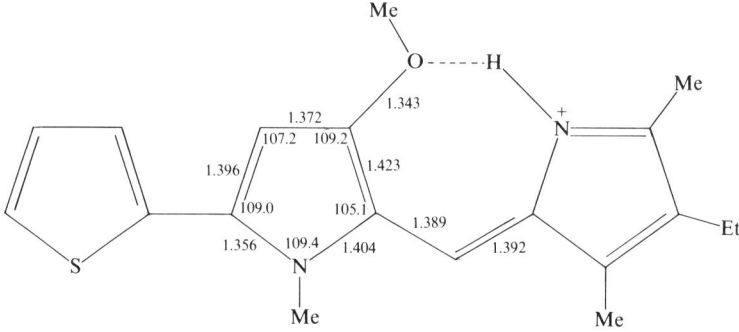

Figure 4.7. Selected bond lengths (Å) and angles (degrees) of prodigiosin analog.

4.8. ACKNOWLEDGMENT

We are most grateful to Dr. J. K. MacLeod (Australian National University) for carrying out the on-line substructure literature searches on 3-hydroxypyrroles and pyrrol-3-ones, which provided the basis for the review presented in this chapter.

4.9. REFERENCES

1. E. Benary and B. Silbermann, *Ber. Deut. Chem. Ges.*, **46**, 1363 (1913).
2. R. A. Jones and G. P. Bean, *The Chemistry of Pyrroles*, Academic Press, London, 1977.
3. R. C. F. Jones and J. M. Patience, *J. Chem. Soc., Perkin Trans. 1*, **1990**, 2350.
4. R. A. Jones Ed., *The Chemistry of Heterocyclic Compounds*, Vol. 48, Pyrroles, Part 1, Wiley, New York, 1990.
5. F. Ledl and E. Schleichter, *Angew. Chem. Int. Ed. Engl.*, **29**, 565 (1990).
6. B. Capon and F. -C. Kwok, *Tetrahedron Lett.*, **27**, 3275 (1986).
7. B. Capon and F. -C. Kwok, *J. Am. Chem. Soc.*, **111**, 5346 (1989).

4.9. References

8. H. Bauer, *Liebigs Ann. Chem.*, **736**, 1 (1970).
9. G. Pfeiffer and H. Bauer, *Liebigs Ann. Chem.*, **1976**, 383.
10. T. Momose, T. Tanaka, T. Yokota, N. Nagamoto, and K. Yamada, *Chem. Pharm. Bull.*, **27**, 1448 (1979).
11. W. Flitsch, K. Hampel, and M. Hohenhorst, *Tetrahedron Lett.*, **28**, 4395 (1987).
12. W. Flitsch and M. Hohenhorst, *Liebigs Ann. Chem.*, **1990**, 397.
13. T. Severin, W. Supp, and G. Manninger, *Chem. Ber.*, **112**, 3013 (1979).
14. J. Patjens, R. Ghaffari-Tabrizi, and P. Margaretha, *Helv. Chim. Acta*, **69**, 905 (1986).
15. B. P. Gusev, E. A. Él'perina, and V. F. Kucherov, *Khim. Geterotsikl. Soedin*, **1971**, 1424.
16. R. R. Schmidt, W. J. W. Mayer, and H. -U. Wagner, *Liebigs Ann. Chem.*, **1973**, 2010.
17. A. Hassner, A. S. Miller, and M. J. Haddadin, *J. Org. Chem.*, **37**, 2682 (1972).
18. C. -S. Chien, T. Kawasaki, and M. Sakamoto, *Chem. Pharm. Bull.*, **33**, 5071 (1985).
19. H. J. Gordon, J. C. Martin, and H. McNab, *J. Chem. Soc., Chem. Commun.*, **1983**, 957.
20. H. McNab and L. C. Monahan, *J. Chem. Soc., Chem. Commun.*, **1985**, 213.
21. C. L. Hickson, E. M. Keith, J. C. Martin, H. McNab, L. C. Monahan, and M. D. Walkinshaw, *J. Chem. Soc., Perkin Trans. 1*, **1986**, 1465.
22. H. McNab and L. C. Monahan, *J. Chem. Soc., Chem. Commun.*, **1987**, 138.
23. H. McNab and L. C. Monahan, *J. Chem. Soc., Perkin Trans. 1*, **1988**, 863.
24. H. McNab and L. C. Monahan, *J. Chem. Soc., Perkin Trans. 1*, **1988**, 869.
25. E. Anklam, R. Ghaffari-Tabrizi, H. Hombrecher, S. Lau, and P. Margaretha, *Helv. Chim. Acta*, **67**, 1402 (1984).
26. G. A. Hunter and H. McNab, unpublished results; G. A. Hunter, Ph.D. thesis, University of Edinburgh, 1990.
27. P. Lorenčak, J. C. Pommelet, J. Chuche, and C. Wentrup, *J. Chem. Soc., Chem. Commun.*, **1986**, 369.
28. H. Dhimane, J. -C. Pommelet, J. Chuche, G. Lhommet, M. G. Richaud, and M. Haddad, *Tetrahedron Lett.*, **26**, 833 (1985).
29. J. -C. Pommelet, H. Dhimane, J. Chuche, J. -P. Célérier, M. Haddad, and G. Lhommet, *J. Org. Chem.*, **53**, 5680 (1988).
30. H. J. Gordon, J. C. Martin, and H. McNab, *J. Chem. Soc., Perkin Trans. 1*, **1984**, 2129.
31. H. Briel, A. Lukosch, and C. Wentrup, *J. Org. Chem.*, **49**, 2772 (1984).
32. J. Davoll, *J. Chem. Soc.*, **1953**, 3802.
33. H. Berner, G. Schulz, and H. Reinshagen, *Monatsh. Chem.*, **109**, 137 (1978).
34. E. R. Talaty, A. R. Clague, M. O. Agho, M. N. Deshpande, P. M. Courtney, D. H. Burger, and E. F. Roberts, *J. Chem. Soc., Chem. Commun.*, **1980**, 889.
35. T. Eicher and J. L. Weber, *Tetrahedron Lett.*, **1974**, 1381.
36. T. Eicher, J. L. Weber, and G. Chatila, *Liebigs Ann. Chem.*, 1203, 1978.
37. O. Widman, *Liebigs Ann. Chem.*, **400**, 86 (1913).
38. H. H. Wasserman and J. B. Brous, *J. Org Chem.*, **19**, 515 (1954).
39. H. H. Wasserman and J. B. Brous, *J. Am. Chem. Soc.*, **76**, 5811 (1954).
40. J. Chauvelier, *Bull. Soc. Chim. Fr.*, **1954**, 734.
41. T. Metler, A. Uchida, and S. I. Miller, *Tetrahedron*, **24**, 4285 (1968).
42. G. Heinisch, W. Holzer, and H. Völlenkle, *Monatsh. Chem.*, **118**, 409 (1987).
43. M. E. Jung and B. E. Love, *J. Chem. Soc., Chem. Commun.*, 1288 (1987).
44. R. Chong and P. S. Clezy, *Aust. J. Chem.*, **20**, 935 (1967).
45. A. Treibs and A. Ohorodnik, *Liebigs Ann. Chem.*, **611**, 139 (1958).

46. G. Pfeiffer and H. Bauer, *Liebigs Ann. Chem.*, **1980**, 564.
47. O. Hromatka, D. Binder, and P. Stanetty, *Monatsh. Chem.*, **106**, 213 (1975).
48. G. Kresze and H. Härtner, *Liebigs Ann. Chem.*, **1973**, 650.
49. G. Albers-Schönberg, B. H. Arison, O. D. Hensens, J. Hirshfield, K. Hoogsteen, E. A. Kaczka, R. E. Rhodes, J. S. Kahan, F. M. Kahan, R. W. Ratcliffe, E. Walton, L. J. Ruswinkle, R. B. Morin, and B. G. Christensen, *J. Am. Chem. Soc.*, **100**, 6491 (1978).
50. D. H. Shih and R. W. Ratcliffe, *J. Med. Chem.*, **24**, 639 (1981).
51. A. G. Brown, D. F. Corbett, A. J. Eglington, and T. T. Howarth, *Tetrahedron*, **39**, 2551 (1983).
52. M. Sletzinger, T. Liu, R. A. Reamer, and I. Shinkai, *Tetrahedron Lett.*, **21**, 4221 (1980).
53. J. S. Davies, *Tetrahedron Lett.*, **23**, 5089 (1982).
54. J. Häusler, *Liebigs Ann. Chem.*, **1981**, 1073.
55. H. H. Wasserman, J. Fukuyama, N. Murugesan, J. van Duzer, L. Lombardo, V. Rotello, and K. McCarthy, *J. Am. Chem. Soc.*, **111**, 371 (1989).
56. R. V. Hoffman, H. -O. Kim, and A. L. Wilson, *J. Org. Chem.*, **55**, 2820 (1990).
57. H. H. Wasserman and C. B. Vu, *Tetrahedron Lett.*, **31**, 5205 (1990).
58. H. H. Wassermann, J. D. Cook, J. M. Fukuyama, and V. M. Rotello, *Tetrahedron Lett.*, **30**, 1721 (1989).
59. H. H. Wasserman and R. M. Amici, *J. Org. Chem.*, **54**, 5843 (1989).
60. H. H. Wasserman, J. D. Cook, and C. B. Vu, *Tetrahedron Lett.*, **31**, 4945 (1990).
61. H. H. Wasserman and L. J. Lombardo, *Tetrahedron Lett.*, **30**, 1725 (1989).
62. O. S. Wolfbeis and H. Junek, *Monatsh. Chem.*, **110**, 1387 (1979).
63. E. Benary and R. Konrad, *Ber. Deut. Chem. Ges.*, **56**, 44 (1923).
64. S. K. Gupta, *Synthesis*, **1975**, 726.
65. W. Madelung and L. Obermann, *Ber. Deut. Chem. Ges.*, **63**, 2870 (1930).
66. E. Campaigne and G. M. Shutske, *J. Heterocycl. Chem.*, **12**, 67 (1975).
67. A. San Feliciano, E. Caballero, J. A. P. Pereira, and P. Puebla, *Tetrahedron*, **45**, 6553 (1989).
68. A. R. Mattocks, *J. Chem. Soc., Perkin Trans. 1*, **1978**, 896.
69. E. Benary and W. Lau, *Ber. Deut. Chem. Ges.*, **56**, 591 (1923).
70. E. von Meyer, *J. Prakt. Chem.*, **90**, 1 (1914).
71. I. T. Kay and N. Punja, *J. Chem. Soc. (C)*, **1970**, 2409.
72. R. Kuhn and G. Osswald, *Chem. Ber.*, **89**, 1423 (1956).
73. R. L. Rosati, L. V. Kapili, P. Morrissey, J. Bordner, and E. Subramanian, *J. Am. Chem. Soc.*, **104**, 4262 (1983).
74. J. P. Bazureau, J. Le Roux, and M. Le Corre, *Tetrahedron Lett.*, **29**, 1921 (1988).
75. T. Momose, T. Tanaka, and T. Yokota, *Heterocycles*, **6**, 1821 (1977).
76. T. Momose, T. Tanaka, T. Yokota, N. Nagamoto, and K. Yamada, *Chem. Pharm. Bull.*, **26**, 2224 (1978).
77. H. Rapoport and K. G. Holden, *J. Am. Chem. Soc.*, **84**, 635 (1962).
78. T. C. -G. Kazembe and D. A. Taylor, *Tetrahedron*, **36**, 2125 (1980).
79. E. Campaigne and G. Shutske, *J. Heterocycl. Chem.*, **11**, 929 (1974).
80. E. Campaigne, G. Shutske, and J. C. Payne, *J. Heterocycl. Chem.*, **14**, 329 (1977).
81. T. Murata, T. Sugawara, and K. Ukawa, *Chem. Pharm. Bull.*, **21**, 2571 (1973).
82. T. Murata, T. Sugawara, and K. Ukara, *Chem. Pharm. Bull.*, **26**, 3080 (1978).
83. E. Winterfeldt and H. -J. Dillinger, *Chem. Ber.*, **99**, 1558 (1966).
84. S. K. Khetan and M. V. George, *Tetrahedron*, **25**, 527 (1969).
85. A. Zinoune, J. -J. Bourguignon, and C. -G. Wermuth, *Heterocycles*, **28**, 1077 (1989).

4.9. References

86. J. P. Freeman and M. J. Haddadin, *J. Org. Chem.*, **45**, 4898 (1980).
87. W. Ried and U. Reiher, *Chem. Ber.*, **120**, 1597 (1987).
88. G. Bianchi, A. Gamba-Invernizzi, and R. Gandolfi, *J. Chem. Soc., Perkin Trans. 1*, **1974**, 1757.
89. D. A. Lightner and L. K. Low, *J. Heterocycl. Chem.*, **12**, 793 (1975).
90. M. Weigele, J. F. Blount, J. P. Tengi, R. C. Czajkowski, and W. Leimgruber, *J. Am. Chem. Soc.*, **94**, 4052 (1972).
91. M. Weigele, S. L. De Bernardo, J. P. Tengi, and W. Leimgruber, *J. Am. Chem. Soc.*, **94**, 5927 (1972).
92. M. Weigele, J. P. Tengi, S. L. De Bernardo, R. Czajkowski, and W. Leimgruber, *J. Org. Chem.*, **41**, 388 (1976).
93. E. Gross and J. Meienhofer, Ed., *The Peptides*, Vol. 4, Academic Press, New York, 1981, Chapters 3 and 4.
94. V. Toome, S. De Bernardo, and M. Weigele, *Tetrahedron*, **31**, 2625 (1975).
95. D. D. Keith, S. De Bernardo, and M. Weigele, *Tetrahedron*, **31**, 2629 (1975).
96. D. D. Keith, J. A. Tortora, K. Ineichen, and W. Leimgruber, *Tetrahedron*, **31**, 2633 (1975).
97. V. Toome and B. Wegrzynski, *Biochem. Biophys. Res. Commun.*, **85**, 1496 (1978).
98. H. H. Wasserman, R. Frechette, T. Oida, and J. M. van Duzer, *J. Org. Chem.*, **54**, 6012 (1989).
99. S. Gelin, *Synthesis*, **1985**, 291.
100. B. Chantegrel and S. Gelin, *J. Heterocycl. Chem.*, **15**, 1215 (1978).
101. F. Dallacker and V. Mues, *Chem. Ber.*, **108**, 569 (1975).
102. H. McNab and L. C. Monahan, *J. Chem. Soc., Perkin Trans. 1*, **1989**, 419.
103. H. McNab and A. D. Macpherson, unpublished work.
104. R. Fuks and H. G. Viehe, *Tetrahedron*, **25**, 5721 (1969).
105. W. Steglich, G. Höfle, W. König, and F. Weygand, *Chem. Ber.*, **101**, 308 (1968).
106. P. C. Srivastava, C. M. Gupta, and A. P. Bhaduri, *Tetrahedron Lett.*, **1971**, 1975.
107. A. Kjaer, *Acta Chem. Scand.*, **7**, 1027 (1953).
108. C. E. Ward, U.S. Pat. 4643762 (1987); *Chem. Abstr.*, **106**, 156271 g (1987).
109. K. Burger, A. Meffert, and A. Gieren, *Liebigs Ann. Chem.*, **1978**, 1037.
110. H. -D. Stachel and K. K. Harigel, *Arch. Pharm.* (*Weinheim*), **302**, 654 (1969).
111. H. -D. Stachel, K. K. Harigel, H. Poschenrieder, and H. Burghard, *J. Heterocycl. Chem.*, **17**, 1195 (1980).
112. H. C. Clark, C. R. C. Milne, and N. C. Payne, *J. Am. Chem. Soc.*, **100**, 1164 (1978).
113. K. -J. Boosen, *Helv. Chim. Acta*, **60**, 1256 (1977).
114. T. Eicher and J. L. Weber, *Tetrahedron Lett.*, **1973**, 1541.
115. J. P. Freeman and M. J. Haddadin, *Tetrahedron Lett.*, **1979**, 4813.
116. M. Takahashi, S. Yamaguchi, and N. Igari, *Chem. Lett.*, **1977**, 1497.
117. M. Takahashi, N. Inaba, H. Kirihara, and S. Watanabe, *Bull. Chem. Soc. Jpn.*, **51**, 3312 (1978).
118. C. Kascheres, A. Kascheres, and P. S. H. Pilli, *J. Org. Chem.*, **45**, 5340 (1980).
119. T. Eicher and D. Krause, *Synthesis*, **1986**, 899.
120. T. Eicher and J. L. Weber, *Tetrahedron Lett.*, **1974**, 3409.
121. T. Eicher, J. L. Weber, and J. Kopf, *Liebigs Ann. Chem.*, **1978**, 1222.
122. T. Eicher, F. Abdesaken, G. Franke, and J. L. Weber, *Tetrahedron Lett.*, **1975**, 3915.
123. T. Eicher and R. Rohde, *Synthesis*, **1985**, 619.
124. T. Eicher and G. Franke, *Liebigs Ann. Chem.*, **1981**, 1337.
125. H. Yoshida, S. Sogame, S. Bando, S. Nakajima, T. Ogata, and K. Matsumoto, *Bull. Chem. Soc. Jpn.*, **56**, 3849 (1983).

126. M. Takahashi, T. Funaki, H. Honda, Y. Yokoyama, and H. Takimoto, *Heterocycles*, **19**, 1921 (1982).
127. J. J. Barr, R. C. Storr, and V. K. Tandon, *J. Chem. Soc., Perkin Trans. 1*, 1147 (1980).
128. H. Yoshida, S. Sogame, Y. Takishita, and T. Ogata, *Bull. Chem. Soc. Jpn.*, **56**, 2438 (1983).
129. K. -H. Klaska, O. Jarchow, T. Eicher, and H. Preut, *Acta Cryst. C*, **35**, 2788 (1979).
130. A. Treibs and A. Ohorodnik, *Liebigs Ann. Chem.*, **611**, 149 (1958).
131. N. Bodor, M. J. S. Dewar, and A. J. Harget, *J. Am. Chem. Soc.*, **92**, 2929 (1970).
132. A. Karpfen, P. Schuster, and H. Berner, *J. Org. Chem.*, **44**, 374 (1979).
133. A. J. Blake, H. McNab, and L. C. Monahan, *J. Chem. Soc., Perkin Trans. 2*, **1988**, 1455.
134. R. S. Atkinson and E. Bullock, *Can. J. Chem.*, **41**, 625 (1963).
135. H. Berner, G. Schulz, and H. Reinshagen, *Monatsh. Chem.*, **108**, 915 (1977).
136. J. Elguero, C. Marzin, A. R. Katritzky, and P. Linda, *Adv. Heterocycl. Chem., Suppl. 1*, **1976**, 239.
137. H. McNab, L. C. Monahan, and J. C. Walton, *J. Chem. Soc., Perkin. Pevkin Trans. 2*, **1988**, 759.
138. G. A. Hunter, H. McNab, L. C. Monahan, and A. J. Blake, *J. Chem. Soc., Perkin Trans. 1*, **1991**, 3245.
139. J. Kao, A. L. Hinde, and L. Radom, *Nouveau J. Chim.*, **3**, 473 (1979).
140. R. Ghaffari-Tabrizi, P. Margaretha, and H. W. Schmalle, *Helv. Chim. Acta*, **67**, 1957 (1984).
141. J. Dabrowski and U. Dabrowska, *Chem. Ber.*, **101**, 2635 (1968).
142. J. Dabrowski and U. Dabrowska, *Chem. Ber.*, **101**, 3392 (1968).
143. M. Beyer, R. Ghaffari-Tabrizi, M. Jung, and P. Margaretha, *Helv. Chim. Acta*, **67**, 1535 (1984).
144. T. J. Batterham, *N.M.R. Spectra of Simple Heterocycles*, Wiley-Interscience, New York 1973.
145. H. McNab and L. C. Monahan, *J. Chem. Soc., Perkin Trans. 2*, **1991**, 1999.
146. P. Walter and T. M. Harris, *J. Org. Chem.*, **43**, 4250 (1978).
147. P. E. Hansen, *Prog. Nucl. Magn. Reson. Spectrosc.*, **14**, 175 (1981).
148. H. McNab and L. C. Monahan, *J. Chem. Soc. Perkin Trans. 2*, **1988**, 1459.
149. P. S. Clezy and N. W. Webb, *Aust. J. Chem.*, **25**, 2217 (1972).
150. T. -L. Ho, *Tetrahedron*, **41**, 3 (1985).
151. A. J. Blake, G. A. Hunter, and H. McNab, *J. Chem. Soc., Chem. Commun.*, **1990**, 734.
152. G. Kresze, M. Morper, and A. Bijev, *Tetrahedron Lett.*, **1977**, 2259.
153. T. Momose, T. Tanaka, T. Yokota, N. Nagamoto, and K. Yamada, *Chem. Pharm. Bull.*, **26**, 3521 (1978).
154. G. Manninger and T. Severin, *Z. Lebensm. Unters. Forsch.*, **161**, 45 (1976).
155. A. J. Blake, H. McNab, and L. C. Monahan, *J. Chem. Soc., Perkin Trans. 2*, **1988**, 1463.
156. H. McNab and K. A. Taylor, unpublished work.
157. H. Bauer, *Chem. Ber.*, **100**, 1704 (1967).
158. T. Wieland and T. Severin, *Z. Lebensm. Unters. Forsch.*, **153**, 201 (1973).
159. R. S. Atkinson and E. Bullock, *Can. J. Chem.*, **42**, 1524 (1964).
160. S. Gelin and C. Deshayes, *Synthesis*, **1982**, 657.
161. J. P. Freeman and M. K. Fettes-Fields, *Heterocycles*, **23**, 1073 (1985).
162. H. Yoshida, S. Bando, S. Nakajima, T. Ogata, and K. Matsumoto, *Bull. Chem. Soc. Jpn.*, **57**, 2677 (1984).
163. J. Patjens and P. Margaretha, *Helv. Chim. Acta*, **72**, 1817 (1989).
164. H. Bauer, *Chem. Ber.*, **104**, 259 (1971).
165. T. Momose, T. Tanaka, T. Yokota, N. Nagamoto, H. Kobayashi, and S. Takano, *Heterocycles*, **15**, 843 (1981).

4.9. References

166. A. J. Blake, H. McNab, and L. C. Monahan, *J. Chem. Soc., Perkin Trans. 1*, **1991**, 701.
167. L. Capuano and T. Tammer, *Chem. Ber.*, **114**, 456 (1981).
168. K. E. Schulte, J. Reisch, and U. Stoess, *Arch. Pharm. (Weinheim)*, **305**, 523 (1972).
169. H. McNab and L. C. Monahan, unpublished results; L. C. Monahan, Ph.D. thesis, University of Edinburgh, 1986.
170. H. Bauer *Angew. Chem. Int. Ed. Engl.*, **7**, 734 (1968).
171. T. Eicher and A. Kruse, *Synthesis*, **1985**, 612.
172. H. H. Inhoffen, N. Schwarz, and K. -P. Heise, *Liebigs Ann. Chem.*, **1973**, 146.
173. W. Flitsch, R. A. Jones, and M. Hohenhorst, *Tetrahedron Lett.*, **28**, 4397 (1987).
174. W. Flitsch and M. Hohenhorst, *Liebigs Ann. Chem.*, **1990**, 449.
175. H. Bauer, *Chem. Ber.*, **101**, 1286 (1968).
176. S. L. Galdino, I. R. Pitta, C. D. Luu, B. Lucena, L. C. Oliveira, and M. Rosalia, *Eur. J. Med. Chem.–Chim. Ther.*, **20**, 439 (1985).
177. S. Gelin and R. Gelin, *J. Org. Chem.*, **44**, 808 (1979).
178. H. G. Viehe, Z. Janousek, R. Merenyi, and L. Stella, *Acc. Chem. Res.*, **18**, 148 (1985) and references cited therein.
179. S. Gelin, *J. Org. Chem.*, **44**, 3053 (1979).
180. R. A. Y. Jones and N. Sadighi, *J. Chem. Soc., Perkin Trans. 1*, **1976**, 2259.
181. G. Pfeiffer, W. Otting, and H. Bauer, *Angew. Chem. Int. Ed. Engl.*, **15**, 52 (1976).
182. R. L. N. Harris, *Aust. J. Chem.*, **23**, 1199 (1970).
183. T. Momose, T. Tanaka, and T. Yokota, *Heterocycles*, **6**, 1827 (1977).
184. K. Schofield, *Heteroaromatic Nitrogen Compounds*, Butterworths, London, 1967, p. 95.
185. R. Ghaffari-Tabrizi and P. Margaretha, *Helv. Chim. Acta*, **65**, 1029 (1982).
186. R. Ghaffari-Tabrizi and P. Margaretha, *Helv. Chim. Acta*, **66**, 1902 (1983).
187. T. Severin and H. Poehlmann, *Chem. Ber.*, **110**, 491 (1977).
188. M. Nagarajan and H. Shechter, *J. Org. Chem.*, **49**, 62 (1984).
189. V. Dragisich, C. K. Murray, B. P. Warner, W. D. Wulff, and D. C. Yang, *J. Am. Chem. Soc.*, **112**, 1251 (1990).
190. R. Aumann and H. Heinen, *Chem. Ber.*, **119**, 3801 (1986).
191. R. Aumann, *Angew. Chem. Int. Ed. Engl.*, **27**, 1456 (1988).
192. K. S. Kochhar and H. W. Pinnick, *J. Org. Chem.*, **49**, 3222 (1984).
193. M. Giles, M. S. Hadley, and T. Gallagher, *J. Chem. Soc., Chem. Commun.*, **1990**, 831.
194. A. A. Nagel and L. A. Vincent, *J. Org. Chem.*, **44**, 2050 (1979).
195. F. Bellamy, J. Streith, and H. Fritz, *Nouveau J. Chim.*, **3**, 115 (1979).
196. F. Bellamy, P. Martz, and J. Streith, *Heterocycles*, **3**, 395 (1975).
197. F. Bellamy and J. Streith, *J. Chem. Res., (S)* **1979**, 18; (M) **1979**, 0101.
198. H. Sawanishi, K. Tajima, and T. Tsuchiya, *Chem. Pharm. Bull.*, **35**, 3175 (1987).
199. H. Rapoport and C. D. Willson, *J. Am. Chem. Soc.*, **84**, 630 (1962).
200. M. Vaultier, R. Danion-Bourgot, D. Danion, J. Hamelin, and R. Carrié, *Comptes Rendus*, **277C**, 1041 (1973).
201. M. Vaultier, R. Danion-Bourgot, D. Danion, J. Hamelin, and R. Carrié, *Bull. Soc. Chim. Fr.*, **1976**, 1537.
202. A. R. Mattocks, *J. Chem. Soc., Perkin Trans. 1*, **1974**, 707.
203. M. Vaultier, R. Danion-Bougot, F. Tonnard, and R. Carrié, *Bull. Soc. Chim. Fr.*, **1985**, 803.
204. L. Bonaccina, P. Mencarelli, and F. Stegel, *J. Org. Chem.*, **44**, 4420 (1979).
205. E. Campaigne and G. M. Shutske, *J. Heterocycl. Chem.*, **13**, 497 (1976).

206. D. L. Boger, R. S. Coleman, J. S. Panek, and D. Yohannes, *J. Org. Chem.*, **49**, 4405 (1984).
207. D. L. Boger and M. Patel, *J. Org. Chem.*, **53**, 1405 (1988).
208. G. S. Coumbarides, J. M. Mercey, and T. P. Toube, *J. Chem. Res. (S)*, **1990**, 151.
209. R. J. Cushley, R. J. Sykes, C.-K. Shaw, and H. H. Wasserman, *Can. J. Chem.*, **53**, 148 (1975).
210. J. M. Mercey and T. P. Toube, *J. Chem. Res.*, (S) **1987**, 62; (M) **1987**, 0680.
211. R. Chong and P. S. Clezy, *Aust. J. Chem.*, **20**, 951 (1967).
212. E. Campaigne and G. M. Shutske, *J. Heterocycl. Chem.*, **12**, 317 (1975).
213. E. Campaigne and G. M. Shutske, *J. Heterocycl. Chem.*, **12**, 1047 (1975).
214. A. J. Blake, G. A. Hunter, and H. McNab, *J. Chem. Res. (S)* **1991**, 316; (M) **1991**, 2885.
215. R. Bonnett, P. Cornell, and A. F. McDonagh, *J. Chem. Soc., Perkin Trans. 1*, **1976**, 794.
216. G. Dattolo, G. Cirrincione, A. M. Almerico, G. Presti, and E. Aiello, *Heterocycles*, **20**, 829 (1983).
217. R. Knorr and R. Huisgen, *Chem. Ber.*, **103**, 2598 (1970).
218. F. Wrede and A. Rothhass, *Z. Physiol. Chem.*, **226**, 95 (1934).
219. N. N. Gerber, A. G. McInnes, D. G. Smith, J. A. Walter, J. L. C. Wright, and L. C. Vining, *Can. J. Chem.*, **56**, 1155 (1978).
220. N. N. Gerber, *Crit. Rev. Microbiol.*, **3**, 469 (1975).
221. D. Brown, D. Griffiths, M. E. Rider, and R. C. Smith, *J. Chem. Soc., Perkin Trans. 1*, **1986**, 455.
222. B. S. Deol, J. R. Alden, J. L. Still, A. V. Robertson, and J. Winkler, *Aust. J. Chem.*, **27**, 2657 (1974).
223. H. H. Wasserman, J. E. McKeon, L. A. Smith, and P. Forgione, *Tetrahedron, Suppl. 8*, Part 2, **1966**, 647.
224. H. H. Wasserman, D. J. Friedland, and D. A. Morrison, *Tetrahedron Lett.*, **1968**, 641.
225. S. Matsunaga, N. Fusetani, and K. Hashimoto, *Experientia*, **42**, 84 (1986).
226. R. Kazlauskas, J. F. Marwood, P. T. Murphy, and R. J. Wells, *Aust. J. Chem.*, **35**, 215 (1982).
227. H. Laatsch and R. H. Thomson, *Tetrahedron Lett.*, **24**, 2701 (1983).
228. N. N. Gerber, *Tetrahedron Lett.*, **24**, 2797 (1983).
229. B. Carte and D. J. Faulkner, *J. Org. Chem.*, **48**, 2314 (1983).
230. H. H. Wasserman, R. J. Sykes, P. Peverada, C. K. Shaw, R. J. Cushley, and S. R. Lipsky, *J. Am. Chem. Soc.*, **95**, 6874 (1973).
231. R. J. Cushley, D. R. Anderson, S. R. Lipsky, R. J. Sykes, and H. H. Wasserman, *J. Am. Chem. Soc.*, **93**, 6284 (1971).
232. H. H. Wasserman, J. E. McKeon, L. Smith, and P. Forgione, *J. Am. Chem. Soc.*, **82**, 506 (1960).
233. H. H. Wasserman, G. C. Rodgers, and D. D. Keith, *Tetrahedron*, **32**, 1851 (1976).
234. H. H. Wasserman, D. D. Keith, and G. C. Rodgers, *Tetrahedron*, **32**, 1855 (1976).
235. H. H. Wasserman and J. M. Fukuyama, *Tetrahedron Lett.*, **25**, 1387 (1984).
236. W. R. Hearn, J. Medina-Castro, and M. K. Elson, *Nature*, **220**, 170 (1968).

Index

1-(1-Acetoxyethyl)pyrrole:
  thermal cleavage of, 168
3-Acetyl-1-benzenesulfonylpyrrole:
  Diels–Alder reaction with 2-methylbutadiene, 95
2-Acetyl-1-methylpyrrole:
  reaction of oxime with acetylene, 244
2-Acetylpyrrole:
  aldol condensation reaction, 34, 35
  oxidation with selenium oxide, 65
  reaction with ethyl trifluoroacetate, 35
  reaction with ethyl oxalate, 35
3-Acetylpyrroles:
  aldol condensation reaction, 34
  oxidation with selenium oxide, 65
  reaction with thallium(III) nitrate, 97, 98
2-Acetylpyrrole-3-carboxylic esters:
  reaction with hydrazine, 57
  reaction with hydroxylamine, 58
3-Acetyl-1-tosylpyrrole:
  haloform reaction on, 66
1-Acylaminopyrroles:
  cleavage of acyl group, 348
4-Acyl-1-methyl-2-vinylpyrroles:
  [4+2]cycloaddition reaction with DMAD, 95
3-Acyloxypyrroles:
  C-acetylation of, 605
  deacylation reaction, 605
  Fries rearrangement, 605
  halogenation of, 605
  hydride cleavage of acyl group, 606
  IR spectra of, 604
  mass spectra of, 604
  NMR spectra of, 604
  synthesis by lead tetracetate oxidation of 2,5-disubstituted pyrroles, 603
  synthesis by benzoyl peroxide oxidation of 2,5-disubstituted pyrroles, 603
  synthesis from 3-diazopyrroles, 603
  synthesis from acetoxybutadienes, 603
  synthesis from oxazolium oxides, 604
3-Acyloxydipyrrolylmethanes:
  synthesis of, 605
Acylpyrroles:
  acid-catalysed migration of acyl group, 96
  Baeyer–Villiger oxidation of, 64
  catalytic reduction of, 73

1-(2-chloroethyl) derivative, 86
Claisen condensation reaction, 35
cleavage of acyl group, 90
conversion into trimethylsilyl enol ether, 84
dithioketals, 90
electrochemical polymerization, 110
from higher plants, 8
from microbial sources, 2
from sponges, 6
Horner reaction, 30
hydride reduction of, 68, 70, 71
in pharmaceuticals, 10
phase-transfer catalysed N-alkylation, 100
N-protection of, 83–86
reaction with Grignard reagents, 22
reaction with hydrazines, 51
reduction with diborane, 71
reduction with zinc amalgam, 75
reductive alkylation of, 21
Reformatzky reaction, 23
resonance structures, 12
1-(tert-butyloxycarbonyl) derivative, 86
1-(p-toluenesulfonyl) derivative, 86
1-(2-trimethylsilyl)ethoxymethyl derivative, 86
Wittig reaction, 24, 30
Wolff-Kishner reduction of, 75
1-Acylpyrroles:
  as acylating agents, 108, 109
  base-catalysed cleavage of 1-acyl group, 83, 84
  electrochemical polymerization, 110
  hydrogenolysis of 1-benzoyl group, 84
  palladium(II) acetate catalysed reactions, 100–101
3-Acylpyrroles:
  halogenation of, 102
Alkoxybipyrroles:
  biosynthesis, 608
  natural occurence, 606
  synthesis of, 609
3-Alkoxydipyrrolylmethenes:
  synthesis of, 599
1-(1-Alkoxyethyl)pyrroles:
  synthesis of, 183–185
3-Alkoxy-2-formylpyrroles:
  synthesis of, 601

3-Alkoxy-5-formylpyrroles:
  synthesis of, 602
3-Alkoxypyrroles:
  aerial oxidation of, 600
  Claisen rearrangement of 3-allyloxy pyrrole, 602
  O-dealkylation of, 595
  diazo-coupling, 597
  halogenation, 596
  IR spectra of, 592
  mass spectra of, 593
  Michael reaction, 599
  NMR spectra of, 593, 594
  oxidation of, 600, 601
  photolytic synthesis from 4-alkoxypyridine-1-oxides, 589
  protonation of, 595
  reaction with benzaldehyde, 598
  reaction with DMAD, 598
  reaction with methoxymethylene derivative of Meldrum's acid, 598
  sulfenylation of, 599
  synthesis of Dibal reduction of 4-alkoxy-pyrrol-2(5H)-ones, 589
  synthesis by ring contraction of alkoxypyridazines, 591
  synthesis from 3-alkoxy-2,5-dihydropyrroles, 590
  synthesis from 3-alkoxy-4,5-dihydropyrroles, 590
  synthesis from alkoxycarbonylaziridines and phosphoranes, 590
  synthesis from 1,4-diketones, 588
  synthesis from ketohydrazones, 588
  synthesis by nucleophilic displacement of a 3-nitro group, 590
  molecular structure, 591
  UV spectra of, 592
  Vilsmeier–Haack formylation, 597
3-Alkoxypyrrole-5-carboxylic acids:
  iodinative decarboxylation of, 596
3-Alkoxypyrrole-2-carboxylic esters:
  base-catalysed hydrolysis of, 602
  hydride reduction of, 601
3-Alkoxypyrrole-4-carboxylic esters:
  acid-catalysed hydrolysis of, 602
  hydride reduction of, 601
3-Alkoxypyrrole-5-carboxylic esters:
  base-catalysed hydrolysis of, 602
  hydride reduction of, 601
  iodination of, 596
1-(2-Alkylthioethyl)pyrroles:
  synthesis of, 185

2-Alkyl-1-vinylpyrroles:
  synthesis of, 136, 138
2-Amino-3-carboxamidopyrrole:
  reaction with thionyl chloride, 484
2-Amino-3-cyanopyrroles:
  acid-catalysed hydrolysis of, 486
  reaction with carbon disulfide, 487
  reaction with DMAD, 226
  reaction with trimethylorthoformate, 487
  synthesis of, 336, 513
"2-Amino-5-hydroxypyrroles:"
  base-catalysed hydrolysis of, 499
  oxidation of, 496
  reaction with hydroxylamine, 499
  tautomerism of, 473
3-Amino-2-hydroxy-2H-pyrroles, 498
1-Aminopyrroles:
  N-acylation of, 483
  N-alkylation of, 483
  analgesic and anaesthetic activity, 508
  antibacterial activity, 509
  N-arylation of, 483
  basicity of, 472
  bond angles, 363
  conversion into 5-azaindolizines, 484
  conversion into 6H-diaza-azulenes, 485
  conversion into pyrazoles, 480, 506
  [4 + 2]cycloaddition reactions, 505–507
  decomposition in mineral acids, 483
  densities, 471
  diazotization of, 485
  heat of combustion, 464
  hydrogenation of, 503
  hypoglycemic activity, 508
  NH stretching frequencies of, 435–446
  ^{13}C NMR spectra of, 404–414
  ^{1}H NMR spectra of, 365–383
  mass spectra of, 460, 462–463
  oxidation of, 495
  photolysis of, 480
  reaction with aryl aldehydes, 483
  reaction with 1,3- and 1,4-dicarbonyl compounds, 484–485
  reaction with diketene, 484
  reaction with DMAD, 505–507
  reaction with isocyanates and isothiocyanates, 483
  reaction with NBS, 495
  reaction with 4-nitrosophenol, 483
  reaction with N-phenylmaleimide, 506
  reaction with pyrylium salts, 485
  reaction with quinones, 484
  refractive index, 471

Schotten–Baumann benzoylation of, 483
stability of, 463
synthesis by N-amination, 348
synthesis from azoalkenes, 325
synthesis from 1,4-dicarbonyl compounds, 315–322
synthesis from 2,5-diethoxytetrahydrofuran, 322
synthesis from 1,4-dihydrazones, 320
synthesis from 1,4-dihydropyridazines, 323
synthesis from epoxyalkynes, 323
thermolysis of, 478
tuberculostatic activity, 508
via Wittig reaction, 324
UV spectra of, 421–425
X-ray diffraction data, 363

2-Aminopyrroles:
acylation of, 486, 487
N-alkylation of, 486
antiarrhythmic activity, 509
anticonvulsant activity, 509
basicity of, 473, 474
bond lengths, 362
bond lengths of protonated species, 362
calculated π-electron populations, 357, 359
calculated stability energies, 356
conformation of amino group, 356, 358, 359
decomposition in mineral acids, 486
diazotization of, 489
hydrogenation of, 504
mass spectra of, 460–461, 465–468
NH stretching frequencies of, 447–454, 459
^{13}C NMR spectra of, 415–417, 424
^{1}H NMR spectra of, 365, 384–394
^{1}H NMR spectra of protonated species, 474
oxidation of, 496
phase-transfer catalysed alkylation of, 486
plant protection agents, 509
protonation of, 473, 474
quantum chemical studies on, 361, 362
reaction with aryl aldehydes, 487
reaction with 1,3-dicarbonyl compounds, 488, 489
reaction with isocyanates, 487
reaction with trimethylorthoformate, 487
stability of, 464
synthesis from acetylenes, 339
synthesis from 2-amino-3-bromotropones, 337
synthesis from α-aminoketones, 335, 336
synthesis from α-amino esters, 339
synthesis from azomethines, 335
synthesis from 2H-azirines esters, 333

synthesis from butadienes, 330
synthesis from 2,5-diamino-3,4-dicyanothiophene, 332
synthesis from (2,2-dichloroalkylidene)imines, 334
synthesis from 1,2-diketones, 336
synthesis from enamines, 337
synthesis from enaminonitriles, 329
synthesis from α-halocarbonyl compounds, 337
synthesis from α-hydroxyketones, 336
synthesis from methylenecyclobuten-1-ones, 333
synthesis from succinonitriles, 331
synthesis from iminopyridazines, 332
synthesis *via* Beckmann rearrangement of pyrrolyl oximes, 349
synthesis *via* Curtius rearrangement of 2-pyrroloylazides, 349
synthesis *via* nucleophilic substitution reactions, 350
synthesis *via* reduction of azo, nitroso, and nitro compounds, 349
synthesis *via* Stobbe condensation, 331
synthons for nucleoside antibiotics, 512
tautomerism of, 359–361, 472
thermolysis of, 481
UV spectra of, 425–430
X-ray diffraction data, 364

3-Aminopyrroles:
aerial oxidation of, 498
N-acylation of, 490, 491
N-alkylation of, 489, 490
analgesic activity, 510
antibacterial activity, 510
antiinflammatory activity, 510
antipyretic activity, 510
N-arylation of, 490
basicity of, 474–477
bond lengths, 362
bond lengths of protonated species, 362
calculated π-electron populations, 357
calculated stability energies, 356
CNS sedatives and myorelaxants, 510
conformation of amino group, 356
diazotization of, 492, 493
from α-aminonitriles, 339, 343
from base-catalysed self condensation of ω-aminoacetophenone, 344
from ring contraction of 4-chloropyrimidines, 340
hypolipidemic activity, 514
mass spectra of, 462, 464, 469–470

3-Aminopyrroles (*Continued*)
  mono-N-methylation of, 490
  NH stretching frequencies of, 455–459
  ^{13}C NMR spectra of, 418–420, 424
  ^{13}C NMR spectra of protonated species, 477
  ^{1}H NMR spectra of, 395–401, 403
  ^{1}H NMR spectra of protonated species, 477
  oxidation of, 497
  quantum chemical studies, 361, 362
  reaction with aryl aldehydes, 491
  reaction with isocyanates and isothiocyanates, 491
  reduction of, 504
  stability of, 471
  synthesis from aspartic acid derivatives, 513
  synthesis from azirines, 341
  synthesis from benzoylpropenes, 340
  synthesis from enamines, 343
  synthesis from enaminonitriles, 340
  synthesis from formylacetonitriles, 342
  synthesis from glycine esters, 343
  synthesis from 3-isoxazolylketoximes, 341
  synthesis from 2-bromopyruvic esters, 344
  synthesis *via* Beckmann rearrangement of pyrrolyl oximes, 351
  synthesis *via* Curtius rearrangement of 3-pyrroloylazides, 351
  synthesis *via* nucleophilic substitution reactions, 353
  synthesis *via* reduction of azo, nitroso, and nitro compounds, 351–353
  tautomerism of, 361, 477
  UV spectra of, 425, 431–433
  X-ray diffraction data, 364
1-Amino-1*H*-pyrrol-3(2*H*)-ones:
  synthesis of, 536
2-Amino-1*H*-pyrrol-3(2*H*)-ones:
  synthesis of, 537, 542
4-Amino-1*H*-pyrrol-3(2*H*)-ones:
  from azopyrrolones, 570
5-Amino-1*H*-pyrrol-3(2*H*)-ones:
  synthesis of, 539
  alkylation of, 559
5-Amino-1*H*-pyrrol-3(2*H*)-one-4-carboxylic esters:
  synthesis of, 540
1-Amino-2-vinylpyrroles:
  synthesis of, 233
Anthelvencin A:
  antibacterial activity, 512
  degradation of, 315
  nematodal control in mice, 512
  total synthesis of, 314
1-Aroylpyrroles:
  reaction with palladium(II) acetate, 99

2-Arylpyrroles:
  synthesis of, 99
2-Aryl-1-vinylpyrroles:
  synthesis of, 137
Azaazulenones:
  synthesis of, 573
Azacyclopentadienones:
  ring expansion of pyridines, 576
  synthesis of, 576
Azadipyrromethenes, 496
Azafulvenes:
  reaction with pyrrolin-2-ones, 230

2-Benzoyl-1-methylpyrrole:
  palladium(II) acetate catalysed reactions, 99
  reaction with 2,3-dimethylbut-2-ene, 234
1-Benzoylpyrrole:
  [4+2]cycloaddition reaction with DMAD, 92
  [4+2]cycloaddition reaction with hexafluorobut-2-yne, 92
  methylenation of, 168
  reaction with maleimides and maleic anhydride, 93
2-Benzylidenepyrrol-3-ones:
  synthesis of, 531
3,3'-Bipyrazoles, 320
1,1'-Bipyrroles:
  synthesis of, 316, 485
2,2'-Bipyrroles:
  synthesis of, 110
  Vilsmeier-Haack formylation of, 20
1,4-Bis(1-vinyl-2-pyrrolyl)benzene:
  protonation of, 178–180
1,4,8,11-Bisimino[14]annulene, 232
3,4-Bis(trifluoromethyl)pyrroles, 92
Z-2-(Buta-1,3-dienyl)pyrrole, 234
1-*tert*-Butyl-3-hydroxy-1,2-dihydropyrrolium cation:
  molecular structure, 563
1-*tert*-Butyl-3-methoxy-2-phenylpyrrole:
  molecular structure, 591
2-*tert*-Butyl-1-vinylpyrrole:
  protonation of, 178

Condensed 1-vinylpyrroles:
  synthesis of, 145–151
Congocidine (see Netropsin)
3-Cyano-2-formylpyrrole:
  reduction with diisobutylaluminum hydride, 72
4-Cyano-3-hydroxypyrroles:
  synthesis of, 534
2-Cyanopyrroles:
  synthesis of, 53

2-Cyanopyrrol-3-ones:
  hydrolysis of, 585
  synthesis of, 536
Cyclic ketoximes:
  reaction with acetylene, 145
[2 + 2]cycloaddition reactions:
  of 2-benzoyl-1-methylpyrrole, 234
  of 2,2-disubstituted 1*H*-pyrrol-3(2*H*)-ones, 586
[4 + 2]Cycloaddition reactions:
  of 3-acetyl-1-benzenesulfonylpyrroles (as a 2π system), 95
  of 4-acyl-1-methyl-2-vinylpyrroles, 95
  of 3-alkoxypyrroles, 598
  of 1-aminopyrroles, 505–507
  of 1-benzoylpyrrole, 92, 93
  of 5-formyl-1-methyl-4-vinylpyrrole, 95
  of 3-hydroxypyrroles, 576
  of pyrrole-1-carboxylic esters, 92, 96
  of 1-substituted pyrroles, 224, 225
  of 1*H*-pyrrol-3(2*H*)-ones, 576
  of vinylindoles, 243, 259
  of 2-vinylpyrroles, 225, 237–242
  of 3-vinylpyrroles, 26, 248–253

Danaidone, 44
Dehydroproline:
  synthesis of, 76
3,4-Diacetylpyrroles:
  reaction with hydrazine, 56
3,3'-Dialkoxy-2,2'-bipyrroles:
  synthesis of, 597, 601
6-Dialkylaminoazafulvenes, 48
Dialkylaminomethylene Meldrum's acid derivatives:
  flash vacuum pyrolysis of, 529, 530
2,3-Dialkyl-1-vinylpyrroles:
  synthesis of, 137
2,3-Dialkenylpyrroles:
  electrocyclic reaction, 263
  synthesis of, 233
2,5-Diamino-3,4-dicyanothiophene:
  conversion into 2-amino-3,4-dicyano-5-mercaptopyrrole, 332
1,2-Diaminopyrroles:
  conversion into pyrrolo[1,2-*b*]-*as*-triazines, 481
  mass spectra of, 461
  NH stretching frequencies of, 459
  ^{1}H NMR spectra of, 402
  synthesis of, 345, 354
  UV spectra of, 434
2,3-Diaminopyrroles:
  ^{1}H NMR spectra of, 403
  synthesis of, 354
  UV spectra of, 434
2,4-Diaminopyrroles:
  N-methylation of, 494
  NH stretching frequencies of, 462
  synthesis of, 346, 355
  tautomerism of, 361, 477, 478
  thermolysis of, 481
2,5-Diaminopyrroles:
  NH stretching frequencies of, 460
  reaction with hydroxylamine, 503
  stability of, 471
  synthesis from succinamidine, 347
  synthesis from succinonitrile, 347
  synthesis *via* Curtius rearrangement of pyrroloylazides, 355
  tautomerism of, 347, 361, 478
  thermolysis of, 482
  UV spectra of, 435
1,4-Dicarbonyl compounds:
  reaction with hydrazines, 315–322
2-Dichloromethylpyrrole-3,5-dicarboxylic esters:
  unusual reaction with potassium hydroxide, 50
5,7*a*-Didehydroheliotridin-3-one, 42
Dieckmann cyclization reactions:
  of enamino esters, 532
  of aminomethylenemalonic esters, 535
Diethyl 1-pyrrolylmethanephosphonate:
  reaction with aldehydes, 165
  synthesis of, 166
Diethyl 1,2,4,5-tetrazine-3,6-dicarboxylate:
  reaction with vinylindoles, 243, 259
2,3-Diformylpyrroles:
  reaction with bis(methoxycarbonylmethyl)sulfide, 38
  reaction with succindinitrile, 36
  reaction with sulfones, 36
  synthesis of, 72
2,4-Diformylpyrroles:
  synthesis of, 65
2,5-Diformylpyrroles:
  borohydride reduction of, 70
  complexes with metal ions, 111
  reaction with 2-aminophenol, 114
  reaction with heteroarylphosphonates, 27
  reaction with hydrazines, 51
  synthesis of, 65
3,4-Diformylpyrroles:
  reaction with cyclohexane-1,4-dione, 16
  reaction with arylamines, 48
  reaction with bis(2-aminoethyl)amine, 56
  reaction with dialkylketones, 33, 249
  reaction with α,ω-diaminoalkanes, 56
  reaction with glycine, 58

3,4-Diformylpyrroles (*Continued*)
   reaction with hydrazine, 56
   reaction with sulfones, 37
Dihydroindoles:
   synthesis from vinylpyrroles, 95
Dihydroindolizines:
   synthesis of, 26
Dihydropyridazines:
   from 1,3-dicarbonyl compounds and semicarbazones, 324
   from 1,4-dicarbonyl compounds, 316
   from 2,5-diethoxytetrahydrofuran, 322
1,2-Dihydro-3*H*-pyrrol-3-ones (see 1*H*-pyrrol-3(2*H*)-one)
3,4-Dihydroxypyrroles:
   reaction with dibromoethane, 559
   reaction with bromochloromethane, 559
   synthesis of, 539
Diimidoporphyrins, 481, 501
Di(3-methoxy-2-pyrrolyl)methanes:
   synthesis of, 596, 598
1,3-Dipolar cycloaddition reactions:
   of DMAD with azirines, 341
   of pyrrole-1-carboxylic esters with β-ketonitriles, 96
   of 2-vinylpyrroles, 244
   of 3-vinylpyrroles, 261
1,1-Di(2-pyrrolyl)alkenes:
   synthesis of, 227
1,2-Di(2-pyrrolyl)ethenes:
   photocyclization, 43
   synthesis of, 223, 236
Dipyrrolylmethenes:
   synthesis of, 19
Dipyrrolyltrimethine dyes:
   synthesis of, 228
Distamycin A:
   analogs of, 306, 513
   antiviral activity, 511
   degradation of, 305
   inhibition of DNA formation, 511
   isomers of, 308
   total synthesis of, 305–307, 309–310
5,5-Disubstituted 2-aminopyrrol-4-ones:
   synthesis from oxazolones, 333
4,5-Disubstituted 2-formylpyrroles:
   synthesis of, 49, 50
1,1-Disubstituted 3-oxopyrrolium salts:
   synthesis of, 531
2,2-Disubstituted 1*H*-pyrrol-3(2*H*)-ones:
   O-acylation of, 561
   bromination of, 566
   [2+2]cycloaddition with alkenes, 586
   ^{1}H NMR spectra of 553–554
   ^{13}C NMR spectra of, 554–555
   ozonolysis of, 582
   photochemical dimerization, 586
   photosensitized migration of 2-allyl group, 587
   reaction with aryldiazonium salts, 569
   reaction with methoxymethylene Meldrum's acid, 571
   synthesis of, 529

Enamino esters:
   Dieckmann cyclization of, 532
5-Ethoxycarbonyl-3*H*-pyrrolizine:
   synthesis of, 216
1-Ethyloctahydroindole, 172
2-Ethylpyrrole:
   dehydrogenation of, 231
Ethyl 5-(2-pyrrolyl)penta-2,4-dienoate:
   synthesis of, 25

Fluorescamine, 538
2-Formylfuran:
   acid-catalysed condensation with pyrroles, 169
2-Formyl-1-methylpyrrole:
   conversion into N,N-dimethyl 1-methylpyrrole-2-carboxamide, 55
   crossed Cannizzaro reaction with formaldehyde, 76
   reaction with benzocycloheptanone, 39
   reaction with N-hydroxy-N'-aminoguanidine, 51
   reaction with lithium aluminum hydride, 67
   silyl enol ether derivative, 21
   synthesis of, 65
5-Formyl-1-methyl-4-vinylpyrrole:
   [4+2]cycloaddition reaction with DMAD, 95
2-Formyl-1-(2-nitrobenzyl)pyrrole:
   catalytic reduction of, 59
2-Formylpyrroles:
   acid-catalysed migration of formyl group, 96
   aldol condensation with 2-acetylpyrrole, 34
   1-(2-chloroethyl) derivative, 86
   cleavage of formyl group, 90
   conversion into 2-cyanopyrrole, 53
   conversion into pyrrolizin-3-one, 36
   dithioketals, 90
   Knoevenagel condensation, 215
   oxidation of, 64
   Perkin reaction, 215
   phase-transfer catalysed N-alkylation of, 100
   pinacol formation, 75
   pyrolysis of dibutyl acetal, 109
   reaction with 2-acetylpyrrole, 214
   reaction with acylglycines, 40

reaction with alkyl hydrogen malonates, 215
reaction with amidic ketals, 38
reaction with arylacetonitriles, 216
reaction with azlactones, 217
reaction with barbituric acid, 217
reaction with benzil and ammonium acetate, 63
reaction with 1,2-diaminonapthalenes, 63
reaction with 2-(dimethylamino)ethyl phenyl ketone, 214
reaction with L-cysteine methyl ester, 112
reaction with ethyl acrylate, 216
reaction with heterocyclic quaternary salts, 218
reaction with hydantoin, 217
reaction with hydrazine, 58
reaction with Meldrum's acid, 217
reaction with methyl cyanoacetate, 215
reaction with methyl ketones, 213–214
reaction with phenylacetyl chloride, 216
reaction with phosphonates, 223
reaction with phosphoranes, 218–224
reaction with Δ^3-pyrrolin-2-ones, 218
reaction with 2-pyrrolylacetic acid, 215
reaction with secondary amines, 48
reaction with succinyl hydrazide, 38
reduction with lithium aluminum hydride, 67
reductive amination of, 71
semicarbazone formation, 51
Stobbe reaction, 216
1-(2-trimethylsilyl)ethoxymethyl derivative, 86
3-Formylpyrroles:
  Knoevenagel reaction, 248
  reaction with nitromethane, 249
  reaction with heterocyclic quaternary salts, 252
  synthesis of, 65
  Wittig reaction, 252
2-Formylpyrrole-3-carboxylic esters:
  reaction with hydrazine, 57
Formylpyrroles:
  acid-catalysed migration of formyl group, 96
  N-acylation of, 83
  Benzoin reaction, 14
  Baeyer-Villiger oxidation of, 64
  Cannizzaro reaction, 76
  catalytic reduction of, 73
  conversion into cyclic acetals, 77
  N-ethylation with triethyl phosphate, 107
  Horner reaction, 24–25, 27
  intramolecular Horner reaction, 106
  Knoevenagel reaction, 30
  C-methylation of, 103
  mixed Benzoin reaction, 15

Perkin reaction, 18
phase-transfer catalysed Horner reaction, 25
N-protection of, 83–86
reaction with alkylamines, 46
reaction with arylamines, 46
reaction with barbituric acid, 32
reaction with but-3-en-2-one, 110
reaction with cyanide ions, 13
reaction with cyanoacetic esters, 31
reaction with cyclopentadienyl anion, 19
reaction with cyclopropylphosphonium salts, 26
reaction with diaminobenzenes, 47
reaction with 2,2-dimethylpropane-1,3-diol, 77
reaction with (heteroarylmethyl)phosphonium salts, 27
reaction with heterocyclic quaternary salts, 21
reaction with hydantoin, 18
reaction with hydrazines, 51, 58
reaction with hydroxylamine, 53
reaction with malononitrile, 77
reaction with Meldrum's acid, 32
reaction with methyl ketones, 15
reaction with 2-methylpentane-2,4-diol, 77
reaction with nitromethane, 17
reaction with oxazolone, 18
reaction with phenylacetonitrile, 17
reaction with rhodamine, 18
reaction with α-unsubstituted pyrroles, 19
reduction with aluminum amalgam, 75
reduction with diborane, 71
reduction with diisobutylaluminum hydride, 72
reductive alkylation of, 21
Stobbe reaction, 15
Wittig reaction, 24, 42
Wolff-Kishner reduction of, 52, 75
2-Formylpyrrolyl anion:
  reaction with 1,2-bis(phenylsulfonyl)ethene, 41
  reaction with vinyltriphenylphosphonium salts, 41
Functionalised ketoximes:
  synthesis of 1-vinylpyrroles from, 151–153
2-(2-Furyl)-1-vinylpyrroles:
  synthesis of, 143
  protonation of, 176
  trifluoroacetylation of, 190

Heptafluoropropylpyrroles:
  synthesis of, 103
Heterocyclic ketoximes
  reaction with acetylene, 145
Horner reaction, 24–25, 27, 165, 223

2-Hydrazonopyrrolones, 569
2-(1-Hydroxyalkyl)pyrroles:
    conversion into 2-vinylpyrroles, 232
    synthesis of, 232
3-Hydroxypyrroles (see also 1H-pyrrol-3(2H)-
        ones):
    acidity of, 578
    O-acylation of, 560
    aerial oxidation of, 579
    C vs O alkylation of, 557
    O-alkylation of, 557
    aldol reaction, with aldehydes, 572
    conversion into carbamates, 561
    [4 + 2]cycloaddition reactions, 576
    N,O-diacylation of, 561
    keto-enol tautomerism of, 543–545
    IR spectra of, 547
    mass spectra of, 556
    Michael reactions, 577, 578
    NMR spectra of, 550–552
    one-electron oxidation of, 581–582
    oxidation of, 578–583
    photochemistry of, 585–588
    reaction with formylpyrroles, 573
    reaction with glyoxal, 573
    reaction with isocyanates, 561
    reaction with methoxymethylene Meldrum's
        acid, 571
    reaction with isatin, 573
    reaction with trimethyl orthoformate, 573
    reduction of, 583–584
    synthesis of, 527–542
    O-trimethylsilyation of, 557
    UV spectra of, 548
3-Hydroxypyrrole-2-carboxylic acids:
    stability of, 584–585
3-Hydroxypyrrole-4-carboxylic acids:
    decarboxylation of, 585
3-Hydroxypyrrole-2-carboxylic esters:
    acid-catalysed hydrolysis of, 584
    C-alkylation of, 559
    O-alkylation of, 558
    O vs N vs C alkylation of, 559
    chlorination of, 567
    N,O-dialkylation of, 558
    hydride reduction of, 583
    hydrogenolysis of benzyl esters, 584
    IR spectra of, 547
    Mannich reaction, 575
    Michael reaction, 577
    photodimerization, 110, 587
    photooxidation of, 581
        reaction with ethyl acrylate, 577
    ring expansion of 568, 570
    synthesis of, 532, 533
    tautomerism of, 545
    UV spectra of, 548
3-Hydroxypyrrole-4-carboxylic esters:
    O-alkylation of, 559
    conversion into pyrrolopyrones, 575
    Friedel-Crafts acylation of, 571
    hydrolysis of, 585
    IR spectra of, 547
    nitrosation of, 568
    oxidative dimerization, 582
    reaction with aryl aldehydes, 574
    reaction with aryl diazonium salts, 570
    reaction with acetylacetone, 574
    reaction with phosphorus oxychloride, 567
    reaction with pyruvic aldehyde, 575
    synthesis of, 533–534
    UV spectra of, 548
    Vilsmeier-Haack formylation of, 571
3-Hydroxypyrrole-5-carboxylic esters:
    reaction with aryl diazonium salts, 570
    reaction with isatin, 573
    synthesis of, 535
3-Hydroxypyrrole-2,4-dicarboxylic esters:
    O-alkylation of, 559
    O vs C alkylation of, 559
    bromination of, 567
    synthesis of, 535
3-Hydroxypyrrole-2,5-dicarboxylic esters:
    O vs N alkylation of, 559
    synthesis of, 536
3-Hydroxypyrrole radicals:
    ESR spectra of, 556
1-Hydroxy-1H-pyrrol-3(2H)-ones:
    synthesis of, 537
2-Hydroxy-1H-pyrrol-3(2H)-ones:
    acyloin reaction, 578
    conversion into 2-methoxypyrrolones, 560
    from hydrogenation of oxazoles, 537
    from 2-methoxy-3(2H)-furanones, 537
    ring-chain tautomerism of, 545–546
    synthesis of, 537–579
2-Hydroxy-1H-pyrrol-3(2H)-one-2-carboxylic
        esters:
    synthesis of, 538
2-Hydroxy-1H-pyrrol-3(2H)-one-4-carboxylic
        esters:
    synthesis of, 538
S-(3-Hydroxy-2-pyrrolyl)isothiouronium salts:
    from 1H-pyrrol-3(2H)-one-4-carboxylic es-
        ters, 583

Imidoporphyrins, 501

β-Ketoesters:
  1-acylpyrroles in the synthesis of, 108–109
Ketoximes:
  reaction with acetylene, 134–154
Kikumycins:
  antibacterial activity, 512
  cytotoxic activity, 512
  degradation of, 312
  total synthesis of, 313

Lexitropsins, 4

Maillard reaction, 9
2-Methoxy-1H-pyrrol-3(2H)-ones:
  synthesis of, 537
Methyl 2-formylpyrrole-1-carboxylate:
  reaction with Grignard reagents, 22
1-Methyl-2-nitropyrrole:
  reaction with 1-methyl-2-cyanomethylpyrrole, 236
1-Methylpyrrole:
  reaction with diphenylacetylene, 226
2-Methylpyrrole-3-carboxylic acids:
  abnormal reaction with thionyl chloride, 98
2-(1-Methyl-2-pyrrolyl)-1-vinylpyrroles:
  synthesis of, 144

Netropsin:
  acid degradation of, 302
  alkaline degradation of, 302
  analogs of, 302, 310, 513
  antibacterial activity, 510
  antitubercular activity, 510
  homolog of, 308
  inhibition of DNA formation, 511
  insecticidal activity, 510
  isomers of, 308
  total synthesis of, 302, 303
  trypanocidal activity, 510
2-(2-Naphthyl)-1-vinylpyrrole:
  synthesis of, 142

Oxazinones:
  via oxidation of pyrrol-3-ones, 580
2-Oximinopyrrolones, 567
6-Oxo(2H)cyclohepta(c)pyrroles:
  cycloaddition reaction with DMAD, 260

Peramine, 59
1-(2-Phenylethyl)pyrrole-2-carbonyl chlorides:
  cyclization of, 45
2-Phenylpyrrolin-3-one:
  synthesis of, 216

1-Phenylpyrrol-3(2H)one:
  molecular structure, 546
E-2-(2-Phenylsulfinylvinyl)pyrroles:
  synthesis of, 235
2-Phenyl-1-vinylpyrrole:
  acid-catalysed reaction with ethanal, 194
  protonation of, 178
  reaction with methanol, 182
  reaction with hydroxylamine hydrochloride, 183
  synthesis of, 135, 137
Porphobilinogen, 60
Porphyrins, 505
Prodigiosin:
  natural occurence, 606
  pK_a, 610
  synthesis of, 19, 89, 227
  X-ray structure of 5-thienyl analog, 610
Prostaglandin:
  heterocyclic analogs, 29, 222
Pyrazoles:
  synthesis from 2-acyl-1,4-diketones, 320
Pyrocolls:
  synthesis of, 90
Pyrroles:
  condensation with ω-(N-methylanilino)propenal, 229
  intramolecular acylation, 103
  photochemical reaction with benzene, 235
  reaction with alkynes, 157–165, 224–226, 253
  reaction with carbonyl compounds, 227–231
  reaction with DMAD, 224–225
  reaction with formic acid, 252
  reaction with oxalyl chloride, 235
  reaction with Z-4-(phenylsulfinyl)-4-penten-2-one, 235
  reaction with squaric acid, 230
  Vilsmeier-Haack reaction, 255
Pyrrolecarboxamides:
  2-acylation of 3-carboxamide, 105
  synthesis of, 54
Pyrrolecarboxanilides:
  synthesis of, 54
Pyrrolecarboxylic acids:
  complexes with metal ions, 111
  esterification of, 78
  iodinative decarboxylation, 89
  phase-transfer catalysed esterification of, 79
  pyrolysis of, 90
  thermal decarboxylation, 87
Pyrrolecarboxylic esters:
  acid-catalysed cleavage of tert-butyl esters, 82
  hydride reduction of, 68, 69
  hydrogenolysis of benzyl esters, 82

Pyrrolecarboxylic esters (*Continued*)
  hydrolysis of, 79–83
  synthesis from trichloroacetylpyrroles, 66
  synthesis from trichloromethylpyrroles, 79
Pyrrole-1-carboxylic esters:
  acid-catalysed cleavage of 1-*tert*-butyloxycarbonyl group, 100
  conversion into 1-acylpyrroles, 108
  [4 + 2]cycloaddition reaction with DMAD, 92
  [4 + 2]cycloaddition reaction with hexafluorobut-2-yne, 92
  [4 + 3]cycloaddition reaction with β-ketonitriles, 96
  pseudo-Gomberg reaction, 99
  reaction with maleimides and maleic anhydride, 93
  reaction with ethoxycarbonyl carbene, 95
  reaction with singlet oxygen, 94
Pyrrole-2-carboxylic esters:
  base-catalysed hydrolysis of, 79
  Mannich reaction, 102
  mercuration of, 100
  reaction with styrene epoxides, 45
Pyrrole-3-carboxylic esters:
  acid-catalysed hydrolysis of, 80
  halogenation of, 102
Pyrrole-3,4-dicarboxylic esters:
  synthesis of, 576
Pyrrolethiocarboxylic esters:
  hydrogenolysis of, 74
2-Pyrrolin-4-ones (*see* 1*H*-pyrrol-3(2*H*)-ones)
3*H*-Pyrrolizines:
  synthesis of, 25–26, 36, 41, 216, 219–220
Pyrrolizin-3-ones:
  synthesis of, 36, 42
Pyrrolobenzoquinones:
  synthesis of, 90, 104–105
Pyrrolobenzazepinones:
  synthesis of, 45
Pyrrolodiazepines, 499, 513
Pyrrolodiazepin-2-ones, 491, 513
Pyrroloindoles, 502
1*H*-Pyrrol-3(2*H*)-ones (*see also* 3-hydroxypyrroles):
  acid-catalysed deuterium exchange reactions, 563
  aerial oxidation of, 579
  autoxidation of, 579
  chlorination of, 567
  [4 + 2]cycloaddition reactions, 576
  diazo coupling reactions, 569
  Friedel-Crafts acylation of, 571
  from cyclopropenones, 540

keto-enol tautomerism of, 543–545
nitrosation of, 567
NMR spectra of, 552–556
mass spectra of, 556
one-electron oxidation of, 581–582
oxidation of, 578–583
oxidative dimerization, 582
photochemistry of, 585–588
O-protonation of, 563
reaction with DMAD, 575–577
reduction of, 583–584
ring expansion to oxazinones, 580
spectra of protonated forms, 563
UV spectra of, 548–549
Pyrrolone-N-oxides:
  hydride reduction of, 541
Pyrrolophthalazinones, 479
Pyrrolopyridines, 488–489, 492, 500, 513
Pyrrolopyridones, 43–44, 62, 487, 504
Pyrrolopyrimidines, 62, 487–488, 490, 500, 513
Pyrrolopyrimidinones, 61, 481, 492, 499
Pyrrolopyrones, 576
3*H*-Pyrrolo[1,2-*a*]pyrroles (*see* 3*H*-pyrrolizines)
Pyrrolothiazines, 487
Pyrrolotriazepines, 495
Pyrrolotriazines, 57, 489, 493, 495, 499
Pyrroloyl chlorides:
  reaction with 2-amino-2-methylpropan-1-ol, 55
2-Pyrroloylferrocenes:
  thermal rearrangement to 3-pyrroloylferrocenes, 97
Pyrroloylhydrazines, 52
2-(2-Pyrrolyl)benzimidazoles, 47
4-(2-Pyrrolyl)but-2-enal:
  conversion into the indole, 105
1-(2-Pyrrolyl)hexan-2,4-dione:
  conversion into the indole, 105
Pyrrolyl hydrazones:
  complexes with metal ions, 115
Pyrrolylmethyleneimines:
  complexes with metal ions, 112–114
Pyrrolyl oximes:
  Beckmann rearrangement of, 53
  reduction of, 53, 60
1-(2-Pyrrolyl)pentan-1,4-diones:
  synthesis of, 110
Pyrrolylpotassium:
  reaction with ethoxyethylenes, 166
1-(2-Pyrrolyl)propan-1-one:
  reaction with ethyl magnesium bromide, 22
4-(2-Pyrrolyl)pyrimidines:
  synthesis of, 40

3-(2-Pyrrolyl)-1,2,4-triazole,
  synthesis of, 63

Sibromycines, 60
Sinanomycin (see Netropsin)
1-Styrylpyrroles:
  synthesis of, 166
2-Styrylpyrroles:
  photocyclisation of, 245
  synthesis of, 28, 218, 219
3-Styrylpyrroles:
  photocyclisation of, 262
  synthesis of, 29, 252
4-Substituted 2-formylpyrroles:
  synthesis of, 48
5-Substituted 2-formylpyrroles:
  synthesis of, 48
1-Substituted 3-hydroxypyrroles:
  synthesis of the flash-vacuum pyrolysis, 529
2-Substituted pyrroles:
  synthesis of, 94

T-1384 (see Netropsin)
TAN-868 (see Kikumycins)
Tautomerism:
  of 2-aminopyrroles, 359–361, 472
  of 3-aminopyrroles, 361, 477
  of azo derivatives of 3-hydroxypyrroles, 569
  of 2,4-diaminopyrroles, 361, 477–478
  of 2,5-diaminopyrroles, 347, 361, 478
  of 3-hydroxypyrroles, 543–545
  of 3-hydroxypyrrole-2-carboxylic esters, 545
  of 2-hydroxy-1H-pyrrol-3(2H)-ones, 545–546
Tetrahydropyridazines, 322
Tetramic acids:
  from 2-hydroxypyrrol-3(2H)-ones, 564
2-(2-Thienyl)-1-vinylpyrroles:
  synthesis of, 144
  trifluoroacetylation of, 190
1-Tosylaminopyrrole:
  thermolysis of, 479
1-(2-Trialkylsilyl)ethylpyrroles:
  synthesis of, 187
Triaminopyrroles:
  reaction with aryl aldehydes, 494
  reaction with DMF, 494
  synthesis of, 347
1,2,4-Triaryl-3-hydroxypyrroles:
  synthesis of, 530
Trichloroacetylpyrroles:
  conversion into pyrrolecarboxamides, 55
  conversion into pyrrolecarboxylic acids, 66
  conversion into pyrrolecarboxylic esters, 66

2-Trifluoroacetyl-1-vinylpyrroles:
  base-catalysed hydrolysis of, 191
  hydrogenation of, 191
  ^{19}F NMR spectra of, 208
  phosphorylation of, 191
  synthesis of, 189
  thiolation of, 191
1,2,5-Trisubstituted-3-hydroxypyrroles:
  synthesis of, 530
Trofimov reaction:
  effect of pressure on, 135
  effect of solvent on, 136
  effect of super-base on, 135
  effect of temperature on, 135
  reaction conditions, 134
  regiospecificity of, 153
  use of vinyl halides in, 155
  use of dihaloalkanes in, 155

1-Ureidopyrroles:
  acid-catalysed hydrolysis of, 483
  acidity of, 472
  NH stretching frequencies of, 435, 441–442
  ^{13}C NMR spectra of, 373–375
  ^{1}H NMR spectra of, 365, 373–376
  Schotten-Baumann benzoylation of, 483
  synthesis of, 317–318, 323, 326–327
  thermolysis of, 479
  UV spectra of, 422–424

N-Vinylation of pyrroles:
  effect of structure on, 160
  effect of super-base on, 158
  effect of temperature on, 159
  one-electron channel mechanism for, 162
1-Vinylpyrroles:
  acid-catalysed dimerisation of, 177
  acid-catalysed hydrolytic cleavage of, 180–183
  basicity of, 201
  biological properties of, 209–212
  charge-transfer complexes with electron-acceptors, 195
  CNDO/2 calculations on, 197
  conformation of, 205
  copolymerisation with 1-vinylindole, 195
  dipole moments of, 198–201
  hydrogenation of, 171–172
  IR spectra of, 204–205
  ^{1}H NMR spectra of, 197, 205–206
  ^{13}C NMR spectra of, 197, 205, 207
  ^{14}N NMR spectra of, 197, 208
  oxidation of, 170
  phosphorylation of, 192–194

1-Vinylpyrroles (*Continued*)
  polymerisation of, 195–197, 212–213
  protonation of, 172–180
  reaction with acetylenic alcohols, 184
  reaction with alcohols and phenols, 183–185
  reaction with hydrogen halides, 174
  reaction with tetracyanoethene, 194
  reaction with thiols, 185–186
  reaction with trialkylsilanes, 186–188
  regiospecific protonation of, 174
  synthesis from alkyl heteroaryl ketoximes, 143
  synthesis from dialkyl ketoximes, 136
  synthesis from ketoximes, 134
  synthesis from pyrrolylpotassium
  synthesis *via* Horner reaction, 165
  trifluoroacetylation of, 188
  UV spectra of, 201–204
2-Vinylpyrroles:
  conformation of, 239
  conversion into indoles, 237–242
  [4+2]cycloaddition reactions, 237–242
  1,3-dipolar addition reactions, 244
  dipole moments of, 264
  intramolecular Diels–Alder reaction, 242
  IR spectra of, 269–271
  NMR spectra of, 271–278
  photocyclisation of, 244–246
  reaction with carbenes, 243
  reaction with diethyl azodicarboxylate, 246
  reaction with DMAD, 225, 237–242
  reaction with ethylamine, 243
  reaction with 4-phenyl-1,2-triazoline-3,5-dione, 247
  synthesis from 2-formylpyrroles, 26, 213–224
  synthesis *via* the Wittig reaction, 218–224
  UV spectra of, 265–269
3-Vinylpyrroles:
  conversion into indoles, 256–261
  [4+2]cycloaddition reactions, 256–261
  1,3-dipolar addition reactions, 261
  ^{1}H NMR spectra of, 271
  photocyclisation of, 262
  reaction with DMAD, 257
  reaction with Erhlich's reagent, 264
  synthesis from 3-(2-aminoethyl)pyrroles, 254
  synthesis from ethylenic ketoximes, 254
  synthesis from 3-formylpyrroles, 26, 248–253
1-Vinylpyrrolium cations:
  NMR spectra of, 172
  reaction with 1-vinylpyrroles, 177
(1-Vinyl-2-pyrrolyl)indoles:
  synthesis of, 145

Wittig reaction:
  in the formation of 3-hydroxypyrroles, 535
  in the formation of vinylpyrroles, 24, 30, 42, 218–224, 252
  in the formation of 1-aminopyrroles, 324